Quantitative Methods of Data Analysis for the Physical Sciences and Engineering

This book provides thorough and comprehensive coverage of most of the new and important quantitative methods of data analysis for college and graduate students and practitioners. In recent years, data analysis methods have exploded alongside advanced computing power, and an understanding of such methods is critical to getting the most out of data and for extracting signal from noise. The book excels in explaining difficult concepts through simple explanations and detailed explanatory illustrations. Most unique is the focus on confidence limits for power spectra and their proper interpretation, something rare or completely missing in other books. Likewise, there is a thorough discussion of how to assess uncertainty via use of Expectancy, and easy-to-apply and -understand Bootstrap method. The book is written so that descriptions of each method are as self-contained as possible. Many examples are presented to clarify interpretations, as are user tips in highlighted boxes.

Douglas G. Martinson is a Lamont Research Professor in the Division of Ocean and Climate Physics at Columbia University's Lamont–Doherty Earth Observatory. As a physical oceanographer who researches the role of polar oceans in global climate, his research involves the collection of a large amount of data and considerable quantitative analysis. He developed the course on Quantitative Methods of Data Analysis as an Adjunct Professor for the Department of Earth and Environmental Sciences at Columbia University, and received an Outstanding Teacher Award in 2004.

Quantitative Methods of Data Analysis for the Physical Sciences and Engineering

DOUGLAS G. MARTINSON
Columbia University

CAMBRIDGE
UNIVERSITY PRESS

University Printing House, Cambridge CB2 8BS, United Kingdom

One Liberty Plaza, 20th Floor, New York, NY 10006, USA

477 Williamstown Road, Port Melbourne, VIC 3207, Australia

314–321, 3rd Floor, Plot 3, Splendor Forum, Jasola District Centre, New Delhi – 110025, India

79 Anson Road, #06–04/06, Singapore 079906

Cambridge University Press is part of the University of Cambridge.

It furthers the University's mission by disseminating knowledge in the pursuit of education, learning, and research at the highest international levels of excellence.

www.cambridge.org
Information on this title: www.cambridge.org/9781107029767
DOI: 10.1017/9781139342568

First published 2018

Printed in the United Kingdom by TJ International Ltd. Padstow Cornwall

A catalogue record for this publication is available from the British Library.

Library of Congress Cataloging-in-Publication Data
Names: Martinson, Douglas G.
Title: Quantitative methods of data analysis for the physical sciences and engineering / Douglas G. Martinson (Columbia University, New York)
Description: Cambridge, United Kingdom ; New York, NY : Cambridge University Press, 2018.
Identifiers: LCCN 2017055413| ISBN 9781107029767 (hbk.) | ISBN 1107029767 (hbk.)
Subjects: LCSH: Statistics. | Physical sciences – Statistical methods. | Engineering – Statistical methods.
Classification: LCC QA276 .M3375 2018 | DDC 519.5–dc23
LC record available at https://lccn.loc.gov/2017055413

ISBN 978-1-107-02976-7 Hardback

Additional resources for this publication at www.cambridge.org/martinson

To the love of my life, my wife Rhonda

Contents

Preface

This book is the outcome of a one-semester graduate class taught in the Department of Earth and Environmental Sciences at Columbia University, although the book could be used over two or even three semesters, if desired. I have taught this class since 1985, having taken over from a departing marine seismologist who had taught the course as one on Fourier analysis, the only topic that computers of the day were capable of performing, because of the development of the Fast Fourier Transform. However, at that time computers were rapidly becoming powerful enough to allow application of methods requiring more power and memory. New methods were sprouting yearly, and as the computers grew faster, previously sluggish methodologies were becoming realizable. At the time I started teaching the course, there were no textbooks (none!) that gave a *thorough* introduction to the primary methods. *Numerical Recipes* – published in the early 1980s – did present a brief overview and the computer code necessary to run nearly every method, and it was a godsend. It occurred to me that my class notes should be converted to a book to fill this void. Over the last 30 years many other books have been published, but in my opinion there is still a need for an introductory-level book that spans a broad number of the most useful techniques. Regardless of its introductory nature, I have tried to give the reader a complete enough understanding to allow him or her to properly apply the methods while avoiding common pitfalls and misunderstandings.

I try to present the methods following a few fundamental themes: e.g., Principle of Maximum Likelihood for deriving optimal methods, and Expectancy for estimating uncertainty. I hope this makes these important themes better understood and the material easier to grasp.

How to Use This Book

This book is designed to fill many needs, according to the level of the student. Some, like myself, see the methods clearly if they understand their complete derivation, while others don't require that detailed understanding. In an effort to satisfy both readerships, I have placed complete derivations in boxes highlighted with 25 percent grayscale: these boxes are optional, and the reader is free, if preferred, to skip the box and go straight to the answer (all equations in derivation boxes are prefaced by "D" – for example, "D5.1"). There are student exercises at the end of each chapter, some of which require computing. I do not present code because it changes so quickly, but I do show some MATLAB code in the solution manual. Data for exercises requiring such can be found at www.cambridge.org/martinson. Most of the examples in the book are taken from the natural sciences, although they are presented so as to be understandable to anyone. Special user tips are included in boxes highlighted with 15 percent grayscale. I have attempted to make each chapter stand on its own (as far as possible), so the reader doesn't need to have read the entire book in order to understand material from previous chapters. This should make the book good for easy use and reference for a particular method.

Read it, practice, and when not sure what road to take, take all possible roads and then determine which is the most appropriate for your particular analysis. Then, maybe present several results explaining the differences, and why you favor the method you choose.

Acknowledgments

As with all books evolving from a course, one must acknowledge the considerable input from students and teaching assistants. As any teacher knows, it is usually the one teaching who learns more than anyone – when first teaching this course there were numerous derivations that only were partly developed, then a "miracle occurred" that skipped some "intuitively obvious" steps to the final result. No such skipped steps occur within this book. Over the years, excellent questions from students that I could not answer on the spot forced me to fill in many aspects of the material. So I offer a heartfelt thanks to those who stumped me in class. In that same vein, I would appreciate hearing about any errors still present in the book. The class has benefited from some incredibly smart and motivated teaching assistants, and many of the exercises appearing at the end of chapters originated from them (special thanks go to Sharon Stammerjohn, Chen Chen, and Darren McKee, among many others). Unfortunately, as I transformed my class notes into a textbook, my wife, Rhonda, became a writer's widow for nearly a year – I can't thank her enough for all the support she has given me. And finally, but not least, thanks to my editor, Matt Lloyd, at Cambridge University Press, who was a constant source of improvements and encouragement!

Finally, that ubiquitous message accompanying all such books: any errors in the book are strictly mine. Oh, and the other statement: any views expressed in this book (and there are many) are entirely mine. Enjoy!

Part I

Fundamentals

Analysis of data requires the use of a broad range of techniques spanning a wide range of complexities. Common to most of these techniques is a fundamental foundation consisting of a nomenclature (not always consistent from one author to the next) as well as a set of mathematical and statistical tools. The purpose of Part I is to define that nomenclature and those basic tools.

For this Part, the *order* in which the observations occur is *not* important. This is in contrast to **sequential data**, which are generically referred to as **time series** (whether they vary with time or not), the subject of Part II. For the latter, the order in which the observations occur *is* important.

Part I is dominated by techniques of classical statistics, such as regression analysis, though some newer techniques such as nonparametric and resampling (bootstrap) statistics represent valuable additions to that traditional arsenal. While the tools of statistics are extensive and span a broad range of approaches, the concepts of *Expectation* and *Maximum Likelihood Estimators* are particularly useful in data analysis and will be stressed throughout. Because resampling statistics offers a significant increase in our processing capabilities, it too will be presented for general analyses.

1 The Nature of Data and Analysis

1.1 Analysis

The Random House College Dictionary defines **analysis** as "the separation of any material or abstract entity into its constituent elements." For our analysis to be meaningful, it is implicit that the data being analyzed contain some "signal" representing the phenomenon of interest (or some aspect of it). Satisfying this, we might attempt to separate the signal from the noise present in the data. Then we can characterize the signal in terms of its robust features and, in the case of complex phenomena, separate the signal even farther into constituents, each of which may afford additional insights regarding the character, behavior or makeup of multiple processes contributing to our single phenomenon.

Mathematically, we often desire to rewrite a data set, y_i, as

$$y_i = a_1\varphi_{1i} + a_2\varphi_{2i} + \ldots \quad + a_n\varphi_{ni}, \tag{1.1}$$

where the constituents of the data are now described by functions or vectors, φ_{ki}. These constituents (or some subset) with the appropriate weights a_k can then be recombined to **synthesize** (reconstruct) the original data y_i (or just the signal portion of it). Typically, the fewer constituent terms you can use to describe the greatest amount of the data is better (presuming that the signal is contained in a few, hopefully understandable, constituents).

Equation (1.1) represents a simple linear foundation from which a large number of the techniques and analysis tools developed in this text will build on.

Caution: Most analysis techniques will produce something satisfying that technique's definition of signal, even when performed on pure noise, so be aware that the analysis result may actually be nothing more than a statistical construct. A proper interpretation of your analysis is possible when multiple pieces of evidence support or refute a hypothesized answer to the question being addressed.

1.2 Data Nomenclature

Data (plural) represent measurements of quantities or of variables (a single data point is a **datum**). The variables being measured are classified as discrete or continuous.

Discrete variables are those having discontinuous or individually distinct possible outcomes. Examples include

1) flipping of a coin or rolling of dice
2) counts of individual items or groups of items
3) categorization or classification of measurements

Continuous variables are those having an uninterrupted range of possible outcomes (i.e., with no breaks). Examples include

1) concentrations of a quantity
2) percentage of an item (such data, forced to a constant sum, sometime require special care and attention)
3) magnitude, such as length, mass, etc.

The data (i.e., the measurements of the variables) are also classified according to how they are recorded:

Analog data are those which have been recorded "continuously," such as by a strip recorder (though, technically, even this is not purely continuous, given a non-instantaneous response time of the recorder).

Discrete (or digital) data are those that have been recorded at discrete intervals. All data, when represented on digital computers, are discrete.

In either case, the data must be discretized before they are analyzed in any computational manner that we will consider.

Regardless of how the data are recorded, a sequential series of data are classified as time series.

A **sequential data series** consists of measurements of a quantity that vary as a function of time *or space*, and *the order in which the measurements occur is important*. In this case, the variable being measured is typically referred to as the **dependent variable**, while the time or space variable is the **independent variable**.

Sequential series are commonly referred to by other names such as **time series, traces, records, data series, spatial series**, etc., though "time series" is the most common. The independent variable need not be restricted to time or space. In many instances, it is desirable to measure a quantity as it varies with some other variable whose order of occurrence is important. Regardless, as long as the order of occurrence of the measurements is important, the name "time series" still is commonly applied.

Time series can represent measurements of either discrete or continuous variables and are recorded in either analog or discrete fashion. However, since time and space vary continuously themselves, the discrete or continuous variables being measured often vary continuously as a function of the independent variables. Therefore, discretization of time series data may involve both the dependent and independent variables.

Some authors distinguish between digital and discrete time series as follows. **Discrete series** are sequential series that are discrete in the independent variable but continuous in the dependent variable. **Digital series**, on the other hand, are series that are discrete in both the dependent and independent variables. I will make no such distinction here.

Multidimensional data are those in which the *dependent variable* varies as a function of two or more independent variables simultaneously. For example, weather (measured by a quantity such as temperature) or seismic activity, both of which vary in space and time.

Multivariate data are those in which there are *multiple dependent variables* varying as a function of a single independent variable. Or, if they are not time series or sequential data, then they simply represent a data set that includes two or more dependent variables. An additional, slightly more restrictive criterion is added to this term as used in probability and statistical applications (Chapter 2).

Real versus Complex Data. "Real" data are what we deal with in the real world, but treating them mathematically as complex numbers (described in more detail in later chapters) affords us the ability to conveniently consider rotation (or phase) of a quantity, as well as offering several additional mathematical conveniences. Therefore, real data will sometimes be organized as if they are complex quantities. Contrary to the name, complex quantities are often considerably easier to deal with than real ones.

Data are further classified according to statistical considerations (e.g., samples, realizations, etc.). These are presented in the next chapter (Probability Theory). Since the analyses presented in this text involve computer manipulation, all data are considered to be discrete.

1.3 Representing Discrete Data and Functions as Vectors

It is helpful to use the most convenient and standard form to represent data. This involves the concept of vectors and matrices. A more detailed summary of the matrix techniques utilized in this text is presented in Appendix 1. Here, only the concept of how one stores discrete data and mathematical functions in vectors is presented.

Typically, data are stored in tables (matrices). For example, if one measures the temperature at noontime on each of m days, at n different locations, the data are stored in a table as shown in Table 1.1:

Table 1.1

	Site 1	Site 2	Site 3	⋯	Site n
Day 1	20.1	23.2	24.8		23.6
Day 2	23.2	23.0	23.6		19.8
Day 3	24.8	23.6	24.2		20.5
⋮					
Day m	23.6	19.8	21.9		19.4

Alternatively, you can store temperatures in columns and different locations in rows (*preferred*) as shown in Table 1.2:

Table 1.2

	Day 1	Day 2	Day 3	⋯	Day m
Site 1	20.1	23.2	24.8		23.6
Site 2	23.2	23.0	23.6		19.8
Site 3	24.8	23.6	24.2		21.9
⋮					
Site n	23.6	19.8	20.5		19.4

Which of the two storage schemes you use is a matter of personal taste, but you must pay close attention to the storage form when performing the mathematical manipulations so that the appropriate values are being manipulated as required. For consistency between matrix and vector operations, the form of Table 1.2 is the form used throughout this text.

Initially, however, we will deal predominantly with single-column vectors of data. In the above examples, this is equivalent to having the temperature measured each noontime at one site only (e.g., the first column of Table 1.1, or the first row in Table 1.2), or the noontime temperature of one particular day at n different sites (e.g., the first column of Table 1.2).

Data storage in organized rows and columns is precisely the method used in a matrix. Indeed, each column of numbers represents a column vector. For simplicity, we will assume all vectors are column vectors and thus will drop the descriptor "column" (row vectors are indicated as the *transpose* of a column vector).

In addition to storing discrete observations or data values in table (matrix) form, the storage of mathematical functions, conveniently expressed as formulas when dealing with continuous data, must be presented at discrete values to represent them in vectors, hence the indexing of the constituent terms, φ_{kj} in equation (1.1), where the k represents *which* constituent term (function or vector), and j, the jth discrete value of the term.

1.4 Data Limits

1.4.1 Domain

Domain represents the spread or extent of the *independent variable* over which the quantity being measured varies. It is usually given as the maximum value of the independent variable minus the minimum value. Because no phenomenon is observed over all time or over all space, data have a limited domain (though the domain may be complete relative to the phenomenon of interest, e.g., the finite Earth's surface).

1.4.2 Range

Range represents the spread or extent over which the *dependent variable* (i.e., the quantity being measured, possibly as a function of time or space) can take on values. You will typically present range as the maximum value of the dependent variable minus the minimum value. No measuring technique can record (or transmit) values that are arbitrarily large or small. The lower limit on very small quantities is often set by the noise level of the measuring instrument.

Dynamic Range (DR) is the actual range over which dependent variables are measured or reproduced. Often this is less than the true range of the variable. You present dynamic range on a logarithmic scale in decibels (dB):[1]

[1] You can use some other form to express this if you are uncomfortable with decibels; just make it clear what your form is.

$$\mathrm{DR} = 10 * \log_{10}\left(\frac{\textit{largest power}}{\textit{smallest (nonzero) power}}\right). \tag{1.2}$$

or

$$\mathrm{DR} = 20 * \log_{10}\left(\frac{|\textit{largest value}|}{|\textit{smallest (nonzero) value}|}\right) \tag{1.3}$$

The first formula (1.2) is used if the data represent a measure of power (a squared quantity such as variance or square of the signal amplitude). Otherwise the second formula (1.3) is used. Use the smallest nonzero value for measurement devices that report zero values.

Since power is a quantity squared, the first formula (1.2) is related to the second (1.3) by

$$\mathrm{DR} = 10 * \log_{10}\left[\left(\frac{|\textit{largest value}|}{|\textit{smallest (nonzero) value}|}\right)^2\right]. \tag{1.4}$$

Therefore, the two formulas yield the same answer, given the appropriate input. This is especially useful for instruments that return measurements proportional to variance or power.

An increment of 10 in DR equates to a factor of 10 in R_p (the power ratio).

$$\mathrm{DR} = 10 * \log_{10}(\mathrm{R}_p), \tag{1.5}$$

so

$$\mathrm{R}_p = 10^{(\mathrm{DR}/10)}. \tag{1.6}$$

E.g., DR = 30 so R_p = 1,000; DR = 40, so R_p = 10,000.

Finally, because of the limited range and domain of data, any data set, say $y(x)$, are constrained by

$$X_S < x < X_L \tag{1.7}$$

and

$$|y(x)| < M, \tag{1.8}$$

where X_S, X_L and M are finite constants. *Such functions are manageable and can always be integrated.* This seemingly esoteric fact proves extremely useful in the practical analysis of data.

1.4.3 Frequency

Most methods of data measurement cannot respond instantly to sudden change. The resulting data are thus said to be **band limited**. That is, they will not contain frequency information higher than that representing the fastest response of the recording device, this is an invaluable constraint for some important analysis techniques, though it can be severely limiting regarding the study of certain high-frequency (rapidly varying) phenomena.

1.5 Data Errors

Data are never perfect. Errors can enter data through experimental design, measurement and collection techniques, assumptions concerning the nature and fidelity of the data, discretization and computational or analysis procedures. In analyzing and interpreting data it is important to attempt to estimate all of the potential errors (either quantitatively or qualitatively). That is, it is important to estimate the uncertainty contained in the measurements and their influence on your interpretation. Too often, errors related to one easily determined component are presented while others are completely ignored. You needn't be fanatic in your attempt to estimate the uncertainty – *scientific progress needn't be held hostage to unreasonable quantification* – rather, it is important to **make an honest assessment to the best of your ability to *estimate* the uncertainties associated with your measurements.** If you can't formally estimate an error, simply say so, then make your best educated guess or give a range for the error.

Box 1.1 Errors: Give Them Their Due

It is not uncommon to present, as the only error in the data, the scatter measured when making replicate measurements from a specific measurement sample (e.g., weighing a sample 100 times). While this is certainly a good estimate for the instrument error (sometimes called the "analytic error," "measurement precision" or "instrument precision"), it does not preclude the presence of a variety of other types of error that are likely present in the data. That is, how much scatter would occur if replicate samples were obtained (not just replicate measurements of the same sample)? Is there systematic bias in the instrument making the measurements? How representative is the sample of the process that you think you're sampling? (This is sometimes a dominant source of error that is completely overlooked.)

It is also important to provide an explanation as to how the estimate of the uncertainty was arrived at so the reader can make their own assessment of the techniques employed.

1.5.1 Instrument Error

Errors (uncertainties) in data are classified according to their source. Under the best circumstances, the quality of the data is predominantly controlled by the capabilities of the recording device. Measurement capabilities are classified according to:

1) **Precision** specifies how well a specific measurement of the same sample can be replicated. In statistical terms, the precision is a measure of the variance (or standard deviation) of the sample.

For example, if a substance was repeatedly weighed 100 times, giving a mean weight of 100 kg but with a scatter about this mean of 0.1 kg, then 0.1 kg would represent the precision of the measurement.

2) **Accuracy** specifies how well a specific measurement actually represents the true value of the measured quantity (often considered in terms of, say, a long-term instrument drift). In statistical terms, accuracy is often reported in terms of **bias**. For example, if a scale repeatedly returns a weight of ~100.0 kg for a substance, but its true weight is 105.3 kg – the mismatch between the measured value and true value reflects the bias of the measurement. So the scale is good to an accuracy of just over 5 kg, or the scale has a bias of ~5 kg.
3) **Resolution** specifies the size of a discrete measurement interval of the recording instrument used in the discretization process. In other words, it indicates how well the instrument (or digitized data) can *resolve* changes in the quantity being measured. For example, if a thermometer only registers changes in temperature of 0.01° C, then it cannot distinguish changes in temperature smaller than this resolution. One would achieve the same resolution if, in the process of digitizing higher-resolution data, all values were rounded off to the nearest 0.01° C.
4) **Response time** specifies how quickly an instrument can respond to a change in the quantity being measured. This will limit the bandwidth (range of frequencies) of a measured time series (discussed in more detail latter).

In general, accuracy *reflects the degree of systematic errors, while* precision *reflects the degree of random errors*. Statistics are well designed to treat the latter (precision), whereas they are not generally designed to address the former (accuracy). Accuracy must be estimated, using whatever means are practical and reasonable, by the person who understands the instrument.

1.5.2 Experimental/Observational Error

The experimental design, sampling program or observational methods may also lead to errors in precision and accuracy.

Precision. Consider estimating the precision of a specific brand of thermometer. If 100 of the thermometers were simultaneously used to measure the temperature of a well-mixed bath of water, the scatter about the mean temperature might typically be presented as the precision of the thermometers. However, this is really the precision of the estimated temperature and it reflects the precision of the experimental design. Any one particular thermometer may have significantly better or worse precision than that suggested by the scatter achieved between 100 different thermometers. Also, the "well-mixed" bath may actually contain temperature gradients to some extent, which will also influence the scatter observed in the measurements. Repeating the above calibration, only this time making 100 replicate measurements using a single thermometer, may include some scatter due to subtle changes in the water bath between replicate measurements. Thus, even that measured precision reflects some combination of instrument precision and experimental scatter.

In this respect, precision errors may well be attributable to a combination of both instrument and experimental error. This combination is responsible for the observed random scatter in replicate measurements, which is often referred to as **noise** in data. Noise can also represent any portion of the data that does not conform with preconceived ideas concerning the nature of the data – recall the expression that "one person's noise is another person's signal."

One of the goals in data analysis is to detect signal in noise or reduce the degree of noise contamination. Noise is sometimes classified according to its contribution relative to some more-stable (or non-fluctuating) component of the observations, referred to as the **signal**.

Signal-to-Noise Ratio (SN) is the common measure for comparing the ratio of signal to noise in a data set. As with dynamic range, you give this ratio on a logarithmic scale in decibels (dB):

$$\mathrm{SN} = 10 * \log_{10}\left(\frac{power\ of\ signal}{power\ of\ noise}\right) \tag{1.9}$$

or

$$\mathrm{SN} = 20 * \log_{10}\left(\frac{|amplitude\ of\ signal|}{|amplitude\ of\ noise|}\right). \tag{1.10}$$

Exactly how one determines the values to insert in the above formulas depends on the data and how they were collected. In some cases, it is appropriate to use the mean value of the data (or measured range, for time-series data) as the amplitude of signal and the (known) instrument error (or precision) as the amplitude of the noise. With time series, the noise may be alternatively estimated by computing the measured scatter in a series of replicate time-series measurements of the same quantity (under the same sampling conditions). The amplitude of signal might then be regarded as the observed range in the average of the replicate time series.

The signal-to-noise ratio is also given as the ratio of the variance (a power) of the signal to the variance of the noise, or it can be written in any other manner that essentially provides some ratio between the variance of the signal and that of the noise.

If one uses the mean of the data as the signal and the standard deviation as the noise, then it is often convenient to present this form of SN (actually, SN^{-1}) as a **coefficient of variation**:[2]

$$V = \frac{standard\ deviation\ of\ data}{mean\ of\ data}. \tag{1.11}$$

V can be presented as percentage or in dB, but its interpretation is conveniently intuitive. That is, in the vicinity of $V = 1$, it is a suggestion that the scatter (noise) in the data is comparable in size to the signal itself. At 200 percent (~3 dB), the scatter is twice the size of the signal. At 25 percent (~ −6 dB), the noise is only one-fourth of the signal amplitude.

[2] It is common to present many statistical quantities ("moments") as coefficients of the moment. For moment μ^k, a coefficient of the moment is given as μ^k/μ^{k-1} (as is the case for V here). This will make more sense after the discussion of moments in Chapter 2, "Probability Theory."

Accuracy. While measurements of a specific sample may be extremely accurate relative to the sample's true value, the sample itself may not be representative of the phenomenon you think you are examining. This reflects a difference between instrument bias and observational (or experimental) bias.

Alternatively, in a previous example where 100 measurements of a substance suggested that it weighs ~100.0 kg, while the true weight was actually 105.3 kg, some of the discrepancy may have been due to a miscalibration of the scale, while the remainder may simply reflect a systematic error in reading the scale (e.g., a parallax problem).

Therefore, as with precision errors, errors in accuracy may reflect a combination of both a bias in the recording instrument as well as in the experiment, sampling program, observational methods, etc.

You must be aware of the various sources of error so that the influence of each can be considered at some level. For example, your lab personnel may have determined that the scale has a bias of 5 kg, but this will not take into account the additional bias introduced into the final measurements due to a systematic error when reading the scale.

More precise definitions of different types of noise will be given for time series in later chapters.

1.5.3 Digital Representation and Computational Errors

One can introduce additional "errors" into the data during discretization or digitization. These errors are minimized by taking care to digitally represent the measurements with the proper number of significant digits.

The number of **significant digits** (or **significant figures**) used in representing the data should reflect their precision – or, if the values are to be subject to arithmetic operations, one digit beyond the precision, to preserve precision during numerical operations.

The number of significant digits is defined according to whether or not the numbers contain a decimal point:

1) If there is no decimal point, the number of significant digits is that which occurs between the leftmost and rightmost nonzero digits. For example, the number 3493800 has 5 significant digits, as do the numbers 34938 and 30008.
2) If there is a decimal point, then the number of significant digits is that which occurs between the leftmost nonzero digit and the rightmost digit (whether it is zero or not). For example, all of the following numbers have three significant digits: 300., 3.10, 0.000570.

Alternatively, errors are introduced during arithmetic operations due to the manner in which numbers are stored on a digital computer (roundoff error), or due to the fact that computers can only deal with discrete values of continuous functions (truncation error).

When reporting results, numbers should be given to no more significant figures than one beyond that representing the precision of the data.

Roundoff error arises when the amount of storage available for storing a number is not large enough to accommodate the number, so the computer simply rounds it off to a size that it can accommodate. For example, at the time of this writing (we just moved into computers from the abacus), a typical (IEEE standard) storage scheme for a floating-point number (i.e., a number that can contain a fractional component) uses 32 bits, of which 1 bit is used to specify the sign of the number, 8 bits to define the exponent, and 23 bits to define the mantissa. Thus the mantissa can contain numbers as large as $2^{23} - 1 = 8388607$, or it can contain approximately 7 significant digits. *Therefore, if during the course of numerical operations the stored sum or product must exceed this number of significant digits, the computer will essentially ignore any such excess.*

For example, if you try to add the number 10.033 to the number 1234567, the computer cannot accommodate the fractional part (the 0.033), since that would require storing 10 significant digits. Consequently, the 0.033 would be ignored and the calculation would continue as if that fraction had never existed. When adding large quantities of numbers, the consequences of ignoring this fraction can become large, even huge, resulting in nonsensical results. Consequently, one must exercise care to minimize such problems (there is a variety of possible solutions, such as increasing the storage capacity using double precision, or altering the manner in which the calculations are made, such as by subtracting-off a large constant before summing numbers that differ by a small amount).

Truncation error arises when the discrete approximation to a continuous function is computed using a limited number of terms in an infinite expansion or using a limited number of points to represent the function. This error can be minimized through the use of appropriate algorithms.

Additional classifications of errors will be considered in the next chapter, as will their statistical assessment and their propagation during arithmetic operations.

Fortunately, most of these latter problems are being reduced continuously as computers use more memory and higher-bit storage schemes, and become fast enough to easily and quickly make routine calculations of complicated functions (bypassing the need for low-order truncation schemes). So, essentially I have just told you one of those "father" stories: "When I was a kid, I used to walk 5 miles to school in blizzards, with hot potatoes in my pockets to keep my hands warm, and the walk was uphill in both directions; you kids are lucky because they invented the automobile and jackets." But, even with large computers, if you have a huge quantity of numbers you can still encounter roundoff error.

1.6 Practical Issues

Analysis means to separate something into components in order to identify, interpret and/or study the underlying structure. In order to do this properly, you should have some idea of what the components are likely to be. Therefore, you must have some sort of model of the data in mind (whether this is a conceptual, physical, intuitive or other type of model is not important); you essentially need some sort of guideline to aid in your analysis beyond the simple statistical classifications. For example, it is not good to take a data set and compute its Fourier series simply because you know about Fourier analysis (or at least soon you will). You need to have some idea as to what to look for in the data. If you can't figure out a technical basis for why you should use one method over another, experiment with several and state how your results are sensitive to the actual form used – this in itself can yield valuable insights regarding the data and/or analysis strategy.

Often, your analysis strategy will evolve as the analysis proceeds and new insights are gained along the way.

In general, we decompose data along a few well-worn, or at least well-understood, paths: (1) decomposition into standard statistical moments and (2) decomposition into a set of functions, where the functions have intuitive, physical, and/or "natural" appeal. This collapses data analysis into a couple of rather general categories that might be described as "how you define moments," "how you fit functions," and then "how you evaluate the results."

Analysis: Take Little for Granted

Consider your model to be no more than a guide – it is important to keep an *open mind* about the data and analysis results, since the data represent the real world, not necessarily the model you are imposing on them. Always consider other interpretations; the potential for bias introduced by, or related to, the analysis techniques employed; errors related to the sampling strategy or collection techniques; and preconceived notions or underlying assumptions influencing any interpretation. You should take very little for granted (this takes strong discipline).

It is equally important not to fall into the common trap well expressed by the quote: "Ignorance of one's data is not to be confused with objectivity."[3] It is not uncommon to see suspect data points left in data sets under analysis simply because the analyst thought it would be "non-objective" to remove such values. Complete understanding of the nature of one's data and the techniques used to collect them can dictate the removal or "correction" of suspect points, which may dramatically improve the results. In fact, such an understanding often leads to the identification of bad data values. Similarly, all techniques being used (methodological and analytical) must be thoroughly understood. Such misunderstandings, sadly, account for a significant number of misinterpretations present in the literature. In the end, you should be your own strongest critic when evaluating your analysis (better to find your own mistakes than to have someone else – especially in a public forum – point them out).

Along those lines, intuition may prove to be a valuable guide, but with statistical matters it requires the strongest self-critical assessment. The most famous case (or perhaps most infamous case) of failed intuition is recounted in a book about the mathematical genius Paul Erdös.[4] This case involves a self-proclaimed genius, Marilyn vos Savant, who wrote a magazine column titled "Ask Marilyn." Vos Savant addressed the following problem in 1990 (based on the popular TV game show "Let's Make a Deal"). There are three doors: behind one is a car, and behind the other two are goats. You choose one door (say, door 1), and then the host, knowing what is behind each door, shows you what is behind one of the other doors (say, door 2), revealing a goat. The host then gives you the choice of sticking with door 1 or switching to the other door (door 3). The question was whether you should switch or not. Vos Savant claimed that if

[3] I heard this great quote from Dr. Warren Prell of Brown University in the early 1980s in response to a particular study he had just evaluated.

[4] Hoffman, P., 1998. *The Man Who Loved Only Numbers*, Hyperion, New York (see Hoffman's chapter 6, "Getting the Goat").

you switch you have a 2-out-of-3 chance of winning, but if you stay with door 1 you only have a 1-in-3 chance. Her response was attacked by a deluge of letters from irate mathematicians who said she was contributing to the poor understanding of math in the country and that she should correct her answer, since the odds were clearly 50:50, whether you switched or not. Turns out, she was correct (you are considerably better off to switch than to stay with your original choice). This example fooled even the best mathematicians at the time (including Erdös).

One other aspect of analysis is critically important: that is, not only do you have to know which tools to apply for any particular desired analysis, and how to apply those tools, but you must also be acutely aware of the specific result desired, and must make sure that the tools being applied and the result obtained are consistent with the particular question being asked. For example, consider the question: "Is a person more likely to be involved in a car accident when driving on a trip more than 25 miles from their house than they are when driving within a close proximity of their house (within 25 miles)?" Some years ago, during a national campaign to get people to wear seatbelts, there was constant quoting of a statistical result that "most accidents occur within 25 miles of home."

At first blush, this statistic may sound like it answers our original question, and that one is actually safer driving on a trip away from home. However, it may be that 99.9 percent of the time the people being surveyed are driving within 25 miles of their house, and only 1 in 1000 times are they on a "trip." If it turns out that in 999 near-home excursions, they average 5 accidents, yet in each of their (rare) trips they always have an accident, the result that most of the accidents occur near home is still true, while the answer to our specific question is that you are absolutely *not* safer driving on a trip, since it appears that you have nearly a 100 percent chance of having an accident, versus only a 0.5 percent chance of having an accident during any one particular outing near your home.

How about this Nobel-prize-winning science? I can prove that gravity is not constant. Take a heavy bag of books and hold them at arm's length with one arm. The longer you hold them, the heavier they get, so obviously gravity is getting stronger. Maybe the study should consider other possible factors, like human physiology.

The main point of these examples is to emphasize how important it is to make sure that your question is well thought out and well articulated and that the analysis tools are suited to addressing the question. The precise manner in which you phrase your particular questions of interest can have a profound affect on the result you obtain. It is all too common to hear the results of an analysis answering a question that was in fact different from the one originally posed. The most common trap in this case is where people find one result after careful analysis, but then extrapolate the answer to a broader question that the analysis did not directly address. *Articulating your questions clearly and succinctly is usually the all-important starting point for any analysis (and research project). It will dictate that the best analysis tools keep the analysis focused and will help immeasurably in explaining the research, analysis and results. Likewise, when deep into your analysis it is always good to re-ask yourself, "What am I trying to accomplish here?"*

One final point: it is common that in answering the original question, your analyses reveal more questions. Don't despair; this is how research is advanced.

2 Probability Theory

2.1 Overview

Most data are subject to uncertainty. This uncertainty leads to noise, or "scatter," in the data. Statistics deals with techniques for collecting and classifying "noisy" data (descriptive statistics) as well as extracting information and drawing valid conclusions from such data (inferential statistics). Providing the stats of a baseball player or results of a census are typical examples of descriptive statistics, while projecting a baseball player's future or drawing conclusions about society based on the census fall under inferential statistics. From my perspective, descriptive statistics usually provides an organized basis from which the information of interest is then extracted and conclusions drawn.

Probability theory provides the mathematical foundation and core upon which statistical techniques are based.

Probability and statistics provide a considerable and powerful set of tools as well as a general nomenclature and philosophy regarding the analysis of data. Typically, this analysis follows a standard sequence:

1) You desire to characterize, understand, display, draw conclusions from or make predictions about some phenomenon of interest (i.e., there is a particular problem that motivates your interest and study).
2) You have collected a set of observations (a *sample*) from an experiment or from nature, representative of this phenomenon. This may be in addition to your study of any associated theory regarding the phenomenon, which may help you formulate the observational program and guide your data analysis.
3) You analyze the sample to provide an estimate of the desired characteristics of the phenomenon and draw conclusions about it or test an existing theory. Hopefully, the analysis also provides a quantifiable level of confidence regarding the validity of the estimated characteristics and conclusions.

Probability theory specifies the manner in which phenomena should be described and establishes the rules regarding the manipulation and quantification of the descriptions. Statistics then uses these rules to define the techniques with which the characteristics of the phenomenon are estimated, how the conclusions are drawn from the sample and how the associated uncertainties are assessed. Whereas all following chapters typically focus on a specific analysis technique, the material of this chapter establishes the methods used in formulating these later techniques and assessing the uncertainties associated with them. It will also be useful in its own right for designing specific analysis techniques, modifying existing ones and customizing observational programs.

2.2 Definitions

A **deterministic variable** is a variable whose values *can* be stated with certainty. It may also be described as the unique response to a known cause.

A **random variable (rv)** is any variable whose exact value *cannot* be stated with certainty; at best, it can be predicted to lie within some range of possible values. Key to the understanding of probability theory and statistics is the concept of a random variable. A random variable is usually denoted by an uppercase letter, such as X.

Data represent measurements of random variables, or of some combination of random variables – at some level, uncertainty unavoidably enters the measurements, so they can never be known with absolute certainty.[1] It is this uncertainty that requires statistical methods to help quantify the level of uncertainty and eventually determine how it influences the conclusions you draw from the data.

An **event** is a possible outcome: a specific value of a random variable, or some combination of outcomes. For example, the raw measurements may be the face value on two tossed dice, while the "events" of interest are simply whether the sum of those values is an odd or even number. So, even though there are 36 individual outcomes possible for this experiment, for the problem of interest, there are only two events. Also, the raw, random variables in this case are the two individual face-values observed for each toss, but the function of random variables is itself a random variable, so you could chose to define the random variable that you will be analyzing as having values of 0 = even, and 1 = odd. In this case, your two events correspond to your two possible outcomes of the random variable. For convenience, you will typically refer to events as particular values of your data (your random variable), though by this it is implicit that the events may in fact be combinations of values of some other random variable or variables. Events are the things of interest; ultimately, in statistics, you are interested in the likelihood of obtaining any particular event.

A **population** or **parent population** is the set of *all* possible events. The population thus represents a well-defined (*all-inclusive*) set of possible outcomes, or values, of the random variable of interest. In statistics you attempt to make inferences about the characteristics and properties of the population from your data. The more complex the phenomenon of interest being studied, the more complex it may be to define the population or events of interest, in which case more care is required in precisely articulating the specific problem being addressed and doing so in a manner that can be classified, which may require some clever thinking in some cases.

[1] For practical purposes, if the level of uncertainty is exceedingly small relative to the precision required for extracting the desired signal, you may just as well treat the random data as deterministic.

A **realization** is the particular event realized in any one experiment or measurement. That is, it is one value drawn from the population: one value of the random variable determined during an experiment or measurement. Realizations are typically denoted by a subscripted lowercase letter – the same letter used in uppercase form to denote the random variable. So, x_i represents the *i*th realization of the rv *X*.

A **sample** is a collection of realizations. It thus represents a subset of a population, obtained by experiment or observation. Your data typically represent a sample (or multiple samples, if from different experiments). It is from analysis of such samples that inferences are made regarding the population. If the sample represents a truly random collection from the population, then the properties of the population can be estimated from the sample.

A **biased sample** is one that does not represent a truly random subset of the population. Bias may be introduced by sampling restrictions (e.g., you can only sample from a restricted region) or when the sample is not actually representative of the target population (the population you are interested in studying), but instead is a sample of a different population.

A **distribution**, in a generic sense, provides an indication of the relative, or absolute, frequency of occurrence of the various events of a random variable. Essentially, it provides the relevant description of the random variable. The distribution can be for the true population or for the sample(s); the latter is an estimate of the population distribution.

Distributions can be presented in any number of very specific forms, each of them serving a specific purpose and having a specific name,[2] as discussed later. Here, the word "distribution" is used in a generic sense when the specific form of the distribution is unimportant – when it *is* important, however, the specific type of distribution will be clearly indicated.

Independent variables are those that have no relationship between them. This concept often proves confusing, since independence manifests in several forms (succinctly described following the definitions of Rodgers, Nicewander and Toothaker, 1984):

1) *linearly independent*, defined for variables ... $w, x, y, \ldots z$:

$$\mathrm{a}_1 w + \mathrm{a}_2 x + \mathrm{a}_3 y + \ldots + \mathrm{a}_n z \neq 0.$$

for any nonzero values of a_i; for nonlinear relationships, independence also requires the variables to be uncorrelated (as defined under 3 in this list).

2) *orthogonal*, defined for variables y and x (subscript indicates specific value of the variable):

$$x_1 y_1 + x_2 y_2 + x_3 y_3 + \ldots + x_n y_n = 0$$

or $\mathbf{XY} = 0$.

[2] Unfortunately, while there is a common set of specific names describing particular forms of a distribution, there is no universally agreed-upon usage of them. Consequently, the names are ambiguous unless accompanied by the particular definition for which they are being applied. I will follow one of the more common nomenclatures.

3) *uncorrelated*, defined for y and x:

$$(x-\bar{x})_1(y-\bar{y})_1 + (x-\bar{x})_2(y-\bar{y})_2 + (x-\bar{x})_3(y-\bar{y})_3 + \ldots + (x-\bar{x})_n(y-\bar{y})_n = 0$$

or Cov[**X**,**Y**] = 0 (defined in §2.6.3).

"Orthogonal" and "uncorrelated" are nearly equivalent, except that the latter is centered (i.e., the means have been removed before the comparison) – one does not imply the other.

2.3 Probability

Most of modern probability theory is built upon axioms (independent statements accepted without proof) originally described by Kolmogorov (1956). These axioms make strong use of the *observation* that in many random systems the relative frequency with which specific events occur is statistically stable, given a large number of trials. In other words, given enough experiments, the values taken on by the random variable seem to show a consistent pattern, as the possible values usually appear with the same relative frequency over and over again. This regularity becomes more apparent, given a large enough number of observations of the phenomenon or repetitions of the experiment.

For example, if you plot the fraction of times you get a heads when you toss a coin, this fraction will hover near 50 percent after the coin has been flipped a large number of times. As seen in Figure 2.1, the fraction of times a heads appears, relative to the number of tosses, is near 50 percent. Furthermore, as the number of tosses increases, the excursions from 50 percent will get smaller and smaller, until ultimately they will appear to converge exactly to 50 percent as the number of tosses approaches infinity. Fortunately, the convergence will become apparent well before you reach an infinite number of tosses.

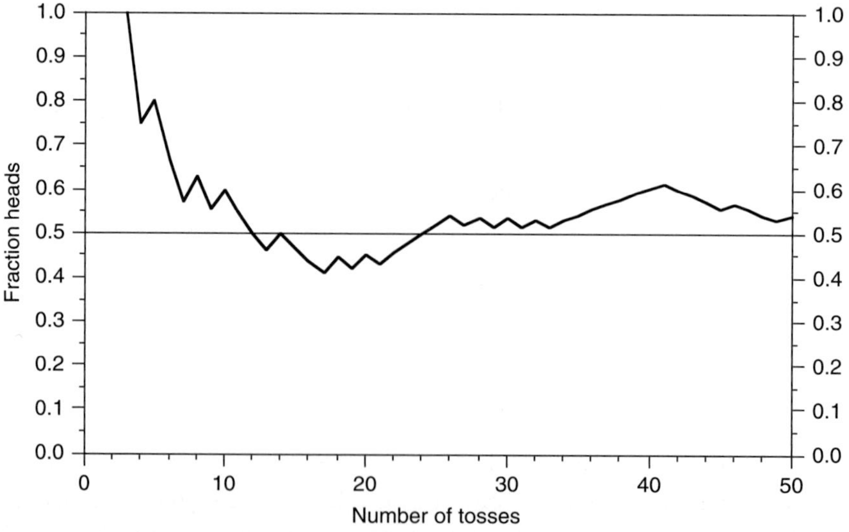

Figure 2.1 Relative frequency of heads received as a function of number of tosses. The fraction is converging toward a stable 50 percent – the probability of getting a heads from flipping a coin.

Similarly, if a single die is thrown many times, a pattern becomes apparent for each of the six possible outcomes. If the die is honest, each value will tend to appear, on average, one-sixth of the time. If the die is not honest, a different, but equally stable pattern will emerge, showing the relative frequency with which each possible outcome is likely to occur on average.

More generally, if you could repeat an experiment numerous times, the probability of getting a particular value, $X = x_i$, is approximated by taking the ratio of the number of times that x_i occurred, $n(x_i)$, to the number of times the experiment was run, N, or, $n(x_i)/N$. The probability, $P\{X = x_i\}$, is the value of this ratio in the limit as the number of experiments approaches infinity or the maximum number of experiments. This observation of a stable frequency-of-occurrence leads directly to the concept of *a unique* **probability**, or *likelihood of occurrence*, for a particular event, x_i, denoted $P\{X = x_i\}$. In the above example for a coin toss, $P\{X = \text{heads}\} = 0.5$.

So, *probability is an ideal, hypothetical constant, reflecting the frequency of occurrence with which you would expect to see a particular event on average, or given a huge number of trials.*

From this, it is clear that the probability of a particular event must be between 0 (the event never occurred) and 1 (the event was the sole outcome in every experiment). This is expressed, for any value, x_i, of the random variable, X, as

$$0 \leq P\{X = x_i\} \leq 1. \tag{2.1}$$

That is, the probability can never be negative, nor can it be greater than 1.

The collection, or *set*, of probabilities for a random variable, regardless of the specific form in which it is presented, represents a **probability distribution** of the random variable. The specific forms of distributions are presented as we continue.

2.4 Univariate Distributions

Univariate data consist of a single random variable (e.g., temperature, weight, etc.), as opposed to **multivariate** data, which consist of multiple random variables.

2.4.1 Discrete Probability Distributions

Probability Mass Function (PMF)

A **probability mass function (PMF)** is a function, $p_X(x_i)$, that gives the unique probability (likelihood of occurrence) for each and every possible value, x_i, of a *discrete* random variable, X. This is shown as:

$$p(x_i) = P\{X = x_i\}. \tag{2.2}$$

This function essentially contains the most fundamental information about a (non-sequential or non-ordered) random variable – as such, you should try to determine it for your data. It provides the distribution of probabilities of the different values of X and

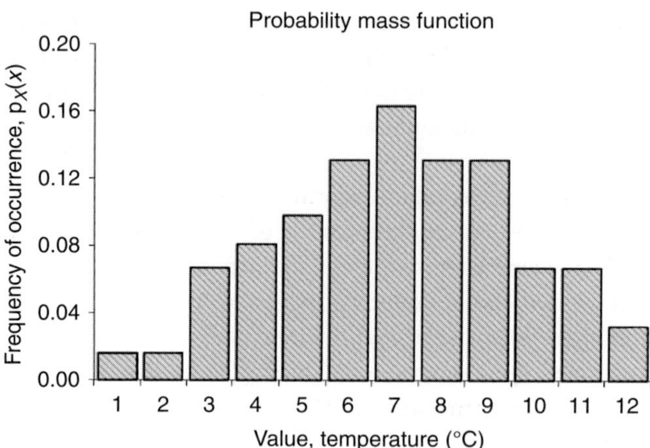

Figure 2.2 Example of a probability mass function. Abscissa shows temperature categories of 1° (that is, for this discrete variable, temperature has been measured at increments (resolution) of 1°.

is also called a **probability function, frequency function**, or simply **distribution**, though this last term will be used here in a more general sense as stated above.

A PMF, or discrete distribution, is conveniently presented in functional form for theoretical or ideal populations and in graphical form (as a histogram) for real populations, as shown in Figure 2.2.

In this graphical form, the attributes of the random variable are clearly visible. For example, from the graph it is clear that the value $x_i = 7$ is that value most likely to occur; on average ~16 out of 100 observations of this phenomenon will have a value of $x_i = 7$, thus it has a probability of ~16 percent. It is also clear that the overall range of X is fairly limited (only 12 possible outcomes), and the probability of observing a value near one of the extremes is an order of magnitude less likely than it is for observing a value located near the center of the distribution (e.g., 1.6 percent probability of observing a value of 1 or 2, versus a 16 percent chance of observing a value of 7).

The explicit values of this random variable's PMF are given as

$$\begin{array}{rcccl} \mathrm{p}_x(1) & = & \mathrm{P}\{X=1\} & = & 0.016 \\ \mathrm{p}_x(2) & = & \mathrm{P}\{X=2\} & = & 0.016 \\ \mathrm{p}_x(3) & = & \mathrm{P}\{X=3\} & = & 0.067 \\ \mathrm{p}_x(4) & = & \mathrm{P}\{X=4\} & = & 0.081 \\ \mathrm{p}_x(5) & = & \mathrm{P}\{X=5\} & = & 0.098 \\ \mathrm{p}_x(6) & = & \mathrm{P}\{X=6\} & = & 0.131 \\ \mathrm{p}_x(7) & = & \mathrm{P}\{X=7\} & = & 0.163 \\ \mathrm{p}_x(8) & = & \mathrm{P}\{X=8\} & = & 0.131 \\ \mathrm{p}_x(9) & = & \mathrm{P}\{X=9\} & = & 0.131 \\ \mathrm{p}_x(10) & = & \mathrm{P}\{X=10\} & = & 0.067 \\ \mathrm{p}_x(11) & = & \mathrm{P}\{X=11\} & = & 0.067 \\ \mathrm{p}_x(12) & = & \mathrm{P}\{X=12\} & = & 0.032. \end{array} \tag{2.3}$$

The probability of $X = 6$ *or* $X = 7$ is given by adding the probabilities of the two individual events, $P\{(X = 6) \text{ or } (X = 7)\} = P\{X = 6\} + P\{X = 7\} = f(6) + f(7) = 0.131 + 0.163 = 0.294$, or ~30 percent. This *additive rule* follows from the fact that the two events are *disjoint* – you cannot possibly get both values at the same time; therefore, the net probability is the sum of the individual probabilities.[3] More generally, this rule can be expanded to include the probability of getting any of a number of events, where the probability is the sum of the probabilities of the individual events. For example, the probability of $X = 2$ or $X = 3$ or $X = 5$ is $p_X(2) + p_X(3) + p_X(5)$. The events do not have to occur in sequential order for this rule to apply.

Cumulative Distribution Function (CDF)

The probability of obtaining a value less than or equal to x_n can be expressed as the **discrete cumulative distribution function (CDF)**, $F_X(x)$, where

$$F_X(x_n) = P\{X \leq x_n\} = \sum_{i=1}^{n} p_x(x_i). \tag{2.4}$$

This function is simply the accumulated sum of probabilities for each possible value of the random variable. Graphically, the two functions are related as shown in Figure 2.3.

Despite the plot of $F_X(x)$ below the cumulative distribution for a *discrete* random variable is not defined for values lying between the distinct possible values of the variable. That is, in the graph below, $F_X(x)$ is only defined at the location of the squares, representing $F_X(x)$ at the possible values of X (e.g., $X = 1, 2, \ldots, 12$), and how the various dots are connected is simply a matter of graphical convenience. Some authors choose to connect them via a staircase structure; others pass horizontal lines through the values with no vertical connection; etc. Here, as is often

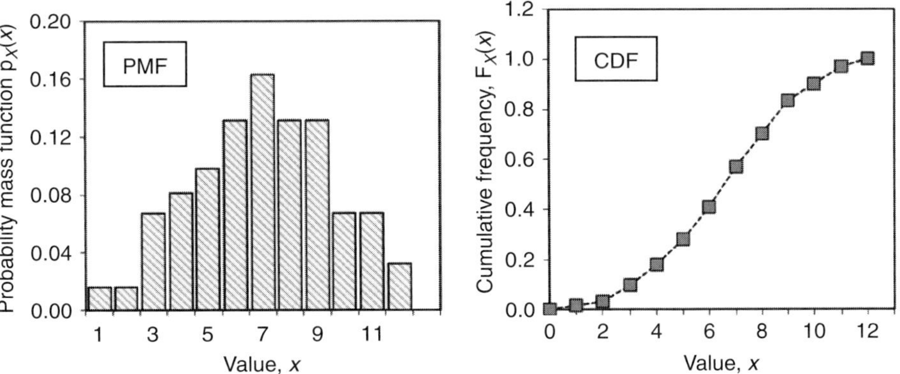

Figure 2.3 Example of previous PMF, and same distribution as a cumulative distribution function (CDF).

[3] This additive rule is actually one of the axioms of probability theory.

done, it is drawn to approximate a continuous curve representative of a continuous random variable.[4]

Taking this concept of cumulative distribution to its limit, it is clear that the probability of getting either $x_i = 1$, or $x_i = 2$, or . . ., $x_i = 12$ – that is, that you get one possible event – is $p_X(1) + p_X(2) + \ldots p_X(12) = 1$, or $P\{X \leq 12\} = 1$; there is 100 percent certainty that you will obtain at least one of the possible values.[5] For example, rolling a single die, you are certain of getting either a 1, 2, 3, 4, 5 or 6.

This rule is written as

$$\sum_{i=1}^{N} p_x(x_i) = 1, \tag{2.5}$$

where N = total number of possible values (including infinite).

2.4.2 Continuous Probability Distributions

Probability Density Function (PDF)

Continuous random variables require different treatment than discrete random variables, especially regarding the nature and interpretation of their probability distributions. In the discrete case, the PMF gives the probability that each individual discrete event will occur in a given (unbiased) sample, or is present in the population. If each event has an equal probability of occurring, the probability for any one event is simply 1/N, where N is the total number of events. Again, a die has six numbers and each one has an equal chance of occurring, thus the probability of any one number occurring is 1/6.

For a continuous random variable such as weight, there are an infinite number of possible events, or particular weights. Even if the weights for the problem of interest are restricted to a small range, there are still an infinite number that can occur between the limits. Thus, the probability of measuring any one *exact* weight is $1/\infty = 0$. That is, *the probability of obtaining any precise value of a continuous random variable is not meaningful.*

In the continuous case, the number of occurrences is not for an individual event, but rather for a range of values, so it is a *relative measure*: number of occurrences/range of values. Therefore, the relative probability of an occurrence is

$$f_X(x) = \frac{P\{x_a < X < x_b\}}{\Delta x}, \tag{2.6}$$

[4] This continuous representation is often done in an attempt to approximate the true continuous nature of a discrete data set. That is, discrete data are sometimes discrete because of observational restrictions associated with the observing instrument, in which case the true phenomenon of interest is continuous, hence the tendency to treat the distribution as if it were continuous.

[5] This is another of the axioms of probability theory. The third axiom is that the probability of getting a particular event is greater than or equal to zero – it is impossible, or it makes no sense, to have a negative probability. This latter axiom is encompassed by the left-hand-side inequality in (2.1). The right-hand-side inequality follows as a theorem from the three axioms.

where the probability is computed as before (number of occurrences for the range divided by the total number of occurrences of all the ranges). In this form, $f_X(x)$ is defined as the **probability density function (PDF)**, or simply **density function**, for a continuous random variable, X.

The PMF and PDF are related in the same way that mass and density are: the former is an absolute entity, the latter is a relative entity (mass/unit volume), converted to the former by multiplying by unit volume. Here the PDF is converted to an absolute probability (which is what the PMF gives directly) *for a range of* x, between x_a and x_b, by multiplying it by that range:

$$f_X(x)\Delta x = P\{x_a < X < x_b\}. \tag{2.7}$$

In other words, $p_X(x_i)$ *and* $f_X(x_i)\Delta x_i$ *are comparable quantities, both providing a probability.*

A good approximation to the true PDF is constructed by making the range, Δx, smaller and smaller, until it is infinitesimal and the steps blend to become a continuous curve (as in Figure 2.4). There, $\Delta x \to dx$, and the probability of drawing a value between any range of x = a to b is still the area under the curve, but now this area is given as

$$\int_a^b f_X(x)dx = P\{x_a < X < x_b\}. \tag{2.8}$$

Alternatively, you could measure the probability of obtaining some weight over a given range, however, such as the range spanned by the resolution of the weighing instrument. For example, Figure 2.5 shows the number of items in a study that weigh from 0–5 g, 5–10 g, 10–15 g, . . ., 75–80 g, or over range intervals, $\Delta x = 5$ g.

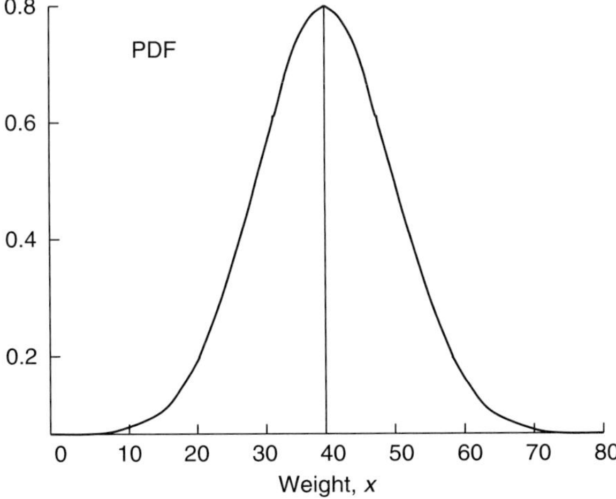

Figure 2.4 Example of a continuous probability density function (PDF). In this probability of any single value, $x_i = 0$, so you must integrate over a range to get a probability of obtaining a weight within the range.

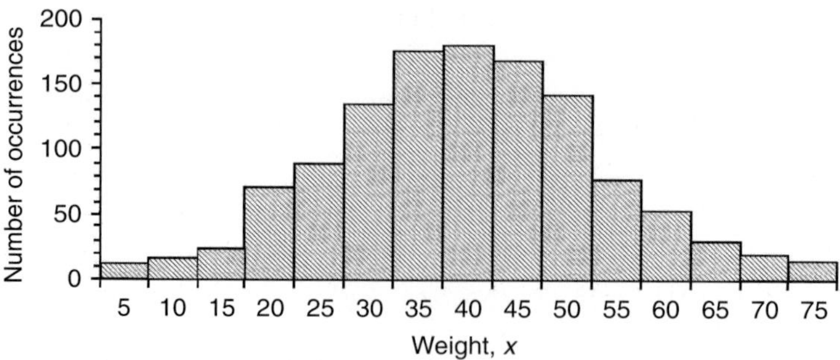

Figure 2.5 Example of a continuous variable categorized into discrete ranges, allowing its distribution to be graphed comparable to a PMF.

This graph looks similar to a PMF for discrete variables, the difference being that discrete events in this "PMF" are now *ranges* of possible values, and instead of showing the fraction of occurrences on the ordinate, the actual number of occurrences for each weight range is given. In the discrete case, the number of occurrences is converted to a frequency, or probability, by dividing by the total number of occurrences.

The PDF possesses the following properties:

$$\int_{a}^{b} f_X(x)dx > 0 \tag{2.9a}$$

$$\int_{-\infty}^{+\infty} f_X(x)dx = 1 \tag{2.9b}$$

$$\int_{a}^{a} f_X(x)dx = P\{X = x_a\} = 0. \tag{2.9c}$$

That is, (2.9a) states that the probability X will take on a value within a range over which it can occur is revealed by the fact that it has a nonzero probability for values in that range; (2.9b) states that the probability that X will take on one of its possible values is 1 (there is no doubt that X will yield some value, so the area under the entire curve, f_X, is equal to one); and (2.9c) states that the probability that X will yield the *exact* value of one unique number is zero (as previously discussed).

Cumulative Distribution Function (CDF)

The cumulative distribution function (CDF), $F_X(x)$, is defined as

$$P\{X \le x\} = F_X(x) \tag{2.10}$$

and it is given, comparable to the discrete case, as

$$F_X(x) = \int_{-\infty}^{x} f_X(x)dx = P\{X \le x\}. \tag{2.11}$$

From this (using a special case of Leibniz's rule),

$$\frac{\mathrm{dF}_X(x)}{\mathrm{d}x} = \mathrm{f}_X(x). \tag{2.12}$$

Therefore, the CDF is simply the area under the density function curve (i.e., the PDF), $\mathrm{f}_X(x)$, from $-\infty$ to x. The real advantage of this CDF function is that it always exists – even in the case where the PDF does not exist. That is, if there are discontinuities in a continuous PDF, the cumulative distribution still exists and is well behaved. The implication here is that *the CDF can be treated identically, whether the random variable it describes is continuous or discrete*. This is particularly advantageous in theoretical developments, since it allows a single general mathematical treatment,[6] and is the reason why many texts deal almost exclusively with it instead of what may seem like a more intuitive PDF or PMF (and why some use the CDF as the generic "distribution").

The cumulative distribution may be quite intuitive and natural for many situations, such as when the key interest is the probability that something not exceed a threshold. This is particularly the case in quality control and safety issues (e.g., what is the probability that an aircraft wing will be subjected to vertical accelerations greater than or equal to some critical limit).[7]

As previously noted, many publications use the names of the various distributions and functions described above interchangeably or non-uniquely. The key point is that all of the above curves express, in one way or another, the probability of obtaining a specific value or range of values for any single observation of the random variable.

Box 2.1 Important Distinction between the PMF and PDF

The foregoing material shows the two most common forms for displaying the information describing the distribution of discrete and continuous non-sequential random variables (via regular and cumulative distributions). However, it is important to remember that while graphs of the discrete probability mass function (PMF) and continuous probability density function (PDF) may look very similar (the latter looking like a smoothed version of the former), there is a significant conceptual difference between the two functions, beyond the fact that one is for discrete and the other is for continuous random variables.

That is, the PMF gives *directly* the probability, or likelihood, of obtaining any particular value of the random variable – this probability can be read directly from the graph. The PDF however, does *not* give this information, even for a range of values! In order to determine the probability that the random value will take on a particular range of values, you must multiply the PDF by the width of the range, Δx, itself. Therefore, the product PDF $\cdot$ Δx (the *area* under the PDF curve) is the quantity that is the continuous equivalent to the discrete PMF, and the PDF is *not* actually a probability distribution (hence, some authors just call it a density function, not a *probability* density function).

[6] This involves a special form of integral called the Riemann–Stieltjes integral.

[7] We all hope that aircraft engineers are using the CDF in the most conservative manner, because typically one does not want to have a window seat on a plane and notice the wing fall off during turbulence.

Box 2.1 (Cont.)

Remember the analogy between mass and density here: mass is a distinct value that can always be given for a particular finite substance (like the PMF). But if the substance is spread out over an infinite surface, you can only give its mass per unit volume (or, in one dimension, per unit length), and thus to get the mass of some segment of the surface, you must multiply the mass/unit-volume by the volume of interest. That is the analog of the PDF, and hence the choice of nomenclature adopted here (mass function versus density function).

The cumulative distribution function (CDF), $F_X(x)$, *is* a probability function for both discrete and continuous random variables. That is, in both cases $F_X(x_i)$ directly gives the probability that your random variable will take on a value less than or equal to x_i, $P\{X \leq x_i\}$.

Notation and the Intervals of Distributions

While the CDFs of the continuous and discrete cases are essentially the same, there is an apparent difference regarding how the probability of an interval is actually computed with respect to the endpoints of the interval between the two cases.

Specifically, for a *continuous* random variable, the probability of observing an event between the limits a and b of a random variable is given by the integral $\int_a^b f_X(x)dx$, which *includes the endpoints of the interval* (by definition of a definite integral). However, such probabilities have the following equivalencies:

$$
\begin{aligned}
P\{X \geq a\} &= P\{X > a\} \\
P\{X \leq b\} &= P\{X < b\} \\
P\{a \leq X < b\} &= P\{a < X < b\}.
\end{aligned}
$$

This is because the integral of a single point – that is, the probability of getting a single precise value of X – is zero.[8] Therefore, whether or not the endpoints are included does not matter, since a single point (of no value) will not increase the integral.

Alternatively, for a discrete random variable, this same probability represents a sum of the discontinuous steps of the discrete CDF, in which case the end intervals are dealt with as follows:

$$
\begin{aligned}
P\{a < X \leq b\} &= F(b) - F(a) \\
&= P\{X \leq b\} - P\{X \leq a\}, \text{so } P\{X = a\} \text{ is eliminated} \\
P\{a \leq X \leq b\} &= F(b) - F(a) + P\{X = a\} \\
P\{a < X < b\} &= F(b) - F(a) - P\{X = b\} \\
P\{a \leq X < b\} &= F(b) - F(a) + P\{X = a\} - P\{X = b\}.
\end{aligned}
$$

Of course, this approach works equally well for the continuous case, since the probabilities of the distinct end points are zero in each case, and thus all four cases are

[8] The actual Riemann integral computation of the probability, when given as an integrable function, between the interval a and b, noted as the closed interval [a,b], ***does*** include the endpoints via the definition of the integral. However, it is common practice in statistics to present this interval probability as the open interval (i.e., not including the endpoints), or (a,b). This is only a presentational difference, since the two values are comparable as shown.

Box 2.1 (Cont.)

identical, as previously suggested. Because of the generality of this latter form, it is the form that should be followed when it matters whether or not the endpoints of an interval are to be included in the computation of the interval's probability from the CDF.

2.5 Multivariate Distributions

Multivariate data (of non-sequential data), unlike univariate data, involve two or more variables (e.g., length, width, depth and weight), and a relationship between them often exists. In that case, the distribution of one random variable is a function of the distribution of the other random variable(s), described by joint probability distributions.

2.5.1 Discrete Joint Probability Distributions

Joint Probability Mass Function (Joint PMF)

Discrete multivariate data are characterized by a **joint probability mass function (joint PMF**, or **JPMF)**, $p(X_1,X_2, \ldots,X_K)$, where

$$p(X_1, X_2, \ldots, X_K) = P\{X_1 = x_{1i} \text{ and } X_2 = x_{2j} \ldots \text{ and } X_K = x_{Kk}\} \qquad (2.13)$$

for all possible events of the K random variables. Note that $P\{X_1 = x_{1i}, X_2 = x_{2i}, \ldots, X_K = x_{Ki}\} = P\{X_1 = x_{1i} \text{ and } X_2 = x_{2\mathrm{i}} \ldots \text{ and } X_K = x_{Ki}\}$.

When $K = 2$ – that is, for two related random variables, or **bivariate data**, $X_{\mathrm{j}}, j = 1,2$, or $X = X_1$ and $Y = X_2$ – this relationship describes the probability that you will draw both x_i and y_i simultaneously. The joint PMF for two identically distributed random variables, in this case obeying a normal distribution (defined later), is shown in Figure 2.6.

The concentric symmetric lines in the center of Figure 2.6 should be considered as the contours of a three-dimensional hill (in this case, resembling a Mayan pyramid) rising out of the page (height equivalent to the probability for that combination of X_1 *and* X_2 occurring), while the curves drawn along each axis reflect **marginal probability PMF**s (the probability of achieving any particular value of X_i, regardless of the value of X_j for that X_i.)

Joint Probability Table

For discrete variables, this joint PMF can be presented in a table, called a **joint probability table**, with columns providing the probabilities of obtaining particular Y values, and rows corresponding to X values.

Y	y_1	y_2	$\cdots$	y_N	Marginal probabilities
X					
x_1	$f(x_1,y_1)$	$f(x_1,y_2)$	$\cdots$	$f(x_1,y_N)$	$p_X(x_1)$
x_2	$f(x_2,y_1)$	$f(x_2,y_2)$	$\cdots$	$f(x_2,y_N)$	$p_X(x_2)$
$\vdots$	$\vdots$	$\vdots$	$\vdots$	$\vdots$	$\vdots$
x_M	$f(x_M,y_1)$	$f(x_M,y_2)$	$\cdots$	$f(x_M,y_N)$	$p_X(x_M)$
Marginal probabilities	$p_Y(y_1)$	$p_Y(y_2)$	$\cdots$	$p_Y(y_N)$	1

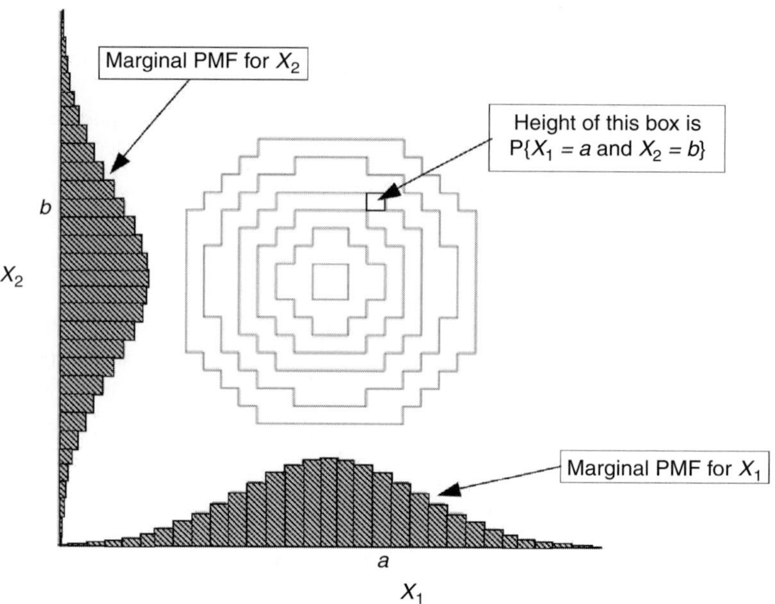

Figure 2.6 Example of a joint probability mass function (joint PMF) for two variables (X_1 and X_2). Contours in the center give the height of the joint PMF in units of probability.

The intersection between a row and column describes the probability of obtaining that specific value of both X and Y. If you were to sum up the probabilities in a single column, the value will give the probability, $p_Y(y_j)$, of obtaining that particular value of Y, regardless of the X values. Similarly, summing up the probabilities in a single row would give the probability, $p_X(x_i)$, of getting that particular X value, regardless of the Y values. Therefore,

$$p_Y(y_j) = \sum_{i=1}^{M} p(x_i, y_j) \tag{2.14a}$$

$$p_X(x_i) = \sum_{j=1}^{N} p(x_i, y_j). \tag{2.14b}$$

These sums of each column and each row (given in the margins of the joint probability table) yield the marginal probability mass functions (marginal PMFs). As with the univariate PMF, a marginal PMF must sum to 1.

Joint Cumulative Distribution Function (Joint CDF)

There is also a **joint cumulative distribution function (joint CDF)**, which gives the probability of getting up to, or including, any particular combination of events. For bivariate data, or $K = 2$,

$$F(x_m, y_n) = P\{X \leq x_m \text{ and } Y \leq y_n\} = \sum_{i=1}^{m} \sum_{j=1}^{n} p(x_i, y_j). \tag{2.15}$$

Y	y_1	y_2	y_3	·	y_n	y_{n+1}	·	y_N	Marginal PMF
X									
x_1	$f(x_1,y_1)$	$f(x_1,y_2)$	$f(x_1,y_1)$	·	$f(x_1,y_n)$	$f(x_1,y_{n+1})$	·	$f(x_1,y_N)$	$p_X(x_1)$
x_2	$f(x_2,y_1)$	$f(x_2,y_2)$	$f(x_2,y_1)$	·	$f(x_2,y_n)$	$f(x_2,y_{n+1})$	·	$f(x_2,y_N)$	$p_X(x_2)$
x_3	$f(x_3,y_1)$	$f(x_3,y_2)$	$f(x_3,y_1)$	·	$f(x_3,y_n)$	$f(x_3,y_{n+1})$	·	$f(x_3,y_N)$	$p_X(x_3)$
$\vdots$	$\vdots$	$\vdots$	$\vdots$	$\vdots$	$\vdots$	$\vdots$	$\vdots$	$\vdots$	$\vdots$
x_m	$f(x_m,y_1)$	$f(x_m,y_2)$	$f(x_m,y_1)$	·	$f(x_m,y_n)$	$f(x_m,y_{n+1})$	·	$f(x_m,y_N)$	$p_X(x_m)$
x_{m+1}	$f(x_{m+1},y_1)$	$f(x_{m+1},y_2)$	$f(x_{m+1},y_1)$	·	$f(x_{m+1},y_n)$	$f(x_{m+1},y_{n+1})$	·	$f(x_{m+1},y_N)$	$p_X(x_{m+1})$
$\vdots$	$\vdots$	$\vdots$	$\vdots$	$\vdots$	$\vdots$	$\vdots$	$\vdots$	$\vdots$	$\vdots$
x_M	$f(x_M,y_1)$	$f(x_M,y_2)$	$f(x_M,y_1)$	·	$f(x_M,y_n)$	$f(x_M,y_{n+1})$	·	$f(x_M,y_N)$	$p_X(x_M)$
Marginal PMF	$p_Y(y_1)$	$p_Y(y_2)$	$p_Y(y_1)$	·	$p_Y(y_n)$	$p_Y(y_{n+1})$	·	$p_Y(y_N)$	**1**

In the joint probability table, this double sum is the sum of all individual cells (each cell containing the probability of obtaining the particular X and Y events that intersect at the cell) within a rectangular area, starting with the upper left cell and ending with the cell in which x_m and y_n intersect (as seen in the table).

The summed probabilities over the entire shaded region represent the joint CDF. Typically though, the joint probability CDF, if presented in tabular form, would show the cumulative sum of each column and row up to each individual table cell.

The sum of all probabilities, as with the univariate case, must equal 1:

$$\sum_{i=1}^{M}\sum_{j=1}^{N} p(x_i, y_j) = 1, \tag{2.16a}$$

or, for the more general case of K random variables (that are related),

$$\sum_{i=1}^{M}\sum_{j=1}^{N}\cdots\sum_{k=1}^{P} p(X_{1i}, X_{2j}, \ldots, X_{Kk}) = 1. \tag{2.16b}$$

That is, the likelihood that you will obtain at least one combination of all of the possible combinations of events is exactly 1, or 100 percent – it can be no more or less. If it were less, you would not have considered all possible combinations of events. It is also true that $p(X_1,X_2, \ldots,X_K) \geq 0$, so that for any possible combination of events, there is at least some likelihood that that combination will occur (otherwise the event isn't possible).

2.5.2 Continuous Joint Probability Distributions

Joint Probability Density Function (Joint PDF)

Continuous multivariate data are described by a **joint probability density function (joint PDF)**, $f(X_1,X_2, \ldots,X_K)$, where

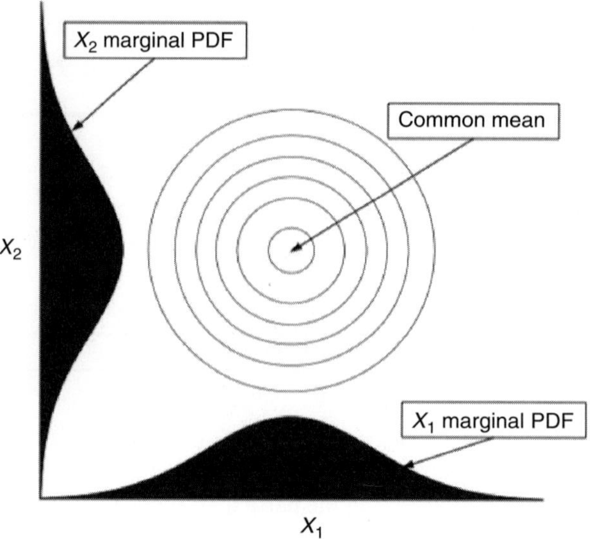

Figure 2.7 Example of a continuous joint probability density function (joint PDF), comparable to the discrete case of Figure 2.6.

$$\int_a^b \int_c^d \cdots \int_u^v \mathrm{f}(x_{1i}, x_{2j}, \ldots, x_{Kk}) \mathrm{d}x_1 \mathrm{d}x_2 \ldots \mathrm{d}x_K = \mathrm{P}\{x_{1a} < X_1 < x_{1b}, x_{2c} < X_2 < x_{2d}, \ldots, x_{Ku} < X_K < x_{Kv}\} \tag{2.17}$$

for all possible events, X_{jk}, and all ranges of the K random variables. That is, as with the univariate case, the probability of getting any one precise combination of events, say x_i and y_j, is zero, so you still have to consider the likelihood of getting a combination of events within a range of possible events.

Graphically, for two identically distributed (related) continuous random variables, X and Y, the joint PDF looks like that shown in Figure 2.7.

The continuous case does not lend itself to construction of a joint probability table, due to the infinite nature of the distributions. However, the concept of marginal distributions is still valid, in which case they represent **marginal probability density functions (marginal PDF)** given for two random variables, as

$$\mathrm{f}_Y(y) = \int_{-\infty}^{+\infty} \mathrm{f}(x, y) \mathrm{d}x \tag{2.18a}$$

$$\mathrm{f}_X(x) = \int_{-\infty}^{+\infty} \mathrm{f}(x, y) \mathrm{d}y, \tag{2.18b}$$

or, for the general case of $K + 1$ related random variables, as

$$\mathrm{f}(x) = \int_{-\infty}^{+\infty}\int_{-\infty}^{+\infty}\ldots\int_{-\infty}^{+\infty}\mathrm{f}(x_1, x_2, \ldots x_K, x_{K+1})\mathrm{d}x_1\mathrm{d}x_2\ldots\mathrm{d}x_K. \quad (2.18c)$$

Joint Cumulative Distribution Function (Joint CDF)

Finally, the **joint cumulative distribution function (joint CDF)**, $\mathrm{F}(X_1,X_2, \ldots,X_K)$, is given as the integral

$$\mathrm{F}(X_1, X_2, \ldots, X_K) = \mathrm{P}\{X_1 \leq x_m, X_2 \leq x_{2n}, \ldots, X_K \leq x_{Kp}\} = \int_{-\infty}^{m}\int_{-\infty}^{n}\ldots\int_{-\infty}^{p}\mathrm{f}(x_1, x_2, \ldots, x_K, x_{K+1})\mathrm{d}x_1\mathrm{d}x_2\ldots\mathrm{d}x_K. \quad (2.19)$$

As with the discrete joint CDF, this integral represents the probability of obtaining any combination of events less than or equal to the values of the random variables specified by the upper integration limit.

Bayesian Statistics

Bayes theorem is given as $\mathrm{P}(A|B) = \dfrac{\mathrm{P}(B|A)}{\mathrm{P}(B)}$, and Bayesian statistics is the employment of this formula for utilizing all relevant information to make a conditional probability estimate. Some feel this approach is better than the Principle of Maximum Likelihood, and some feel it is the only approach. However, the topic warrants an entire book to cover, so it will not be covered here. There are numerous excellent introductions to Bayesian Statistics, such as Eddy (2004) and Lavine (2000), and for objections, Gelman (2008).

2.6 Moments of Random Variables

The following discussion presents some of the fundamental theory and concepts of statistical moments. This discussion assumes you are working with the complete population (or its distribution) of a random variable or set of random variables. In the next chapter you will deal with *estimating* the distribution or its moments from samples. This will introduce some additional considerations that must be taken into account at that time.

2.6.1 General

With the distribution of a random variable, you have the relevant statistical information concerning that random variable, and the details of the analysis will often be dictated to some extent by the nature of that distribution. Consequently, it is worth some effort to attempt to determine the distribution of the random variable(s) of interest. Unfortunately, it is not always possible to determine the complete distribution, or the distribution might be unwieldy and thus awkward or difficult to work with. Regardless, it is helpful to compute the most fundamental characteristics of the distribution in the form of simple constants.

The most basic (and standard) characteristics that are employed to describe a distribution include those related to its central value and spread.

Central Value

The central value is the most "representative" value of the distribution. It can be described in several ways:

Mode (μ_{max})	This is the value of the random variable that has the highest probability or highest frequency of occurrence. For discrete data, $P\{X=\mu_{max}\} \geq P\{X \neq \mu_{max}\}$; for continuous data $f(\mu_{max}) \geq f(X \neq \mu_{max})$.
Median ($\mu_{1/2}$)	This is the value in relation to which half of the possible values are larger and half are smaller: $P(X \leq \mu_{1/2}) = P(X \geq \mu_{1/2}) = 0.5$.
Mean (μ)	This is the algebraic average: conceptually, the balancing point at which a PMF or PDF plot would balance on a pivot point defined for a discrete random variable. as:

$$\mu = \frac{1}{N} \sum_{i=1}^{N} x_i \tag{2.20}$$

For a symmetrical distribution, the three central value indicators will coincide. For a non-symmetrical distribution, the mode will be at the distribution peak and the median will fall somewhere between the mean and mode, as shown in Figure 2.7.

In asymmetrical distributions (Figure 2.8), the usefulness of the mean degrades as the distribution becomes more and more skewed or asymmetrical. Likewise, for multi-modal distributions, a measure of the central value in general is of limited use, and you may desire to describe such a distribution as a linear combination of more than one distribution (among other things). In any case, this is one reason why it is useful to attempt to determine (and examine) a random variable's distribution.

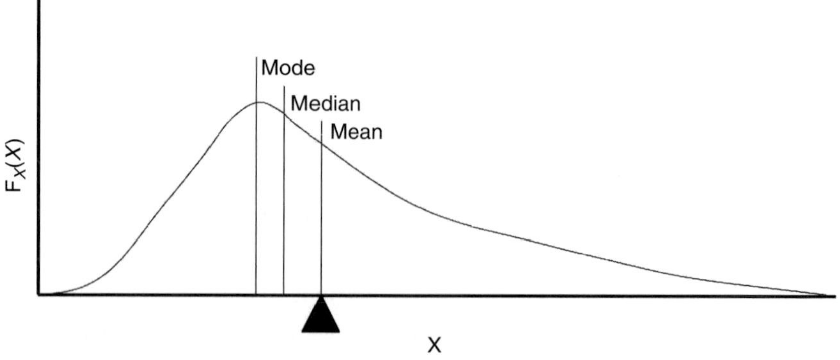

Figure 2.8 Example showing the three most popular versions of the central value of a distribution. The mean value is the one that represents the balancing point of the distribution, shown here by the distribution balancing on a pivot.

Spread (Dispersion)

The spread of a distribution gives an indication of the degree of scatter (noise or error) in the system. A wide distribution implies noisy data. A narrow distribution implies relatively noise-free data – that is, the data are clustered fairly tightly, showing little dispersion or spread, about the central value. The measure of the spread, or dispersion, can, like the central value, be measured in a variety of ways:

Range – a measure of the spread given by the difference between the maximum and minimum values:

$$\text{range} = (x_{max} - x_{min}). \tag{2.21}$$

This measure can also be given by x_{max} and x_{min} directly (instead of as the difference). In either case, this form is often quite useful when it is of more importance to have an idea of the range of your variable(s) than it is to have a single value giving some measure of spread. For example, knowing the maximum and minimum temperatures of an area may be more useful than knowing that there is a 40° temperature spread about the mean.

L1 norm[9] – this is the average of the absolute value of the differences (deviations) about the mean. This is also referred to as the "mean deviation" and is defined, for a discrete random variable, as

$$\text{L}_1 \text{ norm} = \frac{1}{\text{N}} \sum_{i=1}^{\text{N}} |x_i - \mu|. \tag{2.22}$$

Variance (σ^2 or **L2 norm**[10]) – the most common measure of spread is the average of the squares of the deviations about the mean, defined, for a discrete random variable, as

$$\sigma^2 = \frac{1}{\text{N}} \sum_{i=1}^{\text{N}} (x_i - \mu)^2 \tag{2.23}$$

This statistic has units of the variable squared; for convenience, the square root of this quantity, the **standard deviation** (σ), is often used.

2.6.2 Univariate Expectance

As suggested above, it is possible to develop any number of characteristics for describing a distribution, and in fact, these various forms each have distinct advantages that are useful in certain situations (noted where appropriate). Regardless, one standard form

[9] This particular parameter is one case of a broader class of components known as norms (discussed in more detail in a later chapter), and it is useful in many applications in inverse theory, linear programming and other areas. Its definition here, as an estimator of spread, is a specific case of the L1 norm since the normal definition of norm does not include the 1/N scaling.

[10] As with the L1 norm, this is a specific (normalized) case of the L2 norm. As discussed later, the precise form of the normalization will change when estimating this value from a sample.

involves statistical **moments** determined using **expectance** (or **expectancy**) as the **expectance operator**, E[ ].

The number of moments that it takes to completely describe a random variable's population is dependent upon the nature of the distribution, as is demonstrated below. The first two moments provide the central value (specifically, the mean) and spread (specifically, the variance).

Mean (Expected Value) of a Random Variable

The **expected value** represents the probability-weighted mean of a random variable, X. This is the simple *mean* of a random variable and it is also referred to as the **expected measure** or **first moment**. For a discrete random variable, the expected value is given by

$$\mathrm{E}[X] = \mu = \sum_{j=1}^{h} x_j \mathrm{p}_X(x_j). \tag{2.24a}$$

(2.24a) is the formula when the N values of x have been grouped into h categories or classes and $\mathrm{p}_X(x_j)$, the probability, or likelihood of occurrence, of each event or class of events (i.e., the PMF, consisting of h events or classes). If not grouped, then the expected value is the arithmetic average,[11] as

$$\mathrm{E}[X] = \mu = \frac{1}{\mathrm{N}} \sum_{i=1}^{\mathrm{N}} x_i. \tag{2.24b}$$

For a continuous random variable,

$$\mathrm{E}[X] = \mu = \int_{-\infty}^{+\infty} x\mathrm{f}(x)\mathrm{d}x, \tag{2.24c}$$

where f(x) is the standard probability density function (PDF).

Linear Properties of Expectance

As defined previously, the expectance operator provides nothing more than the simplest estimate of a mean quantity. However, as a formal *linear operator*, it obeys a set of simple linear rules that lend it considerable power and versatility and make it user friendly!

Specifically, for the random variables X and Y and the constants a and b, where a and b are *not* random variables,

$$[\mathrm{a}X] = \mathrm{aE}[X] \tag{2.25a}$$

$$\mathrm{E}[X + Y] = \mathrm{E}[X] + \mathrm{E}[Y] \tag{2.25b}$$

$$\mathrm{E}[\mathrm{a} + \mathrm{b}X] = \mathrm{a} + \mathrm{bE}[X]. \tag{2.25c}$$

[11] As you might guess, these two forms (2.24a and b) are identical, the second form simply being the case where the values have not been grouped so they each have the same likelihood of occurring, which is $1/N$.

The last property is that of invariance under any linear transform, and it is a consequence of the first two properties. All of these expectance properties can be derived directly by expansion of the sums or integrals in (2.24).

Expectance of Functions of Random Variables

The expectance operator also provides a means for computing much more than the simple mean of a random variable. Consider the case in which our random variable, Y, is itself a function, $Y(X_j)$, of one or more other random variables, X_j (e.g., variables that are readily observed). The function Y is a random variable because its value contains uncertainty, since the specific values of x_{ji} that go into computing it have uncertainty. You can use the expectance operator to compute the mean of the new random variable, that is, the mean of the function.

For a discrete function, using (2.24a),

$$\mathrm{E}[Y] = \sum_{j=1}^{h} y_j \mathrm{p}_Y(y_j), \tag{2.26a}$$

or, for the continuous case, using (2.24c),

$$\mathrm{E}[Y] = \int_{-\infty}^{+\infty} y \mathrm{f}_Y(y) \mathrm{d}y. \tag{2.26b}$$

As given, these equations work fine, but they assume you know the distribution of the function Y, which you may indeed know. Alternatively, you can bypass this by substituting the explicit form of the function into the relationship so that the sum or integral now becomes a strict function of X alone. In this respect, you could work solely with the original random variable, X, ignoring the distribution f_Y. The linear properties above tell us how this can be done for simple linear function relationships.

The advantage of this is due to the fact that for a one-to-one transformation of Y and X (that is, where there is a unique value of y for each value of x, and vice versa), then it is fairly easy to show that the distribution of Y, $\mathrm{p}_Y(Y)$, can be given directly in terms of the distribution of X, $\mathrm{p}_X(x)$.[12]

Consequently, (2.26) can be rewritten as

$$\mathrm{E}[Y] = \sum_{j=1}^{h} y(x_j) \mathrm{p}_X(x_j) \tag{2.27a}$$

and

$$\mathrm{E}[Y] = \int_{-\infty}^{+\infty} y(x) \mathrm{f}_X(x) \mathrm{d}x. \tag{2.27b}$$

For example, consider the function

[12] It can also be shown that if X is discrete, then $\mathrm{y}(X)$ will also be a discrete function.

$$Y = \mathrm{a} + \mathrm{b}X. \tag{2.28}$$

So, from (2.26a),

$$\begin{aligned} \mathrm{E}[Y] &= \sum_{j=1}^{h} y_j \mathrm{p}_Y(y_j) \\ &= \sum_{j=1}^{h} (\mathrm{a} + \mathrm{b}x_j)\mathrm{p}_X(x_j) \\ &= \sum_{j=1}^{h} \mathrm{a}\mathrm{p}_X(x_j) + \sum_{j=1}^{h} \mathrm{b}x_j\mathrm{p}_X(x_j) \\ &= \mathrm{a} + \mathrm{b}\sum_{j=1}^{h} x_j\mathrm{p}_X(x_j) \\ &= \mathrm{a} + \mathrm{b}\mathrm{E}[\mathrm{X}] \\ &= \mathrm{a} + \mathrm{b}\mu_X. \end{aligned} \tag{2.29}$$

Alternatively, you could have applied the linear property (2.25c) directly to (2.28) to also yield

$$\mathrm{E}[Y] = \mathrm{E}[\mathrm{a} + \mathrm{b}X] = \mathrm{a} + \mathrm{b}\mathrm{E}[X] = \mathrm{a} + \mathrm{b}\mu_X. \tag{2.30}$$

Variance of a Random Variable

Consider the specific function $Y = (X - \mu)^2$. The *expectance of this function gives the variance for X*. For the discrete case,

$$\mathrm{E}[(X - \mu)^2] = \sigma^2 = \sum_{j=1}^{h} (x_j - \mu)^2 \mathrm{p}_X(x_j). \tag{2.31a}$$

For the continuous case,

$$\mathrm{E}[(X - \mu)^2] = \sigma^2 = \int_{-\infty}^{+\infty} (x - \mu)^2 \mathrm{f}_X(x)\mathrm{d}x. \tag{2.31b}$$

Consider an expansion of the left-hand side (**LHS**) of (2.31) using the linear operator relationships of (2.25):

$$\begin{aligned} \sigma^2 &= \mathrm{E}[(X - \mu)^2] \\ &= \mathrm{E}[X^2 - 2\mu X + \mu^2] \\ &= \mathrm{E}[X^2] - 2\mu\mathrm{E}[X] + \mu^2 \\ &= \mathrm{E}[X^2] - 2\mu^2 + \mu^2 \\ &= \mathrm{E}[X^2] - \mu^2. \end{aligned} \tag{2.32a}$$

Thus, the variance (spread) of a random variable is equivalent to the mean of the sum of squares of the random variable minus the random variable's mean squared.

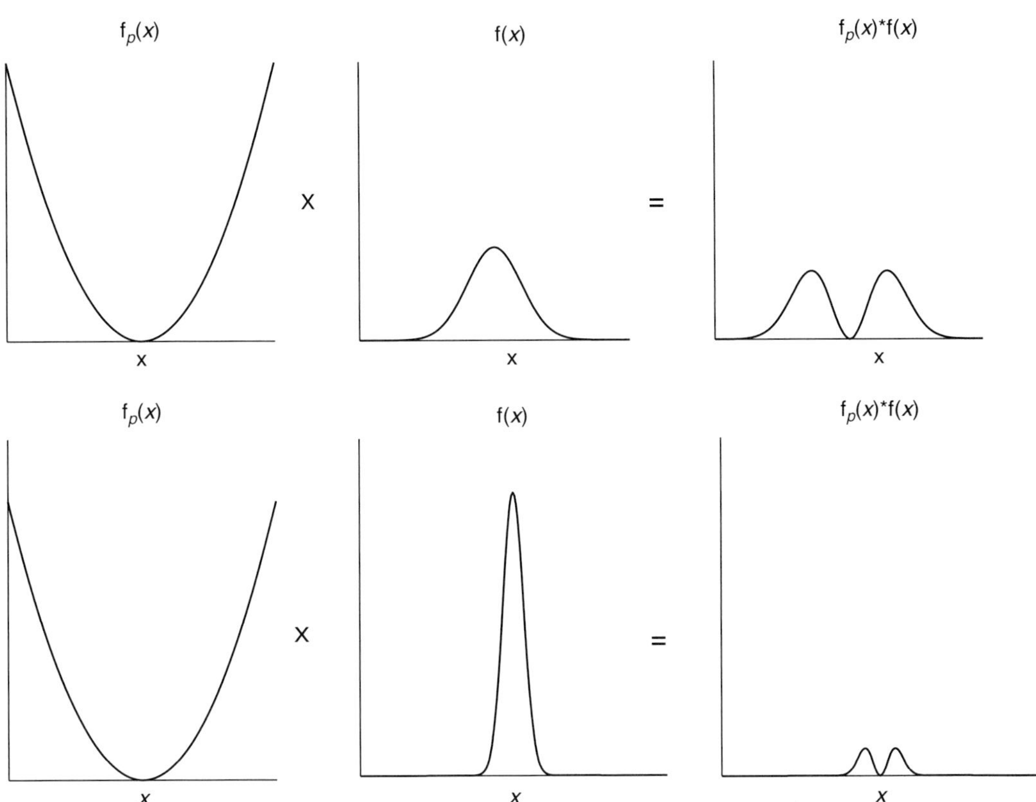

Figure 2.9 Conceptualization of variance (adapted from Menke, 1984) showing that the variance is the integral under the curve of a parabola, $f_P(x) = (x - \mu)^2$, times the distribution.

Notice that μ is nothing more than E[X], so μ^2 = E[X]^2, and (2.32a) can also be written as

$$\begin{aligned}\sigma^2 &= \mathrm{E}[X^2] - \mathrm{E}[X]^2 \\ &= \mathrm{Var}[X].\end{aligned} \tag{2.32b}$$

Because this is such a common and important use of the expectance operator, $E[(X-\mu)^2]$ *is often given directly as Var[X].* Its root is the standard deviation.

The concept of variance is conceptualized by recognizing that it represents the width of a distribution, f(x), after being multiplied by a parabola of the form: $f_p(x) = (x - \mu)^2$.

When multiplied against the distribution f(x), if the distribution is narrow you get a small value for the area under the curve, as shown graphically in Figure 2.9.

And if multiplied against a wide distribution, the area under the curve is large. For asymmetrical distributions – a chi-squared distribution in Figure 2.10 – the variance is still computed the same way, but the distribution of the uncertainty (e.g., standard deviation) is asymmetrical about the mean, eliminating the ability to say μ ± 1.2, requiring something like μ −0.2/+1.0.

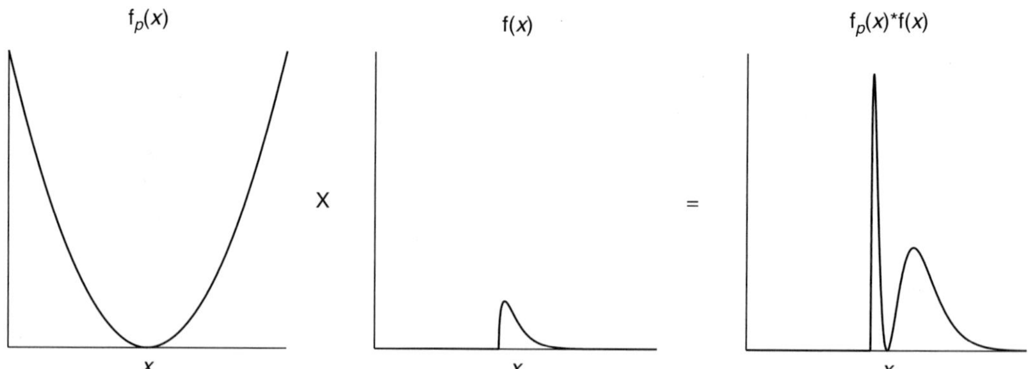

Figure 2.10 Variance for an asymmetrical distribution, leading to an asymmetrical distribution of the variance about the mean.

So, from this, the measure of variance is given as

$$\sigma^2 = \int_{-\infty}^{+\infty} (x - \mu)^2 \mathrm{f}(x)\mathrm{d}x. \tag{2.33}$$

More importantly, you can conceptually see how asymmetrical distributions f(x) will give asymmetrical variance, though the single value of variance will still provide an indication of the overall spread in the distribution (albeit unevenly about the mean).

Variance of a Function of a Random Variable

A *major advantage* of the above definition for variance, using the expectance operator, is that it allows you to determine the variance of a function of the original random variable. This is invaluable for data analysis, since the analysis often involves the manipulation of your data (i.e., converting the random variable to a new form through some mathematical relationship). The above relationship provides a means by which the uncertainty in the new function can be determined via knowledge of the uncertainty (i.e., the variance or spread) of the original raw data.

The variance (uncertainty, as scatter about the mean, or precision) of a function $Y(X)$ is determined as

$$\begin{aligned} \sigma_Y^2 &= \mathrm{Var}[Y] \\ &= \mathrm{E}[(Y - \mathrm{E}[Y])^2] \\ &= \mathrm{E}[Y^2] - 2\mathrm{E}[Y\mathrm{E}[Y]] + \mathrm{E}[Y]^2 \end{aligned} \tag{2.34a}$$

and, since the middle term is the expectance of Y times a constant, the constant being the mean of $Y = \mathrm{E}[Y] = \mu_Y$, the linear property (2.25a) indicates that it is equal to the constant times the expectance, or $\mathrm{E}[Y\,\mathrm{E}[Y]] = \mathrm{E}[Y]\mathrm{E}[Y] = \mathrm{E}[Y]^2$, so (2.34a) reduces to

$$= \mathrm{E}[Y^2] - \mathrm{E}[Y]^2. \tag{2.34b}$$

Consider the specific case where $Y = \mathrm{a}X$. Then,

$$\begin{aligned}\sigma_Y^2 &= \mathrm{Var}[Y] \\ &= \mathrm{E}[Y^2] - \mathrm{E}[Y]^2 \\ &= \mathrm{E}[\mathrm{a}^2X^2] - \mathrm{E}[\mathrm{a}X]^2 \\ &= \mathrm{a}^2\mathrm{E}[X^2] - \mathrm{a}^2\mathrm{E}[X]^2 \\ &= \mathrm{a}^2(\mathrm{E}[X^2] - \mathrm{E}[X]^2).\end{aligned} \tag{2.35a}$$

Therefore,

$$\begin{aligned}&= \mathrm{a}^2\mathrm{Var}[X] \\ &= \mathrm{a}^2\sigma_X^2.\end{aligned} \tag{2.35b}$$

Now consider a slightly different function, $Y = \mathrm{a}X + \mathrm{b}$. Then,

$$\begin{aligned}\sigma_Y^2 &= \mathrm{Var}[Y] \\ &= \mathrm{E}[Y^2] - \mathrm{E}[Y]^2 \\ &= \mathrm{E}[(\mathrm{a}X + \mathrm{b})^2] - \mathrm{E}[\mathrm{a}X + \mathrm{b}]^2.\end{aligned} \tag{2.36a}$$

From the linear properties, you know that $\mathrm{E}[\mathrm{a}X + \mathrm{b}] = \mathrm{a}\mathrm{E}[X] + \mathrm{b}$, so the second term on the RHS is $\mathrm{a}^2\mathrm{E}[X]^2 + 2\mathrm{ab}\mathrm{E}[X] + \mathrm{b}^2$ and the first term on the RHS is $\mathrm{E}[\mathrm{a}^2X^2 + 2\mathrm{ab}X + \mathrm{b}^2]$, so (2.36a) is

$$\begin{aligned}&\mathrm{a}^2\mathrm{E}[X^2] + 2\mathrm{ab}\mathrm{E}[X] + \mathrm{b}^2 - \mathrm{a}^2\mathrm{E}[X]^2 - 2\mathrm{ab}\mathrm{E}[X] - \mathrm{b}^2 \\ &= \mathrm{a}^2\mathrm{E}[X^2] - \mathrm{a}^2\mathrm{E}[X]^2 \\ &= \mathrm{a}^2(\mathrm{E}[X^2] - \mathrm{E}[X]^2).\end{aligned} \tag{2.36b}$$

Therefore,

$$\mathrm{Var}[\mathrm{a}X + \mathrm{b}] = \mathrm{a}^2\mathrm{Var}[X] = \mathrm{a}^2\sigma_X^2. \tag{2.36c}$$

That is, the addition of a constant to a random variable does not influence the spread or uncertainty of the function of the random variable, so $\mathrm{Var}[\mathrm{a}X + \mathrm{b}] = \mathrm{Var}[\mathrm{a}X] = \mathrm{a}^2\mathrm{Var}[X]$.

Notice that the above manipulations are independent of whether the random variable is continuous or discrete. This is because the expectance operator is linear in both cases, and thus the general operator properties, which allow the above manipulations, apply for both.

Higher-Order Moments

Moments represent a special case of expectance and they prove very useful in studying properties of distributions. In some cases, the distribution can be uniquely determined by its moments.

Expectance provides statistical moments of a data set when specific forms of $Y(X)$ are introduced, as shown above for the variance. In general, the ***k*th moment**, or ***k*th moment about the origin**, is defined with $Y(X) = X^k$ (for h classes),

$$\mu_k' = E[X^k] = \sum_{j=1}^{h} x_j^k p_X(x_j), \tag{2.37a}$$

or, without partitioning into classes for N data points,

$$\mu_k' = E[X^k] = \frac{1}{N}\sum_{j=1}^{N} x_j^k. \tag{2.37b}$$

Alternatively, the ***k*th central moment**, or ***k*th moment about the mean**, results when $Y(X) = (X - \mu_X)^k$, so

$$E[(X-\mu)^k] = \sum_{j=1}^{h} (x_j - \mu)^k p_X(x_j). \tag{2.38a}$$

Or, again without partitioning the data,

$$E[(X-\mu)^k] = \frac{1}{N}\sum_{j=1}^{N} (x_j - \mu)^k. \tag{2.38b}$$

Comparable integral formulas exist for the continuous case.

From (2.38), it is clear that *the second central moment defines the variance*. The third moment about the mean that describes the asymmetry in a distribution is called the **skewness**, while the fourth moment about the mean, which describes the degree of "peakedness" of a distribution, is called **kurtosis**.

Most often you work with central moments, but since it is sometimes computationally easier to calculate the moments about the origin, the central moments can be rewritten in terms of the moments about the origin. This involves expansion, as done in (2.32) when showing the expanded form of the variance in terms of the expectance operator. That is, the second moment about the mean (i.e., the variance) was expanded (in (2.32)) to give

$$E[(X-\mu)^2] = \sigma^2 = E[X^2] - E[X]^2. \tag{2.39}$$

This gives the variance in terms of the second moment about the origin (the first term on the RHS) and the square of the first moment about the origin (the second term on the RHS, the mean). A similar approach following the expansions of (2.32) can be applied to any of the higher-order moments as well.

Note that some authors use the notation such that $E^2[X] = (E[x])^2 = E[X]^2$. This notation is nice and often is a bit clearer when working with functions, but it can be confused with another similar notation whereby $E^k[X]$ represents application of the expectance operator to compute the kth moment of the random variable. Therefore, to avoid confusion, I will stick with the notation as shown in (2.34b): $E[X]^2$.

2.6.3 Multivariate Expectance

Mean (Expected Value) of Multivariate Random Variables

A function $Y(X_1,X_2,\ldots,X_K)$ with joint PMF $p(X_1,X_2,\ldots,X_K)$ has an expected value given by the joint PMF-weighted mean of the random variables, $X_1,X_2,\ldots,X_K$:

$$\begin{aligned} E[Y(X_1,X_2,\ldots,X_K)] &= \mu_Y \\ &= \sum_{i=1}^{M}\sum_{j=1}^{N}\cdots\sum_{k=1}^{P} y(x_{1i},x_{2j},\ldots,x_{Kk})p(x_{1i},x_{2j},\ldots,x_{Kk}). \end{aligned} \tag{2.40}$$

In other words, as before, you can evaluate the function $Y(X_j)$ in terms of the joint PMF for the X_j (as opposed to the PMF for Y).

Variance of Multivariate Random Variables

Consider the general linear function $Y(X_j)$, j = 2 (i.e., bivariate data),

$$Y(X_j) = a_1X_1 + a_2X_2, \tag{2.41}$$

where the a_1 and a_2 are constants. For example, Y may be the quantity of interest, say seawater density, which you linearly approximate as a weighted sum of the "raw" observations, X_1 and X_2, temperature and salinity.

You can compute the variance of this function using expectance as done previously with univariate data, starting with those previous results that showed $\text{Var}[X] = E[X^2] - E[X]^2$, so

$$\begin{aligned} \sigma_Y^2 &= \text{Var}[a_1X_1 + a_2X_2] = \text{Var}[Y] = E[Y^2] - E[Y]^2 \\ &= E[Y^2] - E[Y]^2 = E[(a_1X_1 + a_2X_2)^2] - E[(a_1X_1 + a_2X_2)]^2 \\ &= E[(a_1X_1)^2] + 2E[a_1a_2X_1X_2] + E[(a_2X_2)^2] - (E[a_1X_1] + E[a_2X_2])^2 \\ &= a_1^2E[X_1^2] + 2a_1a_2E[X_1X_2] + a_2^2E[X_2^2] \\ &\quad -a_1^2E[X_1]^2 - 2a_1a_2E[X_1]E[X_2] - a_2^2E[X_2]^2. \end{aligned} \tag{2.42a}$$

This expression can be reduced, since you know from (2.35) $\text{Var}[aX] = a^2E[X^2] - a^2E[X]^2$, so several terms in (2.42a) can be combined to give

$$\begin{aligned} &= a_1^2\text{Var}[X_1] + a_2^2\text{Var}[X_2] + 2a_1a_2E[X_1X_2] - 2a_1a_2\mu_1\mu_2 \\ &= a_1^2\sigma_{X_1}^2 + a_2^2\sigma_{X_2}^2 + 2a_1a_2E[X_1X_2] - 2a_1a_2\mu_1\mu_2, \end{aligned} \tag{2.42b}$$

where $\mu_1 = E[X_1]$ and $\mu_2 = E[X_2]$.

The expression in (2.42b) now contains a new term, not present in the previous univariate cases, which involves $E[X_1X_2]$. This term provides information regarding the extent to which the two random variables X_1 and X_2 are dependent upon one another, or how they **covary**. The degree to which random variables covary involves the concept of **covariance**.

Covariance

Covariance is comparable to variance for a single random variable, only in this case the concept extends to include how two or more random variables vary together. Specifically, the covariance between two random variables X_1 and X_2, denoted by $\mathrm{Cov}[X_1X_2]$ or σ_{12}^2, is given as the expected value of the product of their deviations about their means:

$$\begin{aligned}\sigma_{12}^2 &= \mathrm{Cov}[X_1X_2] = \mathrm{E}[(X_1 - \mathrm{E}[X_1])(X_2 - \mathrm{E}[X_2])]\\ &= \mathrm{E}[X_1X_2 - \mathrm{E}[X_1]X_2 - \mathrm{E}[X_2]X_1 + \mathrm{E}[X_1]\mathrm{E}[X_2]]\\ &= \mathrm{E}[X_1X_2 - \mu_1X_2 - \mu_2X_1 + \mu_1\mu_2]\\ &= \mathrm{E}[X_1X_2] - \mu_1\mathrm{E}[X_2] - \mu_2\mathrm{E}[X_1] + \mu_1\mu_2\\ &= \mathrm{E}[X_1X_2] - \mu_1\mu_2 - \mu_2\mu_1 + \mu_1\mu_2\\ &= \mathrm{E}[X_1X_2] - \mu_1\mu_2,\end{aligned} \tag{2.43a}$$

or

$$= \mathrm{E}[X_1X_2] - \mathrm{E}[X_1]\mathrm{E}[X_2]. \tag{2.43b}$$

From this definition, it is seen that (2.42b) can be further reduced to

$$\begin{aligned}\mathrm{Var}[a_1X_1 + a_2X_2] &= a_1^2\mathrm{Var}[X_1] + a_2^2\mathrm{Var}[X_2] + 2a_1a_2\mathrm{Cov}[X_1X_2]\\ &= a_1^2\sigma_{X_1}^2 + a_2^2\sigma_{X_2}^2 + 2a_1a_2\sigma_{12}^2.\end{aligned} \tag{2.44}$$

So, with respect to the previous example of estimating the seawater density, the uncertainty (i.e., the variance) in the computed value is the linear sum, weighted by the coefficients (a_i), of the variance of seawater temperature and salinity values, respectively, and the covariance between the temperature and salinity.

The similarity between variance of a univariate random variable and covariance is plainly seen when considering the case where $X_1 = X_2$. In that case, (2.43) reduces to $\mathrm{E}[X_1{}^2] - \mu^2$, which is the variance of X_1 (recall equation (2.32a)), so

$$\mathrm{Cov}[X_1X_1] = \mathrm{Var}[X_1]. \tag{2.45}$$

Independent Random Variables

If two or more random variables are related in some manner, then their joint distribution conveys this dependence. That is, the probability of getting any one value of a random variable is dependent upon the value of the other random variable, something known as a **conditional probability** (i.e., what is the probability of getting x_{2i}, given that we have x_{1j}, or $\mathrm{P}\{x_{2i}|x_{1j}\}$). If the variables are not related, then they are considered to be **linearly independent**, or simply **independent**, random variables (other statements of independence were presented previously in this chapter). In this case, the distribution of one random variable is given strictly by its own distribution, independent of the distribution of the other random variable(s).

When random variables are independent, then,

$$P\{X_1 = x_{1i}, X_2 = x_{2j}\} = P\{X_1 = x_{1i}\}P\{X_2 = x_{2j}\}, \tag{2.46}$$

or

$$p_{1,2}(x, x) = p(x_{1i})p(x_{2j}) \tag{2.47a}$$

for discrete variables, or

$$f_{1,2}(x, x) = f(x_{1i})f(x_{2j}) \tag{2.47b}$$

for continuous variables. Their joint probability table would show each value of x and y equal to the product of their marginal probabilities. In fact, the joint probability table is unnecessary, since just the marginal PMFs show all that is needed for each variable.

Y	y_1	y_2	$\cdots$	y_N	Marginal probabilities
X					
x_1	$p(x_1)\ p(y_1)$	$p(x_1)\ p(y_2)$	$\cdots$	$p(x_1)\ p(y_N)$	$p(x_1)$
x_2	$p(x_2)\ p(y_1)$	$p(x_2)\ p(y_2)$	$\cdots$	$p(x_2)\ p(y_N)$	$p(x_2)$
$\vdots$	$\vdots$	$\vdots$	$\vdots$	$\vdots$	$\vdots$
x_M	$p(x_M)\ p(y_1)$	$p(x_M)\ p(y_2)$	$\cdots$	$p(x_M)\ p(y_N)$	$p(x_M)$
Marginal probabilities	$p(y_1)$	$p(y_2)$	$\cdots$	$p(y_N)$	1

Independence is also reflected in the cumulative distributions, so

$$F(x_1, x_2) = F_1(x_1)F_2(x_2). \tag{2.48}$$

When random variables are independent, the expectance of their product reduces to the product of their expectances, which is essentially a restatement of (2.46), so

$$E[X_1X_2] = E[X_1]E[X_2] \text{ (if } X_1 \text{ and } X_2 \text{ are independent)}, \tag{2.49}$$

reducing (2.43) to

$$\text{Cov}[X_1X_2] = 0 \text{ (if } X_1 \text{ and } X_2 \text{ are independent)}. \tag{2.50}$$

Therefore, independent random variables do *not* covary. Conversely, if two or more random variables have zero covariance, then they are linearly independent, though as discussed later (when introducing the concept of correlation), it will be shown that they may still be *dependent* in a *nonlinear* fashion.

Following the example of seawater density, in this case, if the seawater temperature and salinity were independent of one another – that is, if the covariance between them were zero (the value of temperature shows no consistent dependency on the salinity and vice versa) – then the variance of seawater density would be the linear sum, weighted by the squared coefficients, of the variance of temperature with the variance of salinity.

Variance of Multivariate Functions of Random Variables

The expectance operator can be applied to functions of multivariate random variables, with the most common application aimed at determining the variance of the function, that is, the *uncertainty* of the function stemming from its dependence on random variables, themselves uncertain – in fact, we use this to determine the uncertainty of results from most of the analysis methods introduced in this text.

Consider the general expression

$$\begin{aligned} Y(X_j) &= \mathrm{a}_1 X_1 + \mathrm{a}_2 X_2 + \mathrm{a}_3 X_3 + \ldots + \mathrm{a}_K X_K \\ &= \sum_{j=1}^{K} \mathrm{a}_j X_j. \end{aligned} \tag{2.51}$$

The variance of $Y(X_j)$ is given as

$$\begin{aligned} \sigma_Y^2 &= \mathrm{Var}\left[\sum_{j=1}^{K} \mathrm{a}_j X_j\right] \\ &= \mathrm{E}\left[\sum_{j=1}^{K} (\mathrm{a}_j X_j)^2\right] - \left\{\mathrm{E}\left[\sum_{j=1}^{K} (\mathrm{a}_j X_j)\right]\right\}^2 \\ &= \mathrm{E}[(\mathrm{a}_1 X_1 + \mathrm{a}_2 X_2 + \ldots + \mathrm{a}_K X_K)^2] - \mathrm{E}[(\mathrm{a}_1 X_1 + \mathrm{a}_2 X_2 + \ldots + \mathrm{a}_K X_K)]^2 \\ &= \sum_{i=1}^{K}\sum_{j=1}^{K} \mathrm{a}_i \mathrm{a}_j \mathrm{Cov}[X_i X_j]. \end{aligned} \tag{2.52a}$$

The covariance between *independent* random variables is zero, so if the X_j in $Y(X_j)$ are independent, all of the terms in (2.52) equal zero *except* when i = j. So, expanding (2.52),

$$\begin{aligned} \sigma_Y^2 = {} & \mathrm{a}_1\mathrm{a}_1\mathrm{Cov}[x_1x_1] + \mathrm{a}_1\mathrm{a}_2\mathrm{Cov}[x_1x_2] + \ldots + \mathrm{a}_1\mathrm{a}_K\mathrm{Cov}[x_1x_K] + \ldots + \\ & \mathrm{a}_2\mathrm{a}_1\mathrm{Cov}[x_2x_1] + \mathrm{a}_2\mathrm{a}_2\mathrm{Cov}[x_2x_2] + \ldots + \mathrm{a}_2\mathrm{a}_K\mathrm{Cov}[x_2x_K] + \ldots + \\ & \mathrm{a}_K\mathrm{a}_1\mathrm{Cov}[x_Kx_1] + \mathrm{a}_K\mathrm{a}_2\mathrm{Cov}[x_Kx_2] + \ldots + \mathrm{a}_K\mathrm{a}_K\mathrm{Cov}[x_Kx_K]. \end{aligned} \tag{2.52b}$$

All terms are eliminated (= 0) except for those where $i = j$, so for *independent* X_j, the variance of $Y(X_j)$ reduces to

$$\sigma_Y^2 = \sum_{j=1}^{K} \mathrm{a}_j^2 \mathrm{Var}[X_j] \tag{2.53a}$$

$$= \sum_{j=1}^{K} \mathrm{a}_j^2 \sigma_j^2. \tag{2.53b}$$

For cases in which all of the independent random variables, X_j, have the same variance, so $\mathrm{Var}[X_j] = \sigma^2$, then (2.53) reduces to

$$\sigma_Y^2 = \sigma^2 \sum_{j=1}^{K} \mathrm{a}_j^2. \tag{2.54}$$

If the function $Y(X_j)$ is given by

$$\begin{aligned} Y(X_j) &= \mathrm{a}_1 X_1 + \mathrm{b}_1 + \mathrm{a}_2 X_2 + \mathrm{b}_2 + \ldots + \mathrm{a}_K X_K + \mathrm{b}_K \\ &= \mathrm{a}_1 X_1 + \mathrm{a}_2 X_2 + \ldots + \mathrm{a}_K X_K + \mathrm{b} \\ &\left[\sum_{j=1}^{K} (\mathrm{a}_j X_j)\right] + \mathrm{b}, \end{aligned} \tag{2.55}$$

where the b_j are constants, and the constant b is the sum of the b_j. The addition of a constant b is comparable to the univariate case of (2.36), where it was shown that an added constant does not influence the variance of a function. Consequently, the variance of (2.55) is the same as that of (2.51), so, again,

$$\sigma_Y^2 = \sum_{i=1}^{K} \sum_{j=1}^{K} \mathrm{a}_i \mathrm{a}_j \mathrm{Cov}[X_i X_j]. \tag{2.56}$$

One of the most valuable aspects of this application of the expectance operator to a function (e.g., a variable that is constructed from other variables or parameters) is that if variability (uncertainty) of the function Y is what is of interest (e.g., climate variability), then the variance of the function as determined from Var[X] reveals its ***dependencies*** on the uncertainty of the parameters on which it depends. Therefore you can determine the sensitivity of the function to these parameters – i.e., which parameters, X_k, contribute most to the uncertainty of Y, the function of interest. This is most useful for interpretation, hypothesis testing and future sampling strategies (e.g., sample with greater precision that parameter causing the biggest error, or define the level of sample precision needed in each parameter to be able to estimate the function to the desired precision).

2.6.4 Moments of Nonlinear Functions of Random Variables

When functions $Y(X_j)$ are not linear in the random variables, simple reductions with the linear expectance operator often cannot be adopted. In these cases, the expectance, used to compute the first two central moments of the nonlinear function (i.e., the mean and variance), can be approximated by expanding the nonlinear function in a Taylor series about the mean and then truncating this expansion to first order, providing a linear approximation to the function. The expectance operator can then be easily applied to this truncated, linear approximation to provide an approximation of the moments.

So, for nonlinear function $Y(X_j)$,

$$Y(X_1, X_2, \ldots, X_K), \tag{2.57}$$

the **Taylor series expansion** of this function about the mean, μ, truncated to first order (that is, after the first derivative term) is given by

$$Y(X_1, X_2, \ldots, X_K) \approx Y(\mu_1, \mu_2, \ldots, \mu_K) + \sum_{j=1}^{K} \frac{\partial Y(X_j)}{\partial X_j}\Big|_{\mu_j} (X_j - \mu_j), \tag{2.58}$$

where the derivatives are evaluated at the μ_j.

Mean of Nonlinear Functions of Random Variables

The expected value of the first-order *approximation* to $\mu(X_j)$ in (2.58) is given as

$$\begin{aligned} Y(X_1, X_2, \ldots, X_K) &\approx Y(\mu_1, \mu_2, \ldots, \mu_K) + \sum_{j=1}^{K} \frac{\partial Y(X_j)}{\partial X_j}\Big|_{\mu_j} (\mathrm{E}[X_j] - \mu_j) \\ &\approx Y(\mu_1, \mu_2, \ldots, \mu_K). \end{aligned} \tag{2.59}$$

The second term on the RHS is eliminated, since the derivatives evaluated at the points μ_j yield constants, in which case the expectance operator operates on the only random variables within the sum, which are $\mathrm{E}[X_j]$. Since, $\mathrm{E}[X_j] = \mu_j$, the difference within the sum, $\mathrm{E}[X_j] - \mu_j = \mu_j - \mu_j$, eliminates the term, leaving only the first term.

Therefore, the mean of a nonlinear multivariate function of a random variable, $Y(X_j)$, is approximated by the function evaluated at the mean of each of the individual random variables, of which $Y(X_j)$ is a function.

Variance of Nonlinear Functions of Random Variables

Equation (2.58) is providing us with a linear approximation of Y of the form

$$\approx \mu_Y + \sum_{j=1}^{K} (\mathrm{a}_j X_j - \mathrm{b}_j), \tag{2.60}$$

where the constants a_j are the derivatives evaluated at the means, $\frac{\partial Y(X_j)}{\partial X_j}\Big|_{\mu_j}$, and b_j are the products of the derivatives (evaluated at the means, the a_j) and the means, μ_j. The sum of the b_j terms with the μ_Y can be lumped into a single constant b, giving

$$\approx \sum_{j=1}^{K} (\mathrm{a}_j X_j) - \mathrm{b}. \tag{2.61}$$

This is of the same form as (2.55), which shows that the variance of this function is given by (2.56), or

$$\begin{aligned}\mathrm{Var}[Y(X_1,X_2,\ldots,X_K)] &= \sigma_Y^2 \approx \sum_{i=1}^{K}\sum_{j=1}^{K} \mathrm{a}_i\mathrm{a}_j\mathrm{Cov}[X_iX_j]\\ \approx \sum_{i=1}^{K}\sum_{j=1}^{K} \frac{\partial Y(X_i)}{\partial X_i}\Big|_{\mu_i} \frac{\partial Y(X_j)}{\partial X_j}\Big|_{\mu_j} &\mathrm{Cov}[X_iX_j]\end{aligned} \tag{2.62}$$

If the X_j are independent, then the above sum is reduced to become equivalent in form with (2.53); if the independent random variables all have the same variance, it reduces to a form equivalent to (2.54).

Moments of Univariate Nonlinear Functions

When $K = 1$ in the preceding nonlinear functions, the moments are reduced to those for the univariate case where the sums are replaced in the above expressions to single terms. For example, consider $K = 1$ and $Y(X) = X^2$. In this case, the mean is given by (2.59), or, specifically,

$$\begin{aligned}\mathrm{E}[X] &= \mu_Y \approx Y(\mu_X)\\ &= \mu_X^2.\end{aligned} \tag{2.63}$$

The variance is given by (2.62):

$$\begin{aligned}\sigma_Y^2 &\approx \sum_{i=1}^{1}\sum_{j=1}^{1} \frac{\partial Y(X_i)}{\partial X_i}\Big|_{\mu_i} \frac{\partial Y(X_j)}{\partial X_j}\Big|_{\mu_j} \mathrm{Cov}[X_iX_j]\\ &\approx Y(\mu_X) + \frac{\partial Y(X_1)}{\partial X_1}\Big|_{\mu_1}^2 \mathrm{Cov}[X_1X_1]\\ &\approx \mu_X^2 + 4\mu_X^2\mathrm{Var}[X]\\ &\approx \mu_X^2 + 4\mu_X^2\sigma_X^2 = \mu_Y + 4\mu_Y\sigma_X^2.\end{aligned} \tag{2.64}$$

The degree of nonlinearity of a function will determine just how well, or how poorly, the simple linear approximation works. From a practical perspective, the approximations do provide reasonable estimates of the lower-order moments. If more precise values are required, more elaborate numerical techniques must be employed.

Box 2.2 Example of First-Order Linear Approximation Error

Consider the simple situation where $X = 2, 3, 4, 5$, so $\mu_X = 3.5$, and $\sigma_x^2 = 1.25$; whereas $X^2 = 4,9,16,25$ has a true mean of 13.5 and true variance of 62.25. The first-order linear approximations to X^2 using (2.63) for the mean and (2.64) for the variance give ${\mu_X}^2 = 12.25$ for the mean and $\mu_Y + 4\mu_Y\sigma_x^2 = 73.5$ for the variance. The mean of this first-order approximation to X^2 is ~91 percent of the true value, while the variance is 18 percent larger than true σ_Y^2.

Moments of Nonlinear Functions of Independent Random Variables

If the random variables of the nonlinear function are *independent*, then the covariance terms in the variance approximation of (2.62) are zero whenever $i \neq j$, so (2.62) reduces to

$$\begin{aligned}\mathrm{Var}[Y(X,X,\ldots,X)] = \sigma_Y^2 &\approx \sum_{j=1}^{K} \frac{\partial Y(X_j)}{\partial X_j}\Big|_{\mu_j}^2 \mathrm{Var}[X_j] \\ &\approx \sum_{j=1}^{K} \left(\frac{\partial Y(X_j)}{\partial X_j}\Big|_{\mu_j}^2 \sigma_j^2 \right).\end{aligned} \tag{2.65}$$

If the variances of the different independent X_j are equivalent, the approximation of (2.65) reduces to

$$\mathrm{Var}[Y(X,X,\ldots,X)] = \sigma_Y^2 \approx \sigma^2 \sum_{j=1}^{K} \frac{\partial Y(X_j)}{\partial X_j}\Big|_{\mu_j}^2. \tag{2.66}$$

Practical Considerations

Much of this text is concerned with presenting the various tools that are useful, if not invaluable (e.g., expectance), in analyzing data. However, it is important to keep in mind the limitations of the tools, and what exactly they are providing. In the case of statistical moments computed via the expectance operator, the moments are providing the characteristics of the distribution of the function $Y(X_j)$ based on the characteristics of the distributions of the individual random variables that go into defining $Y(X_j)$.

Why Use Expectance?

Much of this chapter demonstrates how expectance can be used to compute the lower-order moments (e.g., the mean and variance) of functions of random variables. In data analysis, such functions often arise, since it is common that the quantity of interest is not measured directly, but rather computed as a function of the variables measured or sampled. You might wonder, however, why not simply compute the moments directly from the variability measured in the computed function, a distinct possibility, instead of employing the expectance operator. However, the advantage of expectance is realized by its ability to precisely show how the moments of the *function* are controlled by the moments of the *measured variables*. You can use this information to develop or refine sampling procedures so as to yield the best precision in the function from the most efficient sampling strategy (e.g., focus your sampling strategy so as to provide the best estimate of that random variable to which your function's uncertainty is most sensitive).

Also, expectance provides a means for determining how uncertainty is propagated through different theoretical functional relationships when developing or testing them. Finally, it is not uncommon in the digital age to have so much information that it may simply prove impractical to compute the statistical moments of the function directly. Conversely, large bulky or numerical functions of random variables may also prove unwieldy for the calculation of expectance as well. In such cases, the uncertainty may have to be computed numerically or the expectance may have to be estimated numerically. Despite the advantages of the expectance operator, numerical methods, particularly those involving using the bootstrap or resampling techniques, discussed in the next chapter and in Appendix 2, will often prove more convenient or practical to use. In the end, you should attempt to match the tools and efforts most appropriate for addressing the issues at hand. While expectance is certainly one of the more versatile tools, it too has its limitations.

Working with Expectance

A frequently encountered problem when applying the expectance operator to a formula is experienced when attempting to reduce the expectance results. In general, you should always work to reduce the results to a set of meaningful or easy-to-compute moments – that is, either the lower-order central moments (e.g., μ, σ^2, σ_{12}^2, etc.) or straight moments (e.g., $E[X]$, $E[X^2]$, etc.). Therefore, when computing the expectance, you will typically have terms such as $E[X]$, $E[X^2]$, $E[aX]$, etc. – these should be converted to standard moments. To convert to central moments, recall that $E[X] = \mu$, $E[aX] = aE[X] = a\mu$, $\sigma^2 = E[X^2] - \mu^2$ so $E[X^2] = \sigma^2 + \mu^2$, etc. Similar expansions can be determined for the higher-order moments, such as $E[X^3]$.

By always working to reduce the expectance set to these fundamental moments, the results will always be presented in the terms of the most basic, intuitively meaningful and easy-to-compute statistical moments.

Interpreting the Values of Statistical Moments

It is important to understand how to interpret the actual numerical value that is assigned to the various moments of random variables or functions of random variables. The actual values can only be properly interpreted, in an *absolute* sense, when knowledge of the distribution for the random variable or its function is known (or approximately known).[13]

For example, the mean price of a house in an area may not reflect the information of interest to someone looking for an affordable place to live if 100 hundred houses have a similar price and 5 houses cost 100 times more (producing a highly skewed distribution of cost).[14] Likewise, if the value of the variance for some distribution is 3.5, this in itself means little, though *relative* to the mean it does give some indication of the spread about the mean. For example, if the mean is 1, then the variance of 3.5 suggests that the random variable has a spread so that some significant fraction of the values lie within $\pm(3.5)^{1/2}$ of the mean, and therefore the noise (that is, this spread) is fairly large relative to the signal (the mean). If, on the other hand, the mean is 1000, then this suggests that the scatter in the variable is exceedingly small relative to the signal (well under a fraction of

[13] Clearly, the first moment always gives an indication of the central value of a distribution, and the variance gives an indication of its spread. However, the actual interpretation or practical implication of the mean will be different if the distribution is highly symmetrical versus highly skewed.

[14] You will use parameters that are more robust to skewed distributions in these cases, such as the median or L1 norm.

a percent). However, if you determine that the random variable displaying this variance has an approximate "normal" distribution (described below), then you can add the additional interpretation that over 90 percent of the measurements made for this random variable should lie within $\pm 2[(3.5)^{1/2}]$ of the mean of this function. From this, you can put quantitative estimates on the likelihood of obtaining any particular value of the random variable or functions of it. Likewise, specific values of the higher-order moments can also be interpreted with respect to the general characteristics of the distribution.

For distributions other than normal, which is the most popular and easily conceptualized distribution, you can return to the graphical interpretation presented for (2.33), in which case knowledge of the distribution allows you to see in a relative sense how the variance (or an analogous treatment for higher-order moments) is distributed for a particular distribution. Either way, though, you need to have some idea regarding the distribution in order to appropriately interpret any of the higher-order moments for the random variable in question.

2.7 Common Distributions and Their Moments

A number of ideal or theoretical distributions (sometimes called **probability laws**) have been determined and studied that prove useful in data analysis and statistical methods. The main reason for this is the fact that many natural processes and experiments seem to obey these standard (ideal) distributions, the properties and characteristics of which are now well understood. Furthermore, even if your random variable or variables do not obey any of these distributions, the distributions often describe common *functions* of random variables, regardless of the random variables' original distribution. For example, the exceedingly popular "normal" distribution (discussed later), applies to almost any function that consists of a linear sum of random variables (such as the mean). Since a large number of analysis techniques involve linear summation, the normal distribution is found to govern the distribution of such sums consistently. Consequently, the characteristics of these well-established and well-understood distributions can be directly applied to the analysis products even when the distributions of the raw data are not well known, or were highly irregular and thus otherwise difficult to work with.

Several of the ideal distributions are called **sample-based distributions**. These are distributions that govern the spread of *estimates* of statistical quantities or parameters (i.e., functions of random variables) and, as such, the specific characteristics of the distribution vary with the number of sample points used to make the estimate. For example, if your sample is from a normal distribution but the sample size is small, you are not likely to get a good representation of the normal distribution, and the t-distribution takes this into account, adjusting to what is expected, given a small sample. These particular sample-based distributions are presented in the next chapter after the concept of parameter estimation is presented. Here, some of the more common, non-sample-based distributions are presented.

2.7.1 Normal Distribution

One of the most common and useful continuous distributions frequently encountered in the real world resembles a symmetrical, bell-shaped curve called the **normal**, or **Gaussian**, distribution. The normal distribution describes many natural phenomena. Also, this distribution, as will be seen by the Central Limit Theorem (discussed later), governs the distribution of estimates of the mean from a sample and of linear combinations of variables in general, regardless of the distribution of the random variable from which the mean or sum is generated. In addition to this feature, which makes the normal distribution arise in a multitude of physical situations, the fact that this distribution is completely described by its first two moments makes it clearly the most convenient and useful distribution known.[15]

The Gaussian PDF is given by the following formula:

$$\mathrm{f}(x) = \frac{1}{\sigma\sqrt{2\pi}} \mathrm{e}^{\left[\frac{-(x-\mu)^2}{2\sigma^2}\right]}. \tag{2.67}$$

Inspection of the PDF reveals that the maximum value of the distribution occurs when $x = \mu$, where $\mathrm{f}(x = \mu) = 1/[\sigma(2\pi)^{1/2}]$

From the equation for variance given by the expectance (2.31b), with $\mathrm{f}(x)$ given by (2.67), you can derive the variance of a random variable with this distribution to be

$$\begin{aligned} \mathrm{Var}[X] &= \int_{-\infty}^{+\infty} (x-\mu)^2 \left[\frac{1}{\sigma\sqrt{2\pi}} \mathrm{e}^{\frac{-(x-\mu)^2}{2\sigma^2}}\right] \mathrm{d}x \\ &= \sigma^2. \end{aligned} \tag{2.68}$$

So, the variance for a normally distributed random variable is given by the parameter σ^2, and the standard deviation is σ. Again by inspection of (2.68), it is revealed that if you move one standard deviation, σ, from the mean, μ (i.e., $x = \mu \pm \sigma$), the value of $\mathrm{f}(x)$ will have dropped from $1/[\sigma(2\pi)^{1/2}]$, its value at μ, to $\mathrm{e}^{-1/2}$ of this maximum.

A normal distribution of mean μ and variance σ^2 is indicated using the notation $\mathbf{N(\mu,\sigma^2)}$.

To compute the probability of drawing a value of X that lies in any particular interval of x, say from a to b, you compute the integral of $\mathrm{f}(x)$ over that range, [a, b]. These values have been tabulated already and can be looked up readily. This is fortunate, since there is no analytic solution to this definite integral except for the case with limits from $-\infty$ to $+\infty$. However, to use the tables you must first *standardize* your data so that they take on some standard range – otherwise the tables would have to be calculated for every combination of μ and σ^2.

[15] The higher-order moments of the normal distribution are not all zero; only the odd moments are zero, whereas the even moments are simple functions of the mean and variance.

Standardization

Standardization involves normalizing a random variable (the data, X) so that it has 0 mean and 1 variance ($\sigma^2 = 1$). Standardization puts the random variable into standard normal form. The standardized variable is often called **Z**,[16] where

$$Z = (x - \mu)/\sigma. \tag{2.69}$$

Notice that when $x - \mu = 1\sigma$, $Z = 1$; when $x - \mu = 2\sigma$, $Z = 2$; and more generally, when $x - \mu = n\sigma$, $Z = n$. Thus, the value of Z indicates how many standard deviations that particular value is from the mean.

So, if you take a normally distributed data set, available as a set of numbers representing a random variable, X, compute the mean and variance of the data, and then subtract the mean from each value and divide this difference by the standard deviation, the resultant quotient for each value of X is now the standardized value of the data, or Z. If $Z = 1$, that particular data value is one standard deviation larger than the mean value of the data set. *Whatever the value of Z, that value represents how many standard deviations away from the mean that particular data value lies.*

The probability of drawing a value of Z between the range of [a, b] is determined by integrating (2.68) after substituting Z and dZ for x and dx. That is,

$$x = Z\sigma + \mu \tag{2.70}$$

and

$$\mathrm{d}x = \sigma\, \mathrm{d}Z. \tag{2.71}$$

Integrating (2.68) after substituting (2.71) and (2.72) into the integrand gives the probability of obtaining a value of Z between a and b:

$$\mathrm{P}\{\mathrm{a} \le Z \le \mathrm{b}\} = \frac{1}{\sqrt{2\pi}} \int_{\mathrm{a}}^{\mathrm{b}} \mathrm{e}^{\frac{-Z^2}{2}}\, \mathrm{d}Z. \tag{2.72}$$

This integral is frequently referred to as the **normal-error integral** or **error function** in many branches of mathematics and physics, and it is usually denoted as **erf(x)**.[17]

Once a normally distributed random variable with distribution $N(\mu,\sigma^2)$ has been standardized, the standardized random variable has distribution N(0,1) and the appropriate probability for a specified range can be calculated directly from the tables.

Reading Standard Normal Probability Tables

Typically, tables of the standard Gaussian distribution (and in fact, most distributions) are presented in terms of their cumulative distribution function (CDF). This gives the

[16] The Z-score is used sometimes in computer statistical packages, but rarely anywhere else (to my knowledge).

[17] Many standard equations often reduce to some form of this integral, the values of which are easily found in tables or computed numerically.

probability of getting a value less than or equal to $x = \mathrm{b}$, i.e. it is the integral $\int_{-\infty}^{\mathrm{b}} \mathrm{f}(x)\mathrm{d}x$. Therefore, to determine the probability of obtaining an observation between a and b (i.e., $\mathrm{a} < x \leq \mathrm{b}$; see note earlier in this chapter regarding limits, titled "Notation and the Intervals of Distribution"), you must look up the cumulative probability of both $x = \mathrm{a}$ and $x = \mathrm{b}$ and then subtract $\mathrm{F(b)} - \mathrm{F(a)}$ to get the probability of obtaining a value of x within this particular interval.

Box 2.3 Example Using a Cumulative Distribution Table for Normal Distribution

If the data have a standard deviation of 4 and a mean of 11, what is the probability of selecting a sample with a value less than or equal to 3? First, convert to Z:

$$Z = (3 - 11)/4 = -2.$$

Since the value of Z represents the number of standard deviations about the mean, for this example, the probability of finding a value ≤ 3 or between $-\infty$ and 3 is the probability of finding a value $-\infty \leq Z \leq -2$ or a value ≤ -2 standard deviations from the mean. So in that case, you look up the cumulative probability corresponding to -2 ($= 2.27\%$) and $-\infty$ ($= 0$), so the probability of getting a value ≤ 3 (2σ shy of the mean) is 2.27 percent.

(The conversion to Z-values presented here is not the same as a Z-transform, which is a particular transformation associated with time series, introduced later in the text.)

Some probabilities for a standardized normal distribution are shown in Figure 2.11 (as approximations).

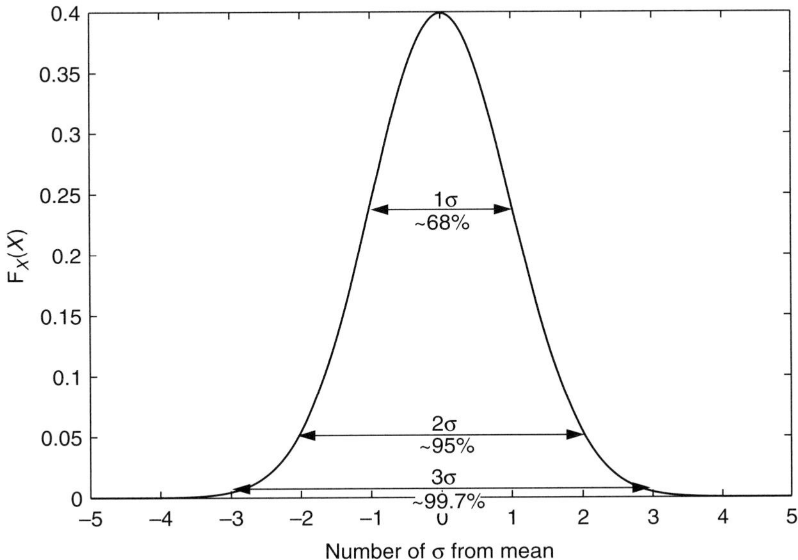

Figure 2.11 Standard Gaussian PDF, with approximate probabilities of drawing a value within 1, 2 and 3 standard deviations of the mean.

So, from Figure 2.11 it is seen that the probability that a single realization (observation or measurement) drawn from a normally distributed random variable will have a value lying within ±1 standard deviation of the true mean is ~68 percent, and within $\pm 2\sigma$ is ~95 percent (the actual number for 95 percent is $\pm 1.96\sigma$). Of course, this also implies that 5 out of 100 realizations will lie outside this interval. The probability is 99.7 percent that a realization will lie within $\pm 3\sigma$ of the mean. That is, on average you would expect 997 out of every 1000 observations to lie within $\pm 3\sigma$ of the mean value, while only 3 would lie outside this range – *but*, 3 out of 1000 *will* likely fall outside that range, and it just may happen that you get one of those three. About 99.99 percent of the observations will fall within $\pm 4\sigma$.

Therefore, if you know the mean and variance of the normally distributed population from which an observation is drawn, you can compute what the actual values of the variables are defining the $\pm 2\sigma$ (i.e., the values between which 95 percent of the observations will lie). This is done by using (2.71):

$$-2 \leq Z \leq +2 \tag{2.73a}$$

and

$$\begin{aligned} &-2 \leq (x - \mu)/\sigma \leq +2 \\ &-2\sigma \leq (x - \mu)/ \leq +2\sigma \\ &\mu - 2\sigma \leq x \leq \mu + 2\sigma. \end{aligned} \tag{2.73b}$$

Therefore,

$$\mathrm{P}\{\mu - 2\sigma \leq x \leq \mu + 2\sigma\} \approx 0.95. \tag{2.74}$$

So, there is approximately a 95 percent probability that other observations (realizations) drawn from the same population will lie within $\mu \pm 2\sigma$ ($\mu \pm 1.96\sigma$, for exactly 95 percent probability).

This interval relationship of (2.74) assumes that you *know* the true mean and variance of your normally distributed random variable (and, in fact, that you *know* that the random variable is normally distributed). In data analysis, you are typically more concerned with *estimating* the mean, variance and distribution in general, from which point you are then interested in the much more difficult question of, given these estimates, what is the probability that the *true* mean lies within the interval of interest? This is essentially the inverse of (2.74) and it is the focus of a statistical discussion for confidence intervals in Chapter 3.

2.7.2 Central Limit Theorem

The widespread use of the normal distribution is in large part due to the consequence of the **Central Limit Theorem**. This theorem is one of the most useful theorems in classical statistics. Its main result states that *the sum of a large number of linearly independent random variables will be approximately normally distributed almost regardless of their individual distributions.*

So, for $Y(X_j) = X_1 + X_2 + \cdots + X_j$, where X_j each have mean μ_j and variance σ_j^2 (despite their individual distributions), as long as $\sigma_j^2 < \infty$ and they are all *approximately the same order of magnitude*, it can be shown that $Y = N(\mu_Y, \sigma_y^2)$.[18] Note that if one of the random variables in the sum has a variance that is many orders of magnitude larger than those of the others, the sum is dominated by that random variable and its distribution, as you might guess.

How "large" the sum must be before it approximates a normal distribution depends on the nature of the individual distributions of the random variables making up the sum. When the component distributions are relatively symmetrical, the sum converges extremely quickly, with as few as say 10 random variables, or even just 4 or 5 for highly symmetrical distributions. For more skewed and irregular distributions, it may require 50 to 100 or more random variables in the sum before it converge close enough to the normal distribution.

The fact that so many random variables in nature are actually the sum of other independent random variables helps explain why the normal distribution is so common in the real world. Also, for random variables that are the product of many different independent random variables, the logarithm of the product can be expected to be normally distributed, since the logarithm of products is equal to the sum of the logs (a random variable whose logarithm follows a normal distribution is distributed according to a log-normal distribution).

In addition to its general importance, the results of the Central Limit Theorem are particularly important when dealing with estimates of the mean of a random variable, since such an estimate consists of a linear sum of realizations of a single random variable, but each of these can be considered an independent random variable (for non-sequential data). In that case, the results of the Central Limit Theorem define the sampling distribution (including a quantitative estimate of its mean and variance) that describes the PDF of an estimate of the mean. This is discussed in more detail in the Chapter 3 when dealing with sample distributions.

Special Case for Summing iid Data

Now consider the special case in which n random variables are being summed to form the mean of the variables, or the random variable Y, and each of the original random variables is *independent and identically distributed* (**iid**), with means μ and variance σ^2. In this case, the Central Limit Theorem provides an even more useful result that is applicable to the evaluation of the uncertainty in means, since it now indicates that not only is Y evenly distributed in the limit as n goes to infinity, but it has variance equal to the original variance divided by the number of random variables contained in the sum. That is,

$$\mu_Y = \mu \tag{2.75}$$

[18] The proof employs the method of moments, or its more general treatment through characteristic functions. For the Cauchy distribution, which does have infinite variance, the sum of Cauchy-distributed random variables results in a Cauchy-distributed sum.

$$\sigma_Y^2 = \sigma^2/n. \tag{2.76}$$

So, $(Y - \mu)/(\sigma^2/n)^{1/2}$ obeys N(0,1) as $n \longrightarrow \infty$.

This result can be shown directly from use of the expectance operator (see Box D2.1) shown by De Moivre in 1733.

The relationship described in Box D2.1 holds even when making weighted sums of random variables, since the weights simply change the distribution of the component variables relative to one another.

Box D2.1 Derivation of Mean and Variance for the Special Case of the Central Limit Theorem via Expectance

Consider the function $Y = \bar{x} = n^{-1}\sum_{i=1}^{n} x_i$, which computes the mean of n identical and independently distributed random variables (here, the notation is such that each random variable x_i is a different realization from the same population X, though the random variables can each be from separate populations as well). The mean of this is computed by the expectance, as

$$\begin{aligned}\mu_Y &= \mathrm{E}[Y] \\ &= \mathrm{E}\left[\frac{1}{n}\sum_{i=1}^{n} x_i\right] \\ &= \frac{1}{n}\sum_{i=1}^{n}\mathrm{E}[x_i] \\ &= \frac{1}{n}\sum_{i=1}^{n}\mu_i,\end{aligned} \tag{D2.1.1a}$$

and for identical means, so $\mu_i = \mu$

$$\begin{aligned}&\frac{1}{n}\sum_{i=1}^{n}\mu \\ &= n\mu/n \\ &= \mu.\end{aligned} \tag{D2.1.1b}$$

The variance of a linear function is

$$\sigma_Y^2 = \mathrm{E}[Y^2] - E[Y]^2 \tag{D2.1.2}$$

so, if $Y = \frac{1}{n}\sum_{i=1}^{n} x_i$, then:

$$\begin{aligned}\sigma_Y^2 &= \mathrm{Var}\left[\frac{1}{n}\sum_{i=1}^{n} x_i\right] \\ &= \mathrm{E}\left[\left(\frac{1}{n}\sum_{i=1}^{n} x_i\right)^2\right] - \left\{\mathrm{E}\left[\frac{1}{n}\sum_{i=1}^{n} x_i\right]\right\}^2\end{aligned}$$

Box D2.1 (Cont.)

$$
\begin{aligned}
&= \frac{1}{n^2}\mathrm{E}[(x_1+x_2+\ldots+x_n)^2] - \frac{1}{n^2}\left\{\sum_{i=1}^{n}\mathrm{E}[x_i]\right\}^2 \\
&= \frac{1}{n^2}\mathrm{E}\left[\sum_{i=1}^{n}\sum_{j=1}^{n}x_i x_j\right] - \frac{1}{n^2}(\mathrm{E}[x_1]+\mathrm{E}[x_2]+\ldots+\mathrm{E}[x_n])^2 \\
&= \frac{1}{n^2}\mathrm{E}\left[\sum_{i=1}^{n}\sum_{j=1}^{n}x_i x_j\right] - \frac{1}{n^2}\sum_{i=1}^{n}\sum_{j=1}^{n}\mathrm{E}[x_i]\mathrm{E}[x_j] \\
&= \frac{1}{n^2}\left\{\sum_{i=1}^{n}\sum_{j=1}^{n}\mathrm{E}[x_i x_j] - \sum_{i=1}^{n}\sum_{j=1}^{n}\mu_i\mu_j\right\} \\
&= \frac{1}{n^2}\sum_{i=1}^{n}\sum_{j=1}^{n}\mathrm{Cov}[x_i x_j], \qquad \text{(D2.1.3)}
\end{aligned}
$$

recalling that $\mathrm{Cov}[XY] = \mathrm{E}[XY] - \mathrm{E}[X]\mathrm{E}[Y] = \mathrm{E}[XY] - \mu_x\mu_y$. The sums in (D2.1.3) can be expanded as

$$
\begin{aligned}
= \frac{1}{n^2}\{&\mathrm{Cov}[x_1x_1] + \mathrm{Cov}[x_1x_2] + \ldots + \mathrm{Cov}[x_1x_n] + \\
&\mathrm{Cov}[x_2x_1] + \mathrm{Cov}[x_2x_2] + \ldots + \mathrm{Cov}[x_2x_n] + \ldots + \\
&\mathrm{Cov}[x_nx_1] + \mathrm{Cov}[x_nx_2] + \ldots + \mathrm{Cov}[x_nx_n]\}. \qquad \text{(D2.1.4)}
\end{aligned}
$$

For each term in which $i \neq j$, $\mathrm{Cov}[x_i x_j] = 0$, since the x are independent of one another – that is, they are independent observations of X. When $i = j$, $\mathrm{Cov}[x_i x_i] = \mathrm{Var}[x_i]$, so the preceding reduces to

$$
\begin{aligned}
&= \frac{1}{n^2}(\mathrm{Cov}[x_1x_1] + \mathrm{Cov}[x_2x_2] + \ldots + \mathrm{Cov}[x_nx_n]) \\
&= \frac{1}{n^2}\sum_{i=1}^{n}\mathrm{Var}[x_i] \qquad \text{(D2.1.5)} \\
&= \frac{1}{n^2}n\sigma^2 \\
&= \sigma^2/n.
\end{aligned}
$$

2.7.3 Binomial (or Bernoulli) Distribution

The binomial distribution is the distribution that describes the probability that any particular event will occur x times (integer number of successes) in N independent trials (or realizations), where the probability of the event occurring in any one trial or realization is given by p. The distribution is given by

$$
\mathrm{P}(x) = \frac{N!}{x!(N-x)!}\mathrm{p}^x(1-\mathrm{p})^{N-x}, \qquad (2.77)
$$

where ! indicates factorial ($N! = N(N-1)(N-2)\ldots 1$ and $0! = 1$). $N - x$ therefore represents the number of times the event will not occur (failures) and $1 - \mathrm{p}$ (typically denoted as q) is the probability that the event will not occur in any one trial.

The ratio in (2.77) is often abbreviated as

$$\binom{N}{x} = \frac{N!}{x!(N-x)!}\mathrm{p}^x\mathrm{q}^{N-x}. \tag{2.78}$$

From this it is seen that the probability of obtaining 5 heads in 8 tosses is

$$\begin{aligned}\binom{8}{5} &= \frac{8!}{5!(3)!}(0.5)^5(0.5)^3 \\ &= \frac{40320}{120(6)}(0.03125)(0.125) = 0.21875.\end{aligned} \tag{2.79}$$

Another case (which will prove useful later) is the probability of getting x successes in x trials. That is, $x = N$. In this case,

$$\begin{aligned}\frac{N!}{N!(N-N)!}\mathrm{p}^N(1-\mathrm{p})^{N-N} &= \frac{1}{0!}\mathrm{p}^N(1-\mathrm{p})^0 \\ &= \mathrm{p}^N.\end{aligned} \tag{2.80}$$

In other words, the probability of getting N successes in a row is equal to the product of the probability of getting one success, N times.

The first two moments of the binomial distribution are given by

$$\mu = N\mathrm{p} \tag{2.81}$$

$$\sigma^2 = N\mathrm{pq}. \tag{2.82}$$

When N is large, the binomial distribution can be approximated by the normal distribution, with the standardized variable given as

$$Z = \frac{X - N\mathrm{p}}{(N\mathrm{pq})^{1/2}}. \tag{2.83}$$

The approximation improves with increasing N and becomes exact when N is infinite. Typically, the approximation is very close even when Np and Nq are greater than 5.

2.7.4 Poisson Distribution

A distribution that is related to the binomial distribution is the Poisson distribution. The distribution, for positive integer values of x as in the binomial distribution, is given by

$$\mathrm{P}(x) = \frac{\lambda^x e^{-\lambda}}{x!}, \tag{2.84}$$

where λ is a positive constant ($\lambda > 0$).

This distribution closely approximates the binomial distribution when the probability of occurrence of an event is close to 0 (so $q = 1 - p \approx 1$), so the occurrence of the event is rare. Rarity is typically realized when $Np < 5$, for which case $\lambda = Np$. As λ increases, the Poisson distribution approaches a normal distribution with standardized variable

$$Z = \frac{X - \lambda}{\lambda^{1/2}}. \tag{2.85}$$

The first two moments of this distribution are

$$\mu = \lambda \tag{2.86}$$

$$\sigma^2 = \lambda. \tag{2.87}$$

2.7.5 Uniform Distribution

Another important distribution is the uniform distribution in which all events have an equal likelihood of occurrence over its entire range (from a to b). The PDF is given as a rectangle whose value is equal to 1/(b − a) over the entire range of values within this range, and 0 outside of this range. The cumulative distribution is a steadily growing ramp.

2.7.6 Exponential Distribution

Finally, a general PDF whose special cases include the binomial, Poisson and a great many other distributions is the exponential distribution (the full suite of special cases are sometimes referred to as members of the exponential family of distributions). This distribution proves useful regarding some of the analysis techniques presented later in the text, at which point this PDF will be discussed in more detail.

2.8 Take-Home Points

1. A random variable is any variable whose exact value cannot be assigned with certainty.
2. The best way to classify a continuous random variable is via a probability density function (PDF), or for discrete data, a probability mass function (PMF, for discrete data).
3. Expectancy (the average) is used to compute the mean. It is indicated by $E[X] = \frac{1}{n}\sum_{i=1}^{n} X_i$.
4. Expectancy in the form of $E[(X - \bar{x})^2] = E[X^2] - E[X]^2$ gives the variance of X, or the formula $y(X)$. ***This is the form that will be applied to your data formulas to estimate the uncertainty in your results***.
5. The ***Central Limit Theorem*** is the most important theorem in statistics. It states that the average of a large sum of independent variables will be distributed as a Gaussian

distribution, no matter what distribution the numbers originally have. For the special case where all variables come from the same distribution (independent identically distributed, iid; not uncommon for data collection), the average will be the true average, and the variance of the sum (e.g., the mean) will be the original variance (σ^2) divided by n, the number in the sum; the square root of that variance is called the ***standard error***; $\sigma_{se} = (\sigma^2/n)^{1/2}$.

2.9 Questions

Pencil and Paper Questions

1. a. What is the difference between a PDF and a PMF?
   b. What is the uncertainty in a fourth-order third-degree polynomial (i.e., what is Var $[y(x)]$)?
   c. Write the kth-order central moment in terms of expectance.
   d. Describe mean, median and mode.

2. For large-scale oceanic and atmospheric circulations, it is often a good approximation to assume that the pressure gradient force balances the Coriolis force (an apparent force due to the earth's rotation). In the ocean, we can measure profiles of salinity and temperature at discrete locations and compute geopotential anomalies. These are basically height anomalies relative to a fixed level that indicate a density gradient (and hence, a pressure gradient) between the two locations, and from these we can estimate what the currents between those two locations are. The dynamic method of calculating geostrophic currents v is discretized as the following function of multiple random variables:

$$v(\mathrm{p}) = \frac{1}{f}\frac{\Delta\varphi_2 - \Delta\varphi_1}{L} + v_{ref}.$$

   Here, $\Delta\varphi_1$ and $\Delta\varphi_2$ are the geopotential anomalies between a level reference surface and surface p at locations 1 and 2, L is the distance between locations 1 and 2, and v_{ref} is the velocity at the reference level. The Coriolis parameter f can be regarded as a known constant. Suppose $\Delta\varphi_1$ and $\Delta\varphi_2$ have uncertainty σ_1 and σ_2 and that L and v_{ref} have uncertainties σ_L and σ_v, respectively. Derive an expression for the uncertainty in v. You should assume that the errors are uncorrelated.
3. Consider the Poisson distribution $P\{X = x\} = \lambda^x e^{-\lambda}/x!$. This distribution describes the probability of a given number of events (x) occurring in an interval of time (t) if the events occur independently of each other, but with a certain average rate ($1/\tau$). It can be used to model, for example, earthquake occurrence. The distribution depends on the single parameter $\lambda = t / \tau$ (≥ 0), and its first two moments are both μ. Suppose you've collected a sample of n measurements x_i (e.g., numbers of earthquakes in a given time interval) from a population known to follow a Poisson distribution. Use the Principle of Maximum Likelihood to estimate λ, as $\hat{\lambda}$. Now

suppose you've measured many independent datasets X_j from the population, supposing the population λ is known. Compute $E[\hat{\lambda}]$ and $Var[\hat{\lambda}]$. Is $\hat{\lambda}$ unbiased? Is it consistent? With one week as the unit of time (t), you've used your data to determine that earthquakes occur with the average rate $1/\hat{\tau} = 2$ per week ($\hat{\lambda} = t/\hat{\tau} = 2$). What is the probability that at least three earthquakes occur during the next week?

4. Give $\sigma_\mu^2 = Var\left[\frac{1}{n}\sum_{i=1}^{n} x_i\right]$ where the x_i (with σ_i^2) are not squared and are independent and identically distributed ($\sigma_i = \sigma$). Give your answer in terms of original σ. Show how you get this answer.

3 Statistics

3.1 Overview

The distribution of a random variable (rv) characterizes the random variable. Because of this, one of the most fundamental goals of statistics is to be able to infer or estimate the *true* distribution of a random variable based on the results of one or more experiments or observational sets – samples of the data. Another important goal of statistics is to provide a quantitative means for assessing the reliability of these inferred properties. Statistics also provides the means for using a sample (i.e., your data) to extract information of interest (e.g., signal from noise, trends, relationships to other variables) and to quantitatively assess how consistent the extracted information is with a particular hypothesis or model.

Most of this chapter deals with *estimation*, how we extract specific information from the data, the backbone of **inferential statistics**.

3.2 Estimation

Methods of inferring information about the distribution or characteristics of a random variable involve the concept of estimation. In particular, you need to estimate the distribution of the random variable itself, its most relevant moments and/or other relevant characteristics from the data. The data represent a *sample* drawn from some population. The term "population" represents the complete phenomenon being studied, and it is characterized by the *true* distribution of the random variable. The random variable (or some combination of random variables) is typically a numerical quantification of the phenomenon being studied and it is represented by the data (the observations or realizations making up the sample) or some functional combination of the data. If properly set up, the data represent measurements drawn from the actual population you are interested in studying; otherwise the results of the analysis will be biased.

While the concepts of population, random variable and sample are simple, in practice you must be very careful to ensure that these various components of the study are well thought out and internally self-consistent. That is, the random variable(s) being observed and/or combined must indeed represent the phenomenon of interest. The population must accurately represent the phenomenon, recognizing that many phenomena of interest are not easily compartmentalized, as they may evolve in time or over space.

These seem like obvious considerations, but the inconsistencies that often arise between random variable, population and sample are usually subtle and easily overlooked, and this can lead to erroneous conclusions.

Box 3.1 Example of a Population

If the phenomenon of interest is the average surface temperature during the month of June in any one location or region on the Earth, then the population consists of all average temperatures for any June that has ever existed (at least under general global climate conditions similar to today's), as well as the set of all average temperatures that ever could exist if the Earth were to be "rerun" an infinite number of times in order to see what average temperatures ultimately occur, and with what relative frequency. The true distribution presents the relative frequency with which any average temperature can occur (specifically, for any small range of temperatures for this continuous random variable). That distribution would describe the (infinite) collection of average temperatures present in the population. However, we must consider whether we will admit into the population any surface temperature within the region, or only those that are within a specific topographic elevation range or specific geographic setting (e.g., open plains, or excluding valleys or concentrated urban centers). If there are bodies of water in the region, will the population include the land-only surface temperature, or the average temperature of the entire region including that of the water?

These are some of the easy considerations. More difficult in this case is the time period over which June temperatures will still be admitted to the population. If you are trying to study modern climate, you certainly do not want to include June temperatures that might occur during an ice age, or during times millions of years ago when the locale of interest was located on a completely different spot of the globe, owing to plate tectonics. Those are obvious exclusions, but what about data collected in this century that may include influences of anthropogenic warming? Whether or not those are admitted into the population depends upon the questions being asked. If you do not admit those, you might define our population based on all *natural* average temperatures (those without the influence of society), against which you could then compare any present day value and use this comparison to help decide whether the values obtained in the past few decades are unreasonable (in a statistical sense) and therefore likely display the effects of a greenhouse warming.

Is temperature the random variable most suited to define the population to characterize the phenomenon of interest, or is heat content better? Perhaps climate is the desired phenomenon; temperature represents only one aspect of climate, when you may actually wish to also know precipitation, evaporation, windiness, cloudiness etc.

If you stick with surface temperature, the sample represents a finite number of average June temperatures that have been measured over some number of years. If you have excluded or admitted specific geographic settings from the

Box 3.1 (Cont.)

population, the sample must also be true to those criteria. From this sample, you wish to produce an estimate of the true distribution of the population of all June average temperatures so that you can assess how representative any particular year is, relative to what to expect on average. You may also wish to know what sort of range of values can be expected; what is the most likely value for any one June; what are the more extreme values you might expect in bad years, hot or cold; are you more likely to deviate from the mean value toward more cold values, more hot values, or approximately equal representation of hot and cold deviations, etc.

The object of statistics is to provide, among other things, guidelines for determining just how unreasonable such values are, or how unreasonable the values would have to be before one felt highly confident that they were showing an actual deviation from a naturally varying month of June (e.g., the influence of society, etc.). These turn out to be very simple questions with extremely difficult to determine, yet important, answers.

A characteristic of the population is called a **parameter**. An *estimate* of a population parameter *and* the formula used to make the estimate is generally referred to as an **estimate, statistic** or **sample statistic**.

Formally, a statistic is a *function* of a random variable. That is, in order to estimate a parameter of the population from a particular sample, you must combine the sample realizations in some manner prescribed by a formula or function. This is an important concept, since the sample statistic is based on the mathematical manipulation of the observations of the random variable, and *thus is itself a random variable.*

For example, the mean of a population, one of the most fundamental statistics, is estimated by adding all of the sample values together and dividing by the sample size. This formula *and* the value it produces are both called a statistic, **estimator** or, in this case, **point estimate** (to be differentiated from interval estimate, as discussed later). It is important to remember that the single value estimated is actually the output of a specific function, and its value can thus change with each new realization added to the sample, reflecting the fact that the estimate itself is a random variable.

Notation

It is relatively standard practice to use Greek letters to represent parameters of the population (i.e., the true value describing some characteristic of the true distribution) and Roman letters to represent the corresponding sample statistic, estimated from the sample (i.e., the *estimate* of the true parameter). Another common notation is to use a Greek letter, such as φ, for the true parameter value, and the same letter with a "hat" to denote the estimate, $\hat{\varphi}$.[1]

[1] Some people only use the "hat" notation to designate "maximum likelihood" statistics (discussed later in this chapter).

I will deviate from that standard when using certain common parameters and their estimates, such as that for population mean, where the parameter is usually given by μ but the estimated value is often given by placing a bar over the rv or by placing it in angle brackets, as $\bar{x}$ and $<x>$. Likewise, variance, given by σ^2, has an estimate usually denoted by s^2. Sometimes the character is subscripted to indicate the rv for which it applies, such as s_x^2, denoting the sample variance of the rv X. Similar changes apply to the common higher-order moments (so, you ask, what else is there?), but I will make it clear whenever a new sample statistic is being introduced. A statistic includes uncertainties that you ultimately wish to assess, whereas the parameter is the real exact value of the population that has no uncertainty whatsoever (and is, unfortunately, unknown). Ideally, the central value of the estimator should lie exactly on the population parameter it is estimating, and the spread of the estimator should be as tight as possible about this central value so the uncertainty in our estimate will be relatively small.

The following definitions more formally encapsulate these ideal needs.

A **consistent statistic** is one such that, the larger the sample size (n), the more accurate and precise is the estimate produced. An **inconsistent statistic** does not provide a more accurate estimate with increasing sample size (the periodogram, which is discussed later, is an example of an important inconsistent statistic).

An **unbiased statistic** is one that produces an estimate equal to the true population parameter value in the limit when the sample size is equal to the entire population (so the entire population was sampled, or, alternatively, an estimator whose average gives the true parameter value); a **biased statistic** does not converge to the true value. Here, *statistic* refers to the formula used to make the estimate, as much as to the value produced. In other words, the formula used to make the estimate is flawed in that the value it produces would not equal the true parameter value even if the entire population had been sampled to estimate it.

The bias, B, of a statistic, $\hat{\alpha}$, estimating a parameter α, is defined as

$$B(\hat{\alpha}) = \alpha - E[\hat{\alpha}]. \tag{3.1}$$

Note that statistics can be designed with minimal variance, but this can lead to a common problem whereby making the variance smaller leads to a larger bias (which may or may not be acceptable, as shown in a later chapter on spectral analysis). Alternatively, a statistic can be designed that minimizes the **mean square error** (**MSE**), where MSE is given as

$$\text{mse}(\hat{\alpha}) = E(\hat{\alpha} - \alpha)^2 = \text{Var}(\hat{\alpha}) + |B(\hat{\alpha})|^2. \tag{3.2}$$

Minimization of MSE provides a balance between minimizing the variance and bias together. If the minimal MSE estimator is achieved with zero bias, the estimator is called a **minimum variance-unbiased estimator**, or **MVUE.**[2]

Restricting the evaluation to the class of all linear estimators, then, for a random sample consisting of n values,

[2] To determine the true MVUE, all estimators must be evaluated (linear or other, and thus its evaluation is more difficult), typically involving what is known as the Cramer Rao inequality, which is a ratio comparing the variance of estimators (not considered here).

$$\hat{\alpha} = a_1x_1 + a_2x_2 + \ldots + a_nx_n$$

$$\hat{\alpha} = \sum_{i=1}^{n} a_i x_i. \tag{3.3a}$$

The mean of this type of estimator is

$$E[\hat{\alpha}] = E\left[\sum_{i=1}^{n} a_i x_i\right] = \sum_{i=1}^{n} a_i E[x_i] = \mu \sum_{i=1}^{n} a_i. \tag{3.3b}$$

As long as $\sum_{i=1}^{n} a_i = 1$, the statistic is unbiased. Minimizing this function provides the **best linear unbiased estimator (BLUE)**.

An **efficient statistic** is a consistent estimator that converges more rapidly than another to the true parameter value as n increases to infinity or to the true, finite population size. The efficiency of one estimator relative to another is often presented as a ratio of the variances for the two estimators (computed by applying the expectance operator on the estimators).

Ideally, it is best to use consistent, unbiased and efficient estimators. In some cases, the non-efficient ones may introduce unacceptably large errors in the estimate. For a given sample size, because of the faster convergence of the efficient estimator, it provides the estimate with a smaller error (i.e., a better estimate for the same size sample). Sometimes, however, it may prove convenient to simply increase the sample size and use a less efficient estimator than to use a more efficient estimator, which may be more computationally difficult with the smaller sample size. Similarly, it is sometimes convenient to use a biased statistic, if the bias is small and the other characteristics of the statistic are advantageous – you decide based on your analysis objectives and the statistics you have available.

3.3 Estimating the Distribution

Since the most complete information regarding a random variable is its distribution, an obvious place to begin an analysis is with an estimation of the distribution based upon the distribution of values within the sample.

A **relative frequency distribution plot** (or **histogram**) is the easiest graphical estimate of the true distribution. To construct this, one divides the data into classes (contiguous or discontiguous, but unique, segments, also called categories, cells, groups, ranges, bins and other similar descriptors). The number of observations (i.e., the frequency of occurrence) in each class is then counted, and the sum is called the **class frequency**. As shown for constructing the PMF in the previous chapter, each class frequency is then divided by the sample size, n, to yield its **relative frequency of occurrence** – an estimate of that class's probability. A plot is constructed in which this number (the relative frequency of occurrence) is plotted against the class value – the former on the ordinate, the latter on the abscissa. For discrete data, each class might represent a unique event, or if there are too many events to be useful, one might construct classes combining the events into logical groupings (e.g., contiguous ranges or neighboring events).

Box 3.2 Real PMF Example

Consider the problem of trying to determine if the Earth over the last several million years has had two stable states – glacial (ice age) and interglacial (~ice free) climates – or if the transitions between these states leads to a continuum of climates. For this, we examine the 5.3 million-year-long record of ice ages (LR04) of Lisiecki and Raymo, 2005. This record is constructed by measuring the relative fraction of "heavy" oxygen isotopes (^{18}O): this heavy isotope requires more energy to evaporate and is more likely to be precipitated out from the atmosphere. The light (more common) isotope (^{16}O) is evaporated and, eventually, precipitated onto land, where it runs off back into the ocean, unless it is trapped in glacial ice and doesn't return to the ocean, leaving the ocean enriched in the heavier isotope (determined by examination of the oxygen isotope composition of the shells of marine microfossil shells, one of the great tools of paleoceanographers). LR04 shows the oxygen isotope values through the last 5.3 million years, as shown in Figure 3.1.

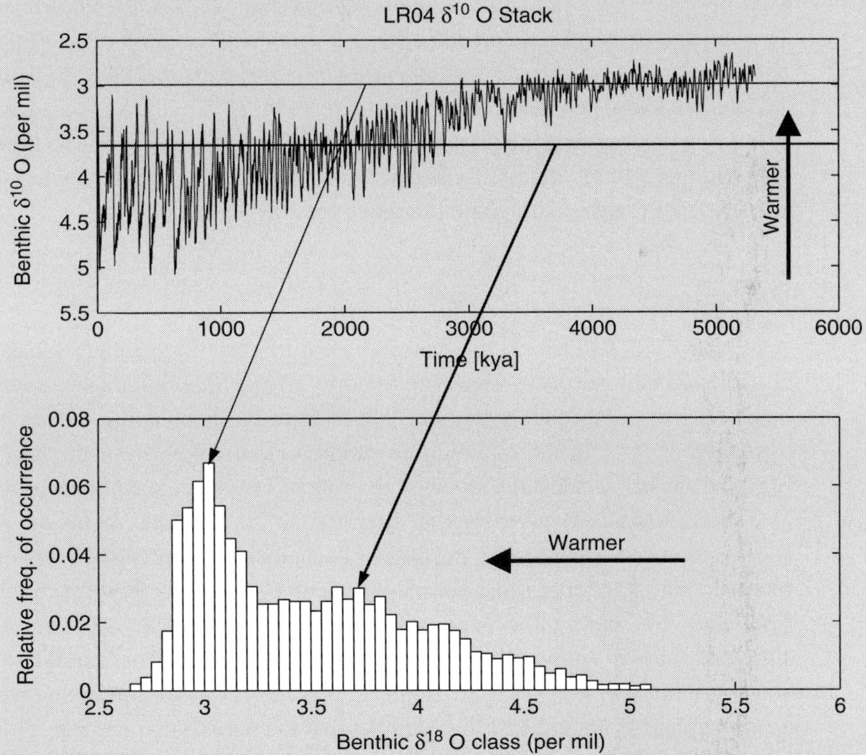

Figure 3.1 The LR04 averaged climate record over the past 5.3 million years and PMF of the values in the climate record, compared in an attempt to determine if (1) there have been two stable (warm and glacial) climate states over this period or if (2) there has been a continuum of climate states (or perhaps some other non-considered arrangement).

Box 3.2 (Cont.)

A PMF of the values hints at a slight tendency toward two peaks, suggesting that perhaps there are two stable modes of climate: cold periods (more global ice) and warm periods.

However, when we look at the original data in the Figure 3.1 upper panel, it is apparent that perhaps the "warmest mode" is a consequence of the overall warmer climate period prior to the last million years. We will account for this later by removing this time-varying mean state. We will develop this particular example throughout the text and continually extract more information via more analyses.

The sample PMF (i.e., histogram for the data sample) plot presents the degree of symmetry of the distribution and whether the sample shows a multi-modal distribution – that is, whether the values tend to group into clusters.

The estimate of the PMF through the relative frequency distribution is one of the most fundamental displays you can make for your samples, and a simple cursory examination of this will often provide the appropriate direction of an entire analysis approach.

Later we show how such estimates can be compared to theoretical ideal distributions in order to determine if the random variable of interest displays one of these well known distributions, which, if so, affords a variety of additional insights owing to the detailed knowledge already existing and tabulated for such distributions.

3.3.1 Outliers

A sample PMF also reveals the presence of **outliers**[3] – anomalous values that occur in classes well beyond those where the majority of the observations lie. Such outliers may represent true, albeit highly unusual values from the population, or they may represent a mistake or error in the sampling procedure or other portion of the analysis treatment (e.g., the analyst accidentally stood on the patient's broken toe while measuring his pulse).

Outliers should receive special attention to see if their inclusion in the data is warranted and to see what the impacts of such aberrant points may have on the analysis results. That is, a certain location may typically experience low precipitation amounts (mm/day), but during the passage of a hurricane it might experience rainfall in a single day that is equivalent to several years' worth of normal rainfall. When looking at a distribution of daily rainfall amounts, this single day of hurricane precipitation will clearly appear as an outlier, lying well beyond the normal distribution. The value is real, but one needs to consider the purpose of the analysis and the underlying issues being addressed to see if this outlier value should be included in the analysis. If a hurricane only passes through the region once in 100 years on average and the primary question is whether the region is appropriate for growing a precipitation-sensitive crop, the unusual

[3] Sometimes more colorfully referred to as "**fliers**," not "fryers," which are usually chickens.

occurrence of a hurricane may be irrelevant to the question and its admittance to the data set may only serve to muddle the analysis and conclusions.

Alternatively, *if it is not clear whether an outlier is real or relevant, you might repeat an analysis with and without it to see what its consequence is* (we show later how such values can exercise strong control on certain analysis techniques and approaches). It is perfectly acceptable to present both results and then state which path you will take, and follow it (periodically you could mention how your conclusions might change if the outlier(s) had been left in or taken out of the analysis). You needn't pretend you know things better than you do – you are just trying to make reasoned deductions based on the available evidence.

3.4 Point Estimates

A distribution itself is somewhat unwieldy mathematically unless you can show that it well approximates one for which a function expression exists. Consequently, it is typically more convenient to deduce the important characteristics or parameters of the true population from your sample. That is, *you wish to reduce the data so that the relevant information is expressed by relatively few numerical values, which provide the most useful information about the process being studied.* However, in order to know which parameters may be most useful (e.g., mode, median or mean) and how to properly interpret them, you must first have an idea of the nature of how the data are distributed (so, start with a sample PMF).

Much of what you wish to know falls under the category of **point estimates** – estimates of particular characteristics or parameters that describe the distribution. Any distribution can be completely described by the appropriate number of parameters. In Chapter 2, when working with the full population we used expectance directly to compute the parameters, but now we are only working with a sample, and we must *estimate* the parameters, and this may as a consequence require subtle alterations to the straight form of the expectance operator.

3.4.1 Estimating the Central Value of a Random Variable

The mean is the most efficient statistic of the various estimates of the central value of a random variable, and typically the one most often employed. Because of the Central Limit Theorem, the mean is often ideal for use even when the parent population has a distribution that is not symmetrical or well behaved. The mean is a statistic that is consistent, unbiased and efficient.

For random variables that are symmetrically distributed, the median also provides a consistent and unbiased estimate of the population mean. However, the ratio of the variance of these two different estimators, the mean and median, shows that the efficiency of the median compared to the mean is 64 percent. Since both estimators are proportional to the number of samples, this indicates that the arithmetic mean will produce an estimate of the mean using only 64 realizations, at a precision that could only be matched by the median using 100 realizations. In other words, the median estimator is wasting 36 percent of the observations – with 36 percent fewer observations, you could have achieved an equally good precision in your estimate of the central value if you had used the arithmetic mean.

On the other hand, the median has certain characteristics that make it particularly "**robust**" in situations involving random variables whose distributions are highly skewed. That is, as demonstrated later, the median is capable of making estimates of the central value that are relatively insensitive to the presence of realizations with values far from the mean (e.g., outliers), whereas an estimate such as the mean is relatively sensitive to the presence of such values. Consequently, the estimated value of the mean may change significantly, given one or more outliers, while the value of the median will change fairly little.

The mean is estimated in a variety of ways. For discrete variables grouped into h classes (representing the h values the discrete variable can take), the estimate, $\bar{x}$ (or $<x>$), or **sample mean**, is given directly from the expectance operator by

$$\bar{x} = \sum_{j=1}^{h} x_j \mathrm{p}_X(x_j), \tag{3.4a}$$

where the probability of each of the h classes of X is given by $\mathrm{p}_X(x_j)$, or for ungrouped data,

$$\bar{x} = \frac{1}{n}\sum_{i=1}^{n} x_i. \tag{3.4b}$$

Equations (3.4a) and (3.4b) are identical, except that the latter works with the individual realizations, each of which has the same probability of occurring. That is, each of n values ($i = 1,2, \ldots, n$) of x_i has probability $\mathrm{p}_X(x_i) = 1/n$. If our sample size is n, we know we have a $1/n$ chance of drawing any realization of X. The former groups the like-values of x_j into a collection of h classes ($j = 1,2, \ldots, h$), the PMF. Each class value has a unique probability of occurring, which is estimated by dividing how many values occurred in the class by the total number of realizations.

You determine if the sample mean is a biased estimator by employing (3.1),

$$\begin{aligned} \mathrm{B}(\bar{x}) &= \mu - \mathrm{E}[\bar{x}] \\ &= \mu - \mathrm{E}\left[\frac{1}{n}\sum_{i=1}^{n} x_i\right] \\ &= \mu - \frac{1}{n}\sum_{i=1}^{n} \mathrm{E}[x_i] \\ &= \mu - \frac{1}{n}\sum_{i=1}^{n} \mu \\ &= \mu - \mu \\ &= 0. \end{aligned} \tag{3.4c}$$

Thus, the expected value of the sample mean – that is, the estimate of the mean you would obtain if the entire population were sampled – is indeed equal to the true mean: no bias.

For continuous variables with range between x_a and x_b (designated a and b, respectively), say $y(x)$, the mean, $\bar{y}$, is calculated as

$$\bar{y} = \frac{1}{x_{range}} \int_{a}^{b} y(x) \mathrm{d}x, \tag{3.5}$$

where $x_{range} = x_b - x_a$. This latter form (3.5), versus (3.4b), is important when computing the mean of a time series that has been sampled at uneven increments of the independent variable.[4]

You can compute a **trimmed mean** by removing the m largest and m smallest values of the sample. This often leads to a more robust calculation of the mean. There are other ways to compute the mean when, for example, the error associated with each value x_i varies with the x_i (discussed later in the book). Instead of the trimmed mean, the median, which is robust without removing the highest and lowest values as stated above, is typically more respected – unless the extreme values are clear outliers, in the sense that good knowledge of understanding of the data indicates that they are completely unrepresentative of the process being measured and consequently can be removed (e.g., if a fly flew into the spectrometer during one of the measurements, that measurement can safely and honestly be removed, unless you are measuring flies).

3.4.2 Estimating the Spread of a Random Variable

Next we wish to estimate the uncertainty or variance in our statistic (in the above case, the sample mean estimator). When estimating variance, the estimate is indicated by s^2, and we might assume that we should follow the standard form of the expectance operator for this second central moment, as shown in the previous chapter. In that case, using expectance for ungrouped data,

$$s^2 = \frac{1}{n} \sum_{i=1}^{n} (x_i - \mu)^2. \tag{3.6}$$

Thus, the sample variance as shown here is computed about the true mean. However, the true mean is typically unknown, and must be estimated itself, as $\bar{x}$, from (3.2). In that case, (3.6) would look like

$$s^2 = \frac{1}{n} \sum_{i=1}^{n} (x_i - \bar{x})^2. \tag{3.7}$$

Because the true mean is not used, it tends to introduce a bias into the estimate as formulated above. This is seen by computing the bias, as shown in the derivation box below that gives a bias of σ^2/n. That bias is eliminated by multiplying the sum by $1/(n-1)$ instead of n. Thus, an unbiased estimate of the variance is produced if the following form is adopted for the estimator:

[4] When computing the integral, you are assigning a functional shape to fill in the gaps between the data points, so you have not really overcome the problem of uneven sampling, but rather have simply transferred it to an area that is frequently out of sight (and thus out of mind). In this case there is no reason not to interpolate the data to even increments.

$$s^2 = \frac{1}{n-1}\sum_{i=1}^{n}(x_i - \bar{x})^2. \tag{3.8}$$

This provides an unbiased estimate, which may be better when the sample size is small, since the bias, σ^2/n, is close to 0 for large sample sizes but may be significantly different from 1 for smaller sample sizes.

Box D3.1 Derivation of the Bias of Sample Variance

The bias of an estimator is reflected in how the expected value of the estimator deviates from the true value of the parameter being estimated (the latter being determined, as always, by applying the expectance operator in which the sample size n = N; the full sample size). So, in order to compute the bias in the variance estimator as defined by (3.7), it is necessary to compute its expected value and compare that to the true value, σ^2.

First consider the expected value itself when n = N, but using the estimated value of the mean, $\bar{x}$, instead of the true mean, σ. In other words, we must compute the expectance of the operator, but the operator will contain an estimate of the mean, not the true mean.

$$s^2 = \frac{1}{n}\sum_{i=1}^{n}(x_i - \bar{x})^2, \tag{D3.1.1}$$

and the expected value of this estimator is

$$\begin{aligned} E[s^2] &= E\left[\frac{1}{n}\sum_{i=1}^{n}(x_i - \bar{x})^2\right] \\ &= E\left[\frac{1}{n}\sum_{i=1}^{n}(x_i^2 - 2x_i\bar{x} - \bar{x}^2)\right] \\ &= E\left[\frac{1}{n}\sum_{i=1}^{n}x_i^2\right] - E\left[\frac{1}{n}\sum_{i=1}^{n}2x_i\bar{x}\right] - E\left[\frac{1}{n}\sum_{i=1}^{n}\bar{x}^2\right]. \end{aligned}$$

Note that you could have taken the expectance operator into the sum and then computed the expectance of each term, as $= \frac{1}{n}\sum_{i=1}^{n}\left(E[x_i^2] - E[2x_i\bar{x}] - E[\bar{x}^2]\right)$. However, since $\bar{x}$ is a random variable (a parameter that is a function of random variables), it cannot come outside of the expectance operator and its value depends on the x_i values, so $\bar{x}$ and x_i covary. This introduces extra work for us in computing $E[2x_i\bar{x}]$. Therefore, *in this case*, it is easier to perform the summation first, producing standard terms and allowing some reductions, and then take the expectance.

Box D3.1 (Cont.)

$$= \mathrm{E}\left[\frac{1}{n}\sum_{i=1}^{n}x_i^2\right] - \mathrm{E}\left[2\bar{x}\frac{1}{n}\sum_{i=1}^{n}x_i\right] + \mathrm{E}[\bar{x}^2]$$

$$= \mathrm{E}\left[\frac{1}{n}\sum_{i=1}^{n}x_i^2\right] - 2\mathrm{E}[\bar{x}\bar{x}] + \mathrm{E}[\bar{x}^2] \qquad \text{(D3.1.2a)}$$

$$= \frac{1}{n}\sum_{i=1}^{n}\mathrm{E}[x_i^2] - \mathrm{E}[\bar{x}^2],$$

since $\mathrm{E}[x^2] - \mu^2 = \sigma^2$, then, $\mathrm{E}[x^2] = \sigma^2 + \mu^2$, so

$$\begin{aligned} &= \frac{1}{n}\sum_{i=1}^{n}(\sigma^2 + \mu^2) - \mathrm{E}[\bar{x}^2] \\ &= \sigma^2 + \mu^2 - \mathrm{E}[\bar{x}^2]. \end{aligned} \qquad \text{(D3.1.2b)}$$

Likewise, $\mathrm{E}[\bar{x}^2] - \mu^2 = \sigma_{\bar{x}}^2$, but $\sigma_{\bar{x}}^2 = \sigma^2/n$ (by the Central Limit Theorem), so (D3.1.2b) is

$$\begin{aligned} &= \sigma^2 + \mu^2 - \sigma^2/n - \mu^2 \\ &= \sigma^2 - \sigma^2/n \\ &= \frac{n-1}{n}\sigma^2. \end{aligned} \qquad \text{(D3.1.2c)}$$

Now the bias can be computed directly from (3.1) and using n in place of the full sample size N:

$$\begin{aligned} \mathrm{B}(s^2) &= \sigma^2 - \mathrm{E}[s^2] \\ &= \sigma^2 - \frac{n-1}{n}\sigma^2 \\ &= \frac{1}{n}\sigma^2. \end{aligned} \qquad \text{(D3.1.3)}$$

Therefore, the estimate of the variance using (3.6) is biased by the amount, σ^2/n, which is relatively minor for large n, but could be important for small n, in which case this fraction can deviate significantly from 0.

Examination of (D3.1.2c) shows that if we multiplied the estimator by $n/(n-1)$, we would get $\mathrm{E}[s^2] = \sigma^2$ and no bias. Doing this, we get

$$\begin{aligned} s^2 &= \frac{n}{n-1}\frac{1}{n}\sum_{i=1}^{n}(x_i - \bar{x})^2 \\ s^2 &= \frac{1}{n-1}\sum_{i=1}^{n}(x_i - \bar{x})^2. \end{aligned} \qquad \text{(D3.1.4)}$$

Thus, the sample variance estimated by (D3.1.4) provides a nonbiased estimator of the true variance. In other words, we divide the sum of the squared deviants by $n - 1$ instead of by n.

Another estimate of the spread may be obtained using the L1 norm, as previously discussed. That particular estimate is not an efficient estimator, and it is not used as much as the more efficient estimator, variance. The L1 norm estimator does play an important role in other robust estimation techniques, but the straight average (i.e., not absolute value) of the deviations about the mean is equal to 0, and therefore of no use as an indicator of spread. Norms in general are discussed in more detail later, in the section on linear regression.

Estimating Covariance of a Random Variable

In practice, covariance, like variance, is a biased estimator if the pure form of expectance is used when the true means of the multivariate X_j are not known, but rather are estimated from the sample. For this parameter, it can be shown in a manner similar to that used for variance that an unbiased estimate results when normalizing by $1/(n-1)$ instead of $1/n$. Thus, the sample covariance, s_{xy}, is estimated by

$$s_{xy} = \frac{1}{n-1}\sum_{i=1}^{n}(x_i - \bar{x})(y_i - \bar{y}). \tag{3.9}$$

You might wonder, why not normalize by $1/(n-2)$, since we are computing two sample means before computing? But since both x and y are measured at the same times or locations (both at index i), they both have n values. If we estimate their means, we only lose one point from the n values. That is, we can estimate one of the x values, given knowledge of the sample mean, and likewise for y, so each loses only one point of the common n.

Box 3.3 Interpreting Expectance

Consider the density of seawater that is dependent upon the temperature, pressure and salinity of the water, though salinity itself is determined from the water's conductivity, which shows a dependence on temperature, so for this variable we might construct a linear function of three random variables using (2.51), where $j = 3$, X_1 = temperature, X_2 = pressure and X_3 = conductivity for any one particular parcel of water. The coefficients a_1, a_2 and a_3 are determined through theory or very well-controlled calibration procedures in a laboratory somewhere (the actual density of seawater is a much more complicated function of these variables than the simple weighted sum presented here, though when the coefficients a_j are determined from a linear regression (described in Chapter 5), this linear form (using salinity directly in place of conductivity) provides a good approximation over rather wide ranges of temperature, salinity and pressure).

The uncertainty in seawater density for some parcel of water is thus computed through (2.56):

Box 3.3 (Cont.)

$$\sigma_Y^2 = \sum_{i=1}^{K}\sum_{j=1}^{K} a_i a_j \mathrm{Cov}[X_i X_j],$$

a weighted sum of the variance of the temperature, pressure and conductivity, and covariance between these variables, for the water sample. The water sample may consist of multiple *in situ* measurements of the water or multiple flasks collected at the same location, possibly at different times, or at the same time – the former predominantly indicating the variation over time at this location and depth, the others predominantly reflecting the variation inherent in the sampling methods, though they can also reveal small scale spatial and temporal fluctuations since the measurements undoubtedly are not at the same precise location each time, even if taken in rapid succession.

In this example, variance (second centered moment) provides an estimate of the uncertainty in the value of the function: the density of seawater based on a selection of water measurements at one location. The variance provides an indication of just how good the estimate of density or pressure is, depending on how much scatter was present in the raw measurements (of temperature, conductivity and pressure for seawater, or temperature and humidity for atmospheric pressure). However, we are ignoring the fact, in this estimate, that this simplistic linear function itself may not be perfect. That is, the linear combination of the raw variables does not necessarily provide a perfect description of the seawater density.[5] We must ultimately take into account the additional uncertainty of the raw variables in order to provide a more complete estimate of the potential error in the final density estimate we are producing – that is, there is uncertainty associated with the scatter present in the raw measurements (which is accounted for via the expectance operator as shown here), and there may be additional uncertainty inherent in the approximation itself, even if the data were perfect. How to estimate this latter error is described in Chapter 5, which discusses fitting curves to data.

In general, the moments of a function provide the statistical information that allows a good understanding of how the function (the quantity of interest, i.e., your analysis product) is distributed: what is its central value; what is the spread of the function about this central value; is the spread asymmetrical (more skewed to one side of the mean than the other); how peaked is the function (values concentrated near the central value fall off rapidly away from this value, or show a similar likelihood for a broad distance from this central value)? From this information, the function's characteristics can be explained and used as if it were a simply a direct random variable itself, remembering that there may be other errors inherent in its estimate that you need to consider, and make your best estimate as to their size.

[5] But it may prove to be excellent over short ranges of density; a comparison (regression) of this linear estimate to true density will assess that.

3.5 Principle of Maximum Likelihood (An Important Principle)

So far, we have discussed a number of individual estimation techniques, but have not considered the best way to consistently combine various observations or samples to produce the overall best estimate of the mean, variance or any other population parameter. The **Principle of Maximum Likelihood** provides one tool for determining a "best" estimate. This principle forms the basis for determining the best estimates of a great many statistical properties of data.

Consider estimates of the amount of sea level rise following the last glacial maximum (~20,000 years ago) as the landlocked glacial ice melts, as indicated from measuring the change in oxygen isotope concentration (as described earlier in this chapter) in benthic foraminifera from several different ocean basins.[6] We may have a collection of measurements from the North Atlantic, the South Pacific, the Western Indian Ocean, etc., with sea level estimates averaged over each ocean basin for a basin-wide average (Figure 3.2). Each estimate therefore represents a mean value, $\bar{x}_i$, for which we have one estimate from each basin (represented by the subscript i), each with its sample standard error, $s_{\bar{x}i}(= s_i/\sqrt{n_i}$, via the Central Limit Theorem).

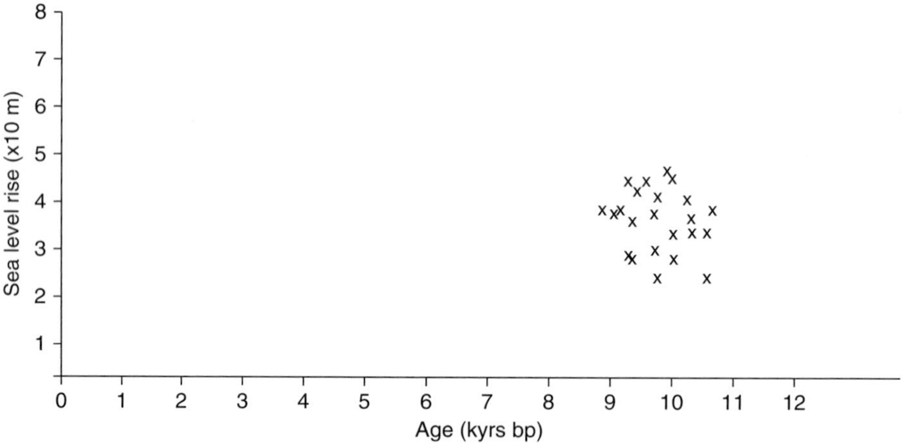

Figure 3.2 Example of a collection of sample means of mean sea level rise at some time in the past (units are thousands of years before present (kyrs bp)). By eye, it is clear that these samples have not been pulled from a population whose mean is 70 or 80 m ~10 kyrs bp, but rather something like ~40 to 50 m. The Principle of Maximum Likelihood formalizes this to find the optimal mean depth by finding that mean most consistent with the values of the sample values (just as we did with our eyes). That will be the population mean that maximizes the likelihood of drawing the samples we did draw.

[6] Actually sea level is more complex than depicted here, since each ocean basin will have a different value of sea level change due to configurations of the ocean basins, their large scale circulation and many other factors, which are ignored here for this simple example.

We now wish to make a "*best*" or "*most representative*" estimate of the global mean sea level rise following the last ice age. The method of maximum likelihood provides the appropriate estimator. This principle states, *we wish to estimate those values of* μ *and* σ *that are the most likely, given knowledge of the distribution from which the samples were drawn*. Fortunately, the Central Limit Theorem tells us that our estimates of the mean values are normally distributed. The method of maximum likelihood can therefore be directly applied.

Suppose we have n independent estimates of the mean, $\bar{x}_\mathrm{i}$. The probability of obtaining those n particular estimates is given by what is called a **likelihood function**, $\mathrm{L}(\bar{x})$. A likelihood function is a probability distribution for a group of observations, expressed in terms of one or more unknown parameters.

We know that the probability of obtaining any two particular values, say $\bar{x}_1$ and $\bar{x}_2$, is given by the joint probability function for these two specific values, i.e., $\mathrm{P}\{X_1 = \bar{x}_1,\ X_2 = \bar{x}_2\}$. For the current problem, we know that the individual means are independent of one another, and in that case the probability of getting $\bar{x}_1$ and $\bar{x}_1$ is the product of the individual probabilities, or $\mathrm{P}\{X_1 = \bar{x}_1\}\ \mathrm{P}\{\ X_2 = \bar{x}_2\}$. The product of probabilities represents the joint probability for any n independent random variables. It can be expressed as a likelihood function:[7]

$$\mathrm{L}(\bar{x}) = \prod_{i=1}^{n} \mathrm{P}(\bar{x}_i) \tag{3.10}$$

In this particular case, you desire the probability of getting the n events that you actually did get. For example, the probability that you will get a heads when flipping an unbiased coin is $\mathrm{P}\{X = \text{heads}\} = 0.5$ (i.e., 50 percent). The probability that you will get five heads in a row when flipping an unbiased coin is $\prod^{5} \mathrm{P}\{X = \text{heads}\} = 0.5 * 0.5 * 0.5 * 0.5 * 0.5 \approx 0.031$ (there is only a 3 percent chance that you will get five heads in a row).

Since the distribution of the sample means is normal (as stated by the Central Limit Theorem), we can substitute the actual form of the normal distribution for $\mathrm{P}(\bar{x}_i)$ in (3.10):

$$\mathrm{L}(\bar{x}) = \prod_{i=1}^{n} \left(\frac{\mathrm{d}\bar{x}}{\sigma_i\sqrt{2\pi}}\right) \mathrm{e}^{\frac{-(\bar{x}_i - \mu')^2}{2\sigma_i^2}}, \tag{3.11}$$

where μ' represents the best estimate (what we want) of the true mean, μ, and σ_i is the true standard deviation, $\sigma_{\bar{\mathrm{X}}}$, for each of the estimates, $\bar{x}_i$ (for σ_i, this assumes that we know σ of the original distribution, and then the true $\sigma_{\bar{\mathrm{X}}}$ follows directly from the Central Limit Theorem; we will discuss later the case in which we do not know the true variance).

[7] The symbol Π (uppercase π) is treated similarly to Σ (uppercase sigma), except that in the case of Π the terms are multiplied together, whereas with Σ they are added. Therefore, Π indicates a product.

This form of the normal PDF in (3.11) differs from how we have previously written it. That is, normally we would write it in terms of the true mean, μ, and standard deviation, σ, but here it was written in terms of known x_i and unknown distribution parameters. Also, it is multiplied by $d\bar{x}$ to convert it to a PMF. This makes it distinctly different from a standard probability law, which gives $P\{x_i|f\}$ (i.e., the probability of obtaining specific x_i, given the probability law f). Instead, it is the "inverse" of such a law: we have the *likelihood* of f, given the values x_i. $L\{f|x_i\}$ is proportional to $P\{x_i|f\}$, and the constant of proportionality is inconsequential in most applications of maximum likelihood (**ML**) estimates. We multiply the PDF by $d\bar{x}$ in (3.10), converting the PDF to a PMF (i.e., to an actual probability), which allows us to compute the likelihood for all of the observed x, and *this term can then be ignored*, as it is absorbed by the unnecessary constant of proportionality.

We wish to determine that estimate of μ, μ′, which maximizes the likelihood of getting the n estimates of $\bar{x}$ that we actually got. That is, we want to estimate the μ that maximizes $L\{f\,|x_i\}$. Conceptually, this is like having 20 values of a mean, each being close to 100, and consequently it is obvious that the true mean is probably close to 100, not close to 30. Here we are formalizing this.

Calculus tells us that the maximum of a function occurs where the derivative goes to zero, so we take the derivative of the above function with respect to μ′, set this derivative to zero and solve. First note that the multiplication of a constant n times is simply the constant raised to the nth power, and the product of n times the exponential term is simply a single exponential with the exponents summed. Applied to (3.11),

$$L(\bar{x}) = \left(\frac{d\bar{x}}{\sigma_i\sqrt{2\pi}}\right) e^{\frac{1}{2}\sum_{i=1}^{n}\frac{-(\bar{x}_i-\mu')^2}{\sigma_i^2}} \tag{3.12a}$$

for constant σ_i, whereas for variable σ_i,

$$L(\bar{x}) = \left[\prod_{i=1}^{n}\left(\frac{d\bar{x}}{\sigma_i\sqrt{2\pi}}\right)\right] e^{\frac{1}{2}\sum_{i=1}^{n}\frac{-(\bar{x}_i-\mu')^2}{\sigma_i^2}}. \tag{3.12b}$$

Inspection of (3.12) indicates that the maximum of the function occurs where the exponent (a negative number) is minimum. So, we need only minimize the exponent term (the −1/2 being inconsequential):

$$-\frac{1}{2}\sum_{i=1}^{n}\left(\frac{(\bar{x}_i-\mu')}{\sigma_i}\right)^2 = \sum_{i=1}^{n}\left(\frac{(\bar{x}_i-\mu')^2}{\sigma_i^2}\right). \tag{3.13}$$

Setting (3.13) to zero and solving for μ', for the case of constant μ_i, yields

$$\mu' = \frac{1}{n}\sum_{i=1}^{n}\bar{x}_i. \tag{3.14a}$$

Therefore, the ML method yields the standard form for the sample mean for normally distributed X.

For the case with variable σ_i (i.e., where the σ_i are different for each of the $\bar{x}$), setting (3.11) to zero and solving for μ' yields

$$\mu' = \frac{\sum_{i=1}^{n}\left(\bar{x}_i/\sigma_i^2\right)}{\sum_{i=1}^{n}\left(1/\sigma_i^2\right)}. \tag{3.14b}$$

Therefore, in our example of sea level estimates, the "best" estimate for the mean is given by (3.14). This is all possible because the Central Limit Theorem tells us that the means are normally distributed. However, the approach works equally well for parameters in distributions that are not normal, since this principle provides an analytic form from which to estimate the optimal parameter values. Of course, the method *does* require knowledge of the distribution, so this does represent its biggest limitation (the Central Limit Theorem and chi-squared notwithstanding).

If we do not know the true variance, we can determine the optimal manner with which to estimate it in a similar way. Sampling from a Gaussian distribution with known mean, σ, again we write the likelihood function, and this case, minimize L with respect to s, as

$$\mathrm{L}(x_i) = \prod_{i=1}^{n}\left[\left(\frac{\mathrm{d}x}{s\sqrt{2\pi}}\right)\mathrm{e}^{\frac{-(x_i-\mu')^2}{2s_i^2}}\right], \tag{3.15a}$$

but with varying σ

$$= \left(\frac{\mathrm{d}x}{s\sqrt{2\pi}}\right)^n \mathrm{e}^{\frac{1}{2}\sum_{i=1}^{n}\frac{-(x_i-\mu)^2}{s_i^2}}.$$

Here it is convenient (but not necessary) to take the log of L before differentiating (using the entire expression for L, since σ occurs in both terms)

$$\ln[\mathrm{L}(x_i)] = \ln\left[\left(\frac{\mathrm{d}x}{s\sqrt{2\pi}}\right)^n \mathrm{e}^{\frac{1}{2}\sum_{i=1}^{n}\frac{-(x_i-\mu)^2}{\sigma_i^2}}\right],$$

setting $a = (2\pi)^{1/2}$ and $b = (x_i - \mu)$

$$= \ln(as)^{-n} + \ln\left(e^{\frac{1}{2}\sum_{i=1}^{n}\frac{b^2}{s^2}}\right)$$

$$= -n\ln(as) - \frac{1}{2}\sum_{i=1}^{n}\frac{b^2}{s^2}$$

$$= -n\frac{d\ln(as)}{das}\frac{das}{ds} - \frac{1}{2}\sum^{n}\frac{d}{ds}b^2s^{-2}$$

$$= -n\frac{a}{as} + \sum^{n}\frac{b^2}{s^3}.$$

So,

$$s^2 = \frac{1}{n}\sum^{n}(x_i - \mu)^2. \quad (3.15b)$$

As with the mean, we see that the optimal way to compute the sample variance, given that value most consistent with the observations we did indeed sample, is to combine them as we have for the standard definition of variance (note that we are not dividing the sum by $n - 1$, because in this example, we said we know the true mean).

Maximum likelihood estimators represent sufficient statistics – all of the information concerning the parameter being estimated has been extracted from the observations. The observations (the sample) can offer absolutely no more information regarding the parameter being estimated after having estimated it using a sufficient statistic. Maximum likelihood estimators also provide the estimates with the smallest mean square error. Unfortunately, these estimators may be biased, in which case one may wish to perform an evaluation of the bias, as was done previously for variance, and introduce a small normalizing factor to eliminate the bias. However, it will be shown later that it is sometimes worthwhile to use the maximum likelihood estimator directly, accepting the bias, as it can still provide an overall "best" estimator in a specific sense that may be more important than a small bias.

3.6 Interval Estimates

One of the most useful contributions of statistics is its ability to provide quantitative estimates of the uncertainty associated with any quantity estimated from a data set. This leads to the concept of interval estimates. So, while point estimates represent estimates of the sampled population's distribution characteristics, *interval estimates provide estimates of the probability that the parameter of interest lies within a particular interval of the random variable.* Interval estimates allow one to make statements regarding the reliability of an estimate and are most commonly presented in the form of confidence intervals.

3.6.1 Confidence Intervals

A **confidence interval** is an interval for which there is an estimable degree of confidence that the value of a parameter being estimated lies within the interval. That is, besides estimating the value of the parameter, it is often necessary to estimate a range for the parameter. Key to this concept is the fact that the range, as was the case for the point estimate, is now an estimate, and that range itself is thus subject to uncertainty.

Specifically, for a random variable, X, and some characteristic of its distribution described by parameter θ, a confidence interval describes an interval bounded by a lower limit, $\hat{\theta}_L$, and upper limit, $\hat{\theta}_U$, where

$$P\{\hat{\theta}_L \leq \theta \leq \hat{\theta}_U\} = 1 - \alpha, \tag{3.16}$$

where α is the **level of significance** (discussed in more detail later), $\hat{\theta}_L$ and $\hat{\theta}_U$ are the **confidence limits** (or **fiducial limits**), which are themselves statistics, and $(1 - \alpha)$ 100 percent is the **confidence level** for any specific value of α. Note that the choice of notation for the confidence limits was chosen to reflect that the limits are statistics – they are estimates based on the realizations in any one sample.

So, if $\alpha = 0.05$ (5 percent), then the confidence level is 95 percent, and there is a 95 percent probability that the true parameter value θ lies within the confidence interval. The typical nomenclature is such that a confidence interval reflecting a 95 percent probability would be called a 95 percent confidence interval (in this example), to distinguish it from confidence intervals based on different confidence levels, say a 99 percent confidence level.

The purpose of such intervals is to allow assessment regarding the precision of the estimate made for any particular population parameter. For example, if you estimate the mean for a population based on a sample consisting of 1000 observations ($n = 1000$) using the standard formula for arithmetic mean (3.4), there is no information provided regarding how good (how *precise*) that estimate is. How much better is it than an estimate of the mean based on a sample consisting of 100 observations? The confidence interval, in association with the results of the Central Limit Theorem, allows you to present a quantitative estimate for the precision of your estimate of the mean, as seen below.

Confidence Interval for the Sample Mean

As previously mentioned, one of the great consequences of the Central Limit Theorem is realized through its impact on estimates of the mean. This is because the Central Limit Theorem indicates that a "large" sum of independent random variables is approximately normally distributed. A special result occurs if we sum a set of independent, identically distributed (**iid**) random variables with mean μ and variance σ^2. In that case, the mean and variance of the summed random variables, $Y(X_j)$, μ_Y and σ_Y^2, respectively, are given as

$$\mu_Y = \mu \tag{3.17}$$

$$\sigma_Y^2 = \sigma^2/n\,. \tag{3.18}$$

This result follows directly from the expectance operator as shown in Chapter 2.

The sample mean, $\bar{x}$, is computed according to the sum given by (3.4b). In that form, each individual realization in the sum has the identical probability ($1/n$) and each realization is independent (i.e., the individual realizations do not depend on the previous realizations except in the special case of sampling without replacement, which is noted below). Therefore, the computation of a sample mean satisfies the criterion of this special case and has a normal distribution of μ and σ^2/n, where μ and σ^2 are the mean and variance of the original population, X, from which the sample was drawn. So, $\bar{x} \approx \mathrm{N}(\mu_{\bar{x}}, \sigma^2_{\bar{x}})$ where

$$\mu_{\bar{x}} = \mu \tag{3.19}$$

$$\sigma^2_{\bar{x}} = \sigma^2/n, \tag{3.20}$$

or, in terms of the standard deviation, for the variance reduced by $n^{-1/2}$, the **standard error** is

$$\sigma_{\bar{x}} = \sigma/\sqrt{n}. \tag{3.21}$$

As the sample size increases, the distribution of $\bar{x} \rightarrow \mathrm{N}(\mu, 0)$ as $n \rightarrow \infty$ (or as $n \rightarrow \mathrm{N}$ for a finite size population of size N). That is, there is no spread in the distribution of the mean, since the estimate will be perfect if the entire population is used to estimate it.

For small random samples *without replacement* that is, where the observations are removed from the population when they are drawn and the sample size, n, represents an appreciable proportion of the total population size, N (say, $n > 0.1{*}\mathrm{N}$) then (3.20) is modified somewhat to

$$\sigma^2_{\bar{x}} = \left(\sigma^2/n\right)[(\mathrm{N} - n)/(n-1)], \tag{3.22}$$

where $(\mathrm{N}-n)/(n-1)$ is a *finite population correction factor*. For example, when sampling the chemical composition of a meteorite, there is only a finite amount of material in total, and each sample drawn from the population may significantly deplete the remainder of the population of its potential realizations.

Box D3.2 Lower Moments for the Special Case of the Central Limit Theorem

Consider the function of n independent, identically distributed (iid) random variables used to compute the sample mean $\bar{x} = \frac{1}{n}\sum_{i=1}^{n} x_i$. The mean of this function is given via the expectance operator:

$$\begin{aligned} \mu_{\bar{x}} &= \mathrm{E}[\bar{x}] \\ &= \mathrm{E}\left[\frac{1}{n}\sum_{i=1}^{n} x_i\right] \\ &= \frac{1}{n}\sum_{i=1}^{n}\mathrm{E}[x_i]. \end{aligned} \tag{D3.2.1a}$$

Because each of the x_i are identically distributed, they have the same mean, μ, so

Box D3.2 (Cont.)

$$= \frac{1}{n}\sum_{i=1}^{n}\mu \qquad \text{(D3.2.1b)}$$
$$= \mu.$$

Now consider the variance of the function $\bar{x}$:

$$\sigma_{\bar{x}}^2 = \mathrm{Var}\left[\frac{1}{n}\sum_{i=1}^{n}x_i\right]. \qquad \text{(D3.2.2a)}$$

This form is identical to that already presented in (2.51). For independent random variables with the same variance (satisfied in this case where the x_j are identically distributed, so that the variance is the same for each x_j), the variance reduced to (2.54), or

$$= \sigma^2\sum_{i=1}^{n}a_i^2. \qquad \text{(D3.2.2b)}$$

In this case, the a_j are all equal, i.e., $a_j = 1/n$, so (D3.2.2b) further reduces to

$$= \frac{1}{n^2}n\sigma^2 \qquad \text{(D3.2.2c)}$$
$$\sigma_{\bar{x}}^2 = \sigma^2/n.$$

Now consider the situation where we draw a sample of size n from a population and compute the sample mean from the sample, giving, $\bar{x}_1$. We then repeat this procedure a great number of times, where with each sample we compute a new mean, $\bar{x}_j$. We can then compute a distribution of these means, which is called the **sampling distribution of the mean**. That is, we can now generate a distribution that describes the probability of obtaining any one particular estimate (or range of estimates for continuous data) of the mean based on a random sample of size n. However, because of the Central Limit Theorem, we do not need to do this, since we know that this sampling distribution of the mean is $N(\mu_{\bar{x}}, \sigma_{\bar{x}}^2)$, where the variance grows smaller as the sample size from which the sample mean is computed grows larger.

Because we know the distribution of a sample mean, assuming knowledge of the variance of the original random variable, σ, we can compute a confidence interval within which we have a known probability that the true mean of the sample mean will lie.

For example, we know for a normal distribution that 95 percent of the values lie within $\pm 1.96\sigma$ of the mean, so:

$$P\{\mu - 1.96\sigma_{\bar{x}} \leq \bar{x} \leq \mu + 1.96\sigma_{\bar{x}}\} = 0.95 \quad (3.23a)$$

or

$$P\{\mu - 1.96\sigma/\sqrt{n} \leq \bar{x} \leq \mu + 1.96\sigma/\sqrt{n}\} = 0.95 \quad (3.23b)$$

or

$$\bar{x} \pm 1.96\sigma/\sqrt{n}\,. \quad (3.23c)$$

If we standardize our data prior to computing the mean, we get $\bar{x}_Z \approx N(0, 1/n)$. In that case,

$$P\{\mu - 1.96/\sqrt{n} \leq \bar{x}_Z \leq \mu + 1.96/\sqrt{n}\} = 0.95. \quad (3.24)$$

This example assumes we know σ; the result is similar but with a slightly broader spread when we must estimate σ, as shown later. In that case, we will use the student t-distribution instead of the Gaussian distribution to determine how many standard deviations from the mean the desired confidence interval spans (the t-distribution being similar to the Gaussian, except that it accounts for the small sample size, so instead of 1.96σ containing 95 percent of the rv it will be a broader range and a function of how many data points are in the sample).

Interpreting the Variance of the Sample Mean

It is important to differentiate between σ and $\sigma\bar{x}$. The former represents the scatter present in a population. That is, for a given sample, drawn from the population, σ provides an indication of how much scatter there is among the various realizations making up the population (and thus the average sample). On the other hand, the $\sigma\bar{x}$ represents the amount of scatter that would occur between numerous estimates of $\bar{x}$ made from a variety of different samples (collected from the same population).

That is, if you collected many samples and then computed $\bar{x}$ for each sample, you would have a large number of $\bar{x}$ estimates. You could then lump all of these estimates of $\bar{x}$ together and look at how they are distributed and the amount of scatter between them. What you would find is that the $\bar{x}$ have a distribution that is Normal with mean μ and standard deviation $\sigma_{\bar{x}}$.

Given knowledge of n, you can always compute the sample size required to achieve any desired precision (say, $\sigma_{\bar{x}r}$) in your estimate of the mean by rearranging (3.20):

$$n = (\sigma/\sigma_{\bar{x}r})^2. \quad (3.25)$$

That is, the value of n can be determined so that the confidence interval for the sample mean is small enough to answer the underlying question(s) that motivated the sampling in the first place.

Box 3.4 Example of Probability of Standardized Variables

Consider a normal population, $N(\mu,4)$ (i.e., $\sigma^2 = 4$, so $\sigma = 2$). If the estimate of the mean, $\bar{x} = 8$, is computed from a sample of 100 observations, the sample mean has a standard error $\sigma_{\bar{x}} = \sigma/\sqrt{n} = 0.2$. Computation of the ~95 percent confidence limits is facilitated by working with the standardized sample means. In that case, the 95 percent confidence interval is defined as

$$-1.96 \leq Z \leq 1.96 \tag{3.26}$$

and

$$\mathrm{P}\{-1.96 \leq Z \leq 1.96\} = 95\%. \tag{3.27}$$

Since

$$Z = (\bar{x} - \mu)/\sigma_{\bar{x}}, \tag{3.28a}$$

then

$$-1.96 \leq (\bar{x} - \mu)/\sigma_{\bar{x}} \leq 1.96 \tag{3.28b}$$

$$\bar{x} - 1.96\sigma_{\bar{x}} \leq \mu \leq \bar{x} + 1.96\sigma_{\bar{x}}, \tag{3.28c}$$

or, specifically,

$$7.6 \leq \mu \leq 8.4. \tag{3.28d}$$

Therefore, there is a 95 percent probability that the true mean lies between the values of 7.6 and 8.4, i.e., $\mathrm{P}\{7.6 \leq \mu \leq 8.4\} = 0.95$, or 95 percent probability that the true mean lies within 8 ± 0.39.

If we wished to improve the accuracy of our estimate of the mean by a factor of 2 (i.e., from $s_{\bar{x}} = 0.2$ to $s_{\bar{x}} = 0.1$), we could do this by increasing the sample size to n_{new}:

$$s_{\bar{x}} = \left(\sigma^2/100\right)^{1/2} = 2/10 = 0.2$$

$$s_{\bar{x}}^{new} = 0.1 = \left(\frac{\sigma^2}{n_{new}}\right)^{1/2} \tag{3.29}$$

$$(n_{new})^{1/2} = \frac{\sigma}{0.1} = \frac{2}{0.1} = 20$$

$$n_{new} = 400 \tag{3.30}$$

Similar intervals, given knowledge of specific estimators, allow us to construct confidence intervals for them as well, though few of them are as generally achieved as is the case for the mean. For example, the variance will have a well-known distribution in the case where the underlying random variable is normally distributed to begin with, which is useful, owing to the Central Limit Theorem, but otherwise certainly more restrictive. These intervals are introduced later, as they arise.

3.6.2 Resampling Statistics (Bootstrap)

There is a relatively new means of estimating uncertainties that has proliferated because of the power of computers. This is presented as "resampling" statistics by Simon (1991), or the bootstrap method. Efron (1981) provides a rigorous foundation for the bootstrap, though the theoretical development is a bit difficult to follow for the casual user. Simon has presented the general concept as resampling statistics and suggests in his book that the technique is ideal for teaching most of statistics, as it tends to demystify statistics and probability, replacing the "black magic" formulations with an intuitively obvious approach.[8]

The basic concept of this technique is similar to that of maximum likelihood, where, in that case (knowledge of the PDF), one could use the method of maximum likelihood to determine the best estimates of the population parameters in the sense that they would be those parameters most likely to yield the sample we did indeed draw. A logical extension of this is to ignore the mathematical form of the distribution, but rather use the distribution of the sample as representative of the true population and to then repeat your own sampling from the sample population multiple times to determine the likelihood of achieving any particular statistic or test result. You can do this multiple times, to compute, say, the likelihood of achieving a particular ratio or other result by drawing multiple random samples from your distribution and then computing the quantity of interest for each sample drawn, and then from those, computing the PMF of the quantity and using that to estimate the likelihood of achieving the answer you indeed got. I will provide some examples of this later, particularly with time series analysis.

3.7 Hypothesis Testing

Another component of statistics besides estimation is addressing how consistent specific estimates are with hypothetical or modeled values expected for the phenomenon of interest. That is, often we wish to draw conclusions about some population based on a careful study of a set of observations drawn from that population. This is the area of statistical inference that concerns itself with making an assertion, assumption or general statement (a **statistical hypothesis**) about the population.

[8] Simon notes that he started this work in 1969, while Efron studied it independently, calling it "bootstrap," in 1979.

For example, we may want to know if some material is a specific substance, and the only way to identify the substance is through knowledge of its density. To test this, we have to compare the density measured from our sample to that expected for the known substance, and then draw some conclusion as to whether it is or is not the substance of interest.

While the reliability with which a property of the population is estimated is indicated through use of confidence intervals, the likelihood of a hypothesis being correct is assessed through the methods of **hypothesis testing.** Hypothesis testing usually involves estimated parameters and their confidence intervals.

Suppose we have just computed a "best" estimate (mean) and related error (variance) for the velocity measured, via a logging tool,[9] at several depth positions in a suspected sandstone unit drilled at a potential oil-bearing location. Geophysical measurements collected over many years and including all types of sandstone have produced a fairly good representation of the velocity distribution expected from sandstone. The distribution is normal, with mean μ_{ss}, and standard error σ_{ss}. We now wish to compare our "best" velocity estimate determined in the field, $\bar{x}$, with sample variance, s^2, to the normal distribution describing the range of velocity values expected for sandstone. This is where hypothesis testing comes in – it plays a major role in classical statistics and is designed to help us evaluate whether the values we measure are consistent with what we expect or with some hypothesized value or distribution.

In general, the testing of hypotheses in a statistical sense involves the following steps:

1) comparison of a sample-based estimate (e.g., a statistic such as the mean or variance) or sample distribution to a hypothesized (or known) value of the true parameter value or distribution
2) using a test statistic to make the comparison; this statistic, computed from the sample statistic value, will have a known distribution from which you can then evaluate, if the observed value is reasonably close to the hypothesized one
3) determining a level of significance that specifies the probability that you will reject the null hypothesis (the hypothesis that we hope fails the test – e.g., it is not a sandstone), when it is in fact true
4) computing the test statistic and comparing the resulting value to the appropriate distribution table, then either (1), if the value is within the specified range of acceptable answers, the hypothesis is accepted, if not it is rejected; or (2) determine the level of significance required to make the hypothesis acceptable and then decide whether to accept the hypothesis given that level of significance

The second option of step 4 is rarely, if ever, promoted in the statistical literature, since it opens the opportunity for undermining the objectivity of the test. That is, some consider that the test is meaningless if one ultimately decides to accept or reject it after the result is known. It can only be objective if you define the acceptable limit in advance of seeing the

[9] A logging tool is not a chainsaw, but rather any instrument that can be lowered down through the hole just drilled to measure various properties as it is lowered and transmit the data back to the surface via the wire used to lower it.

results, and then accept or reject according to that *a priori* criterion. However, in science our advancing a theory or hypothesis is rarely that black-and-white. Far too often, one sees an investigator reject a perfectly plausible theory because it did not pass the test at a 5 percent level of significance, whereas it may have passed the test at a 7 percent level. Therefore, in science you might typically consider the probability at which the hypothesis would pass the test, and then assess whether or not this probability warrants additional consideration or investigation of the theory.

Most statistical tests are based upon acceptance or rejection of the null hypothesis. The null hypothesis (H_0) is one that represents the answer of no change, or no difference, or simply status quo (e.g., there is no trend in the data, so the slope is not different from 0). It is stated so as to represent an explicit relationship between the sample-based estimate and some parameter that is the simplest (pessimistic) value (i.e., no, your data do not show an unusual mean value representing a new material; it is just some slight variant of the more common material). Typically, we are hoping to reject the null hypothesis – that is, *we hope to establish that the parameter being tested is not that of the null hypothesis.*

So, if one is comparing means, the null hypothesis is *explicit*, as

$$H_0 : \mu_1 = \mu_0 \tag{3.31}$$

states that the null hypothesis assumes that the mean of the sample (μ_1) is the same as that of the undesired, but simplest explanation, distribution (μ_0), which our sample may have come from, but which we hope it did not come from (e.g., the distribution of the density of pyrite, when we are hoping we have gold, which has a different distribution).

The alternate hypothesis (H_1), the one we desire, is the more general alternative:

$$H_1 : \mu_1 \neq \mu_0. \tag{3.32}$$

Since H_0 and H_1 are mutually exclusive and all inclusive, they cover all possibilities and one is true while the other is false. Therefore, one of the following outcomes must occur:

	Null Hypothesis Is True	Null Hypothesis Is False
Accept hypothesis	Correct decision	Type II error ($P = \beta$)
Reject hypothesis	Type I error ($P = \alpha$)	Correct decision

Since the null hypothesis is explicit and the alternate (H_1) is general, the probability of committing a Type I error ($= \alpha$) can be specified explicitly. In cases where both H_0 and H_1 are explicit, then, both α and β are often dependent and can be specified. Otherwise, we can only specify the probability of rejecting the null hypothesis when it is true – we cannot make any inferences about the degree of probability of accepting H_0 when it is in fact false. This is because we know the characteristics of the population against which we are making our comparison and therefore we can specify the probability of obtaining a sampled value within some range, but we know nothing about any other population from which the sample may actually have been drawn – it may have characteristics that are very similar, possibly identical, to those that are being examined.

So, we can specify a range for which an occurrence of the sampled value would be highly *unlikely* to occur if it were in fact from the null-hypothesized population (i.e., a range for which the value would be extremely unlikely, given knowledge of the hypothesized distribution or parameter). This range of values represents the **critical range** or **critical region**, and the probability that the sample statistic would actually be in the critical range if drawn from the hypothesized population has a known (pre-specified) probability. This probability, α, is known as the level of significance (see later discussion). This is the same α that was previously introduced when discussing interval estimates.

If our sampled value indeed falls within the critical range, we would state that it is highly unlikely (with a probability given by α), though not impossible, that the sampled value was drawn from the hypothetical population. We cannot, however, say anything about the likelihood that, if the value was within an acceptable range, it was actually *not* from the hypothesized population. So, if the value is within our acceptable range, we accept the hypothesis – that is, we conclude that the sample is indeed from the hypothetical population even though there is some unknown probability that is *not* from the hypothetical distribution. Regarding this latter possibility, it may have been drawn from a population that had very similar characteristics to those of the hypothetical one.

In this respect, the nature of statistical testing is only one of rejection or one of consistency – *in no way does it confirm a hypothesis*, it only will indicate whether an observation is not significantly different from some hypothetical value. This, of course, does not indicate that if it is not significantly different, it is therefore the same. It simply states that we have no reason to reject the null hypothesis and that the result is consistent with the hypothesis (at some specified level of significance).

It has been argued that the null hypothesis should always be "simpler" than any alternative hypothesis. This is based on **Occam's razor**, which implies that a simple hypothesis is *a priori* more likely than a more complicated hypothesis.[10]

3.7.1 Level of Significance (α; or Alternatively, p)

The **level of significance** is the probability of committing a Type I error ($= \alpha$) – that is, rejecting the null hypothesis when it is actually true. The value $(1 - \alpha)$, as previously stated, represents the probability at which a sample statistic could be expected to deviate from the hypothetical value as a result of simple scatter in the values, and still be sufficiently close to allow acceptance of the null hypothesis. Typically, one might choose a value of 0.05 or 0.01 for α. These values arise from convention, but the actual level of significance should be chosen according to criteria relevant to the particular problem and the relative consequences of the chosen value.

[10] This argument was put forth strongly by Harold Jeffreys in his book, *Theory of Probability*, 1998, which advocated Bayesian methods – those that concern conditional probabilities. Another conditional probability rule was put forth by Occam's evil brother Murphy, whereby that result most undesired is that result most likely to happen (e.g., jelly side of toast lands face-down on white carpet), now known as Murphy's Law.

For example, the null hypothesis may be such that it is actually preferable to risk rejection of the hypothesis when it is true (this is most common in those tests where both α and β can be assessed). Also, you may want to only risk rejecting the hypothesis when it is true in an extremely few instances (like 1 in a billion), if safety is involved.

If $\alpha = 0.05$, then the null hypothesis is rejected if the sampled value lies outside of the 95 percent confidence level of the true value, based upon knowledge of the true population's mean and variance. As a consequence, 5 out of 100 times we expect to reject the null hypothesis even though it is true, since 5 out of 100 times we actually *expect* a value to lie outside the 95 percent confidence limit as a simple consequence of natural scatter in the data. This is the risk we accept by specifying α = 5 percent. Specifying α = 1 percent states that we reject the null hypothesis, on average, 1 out of 100 times when it is in fact true, because the natural scatter is such that 1 value in 100 will lie in this outer region.

Alternatively, for your null hypothesis to be true, what is the probability of getting the value you did get, which is called the **p value**? If you are comparing means, then, and your sample mean is $\bar{x}$ and the null hypothesis says it is μ, then the p-value is $P\{X = \bar{x}, \mu\}$. That is the probability of getting $\bar{x}$ when the true mean is μ. For example, you are looking at the rate of change in temperature over a two-decade period and find it is $a_0 = 1°/yr$ and the null hypothesis is that there is no trend, $a_0 = 0$. Given the proper test (in this case, likely a normal distribution test), what is the probability of getting a value as large (1°/yr) as you did get if in fact the trend was 0? Well, the smaller the value, the more unlikely it is that the null hypothesis is true. If $p = 0.005$, this means that only 1 in 2000 times would you expect such a value if the null hypothesis is actually zero. So, you reject H_0 at $p = 0.005$. Alternatively, you could claim that you reject the null hypothesis for $\alpha = 0.005$. Formally, you would have set a significance level (α) for rejecting the null hypothesis, and if $p \leq \alpha$, you reject the null hypothesis.

The usefulness of the p statistic is best summed up by Stanford statistician R. G. Miller (1986): " . . . I don't remember ever having fixed α and having tested a hypothesis. Instead I report the p-value, which is the probability under the null hypothesis of obtaining a result equal to, or more extreme than, the observed . . . The smaller p is, the less likely one feels the null hypothesis can be true."

Originally the testing would be as follows: pick an α (a probability) that you are willing to risk for rejecting H_0 when it is true, then compute p to see just how unlikely it would be to get such a value if H_0 were true. This p-value is excellent because it shows just how good your reason is for rejecting the null hypothesis. For example, you will reject at $\alpha = 0.01$, but you find p is actually 0.001 (i.e., your α was way more conservative than it needed to be).

Unfortunately, the p-value seems to be one of the more commonly misused (misunderstood and thus misrepresented) statistics. Therefore, if you do use the p-value, you should make it clear how you are using it: rejecting the null hypothesis because only with a probability of p would you ever expect to get a value as high (or low according to the particular test you use) as you did indeed get, if H_0 were indeed true. So, the lower the value of p, the less likely it is that H_0 is true.

So, reiterating the hypothesis testing procedure:

1) Specify a test statistic, which is used to make the comparison between our sample population and the hypothetical population.
2) Specify a level of significance – this choice should be based on as much additional information as possible and the acceptable level for rejection of the null hypothesis when it is in fact true (be realistic here – don't just follow some convention, like 5 percent, if it is not warranted; *all tests are significant*, it just depends at what level, and the beauty of the p-value is that it states exactly what value).
3) Determine the limiting value(s) of the test statistic that correspond to the level of significance. This determines the critical region of the test. If the test statistic yields a value that lies within the critical region, you conclude that the sample statistic is significantly different from the population parameter hypothesized to be the same and you reject the null hypothesis.

If the test statistic does not fall within the critical region, then we accept the null hypothesis – or more accurately, we *fail to reject* the null hypothesis – and conclude that the sample statistic is not inconsistent with that status quo hypothesis (in other words, we did not deviate sufficiently from the expected value of the null hypothesis to allow us to comfortably accept our alternate hypothesis (the one we hoped to accept).

3.7.2 Testing Normal Distribution Means

Consider the previous example of comparing the sample mean seismic velocity of a *suspected* sandstone ($\bar{x}$) to that of the "known" mean velocity of sandstone (μ_{ss}). Because of the Central Limit Theorem, we know that the distribution of the sample mean is approximately Normal, and if it is drawn from the true population (i.e., from sandstone) it will have mean μ_{ss} with spread given by the standard error, $\sigma_{\overline{X}} = \sigma_{ss} n^{-1/2}$. Therefore, we expect that ~95 percent of the *mean* velocities from samples of sandstone will have values within 1.96 standard deviations of the true mean velocity. Thus the appropriate (at least, the most obvious) test statistic is the standardized variable Z (i.e., it is that statistic with known distribution for the estimate of interest):

$$Z = (\bar{x} - \mu_{ss})/\sigma_{ss} n^{-1/2}. \tag{3.33}$$

By computing Z, we see how far the sample mean is from the true mean, and given a level of significance, we can decide whether the sample mean lies *acceptably* close enough to the true mean – that is, close enough for us to accept the hypothesis that the sample mean is consistent with that of a sandstone.

We choose (for convenience only) a level of significance of 5 percent, so

$$\alpha = 0.05 \tag{3.34a}$$

$$Z_{0.05} = \pm 1.96 \text{ (critical values)} \tag{3.34b}$$

$$H_0 : \bar{x} = \mu_{ss} \tag{3.34c}$$

$$H_1 : \bar{x} \neq \mu_{ss}. \tag{3.34d}$$

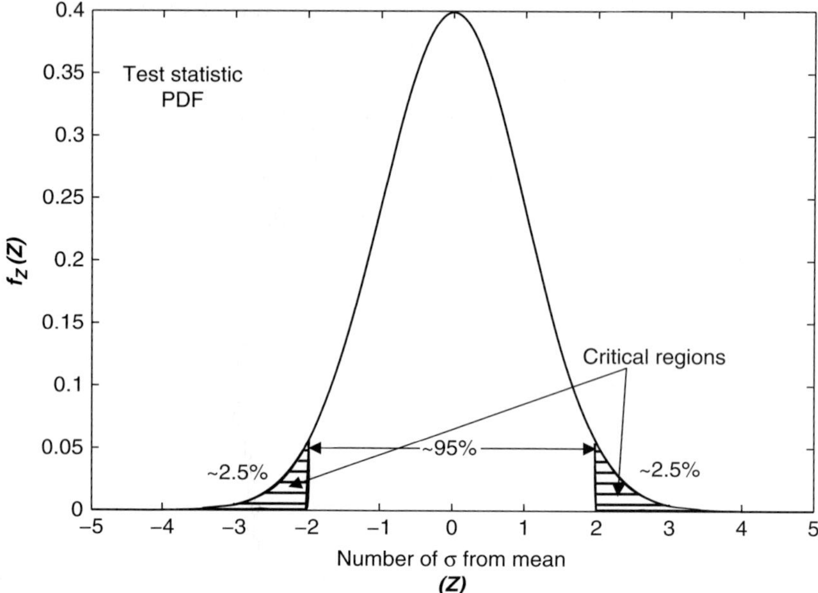

Figure 3.3 Standardized Normal PDF, with critical regions for an $\alpha = 0.05$ indicated. If our sample mean falls into the critical regions, we reject the null hypothesis (that this PDF describes our sample), recognizing that we have a 5 percent chance of rejecting this when it is indeed true.

The value $Z_{0.05} = \pm 1.96$ indicates that, as previously established for a normal distribution, 95 percent of the distribution lies within ± 1.96 standard deviations of the mean of the population. This corresponds to $Z = \pm 1.96$ because Z is the number of standard deviations from the mean. Furthermore, this indicates that the remaining 5 percent of the normal distribution lies in the range of $|Z| > 1.96$, which thus defines the critical region of the test as shown in Figure 3.3.

We then plug our determined value of $\bar{x}$ into the equation for the Z statistic (3.31). *For this test we must assume knowledge of the true mean and variance of the population.*[11] We use the variance of the population reduced by n^{-1}, instead of the variance computed from the sample, though it is acceptable for large enough sample sizes to use the sample based estimate, s, for σ. If the value of Z lies inside the critical region (i.e., if $|Z| > 1.96$), we conclude that the measured velocity is significantly different (at $\alpha = 0.05$) from that expected for a sandstone, and our measured unit is likely some other rock type. However, we make this rejection accepting the fact that there is a 5 percent chance that we are wrong in this conclusion, since 1 in 20 samples (on average) will in fact yield a mean larger than 1.96 standard deviations from the mean.

On the other hand, if the value of Z does not lie in the critical region, then we conclude that the measured velocity is consistent with that expected for a sandstone, at the 5 percent level of significance, or at the 95 percent confidence level. Unfortunately,

[11] These can be very restrictive criteria, and in many cases knowledge of the true mean and variance of the underlying population is not known. How to accommodate these cases is dealt with later.

we cannot state that it *is* a sandstone, because there is nothing to indicate that it couldn't be a rock type that has velocities similar (or identical) to those of sandstone. Therefore, even if the velocity was estimated from a huge sample and gave a mean nearly identical to that of the population mean, we still cannot rule out the possibility that the measured velocities are actually those of another rock type.

This leads to one of the greatest misstated comments in a statistical sense. People too often state (for this example) that, since the test was accepted at the 5 percent level of significance, there is a 95 percent probability that the unit is a sandstone. This is absolutely *not the case*, as just described. The proper statement is that there is no indication (at the 5 percent level of significance) that the measured mean is different from that of a sandstone. Or, if the test were rejected, the statement should be that there is a probability of < 5 percent that the rock type is indeed a sandstone (as determined from its seismic velocity). This also assumes that the sample was collected in a random manner.

If we had information that suggested that the measured unit was either a sandstone or a shale (e.g., from drill core samples), then the hypotheses would have been

$$H_0 : \overline{x} = \mu_{ss} \tag{3.35}$$

$$H_1 : \overline{x} = \mu_{sh}. \tag{3.36}$$

In this case, both hypotheses are explicit and we can therefore specify the probability of committing a Type I error (α) as well as a Type II error (β).

When the sample size is small or the variance of the known population is being estimated from the sample itself, we must take the uncertainty introduced by this into account. In such a case we no longer use the Z statistic, but instead must use the student t test. The t-distribution, like most other test statistics, varies as a function of sample size, which is expressed in terms of degrees of freedom.

3.7.3 Degrees of Freedom

Degrees of freedom (d.f. or **v)** is an important concept in statistics as well as mathematics, modeling and physics. In statistics, degrees of freedom represent the number of *independent* observations used to estimate a population parameter. Effectively, they may be thought of as representing the number of independent observations in the sample (n) minus the number of parameters already estimated from these n observations (n_p):

$$v = n - n_p \tag{3.37}$$

For example, if we are computing the mean of a sample of size $n = 100$, then there are 100 independent observations that will determine the value of $\overline{x}$,

$$\overline{x} = \frac{1}{n}\sum_{i=1}^{n} x_i \tag{3.38a}$$

or

$$n\bar{x} = x_1 + x_2 + \ldots + x_n, \tag{3.38b}$$

so the n independent values of X completely determine the value of $\bar{x}$.

For computing the sample variance,

$$s^2 = \frac{1}{n-1}\sum_{i=1}^{n}\left(x_i - \frac{1}{n}\sum_{i=1}^{n}x_i\right)^2 = \frac{1}{n-1}\sum_{i=1}^{n}(x_i - \bar{x})^2. \tag{3.39}$$

The value of s^2 is dependent upon the n values of X as well as on the value of $\bar{x}$. But, from (3.38b),

$$x_n = n\bar{x} - (x_1 + x_2 + \ldots + x_{n-1}). \tag{3.40}$$

From this it is seen that, given $\bar{x}$ and $n - 1$ values of X, the nth value of X is completely determined. This means that there is one less d.f. in this equation owing to the information already provided by knowledge of $\bar{x}$. Thus, this equation has $n - 1$ degrees of freedom, which is *conceptually* why the sum of squares in (3.37) for variance is divided by $n - 1$ to prevent bias instead of by n (though formally, this was demonstrated through the expectance operator, as previously shown).

The numerous sampling distributions that we will be dealing with are almost all a function of the degrees of freedom in the sample and statistic. This allows all of the quantities to be normalized to a common ground – that is, what was the number of degrees of freedom (independent observations) of the system? This allows us, for example, to compare two different estimates of a mean and standard deviation even though they may have been computed from samples of grossly different sizes. Intuitively, one can imagine that a statistic computed from fewer observations is not as reliable as one computed from considerably more observations. This is indeed a fact, determined from both theory and experimentation, and it is taken into account in determining the distributions of the various test statistics we will use.

3.7.4 Practical Considerations (Data Trolling)

In many sectors of society, one can unambiguously and *a priori* choose a level of significance for which a test's failure is unacceptable (e.g., safety issues, manufacturing issues, both where lives or costs can be compared to quality efforts/costs) – but in science, this is often not the case. Like many other aspects of science, even the acceptable level of significance is not clear in terms of what is or is not tolerable. One must remain painfully aware of this, as it is often inexcusable to simply dismiss an entire hypothesis because some component failed the null hypothesis at 6 percent level because the test stated that only 5 percent passage was acceptable. In complex science situations, we might be satisfied with passing at the 40 percent level. From my perspective, it is often better to note the level of significance at which the test is passed, and then decide if this is good enough to continue the investigation (statisticians call this **data trolling** and worry that one might not be honest in doing this; so, *be honest* – you can only hurt yourself and science if you force something that doesn't warrant it). If the level of significance *is* good enough, in your mind, to continue, present this level and go on with your overall evaluation of the hypotheses. Ultimately it will be the overall balance of the evidence that matters, not the pass/fail of a single piece of evidence. Perhaps you can come up with a novel test where the null hypothesis states the likelihood of a successful hypothesis, given the level at which the various subcomponents passed (or failed) their individual null hypotheses.

One other cautionary remark: statistical tests are often looked to as the definitive final word, leading to that strongly desired black-or-white answer. This is understandable, since such tests probably assess the consistency of the data with your hypothesis as well as anything could. Unfortunately, it is far too commonly used where the test was not designed properly (e.g., not completely or carefully enough, etc.), and then the answer is provided and followed as if it were the definitive answer – indeed, it is some form of a definitive answer, but all too often not for the question that the investigator or reporter is thinking it is for. So, it is critically important that the tests be fully thought through and designed to most closely test the hypothesis being tested or to answer the question being asked, so that the conclusion of the test will be relevant to the conclusions being drawn and presented.

Box 3.5 Example of the Difficulty of Properly Posing a Hypothesis

Consider a case where statistics show that 95 percent of the people killed in auto accidents were wearing their seatbelts, leading to a conclusion that seatbelts are ineffective – obviously, if 99 percent of the people in accidents had their seatbelts on, as they should, then this test really doesn't say anything about seatbelts. We need a test that would somehow not only differentiate between the fraction of deaths for people wearing or not wearing seatbelts, but we would also need to somehow normalize the comparison so that we were not comparing accidents in which the non-seatbelt wearers were just backing up in parking lots (for which reason they may not have been wearing seatbelts, and for which the accident itself is not likely to be serious).

These are obvious examples, but one can imagine that it is easy to get more and more complex with the difficulty of the problem itself, and in science this complexity is often overwhelming, so tests can only test part of the problem, a further reason why making a definitive acceptance or rejection based on a single α is so dangerous, even though it sounds very responsible. In fact, for many of our scientific issues, it is difficult to make a clearly articulated hypothesis that lends itself to such testing. So, not only does significance level and its interpretation mean something – the test design is equally critical. My (perhaps wrong) impression is that most of the abuses I see are due to a conclusion that is consistent with the statistic, but inconsistent with what the test actually tested. Perhaps the bottom line, here, is to not be so impressed with a statistical test that you allow it to blind (or, more often, intimidate) you from examining more carefully the value of the test and its result, itself.

3.8 Sample-Based Distributions

Most test statistics we will employ have distributions that depend on the sample size, or, more accurately, on the degrees of freedom. Such distributions are called **sample-based distributions**. They serve a fundamental role in small sample theory, which represents a very important part of inferential statistics since, as pointed out by R. A. Fisher, for large sample sizes, most of the answers to statistical questions are obvious – it is only when dealing with the uncertainties associated with small sample sizes that statistical problems are most likely to arise and render decision-making difficult.

These sample-based distributions are introduced here; more detailed discussion is given in context for later applications.

3.8.1 t-Distribution

The t-distribution, a sample-based distribution, was originally designed to take into account the uncertainties introduced when *estimating* a population characteristic in place of the true population parameter – particularly the variance. For example, for computing the spread in sample means, the statistic Z assumes knowledge of σ,

$$Z = \frac{\bar{x} - \mu}{\sigma/\sqrt{n}}. \tag{3.41}$$

The comparable t statistic does not,

$$t = \frac{\bar{x} - \mu}{s/\sqrt{n}}, \tag{3.42}$$

as this distribution is based on the sample based estimate, s, instead of the infrequently known true value, σ.

Obviously, with increasing sample size, n, you would expect that the estimate of s would more closely approximate the true value of σ. Therefore, the t-distribution is shaped like the normal distribution, but its spread is a function of the sample size n. The smaller the n, the broader the normal shaped distribution of t, and the larger the n, the more the t-distribution approaches a standard normal distribution. In fact, for large n ($\gtrsim 30$), the t-distribution differs very little from that for the Z statistic, and therefore one can justifiably use the Z statistic instead. For smaller samples, though, this is not the case.

The sample dependence of the t-distribution is based on the degrees of freedom, given as

$$\nu = n - 1 \tag{3.43}$$

for the formulation of the t statistic, as given in (3.39). This measure of ν reflects the fact that one of the n observations making up the sample has been "used" in making the estimate of the sample standard deviation, s, which uses the value of $\bar{x}$. Note that μ does not "use" a degree of freedom, since it is a known or assumed value (it is not based upon the observations, but rather upon the hypothesized distribution being tested).

The **t test** or **Student's t test** is used most frequently to (1) compare a sample mean to a hypothetical mean or (2) compare two sample means, when sample sizes are small (say, $n \leq 25$). For the former, the test is performed as done previously with the Z statistic. For the latter, the test statistic takes the form

$$\mathrm{t} = \frac{\bar{x}_1 - \bar{x}_2}{s_\mathrm{p}(1/n_1 + 1/n_2)^{1/2}}, \tag{3.44a}$$

where

$$s_\mathrm{p} = \left[\frac{(n_1 - 1)s_1^2 + (n_2 - 1)s_2^2}{n}\right]^{1/2} \tag{3.44b}$$

and s_p = pooled variance and $\nu = n_1 + n_2 - 2$ = degrees of freedom. The sample variances (s_1^2) should be calculated by dividing by $(n - 1)$ as opposed to n, and (3.44b) assumes that the variance for each population is the same. If this latter assumption does not hold, then (3.44b) can be modified, though the modification is of limited value since it only holds for fairly large sample sizes, and the advantage of this distribution is for its use with small sample sizes.

For this test, the hypotheses are stated as

$$\mathrm{H}_0 : \mu_1 = \mu_2 \tag{3.45a}$$

$$\mathrm{H}_1 : \mu_1 \neq \mu_2. \tag{3.45b}$$

3.8.2 Chi-Squared (χ^2) Distribution

The chi-squared distribution arises naturally because it represents the distribution of a sum of normally distributed variables that have been squared (e.g., a variance). A chi-squared distribution with degrees of freedom, ν, is indicated by χ^2_ν. This distribution will play a major role in determining the variance associated with spectral estimates later on.

The distribution is also useful in evaluating whether a sample distribution has a shape similar to that of a hypothetical distribution. This test requires evaluation of the **index of dispersion**, given as

$$\chi^2 = \sum_{i=1}^{k} \frac{(o_i - e_i)^2}{e_i}, \tag{3.46}$$

where the observations have been divided up into k classes (as for a PMF) and their frequency of occurrence computed for each class (there should be at least one observed value in each class, for best results); o_i = frequency of occurrence in each class

$\left(\sum_{i=1}^{k} o_i = n\right)$; and e_i = expected frequency of occurrence in each class (from hypothetical distribution). The number of degrees of freedom, ν, is given by

$$\nu = k - 1 - n_{\mathrm{p}}, \tag{3.47}$$

where n_{p} is the number of population parameters required in computing the distribution shape before classifying the observations (e.g., if we are comparing the sample distribution to the standard normal distribution, then we require knowledge of the mean and variance in order to standardize the sample and compute the frequency in each standardized class); there is always one constraint that the sum of frequencies must equal the sample size (therefore, given knowledge of $k - 1$ class frequencies, we can always predict the frequency of occurrence for the last class).

Conceptually, with fewer classes (e.g., $k = 3$), it will be relatively easy for the observed frequencies to match the predicted frequencies and, consequently, produce small values of the index of dispersion (the smaller the index, the better the overall match). Therefore, for small values of k (which leads to small ν), most of the χ^2 distribution clusters are near zero (zero is a perfect match). As the number of classes increases, the likelihood of obtaining a perfect match becomes increasingly difficult, and consequently the χ^2 distribution becomes more symmetrical, ultimately approaching a Normal distribution for large degrees of freedom ($\nu \geq 100$).[12] For that case, compute the Z statistic using

$$Z = (2\chi^2)^2 - (2\nu - 1)^2. \tag{3.48}$$

The development of a chi-squared distribution helps us derive its fundamental moments. It commonly arises from the sum of n squared values of a normally distributed random variable. Consider a set of standardized normally distributed Z, so Z = N(0,1):

$$Y = \sum_{i=1}^{n} z_i^2. \tag{3.49}$$

This variable Y can be shown to have a *chi-squared distribution*, χ^2_ν, with $\nu = n$ degrees of freedom. This should not be too surprising, given the similarity of the form of (3.49) to the chi-squared goodness-of-fit statistic.

[12] Yes, the Central Limit Theorem still holds: even when adding squared normal deviates, add enough and you will have the normal distribution as expected (so these two facts, (1) add summed normal deviates squared and you have a chi-squared distribution, and (2) add any deviates, squared or not, and you will get a normal distribution, are not contradictory). Chi-squared simply states that, oh, by the way, on the way to getting a normal distribution you will have a well-defined other distribution.

A χ^2_ν distribution with $\nu = n$ degrees of freedom, the sum of n-squared Normal deviates, has a mean of n and a variance of $2n$ for the case of $Z = \mathrm{N}(0,1)$. This is easily shown by examining the expectance of the sum in (3.49). By definition,

$$\mathrm{E}[\mathrm{Z}] = 0. \tag{3.50}$$

Thus,

$$\begin{aligned} \mathrm{E}[Y] &= \mathrm{E}\left[\sum_{i=1}^{n} z_i^2\right] \\ &= \sum_{i=1}^{n} \mathrm{E}[z_i^2], \end{aligned} \tag{3.51a}$$

but recall that the variance of a random variable can be given as $\mathrm{Var}[\mathrm{Z}] = \mathrm{E}[\mathrm{Z}^2] - \mu_z^2$, so $\mathrm{E}[\mathrm{Z}^2] = \mathrm{Var}[\mathrm{Z}] + \mu_z^2$, and (3.51a) becomes

$$= \sum_{i=1}^{n} (\sigma_z^2 + \mu_z^2), \tag{3.51b}$$

where $\mu_Z = 0$ and $\sigma_Z^2 = 1$ (thus $\mathrm{E}[\mathrm{Z}^2] = 1$), so

$$\begin{aligned} &= \sum_{i=1}^{n} 1 \\ &= n. \end{aligned} \tag{3.51c}$$

By a similar, though more difficult approach,[13] $\mathrm{Var}[Y] = 2n$.

For the case in which Z does not have zero mean and unit variance, we must compensate, as always, for the loss in degrees of freedom associated with estimating the mean and variance in order to standardize Z, in which case the mean and variance of Y are generalized by $\mathrm{E}[Y] = \nu$ and $\mathrm{Var}[Y] = 2\nu$, where ν is the degrees of freedom ($n - 2$, if both the mean and variance were estimated for Z).[14]

Further, when the normally distributed random variables, z_i, are *not* standardized and have a distribution $\mathrm{N}(0,\sigma^2)$, then the mean and variance are $\nu\sigma^2$ and $2\nu\sigma^4$, respectively. This is easily seen in (3.51b), where we would no longer replace the σ^2 with 1, resulting in $n\sigma_Z^2$, or $\nu\sigma_Z^2$ in the general case.

For the simple case where $\nu = 2$, this represents a special case of the χ^2_ν distribution known as the **exponential distribution** (both the chi-squared and exponential distributions are special cases of the **gamma distribution**).

Consider the functional form of the general χ^2_ν distribution for a random variable x,

[13] This is most easily done by actually inserting the mathematical function for the chi-squared distribution (which is what this sum of squared normal deviates produces) into the expectance operator and solving analytically. Actually, it is often done for the more general case of the gamma distribution, of which the chi-squared distribution is a special case.

[14] Note that if the z_i are not independent, their squared sum is still a chi-squared distribution, but the degrees of freedom are reduced to reflect the shared information contained by the correlation between the variables being summed.

$$f_{\chi_v^2}(x) = \frac{x^{[(v/2)-1]}}{2^{v/2}\Gamma(v/2)} e^{(-x/2)}, \tag{3.52}$$

where Γ(v/2) is the **gamma function** given by

$$\Gamma(v/2) = \int_0^\infty e^{-t} t^{[(v/2)-1]} dt. \tag{3.53}$$

For v = 2, Γ(1) = 1 and

$$f_{\chi_2^2} = \frac{1}{2} e^{(-x/2)}. \tag{3.54}$$

Hence, the χ_v^2 distribution does take on an exponential distribution shape for v = 2.

3.8.3 F Distribution

The F distribution arises by comparing the ratio of all pairs of sample variance arising from random samples from a Normal population. That is,

$$F = \left(s_1^2/\sigma_1^2\right)/\left(s_2^2/\sigma_2^2\right), \tag{3.55}$$

where $s_1^2 \geq s_2^2$. Therefore, you can use this distribution to determine if the sampled populations have similar variances by

$$P\left\{\left(s_2^2/s_1^2\right)F\left(v_1, v_2, \alpha/2\right) \leq \sigma_2^2/\sigma_1^2 \leq \left(s_2^2/s_1^2\right)F\left(v_1, v_2, 1-\alpha/2\right)\right\} = 1-\alpha \tag{3.56}$$

3.9 Take-Home Points

1. The *population* represents the *true* distribution of a random variable, and your sample allows you to *estimate* the population, so the sampling must be carefully planned to represent the true population being studied.
2. An estimate of a population characteristic (termed a parameter) is itself a random variable, and with each new realization added to the sample, that parameter estimate can change, reflecting this fact.
3. Level of significance must be given when stating that something is or is not statistically significant. There is nothing magic about 5 percent significance (in fact, in many cases in nature, 5 percent is a ridiculously high standard that would be difficult to achieve in all but the most insanely simple cases). If NOT significant, give α at which it fails (dropping a hypothesis because it is significant at α = 0.07 instead of 0.05 may not make any sense).
4. The *Principle of Maximum Likelihood* is a means of estimating population parameters, given knowledge of the population distribution, because it finds those

parameters of the distribution that are most consistent with the values actually attained in the sample.

5. Confidence intervals are intervals with a known confidence of containing the true population parameter value, based on the sample estimate.
6. Confidence intervals can be constructed using the *bootstrap* method, whereby you will use the distribution of your sample as representative of the population, and redraw new samples from it to establish the likelihood of getting the statistical result you actually did get.
7. The null hypothesis used in a hypothesis test should be either the one of "no change" or the easiest hypothesis; in either case, the one you hope to reject. For example, the null hypothesis is that your data have no slope (but you hope it does have slope).

3.10 Questions

Pencil and Paper Questions

1. a. Give $\text{Var}\left[\frac{1}{n}\sum_{i=1}^{n} x_i\right]$ in terms of standard central moments and X, assuming nothing about X. Show the steps of your expansions and reductions.
   b. What is the result if the x_i are independent and identically distributed (iid)?

2. a. Show the bias of a sample variance, when computed as follows:

$$s^2 = \frac{1}{n}\sum_{i=1}^{n}(x_i - \overline{x})^2.$$

   b. How can you compute the sample variance to eliminate the bias (show the derivation).

3. Are the following statements true (explain answer)?

   a. The level of significance for hypothesis testing should always be 5 percent.
   b. $H_0 = x =$ sandstone; $H_1 = x \neq$ sandstone; $\nu = 0.02$.
   Test shows that the statistic does not fail (no reason to reject the null hypothesis), then we conclude that the rock is a sandstone at 98 percent confidence.

Part II

Fitting Curves to Data

Fitting curves to data, or **curve fitting**, deals with fitting a function (i.e., a curve) to a set of discrete data points. It is useful for estimating the values of unsampled points, reducing a bulky collection of data to a useful and manageable form, examining the relationship between variables, comparing theoretical predictions to data (e.g., fitting data to a theoretical formula) and determining model coefficients, among other things.

Curve fitting is readily divided into two categories: (1) exact fitting (interpolation) and (2) "smoothed" fitting (optimization). The first of these arises naturally when dealing with continuous data sampled at a discrete set of points; you may later need to estimate the value of points not sampled. The second category is common because ubiquitous noise can hide or distort the underlying signal in sampled data or introduce disagreement or scatter between repetitive samplings of the function. In that case, we wish to obtain an estimate of the underlying signal that is otherwise disguised by the noise. Such an estimate represents a "smoothing" of the data, since it attempts to present the function stripped of the irregularities thought to be attributable to the noise. Smoothed curve fits also arise for data that represent more than one process, and the smoothed fit may represent a subcomponent of interest.

In this Part, interpolation and that subset of smoothing problems involving optimization (including regression analysis) are considered. The other fundamental smoothing technique involves the concept of *filtering* (technically equivalent to *running averages* with various weighting schemes, or to other forms of synthesizing the data from its analyzed components while omitting specific components thought to be noise). Filtering is considered in later chapters that involve time series analysis, where the necessary tools of those techniques are developed.

4 Interpolation

4.1 Overview

Interpolation is the process of estimating values of a discretely sampled function, $y(x)$, *between* observed values of the function. This is in contrast to **extrapolation**, which involves estimating the function *outside* the interval for which the data were sampled. Here, interpolation takes on the more restricted formal "mathematical" definition in which a function, I(x), is used to estimate the values of $y(x)$ at any arbitrary value of x, where I(x) *is subject to the constraint that it agrees exactly with the values of $y(x)$ at the x positions for which the function is known*. That is, interpolation involves the *exact fitting* of a set of discrete points, y_i. Extrapolation is treated as a special case within the general category of interpolation.

This "exact fit" constraint is consistent with the earliest definitions of interpolation and is still practical in a variety of situations. However, with the advent of faster computers with more memory, there has been an explosion in the number of practical interpolation schemes that are not restricted by having to exactly fit the data points. That is, they produce smoothed fits to the data, presumably fitting the signal and not the noise present in the exact values sampled. Interpolation doesn't rank among the sublime data analysis tools.[1] Consequently, it is a subject often ignored, yet it probably represents one of the most widely used (and misused) analysis tools. It often enters into an analysis through stealth, when, for example, you do a numerical integration (which must perform some sort of interpolation) or when you are simply converting your data to evenly spaced intervals and don't even think about the implications of doing so via interpolation. Interpolation is not bad, but still must be understood when and where it is applied.

4.1.1 What Is Involved

Interpolant is the name given to the approximating function, I(x). For I(x) to represent an interpolant, it must take on the *observed* values of the function $y(x)$ at the points, x_i. So,

[1] The relatively new **optimal interpolation** schemes should rank with the sublime tools. I will not cover optimal interpolation here, but will give a brief discussion after introducing Empirical Orthogonal Function analysis (Chapter 15), which lies at the heart of some of these methods.

$$I(x_i) = y(x_i) = y_i, \tag{4.1}$$

and $y(x_i)$ or y_i represents a discrete (sampled) representation of $y(x)$.

The most fundamental problem of interpolation (and all curve fitting), in practice, is that the process of fitting a function to a set of discrete points is *non-unique*. Consequently, this operation requires additional information about the data being fit in order to best estimate the true nature of $y(x)$. Specifically, you need to

1) define the concept of fit required (e.g., exact fit or smoothed fit; degree of smoothness; bounds; etc.)
2) determine what type of function is most appropriate for the data being fit

The concept of fit is dictated by one's understanding of the data and the purpose of the interpolation. For example, if the goal of the interpolation is to provide an evenly spaced time series to be analyzed for its frequency content by spectral analysis techniques, then, unless the true form of $y(x)$ is known, it is most sensible to choose an interpolant that preserves the frequency content of the discrete samples. That is, you don't want to introduce new frequency content as a consequence of the interpolation process itself, unless warranted by other information about the nature of $y(x)$. Therefore, you must choose an interpolant that satisfies this requirement (i.e., one using sines and cosines, in this example).

Similarly, if your data points represent a process that is known to vary in a smooth manner, then you want a "smooth" interpolant to capture this quality (e.g., a spline). You must carefully consider all information about the data and the analysis goals to properly define the concept of fit. After this, choosing or developing the appropriate interpolant becomes a technical detail.

Box 4.1 To Interpolate or Not to Interpolate

Interpolation is sometimes frowned upon, in that the user is effectively estimating the values of the data at locations other than those for which the observations were collected. Therefore, it is often considered that a certain sacrifice of objectivity has been made when interpolating. Certainly, there are instances where interpolation is unnecessary and the above is true. However, there is an entire class of situations where such is not the case, and considerable computational effort is often wasted in needlessly avoiding interpolation.

For example, consider a sequence y_i that is about to undergo analysis, which has been measured at uneven increments of x (that is, Δx is not constant). You now have two options. The first involves interpolating the series to one, in which the y_i are at evenly spaced increments of x. This offers the advantage that the most common analysis techniques are often based on algorithms that assume, or in fact demand, that the data are evenly spaced in x (i.e., that $\Delta x = \text{constant}$). Alternatively, you can leave the data in the original form and use (or develop) less common algorithms (usually more complicated and computationally slower) that perform the same analyses but do not assume a constant interval Δx.

Your choice might justifiably vary from case to case, but the reasoning employed to decide which track to follow should not be corrupted by a misunderstanding of the consequence of interpolation. Specifically, if the analysis involves any sort of integration

Box 4.1 (cont.)

of the unevenly spaced function (as many do), this too will make assumptions involving the shape of the function between the observed points. Therefore, you might achieve a considerable computational savings by first interpolating the data to an evenly spaced series using an interpolant whose form is consistent with what you feel is appropriate for the data (as opposed to that used by the integration, which may not be). Then use a form of the integration that takes advantage of evenly spaced values to achieve computational efficiency, often a simple sum.

In general, *being aware of how the various analysis techniques work can lead you to the most accurate and internally consistent results.*

4.1.2 Interpolant Types

Now consider the different types of interpolants, their general development and application concepts. Interpolants generally fall into one of two categories: (1) piecewise continuous or (2) continuous.

Piecewise Continuous Interpolants

The piecewise continuous class of interpolants involves breaking the discrete points into adjoining (contiguous) intervals (groupings) of x, and then defining the interpolant for each interval separately. The individual interpolants of neighboring intervals, $I_1(x)$, $I_2(x)$, etc. (usually of the same functional form) connect sequentially at the common interval boundary (at points called **breakpoints, nodes** or **knots**) in a manner so as to avoid any jump in I(x) at these interval boundaries. In this manner, the connected interpolants form a continuous curve. The word "piecewise" indicates that the interpolant is not continuous in all of its derivatives at the node points (this dictates the "smoothness" of the overall curve).

For example, consider Figure 4.1, in which every three data points form an interval, i, which is fit using an interpolant, $I_i(x)$, where x is nonzero in the *i*th interval only – the interval that includes limiting values at the nodes.[2]

Continuous Interpolant

Continuous interpolants involve the fitting of a single function over the entire range of the discrete data points. This requires combining *n* terms of a **basis**, or **characteristic function**, to exactly fit the *n* discrete points. Conceptually, a continuous basis is a set of linearly independent functions of similar form. For example, a polynomial basis is made of independent functions of the form x^n (e.g., x^0, x^1, x^2, x^3, . . .); a sinusoid basis is made of independent functions of the form $\sin\left(\frac{n2\pi x}{x_{range}}\right)$, that is, $\sin\left(\frac{2\pi x}{x_{range}}\right)$, $\sin\left(\frac{4\pi x}{x_{range}}\right)$, $\sin\left(\frac{6\pi x}{x_{range}}\right)$, etc.

A basis is more easily defined for vector spaces, where it represents a set of linearly independent vectors such that every vector in the same dimensional (length of the

[2] The nodes (interval endpoints) need not coincide with data points.

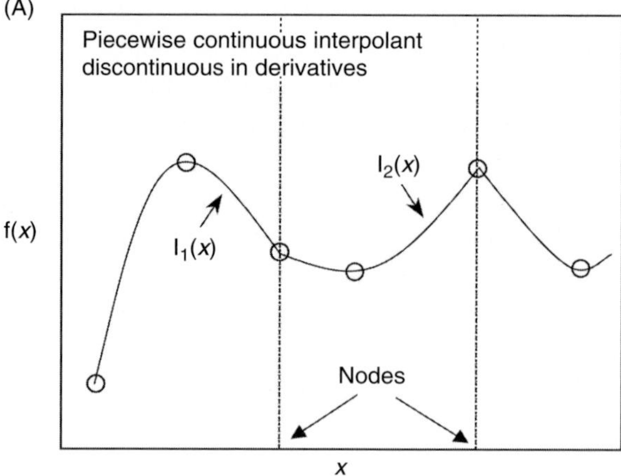

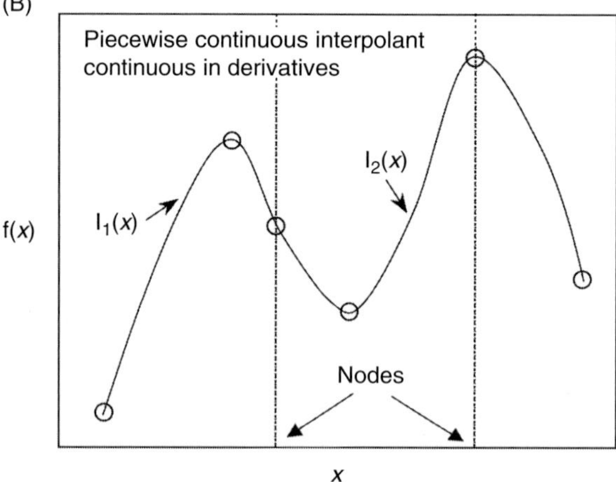

Figure 4.1 **(A)** Schematic of piecewise continuous interpolants where the curve is discontinuous in derivatives, and **(B)** where it is continuous in derivatives. "Smoothness" of interpolant increases with continuity of higher derivatives (though increases are mostly imperceptible after continuity of the second derivative).

vector) space can be represented by some finite linear combination of vectors of the basis. So if the basis consists of a set of N vectors $\mathbf{b}_i$, then every vector $\mathbf{y}$ in the vector space can be uniquely expressed in the form

$$\mathbf{y} = \sum_{i=1}^{\leq N} a_i \mathbf{b}_i. \tag{4.2}$$

The most intuitive example is the case in which N = 3, where the three basis vectors correspond to the standard three spatial dimensions, x, y and z. Vectors in these three directions can be combined through different weightings to produce any vector in the 3-dimensional world, regardless of whether or not they lie precisely along one of these primary axes (e.g., some vector that lies at a 60° angle to each of the primary axes). So, if the system we are studying is a geographic location, any location can be described by a linear combination of fixed-length vectors, where one lies along the east-west direction, one along the north-south direction and one in the vertical.[3]

N represents the **dimension** of the vector space that is also known as the *degrees of freedom* of the space. There may be any number of bases with which to uniquely represent the vector space, but such bases have the same dimension. In physical problems, "degrees of freedom" analogously represent the minimal number of independent "processes" that need to be considered in order to fully describe the phenomenon of interest. One goal of analysis of a physical system is often determination of the number of degrees of freedom (or dimension) of a system – that is, what is the minimal set that must be resolved, observed or monitored in order to completely describe the system being studied.

Here, the basis represents the minimal number of independent functions that can be linearly combined to form any functional shape. There are typically an infinite number of bases in such problems, but the dimension will always be the same. For example, in the previous example in which N = 3, the basis vectors, instead of lying along the standard x, *y* and *z* axes, may lie at angles to the x, y and z axes (e.g., in the NE-SW, NW-SE and off-vertical directions). In either case, starting with these three independent vectors, we can then describe any other vector in the 3-dimensional geographic world.

4.1.3 Interpolation Schemes

Contrasts between piecewise and continuous interpolants mainly center on the concepts of global versus local interpolation schemes and ease of use. Regarding the two schemes, consider the differences in the piecewise and continuous interpolants that pass through the same set of points, as shown in Figure 4.2.

The sharp gradient created by the first two data points relative to the next two data points introduces no undesirable effects in the piecewise interpolant (local scheme) on the left of Figure 4.2, other than a relatively sharp (non-smooth) break in the curve. The global nature of the continuous interpolant on the right leads to a much smoother fit,

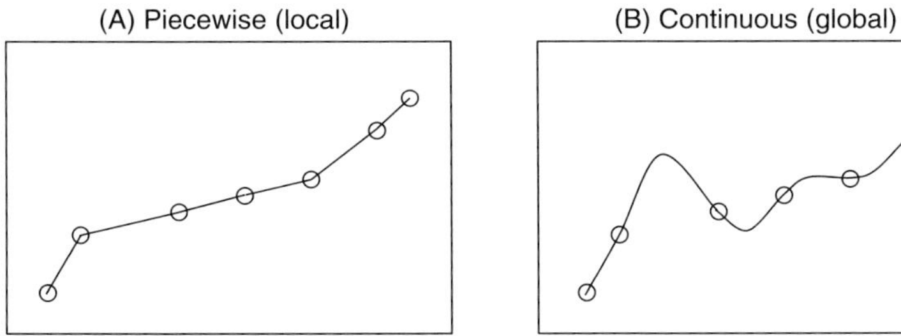

Figure 4.2 **(A)** Piecewise and **(B)** global interpolants, demonstrating the discontinuity of derivatives in piecewise and the continuity of derivatives in global, but leading to overshoot.

[3] See Appendix 1 on Matrix Algebra for a discussion of basis and vector length. Ideally, we would use unit vectors as our basis, but this is not a requirement.

but the cost of "smoothness" is that the sharp gradient leads to "overshoot" (**oscillation**, or **disturbance**) in the fit between the next pairs of points. To preserve smoothness, the gentle and continuous change in slope of the curve, this initial disturbance then introduces an overshoot in the next pair of points. In this manner the initial, or local, disturbance is propagated over much or even the entire length of the interpolant.

In general,

1) Continuous interpolants are smooth and continuous over all derivatives, but involve **global schemes** in which local disturbances can be propagated through the interpolant, resulting in undesirable oscillatory behavior. Because of this propagation, *global schemes are sensitive to the end conditions utilized.*
2) Piecewise interpolants can overcome the global scheme problem of disturbance propagation by employing **local schemes**. These local schemes typically avoid "disturbances" completely, since the interpolant is only sensitive to the local points bounding, and within, an interval – or, if disturbances are generated (e.g., if an interval contains interior points), they are isolated to the interval of origin. *Local schemes, however, are not smooth* because they do not take into account the change in slope of the interpolant that will take place in the adjoining intervals.
3) Piecewise interpolants using global schemes have been developed as a "best" compromise between the smooth, global continuous interpolants and "stable," local piecewise interpolants. These are continuous in some derivatives and thus smooth – in fact, the "smoothest" interpolant in a restricted definition (one with minimum curvature), a natural cubic spline, is a piecewise interpolant. However, they will propagate disturbances like continuous interpolants. The compromise is usually made that the smoothness is reduced somewhat at the cost of reducing the degree of disturbance propagation. That is, they can give a relatively smooth curve with a minimal amount of undesirable oscillatory behavior. In fact, the rate at which a disturbance is attenuated away from the disturbance source represents a major attribute distinguishing the various piecewise global schemes.
4) For both types of interpolant, it is relatively easy to establish the mathematical solutions to the unknown coefficients of the problem, but in practice, the actual numerical implementation may be considerably different. See Press et al. (1986) for examples in order to see some of the more efficient and stable schemes that are best used in practice to carry out some of the interpolants described here.

4.2 Piecewise Continuous Interpolants

Here it is convenient to formalize the simple linear interpolant in order to demonstrate the fundamental philosophy underlying the construction of all piecewise continuous schemes. However, while the mathematical development is general, it will often prove inefficient or impractical to solve the resulting system of equations directly. Typically, such schemes are numerically unstable or grossly inefficient, and consequently it is more practical to utilize well-tested pre-existing algorithms.[4]

4.2.1 Piecewise Linear Interpolant

A piecewise linear interpolant is the most common and simplest interpolant to work with, and it demonstrates nicely the general solution techniques of a local piecewise interpolant. Recognize that *"linear" interpolant means that the interpolant is linear in the coefficients; it does not mean that the interpolant is a straight line.*

The linear interpolant is a piecewise polynomial interpolant of degree 1, order 2, where[5]

degree = value of largest exponent present in the polynomial
order = number of distinct x^n terms

So, for example,

$$I(x) = a_0 + a_1 x + a_2 x^2 \tag{4.3}$$

is a second-degree, third-order polynomial (linear in the three a_i coefficients). A second-degree equation (i.e., one with the highest degree term x^2) is often referred to as a **quadratic** equation, regardless of the order; that is, whether the x^0 and x^1 terms are present or not. A third-degree equation is typically referred to as a **cubic** equation.

Fitting a piecewise first-degree, first-order interpolant to a set of n points, y_i (where $y(x_i) = y_i$), results in an interpolant where the data points are connected by straight lines. This provides estimates for all the values of f(x), not just for the data points (x_i, y_i).

For the interpolant of Figure 4.3, the following applies:

1) ξ_i, the breakpoints or nodes, correspond to the x_i at which $y(x)$ is sampled (i.e., the actual data points).
2) $\xi_1 < \xi_2 < \xi_3 < \ldots$
3) The y_i have no restrictions (e.g., they needn't increase monotonically).
4) The L_i are the discrete linear segments connected at the breakpoints and interpolate the function $y(x_i)$ at y_i, $i = 1, 2, \ldots, n$.

So, the piecewise linear interpolant is given by

$$L_i(x_j) = \sum_{j=1}^{n-1} L_i(x_j), \tag{4.4}$$

[4] The book *Numerical Recipes* by Press et al. (1986) contains a great many such schemes, as do standard math libraries such as IMSL or MATLAB.

[5] Some authors use *order* and *degree* differently, and some assume that *order* is always equal to *degree* + 1 (a restriction we need not impose).

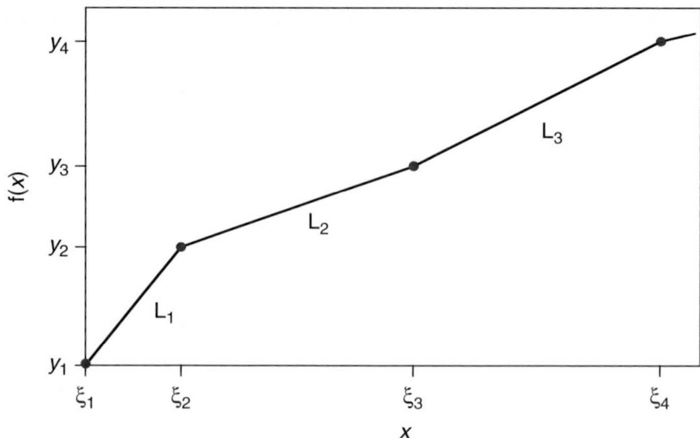

Figure 4.3 Example of the simplest piecewise linear interpolant, fitting one straight line per every two data points.

where

$$L_i(x) = \begin{cases} a_{i,0} + a_{i,1}(x - \xi_i) & \xi_i \le x \le \xi_{i+1} \quad (4.5a) \\ 0 & \text{elsewhere} \quad (4.5b) \end{cases}$$

and $i = 1,2,\ldots,n-1$, corresponds to the $n-1$ intervals formed by the n discrete data points.

Written as a system,

$$\begin{aligned} L_1(x) &= a_{i,0} + a_{i,1}(x - \xi_1) && (\xi_1 \le x \le \xi_2) \\ L_2(x) &= a_{2,0} + a_{2,1}(x - \xi_1) && \xi_2 \le x \le \xi_3 \\ &\vdots \\ L_{n-1}(x) &= a_{(n-1),0} + a_{(n-1),1}(x - \xi_{(n-1)}) && \xi_{(n-1)} \le x \le \xi_n. \end{aligned} \tag{4.6}$$

Each segment of this system contains two unknown coefficients, a_0 and a_1. Therefore, if the values of the $a_{i,j}$ coefficients were known, the system would be well posed, allowing solution. As it stands, since the values of $a_{i,j}$ are not known, the system requires another $2(n-1)$ equations (two equations for each of $n-1$ segments) to provide as many equations as unknowns.

The additional equations are given by satisfying the two most fundamental interpolation criteria. Note that more elaborate interpolants will have more degrees of freedom, allowing us to satisfy not only the fundamental criteria (given later), but other desirable criteria or constraints as well. This allows us to build interpolants with specific qualities.

The two most fundamental interpolation criteria are given by the following constraints:

The **interpolation condition** states that in order for I(x) to be a true interpolant, it must pass exactly through the observed values of $y(x)$ (i.e., the y_i).

The **continuity condition** states that the interpolants of each interval must connect at the adjoining interval limits (at the node or breakpoints) to make a single continuous curve over the entire set of points. This is also known as the **collocative condition.**[6]

In equation form, these constraints are written as

$$\mathrm{L}_i(\xi_i) = y_i \tag{4.7}$$

$$\mathrm{L}_i(\xi_{i+1}) = y_{i+1}. \tag{4.8}$$

These provide the required $2(n-1)$ additional constraints (recall that $i = 1, 2, \ldots, n-1$, representing the $n-1$ linear segments). Therefore, (4.7) says that the beginning of each segment must pass through the segment's initial node, and (4.8) says the end of each segment must pass through the segment's upper node. These also satisfy the collocative condition, since (4.8) says the former segment captures node $i + 1$, while (4.7) states that the next segment also hits that node, as i.

Box 4.2 Computation of Piecewise Coefficient Values

Substituting constraint (4.7) into the equation for $\mathrm{L}_i(\mathrm{x})$, equation (4.5a), gives

$$\mathrm{L}_i(\mathrm{x} = \xi_i) = y_i = a_{i,0} + a_{i,1}(\xi_i - \xi_i), \tag{4.9}$$

so

$$a_{i,0} = y_i, \tag{4.10}$$

or the $a_{\mathrm{i},1}$ coefficients are simply equal to the values y_i that begin each interval.

Rewriting constraint (4.7) in terms of $i + 1$ and combining this with constraint (4.8) to eliminate y_{i+1} provides a more explicit form of the continuity condition:

$$\mathrm{L}_i(\xi_{i+1}) = \mathrm{L}_{i+1}(\xi_{i+1}). \tag{4.11}$$

Writing (4.11) in terms of $\mathrm{L}_i(x)$ in (4.5a) then gives

$$a_{i,0} + a_{i,1}(\xi_{i+1} - \xi_1) = a_{i+1,0} + a_{i+1,1}(\xi_{i+1} - \xi_{1+1}). \tag{4.12}$$

Rearranging this for $a_{i,1}$ (note that the second term on the RHS drops out) gives

$$a_i = \frac{a_{i+1} - y_i}{\xi_{i+1} - \xi_1}. \tag{4.13}$$

From (4.10), we can substitute y_i for the $a_{i,1}$ terms, giving

[6] *Collocation* is placement side by side, or in this case, end to end.

Box 4.2 (cont.)

$$a_{i,1} = \frac{y_{i+1} - y_i}{\Delta\xi_1}, \tag{4.14}$$

where $x = \mathbf{A}^{-1}\mathbf{b}$. This defines the slope of the segment.

So, we now have the forms of the unknown $a_{i,0}$ (4.10) and $a_{i,1}$ (4.14), which can be substituted directly into equation (4.5a), defining the linear pieces:

$$\begin{aligned} L_i(x) &= y_i + \frac{y_{i+1} - y_i}{\Delta\xi_i}(x - \xi_i) \\ &= y_i + y_{i+1}\frac{x - \xi_i}{\Delta\xi_i} - y_i\frac{x - \xi_i}{\Delta\xi_i} \\ &= y_i\left(1 - \frac{x - \xi_i}{\Delta\xi_i}\right) + y_{i+1}\frac{x - \xi_i}{\Delta\xi_i}. \end{aligned} \tag{4.15a}$$

Since $\xi_i = \xi_{i+1} - \Delta\xi_i$, (4.15) can be rewritten as

$$L_i(x) = y_i\left(1 - \frac{x - \xi_i + \Delta\xi_i}{\Delta\xi_i}\right) + y_{i+1}\frac{x - \xi_i}{\Delta\xi_i}, \tag{4.15b}$$

allowing a symmetrical form, amenable for computation, for each segment i.

That system can be rearranged to yield the following for the piecewise linear interpolant:

$$L_i(x) = y_i\left(\frac{\xi_{i+1} - x}{\Delta\xi_i}\right) + y_{i+1}\left(\frac{x - \xi_i}{\Delta\xi_i}\right). \tag{4.16}$$

Equation (4.16) provides the general form from which values of the interpolating function L(x) can be computed for any interval of ξ and any value of x. This interpolant is not smooth and it is discontinuous for all derivatives. It is therefore not suited for approximating most functions that arise in physical problems. It is, however, reasonable for providing a rough idea of the shape of the data curve. We can overcome many of the shortcomings of a linear interpolant with spline interpolants.

4.2.2 Cubic Spline Interpolant

The cubic spline interpolant is a particular piecewise polynomial interpolant of degree 3, order 4. Relative to the linear interpolant, this has two extra degrees of freedom per interval (due to the second- and third-degree terms), allowing us to fit two more data points per segment or add two more constraints to the interpolant (or one more data point and one more constraint). With the cubic spline, the two additional constraints are used to specify the continuity of the first two derivatives across the breakpoints. This makes the function smooth, but also makes it a more global scheme.

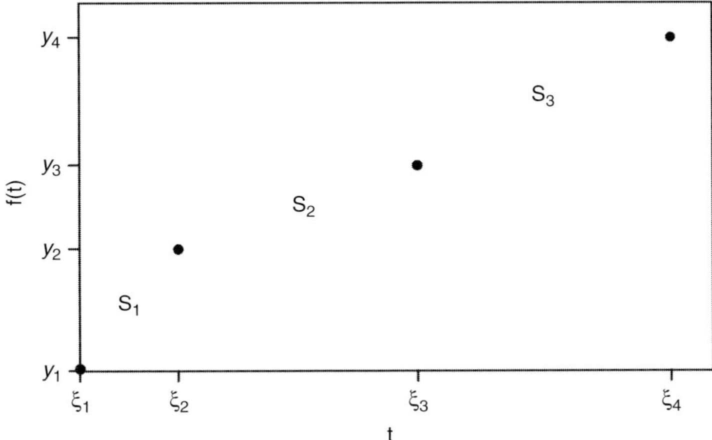

Figure 4.4 Segments to be fit with cubic spline. Fitting only pairs of points as with the straight line fit allows the addition of continuity of first two derivatives across the collocation nodes for a smooth piecewise fit (but allowing extreme overshoot). The overshoot problem will be corrected by making only the first derivative continuous and setting a second condition that defines the second derivative (a "deficient" spline).

The cubic spline interpolant, S(t), has the following properties:

1) S(t) satisfies the interpolation condition.
2) S(t) is a cubic polynomial over each interval.
3) S(t) is continuous, as are S′(t) and S″(t).

Each segment of S(t) as $S_i(t)$ is defined as in Figure 4.4, so for n data points,

$$S_i(t_j) = \sum_{j=1}^{n-1} S_i(t_j) \tag{4.17}$$

$$S_i(t) = \begin{cases} a_{i0} + a_{i1}(t-\xi_i) + a_{i2}(t-\xi_i)^2 + a_{i3}(t-\xi_i)^3 & \xi_i \le t \le \xi_{i+1} \\ 0 & \text{elsewhere} \end{cases}. \tag{4.18}$$

This system requires an additional $4(n-1)$ constraints (to constrain the 4 degrees of freedom due to the $a_{i,j}$, where $j = 1,2,3,4$, for each of the $n-1$ segments).

Interpolation condition and continuity of S(t) is defined in the same manner as for the straight-line interpolant:

$$S(\xi_i) = y_i \tag{4.19a}$$

$$S_i(\xi_{i+1}) = y_{i+1}. \tag{4.19b}$$

These add $2(n-1)$ equations (two constraints for each segment), but we still require an additional $2(n-1)$ equations.

Continuity of the first derivative of S(t), written as S′(t), is defined as

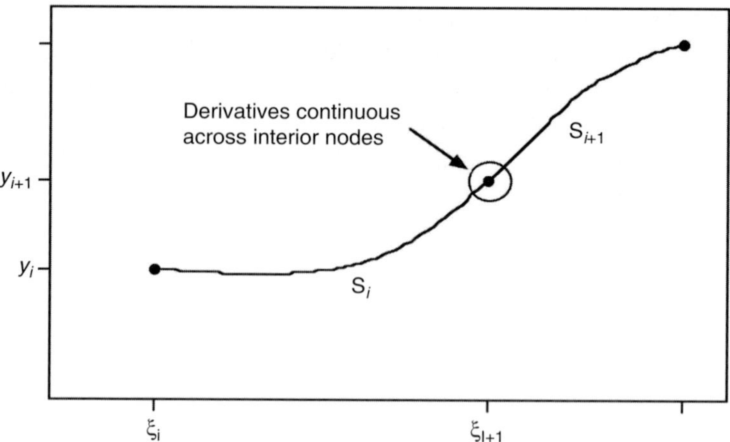

Figure 4.5 Location of nodes defining the end of the cubic polynomial segments.

$$S'_i(\xi_{i+1}) = s'_i \tag{4.20a}$$

$$S'_{i+1}(\xi_{i+1}) = s'_i. \tag{4.20b}$$

This condition states that the first derivative of $S_i(t)$ evaluated at the point defining the end of the interval, ξ_{i+1}, is the same as the derivative of the next segment, $S_{i+1}(t)$, evaluated at the same point, ξ_{i+1}, which for this next segment represents the first point of the interval, as seen in Figure 4.5.

Equations (4.20) add $2(n-2)$ constraints (not $2(n-1)$ constraints, because they only apply to the $n-2$ interior points, not to the $n-1$ segments). They also *introduce* $n-2$ additional degrees of freedom in the s_i'. So, unless the s_i' are specified (which they are in numerous other types of spline interpolants), we still require an additional n equations. That is, we have $5(n-1)$ original unknowns (the unknown $a_{i,j}$ and the unknown value of t) + $n-2$ more added here = $6n-7$ unknowns, versus the $n-1$ original equations + $2(n-1)$ from the interpolation and continuity conditions + $2(n-2)$ equations added here = $5n-7$. So, $6n-7-(5n-7)=n$ equations still required.

Note that the continuity condition (4.20a&b) could have been combined and written as $S_i'(\xi_{i+1}) = S_{i+1}'(\xi_{i+1})$ with the same net gain in number of equations. It is written as given in (4.20) to show a more general form in which the specification of the actual derivative values can be made to achieve certain other desirable splines (as will be discussed more later).

Continuity of the second derivative of S(t), written as S″(t), is used in cubic splines to provide additional smoothness. This is given as

$$S''_i(\xi_{i+1}) = s''_i \tag{4.21a}$$

$$S''_{i+1}(\xi_{i+1}) = s''_i. \tag{4.21b}$$

This constraint adds an additional $2(n-2)$ equations (two constraints for each of the $n-2$ interior points), but introduces an additional $n-2$ unknowns in the S_i''. So, by adding $2(n-2)$ additional equations and $n-2$ additional unknown, we achieve a net gain of $n-2$ equations toward the required n.

The **final two constraints** are applied to the ends of the data set and can be anything, but most often they are

$$S'(\xi_1) \text{ and } S'(\xi_n) = \text{specified} \tag{4.22a}$$

or

$$S''(\xi_1) \text{ and } S''(\xi_n) = 0. \tag{4.22b}$$

Either of these specifications will close the system. The first constraint (4.22a) is most useful for local schemes only (when the values of the s_i' are specified) or if information exists that allows their specification. The second condition is more common, and its use results in the construction of a special type of cubic spline known as a **natural cubic spline**. This condition simply indicates that the slope of the spline across the ends of the data set does not change (i.e., no curvature), but instead continues on beyond the data at the same slope it had at the ends (consistent with a draftsman spline).

Box D4.1 Construction of a Cubic Spline

The development just described shows the setup of a closed system for the cubic spline that can be solved in some traditional (though very cumbersome) manner of elimination. However, for piecewise polynomials in general, after constructing the conceptual setup shown above to ensure that the system closes, the following approach could be followed to produce a computationally simple algorithm – though for these, using trusted software is probably best.

Since each polynomial piece is cubic, we know the interpolant is linear in the second derivative, as shown in Figure D4.1.1.

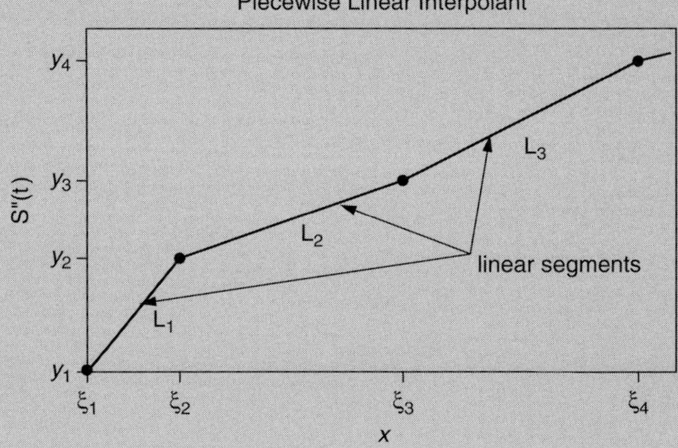

Figure D4.1.1 After taking the second derivative, each cubic segment is now a line slope.

Box D4.1 (cont.)

So, the second derivative of each segment is

$$\frac{\partial^2 S_i(\mathrm{t})}{\partial \mathrm{t}^2} = S_i''(\mathrm{t}) = \frac{\partial^2}{\partial \mathrm{t}^2}\left[a_{i,0} + a_{i,1}(\mathrm{t}-\xi_i) + a_{i,2}(\mathrm{t}-\xi_i)^2 + a_{i,3}(\mathrm{t}-\xi_i)^3\right]$$
$$= \frac{\partial^2}{\partial \mathrm{t}}\left[a_{i,1} + 2a_{i,2}(\mathrm{t}-\xi_i) + 3a_{i,3}(\mathrm{t}-\xi_i)^2\right] \quad \text{(D4.1.1)}$$
$$= 2a_{i,2} + 6a_{i,3}(\mathrm{t}-\xi_i),$$

where each segment is linear (or in standard linear nomenclature: intercept = $2a_{1,2}$ and slope = $6a_{\mathrm{i},3}$).

Therefore, continuity of the second derivatives is achieved by treating the $S_i''(\mathrm{t})$ segments in the same manner as was done for the $L_\mathrm{i}(\mathrm{t})$ segments of the piecewise linear interpolant. In other words, fitting the s_i' with piecewise linear segments, $L_i(\mathrm{t}) = S_i''(\mathrm{t})$, gives, from (4.16),

$$S_i''(\mathrm{t}) = s_i''\frac{(\xi_{1+1}-\mathrm{t})}{\Delta\xi_i} - s_{i+1}''\frac{(\mathrm{t}-\xi_1)}{\Delta\xi_i}. \quad \text{(D4.1.2)}$$

This linearly interpolates the points S_i'', but these S_i'' are unknowns. Note that for the previous linear interpolation discussion, these values (corresponding to the y_i) were known and satisfied the interpolation condition. Here, the interpolation condition merely states that the $a_{i,0}$ coefficients are equal to the unknown values of the second derivatives, the s_i', at the breakpoint positions.

To get the actual $S_i(\mathrm{t})$ segments, we simply integrate $S_i''(\mathrm{t})$ twice. Integrating once gives $S_i'(\mathrm{t})$. So,

$$\int S_i''(\mathrm{t})\mathrm{dt} = \int S_i''\frac{(\xi_{i+1}-\mathrm{t})}{\Delta\xi_i}\mathrm{dt} - \int S_{i+1}''\frac{(\mathrm{t}-\xi_i)}{\Delta\xi_i}\mathrm{dt}. \quad \text{(D4.1.3)}$$

Defining

$$\mathrm{u} = \xi_{i+1} - \mathrm{t};\ \mathrm{du} = -\mathrm{dt} \quad \text{(D4.1.4a)}$$

$$\mathrm{v} = \mathrm{t} - \xi_i;\ \mathrm{dv} = \mathrm{dt} \quad \text{(D4.1.4b)}$$

and substituting these into (D4.1.3) and rearranging gives

$$S_i'(\mathrm{t}) = \frac{S_{i+1}''}{\Delta\xi_i}\int \mathrm{v\,dv} - \frac{S_i''}{\Delta\xi_i}\int \mathrm{u\,du}, \quad \text{(D4.1.5)}$$

which is easily integrated to give

$$S_i'(\mathrm{t}) = \frac{S_{i+1}''}{2\Delta\xi_i}(\mathrm{t}-\xi_i)^2 + \mathrm{C} - \left[\frac{S_i''}{2\Delta\xi_i}(\xi_{i+1}-\mathrm{t})^2 + \mathrm{D}\right]. \quad \text{(D4.1.6)}$$

Integrating again for $S_i(\mathrm{t})$ gives

Box D4.1 (cont.)

$$\int S_i'(\mathrm{t})\mathrm{dt} = \int\left[\frac{s''_{i+1}}{2\Delta\xi_i}(\mathrm{t}-\xi_i)^2+\mathrm{C}\right]\mathrm{dt} - \int\left[\frac{s''_i}{2\Delta\xi_i}(\xi_{i+1}-\mathrm{t})^2+\mathrm{D}\right]\mathrm{dt}$$
$$= \frac{s''_{i+1}}{2\Delta\xi_i}\int \mathrm{v}^2\mathrm{dv} + \mathrm{C}\int \mathrm{dv} + \frac{s''_i}{2\Delta\xi_i}(\xi_{i+1}-\mathrm{t})^2\int \mathrm{u}^2\mathrm{du} + \mathrm{C}\int \mathrm{du}$$
$$\mathrm{S}_i(\mathrm{t}) = \frac{\mathrm{s}''_{i+1}}{6\Delta\xi}(\mathrm{t}-\xi_i)^3 + \frac{\mathrm{s}''_i}{6\Delta\xi}(\xi_{i+1}-\mathrm{t})^3 + \mathrm{C}(\mathrm{t}-\xi_i) + \mathrm{D}(\xi_{i+1}-\mathrm{t}). \tag{D4.1.7}$$

Interpolation constraints and continuity are used to solve for the unknown constants C and D. So

$$\mathrm{S}_i(\xi_i) = y_i \tag{D4.1.8a}$$
$$\mathrm{S}_i(\xi_{i+1}) = y_{i+1} \tag{D4.1.8b}$$

$$S_i(\xi_i) = y_i = \frac{s''_{i+1}}{6\Delta\xi_i}(\xi_i-\xi_i)^3 + \frac{s''_i}{6\Delta\xi_i}(\xi_{i+1}-\xi_i)^3 + \mathrm{C}(\xi_i-\xi_i) + (\xi_{i+1}-\xi_i)$$
$$= \frac{s''_i\Delta\xi_i^2}{6} + \mathrm{D}\Delta\xi_i \tag{D4.1.9}$$

This is solved for D:

$$\mathrm{D} = \left(y_i - \frac{s''_i\Delta\xi_i^2}{6}\right)\Delta\xi_i^{i-1} = \frac{y_i}{\Delta\xi_i} = \frac{s''_i\Delta\xi_i}{6}. \tag{D4.1.10}$$

Similarly, using constraint (D4.1.8b) in (D4.1.7) gives C,

$$s_i(\xi_{i+1}) = y_{i+1} = \frac{s''_{i+1}}{6\Delta\xi_i}\Delta\xi^3 + \mathrm{C}\Delta\xi_i, \tag{D4.1.11}$$

which is solved to give C,

$$\mathrm{C} = \left(y_{i+1} - \frac{s''_{i+1}\Delta\xi_i^2}{6}\right)\Delta\xi_i^{-1} = \frac{y_{i+1}}{\Delta\xi_i} - \frac{s''_i\Delta\xi_i^2}{6\Delta\xi_i}. \tag{D4.1.12}$$

Substituting equations (D4.1.10) and (D4.1.12) for D and C back into equation (D4.1.7) for $S_i(t)$ gives (relative to equation (D4.1.7), the s_i'' and s_{i+1}'' terms here are in reverse order)

$$S_i(\mathrm{t}) = \frac{s''_i}{6\Delta\xi_i}(\xi_{i+1}-\mathrm{t})^3 + \frac{s''_{i+1}}{6\Delta\xi_i}(\mathrm{t}-\xi_i)^3 + \left(\frac{y_{i+1}}{\Delta\xi_i} - \frac{s''_{i+1}}{6\Delta}\right)(\mathrm{t}-\xi_i)$$
$$+\left(\frac{y_i}{\Delta\xi_i} - \frac{s''_i\Delta\xi_i}{6}\right)(\xi_{i+1}-\mathrm{t}) \tag{D4.1.13}$$

Continuity of first derivative must still be added. This is done by taking the derivative of (D4.1.13) above,

Box D4.1 (cont.)

$$S_i'(\mathrm{t}) = \frac{\partial S(\mathrm{t})}{\partial \mathrm{t}} = S_i(\mathrm{t}) + \frac{s_i''}{2\Delta\xi_i}(\xi_{i+1} - \mathrm{t})^2 + \frac{s_{i+1}''}{2\Delta\xi_i}(\mathrm{t} - \xi_i)^2 + \frac{y_{i+1}}{\Delta\xi_i} - \frac{s_{i+1}''\Delta\xi_i}{6} - \frac{y_i}{\Delta\xi_i} + \frac{s_i''\Delta\xi_i}{6},$$

or, rearranging,

$$S_i'(\mathrm{t}) + \frac{s_i''}{2\Delta\xi_i}(\xi_{i+1} - \mathrm{t})^2 + \frac{s_{i+1}''}{2\Delta\xi_i}(\mathrm{t} - \xi_i)^2 + \frac{y_{i+1} - y_i}{\Delta\xi_i} + \frac{\Delta\xi_i}{6}(s_{i+1}'' - s_i''). \quad \text{(D4.1.14)}$$

The first derivative continuity condition is given as

$$\mathrm{S}_{i-1}'(\xi_i) = \mathrm{S}_i'(\xi_i). \quad \text{(D4.1.15)}$$

For symmetry reasons (these are identical to $\mathrm{S}_i'(\xi_{i+1}) = \mathrm{S}_{i+1}'(\xi_{i+1})$ and can now be applied to equation (D4.1.14) in anticipation of the different values of t), first define

$$\Delta\xi_{i-1}^2 = (\xi_i - \xi_{i-1})^2 \quad \text{(D4.1.16)}$$

$$\Delta\xi^2 = (\xi_{i-1} - \xi)^2 \quad \text{(D4.1.17)}$$

Substituting these into (D4.1.14) and rearranging gives

$$\frac{s_{i-1}''\Delta\xi_{i-1}}{6} + \frac{s_i''\Delta\xi_{i-1}}{2} - \frac{s_i''\Delta\xi_{i-1}}{6} + \frac{s_i''\Delta\xi_i}{2} - \frac{s_i''\Delta\xi_i}{6} + \frac{s_{i+1}''\Delta\xi_i}{6} = \frac{\Delta y_i}{\Delta\xi_i} - \frac{\Delta y_{i-1}}{\Delta\xi_{i-1}}. \quad \text{(D4.1.18)}$$

Finally, defining

$$\alpha_i = 2(\Delta\xi_{i-1} + \Delta\xi_i) \quad \text{(D4.1.19a)}$$

$$\beta_i = 6\left(\frac{\Delta y_i}{\Delta\xi_i} - \frac{\Delta y_{i-1}}{\Delta\xi_{i-1}}\right) \quad \text{(D4.1.19b)}$$

and substituting these into (D4.1.18) and rearranging gives

$$s_{i-1}''\Delta\xi_{i-1} + \alpha_1 s_i'' + s_{i+1}''\Delta\xi_i = \beta_i. \quad \text{(D4.1.20)}$$

This equation describes a closed system of equations that can be put into matrix form as

$$\begin{bmatrix} \alpha_2 & \Delta\xi_2 & 0 & 0 & \dots & 0 \\ \Delta\xi_2 & \alpha_3 & \Delta\xi_3 & & \dots & 0 \\ 0 & \Delta\xi_3 & \alpha_4 & \Delta\xi_4 & & 0 \\ \vdots & & & \ddots & & \\ 0 & 0 & & \Delta\xi_{n-3} & \alpha_{n-2} & \Delta\xi_{n-2} \\ 0 & 0 & \dots & & \Delta\xi_{n-2} & \alpha_{n-1} \end{bmatrix} \begin{bmatrix} s_2'' \\ s_3'' \\ \vdots \\ s_{n-1}'' \end{bmatrix} = \begin{bmatrix} \beta_2 - s_1''\Delta\xi_1 \\ \beta_3 \\ \vdots \\ \beta_{n-2} \\ \beta_{-1} - s_n''\Delta\xi_{n-1} \end{bmatrix}. \quad \text{(D4.1.21)}$$

Box D4.1 (cont.)

Note that the unknown column vector (containing the s_i'') does not contain s_i'' or s_n''. Rather, these two values are included in the known vector on the RHS, reflecting the fact that these were specified in the final two constraints closing the system (recall that for the natural spline, $s_i'' = s_n'' = 0$, so these two terms would even drop out of this RHS vector, leaving only the β_i).

This matrix equation represents a **tridiagonal system** that is solved by a simple recursion formula (working from the bottom up, using standard software for this common type of system). The solution vector, containing the values of the s_i'', is then used in (D4.1.13) to completely describe the interpolant and allow the calculation of $S_i(t)$ for any value of t.

Properties of Natural Cubic Spline

The natural cubic spline constructed and solved above has some very nice properties, and as a consequence represents one of the most popular and widespread interpolants, *but it can display some terrible properties and requires care before it can be blindly applied.* In particular,

1) It has minimum curvature (for all twice-differentiable functions), where curvature is defined here approximately as $|S''(t)|$ and the natural cubic spline minimizes the function:

$$\int \left[\frac{\partial^2 S(t)}{\partial t^2}\right]^2 dt. \tag{4.23}$$

This also equates to minimizing the "strain energy," since

$$\frac{\partial^2 S(t)}{\partial t^2} \propto \text{ strain energy}, \tag{4.24}$$

though this too is a restricted case, since the effect of the first derivative in the strain energy is being ignored.

If we minimize internal strain energy, the natural cubic spline closely approximates a draftsman's spline. In fact, the spline was originally derived by minimizing the strain energy in order to best approximate the draftsman's spline. The term "natural" cubic spline simply reflects the fact that a draftsman's spline will flatten to zero curvature at its extremities (i.e., the second derivative goes to zero at the endpoints).

Some people translate the minimum curvature property by saying that the natural cubic spline is the smoothest of any interpolant. Like most such claims, this is only true in a restricted sense – that is, if one wishes to define "smoothness" in terms of the curvature as described above (which, indeed, is a pretty good definition).

2) When

$$\Delta\tau_i \to 0 \text{ as } n \to \infty$$
$$S(t) \to f(t) \text{ and } S'(t) \to f'(t) \tag{4.25}$$

Therefore, the natural cubic spline is excellent for approximating both f(t) *and* its derivative f′(t). Continuous polynomial interpolants *do not* converge in this manner, since their oscillatory behavior tends to exaggerate the difference between S′(t) and f′(t).

3) The scheme just described has global characteristics because of the continuity of the first two derivatives, so the *natural cubic spline is sensitive to end conditions and steep gradients*. This represents its main drawback, since steep gradients in the data may introduce large (in some cases, enormous) oscillations or overshoot (this can be minimized by going to a "deficient" spline, as discussed later). However, in my experience, such overshoot occurs only rarely, so go ahead and apply, but check curve fit to make sure you are not suffering from the extreme overshoot.

4.2.3 Additional Types of Splines

Numerous other types of splines exist that attempt to maintain the advantageous properties of the natural cubic spline while minimizing the sensitivity to oscillatory behavior. This involves a tradeoff, as mentioned previously: smoothness versus stability.

Smoothness

Smoothness is defined by the continuity of derivatives as well as the amount of curvature. So, the more continuous a function is in its derivatives, the smoother it is. This reflects the fact that derivatives are an indication of the rate of change of a function, which to the eye must change in a continuous manner if it is to appear smooth. For practical purposes, a function that is continuous over the first two derivatives is usually as smooth to the eye as a function that is continuous over higher derivatives.

A function with less overall curvature will also appear smoother than one with more curvature (assuming both are continuous to the same degree in their derivatives). Curvature reflects the rate at which a function "turns." The slower it turns, the smoother it looks and the smaller the curvature. Curvature, κ, is specified by

$$\kappa = \frac{f'(x)}{\left[1 + f'(x)^2\right]^{3/2}} \approx \left| f''(x) \right|. \tag{4.26}$$

So, while the natural cubic spline is continuous over the first and second derivatives and minimizes the curvature (in an approximate sense), the minimal curvature property forces the function to respond very slowly to rapid changes in slope in the data points (Figure 4.6). This is directly responsible for the introduction of large "overshooting" in regions of rapid slope change that can lead to the undesirable oscillatory behavior.

In those cases where the fit experiences unacceptably large overshoot, there are other splines that overcome this problem by specifying a functional form of the first

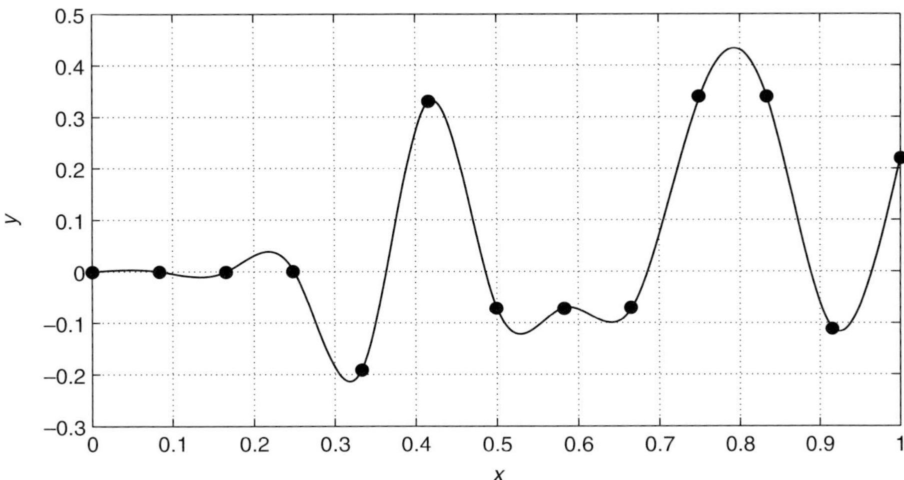

Figure 4.6 Example of a natural cubic spline fit to a data set with some steep gradients near the end, but the fit appears to be excellent with minimal overshoot.

derivatives, $s'(\xi_i)$, at the knots. This makes the interpolant less smooth, since it is no longer continuous in its second derivative, but at the same time, by increasing the rate of curvature relative to the natural cubic spline, the function can respond sooner to rapid slope changes and minimize the overshoot.

Except for the first two names, which simply identify classes of splines, the splines listed here follow the above philosophy in one form or another:

A) Cardinal splines are splines with evenly spaced knots.

B) "Deficient" splines (or splines with multiple knots) make up the general class of splines having less than maximum smoothness (not continuous in the higher derivatives). Many of these specify the value or the first derivative in a piecewise cubic polynomial. For example, here are some of the most popular.

1) **Akima's spline** is a very popular deficient spline – it provides a local scheme and does not introduce oscillations associated with sharp-end gradients by specifying the first derivatives at the knots as

$$S_i' = \frac{\left|\frac{y_{i+2}+y_i-2y_{i+1}}{\Delta\tau_i}\right|\left(\frac{y_i-y_{i-1}}{\Delta\tau_i}\right)+\left|\frac{y_i+y_{i-2}-2y_{i+1}}{\Delta\tau_i}\right|\left(\frac{y_{i+1}-y_i}{\Delta\tau_i}\right)}{\left|\frac{y_{i+2}+y_i-2y_{i+1}}{\Delta\tau_i}\right|+\left|\frac{y_i+y_{i-2}-2y_{i+1}}{\Delta\tau_i}\right|}. \quad (4.27)$$

You are left with specifying s_i' and s_n' to close the system, which must be done carefully.

2) **Cubic Hermite spline**: All $s_i' = f'(\tau_i)$ (so these derivatives at the knots must be known – *great,* if you actually know the derivatives).
3) **Cubic Bessel spline**: the s_i' are made to agree with the slope of $f_p(\tau_i)$, where $f_p(t)$ is computed by fitting three points fit with a parabola centered at τ_i.
4) **Taut splines** control undesirable oscillations by using multiple knots.
5) **Splines in tension** use the basis $1,\ x,\ e^{px}, e^{-px}$.
6) **Continuous basis (β) splines**: use different-order polynomials designed as continuous bases (e.g., parabolic, quartic, etc.).

4.3 Continuous Interpolants

4.3.1 Continuous Polynomial Interpolation

The most common continuous interpolant involves the power basis

$$1\ x\ x^2\ x^3\ \ldots\ \ x^n.$$

These form the foundation for continuous polynomial interpolation in which a set of n data points are fit by adding n terms of the above power functions; the coefficients of which must be determined.

Intuitively, this interpolant may seem advantageous because of its similarity in form to a Taylor series – however, it is generally a *very poor interpolant* because of its strong oscillatory behavior. It also tends toward $a_n x^n$ for large values of $|x|$, and is as a consequence a terrible function for extrapolation (Figure 4.7). If you have control over the sampling points, you can take advantage of function approximation theory, which deals with the best way to approximate a function. In this

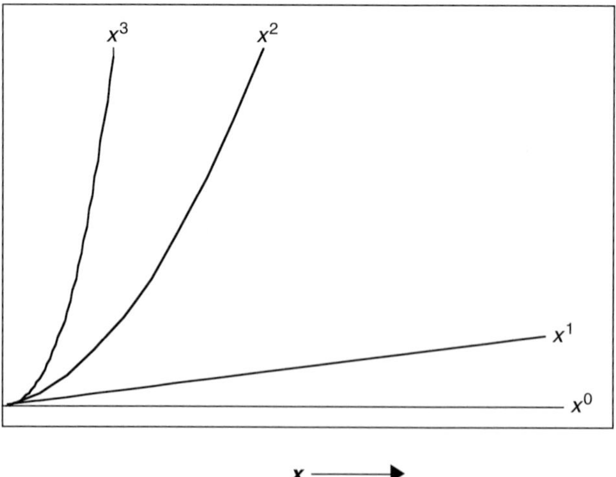

Figure 4.7 Larger powers lead to faster growth, dominating interpolants, and even more so, extrapolation, and fueling strong oscillatory behavior.

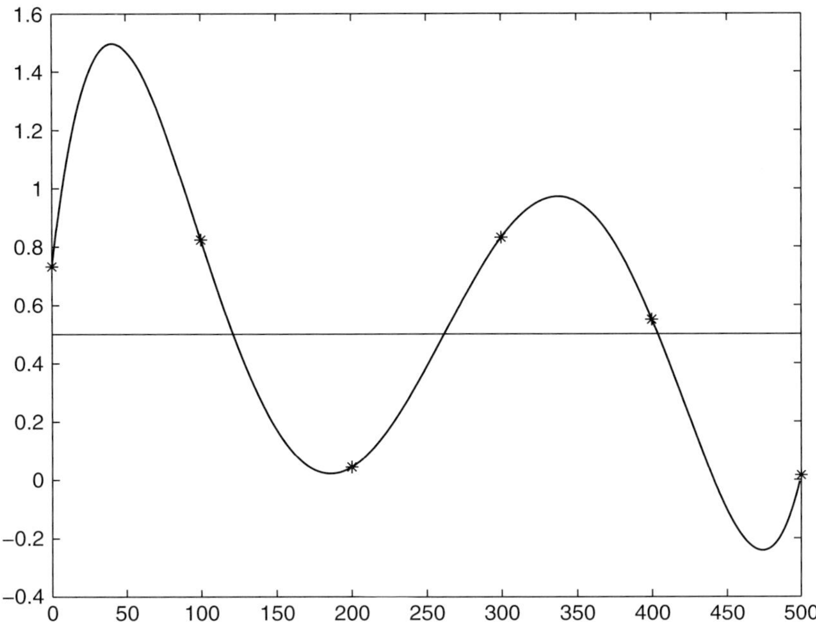

Figure 4.8 Example of overshoot for a polynomial interpolant, fitting a straight line affected by noise.

case, it is best to obtain a concentration of points near the far end, where the x values are largest.

The classic example of the undesirable oscillatory behavior is demonstrated by fitting an nth order polynomial to n data points that nearly, but not quite, define a straight line. The polynomial interpolant will exactly pass through each data point (as required by the interpolation condition), but does so as shown in Figure 4.8.

Regardless of its negative aspects, it serves as a simple model for most continuous interpolation constructions.

A continuous interpolant system is closed by simply adding more terms of higher degree. So, a solvable system has n equations (one for each data point) requiring an nth-order polynomial (n−1 degree).

The **interpolation condition**, as always, states that the polynomial interpolant, $P(x)$, must exactly fit each of the data points at x_i, so

$$P(x_i) = y(x_i) = y_i, \tag{4.28}$$

where

$$P(x_i) = a_0 + a_1(x) + a_2(x)^2 + \ldots + a_{n-1}(x)^{n-1}, \tag{4.29}$$

and the closed system is given by

$$
\begin{aligned}
a_0 + a_1 x_1 + a_2 x_1^2 + \ldots + a_{n-1} x_1^{n-1} &= y_1 \\
a_0 + a_1 x_2 + a_2 x_2^2 + \ldots + a_{n-1} x_2^{n-1} &= y_2 \\
&\vdots \\
a_0 + a_1 x_n + a_2 x_n^2 + \ldots + a_{n-1} x_n^{n-1} &= y_n.
\end{aligned} \tag{4.30}
$$

This provides n equations in n unknowns (the x_i values are not unknowns here, but instead are specific values of x, where the values of y_i are observed).

In matrix form,

$$
\begin{bmatrix} 1 & x_1 & x_1^2 & \ldots & x_1^{n-1} \\ 1 & x_2 & x_2^2 & & x_2^{n-1} \\ 1 & \vdots & & & \vdots \\ 1 & x_n & x_n^2 & \ldots & x_n^{n-1} \end{bmatrix} \begin{bmatrix} a_0 \\ a_1 \\ \vdots \\ a_{n-1} \end{bmatrix} = \begin{bmatrix} y_1 \\ y_2 \\ \vdots \\ y_n \end{bmatrix}. \tag{4.31a}
$$

This system in (4.31a) is written more succinctly in matrix form as

$$
\mathbf{A}x = \mathbf{b}. \tag{4.31b}
$$

For this particular system, matrix **A** (often called a **Vandermonde matrix**) is square ($n \times n$) and of *full rank*,[7] n (by definition, the various functional forms of the basis must be linearly independent, as is the case for a full rank matrix). This equation is easily solved for the unknown coefficients a_i (in the **x** vector):

$$
\mathbf{A}^{-1}\mathbf{A}x = \mathbf{A}^{-1}\mathbf{b} \tag{4.32}
$$

$$
\mathbf{x} = \mathbf{A}^{-1}\mathbf{b}. \tag{4.33}
$$

Once solved for the coefficients, a_i, their values can be inserted into (4.29) to define completely the interpolant P(x) for any desired value of x. However, in practice, the Vandermonde matrix can be highly *ill-conditioned* (see Appendix 1), leading to a numerically unstable solution. Consequently, except for relatively small samples, the solution of this matrix is not the best way to proceed with solving this interpolant. Instead, a special numerical form based on a different form of the polynomial is typically used. In fact, the coefficients are rarely computed directly, unless they are needed explicitly to generate the function analytically. Rather, the interpolant is usually constructed directly.

4.4 Take-Home Points

1. Interpolation (official mathematical definition) means fitting a curve to data so that the curve exactly reproduces the original data points (optimal interpolation, mentioned later, allows a smoothed fit, not reproduction of each point exactly).
2. The curve chosen to interpolate is the interpolant.
3. Care is required when choosing your interpolant because some curves will reproduce every point at a severe price (e.g., significant overshoot, extensive wiggles).

[7] See Appendix 1 on Matrix Algebra for a definition of "full rank."

4.5 Questions

Pencil and Paper Questions

1. Please define the following with a brief discussion:

   a. piecewise interpolation and continuous interpolation
   b. degree and order of a polynomial
   c. cubic spline interpolation

2. Show the matrix solution to $\mathbf{A}x = \mathbf{b}$ for the interpolation problem.
3. What is the difference between interpolation and extrapolation?
4. You have a data set that physically must be smooth, but has very sharp discontinuities. What is the best interpolant for this data set?
5. a. For interpolation of noisy data, describe each interpolant and give its pros and cons:
      1. a polynomial
      2. a natural cubic spline
      3. any orthogonal basis (include an example of such)

   b. What is the very best interpolant for any occasion? (Explain why.)

5 Smoothed Curve Fitting

5.1 Overview

Curve fitting, as described in the previous chapter, involves fitting a curve to a set of data. If we desire to fit the curve exactly through the data points, we are labeling that interpolation (though there are actually a great number of interpolation methods, such as optimal interpolation, where we seek to interpolate a set of data via smoothed, as opposed to exact, curve fits). The other case, smoothed fitting, involves the acknowledgment of noise or scatter in the data that prevents or discourages the fitting of a curve so that it passes exactly through each of the data points. Smoothed curve fitting is the focus of this chapter, and it is probably one of the most ubiquitous and useful statistical tools in existence. This importance is due to the underlying theme of **optimization**, presented in this chapter in the form of regression, though touched upon earlier in the discussion of the Principle of Maximum Likelihood.

Regression[1] is a special case, though the most common form of smoothed curve fitting. It involves quantifying the relationship between two or more variables, sometimes with the ultimate goal of allowing a prediction of one of the variables (to some determined level), given knowledge of the other(s). We will deal with the specific details of regression and its various related tools: calibration, inverse regression and correlation, after establishing the underlying mathematical/statistical concepts involved in the broader class of smoothed curve fitting first.

5.2 Introduction

Curve fitting, seeks to find a functional relationship, $Y = f(x)$, as was the case for interpolation.[2] In this case, however, the curve *passes through the data in such a manner*

[1] The "calibration" problem and "inverse regression" are important but specific aspects of the regression problem that are discussed later in this chapter.

[2] The functional form of this relationship is that for interpolation problems, but here the y and x variables are presented as random variables, Y and X. This acknowledges the fact that the Y and X observations of our sample contain uncertainty; hence, the desire to use a smoothed-curve fit instead of an exact fit, as is the case in interpolation (in which any uncertainty is ignored).

that it produces a "best" fit between the curve and data – it is not constrained to pass exactly through any of the data points, the (x_i,y_i) pairs of observations, as is the case for interpolation.

The most common definition of "best" fit curve is that curve which passes through the data so as to minimize the sum of the squared residual (error) between the predicted curve and the data.[3] In fact, while this is certainly a good choice in many situations, it is not always the most appropriate. We will also consider other definitions of "best" fit line and describe their advantages and disadvantages.

Smoothed curve fitting involves several steps:

1) Determine the functional form of the curve to be fit to the data.
2) Determine the definition of "best" fit.
3) Determine those parameter values providing the "best" fit curve.
4) Determine the uncertainty associated with the optimal parameter values.
5) Assess the fit of the curve (Does the curve do an adequate job of describing the data?) and thus, how appropriate the curve is.

5.3 Functional Form of the Curve

The functional form of the curve to be fit to the data must either be specified *a priori* by additional information concerning the data (as was the case for interpolation) or by inspection through use of a scatter plot or other graphical form. The former assumes knowledge of the process governing the data and some phenomenological law describing the process. The functional form of such a law represents the functional form of the curve to be fit to the data.

For example, simple boundary layer theory suggests that the drag (τ) of a fluid against a boundary is proportional to the square of the fluid speed (v^2), so $\tau \propto v^2$, or $\tau = cv^2$. The constant of proportionality, however, is not easily predictable for many cases of interest (e.g., the drag of the wind on the ocean or land). But, $\tau = cv^2$, so $\tau = cx + b$ where x ($= v^2$) and $b = 0$. Therefore, a plot of the measured τ against x ($= v^2$) should show a linear form with slope c. A fit of a linear curve to the data that passes through the origin should give the "best," or most representative estimate of c (the slope of the best fit curve), which takes into account the natural scatter expected in the data. For more complicated boundary layer theory, we find that the functional form of the curve that describes the relationship is given by $\tau = cv^b$, where both c and b are unknown parameter values that must be estimated from the analysis.

In the absence of outside information specifying the functional form of the curve, one typically must resort to visual inspection. For data that display a smooth trend with increasing growth or decay, you can try plotting some simple transformations of the data (log, log-log or power) to find the appropriate shape. Log transforms are particularly

[3] Hamming (1989) points out that this definition of "best" fit is often believed to be the "right one" because "the mathematicians believe it to be a physical principle, while the physicists believe it to be a mathematical principle."

good, since the constant ratios of raw values produce logs of constant differences, so the log converts ratios and products to simple differences and sums.

Some functional forms that are frequently useful are

$$\text{polynomial}: \quad y = a + bx + cx^2 + \ldots + nx^n \tag{5.1a}$$

$$\text{power law(geometric)}: \quad y = ax^b \tag{5.1b}$$

$$\text{general power law}: \quad y = a + bx^c \tag{5.1c}$$

$$\text{exponential}: \quad y = ae^x \tag{5.1d}$$

$$\text{general exponential}: \quad y = a + bc^{dx} \tag{5.1e}$$

$$\text{hyperbola}: \quad y = a/(b + cx) \tag{5.1f}$$

$$\text{Butterworth(logistic)}: \quad y = a/(bc^x + d), \tag{5.1g}$$

where the letters a through d in the formula represent constants. Note that the polynomial curve (5.1a) is most commonly restricted to $n \leq 4$ (quartic). In its various degrees, it represents fitting a simple constant ($n = 0$), a linear fit ($n = 1$), a quadratic fit ($n = 2$) and cubic fit ($n = 3$). For higher-order fits, it typically produces undesirable oscillatory behavior and can display extreme sensitivity (e.g., a subtle change in a data value can lead to a large change in the nature of the fitted curve).

One can also fit such curves to contiguous segments of the data, as done for piecewise continuous interpolation, only in this case the fit over each segment would be a "best" fit as opposed to an interpolation that passes through each point.

5.4 Defining "Best" Fit

How you define a "best" fit curve is dependent upon (1) the nature of the problem, (2) the manner in which the error between the curve and data is defined and (3) the nature of the data being fit.

5.4.1 Nature of the Problem

Here, the "nature" of the problem refers to whether one is interested in (1) producing a best-fit smooth curve to the data or (2) performing a regression or calibration. For the latter, we seek to determine how one variable can be best predicted, given knowledge of the other(s) upon which it shows some dependency. For both problems, the underlying mathematics and statistical analysis are similar. The primary difference will come in the manner as to how we determine, for the fitted function $y(x)$, which variable will serve as y and which will serve as x.

For the remainder of this chapter, we will deal with the straightforward problem of simple curve fitting. In the next chapter we will introduce the additional considerations that must be addressed in the regression/calibration problem (e.g., weighting the values being fit, assigning constraints, etc.).

5.4.2 Defining Error

Defining the error requires two considerations:

1) Where is the error, ε, distributed in the data? That is, is the noise predominantly in the *Y* values, the *X* values or both?
2) How should the total error, e, be quantified? That is, what quantity are we trying to minimize (e.g., the sum of the errors, the sum of the errors squared, etc.)?

Error Distribution

Consider the common situation in which *X* is a variable, such as time or space, that is known very precisely relative to the *Y* values (measured at each x_i position). In this situation, the scatter in the data lies predominantly in the *Y* values. Consequently, the individual errors, ε_i, that should be minimized when fitting a curve are defined as the difference between the observed and curve-predicted *Y* values at each x_i position.

In this case of fitting *Y* on *X*, *Y* represents the **dependent variable** and *X* represents the **independent variable**. The errors, ε_i, are shown graphically in Figure 5.1.

Conversely, if the errors are predominantly contained in the *X* values, then one would minimize the error in the *X* direction and fit *X* on *Y*. This leads to *X* being the dependent

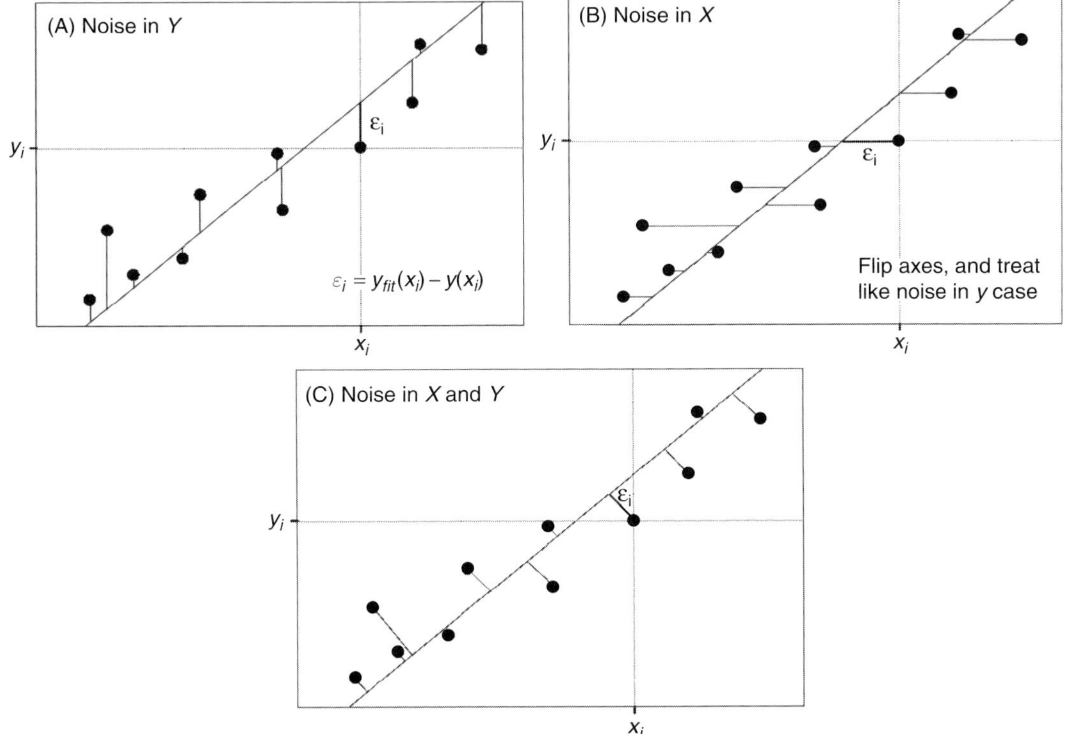

Figure 5.1 Distribution of errors in the data. **A**: errors in *Y* (fitting *Y* on *x*), **B**: errors in *X* (fitting *X* on *Y*) and **C**: errors in both *X* and *Y* (orthogonal fit).

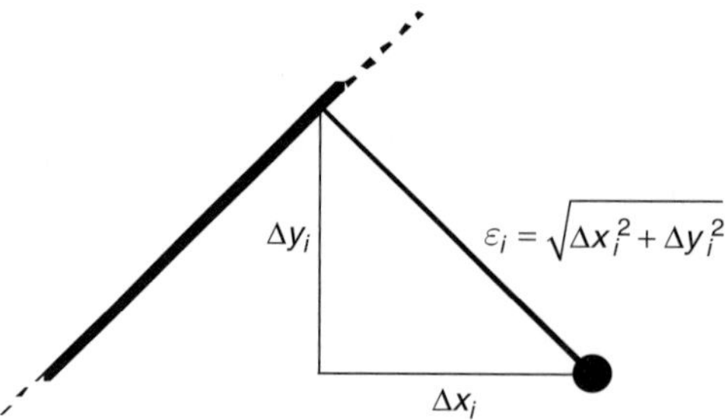

Figure 5.2 Definition of error for orthogonal fit, where uncertainty is in both x and y.

variable and Y being the independent variable. The errors, in this case, are distributed as in Figure 5.1B.

The last possible situation arises when the error is of similar magnitude in both the X and Y. In this case, the individual errors to be minimized represent some combination of mismatch in both the X and Y directions. This leads to **orthogonal fitting**. In the case of orthogonal fits, the errors, ε_i, are not a simple difference but instead are given as shown in Figure 5.2.

Computationally, orthogonal regression is the most difficult, except for the special case of fitting a straight line to a data set. In that case, orthogonal regression is straightforward.

The fitting of Y on X and of X on Y are reduced to being the same problem, the only difference being that in the former, Y is the dependent variable, X the independent, while in the latter, the opposite is true. Therefore, to fit X on Y, simply reverse the variables (flip the axes).

Note that the independent variable is assumed to have very little uncertainty, *at least relative to that in the dependent variable*. Therefore, it is treated as deterministic – not a random variable. Only when both X and Y are random variables, requiring orthogonal fitting, do the terms "dependent variable" and "independent variable" lose their obvious appeal. In this case, the form of the orthogonal fit can be either $Y = f(X)$ or $X = f(Y)$.

The actual best-fit curve will differ according to which error is minimized (i.e., fitting Y on X or X on Y or orthogonal). You should therefore consider the sources of errors carefully.

Error Quantification

The way in which the individual errors are combined to form a total error, e, to be minimized often involves the concept of a norm.

Norm is the (*scalar*) measure chosen to represent the length, magnitude, or size of a vector.[4] For the curve-fitting problem, the elements of the error vector represent

[4] Norms can also be defined for matrices and functions (the former involving the sum over all rows and columns; the latter involving integration in place of summation). Norms can be defined in terms of a weighting function as well. In this case, the form is similar to the standard form, except that the sum or integral is over the product of the vector (function or matrix) with a weighting vector (function or matrix). This represents a more general definition. As defined in the text here, the L_n norms can be considered to have a weighting vector whose elements are unity.

each of the individual errors, ε_i, and the vector norm represents a measure of the total error, e.

A norm is represented by double, or sometimes – rarely – single bars. The most common class of norms, **L_n norms**, are given in terms of a vector **a** as

$$L_n \text{ norm} = \|\mathrm{a}\|_n = \left[\sum_{i=1}^{n} \left|\mathrm{a}_i^n\right|\right]^{1/n}, \tag{5.2}$$

where the power of n defines the specific properties of the norm and the a_i are the elements of the vector. Note that the 1/n power always leaves the norm units in the same units as the error is presented in.

All norms are characterized by the following properties (where bold symbols are same-dimension vectors and c is a scalar constant):

$$\|\mathbf{a}\| \geq 0 \tag{5.3a}$$

$$\|\mathbf{ca}\| = \mathrm{c}\|\mathbf{a}\| \tag{5.3b}$$

$$\|\mathbf{a} + \mathbf{b}\| \leq \|\mathbf{a}\| + \|\mathbf{b}\| \tag{5.3c}$$

$$\|\mathbf{ab}\| \leq \|\mathbf{a}\|\|\mathbf{b}\|. \tag{5.3d}$$

Some specific norm properties vary with the value of n in (5.2). The L_1 norm (i.e., $n = 1$) is simply the sum of the absolute value of the errors and is often called a **robust norm**. The L_2 norm is often called the **Euclidean norm**. If no qualifier is given to the word norm, it most often refers to the L_2 norm.

Another commonly used norm defines the magnitude of the vector as equivalent to the magnitude of the largest element within the vector. This norm is called the L_∞ norm, and is given by

$$L_\infty \text{norm} = \|\mathrm{a}\|_\infty = max|\mathrm{a}_i|. \tag{5.4}$$

Minimization of this norm is known as the **minimax principle** because it **mini**-mizes the **max**imum error (expansion in Chebychev polynomials minimizes this norm).

The particular norm chosen to be minimized during curve-fitting procedures should reflect the nature of the error or scatter contained in the data. This involves examining the data for outliers and determining if the data are asymmetrically distributed. Consider Figure 5.3 (adapted from Menke, 1984) that shows straight lines fit to a data set that contains a single outlier. The various lines represent the best fit as defined by minimizing the L_1, L_2 and L_∞ norms.

The L_1 norm represents the most **robust** of the norms – that is, it is the least sensitive to the influence of the outlier. This is because the L_1 norm is simply the sum of absolute values of the errors. A large error increases the size of the L_1 norm, but even a small adjustment of the straight line toward the outlier (in an attempt to reduce the associated large error between

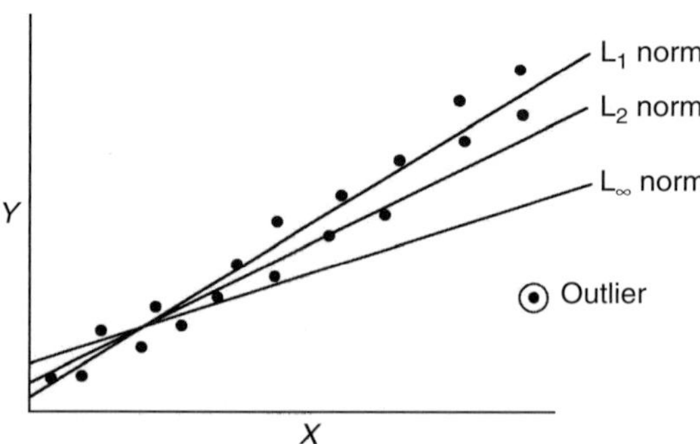

Figure 5.3 Schematic showing influence of a single outlier on some common norms.

the line and outlier) immediately increases the size of all the other errors in the sum. This small increase in all other individual errors quickly accumulates in the sum. Therefore, to obtain a minimal L_1 norm, the line cannot make much of an adjustment to reduce the error associated with the outlier. Consequently, the L_1 norm is robust with respect to outliers.

The L_2 norm represents the sum of the *squared* errors. The influence of a disproportionately large error is tremendous within the L_2 norm because the errors are squared. Consequently, the fitted line can afford to make some adjustment toward the outlier in order to decrease this single error before the incremental increases in errors elsewhere accumulate to a significant size. The L_2 norm is thus sensitive to outliers, and the fitted curves will show some pull due toward them. For this reason, the L_2 norm is best for data that do not contain outliers and which show a symmetrical distribution. The L_2 norm has natural appeal because of its similarity in form to variance. In fact, *by minimizing the L_2 norm, we are finding that curve about which the data show the minimum variance (or root mean square, rms, scatter)*.

With higher norms (i.e., with larger n), the influence of outliers grows even stronger. In the limit, the L_∞ norm shows the strongest influence. The L_∞ norm finds that curve which minimizes the largest single error (sometimes called the *supremum*) between the data and curve. So, for a single outlier as shown above, the line can't completely eliminate the error between the curve and the outlier, because that would introduce a larger error elsewhere. Instead, it must compromise so that the largest single error that exists after the curve is fit is as small as possible. Minimization of this norm is ideal for issues of safety or whenever a threshold most be avoided. In that case, one doesn't care about the size of the individual errors, only that the largest error we expect will be of minimal size.

So, the manner in which the total error is quantified is determined by defining a norm, with the norm chosen to reflect the nature of the noise in the data. If the data obey a Gaussian distribution, there is a strong likelihood that the observations will be distributed *symmetrically* about the true curve position (most of the observations will fall within a couple of standard deviations from the true position). In this case, minimization of the L_2 norm provides the best choice (proven below). For data showing a skewed distribution,[5] we *expect* some of the observations to lie a significant distance from the true curve position, and we may not want such points (outliers) to impart too strong an influence on the fitted curve.[6] For those, the L_1 norm represents a better choice (though another measure based upon the median may prove more useful instead). Finally, if the relevance of distant points (outliers) is significant, then a norm that is more sensitive to outlier points may be appropriate. Obviously, the larger the power n, the more influence the outlier will have on the fit. In such cases, it may be worth investing some effort into determining the distribution about the curve so that Maximum Likelihood could be used to properly account for the scatter. Points that impart a large influence on the fit are called **leverage points**.

5.4.3 Nature of Data: Influence on Defining Best Fit

As stated at the beginning of this section, the best fit is dependent upon how the error between the observations and the fitted curve is defined and the nature of the data being fit. This latter point is best presented by consideration of the following four cases.

Case A (Standard Fit)

In case A (Figure 5.4), all of the data values are known to approximately the same degree of precision and the general range of scatter between values is relatively symmetrical. For this case, the question of defining "best" fit is reduced to deciding the most appropriate manner with which to define the error, ε, and the norm. The error, ε, is decided upon as discussed previously (for the case shown here, the error is contained within the Y values). Selection of the appropriate norm is dependent upon the distribution of the

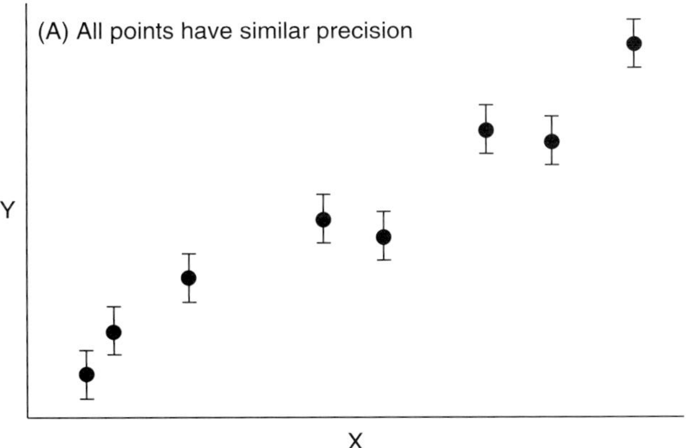

Figure 5.4 Standard fit, with all data known with comparable precision.

[5] This situation can arise for samples taken from a population whose distribution is skewed, or it can result from a small sample size taken from a symmetrically distributed population, but the small sample size does not provide enough observations to yield a balanced (symmetrical) distribution.

[6] Ideally, we would obtain the best-fit line using knowledge of the distribution, if available, in which case "outliers" may in fact be expected and taken into account through the PDF or PMF of the data being fit. This will involve maximum likelihood regression (discussed later).

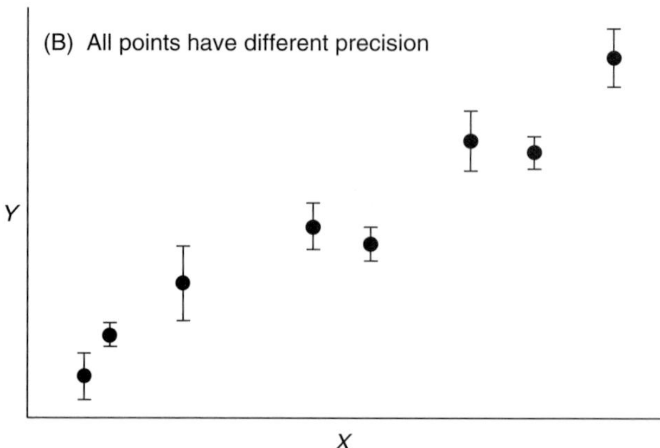

Figure 5.5 Weighted fit, where data are known to different precisions.

scatter in the observations. If the scatter is approximately normal (e.g., if the values being fit are mean values), this case leads to the L_2 norm and the classic **method of least squares** (discussed below).

Case B (Weighted Fit)

In case B (Figure 5.5), the data values are known with different degrees of precision. Naturally, we expect that an observation that is known to extremely high precision should carry more weight in fitting the curve than an observation that is known with considerably less precision. Consequently, this case requires that each data point be given a *weight* so that the "best" fit line reflects this variability. This requires use of a weighted fitting scheme (most commonly, **weighted least squares**).

Case C (Constrained Fit)

In case C (Figure 5.6), we have particular constraints that must be obeyed by the curve, such as a requirement that the curve pass exactly through specific values. In such cases, either we put infinite weight on the point or points that must be fit or we utilize a method that accommodates any number and variety of equality or inequality constraints (most commonly, the method of **constrained least squares**).

Case D (Robust Fit)

In case D (Figure 5.7), the data (regardless of their precision) contain one or more observations that lie well away from the other values (i.e., there are potential outliers present in the data; Figure 5.7a). Or, alternatively, the scatter between observations may be very asymmetrical. For example, there may be a fairly well defined central trend, but a considerable scatter of points above the trend and only a minor amount of scatter below the trend (Figure 5.7b). In both cases, the "best" fit line may be that which reflects the *dominant* trend of the data. Thus, either the outlier points are

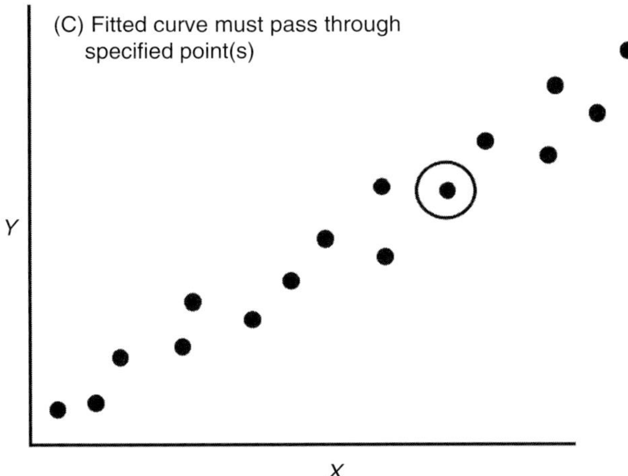

Figure 5.6 Linear equality constraint; best fit must pass through one or more specific points.

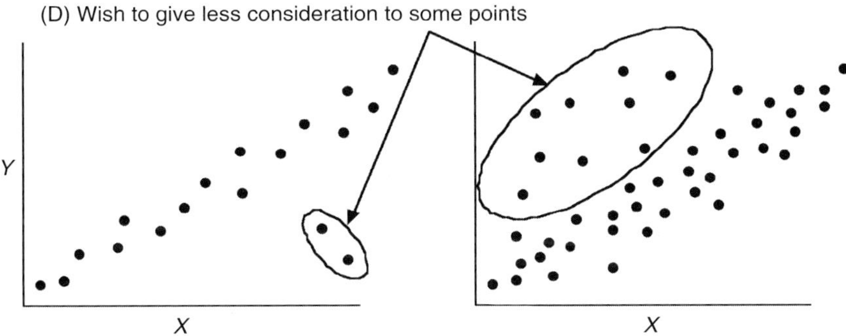

Figure 5.7 Robust fit, giving less consideration to specific points (e.g., outliers).

weighted significantly less than the other points or we utilize a **norm** designed to handle these situations (as discussed previously). In particular, we desire a norm that is relatively insensitive to large individual errors (i.e., a norm that is robust), such as the L_1 norm. This represents **robust fitting**.

Note that there are a tremendous number of other robust fitting schemes for data that are strongly skewed or subject to large outliers. Probably the best of these is known as the **Least Median of Squares (LMS)**, which we will not discuss. The disadvantage of most of these other techniques is that their actual implementation is usually awkward and less efficient (and less elegant) than working directly with an L_n norm. Also, a detailed understanding of their properties and errors has not always been fully developed. Regardless, they can work very well in practice *when warranted*, and their properties (e.g., errors) can be assessed via resampling techniques.

The remainder of this chapter deals with standard curve fitting – the three other special cases presented here (weighted, constrained and robust fits) are discussed in Chapter 6.

5.5 Determining Parameter Values for a Best-Fit Curve

How the parameter values are determined to produce the curve that is best fit to the observations depends upon how the error is quantified and what type of fit is used (i.e., upon the decisions that were made based on the preceding discussion).

5.5.1 Standard Curve Fitting

The **Method of Least Squares** (**LS**, or **LLS** for linear least squares) is the most popular and convenient method for computing standard (and weighted) regressions. This method produces a fit of a specified curve to a set of data points that minimizes the L_2 norm. That is, it minimizes the sum of the squared residuals (errors) between the fitted curve and data – hence the name "least squares." This is the method originally developed independently by both Legendre (in 1806) and Gauss (in 1809) as a method for consistently handling surplus measurements.[7] The method is easily demonstrated by fitting a straight line to a data set.

LS Fitting of a Straight Line

Consider fitting a single "best" linear slope to n data points, as in Figure 5.8.

Here, Figure 5.8 shows a simple function $Y = f(X)$ or a **scatter plot** in which two functions $Y(t)$ and $X(t)$ are plotted against one another at common values of t, so $f[X(t)]$ (this is called a functional – a function of a function).

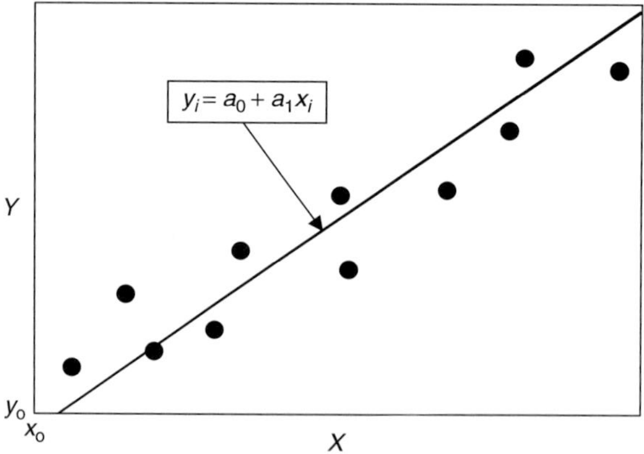

Figure 5.8 Schematic of best straight-line fit.

[7] Legendre published his paper first, then Gauss came along and said he had already developed the technique earlier but hadn't bothered to publish it – what a sport (though Gauss is given credit, so apparently he was able to back up his claim convincingly).

In either case, we treat Y as a function of X (or X as a function of Y if we want to fit X on Y). We then wish to fit to these data a line of the form

$$y_i = a_1 + a_2(X_i - X_0). \tag{5.5a}$$

This equation has two degrees of freedom in the a_1 and a_2 coefficients.

Note that x_0 simply represents any specified value of X, typically the minimum measured X value or the sample mean of X, $\bar{x}$. Alternatively, x_0 is often set equal to zero. In this latter case, the x_0 term drops out, giving

$$y_i = a_1 + a_2 X_i. \tag{5.5b}$$

The advantage of this latter form is twofold. First, the final form of the least squares solution can be written in terms of standard statistical moments (mean and covariance). Second, for numerical considerations when the range of X is a very small fraction of a typical X value (e.g., if X varied from 1×10^7 to 1.000005×10^7), use of the latter form using statistical moments can help avoid roundoff problems. Therefore, the second form (that of equation (5.5b)) will be used here; x_i can be replaced with $x_i - x_0$ throughout the following equations if desired.

To produce the best fit line, we must determine values for a_1 and a_2 that produce a line that minimizes the sum of the squared errors. In this case, we assume that the errors are all contained within Y, so Y is the dependent variable and X will be treated as a deterministic variable,[8] x. The overall error, e, is defined as

$$\mathrm{e} = \sum_{i=1}^{n} (y_{computed} - y_{observed})_i^2 = \sum_{i=1}^{n} \varepsilon_i^2. \tag{5.6}$$

This is the quantity (e) that must be minimized by the values of a_1 and a_2. Note that minimizing the $\sum_{i=1}^{n} \varepsilon_i^2$ also minimizes $\left(\sum_{i=1}^{n} \varepsilon_i^2\right)^{1/n}$, so minimizing (5.6) is identical to minimizing the L_2 norm proper. It also minimizes e/n, which is the formal definition of the rms error.

Now consider the general concept of the approach. We wish to find those values of a_1 and a_2 that produce a straight line that passes through the data points so that the sum of the squared errors (the difference between the curve and each data point) is minimal. In the Figure 5.9, there are three straight lines passing through the data. Examination of the errors at the three data points that have been circled reveals (for arbitrary units): for f1, the first error, ε_1 (lower left of plot) = 6, $\varepsilon_2 = 5$ and ε_3 (upper right) = 2; for f2, $\varepsilon_1 = 3$, $\varepsilon_2 = 3$ and $\varepsilon_3 = 4$; and for f3, $\varepsilon_1 = 2$, $\varepsilon_2 = 3$ and $\varepsilon_3 = 6$.

If we were to square and add up these three errors for each line, e1 (the sum of the three squared errors for f1) = 65; e2 = 34 and e3 = 49. Therefore, from examination of only three points (the answer may change after all the errors are tallied), the f2 curve

[8] Recall that the uppercase notation for X implies a random variable; thus, our conversion to the lowercase form in this case.

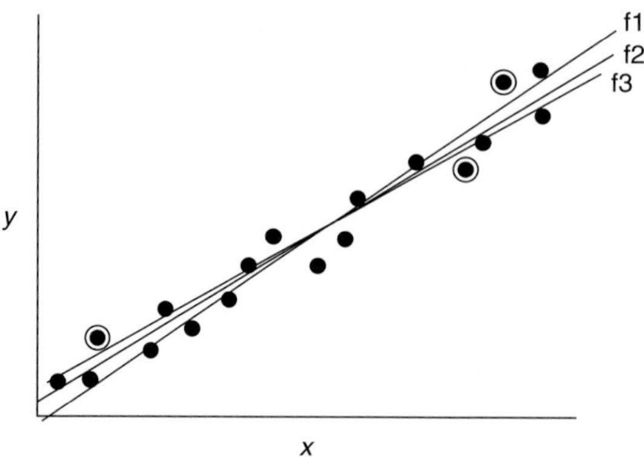

Figure 5.9 Schematic of assessing errors of three specific points.

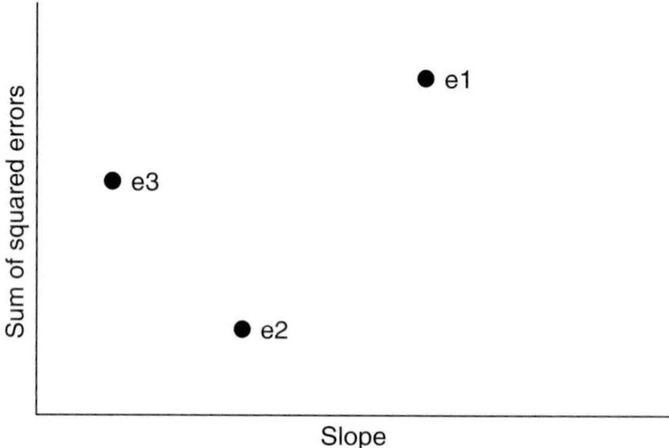

Figure 5.10 Examining error associated with the three data points in Figure 5.9.

seems to have the best fit of the three, though it is not guaranteed to have the best fit of any straight-line curve.

In order to determine the best line overall, consider the fact that the three fitted lines we have discussed differ only in the value of their a_1 and a_2 coefficients (i.e., in their intercepts and slopes). Therefore, we could make a plot of the value of the summed squared errors, the e1, e2 and e3, against the a_1 and a_2 values of each curve, Figure 5.10. This is a three-dimensional bowl (a paraboloid). Consider looking at a slice through this bowl, or, the e1, e2 and e3 versus the a_2 for f1, f2 and f3.

These values form a curve (in this case, a parabola) that shows $e(a_2)$. We wish to find that slope, a_2, at the minimum of this function of squared errors.

The minimum is determined by computing the derivative of the function with respect to the slope. The error function is given by (5.6). Therefore, we take the derivative of (5.6) with respect to a_2 and set this derivative equal to zero, then solve for the value of a_2 that satisfies this condition. That produces the slope of the line that produces the overall minimal error – the "least-squared error" fit.

Because the lines are actually a function of both the intercept and the slope, we must actually compute the similar derivative for the intercept, a_1, and solve for that value at the same time we solve for the a_2 value in order to find the minimum point on the paraboloid described by the function $e(a_1, a_2)$.

Before minimizing (5.6), consider the system we are now dealing with. In the absence of noise, for each observation y_i at x_i,

$$\begin{aligned} a_1 + a_2 x_1 &= y_1 \\ a_1 + a_2 x_2 &= y_2 \\ &\vdots \\ a_1 + a_2 x_n &= y_n, \end{aligned} \tag{5.7a}$$

or, more succinctly,

$$a_1 + a_2 x_i = y_i. \tag{5.7b}$$

This system has many more equations than unknowns. That is, we have n equations (one for each observation at location x_i) versus only two unknowns (two in the degrees of freedom: a_1 and a_2; the x_i and y_i are known from the data). Such a system is **overdetermined,** and there exists no unique solution (unless all the y_i's happen to lie exactly on a single line, in which case any two equations uniquely determine a_1 and a_2).

Intuitively, the non-uniqueness in (5.7) is reflecting the fact that a line is defined by two points, so if we have $n/2$ pairs of points, it is likely that each of these pairs defines a different line. In fact, there can be as many lines as there are permutations of pairs of points $= n(n-1)/2$. This is what we expect because of the noise in the data. If this noise were not present, then we hope that all of the values would indeed lie on a single straight line – it is that line we are hoping to find.

In acknowledgment of the noise, we can modify each equation to consider the error in the fit at each point. So,

$$\begin{aligned} a_1 + a_2 x_1 - y_1 &= \varepsilon_1 \\ a_1 + a_2 x_2 - y_2 &= \varepsilon_2 \\ &\vdots \\ \underbrace{a_1 + a_2 x_n}_{computed} \underbrace{-y_n}_{observed} &\underbrace{= \varepsilon_n}_{error}, \end{aligned} \tag{5.8}$$

where the ε_i represents the error or mismatch at each point i. This error is the difference between the computed y value at the specific point $(= a_1 + a_2 x_i)$ and the observed value, y_i, at that same point. This is equivalent to the error defined in the previous figures showing the regression of Y on X and defined by the individual terms in equation (5.6). By allowing a mismatch, we are also stating that we know we can't satisfy the interpolation condition by having a curve pass through each point y_i exactly (as we did when interpolating).

So, we can now rewrite equation (5.6), defining the error explicitly as

$$e = \sum_{i=1}^{n} \varepsilon_i^2 = \sum_{i=1}^{n} [(a_1 + a_2 x_i) - y_i]^2. \tag{5.9}$$

The summation in equation (5.9) must be minimized by our choice of a_1 and a_2. So, e(a_1,a_2) and the minimum of this function (with respect to the two unknown coefficients) can be determined using basic calculus, which tells us that the minimum occurs where its derivatives go to zero:

$$\frac{\partial e(a_1, a_2)}{\partial a_1} = \frac{\partial e(a_1, a_2)}{\partial a_2} = 0. \tag{5.10}$$

The solutions for a_1 and a_2, obtained by solving equation (5.10) (see boxed details that follow) are given in terms of the standard statistical moments of X and Y (recognizing that X is not actually considered a random variable, but the moments are computed using the standard statistical formula, allowing shorthand notation for the operations required on x). Specifically,

$$a_1 = \overline{y} - a_2 \overline{x} \tag{5.11a}$$

$$a_2 = \frac{s_{xy}}{s_x^2}, \tag{5.11b}$$

where $s_{xy} = (n-1)^{-1} \sum_{i=1}^{n} (y_i - \overline{y})(x_i - \overline{x})$ is the sample covariance between X and Y (the estimate of the covariance in the standard manner) and s_x^2 is the "sample variance" of x. The values for the a_1 and a_2 constants are then substituted into equation (5.7) for a_1 and a_2, producing the best fit line to the data.

Note that the solution for a_1 allows us to rewrite the original equation (5.7b) for the best fit line as

$$y_i = \overline{y} - a_2 \overline{x} + a_2 x_i \tag{5.11c}$$

or

$$= \overline{y} + a_2 (x_i - \overline{x}) \tag{5.11d}$$

or

$$y_i - \overline{y} = a_2 (x_i - \overline{x}), \tag{5.11e}$$

showing that the perturbations of X and Y about their means are proportional and the constant of proportionality is given by the slope, a_2. It also shows that if $x_i = \overline{x}$, then $y_i = \overline{y}$, indicating that the best-fit line passes through the pair $\overline{y}, \overline{x}$.

You may be tempted to estimate a_2 by the average of $(y_i - \overline{y})/(x_i - \overline{x})$ from (5.11e), but note that this would give you the average slope passing through the spread of data points, not the *optimal* fit that passes most closely to the data (quantified by minimum squared error between the fit and data points).

Box D5.1 Solution of Equation (5.10)

The derivatives of equation (5.10) are explicitly given by

$$\begin{aligned}
\frac{\partial \mathrm{e}}{\partial a_1} &= \frac{\partial}{\partial a_1}\left(\sum_{i=1}^{n} \mathrm{e}_i^2\right) \\
&= \frac{\partial}{\partial a_1}\left(\sum_{i=1}^{n}[(a_1 + a_2 x_i) - y_i]^2\right) \\
&= 2\sum_{i=1}^{n}(a_1 + a_2 x_i - y_i) \\
&= 0
\end{aligned} \tag{D5.1.1a}$$

$$\begin{aligned}
\frac{\partial \mathrm{e}}{\partial a_2} &= \frac{\partial}{\partial a_2}\left(\sum_{i=1}^{n}[(a_1 + a_2 x_i) - y_i]^2\right) \\
&= 2\sum_{i=1}^{n}(a_1 + a_2 x_i - y_i)x_i \\
&= 0.
\end{aligned} \tag{D5.1.1b}$$

These two equations can now be expanded and rearranged, forming what are known as the **normal equations**:

$$\sum_{i=1}^{n} a_1 + \sum_{i=1}^{n} a_2 x_i = \sum_{i=1}^{n} y_i \tag{D5.1.2a}$$

$$\sum_{i=1}^{n} a_1 x_1 + \sum_{i=1}^{n} a_2 x_i^2 = \sum_{i=1}^{n} y_i x_i. \tag{D5.1.2b}$$

The normal equations provide a system of two equations in two unknowns that can be uniquely solved.[9] That is, we have one independent normal equation for each unknown.

Rearranging further gives

$$n a_1 + a_2 \sum_{i=1}^{n} x_i = \sum_{i=1}^{n} y_i \tag{D5.1.3a}$$

$$a_1 \sum_{i=1}^{n} x_i + a_2 \sum_{i=1}^{n} x_i^2 = \sum_{i=1}^{n} y_i x_i. \tag{D5.1.3b}$$

Here, all sums involve known values that sum to simple constants.

[9] Note: we know this is true because we have defined "curve" as a series of basis function constituent terms. By definition, the terms of a basis are independent; thus, the derivatives of each term are also independent. In equation (D5.1.2), we have one equation for each basis function; thus, each equation is independent, and therefore, by definition, (D5.1.2) is of full rank and invertible. Also see Strang for explanation directly in terms of linear algebra.

Box D5.1 (Cont.)

Solving for a_1 and a_2 proceeds in the standard manner using elimination. Rearranging (D5.1.3a) in terms of a_1 gives

$$a_1 = \frac{1}{n}\sum_{i=1}^{n} y_i - \frac{a_2}{n}\sum_{i=1}^{n} x_i, \tag{D5.1.4a}$$

which is seen by inspection to be equivalent to

$$a_1 = \bar{y} - a_2\bar{x}. \tag{D5.1.4b}$$

Replacing a_1 in equation (D5.1.3b) with (5.14b) gives

$$(\bar{y} - a_2\bar{x})\sum_{i=1}^{n} x_i + a_2\sum_{i=1}^{n} x_i^2 = \sum_{i=1}^{n} y_i x_i, \tag{D5.1.5}$$

which can be rearranged in terms of a_2 as

$$a_2 = \frac{\sum_{i=1}^{n} y_i x_i - \bar{y}\sum_{i=1}^{n} x_i}{\sum_{i=1}^{n} x_i^2 - \bar{x}\sum_{i=1}^{n} x_i}. \tag{D5.1.6}$$

Now consider the terms in the denominator of equation (D5.1.6),

$$\begin{aligned}\sum_{i=1}^{n} x_i^2 - \bar{x}\sum_{i=1}^{n} x_1 &= \sum_{i=1}^{n} x_i^2 - n\bar{x}^2 \\ &= \sum_{i=1}^{n} (x_i - \bar{x})^2 \\ &= (n-1)s_x^2\end{aligned} \tag{D5.1.7}$$

where s_x^2 is the sample variance of X (the estimate of the variance of X in the standard manner).[10]

Similarly for the numerator of equation (D5.1.6),

[10] The manipulation in equation (D5.1.7) is most easily seen by working in the opposite direction – that is, start with the standard formula for estimating the variance of X (i.e., computing the sample variance), then multiply it by $(n - 1)$ and expand it to get the original form of the denominator. This not-necessarily obvious reduction reflects a common practice, where any time we have sums of a variable or sums of its square (or higher degrees), it is often convenient to rewrite it in terms of its relevant moments. With this goal in mind, one can thus often work backwards from the relevant moments to the raw sums. In this case, having the sum of squares leads us to rewrite it in terms of the second central moment, thus guiding the direction the reduction should take.

Box D5.1 (Cont.)

$$\begin{aligned}\sum_{i=1}^{n} y_i x_i - \bar{y}\sum_{i=1}^{n} x_1 &= \sum_{i=1}^{n} y_i x_i - \bar{y}(n\bar{x}) \\ &= \sum_{i=1}^{n}(y_i - \bar{y})(x_i - \bar{x}) \\ &= (n-1)s_{xy},\end{aligned} \tag{D5.1.8}$$

where s_{xy} is the sample covariance between X and Y (the estimate of the covariance in the standard manner).[11] Recall that for both (D5.1.7) and (D5.1.8), the sample moments are actually not random variables, since x is not treated as a random variable.

Substituting (D5.1.7) and (D5.1.8) into equation (D5.1.6) gives the reduced form of a_2:

$$a_2 = \frac{s_{xy}}{s_x^2}. \tag{D5.1.9}$$

Alternatively, if the original form of the straight line was written using $x_i - x_0$, as in (5.5a), then the solutions for a_1 and a_2 are given by the bulky, but still simple, form of

$$a_1 = \bar{y} - \frac{\frac{1}{n}\left\{\sum_{i=1}^{n}(x_i - x_0)\sum_{i=1}^{n} y_i(x_i - x_0) - \bar{y}\left[\sum_{i=1}^{n}(x_i - x_0)\right]^2\right\}}{\sum_{i=1}^{n}(x_i - x_0)^2 - \frac{1}{n}\left[\sum_{i=1}^{n}(x_i - x_0)\right]^2} \tag{D5.1.10a}$$

$$a_2 = \frac{\sum_{i=1}^{n} y_i(x_i - x_0) - \bar{y}\sum_{i=1}^{n}(x_i - x_0)}{\sum_{i=1}^{n}(x_i - x_0)^2 - \frac{1}{n}\left[\sum_{i=1}^{n}(x_i - x_0)\right]^2}. \tag{D5.1.10b}$$

If the data had the scatter contained within the X values instead of the Y values, requiring us to regress X on Y, you follow the identical procedure except that you swap all of the X values with the Y values in the above equations. If both the X and Y contained errors, then in order to minimize the orthogonal error, we would have redefined the error, e, as $[(x_{\text{computed}} - x_{\text{observed}})^2 + (y_{\text{computed}} - y_{\text{observed}})^2]^{1/2}$, substituted that into (5.6), and

[11] The manipulation of equation (D5.1.8) is also most easily seen by working in the opposite direction.

minimized. Given this, the solution procedure would then proceed in an analogous manner.[12]

General Linear Least Squares

Now consider fitting a set of n data points with m terms of any continuous basis. This can be shown, for any (x_i, y_i) pair, as

$$y_i = \sum_{i=1}^{n}\sum_{j=1}^{n} a_j\varphi_j(x_i) = \sum_{i=1}^{n}(a_1\varphi_{1i} + a_2\varphi_{2i} + \ldots + a_m\varphi_{mi}), \tag{5.12}$$

where the φ_j represent the terms of the chosen basis. For the previous example of a straight line fit, the first two terms of a power basis were used for which $\varphi_1 = x^0$ and $\varphi_2 = x^1$.

Box 5.1 Example of Basis Constituent Functions

The (independent) terms, or functional forms, of a polynomial (or power) basis and a Fourier sinusoid series basis are given as

Polynomial basis	Fourier series basis
$\varphi_1 = x^0$	$\varphi_1 = \sin\frac{2\pi x}{T}$
$\varphi_2 = x^1$	$\varphi_2 = \cos\frac{2\pi x}{T}$
$\varphi_3 = x^2$	$\varphi_3 = \sin\frac{4\pi x}{T}$
$\vdots$	$\vdots$
$\varphi_m = x^{m-1}$	$\varphi_m = \cos\frac{2m\pi x}{T}$.

Recall that if the inner sum of (5.12) was summed to n instead of m, the system for n data points would be n equations in n unknowns and reduce exactly to the interpolation problem. That is, the fitted curve would pass through each data point exactly. In the present case, $m < n$, so the problem is overdetermined, resulting in a smoothed fit.

So, we desire to fit an equation (consisting of m independent terms) of the form

$$y_i = a_1\varphi_{1i} + a_2\varphi_{2i} + \ldots + a_m\varphi_{mi} \tag{5.13}$$

to a data set of n data points by minimizing e (again, we only explicitly consider the case of regressing Y on X; for X on Y, simply switch the X with the Y and vice versa).

Allowing an error at each observation y_i gives[13] (dropping the subscript i from the bases)

[12] Actually, we simplify the procedure by using a matrix approach, which leads to a solution of great simplicity, described later.

[13] Later we will consider two different forms of this and related equations: (1) using matrix notation and (2) using statistical notation. The first is for ease of solution, the second for ease of error assessment in the solution.

$$a_1\varphi_1 + a_2\varphi_2 + \ldots + a_m\varphi_m - y_i = \varepsilon_i, \tag{5.14a}$$

where the φ_i are evaluated at x_i where the y_i are located, or:

$$\sum_{j=1}^{m} a_j\varphi_j(x_i) - y_i = \varepsilon_i. \tag{5.14b}$$

The individual errors, ε_i, are then combined to produce the sum of the squared errors,

$$\begin{aligned} e &= \sum_{i=1}^{n} \varepsilon_i^2 \\ &= \sum_{i=1}^{n} [a_1\varphi_1(x_i) + a_2\varphi_2(x_i) + \ldots a_m\varphi_m(x_i) - y_i]^2, \end{aligned} \tag{5.15}$$

where the ε_i are explicitly given as

$$\begin{aligned} a_1\varphi_1(x_i) + a_2\varphi_2(x_i) + \ldots a_m\varphi_m(x_i) - y_1 &= \varepsilon_1 \\ a_1\varphi_1(x_i) + a_2\varphi_2(x_i) + \ldots a_m\varphi_m(x_i) - y_2 &= \varepsilon_2 \\ &\vdots \\ a_1\varphi_1(x_n) + a_2\varphi_2(x_n) + \ldots a_m\varphi_m(x_n) - y_n &= \varepsilon_n. \end{aligned} \tag{5.16}$$

To minimize the sum of the squared residuals, e,

$$\frac{\partial e(a_j)}{\partial a_j} = 0. \tag{5.17}$$

As was the case for the simple straight-line fit, carrying out the derivatives of (5.17) and rearranging leads to a set of *normal equations* that provide a closed system of m independent equations (one for each coefficient corresponding to the jth basis function) in the m unknown coefficients (see boxed details below). However, in this case, the manipulation of the equations required by elimination techniques to solve for the individual coefficients can quickly become unwieldy. Therefore, we must take advantage of the convenience of matrix manipulation for systems of equations. This provides a succinct expression of these otherwise bulky equations, allowing a surprisingly simple solution procedure.

This provides a system of m equations in m unknowns that can be manipulated by standard elimination methods to obtain the explicit form of any particular coefficient. However, as seen by the size of the expressions for a_1 and a_2 in the case where $m = 2$ using the power basis (the straight-line fit) given in equations (D5.1.10), such a system gets extremely awkward to manipulate as m gets larger than 2.

Matrix Form

The matrix form of the system describing a straight-line fit in equation (5.7) is given as

Box D5.2 Solution of Equation (5.17)

The derivatives of equation (5.17) are explicitly given by

$$\begin{aligned}\frac{\partial e}{\partial a_1} &= \frac{\partial e}{\partial a_1}\sum_{i=1}^{n}[a_1\varphi_1(x_i) + \ldots a_m\varphi_m(x_i) - y_i]^2 \\ &= 2\sum_{i=1}^{n}[a_1\varphi_1(x_i) + a_2\varphi_2(x_i) + \ldots a_m\varphi_m(x_i) - y_i]\varphi_1(x_i) \\ &= 0\end{aligned} \tag{D5.2.1a}$$

$$\begin{aligned}\frac{\partial e}{\partial a_2} &= \frac{\partial e}{\partial a_2}\sum_{i=1}^{n}[a_1\varphi_1(x_i) + \ldots a_m\varphi_m(x_i) - y_i]^2 \\ &= 2\sum_{i=i}^{n}[a_i\varphi_i(x_i) + a_2\varphi_2(x_i) + \ldots a_m\varphi_m(x_i) - y_i]\varphi_2(x_i) \\ &= 0\end{aligned} \tag{D5.2.1b}$$

$$\vdots$$

$$\begin{aligned}\frac{\partial e}{\partial a_m} &= \frac{\partial e}{\partial a_m}\sum_{i=1}^{n}[a_1\varphi_1(x_i) + a_2\varphi_2(x_i) + \ldots a_m\varphi_m(x_i) - y_i]^2 \\ &= 2\sum_{i=1}^{n}[a_1\varphi_1(x_i) + a_2\varphi_2(x_i) + \ldots a_m\varphi_m(x_i) - y_i]\varphi_m(x_i) \\ &= 0,\end{aligned} \tag{D5.2.1c}$$

or

$$\begin{aligned}\frac{\partial e}{\partial a_j} &= \sum_{i=1}^{n}[a_1\varphi_1(x_i) + a_2\varphi_2(x_i) + \ldots a_m\varphi_m(x_i) - y_i]\varphi_j(x_i) \\ &= 0,\end{aligned} \tag{D5.2.1d}$$

which specifies the general form of the normal equations (the factor of 2 has been divided out).

Rearranging these normal equations gives

$$\begin{aligned}&a_1\sum_{i=1}^{n}\varphi_1^2(x_i) + a_2\sum_{i=1}^{n}\varphi_2(x_i)\varphi_1(x_i) + \ldots + a_m\sum_{i=1}^{n}\varphi_m(x_i)\varphi_1(x_i) = \sum_{i=1}^{n}y_i\varphi_1(x_i) \\ &a_1\sum_{i=1}^{n}\varphi_1(x_i)\varphi_2(x_i) + a_2\sum_{i=1}^{n}\varphi_2^2(x_i) + \ldots + a_m\sum_{i=1}^{n}\varphi_m(x_i)\varphi_2(x_i) = \sum_{i=1}^{n}y_i\varphi_2(x_i) \\ &\vdots \\ &a_1\sum_{i=1}^{n}\varphi_1(x_i)\varphi_m(x_i) + a_2\sum_{i=1}^{n}\varphi_2(x_i)\varphi_m(x_i) + \ldots + a_m\sum_{i=1}^{n}\varphi_m^2(x_i) = \sum_{i=1}^{n}y_i\varphi_m(x_i),\end{aligned} \tag{D5.2.2a}$$

Box D5.2 (Cont.)

or, for each a_j derivative,

$$a_1\sum_{i=1}^{n}\varphi_1(x_i)\varphi_j(x_i) + a_2\sum_{i=1}^{n}\varphi_2(x_i)\varphi_j(x_i) + \ldots + a_m\sum_{i=1}^{n}\varphi_m(x_i)\varphi_j(x_i) = \sum_{i=1}^{n}y_i\varphi_j(x_i) \quad \text{(D5.2.2b)}$$

$$\begin{bmatrix} 1 & x_1 \\ 1 & x_1 \\ \vdots & \\ 1 & x_n \end{bmatrix}\begin{bmatrix} a_1 \\ a_2 \end{bmatrix} = \begin{bmatrix} y_1 \\ y_2 \\ \vdots \\ y_n \end{bmatrix}. \quad (5.18)$$

The matrix form of the more general system describing the fit of any function in equation (5.20) is given as

$$\begin{bmatrix} \varphi_1(x_1)\varphi_2(x_1)\ldots\varphi_m(x_1) \\ \varphi_1(x_2)\varphi_2(x_2)\ldots\varphi_m(x_2) \\ \vdots \\ \varphi_1(x_n)\varphi_2(x_n)\ldots\varphi_m(x_n) \end{bmatrix}\begin{bmatrix} a_1 \\ a_2 \\ \\ a_m \end{bmatrix} = \begin{bmatrix} y_1 \\ y_2 \\ \vdots \\ y_n \end{bmatrix}. \quad (5.19)$$

Symbolically,[14] both (5.18) and (5.19) are given as

$$\mathbf{Ax} = \mathbf{b}. \quad (5.20)$$

Matrix **A** is order $n \times m$, so as long as $n > m$ – that is, there are more data points (n) than terms of the basis function (m) and **A** is a non-square matrix and has no inverse. In other words, the matrix equation cannot be inverted and solved. This reflects the non-unique character of the overdetermined system – there are too many solutions possible, which is inconsistent with the (square matrix) system that allows only one unique solution; that is, one set of coefficients determining one specific curve that passes through all n points.

[14] It is important to avoid confusion here by noticing that in common matrix notion, the system "knowns" are in the matrix labeled **A** and the unknowns in a vector labeled **x**. For standard systems of equations (where the contents of **A** and **x** are known and we are solving for **b**), this reflects the fact that the **A** matrix contains the system coefficients, $a_{i,j}$ (and is therefore called the coefficient matrix), while the unknowns in vector **x** are the known x_i positions of the system for which a specific value of y_i will be determined. Often, however, the coefficients represent the unknowns and are placed in the **x** unknown vector, while the x_i are placed in the **A** coefficient matrix (since they are acting as known "coefficients," in this case). Therefore, it is prudent to avoid the temptation to assume that the letters defining the matrix are consistent with the letters symbolically used within it. It is also wise to not blindly assume that the coefficient matrix contains the equation coefficients as defined for the situation here.

The **A** matrix goes by a number of names. It is sometimes referred to as the **design** matrix, the **data kernel** matrix (given its analogy to the kernel in an integral transform equation) or, from group theory, **representation** matrix. I will use "data kernel matrix" here.[15]

If $m > n$, the system is **underdetermined** (more unknowns than equations). This is a common case in discrete inverse theory, which requires some additional $(n - m)$ conditions in order to *close* the system and thus allow a unique solution. If $m = n$ (and of full rank, each data point unique in this case), the problem is **well posed**. This was the case for interpolation: there were as many (independent) equations as unknowns, allowing a unique solution defining one curve that passed exactly through all of the points.

Solution of equation (5.20) proceeds in a manner analogous to that used previously when the system was written in system form. That is, we must first acknowledge that noise exists in the data, which is why we cannot expect a single curve to pass precisely through all of the data points. We do this by admitting an error at each data point,

$$\mathbf{Ax} - \mathbf{b} = \mathbf{e}, \tag{5.21}$$

where the vector $[\varepsilon_1 \varepsilon_2 \varepsilon_3 \ldots \varepsilon_n]^{\mathrm{T}}$ for each individual error, ε_i, is defined for each observation y_i as in equation (5.8) or (5.16).

The sum of the squared errors, given by the *scalar* (i.e., non-vector, or single number) e, is now defined by the vector product

$$\mathrm{e} = \mathbf{e}^{\mathrm{T}}\mathbf{e}. \tag{5.22}$$

Substituting (5.21) into (5.22) gives the dependence of the error e on the unknown coefficients a_i, which are stored in the **x** vector, so

$$\mathrm{e} = (\mathbf{Ax} - \mathbf{b})^{\mathrm{T}}(\mathbf{Ax} - \mathbf{b}). \tag{5.23}$$

We now solve for vector **x**, containing the unknown a_i, by taking the derivatives of e with respect to the a_i and setting these equal to 0:

$$\frac{\partial \mathrm{e}(\mathbf{x})}{\partial \mathbf{x}} = \frac{\partial}{\partial \mathbf{x}}\left[(\mathbf{Ax} - \mathbf{b})^{\mathrm{T}}(\mathbf{Ax} - \mathbf{b})\right] = 0. \tag{5.24}$$

The solution to this equation (see boxed details that follow) yields

$$(\mathbf{A}^{\mathrm{T}}\mathbf{A})\mathbf{x} = \mathbf{A}^{\mathrm{T}}\mathbf{b}, \tag{5.25}$$

which is solved directly to give

$$\mathbf{x} = (\mathbf{A}^{\mathrm{T}}\mathbf{A})^{-1}\mathbf{A}^{\mathrm{T}}\mathbf{b} = \mathbf{A}^{+}\mathbf{b}. \tag{5.26}$$

[15] Actually, I'll usually just call it the A matrix and forgo the fancier "technical" names.

That is, the unknown coefficients to the best-fit curve are given by the product of the $\mathbf{A}^{\mathrm{T}}\mathbf{A}$ inverse with $\mathbf{A}^{\mathrm{T}}\mathbf{b}$. The solution is often obtained most stably through use of the SVD generated pseudo-inverse ($\mathbf{A}^{+} = \mathbf{V}\boldsymbol{\Sigma}^{+}\mathbf{U}^{\mathrm{T}}$, where the SVD decomposition of $\mathbf{A} = \mathbf{V}\boldsymbol{\Sigma}\mathbf{U}^{\mathrm{T}}$, $\boldsymbol{\Sigma}$ is a diagonal matrix containing the singular values, σ_j, and Σ^{+} is the inverse of $\boldsymbol{\Sigma}$ (elements $1/\sigma_j$), with zeroes along the diagonal beyond the nonzero singular values so that this matrix is conformal with the $\mathbf{U}$ and $\mathbf{V}^{\mathrm{T}}$).[16]

As shown in the boxed details below, equation (5.25) is equal to the normal equations arrived at previously by the rearrangement of the derivative equations after solving them and setting them equal to zero.

For example, for a straight line fit, the matrices of equation (5.25) contain

$$\begin{bmatrix} n\sum_{i=1}^{n}(x_i - x_0)^2 & 0 \\ \sum_{i=1}^{n}(x_i - x_0) & \sum_{i=1}^{n}(x_i - x_0)^2 \end{bmatrix} \begin{bmatrix} a_1 \\ a_2 \end{bmatrix} = \begin{bmatrix} \sum_{i=1}^{n} y_i \\ \sum_{i=1}^{n} y_i(x_i - x_0)^2 \end{bmatrix}. \tag{5.27}$$

The two rows of this matrix product coincide with the two normal equations in (D5.1.2). As stated previously, since the basis used contains independent constituent terms, and since differentiation preserves this independence, the matrix product ($\mathbf{A}^{\mathrm{T}}\mathbf{A}$) is square and of full rank – i.e., it is invertible. Therefore, (5.25) is solved by computing this inverse and multiplying both sides by it (which will cancel $\mathbf{A}^{\mathrm{T}}\mathbf{A}$ on the left side), giving (5.26). Furthermore, this holds true whether we are solving for just two coefficients of a straight-line curve or 20 coefficients to a much more complicated curve, assuming that there are always more data points than unknown coefficients.

Notice that the above system solution (5.26) holds, regardless of the actual basis chosen. This is because the system is linear with respect to the coefficients. Even highly nonlinear basis functions can be fit to the data, since their form does not influence the approach. In this respect, *the term "linear" least squares does not imply that a geometrically linear curve (i.e., a straight line) is being fit to the data – it means that the technique is linear in the coefficients.*

When fitting curves that are not linear in the unknown coefficients, we require nonlinear least-squares techniques. However, some nonlinear forms, such as $y = \mathrm{e}^{(a_1 + a_2 x)}$, are **intrinsically linear**, since they can be transformed to the standard linear form (in this case by taking the log).

The simplicity of this solution is tremendous and it allows a least-squares fit of any desired basis to the data. However, potential computer problems may be periodically encountered when inverting a matrix as required here. Consequently,

[16] See Appendix 1 on Matrix Algebra, which discusses SVD in more detail.

Box D5.3 Solution of Equation (5.24)

The solution can be shown in two ways: (1) by direct manipulation of the matrices using the rules of matrix algebra and (2) by tracking the individual operations on the rows and columns of each matrix as an organized system. The first form takes advantage of the convenience of matrix algebra in manipulating large systems. The second helps clarify these symbolic operations for those who are less comfortable with the pure symbolic manipulation of matrices.

Solution by Symbolic Manipulation

Equation (5.24) is

$$\frac{\partial}{\partial \mathbf{x}}\left[(\mathbf{Ax}-\mathbf{b})^{\mathrm{T}}(\mathbf{Ax}-\mathbf{b})\right]=0. \tag{D5.3.1}$$

The left-hand side of this can be expanded, giving

$$\begin{aligned}&\frac{\partial}{\partial \mathbf{x}}\left[(\mathbf{Ax})^{\mathrm{T}}\mathbf{Ax}-(\mathbf{Ax})^{\mathrm{T}}\mathbf{b}-\mathbf{b}^{\mathrm{T}}\mathbf{Ax}+\mathbf{b}^{\mathrm{T}}\mathbf{b}\right]\\&=\frac{\partial}{\partial \mathbf{x}}\left[(\mathbf{x}^{\mathrm{T}}\mathbf{A}^{\mathrm{T}}\mathbf{Ax}-\mathbf{x}^{\mathrm{T}}\mathbf{A}^{\mathrm{T}}\mathbf{b}-\mathbf{b}^{\mathrm{T}}\mathbf{Ax}+\mathbf{b}^{\mathrm{T}}\mathbf{b})\right],\end{aligned} \tag{D5.3.2}$$

where the reversal rule for transposed products has been applied to the right-hand-side terms.

For convenience, call $\mathbf{A}^{\mathrm{T}}\mathbf{A}=\mathbf{N}$ (in anticipation that this product will represent the primary components of the normal equations). Then, differentiating the terms of the sum in (D5.3.2) gives

$$\frac{\partial}{\partial \mathbf{x}}\mathbf{x}^{\mathrm{T}}\mathbf{Nx}-\frac{\partial}{\partial \mathbf{x}}\mathbf{x}^{\mathrm{T}}\mathbf{A}^{\mathrm{T}}\mathbf{b}-\frac{\partial}{\partial \mathbf{x}}\mathbf{b}^{\mathrm{T}}\mathbf{Ax}+\frac{\partial}{\partial \mathbf{x}}\mathbf{b}^{\mathrm{T}}\mathbf{b}=0. \tag{D5.3.3}$$

The convention for vector differentiation of products is:

$$\frac{\partial}{\partial \mathbf{x}}(\mathbf{Ax})\equiv\frac{\partial}{\partial \mathbf{x}}(\mathbf{Ax})^{\mathrm{T}}=\frac{\partial}{\partial \mathbf{x}}\mathbf{x}^{\mathrm{T}}\mathbf{A}^{\mathrm{T}}=\mathbf{A}^{\mathrm{T}}. \tag{D5.3.4a}$$

Therefore, the first term on the left-hand-side of (D5.3.3) can be rewritten as:

$$\frac{\partial}{\partial \mathbf{x}}(\mathbf{x}^{\mathrm{T}}\mathbf{Nx})\equiv\frac{\partial}{\partial \mathbf{x}}(\mathbf{x}^{\mathrm{T}}\mathbf{Nx})^{\mathrm{T}}=\frac{\partial}{\partial \mathbf{x}}(\mathbf{x}^{\mathrm{T}}\mathbf{N}^{\mathrm{T}}\mathbf{x}). \tag{D5.3.4b}$$

while this solution frequently works well, a more robust solution procedure is used in practice, as this one may fail for computational reasons attributed to numerically induced ill-conditioning (see Appendix 1 for discussion).

Box D5.3 (Cont.)

This can be used with the principle of differentiation of parts to give:[17]

$$\frac{\partial}{\partial \mathbf{x}}(\mathbf{x}^{\mathrm{T}}\mathbf{N}\mathbf{x}) = \frac{\partial}{\partial \mathbf{x}}\mathbf{x}^{\mathrm{T}}(\mathbf{N}\mathbf{x}) + \frac{\partial}{\partial \mathbf{x}}\mathbf{x}^{\mathrm{T}}(\mathbf{N}^{\mathrm{T}}\mathbf{x}). \quad \text{(D5.3.4c)}$$

$$= \mathbf{N}\mathbf{x} + \mathbf{N}^{\mathrm{T}}\mathbf{x}. \quad \text{(D5.3.5)}$$

So, (D5.3.3) can be reduced to

$$\mathbf{N}\mathbf{x} + \mathbf{N}^{\mathrm{T}}\mathbf{x} - \mathbf{A}^{\mathrm{T}}\mathbf{b} - \mathbf{b}^{\mathrm{T}}\mathbf{A} = 0. \quad \text{(D5.3.6)}$$

Note that $\mathbf{N} = \mathbf{A}^{\mathrm{T}}\mathbf{A}$ and $\mathbf{N}^{\mathrm{T}} = (\mathbf{A}^{\mathrm{T}}\mathbf{A})^{\mathrm{T}} = \mathbf{A}^{\mathrm{T}}\mathbf{A}$, so $\mathbf{N}\mathbf{x} + \mathbf{N}^{\mathrm{T}}\mathbf{x} = (\mathbf{N} + \mathbf{N}^{\mathrm{T}})\mathbf{x} = 2\mathbf{A}^{\mathrm{T}}\mathbf{A}\mathbf{x}\mathbf{N}\mathbf{x} + \mathbf{N}^{\mathrm{T}}\mathbf{x} = (\mathbf{N} + \mathbf{N}^{\mathrm{T}})\mathbf{x} = 2\mathbf{A}^{\mathrm{T}}\mathbf{A}\mathbf{x}$. Similarly, for the vector product, $\mathbf{A}^{\mathrm{T}}\mathbf{b} = \mathbf{b}^{\mathrm{T}}\mathbf{A}$, so $\mathbf{A}^{\mathrm{T}}\mathbf{b} + \mathbf{b}^{\mathrm{T}}\mathbf{A} = 2\mathbf{A}^{\mathrm{T}}\mathbf{b}$. Therefore, dividing out the 2 and rearranging, (D5.3.6) can be written as

$$\mathbf{A}^{\mathrm{T}}\mathbf{A}\mathbf{x} = \mathbf{A}^{\mathrm{T}}\mathbf{b} \quad \text{(D5.3.7a)}$$

or

$$\mathbf{N}\mathbf{x} = \mathbf{A}^{\mathrm{T}}\mathbf{b}. \quad \text{(D5.3.7b)}$$

The matrix product $\mathbf{A}^{\mathrm{T}}\mathbf{A}$, or matrix $\mathbf{N}$, is square and of full rank (owing to the fact that the constituent terms of the basis function used in system (5.20) are independent and each of the observations is distinct). Therefore, equation (D5.3.7) can be solved in the standard manner: pre-multiply both sides by the inverse of $\mathbf{N}$, $\mathbf{N}^{-1}$, so

$$\mathbf{x} = \mathbf{N}^{-1}\mathbf{A}^{\mathrm{T}}\mathbf{b} \quad \text{(D5.3.8a)}$$

or

$$\mathbf{x} = (\mathbf{A}^{\mathrm{T}}\mathbf{A})^{-1}\mathbf{A}^{\mathrm{T}}\mathbf{b}. \quad \text{(D5.3.8b)}$$

[17] This rule can be somewhat confusing. To help, set $\mathbf{x}^{\mathrm{T}} = \mathbf{y}$, then rewrite $\mathbf{x}^{\mathrm{T}}\mathbf{N}\mathbf{x} = \mathbf{y}\mathbf{N}\mathbf{x}$. Now, consider differentiation by parts for scalars: $\frac{\partial}{\partial x}(\mathbf{y}\mathbf{N}\mathbf{x}) = \mathbf{N}\mathbf{x}\frac{\partial}{\partial x}\mathbf{y} + \mathbf{y}\mathbf{N}\frac{\partial}{\partial x}\mathbf{x}$. However, for vectors, the order of operations is important. Therefore, we take account of the equivalency given in (D5.3.3), which states that the derivatives of $\mathbf{x}^{\mathrm{T}}\mathbf{N}\mathbf{x}$ and $\mathbf{x}^{\mathrm{T}}\mathbf{N}^{\mathrm{T}}\mathbf{x}$ are equal. We use the first form to treat $\mathbf{N}\mathbf{x}$ as being independent of $\mathbf{x}$, so we form the counterpart to $\mathbf{N}\mathbf{x}\frac{\partial}{\partial x}\mathbf{y}$ as $\left(\frac{\partial}{\partial x}\mathbf{x}^{\mathrm{T}}\right)\mathbf{A} = \mathbf{A}$, where $\mathbf{A} = \mathbf{N}\mathbf{x}$. Having done this, we must now treat $\mathbf{x}^{\mathrm{T}}\mathbf{N}$ as being independent of x, which we do by using the second form, in which case the $\mathbf{x}^{\mathrm{T}}\mathbf{N}$ has been transposed and reversed to $\mathbf{N}^{\mathrm{T}}\mathbf{x}$. With this, the derivative can again be written in the proper form for vector differentiation, where the counterpart to $\mathbf{y}\mathbf{N}\frac{\partial}{\partial x}\mathbf{x}$ is $\left(\frac{\partial}{\partial x}\mathbf{x}^{\mathrm{T}}\right)\mathbf{B} = \mathbf{B}$, where $\mathbf{B} = \mathbf{N}^{\mathrm{T}}\mathbf{x}$. In this manner, we have accomplished the product expansion while satisfying the convention for the proper form of vector differentiation.

Box D5.3 (Cont.)

Solution by Column and Row Manipulation

Again consider the system of equations, in matrix form, that describe the error or mismatch between the computed positions of the fit curve and actual observed positions, given any general form of curve to be fit,

$$\begin{bmatrix} \varphi_{11} & \varphi_{12} & & \varphi_{1m} \\ \varphi_{21} & \varphi_{22} & \cdots & \varphi_{2m} \\ \vdots & & \ddots & \\ \varphi_{n1} & \varphi_{n2} & & \varphi_{nm} \end{bmatrix} \begin{bmatrix} a_1 \\ a_2 \\ \vdots \\ a_m \end{bmatrix} - \begin{bmatrix} y_1 \\ y_2 \\ \vdots \\ y_n \end{bmatrix} = \begin{bmatrix} \varepsilon_1 \\ \varepsilon_2 \\ \vdots \\ \varepsilon_n \end{bmatrix}, \tag{D5.3.9a}$$

or, symbolically,

$$\mathrm{A}x - \mathrm{b} = \mathrm{e}. \tag{D5.3.9b}$$

As previously, it is assumed that $n > m$ (there are more data points than unknown coefficients), where $n \times m$ is the order of matrix **A**. Here the jth basis term evaluated at the ith point is indicated by φ_{ij}, compared to the previous notation, where this was given as $\varphi_j(x_i)$.

Alternatively, the **b** vector can be directly combined with the **A** matrix to give

$$\begin{bmatrix} \varphi_{11} & \varphi_{12} & & \varphi_{1m} & -y_1 \\ \varphi_{21} & \varphi_{22} & \cdots & \varphi_{2m} & -y_2 \\ & \vdots & & & \vdots \\ \varphi_{n1} & \varphi_{n2} & & \varphi_{nm} & -y_n \end{bmatrix} \begin{bmatrix} a_1 \\ a_2 \\ \vdots \\ a_m \\ 1 \end{bmatrix} = \begin{bmatrix} \varepsilon_1 \\ \varepsilon_2 \\ \vdots \\ \varepsilon_n \end{bmatrix}, \tag{D5.3.10a}$$

which is written as

$$\mathbf{Cx} = \mathrm{e}, \tag{D5.3.10b}$$

where **C** is the expanded **A** matrix containing the **b** vector, and the unknown vector **X** contains an additional row relative to **x**, which contains the element 1. The additional column in **C** makes it of order $n\times(m+1)$, and the additional row in **X** makes its size $m + 1$, which is conformable with the **C** matrix.

This contracted matrix form can be partitioned into submatrices corresponding to the original matrices of (D5.3.9), as

$$[\mathbf{A} - \mathbf{b}]\begin{bmatrix} \mathbf{x} \\ 1 \end{bmatrix} = [\mathbf{e}]. \tag{D5.3.11}$$

So, the $[\mathbf{A} - \mathbf{b}]$ matrix is the partitioned form of the **C** matrix and the $[\mathbf{x}\ \mathbf{1}]^{\mathrm{T}}$ column vector is the partitioned form of the **X** vector.

Both matrix forms (D5.3.10) and (D5.3.11) describe the system in (5.16), which is written with the shorthand notation as

Box D5.3 (Cont.)

$$\begin{aligned} a_1\varphi_{11} + a_2\varphi_{12} + \ldots + a_m\varphi_{1m} - y_1 &= \varepsilon_1 \\ a_1\varphi_{21} + a_2\varphi_{22} + \ldots + a_m\varphi_{2m} - y_2 &= \varepsilon_2 \\ &\vdots \\ a_1\varphi_{n1} + a_2\varphi_{n2} + \ldots + a_m\varphi_{nm} - y_n &= \varepsilon_n. \end{aligned} \tag{D5.3.12}$$

From this, it is clear that the ith error, ε_i (i.e., the error evaluated at the position x_i), is the dot product

$$\mathrm{C}_i\mathrm{x} = \mathrm{e}_i = [\varphi_{i1}\ \varphi_{i2}\ \cdots\ \varphi_{im}\ \ -y_i]\begin{bmatrix} a_1 \\ a_2 \\ \vdots \\ a_m \\ 1 \end{bmatrix}, \tag{D5.3.13}$$

and the sum of squared errors, e, to be minimized is

$$\mathrm{e} = \mathrm{e}^{\mathrm{T}}\mathrm{e}. \tag{D5.3.14}$$

That is, the summation (5.9), defining the sum of the squared errors, is given as $\mathbf{e}^{\mathrm{T}}\mathbf{e}$, where **e** is the column vector containing the ε_i values.

Substituting (D5.3.13) into (D5.3.14) for **e** gives

$$\mathrm{e} = \mathrm{e}^{\mathrm{T}}\mathrm{e} = (\mathrm{Cx})^{\mathrm{T}}\mathrm{Cx} = \mathrm{x}^{\mathrm{T}}\mathrm{C}^{\mathrm{T}}\mathrm{Cx} \tag{D5.3.15}$$

(recall the reversal rule for the transpose of a matrix product). In the previous symbolic manipulation, this equation corresponds to (5.23), where the matrices were not partitioned into the convenient submatrices as done here.

Defining

$$\mathrm{R} = \mathrm{C}^{\mathrm{T}}\mathrm{C} \tag{D5.3.16a}$$

makes **R** a *square symmetric matrix* of order $m + 1$ given by the matrix product (multiplied out explicitly later)

$$\begin{bmatrix} \varphi_{11} & \varphi_{21} & \varphi_{n1} \\ \varphi_{12} & \varphi_{22}\cdots & \varphi_{n2} \\ \vdots & \ddots & \\ \varphi_{1m} & \varphi_{2m} & \varphi_{nm} \\ -y_1 & -y_2 & -y_n \end{bmatrix} \begin{bmatrix} \varphi_{11} & \varphi_{12} & \varphi_{1m} & -y_1 \\ \varphi_{21} & \varphi_{22}\cdots & \varphi_{2m} & -y_2 \\ \vdots & \ddots & & \vdots \\ \varphi_{n1} & \varphi_{n2} & \varphi_{nm} & -y_n \end{bmatrix} \tag{D5.3.16b}$$

and

$$\mathrm{e} = \mathbf{X}^{\mathrm{T}}\mathbf{R}\mathbf{X}. \tag{D5.3.17}$$

Box D5.3 (Cont.)

To minimize e, we set $\partial e(a_j)/\partial a_j = 0$ (recall the convention for vector differentiation given by (D5.3.4)). Employing the reversal rule for transposed products with differentiation by parts (remember that the **X** vector contains the a_j coefficients) gives

$$\frac{\partial}{\partial a_j}(\mathbf{X}^T\mathbf{R}\mathbf{X}) = \frac{\partial \mathbf{X}^T}{\partial a_j}(\mathbf{R}\mathbf{X}) + \frac{\partial \mathbf{X}^T}{\partial a_j}(\mathbf{R}^T\mathbf{X}) = 0. \quad \text{(D5.3.18)}$$

So, as an example, consider the second coefficient for which

$$\frac{\partial \mathbf{X}^T}{\partial a_2} = [0 \ 1 \ 0 \ \dots \ 0]. \quad \text{(D5.3.19)}$$

So, from (D5.3.18), for this second element,

$$\frac{\partial e(a_j)}{\partial a_2} = \begin{bmatrix} 0 & 1 & 0 & \dots & 0 \end{bmatrix} \mathbf{R} \begin{bmatrix} a_1 \\ a_2 \\ \vdots \\ a_m \\ 1 \end{bmatrix} + \begin{bmatrix} 0 & 1 & 0 & \dots & 0 \end{bmatrix} \mathbf{R}^T \begin{bmatrix} a_1 \\ a_2 \\ \vdots \\ a_m \\ 1 \end{bmatrix}. \quad \text{(D5.3.20)}$$

For all a_j, the derivative forms a matrix (similar to the addition rule in standard calculus),

$$\dot{\mathbf{X}}^T = \frac{\partial \mathbf{X}^T}{\partial a_j} = \begin{bmatrix} 1 & 0 & 0 & & & 0 \\ 0 & 1 & 0 & \dots & & 0 \\ & \vdots & & & & \\ 0 & 0 & 0 & & 1 & 0 & 0 \\ 0 & 0 & 0 & \dots & 0 & 1 & 0 \end{bmatrix} \quad \text{(D5.3.21)}$$

so

$$\frac{\partial e(a_j)}{\partial a_j} = \dot{\mathbf{X}}^T\mathbf{R}\mathbf{X} + \dot{\mathbf{X}}^T\mathbf{R}^T\mathbf{X} = 0. \quad \text{(D5.3.22)}$$

Now consider the form of $\mathbf{R}$ $(= \mathbf{C}^T\mathbf{C})$:

$$\begin{bmatrix} \varphi_{1,1} & \varphi_{1,2} & \cdots & \varphi_{1,m} & -y_1 \\ \varphi_{2,1} & \varphi_{2,2} & \cdots & \varphi_{2,m} & -y_2 \\ \vdots & & \ddots & & \\ \varphi_{n,1} & \varphi_{n,2} & \cdots & \varphi_{n,m} & -y_n \end{bmatrix}$$

Box D5.3 (Cont.)

$$\begin{bmatrix} \varphi_{1,1} & \varphi_{2,1} & \cdots & \varphi_{n,1} \\ \varphi_{1,2} & \varphi_{2,2} & \cdots & \varphi_{n,2} \\ \vdots & & \ddots & \\ \varphi_{1,m} & \varphi_{2,m} & & \varphi_{n,m} \\ -y_1 & -y_2 & \cdots & -y_n \end{bmatrix} \begin{bmatrix} \sum_{i=1}^{n}\varphi_{i,1}^2 & \sum_{i=1}^{n}\varphi_{i,1}\varphi_{i,2} & \cdots & \sum_{i=1}^{n}\varphi_{i,1}\varphi_{i,m} & -\sum_{i=1}^{n}\varphi_{i,1}y_i \\ \sum_{i=1}^{n}\varphi_{i,2}\varphi_{i,1} & \sum_{i=1}^{n}\varphi_{i,2}^2 & \cdots & \sum_{i=1}^{n}\varphi_{i,2}\varphi_{i,m} & -\sum_{i=1}^{n}\varphi_{i,2}y_i \\ \vdots & & \ddots & & \\ -\sum_{i=1}^{n}y_i\varphi_{i,1} & -\sum_{i=1}^{n}y_i\varphi_{i,2} & \cdots & -\sum_{i=1}^{n}y_i\varphi_{i,m} & \sum_{i=1}^{n}y_i^2 \end{bmatrix}. \tag{D5.3.23a}$$

In partitioned form, this matrix product is seen as

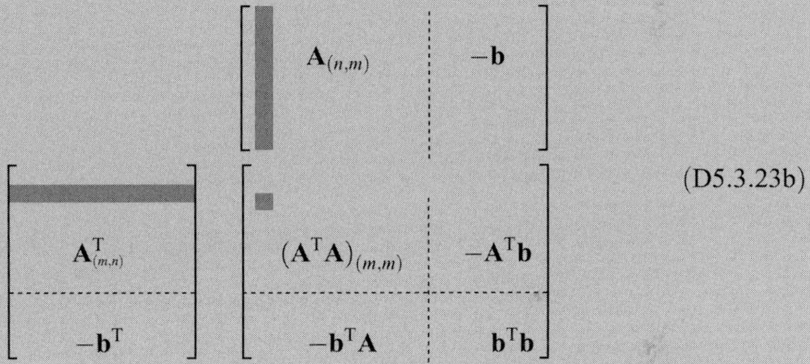

(D5.3.23b)

Notice that $\mathbf{A}^T\mathbf{A}$ = matrix $\mathbf{N}$ [= (5.28d) from the normal equations) and $-\mathbf{A}^T\mathbf{b} = (-\mathbf{b}^T\mathbf{A})^T$ = matrix $\mathbf{B}$, also from (5.28d). Recall that $\mathbf{N}$ and $\mathbf{B}$ were the two known matrices when solving the normal equations in matrix form previously.

Because $\mathbf{R}$ is symmetrical, $\mathbf{R} = \mathbf{R}^T$, which allows simplification of (D5.3.22) describing the minimization criteria

$$\dot{\mathbf{X}}^T\mathbf{R}\mathbf{X} + \dot{\mathbf{X}}^T\mathbf{R}^T\mathbf{X} = 2\dot{\mathbf{X}}^T\mathbf{R}\mathbf{X} = 0. \tag{D5.3.24}$$

Thus, we need only satisfy

$$\dot{\mathbf{X}}^T\mathbf{R}\mathbf{X} = 0 \tag{D5.3.25}$$

after dividing out the 2.

So consider $\dot{\mathbf{X}}^T\mathbf{R}$ in partitioned matrix form:

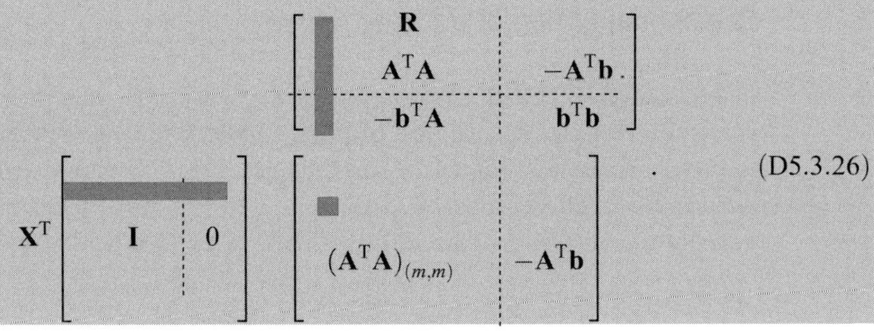

(D5.3.26)

Box D5.3 (Cont.)

Note that the $-\mathbf{b}^T\mathbf{A}$ row vector in $\mathbf{R}$ is always multiplied by the null vector in $\mathbf{X}^T$, so its influence is lost in the product. This is also true of the submatrix (single element) $\mathbf{b}^T\mathbf{b}$ in $\mathbf{R}$.

Completing the matrix product $(\mathbf{X}^T\mathbf{R})\mathbf{X}$,

$$\left[\begin{array}{c:c} \mathbf{A}^T\mathbf{A} & -\mathbf{A}^T\mathbf{b} \end{array}\right] \left[\begin{array}{c} \mathbf{x} \\ \hdashline 1 \end{array}\right] \qquad \text{(D5.3.27a)}$$

$$\begin{bmatrix} \mathbf{A}^T\mathbf{A}\mathbf{x} \\ -\mathbf{A}^T\mathbf{b} \end{bmatrix},$$

or

$$(\dot{\mathbf{X}}^T\mathbf{R})\mathbf{X} = \mathbf{A}^T\mathbf{A}\mathbf{x} - \mathbf{A}^T\mathbf{b} = 0, \qquad \text{(D5.3.27b)}$$

so

$$\mathbf{A}^T\mathbf{A}\mathbf{x} = \mathbf{A}^T\mathbf{b}. \qquad \text{(D5.3.27c)}$$

Recall that $\mathbf{A}^T\mathbf{A} = \mathbf{N}$ and $\mathbf{A}^T\mathbf{b} = \mathbf{B}$. Consequently, we now have the same matrix equation as we have previously solved from the normal equations, only this time it is written in terms of the original system of equations. Its solution procedure is the same as always,

$$[\mathbf{A}^T\mathbf{A}]^{-1}\mathbf{A}^T\mathbf{A}\mathbf{x} = [\mathbf{A}^T\mathbf{A}]^{-1}\mathbf{A}^T\mathbf{b}, \qquad \text{(D5.3.28)}$$

so

$$\mathbf{x} = [\mathbf{A}^T\mathbf{A}]^{-1}\mathbf{A}^T\mathbf{b} \qquad \text{(D5.3.29)}$$

or

$$= \mathbf{N}^{-1}\mathbf{B}, \qquad \text{(D5.3.30)}$$

as derived previously as explicit differentiation.

5.6 Orthogonal Fitting of a Straight Line

Orthogonal curve fitting is required when we wish to minimize the error defined as $(\Delta x^2+\Delta y^2)^{1/2}$. In the special case of fitting a straight line to the data, the solution can be readily obtained via any of several analytic approaches. One approach that is relatively simple requires that we reconsider the problem in terms of variance about a rotated coordinate system. In particular, it is seen that the error, as defined here, is essentially equivalent to minimizing the variance about the best orthogonal fit line. Thus we can search for that rotation angle, Θ, of the original coordinate system that minimizes this "rotated" variance.

If we rotate the coordinate system by an angle θ, then the original Δx, Δy variables take on new values, $\Delta\xi$, $\Delta\psi$, in the rotated framework, given by

$$\Delta\xi_i(\theta) = \Delta x_i \cos\theta + \Delta y_i \sin\theta \tag{5.28}$$

$$\Delta\psi_i(\theta) = -\Delta x_i \sin\theta + \Delta y_i \cos\theta, \tag{5.29}$$

where these two axes are perpendicular, separated by 90°. Therefore, we can now consider the sample variance, s_ξ^2, about the axis parallel to ξ, which is given as

$$\begin{aligned} s_\xi^2 &= \frac{1}{n-1}\sum_{i=1}^{n}(\Delta x_i \cos\theta + \Delta y_i \sin\theta)^2 \\ &= \frac{1}{n-1}\sum_{i=1}^{n}(\Delta x_i^2 \cos^2\theta + \Delta x_i \Delta y_i \cos^2\theta \sin^2\theta + \Delta y_i^2 \sin^2\theta) \\ &= s_{xx} \cos^2\theta + s_{xy} \cos^2\theta \sin 2\theta + s_{yy} \sin^2\theta. \end{aligned} \tag{5.30}$$

Minimizing this with respect to θ gives

$$\frac{\partial e(\theta)}{\partial\theta} = (s_{yy} - s_{xx})\sin 2\theta + 2s_{xy} \cos 2\theta = 0, \tag{5.31}$$

which is solved by

$$\tan^{-1} 2\hat{\theta} = 2s_{xy}/(s_{xx} - s_{yy}). \tag{5.32}$$

We wish to find the minimum of the variance, which is given for the case in which the second derivative is positive, or

$$\frac{\partial^2 e(\theta)}{\partial\theta^2} = -4s_{xy} \sin 2\hat{\theta} > 0. \tag{5.33}$$

This degree of rotation, $\hat{\theta}$, provides the slope of the line producing the orthogonal fit. As with standard regression, this line passes through the original mean of the x and y values (which allows a computation of the intercept as before). An easier solution for angle $\hat{\theta}$ is obtained through Empirical Orthogonal Function (EOF) Analysis, the topic of Chapter 15, where this means of finding the orthogonal best-fit straight line is presented (something that can be applied without detailed knowledge of the EOF methodology).

5.7 Assessing Uncertainty in Optimal Parameter Values

Having now computed the parameter values (the a_i) that produce the "best" fitting curve to the data, at least for standard least squares methods, we must assess the uncertainty in these optimal parameter values. That is, the best-fit parameter values, which are themselves random variables, contain some scatter or uncertainty, reflecting the uncertainty in the original data used to estimate them. We now need to assess just how large this uncertainty is. So, if we find for a straight-line fit that the best fit slope is 2.1, we wish to know if values such as 2.0 or 1.5 would produce significantly worse fits, or fits that are essentially as good as that obtained for a slope of 2.1; is there enough scatter in the original data that a slope of 2.0 might be almost indistinguishable from a slope of 2.1 in an optimal statistical sense, or is the scatter such that a slope of 2.0 would produce a significantly poorer fit of the "best" fit line?

We assess this uncertainty in the same manner that we use to assess the uncertainty in any random variable: by using the expectance operator. For a straight-line curve fit, the slope parameter, a_2, is a function of a random variable Y given by (D5.1.6), where Y is the dependent variable that contains the uncertainty, and X, you will recall, is the independent variable, which is *not* a random variable. Therefore, the uncertainty in the random variable a_2 is given by its variance, computed as

$$\sigma_{a_2}^2 = \mathrm{Var}\left[\frac{s_{xy}}{s_x^2}\right] \tag{5.34a}$$

or

$$= \mathrm{Var}\left[\frac{\sum_{i=1}^{n} y_i x_i - \bar{y}\sum_{i=1}^{n} x_i}{\sum_{i=1}^{n} x_i^2 - \bar{x}\sum_{i=1}^{n} x_i}\right]. \tag{5.34b}$$

Equation (5.34) reduces (see boxed details below) to

$$\sigma_{a_2}^2 = \frac{\sigma_y^2}{(n-1)s_x^2}. \tag{5.34c}$$

Likewise, the variance of a_1 is (see boxed details below)

$$\sigma_{a_1}^2 \approx \left[\frac{\sum_{j=1}^{n} x_j^2}{n^2}\right]\frac{\sigma_Y^2}{s_X^2}. \tag{5.35}$$

One can now estimate the uncertainty in the predicted value of Y, given any value of X, allowing construction of a confidence interval[18] about the predicted Y curve. The variance about any predicted point y_{pi}, at any point x_i is given as[19]

$$\mathrm{Var}[Y_{pi}] = \sigma_Y^2\left[\frac{1}{n} + \frac{(x_i - \bar{x})^2}{(n-1)s_x^2}\right], \tag{5.36a}$$

[18] Technically, a confidence interval is an interval about a parameter prediction, and thus, in the case of an interval about the predicted curve for Y, the interval is known as a "**prediction interval.**"

[19] We will return to this prediction interval after we establish some statistical background on this problem in a later chapter.

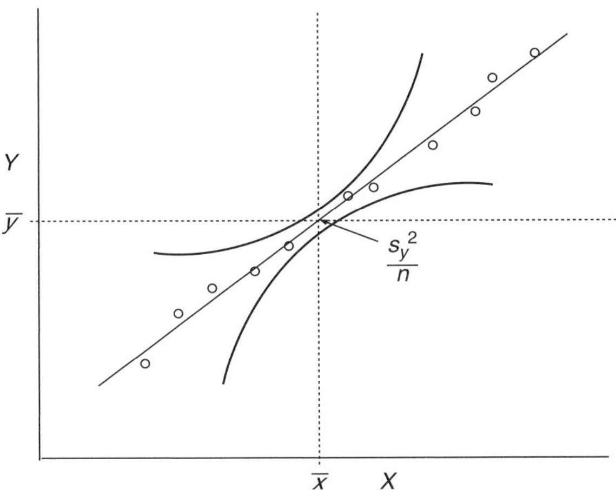

Figure 5.11 Schematic of 95 percent uncertainty limits about the best-fit straight line. Also showing that the best-fit straight line will pass through the mean of x and y, which leads to the uncertainty (variance) at that point to be equal to the Central Limit Theorem (standard error of the mean).

or its sample estimate,

$$s_{y_{pi}}^2 = s_\varepsilon^2 \left[\frac{1}{n} + \frac{(x_i - \overline{x})^2}{(n-1)s_x^2} \right], \tag{5.36b}$$

where s_ε^2 is the sample residual variance about the best-fit line, as described in the boxed derivation of the slope for (D5.4.6).

If you were to draw a region about the best-fit line that encompassed the 95 percent confidence region (or some other selected region), it would look approximately like that shown in Figure 5.11.

Note, when $x_i = \overline{x}$, that $y_i = \overline{y}$ and the estimated uncertainty is at its minimum as $s_{y_{pi}}^2 = s_\varepsilon^2/n$, which reflects the Central Limit Theorem at that point. The uncertainty then increases away (exaggerated in the figure) from this pair of points $(\overline{x}, \overline{y})$, since there is increased uncertainty in the fitted line with respect to how well it represents the true line. When plotted for the actual domain in X, one often will not see the parabolic drift of the prediction interval from the best-fit line (sometimes one has to move to X values that are much larger than the domain before that becomes visually apparent).

The expectance operator can be used to determine the uncertainty in the optimal parameters of any curve fit, including those for constrained regression, in a similar manner to that computed above – see D5.6 and D5.7 for general solutions.

Box D5.4 Derivation of the Variance of the LS Slope and Intercept for Straight-Line Fit

Slope

$$\sigma_{a_2}^2 = \mathrm{Var}\left[\frac{s_{xy}}{s_{\mathrm{x}}^2}\right] \tag{D5.4.1a}$$

$$= \mathrm{Var}\left[\frac{\sum_{i=1}^{\mathrm{n}} y_i \mathrm{x}_i - n\bar{y}\bar{\mathrm{x}}}{(n-1)s_{\mathrm{x}}^2}\right] \tag{D5.4.1b}$$

$$= \mathrm{Var}\left[\mathrm{A}\sum_{i=1}^{n} y_i \mathrm{x}_i - \mathrm{A}n\bar{y}\bar{\mathrm{x}}\right],$$

where $\mathrm{A} = 1/(n-1)s_x^2$ reflecting the fact that X *is not a random variable* and can easily be isolated in the expectance operator.

Expansion of the sum gives

$$= \mathrm{Var}[\mathrm{A}(y_1x_1 + y_2x_2 + \ldots + y\mathrm{n}x_{\mathrm{n}} - \bar{\mathrm{x}}\bar{y}_1 - \bar{\mathrm{x}}\bar{y}_2 - \ldots - \bar{\mathrm{x}}\bar{y}_{\mathrm{n}})]$$

$$= \mathrm{Var}\left[\sum_{i=1}^{\mathrm{n}} \alpha_i y_i\right], \tag{D5.4.1c}$$

where $\alpha_{\mathrm{i}} = (\mathrm{x}_{\mathrm{i}} - \bar{\mathrm{x}})/[(n-1)s_{\mathrm{x}}^2]$.

Recall from (2.51) that

$$\mathrm{Var}\left[\sum_{j=1}^{\mathrm{k}} a_j \mathrm{X}_j\right] = \sum_{i=1}^{\mathrm{k}}\sum_{j=1}^{\mathrm{k}} a_i a_j \mathrm{Cov}[\mathrm{X}_i, \mathrm{X}_j], \tag{D5.4.2}$$

and if the X_j are independent, the covariance is zero when $i \neq j$, reducing this to

$$\mathrm{Var}\left[\sum_{j=1}^{\mathrm{k}} a_{\mathrm{j}} X_j\right] = \sum_{j=1}^{\mathrm{k}} a_j^2 \mathrm{Var}[\mathrm{X}_j]. \tag{D5.4.3}$$

Thus, (D5.4.3) is in the exact form of (D5.4.1c), since the individual observations of Y are independent, so $s_{a_2}^2$ is given by

$$\sigma_{a_2}^2 = \sum_{i=1}^{\mathrm{n}} \alpha_i^2 \mathrm{Var}[\mathrm{Y}_i]$$

$$= \frac{\sum_{i=1}^{\mathrm{n}} \sigma_{\mathrm{y_i}}^2 (x_{\mathrm{i}} - \bar{x})}{(\mathrm{n}-1)^2 s_x^4}. \tag{D5.4.4}$$

Box D5.4 (Cont.)

This formula provides the uncertainty, as a variance, in the a_2 parameter of a straight-line fit to the y_i data points. For standard regression, in which all of the y_i have the same variance, σ_y^2, then (D5.4.4) reduces as[20]

$$\sigma_{a_2}^2 = \sum_{i=1}^{n} \alpha_i^2 \mathrm{Var}[Y_i] = \frac{\sigma_y^2 \sum_{i=1}^{n} (x_i - \bar{x})}{(n-1)^2 s_x^4} = \frac{\sigma_y^2}{(n-1)s_x^2}. \tag{D5.4.5}$$

While this formula appears rather straightforward, there is actually a slight complication. That is, previously we have assumed that the Y are from a single population with variance σ^2, so $\mathrm{Var}[y_1] = \mathrm{Var}[y_2] = \sigma^2$. Now, however, we are assuming that each value y_i varies about a mean that itself varies with X as defined by our best-fit line. So, $\mathrm{Var}[Y_i]$, or σ_Y^2 in (D5.4.5), is the variance about $Y = \bar{y} + a_2(x_i - \bar{x})$, where a_2 is the true slope of the best-fit line. Thus we estimate σ_Y^2 with s_Y^2 as

$$s_Y^2 = \frac{1}{n-1} \sum_{i=1}^{n} (y_i - \bar{y})_i^2 \tag{D5.4.6}$$

where this estimate is referred to as the *sample variance about the best-fit line*. Its root value or *sample standard deviation* will be referred to as the **rms (root mean squared) error**. This is an estimate of the scatter (variance) in each of the sample points $= \sigma_\varepsilon^2$.

As previously mentioned, the s_x is not actually a measure of uncertainty in the x_i, since these are not random variables. Rather, it reflects the spread in the values of x_i measured in the convenient and standard form of a variance.

Intercept

For the intercept uncertainty, we use (5.11a):

$$\begin{aligned}
\sigma_{a_1}^2 &= \mathrm{Var}[\bar{y} - a_2\bar{x}] \\
&= E[(\bar{y} - a_2\bar{x})^2] - E[\bar{y} - a_2\bar{x}]^2 \\
&= E[\bar{y}^2 - 2a_2\bar{x}\bar{y} + (a_2\bar{x})^2] - E[\bar{y}^2] + 2\bar{x}E[a_2]E[\bar{y}] - E[a_2\bar{x}]^2 \\
&= E[\bar{y}^2] - 2\bar{x}E[a_2\bar{y}] + \bar{x}^2E[a_2^2] - E[\bar{y}^2] + 2\bar{x}E[a_2]E[\bar{y}] - \bar{x}^2E[a_2]^2.
\end{aligned}$$

Recall that $\bar{x}$ comes outside of the expectance operator because in this case, X is not a random variable – we only use $\bar{x}$ as a convenient notation for the averaged sum of this

[20] This will not be the case for weighted regression, in which the variance will be given by (D5.4.4).

Box D5.4 (Cont.)

deterministic variable. If X was a random variable, then its mean would also be a random variable and thus an expected value, which (recall from Chapter 2) is μ.

$$
\begin{aligned}
&= \mathrm{E}[\bar{y}^2] - 2\bar{x}(\mathrm{Cov}[a_2\bar{y}] + \mathrm{E}[a_2]\mathrm{E}[\bar{y}]) + \bar{x}^2\mathrm{E}[a_2^2] - \mathrm{E}[\bar{y}]^2 \\
&\quad + 2\bar{x}\mathrm{E}[a_2]\mathrm{E}[\bar{y}] - \bar{x}^2\mathrm{E}[a_2]^2 \\
&= \mathrm{Var}[\bar{y}] + \bar{x}^2\mathrm{Var}[a_2]^2 - 2\bar{x}\mathrm{Cov}[a_2\bar{y}] \qquad \text{(D5.4.7)}
\end{aligned}
$$

(D5.4.7) is now easily reduced, since we know the variance of $\bar{y}$ (given by the Central Limit Theorem) is σ_Y^2/n; there is no covariance between a_2 and $\bar{y}$, so the last term (the covariance term) is eliminated; and we previously computed the variance of a_2, which is given in (5.34c), so:

$$
\begin{aligned}
&= \frac{\sigma_Y^2}{n} + \bar{x}^2 \frac{\sigma_y^2}{(n-1)s_x^2} \\
&= \sigma_Y^2\left[\frac{1}{n} + \frac{\bar{x}^2}{\sum_{j=1}^{n}(x_j-\bar{x})^2}\right] \\
&= \sigma_Y^2\left[\frac{\sum_{j=1}^{n}(x_j-\bar{x})^2}{n\sum_{j=1}^{n}(x_j-\bar{x})^2} + \frac{n\bar{x}^2}{n\sum_{j=1}^{n}(x_j-\bar{x})^2}\right] \\
&= \sigma_Y^2\left[\frac{\sum_{j=1}^{n}(x_j-\bar{x})^2 + n\bar{x}^2}{n\sum_{j=1}^{n}(x_j-\bar{x})^2}\right] \qquad \text{(D5.4.8a)} \\
&= \sigma_Y^2\left[\frac{\sum_{j=1}^{n}x_j^2 - 2n\bar{x}^2 + n\bar{x}^2 + n\bar{x}^2}{n\sum_{j=1}^{n}(x_j-\bar{x})^2}\right] \\
&= \sigma_Y^2\left[\frac{\sum_{j=1}^{n}x_j^2}{n\sum_{j=1}^{n}(x_j-\bar{x})^2}\right] \\
&= \sigma_Y^2\left[\frac{\sum_{j=1}^{n}x_j^2}{n(n-1)s_x^2}\right],
\end{aligned}
$$

Box D5.4 (Cont.)

or, if we approximate $n-1$ as n, introducing a small error when n is large, this reduces to

$$\approx \sigma_Y^2 \left[\frac{\sum_{j=1}^{n} x_j^2}{n^2 s_x^2} \right], \tag{D5.4.8b}$$

so

$$\approx \sigma_{a_1}^2 \left[\frac{\sum_{j=1}^{n} x_j^2}{n^2} \right] \frac{\sigma_Y^2}{s_x^2}. \tag{D5.4.8c}$$

Box D5.5 Derivation of the Variance of Predicted Y Value for Straight-Line Fit

We compute the uncertainty in the predicted values of y using our best-fit parameters in the same manner as we used for the slope and intercept using the expectance operator. For convenience, we work with the form of the best fit as

$$Y_{pi} = \bar{y} + a_2(x_i - \bar{x}), \tag{D5.5.1a}$$

where y_{pi} is any predicted value of y at position x_i. To compute the uncertainty in each predicted value of y_{pi},

$$\begin{aligned}
\mathrm{Var}[Y_{pi}] &= \mathrm{Var}[\bar{y} + a_2(x_i - \bar{x})] \\
&= E[\{\bar{y} + a_2(x_i - \bar{x})\}^2] - E[\bar{y} + a_2(x_i - \bar{x})]^2 \\
&= E[\{\bar{y} + a_2(x_i - \bar{x})\}^2] - E[\bar{y} + a_2(x_i - \bar{x})]E[\bar{y} + a_2(x_i - \bar{x})] \\
&= E[\{\bar{y} + a_2(x_i - \bar{x})\}^2] - \{E[\bar{y}] + E[a_2(x_i - \bar{x})]\}\{E[\bar{y}] + E[a_2(x_i - \bar{x})]\} \\
&= E[\{\bar{y} + a_2(x_i - \bar{x})\}^2] - \{E[\bar{y}] + (x_i - \bar{x})E[a_2]\}\{E[\bar{y}] + (x_i - \bar{x})E[a_2]\} \\
&= E[\{\bar{y}^2 + 2a_2\bar{y}(x_i - \bar{x}) + \{a_2(x_i - \bar{x})\}^2] - E[\bar{y}]^2 - 2(x_i - \bar{x})E[a_2]E[\bar{y}] \\
&\quad -(x_i - \bar{x})^2 E[a_2]^2 \\
&= E[\bar{y}^2] + 2(x_i - \bar{x})E[a_2\bar{y}] + (x_i - \bar{x})^2 E[a_2^2] - E[\bar{y}]^2 - 2(x_i - \bar{x})E[a_2]E[\bar{y}] \\
&\quad -(x_i - \bar{x})^2 E[a_2]^2 \\
&= \mathrm{Var}[\bar{y}] + (x_i - \bar{x})^2 \mathrm{Var}[a_2] + 2(x_i - \bar{x})\mathrm{Cov}[a_2\bar{y}]
\end{aligned} \tag{D5.5.1b}$$

Box D5.5 (Cont.)

$$\sigma^2_{y_{pi}} = \frac{\sigma^2_Y}{n} + (x_i - \bar{x})^2 \frac{\sigma^2_Y}{(n-1)s^2_X}$$
$$= \sigma^2_Y \left[\frac{1}{n} + \frac{(x_i - \bar{x})^2}{(n-1)s^2_X}\right] \quad \text{(D5.5.2a)}$$

or

$$= \sigma^2_Y \left[\frac{1}{n} + \frac{(x_i - \bar{x})^2}{\sum_{j=1}^{n}(x_j - \bar{x})^2}\right] \quad \text{(D5.5.2b)}$$

If we repeat this exercise but apply the expectance operator directly to the original regression curve, $y_{pk} = a_1 + a_2 x_k$ (i.e., we do not make the substitution as shown in (D5.5.1a)), we get

$$\begin{aligned}
\mathrm{Var}[Y_{pi}] &= \mathrm{Var}[a_1 + a_2 x_i] \\
&= E[(a_1 + a_2 x_i)^2] - E[a_1 + a_2 x_i]^2 \\
&= E[a_1^2 + 2a_1 a_2 x_i + a_2^2 x_i^2] - E[a_1]^2 - 2x_i E[a_1]E[a_2] - E[a_2 x_i]^2 \\
&= E[a_1^2] + 2x_i E[a_1 a_2] + E[a_2^2 x_i^2] - E[a_1]^2 - 2x_i E[a_1]E[a_2] - E[a_2 x_i]^2 \\
&= E[a_1^2] - E[a_1]^2 + x_i^2 E[a_2^2] - x_i^2 E[a_2]^2 + 2x_i E[a_1 a_2] - 2x_i E[a_1]E[a_2] \\
&= \mathrm{Var}[a_1] + x_i^2 \mathrm{Var}[a_2] + 2x_i \mathrm{Cov}[a_1 a_2].
\end{aligned} \quad \text{(D5.5.3a)}$$

Now consider the covariance between a_1 and a_2,

$$\mathrm{Cov}[a_1 a_2] = E[(a_1 - E[a_1])(a_2 - E[a_2])] = \mathrm{Cov}\left[\bar{y} - a_2\bar{x}, \frac{s_{xy}}{s^2_x}\right] = -\bar{x}s^2_{a_2}, \quad \text{(D5.5.3b)}$$

so, continuing expanding (D5.5.3a), for $\mathrm{Var}[Y_{pi}]$,

$$= s^2_{a_1} + x_i^2 s^2_{a_2} - 2x_i\bar{x}s^2_{a_2} = s^2_{a_1} - 2\bar{x}s^2_{a_2} + s^2_{a_2}(x_i - \bar{x})^2, \quad \text{(D5.5.3c)}$$

which is seen by a bit of algebra.

Box D5.6 Uncertainty in the Regression Coefficients in a General Regression

Consider the uncertainty in the regression coefficients of a general curve fit (i.e., not restricted to the straight-line case). As was necessary for solving for the regression coefficients, here we must resort to matrix manipulations to determine the uncertainties.

In particular, we know that the least-squares fit regression coefficients, i.e., the estimates of the true coefficient values, are given by

$$\mathbf{x} = (\mathbf{A}^T\mathbf{A})^{-1}\mathbf{A}^T\mathbf{b}. \tag{D5.6.1}$$

We wish to find the uncertainty in the coefficients of vector $\mathbf{x}$, as given by its covariance matrix, Var[$\mathbf{x}$]. So,

$$\mathrm{Var}[\mathbf{x}] = \mathrm{Var}[(\mathbf{A}^T\mathbf{A})^{-1}\mathbf{A}^T\mathbf{b}]. \tag{D5.6.2}$$

The rules of manipulating a random matrix or product and sums of random matrices are given in Appendix 1, section A.8. There it is shown that (D5.6.2) can be reduced as follows. First, define

$$\mathbf{C} = (\mathbf{A}^T\mathbf{A})^{-1}\mathbf{A}^T, \tag{D5.6.3}$$

where $\mathbf{C}$ is not a random matrix. Then

$$\begin{aligned} \mathrm{Var}[\mathbf{x}] &= \mathrm{Var}[\mathbf{Cb}] \\ &= \mathbf{C}\mathrm{Var}[\mathbf{b}]\mathbf{C}^T \\ &= \mathbf{C}\boldsymbol{\Sigma}_\mathbf{b}\mathbf{C}^T \\ &= (\mathbf{A}^T\mathbf{A})^{-1}\mathbf{A}^T\boldsymbol{\Sigma}_\mathbf{b}\mathbf{A}(\mathbf{A}^T\mathbf{A})^{-1}, \end{aligned} \tag{D5.6.4}$$

where $\boldsymbol{\Sigma}_\mathbf{b}$ is the covariance matrix of the observations, the y_i, contained in the $\mathbf{b}$ vector. Thus the uncertainty in the regression coefficients, the a_i, in the general regression curve of (5.19), are given by (D5.6.4). Note that in this relationship, the final inverse term $(\mathbf{A}^T\mathbf{A})^{-1}$ is actually $(\mathbf{A}^T\mathbf{A})^{-T}$, but since $\mathbf{A}^T\mathbf{A}$ is a square symmetrical matrix, $\mathbf{N} = \mathbf{N}^T$ so $(\mathbf{A}^T\mathbf{A})^{-T} = (\mathbf{A}^T\mathbf{A})^{-1}$.

Box D5.7 Uncertainty in the Prediction of a Single Point in a General Regression

We compute the uncertainty in the predicted values of y for the generalized best-fit (in a least-squares sense) curve using the expectance operator on the vector products (see Appendix A1, section A.8 for matrix operation details).

First, consider the general form of the least-squares curve fit for an mth-order fit (using m constituent terms of a basis),

$$\mathrm{y_{pi}} = \mathrm{a_1}\varphi_1 + \mathrm{a_2}\varphi_2 + \mathrm{a_3}\varphi_3 + \cdots + \mathrm{a_m}\varphi_\mathrm{m}, \tag{D5.7.1}$$

Box D5.7 (Cont.)

which is written in vector form as

$$\hat{\mathbf{Y}} = \mathbf{A}\hat{\mathbf{x}}, \tag{D5.7.2}$$

where the vector $\hat{\mathbf{Y}}$ contains the y_{pi} (each row of $\hat{\mathbf{Y}}$ contains one value of y predicted at the point x_i), the **A** matrix contains the individual terms φ_i, evaluated at the points x_i (e.g., first row contains the m φ_i terms evaluated at x_1, second row contains φ_i evaluated at x_2, etc.) and the estimated coefficients, a_j yielding the best-fit curve are contained in $\hat{\mathbf{x}}$. So,

$$\begin{bmatrix} y_{p1} \\ y_{p2} \\ y_{p3} \\ \vdots \\ y_{pn} \end{bmatrix} = \begin{bmatrix} \varphi_1(x_1) & \varphi_2(x_1) & \dots & \varphi_m(x_1) \\ \varphi_1(x_2) & \varphi_2(x_2) & \dots & \varphi_m(x_2) \\ \varphi_1(x_3) & \varphi_2(x_3) & \dots & \varphi_m(x_3) \\ & & \vdots & \\ \varphi_1(x_n) & \varphi_2(x_n) & \dots & \varphi_m(x_n) \end{bmatrix} \begin{bmatrix} a_1 \\ a_2 \\ a_3 \\ \vdots \\ a_m \end{bmatrix} \tag{D5.7.3}$$

We desire to estimate the uncertainty in any predicted y_{pi} value contained in $\hat{\mathbf{Y}}$, so

$$\mathrm{Var}[\hat{\mathbf{Y}}] = \mathrm{Var}[\mathbf{Ax}] = \mathrm{E}[(\mathbf{Ax} - \mathrm{E}[\mathbf{Ax}])(\mathbf{Ax} - \mathrm{E}[\mathbf{Ax}])^{\mathrm{T}}]. \tag{D5.7.4a}$$

Or at an individual point k, so

$$\mathrm{Var}[\hat{\mathbf{Y}}_k] = \mathrm{Var}[a_k^{\mathrm{T}}\mathbf{x}] = \mathrm{E}[(a_k^{\mathrm{T}}\mathbf{x} - \mathrm{E}[a_k^{\mathrm{T}}\mathbf{x}])(a_k^{\mathrm{T}}\mathbf{x} - \mathrm{E}[a_k^{\mathrm{T}}\mathbf{x}])^{\mathrm{T}}], \tag{D5.7.4b}$$

where

$$a_k^{\mathrm{T}} = [\varphi_1(x_k), \varphi_2(x_k), \dots, \varphi_m(x_k)], \tag{D5.7.5}$$

that is, this is the row vector for the kth data point for which the value of y is being predicted (from matrix **A**).

The full form in (D5.7.4a) is reduced from (A1.74), since vector **A** is not a random matrix, so

$$= \mathbf{A}\boldsymbol{\Sigma}_{\mathbf{x}}\mathbf{A}^{\mathrm{T}}, \tag{D5.7.6a}$$

and the case for a single row, given in (D5.7.4b) reduces to

$$= \mathbf{a}_k^{\mathrm{T}}\boldsymbol{\Sigma}_{\mathbf{x}}\mathbf{a}_k, \tag{D5.7.6b}$$

where $\boldsymbol{\Sigma}_{\mathbf{x}}$ is the covariance matrix of the regression coefficients vector **x**,

$$\boldsymbol{\Sigma}_{\mathbf{x}} = \begin{bmatrix} \sigma_{11} & \sigma_{12} & \dots & \sigma_{1m} \\ \sigma_{21} & \sigma_{22} & \dots & \sigma_{2m} \\ \sigma_{31} & \sigma_{32} & \dots & \sigma_{3m} \\ & & \vdots & \\ \sigma_{m1} & \sigma_{m2} & \dots & \sigma_{mm} \end{bmatrix},$$

where variance σ_{ii} represents the variance of the a_i regression coefficient, which is estimated from $s_{a_i}^2$.

Box D5.7 (Cont.)

Now consider the case of a straight-line fit where $m = 2$; that is, there are only two constituent terms of a polynomial basis. So

$$\hat{\mathbf{Y}} = \mathbf{A}\hat{\mathbf{x}} \tag{D5.7.8a}$$

looks like

$$\begin{bmatrix} y_{p1} \\ y_{p2} \\ y_{p3} \\ \vdots \\ y_{pn} \end{bmatrix} = \begin{bmatrix} 1 & x_1 \\ 1 & x_2 \\ 1 & x_3 \\ \vdots & \\ 1 & x_n \end{bmatrix} \begin{bmatrix} a_1 \\ a_2 \end{bmatrix} \tag{D5.7.8b}$$

and (recall that the **A** matrix is not a random matrix but rather a deterministic matrix, so equation A1.78c applies).

$$\mathrm{Var}[\hat{\mathbf{Y}}] = \mathbf{A}\boldsymbol{\Sigma}_{\mathbf{x}}\mathbf{A}^{\mathrm{T}}. \tag{D5.7.9}$$

This requires an estimate of $\boldsymbol{\Sigma}_{\mathbf{x}}$, $\hat{\boldsymbol{\Sigma}}_{\mathbf{x}}$, given as

$$\hat{\boldsymbol{\Sigma}}_{\mathbf{x}} = \begin{bmatrix} s_{a_1}^2 & -\bar{x}s_{a_2}^2 \\ -\bar{x}s_{a_2}^2 & s_{a_2}^2 \end{bmatrix} \tag{D5.7.10}$$

where $s_{a_1}^2$ is given in (D5.4.8c) and $s_{a_2}^2$ is given in (D5.4.5), and the covariance terms given in (D5.5.3b).

Thus, (D5.7.9) is estimated via the following matrix operation:

$$\mathbf{A}\boldsymbol{\Sigma}_{\mathrm{x}}\mathbf{A}^{\mathrm{T}} = \begin{bmatrix} 1 & x_1 \\ 1 & x_2 \\ 1 & x_3 \\ \vdots & \\ 1 & x_n \end{bmatrix} \begin{bmatrix} s_{a_1}^2 & -\bar{x}s,_{a_2}^2 \\ -\bar{x}s_{a_2}^2 & s_{a_2}^2 \end{bmatrix} \begin{bmatrix} 1 & 1 & 1 & \cdots & 1 \\ x_1 & x_2 & x_3 & & x_n \end{bmatrix}$$

$$= \begin{bmatrix} s_{a_1}^2 & -x_1\bar{x}s_{a_2}^2 \\ s_{a_1}^2 & -x_2\bar{x}s_{a_2}^2 \\ s_{a_1}^2 & -x_3\bar{x}s_{a_2}^2 \\ \vdots & \\ s_{a_1}^2 & -x_n\bar{x}s_{a_2}^2 \end{bmatrix} \begin{bmatrix} 1 & 1 & 1 & \cdots & 1 \\ x_1 & x_2 & x_3 & & x_n \end{bmatrix}$$

$$= \begin{bmatrix} s_{a_1}^2 - x_1^2\bar{x}s_{a_2}^2 & s_{a_1}^2 - x_1x_2\bar{x}s_{a_2}^2 & s_{a_1}^2 - x_1x_3\bar{x}s_{a_2}^2 & \cdots & s_{a_1}^2 - x_1x_n\bar{x}s_{a_2}^2 \\ s_{a_1}^2 - x_2x_1\bar{x}s_{a_2}^2 & s_{a_1}^2 - x_2^2\bar{x}s_{a_2}^2 & s_{a_1}^2 - x_2x_3\bar{x}s_{a_2}^2 & \cdots & s_{a_1}^2 - x_2x_n\bar{x}s_{a_2}^2 \\ s_{a_1}^2 - x_3x_1\bar{x}s_{a_2}^2 & s_{a_1}^2 - x_3x_2\bar{x}s_{a_2}^2 & s_{a_1}^2 - x_3^2\bar{x}s_{a_2}^2 & \cdots & s_{a_1}^2 - x_3x_n\bar{x}s_{a_2}^2 \\ & & & \vdots & \\ s_{a_1}^2 - x_nx_1\bar{x}s_{a_2}^2 & s_{a_1}^2 - x_nx_2\bar{x}s_{a_2}^2 & s_{a_1}^2 - x_nx_3\bar{x}s_{a_2}^2 & \cdots & s_{a_1}^2 - x_n^2\bar{x}s_{a_2}^2 \end{bmatrix} \tag{D5.7.11}$$

Box D5.7 (Cont.)

This is the covariance matrix that shows the uncertainty in any one particular estimate of y, given on the diagonal elements, whereas it also shows the covariability between pairs of estimated y values, as given by the off-diagonal elements.

To see this more clearly, now consider examining the error in a predicted value of Y at, say, the kth point. In that case,[21]

$$\hat{Y}_k = a_k^{\mathrm{T}}\hat{\mathbf{x}}. \tag{D5.7.12a}$$

For a straight-line fit, this looks like

$$y_{pk} = [1 \quad x_k]\begin{bmatrix} a_1 \\ a_2 \end{bmatrix} \tag{D5.7.12b}$$

and

$$\mathrm{Var}[\hat{Y}_k] = a_k^{\mathrm{T}}\mathbf{\Sigma}_X a_k. \tag{D5.7.13}$$

This is estimated using the sample covariance matrix, $\hat{\mathbf{\Sigma}}_x$, from (D5.7.10), so that (D5.7.13) expands as

$$\begin{aligned} \mathrm{a}_k^{\mathrm{T}}\mathbf{\Sigma}_X a_k &= [1 \quad \mathrm{x}_k]\begin{bmatrix} s_{a_1}^2 & 0 \\ 0 & s_{a_2}^2 \end{bmatrix}\begin{bmatrix} 1 \\ \mathrm{x}_k \end{bmatrix} \\ &= \begin{bmatrix} s_{a_1}^2 & \mathrm{x}_k s_{a_2}^2 \end{bmatrix}\begin{bmatrix} 1 \\ \mathrm{x}_k \end{bmatrix} \\ &= s_{a_1}^2 + \mathrm{x}_k^2 s_{a_2}^2 \end{aligned} \tag{D5.7.14}$$

This is seen to be equivalent to any diagonal element in (D5.7.11), showing that those elements represent the uncertainty in any one estimated value of $\mathbf{Y}$. It is also equivalent to the previous uncertainty estimated for $\mathbf{Y}$ in (D5.4.3b), where the error was predicted by applying the expectance operator directly to the y_{pi} equation. However, (D5.7.11), derived here via matrix manipulation, gives the more general form of uncertainty in the fit for any linear regression curve.

[21] This seems inconsistent with equation (D5.8.12b), until you recall that for $\mathbf{Ax} = \mathbf{b}$, the $\mathbf{A}$ matrix contains the knowns (the x_i values) and the $\mathbf{x}$ vector contains the unknowns, the a_i coefficients.

5.7.1 Significance of Best-Fit Parameters

Given the assessment of the uncertainty in the optimal regression parameters, the a_1 and a_2 in the case of a straight-line fit, we can now evaluate whether the slope and/or intercepts are significantly different from 0.

Specifically, the regression parameters, obtained by the method of least squares, are normally distributed (shown in the next chapter) and thus we know that for a level of significance of 0.05,

$$P\{-1.96 \leq (a_2 - \beta_1)/s_{a_2} \leq 1.96\} = 95\%, \tag{5.37}$$

where β_1 is the true slope that we are attempting to estimate.

If we knew σ_{a_2} our test statistic would be the Z statistic, $(a_2 - \beta_1)/\sigma_{a_2}$, or a standardized form of the Gaussian. However, since we do not know the true σ_{a_2}, we must estimate it from our sample variance, s_{a_2}. This is okay, but as a consequence, we now use the t statistic, $(a_2 - \beta_1)/s_{a_2}$, which follows the t-distribution (see Chapter 3). This is a Gaussian-like distribution whose actual spread is dependent upon the number of degrees of freedom used to estimate the distribution, which in this case is $n - 2$ (we lose two degrees of freedom in estimating the mean and variance of t).

From a table of t-distributions, we determine to some level of significance the likelihood that the true slope of the data lies within some interval,

$$P\{\hat{\theta}_L \leq t \leq \hat{\theta}_U\} = 1 - \alpha, \tag{5.38}$$

where $\hat{\theta}_L$ and $\hat{\theta}_U$ are the values of the lower and upper limits of the t-distribution (for the number of degrees of freedom) defining the level of significance desired.

To gauge the likelihood that the estimated slope (or other parameter) is *not* significantly different from 0 (the *null hypothesis*), within some specified level of significance, we first find those limits of the t-distribution, given by $\hat{\theta}_L$ and $\hat{\theta}_U$, that contain the desired percentage of a t-distribution. Then, compute the *t* statistic:

$$t = \frac{(a_2 - 0)}{s_{a_2}} = \frac{a_2}{s_{a_2}}. \tag{5.39}$$

Finally, insert the result of (5.39) into (5.38). If t lies within the limits, then there is a 100 $(1 - \alpha)$ percent chance that the value of a_2 is not significantly different from 0. If t lies outside the specified range, then a_2 may still in fact be 0, but only 100α of 100 times on average will this prove to be the case.

For example, if you fit a straight line to 10 data points, then for $\alpha = 0.05$ (i.e., a 95 percent confidence interval) the t-distribution with 8 degrees of freedom shows that $\hat{\theta}_L = -2.306$ and $\hat{\theta}_U = 2.306$. If the value of a_2/s_{a2} falls between these two limits, then your slope is not significantly different from 0 at the 95 percent confidence interval. If it falls outside of these limits, then you will conclude that the slope is significantly different from zero at the 5 percent level of significance, meaning that 5 out of 100 times you may still expect to be incorrect in this conclusion simply due to the random noise in your 10 data points.

Alternatively, you can apply the bootstrap by repeatedly sampling from a noise distribution (consistent with the distribution of the noise about your best-fit line) and estimate the regression coefficients for each of these, then from the distribution of the regression coefficient values, and so determine the probability of getting the values you got for your data (i.e., what is the likelihood of getting a value as high as you did, when the true value was actually the mean values achieved from the noise PDF).

5.8 Assessing the Fit of the Best-Fit Curve

Assessing the fit of the best-fit curve involves two main steps:

1) determining if the curve itself is representative of the data
2) once the first criterion is satisfactorily achieved, assessing how well the curve describes the data and determining the degree to which the dependent variable can be estimated, given knowledge of the independent variable

5.8.1 Appropriateness of the Curve

In practice, the generally acceptable (and clearly the easiest) method for deciding if the curve fit to the data is appropriate is to plot the residuals and look for systematic variations within them. That is, if the fitted curve actually describes the data, then the residuals should be randomly distributed – they should display no systematic variation or shape. If the residuals do show a systematic variation, and they represent a nontrivial fraction of the total variance of the data, then, unless that variation can be clearly assigned to a component of the sample that is being treated as noise, a new curve should be used to better fit the data. In this case, one of three options is usually most useful:

1) If the curve fitting the data is constructed from the sum of one or more constituent functions of a basis (e.g., the first two terms of a polynomial, or of a sine basis), one can continue adding in higher-order terms of the basis until the residuals show random character or their fraction of the total variance becomes negligible (you must exercise caution here, since the rms $\Rightarrow 0$ as $m \Rightarrow n$ (we will deal with this later).
2) Choose a more appropriate curve (i.e., a different functional form, or different basis representation).
3) Fit the residuals themselves with another (different) curve. After fitting the residuals, the residuals from that fit should be examined and fit with another curve if any systematic variation is apparent in them. This procedure can be continued until the residuals of the residual-fits show random character or negligible variance. After the residuals, or residuals of residuals, etc., have been fit, the individual functions can be recombined to give a composite function fitting the data.

For example, if the data, $y(x)$, are made up of two composite curves, $f_1(x)$ and $f_2(x)$, so that $y(x) = f_1(x) + f_2(x)$, then, if the original curve fit captures $f_1(x)$, the residuals are as a result given as $y(x) - f_1(x) = f_2(x)$ (plus noise for each function) and the curve fit to the residuals should capture $f_2(x)$. This works fine in many cases, though care must be exercised if one or both of the functions are not linear.

Tukey suggests the latter approach, since it tends to build the fit piecemeal and may give insights into the underlying structure of the data. Methods of quantifying whether a set of residuals is randomly distributed or not will be presented in a later chapter dealing with time series analysis.

In addition to the "qualitative" assessment suggested above, Draper and Smith (1981) suggest several quantitative methods for determining if the curve chosen to fit the data is appropriate. In general, they all attempt to quantify whether there is systematic variation in the residuals (the errors between the observed values of y_i and the predicted values based on the best-fit curve), though it is not clear whether they are actually any better than the simple qualitative methods we have already discussed.

5.8.2 Quality of Curve Fit

Scatter about the Fitted Curve

Given randomly distributed residuals, one can derive some satisfaction in knowing that the best fit curve is consistent in shape with the distribution of the sample. The next step of assessing the fit of the curve deals with determining the degree of scatter in the data about the curve. That is, do the data lie tightly scattered about the curve, or are they widely distributed about the curve? Clearly, this scatter is similar in concept to variance, which asks the question: how are the data scattered about the *mean* of a distribution? In this case, however, the question is a slight variant of the form: how are the data scattered about the *curve*, instead of a fixed mean?

This question is answered through the concept of **residual variance**, or its root, the **root mean squared error (rms)**, which is the means of defining this scatter about a curve. The residual variance is estimated by

$$s_{\rm rms}^2 = \frac{1}{\nu}\sum_{i=1}^{n}\left[y_i - y_{pi}\right]^2, \tag{5.40}$$

where y_{pi} are the values of y at each position x_i, predicted by the fitted curve, and ν is the number of degrees of freedom (i.e., the number of data points minus the number of parameters that were estimated in order to compute the best-fit curve).

This value, $s_{\rm rms}^2$, is seen to be similar in form to variance, though it is defined about the fitted curve instead of the mean. For the case of $y_{pi} = \bar{y}$, (5.40) reduces precisely to the definition of sample variance (where $\nu = n - 1$, since one degree of freedom has been lost in estimating the mean). In actuality, it is an estimate of the variance of the error, ε, depicted as σ_ε^2, which is an estimate of the scatter in Y given knowledge of X.

As with sample variance, here we divide by the degrees of freedom to produce an unbiased estimate of the spread about the curve. For the case of a straight-line fit to the data, the n data points are used to estimate a slope and intercept (i.e., two curve parameters, a_1 and a_2) so the number of degrees of freedom is $n - 2$.

As with standard deviation, the rms error, σ_ε or s_{rms} (the root of the residual variance), is in the same units as the original data and is thus convenient to use. The smaller the rms scatter, the less the scatter and the "tighter" the fit. For a straight-line fit, one could consider the ratio, in the form of a signal-to-noise ratio, of the rms scatter to the total range in Y, or the range of Y that defines the size of the "signal" of interest. For other curve shapes, one might choose some other appropriate scale defining the magnitude of "signal" described by the fitted curve (say the amplitude of a sinusoid fit to the data) and compare that to the rms scatter to assess the relative magnitude of the scatter in the data about the signal described by the fitted curve. However, if one adds enough terms to the curve being fit (e.g., applying an nth order polynomial to n data points, the fit will be exact and the rms error zero. It is important to recognize a trade-off whereby the more terms you add into the fitting curve, the more likely you are to begin fitting the noise in the data. Typically this trade-off is evaluated through an index where you are rewarded for decreasing the rms value of the fitted curve, but penalized for adding more constituent terms of a basis function. We will evaluate such indices explicitly later, in a variety of forms.

Now consider our previous example of determining if Earth's climate shows two distinct stable modes (glacial and interglacial) or some continuum. We previously estimated this by examining the PDF of the LR04 5.4 MY climate record, which showed hints of bimodality, but some of that was possibly reflecting the fact that the climate prior to 1 MY seemed overall to have been warmer in general. Thus, we imagine that the background climate state may have changed, and we wish to eliminate any trend in the curve to see if the variability about this background is bimodal. For this, we fit a curve to LR04 in an attempt to identify and then remove any trend. Visual inspection of the time series suggests that the series may show a simple linear trend, but it tends to flatten back in time, perhaps suggesting a simple quadratic. We fit both to the series and then examine the rms and residuals to see if one is obviously better than the other, as shown in Figure 5.12.

For this example, both fits show little systematic variability in the residuals until just prior to 5 MY ago, where the residuals shift to higher values systematically. As for quantitative measures, both show a similar degree of rms fit. These are similar enough that it is not clear if one is more justified than the other.

Regarding the PMFs of the de-trended series, these are shown in Figure 5.13.

So, we now see that both de-trended series are nearly unimodal, but both do contain a small second mode in the lowest values, perhaps reflecting the "hump" not removed from the trends near the oldest portion of the series. So we now face the interesting task of either deciding that the small warm mode is real or going to greater lengths to fit a trend that yields a residual with a mean of zero, and no systematic variation from that over some portion of time. Of course, given that we still have a considerable box of tools yet to evaluate, we will continue applying new tools as they become available (and appear

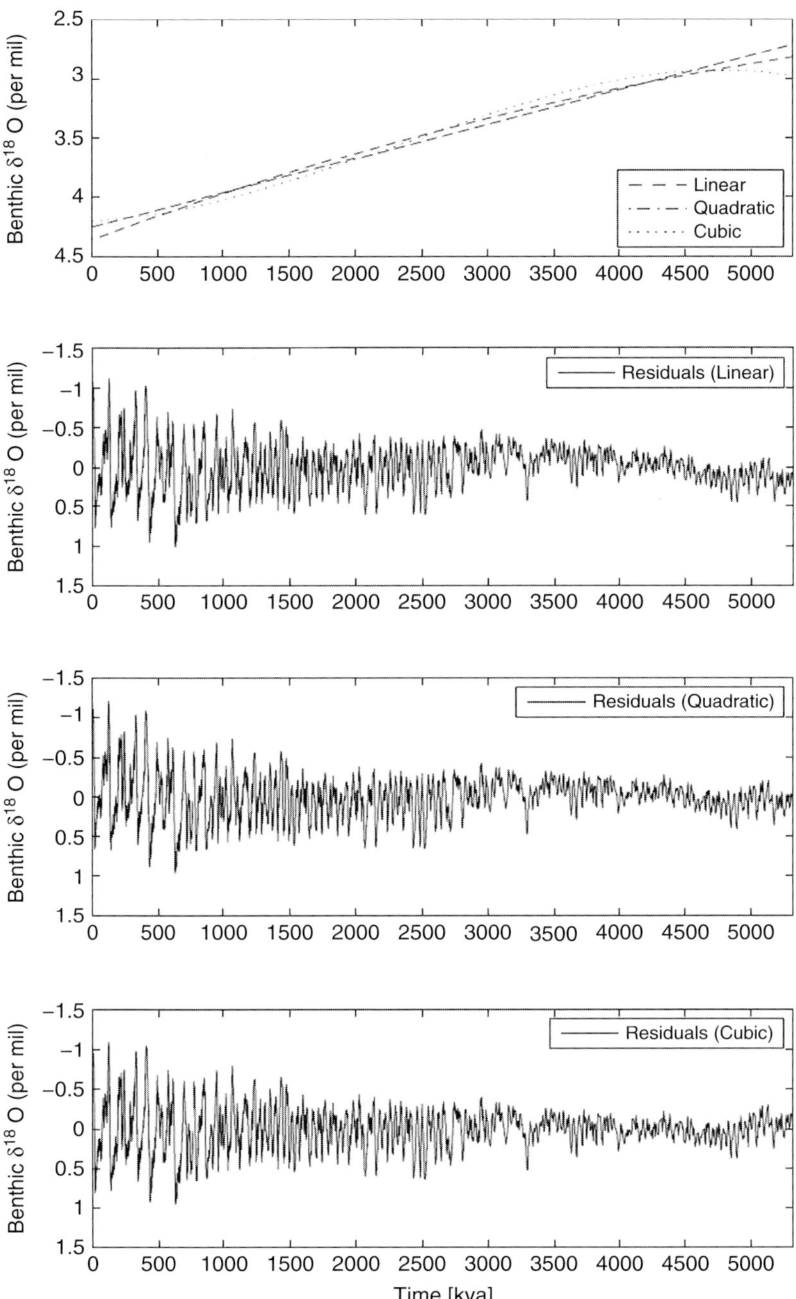

Figure 5.12 Fit of straight-line, quadratic and cubic modes to the LR04 time series, and their residuals about those best fits.

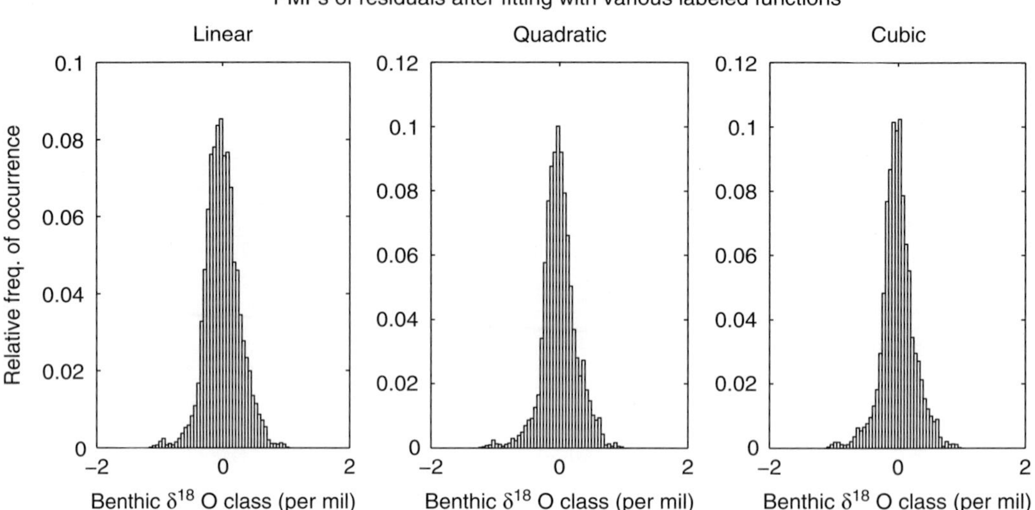

Figure 5.13 PMFs of the residual series in Figure 5.11.

relevant to this problem). Then, based on all of the analyses, we will make a decision regarding the bimodality of this series.

5.9 Take-Home Points

1. Smoothed-curve fitting requires knowledge of where the error lies in the data (in the y values or the *x* values, or both). The first two follow the same procedure, simply swapping *x* for *y*; the third requires orthogonal curve fitting.
2. If the data have a Gaussian distribution, then the method of least squares produces the "best" fit, as defined by providing a curve with the minimal variance (in form of rms) about the fit line. This fit is easily produced via matrix methods using $\mathbf{Ax} = \mathbf{b}$; $\mathbf{x} = (\mathbf{A}^T\mathbf{A})^{-1}\mathbf{A}^T\mathbf{b}$; where $(\mathbf{A}^T\mathbf{A})^{-1}\mathbf{A}^T = \mathbf{C}$, the pseudo inverse.

5.10 Questions

Pencil and Paper Questions

1. Show the matrix solution for the following problems (using $\mathbf{Ax} = \mathbf{b}$) and the contents of **A, x** and **b** for a quadratic polynomial:

   a. least squares
   b. uncertainty of optimal coefficients
   c. uncertainty of estimated y values using optimal coefficients

Computer-Based Questions

2. For *x*, *y* observations (1, 1), (3, 2), (5, 4), (7, 6), (9, 12),

   a. Compute the slope and intercept of the best-fit straight line (assume observations are normally distributed, but state why this matters).

b. Compute the uncertainty in the optimal fit ($\mathbf{x}$), via $\mathrm{Var}(\mathbf{X}) = \mathrm{Var}(\mathbf{Cb}) = \mathbf{C\Sigma_b C^T}$, where the diagonal elements are the variance of the same row $\mathbf{x}$ elements, and the off-diagonal elements are the covariance between the various $\mathbf{x}$ elements.
c. What is the covariance between the various $\mathbf{x}$ elements if the basis being fit is orthogonal?
d. Compute the uncertainty in the prediction of a single point, k, via the regression (i.e., via $\hat{\mathbf{Y}} = \mathbf{Ax}$) as $\mathrm{Var}[\hat{\mathbf{Y}}_\mathrm{k}] = \mathbf{a}_k^T \mathbf{\Sigma_x a}_k$.
e. Assess the best-fit curve by examining nature of the residuals about the curve. Ideally those residuals will show no systematic pattern, but rather white noise (suggesting you have successfully captured the signal, leaving only the noise). If residuals do show a pattern, fit that pattern, and keep doing this until the residuals do not show any pattern. Then combine the various fits into a single function.

6 Special Curve Fitting

6.1 Overview

As discussed in the previous chapter, the best fit of a smooth curve to a data set is dependent upon how the error between the observations and fitted curve is defined *and* the nature of the data being fit. This latter condition resulted in four different types of smoothed curve fitting: (1) *standard fits*, when all data points have a comparable precision or error; (2) *weighted fits*, when the data points have different degrees of precision or errors; (3) *constrained fits*, when specific conditions must be satisfied by the fit curve; and (4) *robust fits*, when wishing to moderate the influence of one or more data points that do not lie within the general vicinity of the other points.

Standard curve fitting, using the method of least squares, was presented in the previous chapter. This chapter presents the appropriate technique for the other three types of curve fit (weighted, constrained and robust).

6.2 Weighted Curve Fits

Conceptually, the approach to be used for weighted curve fitting can be developed through consideration of standard fitting as follows. Consider the situation where some observations are considered or known to be more reliable than others (i.e., different precisions for different observations). In this case it may be desirable to weight those well-known observations more heavily in the calculation of a best-fit line. If, for example, the precision of the third observation is twice as good as that of all the other points, you might consider adding a "weight" to the corresponding error at that location by some amount, say,

$$\mathbf{e} = \begin{bmatrix} \varepsilon_1 \\ \varepsilon_2 \\ 2\varepsilon_3 \\ \vdots \\ \varepsilon_n \end{bmatrix}. \tag{6.1}$$

This weight leads to an effectively larger error at the third point (the squared error is four times bigger when fitting the curve by the method of least squares). Consequently, this requires that the fitted curve be much closer to this observational point relative to the others in order to reduce its contribution to the total error.

This provides a conceptual idea of how to weight the fit, but not a precise manner as to how to weight it. For example, in the above case, do we want ε_3 increased by a factor of 2 or by $2^{1/2}$. Increasing it by 2 is equivalent to having two data points fall at this exact same location, so that when the system of equations is written out the two equations contribute twice the error at this one location of x. On the other hand, increasing it by $2^{1/2}$ means that when the curve minimizes the sum of the *squared* errors, this data point will contribute exactly twice as much to the curve fit as the other data points.

Principle of Maximum Likelihood (ML)

The principle of maximum likelihood (ML) reveals how to quantify the weights. First, consider the case in which *the scatter associated with each data point follows a Gaussian distribution*. Such a situation arises naturally when each of the y_i represents the mean of several observations. As indicated by the Central Limit Theorem, we know the observations (means) are Normal, $\mathbf{N}(\mu_i, \sigma_i^2/n_i)$, where the n_i are the number of observations used in the average for the y_i at each x_i location.

So, if the *true* relationship between *Y* and x is described by a straight line, the scatter, in a single *Y* value, is shown in Figure 6.1.

Or, for multiple data points (each being a mean value), each point displays a Gaussian distribution (as in Figure 6.2), and the Principle of Maximum Likelihood uses that information to find the best fit to those distributions.

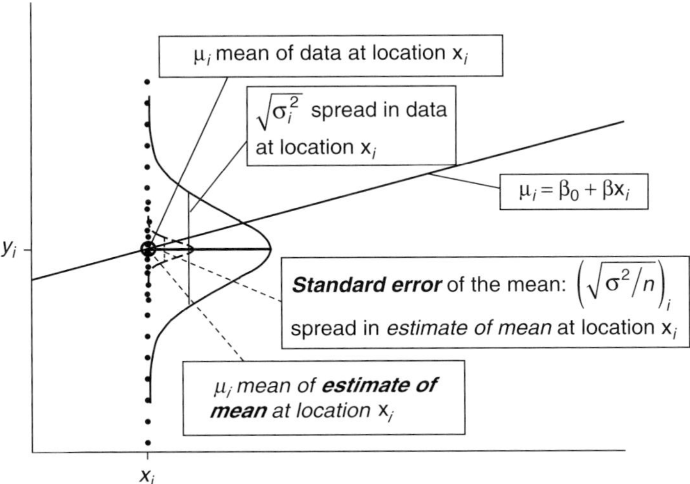

Figure 6.1 Ideal case, where enough data result in the mean lying on the true mean. Important here is to see how the spread in the data (σ) is not the spread in the estimate of the mean, which is reduced by $n^{-1/2}$ due to the Central Limit Theorem. It is this latter spread, the standard error of the mean, that defines the precision of the estimate of the mean used for establishing the weights in a weighted least-squares fit.

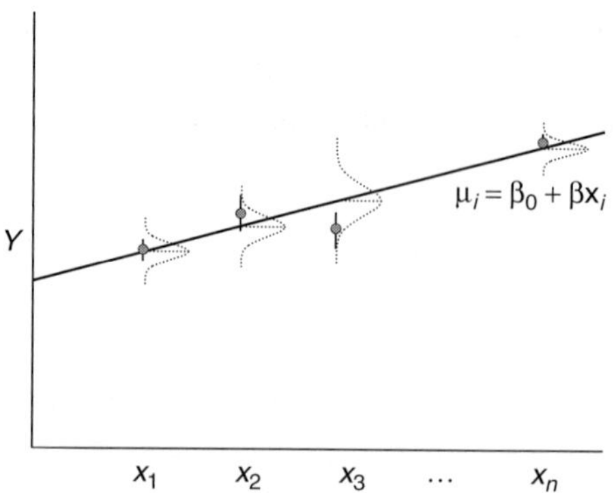

Figure 6.2 Four sample data points, each being the average of some number of values collected at the same x location (or time). This makes each datum, via the Central Limit Theorem, show a Gaussian distribution with spread σ^2/n. This is the information the principle of maximum likelihood uses to formulate the method of least squares for a best-fit curve.

In other words, there is a normal distribution associated with each (y_i, x_i) position on the line that describes the probability of sampling any single observational value, y_i, at that x_i position. The mean of each normal distribution is given by the true (population) line described by

$$\mu_i = \beta_0 + \beta_1 x_i, \tag{6.2}$$

where the β_j terms represent the *true* population parameters of the linear slope, which in turn describes how the mean of Y changes with the values of x.

The likelihood of drawing the n observations that we actually drew is given by the likelihood function, $L(y_i)$, involving the Gaussian PDF, as before,[1]

$$L(y_i) = \sum_{i=1}^{n} \left(\frac{1}{\sigma_i \sqrt{2\pi}} \right) e^{\dfrac{-(y_i - \hat{\mu}_i)^2}{2\sigma_i^2}}, \tag{6.3a}$$

where $\hat{\mu}_i$ represent the best estimates of the mean at each x_i position. As before, this is rewritten by moving the exponential outside of the product and summing the exponent over the data points:

$$L(y_i) = \left[\prod_{i=1}^{n} \left(\frac{1}{\sigma_i \sqrt{2\pi}} \right) \right] e^{-\dfrac{1}{2} \sum_{i=1}^{n} \dfrac{(y_i - \hat{\mu}_i)^2}{2\sigma_i^2}}. \tag{6.3b}$$

Examination of (6.3) shows it to be of the same fundamental form used previously when determining the optimal manner with which to combine observations to estimate the

[1] Chapter 3, on statistics.

mean and variance of a sample drawn from a normal population. The only difference here is that *the true mean, μ, varies as a function of x*, as given by (6.2).

Substituting the functional relationship for μ_i given in (6.2) into (6.3b) for μ_i gives

$$\mathrm{L}(y_i) = \left[\prod_{i=1}^{n}\left(\frac{1}{\sigma_i\sqrt{2\pi}}\right)\right] \mathrm{e}^{-\frac{1}{2}\sum_{i=1}^{n}\frac{[y_i-(\hat{\beta}_0+\hat{\beta}_1 \mathrm{x}_i)]^2}{2\sigma_i^2}}, \tag{6.4}$$

where the $\hat{\beta}_j$ represent the best estimates for the true parameters: intercept, $\beta_{0\,(j=0)}$, and slope, $\beta_1\,(j=1)$.

You need to solve for those values of $\hat{\beta}_j$ that produce the highest likelihood of sampling the values of y_i that we actually sampled. In other words, we assume that our observations are representative of the population from which they were drawn (hopefully, the target population).

The highest likelihood, i.e., the maximum of $\mathrm{L}(y_i)$, occurs where the exponent sum in (6.4) is minimum. So you must find those $\hat{\beta}_j$ that minimize the two normal equations (j =1,2):

$$\sum_{i=1}^{n}\frac{d}{d\hat{\beta}_j}\left[\frac{y_i-(\hat{\beta}_0+\hat{\beta}_1 \mathrm{x}_i)}{\sigma_i}\right]^2. \tag{6.5}$$

Or, for the more general case of any form of curve (in the earlier notation the unknown parameters, the $\hat{\beta}_j$, are designated by a_j), you want those coefficient values that satisfy these m normal equations (one equation for each $\hat{\beta}_j$, $j = 1,2, \ldots m$),

$$\sum_{i=1}^{n}\frac{d}{d\hat{\beta}_j}\left[\frac{y_i-(\hat{\beta}_0\varphi_{i0}+\hat{\beta}_1\varphi_{i1}+\ldots+\hat{\beta}_m\varphi_{im})}{\sigma_i}\right]^2 = 0, \tag{6.6}$$

where $\varphi_{ij} = \varphi_j(\mathrm{x}_i)$ in our previous notation, indicating the jth constituent term of the basis function evaluated at the point x_i.

Computing the derivatives with respect to the $\hat{\beta}_j$ coefficients and setting them equal to zero gives the normal equations

$$\sum_{i=1}^{n}\left\{\left[\frac{y_i-(\hat{\beta}_0\varphi_{i0}+\hat{\beta}_1\varphi_{i1}+\ldots+\hat{\beta}_m\varphi_{im})}{\sigma_i}\right]\frac{\varphi_{ij}}{\sigma_i}\right\} = 0, \tag{6.7a}$$

or, rearranged,

$$\hat{\beta}_0\sum_{i=1}^{n}\frac{\varphi_{i0}\varphi_{ij}}{\sigma_i^2}+\hat{\beta}_1\sum_{i=1}^{n}\frac{\varphi_{i1}\varphi_{ij}}{\sigma_i^2}+\ldots+\hat{\beta}_m\sum_{i=1}^{n}\frac{\varphi_{im}\varphi_{ij}}{\sigma_i^2} = \sum_{i=1}^{n}\frac{y_i\varphi_{ij}}{\sigma_i^2}, \tag{6.7b}$$

giving one equation for each j (of m) unknown $\hat{\beta}_j$.

Equation (6.7) is identical to the general form of the normal equations given in (D5.2.2), with the difference that *each sum here is divided by the different variances* (σ_i^2, estimated by s_i^2). *If the* σ_i *are the same for all data points, they cancel out in* (6.7) *and the above equation is identical to the previous normal equations for the standard least squares problem (where all the precisions were equal to begin with).*

6.2.1 Implications of the Method of Least Squares

The application of the principle of maximum likelihood to the curve-fitting problem reveals the following: *when we minimize the rms error about a smooth fitted curve employing the method of least squares, we are actually assuming that the scatter in the data obeys a Gaussian distribution with the same variance throughout the sample. If the data points obey a Gaussian distribution but with different precisions (different variances), then you minimize by scaling each sum in the standard normal equations by* $1/\sigma_i^2$*, as revealed by* (6.7b). *If the data have scatter (noise) that follows some other distribution, then minimize the error with respect to the appropriate distribution via the principle of maximum likelihood, as done here.*

The principle of maximum likelihood also gives a more conceptual indication of exactly what we are doing when we perform a least-squares fit (weighted or not). Specifically, it is clear that *we are attempting to determine those coefficients for that functional form that describes how the mean of Y varies as a function of* x *most consistent with the data themselves (i.e., if your data have a spread from 5 to 13, you know the mean is not going to be 20, and for Maximum Likelihood Estimators (MLE) you are finding that mean most consistent with the values you sampled)*. From this perspective, it is clear that *the root mean square error,* σ_ε*, is actually a measure of the spread in Y about its changing mean*. In other words, once we subtract out the fitted curve, we have effectively normalized *Y* so that its spread (about that curve) reflects its true distribution about its mean. Thus, rms is essentially a measure of the standard deviation once we have corrected for the complication that the mean of *Y* varies with x. In essence, rms represents the scatter in *Y*, while the fitted curve is a measure of how the mean varies with x.

6.2.2 Matrix Form of Weighted Fits

The actual form of the weights and practical implementation aspects for weighted curve fitting becomes obvious in matrix form.

We wish to minimize

$$e = \sum_{i=1}^{n} (w_i \varepsilon_i)^2, \tag{6.8}$$

where each of the individual errors, ε_i, is to be weighted by w_i. Rewriting ε_i explicitly in terms of the fitted curve and data points,

$$\mathrm{e} = \sum_{i=1}^{n} \left\{ w_i \left[y_i - (a_1\varphi_{i1} + a_2\varphi_{i2} + \ldots + a_m\varphi_{im}) \right] \right\}^2. \tag{6.9}$$

Equation (6.9) can be written in matrix notation as

$$\mathrm{e} = (\mathbf{we})^{\mathrm{T}}(\mathbf{we}) = \mathbf{e}^{\mathrm{T}}\mathbf{w}^{\mathrm{T}}\mathbf{we}, \tag{6.10}$$

where

$$\mathbf{e} = \begin{bmatrix} \varepsilon_1 \\ \varepsilon_2 \\ \varepsilon_3 \\ \vdots \\ \varepsilon_n \end{bmatrix} = \begin{bmatrix} y_1 - (a_1\varphi_{11} + a_1\varphi_{12} + \ldots + a_m\varphi_{1m}) \\ y_2 - (a_1\varphi_{21} + a_2\varphi_{22} + \ldots + a_m\varphi_{2m}) \\ y_3 - (a_1\varphi_{31} + a_2\varphi_{32} + \ldots + a_m\varphi_{3m}) \\ \vdots \\ y_n - (a_1\varphi_{n1} + a_2\varphi_{n2} + \ldots + a_m\varphi_{nm}) \end{bmatrix} \tag{6.11a}$$

and

$$\mathbf{w} = \mathbf{w}^T = \begin{bmatrix} 1/\sigma_1 & & & \\ & 1/\sigma_2 & & \\ & & \ddots & \\ & & & 1/\sigma_n \end{bmatrix}. \tag{6.11b}$$

Pre-multiplication by a diagonal matrix scales the *rows* of a matrix or vector, and the transpose of a diagonal matrix is equal to its non-transposed form (i.e., $\mathbf{w}^{\mathrm{T}} = \mathbf{w}$). The individual weights have been square-rooted in anticipation of the squaring operation in the product $\mathbf{w}^{\mathrm{T}}\mathbf{w}$, later. That is, the individual weights populating the weighting vector, $\mathbf{w}_i^{1/2} = 1/\sigma_i$, or one over the standard deviations (precisions) of the data, σ_i, estimated by s_i in most cases.

The definition of the individual errors or mismatch between observations and curve has not changed relative to the standard least-squares approach. As before, we substitute $\mathbf{Ax} - \mathbf{b}$ for $\mathbf{e}$ in (6.10), giving

$$\mathbf{e} = (\mathbf{A}\mathrm{x} - \mathbf{b})^{\mathrm{T}}\mathbf{w}^{\mathrm{T}}\mathbf{w}(\mathbf{A}\mathrm{x} - \mathbf{b}), \tag{6.12a}$$

or

$$\mathbf{e} = (\mathbf{A}\mathrm{x} - \mathbf{b})^{\mathrm{T}}\mathbf{W}(\mathbf{A}\mathrm{x} - \mathbf{b}), \tag{6.12b}$$

where $\mathbf{W} = \mathbf{w}^{\mathrm{T}}\mathbf{w}$, which is a diagonal matrix in which the elements are now $1/\sigma_{\mathrm{i}}^2$ (as opposed to $1/\sigma_i$).

You can solve for the $\mathbf{x}$ vector (the unknown a_j) by taking the derivative of e with respect to the a_j and setting these equal to 0:

$$\frac{\partial \mathbf{e}(\mathrm{x})}{\partial \mathrm{x}} = \frac{\partial}{\partial \mathrm{x}}\left[(\mathbf{A}\mathrm{x} - \mathbf{b})^{\mathrm{T}}\mathbf{W}(\mathbf{A}\mathrm{x} - \mathbf{b})\right] = 0. \tag{6.13}$$

The solution procedure follows an identical path to that taken for the matrix solution of the standard least-squares solution (see boxed details following), with the additional

$\mathbf{W}$ matrix carried along in this case. As before, $\mathbf{A}^T\mathbf{W}\mathbf{A}$ is square and of full rank (it is the standard normal equation, with each sum divided by the squared weights, σ_i^2). Therefore, this matrix equation is easily solved in the usual manner, giving

$$[\mathbf{A}^T\mathbf{W}\mathbf{A}]^{-1}\mathbf{A}^T\mathbf{W}\mathbf{A}\mathbf{x} = [\mathbf{A}^T\mathbf{W}\mathbf{A}]^{-1}\mathbf{A}^T\mathbf{W}\mathbf{b}, \tag{6.14}$$

so

$$\mathbf{x} = [\mathbf{A}^T\mathbf{W}\mathbf{A}]^{-1}\mathbf{A}^T\mathbf{W}\mathbf{b}, \tag{6.15a}$$

or

$$\mathbf{x} = [\mathbf{A}^T\mathbf{w}^T\mathbf{w}\mathbf{A}]^{-1}\mathbf{A}^T\mathbf{w}^T\mathbf{w}\mathbf{b}. \tag{6.15b}$$

This gives $\mathbf{x}$, whose elements, the $\hat{\beta}_j$ or a_j previously, provide the weighted least-squares fit coefficients to the n data points. Form (6.15b) shows that this solution is essentially identical to the general least-squares solution, with the exception that here the $\mathbf{A}$ matrix and $\mathbf{b}$ column vectors are pre-multiplied by the diagonal matrix containing the square root of the weights ($1/\sigma_i$) for each data point, so instead of $\mathbf{Ax} = \mathbf{b}$, we start with $\mathbf{wAx} = \mathbf{wb}$. That is, *each element in each row of A and b is divided by its precision (standard deviation) or multiplied by the appropriate* $1/\sigma_i$ *(or the estimates* $1/s_i$*). Then, standard curve fitting is employed to give the curve that minimizes the rms error, weighted according to the different precisions of the various data points. This even works for weighted orthogonal curve fitting.*

Uncertainty in Weighted Best-Fit Parameters

Consistent with unweighted fits, the error in the best fit coefficients (in the $\mathbf{x}$ vector) are given as

$$\begin{aligned} \mathrm{Var}[x] &= \mathrm{Var}[(\mathbf{A}^T\mathbf{w}^T\mathbf{w}\mathbf{A})^{-1}\mathbf{A}^T\mathbf{w}^T\mathbf{w}\mathbf{b}] && (6.16a)\\ &= \mathrm{Var}[\mathbf{Cb}], && (6.16b) \end{aligned}$$

where the pseudo-inverse, $\mathbf{C} = (\mathbf{A}^T\mathbf{w}^T\mathbf{w}\mathbf{A})^{-1}\mathbf{A}^T\mathbf{w}^T\mathbf{w}$, is not a random matrix. Then

$$\begin{aligned} \mathrm{Var}[x] &= \mathbf{C}\mathrm{Var}[\mathbf{b}]\mathbf{C}^T\\ &= \mathbf{C}\Sigma_\mathbf{b}\mathbf{C}^T\\ &= [\mathbf{A}^T\mathbf{w}^T\mathbf{w}\mathbf{A}]^{-1}\mathbf{A}^T\mathbf{w}^T\mathbf{w}\Sigma_\mathbf{b}\mathbf{w}^T\mathbf{w}\mathbf{A}[\mathbf{A}^T\mathbf{w}^T\mathbf{w}\mathbf{A}]^{-1}, \end{aligned} \tag{6.16c}$$

where $\mathbf{\Sigma_b}$ is the covariance matrix of the observations, the y_i, contained in the $\mathbf{b}$ vector. In the above relationship, the final inverse term $(\mathbf{A}^T\mathbf{w}^T\mathbf{w}\mathbf{A})^{-1}$ is actually $(\mathbf{A}^T\mathbf{w}^T\mathbf{w}\mathbf{A})^{-T}$, but since $\mathbf{A}^T\mathbf{w}^T\mathbf{w}\mathbf{A}$ is square symmetrical, $\mathbf{N} = \mathbf{N}^T$ so $(\mathbf{A}^T\mathbf{w}^T\mathbf{w}\mathbf{A})^{-T} = (\mathbf{A}^T\mathbf{w}^T\mathbf{w}\mathbf{A})^{-1}$. If all rows, i, in $\mathbf{A}$ and $\mathbf{b}$ are multiplied by $1/s_i$ prior to solving, then the solution to this weighted fit and the uncertainty in the optimum coefficients, a_j, is the same as for the standard fit.

Some Precautions when Using Weighted Fits

Some care must be used with weighted curve fits. For example, consider fitting a curve to a set of points made from two different data sets representing measurements made at Lab1 and Lab2. The errors reported from Lab1 represent only instrument precision (or precision from replication of analyses), while Lab2 reports errors reflecting instrument precision as well as the error due to all other sources (which are also present in the Lab1 data, but are not reported). In this case, the precisions reported for the Lab1 data points should not be weighted in the same manner as those from Lab2, since they represent two different quantities. In other words, if a weighting scheme is to be used, the errors reported for each observation (or y_i value) should be representative of the same processes responsible for introducing the noise.

Also, there are some circumstances in which a weighted fit is inconsistent with the nature of the errors and, say, a constrained fit (discussed below). For example, consider a data set in which the errors are correlated with the values x_i, such that the smaller the x_i, the smaller the errors (typical of measurements in which the measuring instrument shows degradation in precision with larger values of the variable). If these weights are used with a constraint that forces the curve to pass through zero, the data values at large x_i will make little or no contribution to the curve fit, since the fit is heavily weighted by the smaller errors near $x_i = 0$ and the constraint that the curve must pass through zero. In some instances, the constraint or errors may not be appropriate; consequently, considerable care must be taken to understand the implications of such a fit. The easiest way to build this understanding is through experimentation – try a number of different types of fits and carefully consider the implications of the differences between different fits. Alternatively, one can reproduce the system shown earlier, but consider the implications when the errors are correlated to x_i.

6.3 Constrained Fits

Consider the case where you desire to fit a curve to a set of data for which some data points are known exactly – for example, the situation where we have measured one or more points with nearly perfect precision using some extremely high-precision instrument, or the situation where we know from an understanding of the phenomenon that the dependent variable must be equal to some specific value when the independent variable takes on some special value. A common case occurs when the dependent variable is zero and the independent one is also zero (e.g., there is no surface stress when the winds are zero).[2] In these situations, we know that the fitted curve must pass through the special points. More generally, *this is solving an overdetermined set of equations subject to specified* **equality constraints**.

[2] Draper and Smith (1981) discuss this special case of curve fitting through the origin (constraining a curve to pass through zero) a fair amount – they and others suggest not doing it, but instead sticking with standard curve fitting, allowing for the intercept term to be included.

One way to achieve the desired fit is to include the constraint equations as extra rows in the $\mathbf{Ax} = \mathbf{b}$ equation and then apply very large weights to the constraint equations. The errors of the constraints are therefore forced to almost zero at the expense of larger errors elsewhere. That is, the local errors at the constraint positions must be reduced to very small values, since multiplying the large weights times the errors artificially elevates their influence on the total error. This approach, while seemingly inelegant, is completely legitimate and is consistent with the principle of maximum likelihood.[3] That is, if a point is to be fit exactly, it is equivalent to assigning $\sigma_i = 0$ (perfect precision). This corresponds to a weight of $1/\sigma_i^2 = \infty$. Of course, an infinite weight produces an exact fit, since only an error of zero contributes less than an infinite error to the total error.

6.3.1 Solution via Substitution

For simple cases, it may be possible to rewrite the original problem to a basic form that includes the effects of the equality constraints.

Consider the problem

$$\mathrm{f}(a_0,\ a_1,\ a_2) \rightarrow \text{minimized}, \tag{6.17a}$$

where the function is typically the sum of the squared errors, which is a function of the unknown curve coefficients, a_0, a_1, a_2. This function is now subject to the *equality* constraints:

$$\mathrm{g}(a_0,\ a_1,\ a_2) = 0 \tag{6.17b}$$

$$\mathrm{h}(a_0,\ a_1,\ a_2) = 0. \tag{6.17c}$$

(*In*equality constraints involve: $<$, $>$, $\leq$, or $\geq$ in place of $=$; they are typically more difficult to treat and are not covered here.)

One obvious approach would be use (6.16b) and (6.16c) to eliminate two of the variables from (6.16a) and then minimize the new function of only one variable subject to no constraints.

Constrained Straight-Line Fit

Consider the specific case of fitting a straight line to a set of n data points with the constraint that the curve must pass exactly through a point at $(\mathrm{x}_a, \mathrm{y}_a)$. For example, if the line must pass through the pair (4,10), then the best fit curve must pass exactly through the points $\mathrm{x}_a = 4$ and $\mathrm{y}_a = 10$.

Using the above notation in (6.16) for the straight line, $y_i = a_0 + a_1 \mathrm{x}_i$,

[3] Note that when using "canned" computer programs, one can have problems in the assessment of uncertainties. If, for example, the weights are added by replicating the point for which the curve is desired to fit exactly, then the associated confidence intervals computed will not be consistent with those for the actual data. This is because the program will not realize that the added rows are artificial and do not represent real observations.

$$\mathrm{f}(a_0,\ a_1) = \mathrm{e} = \sum_{i=1}^{n}[y_i - (a_0 + a_1\mathrm{x}_i)]^2 \tag{6.18a}$$

$$\mathrm{g}(a_0,\ a_1) = y_a = a_0 + a_1\mathrm{x}_a, \tag{6.18b}$$

where $\mathrm{g}(a_0, a_1)$ is the constraint stating that, at position $\mathrm{x}_i = \mathrm{x}_a$, the value of y_i must equal y_a. That is, the curve must pass through the pair (x_a, y_a).

There is a fundamental difference between these two equations, (6.17) and (6.18), since the latter, the constraint, is a constant, not a variable. That is, (6.18) states how to compute the one and only value of the constant, y_a, given its relationship to the specific point of x, $\mathrm{x_a}$.

The facts that there are only two variables and that they are related by simple linear relationships through (6.17) and (6.18) make this a simple system that lends itself to solution by substitution. Specifically, we can rearrange (6.18) to solve for a_0 in terms of a_1:

$$a_0 = y_a - a_1\mathrm{x}_a, \tag{6.19}$$

which is substituted into (6.17) and rearranged to give

$$\mathrm{e} = \sum_{i=1}^{n}[y_i - y_a - a_1(\mathrm{x}_i - \mathrm{x}_a)]^2. \tag{6.20}$$

Now, proceeding as usual, with the constraint already accounted for in (6.20),

$$\mathrm{e} = \mathbf{e}^{\mathrm{T}}\mathbf{e} \tag{6.21a}$$

$$\mathrm{e} = (\mathbf{A}\mathrm{x} - \mathbf{b})^{\mathrm{T}}(\mathbf{A}\mathrm{x} - \mathbf{b}), \tag{6.21b}$$

where $\mathbf{Ax} = \mathbf{b}$ is given as

$$\begin{bmatrix} (\mathrm{x}_1 - \mathrm{x}_a) \\ (\mathrm{x}_2 - \mathrm{x}_a) \\ \vdots \\ (\mathrm{x}_n - \mathrm{x}_a) \end{bmatrix} \begin{bmatrix} a_1 \\ 1 \end{bmatrix} = \begin{bmatrix} (y_1 - y_a) \\ (y_2 - y_a) \\ \vdots \\ (y_n - y_a) \end{bmatrix}. \tag{6.21c}$$

This system is solved, minimized, for the least-squares best slope, a_1, in the usual manner:

$$\mathrm{x} = [\mathbf{A}^{\mathrm{T}}\mathbf{A}]^{-1}\mathbf{A}^{\mathrm{T}}\mathbf{b}. \tag{6.21d}$$

In this simple case, the solution can also be computed by hand, since we are only dealing with vectors, where all vector products are simple *scalars* (that is, single numbers, or matrices or vectors consisting of a single element).

Specifically, from (6.21d),

$$\mathbf{A}^{\mathrm{T}}\mathbf{A} = \sum_{i=1}^{n}(\mathrm{x}_i - \mathrm{x}_a)^2 \tag{6.22}$$

and

$$\mathbf{A}^{\mathrm{T}}\mathbf{b} = \sum_{i=1}^{n}(\mathrm{x}_i - \mathrm{x}_a)(y_i - y_a). \tag{6.23}$$

Because $\mathbf{A}^{\mathrm{T}}\mathbf{A} = \mathrm{c}$ is a scalar, its inverse, $[\mathbf{A}^{\mathrm{T}}\mathbf{A}]^{-1} = 1/\mathrm{c}$. Also, $\mathbf{x}$ contains only a single element, a_1, so $\mathbf{x} = \mathrm{x} = a_1$ (also a scalar). Therefore, the solution $\mathrm{x} = [\mathbf{A}^{\mathrm{T}}\mathbf{A}]^{-1}\mathbf{A}^{\mathrm{T}}\mathbf{b}$ is nothing more than a ratio of scalars,

$$a_1 = \frac{\sum_{i=1}^{n}(\mathrm{x}_i - \mathrm{x}_a)(y_i - y_a)}{\sum_{i=1}^{n}(\mathrm{x}_i - \mathrm{x}_a)^2}, \tag{6.24}$$

and back to (6.19),

$$a_0 = y_a - a_1\mathrm{x}_a.$$

Thus, the solution of a constrained straight-line fit is exceptionally easy and manageable in this manner. However, such an approach is only possible when the curve and constraining functions are very simple, as they were in this case. More typically, such an easy approach is unattainable and we must resort to a more general (and easier) method.

6.3.2 Method of Lagrange Multipliers

The **method of Lagrange multipliers** offers a general approach for solving nearly any linear-constrained curve-fitting problem.[4] It can be shown to be identical to that of weighted curve fitting by a theoretical examination of the weighted approach when the relevant weight $w_i \to \infty$.

Consider the case of a curve of the form

$$y_i = \sum_{j=0}^{m} a_j\varphi_j(\mathrm{x}_i), \tag{6.26a}$$

where the φ_j are constituent terms of a basis used to construct the desired curve. The equation is linear in the a_j coefficients, and the φ_j are independent. However, if we add one or more constraints of the form

[4] The constraint equations must be differentiable for this method to be used.

$$\theta_k = \mathrm{f}(a_0, a_1, \ldots, a_m) = c_k, \tag{6.26b}$$

where the c_k are constants, the a_j coefficients are forced into being dependent.

For example, consider a constraint of the form $a_0 + a_1 x_0 = y_0$. Because the two terms must sum to y_0 at the point x_0, then a_0 must be a function of a_1. That is, the curve at x_0 must pass through the point y_0, which *constrains* the values of the coefficients, making them dependent upon one another. Given a slope, in order to pass through this desired specific pair of points, there can only be one intercept that will cause this to happen. Similarly, if we specified the intercept, there can only be one slope allowing this intersection to happen. Therefore, given knowledge of the constraint and thus one of the coefficients (*a priori* – i.e., before solving for the coefficients that minimize the error), the other coefficient can be solved as $a_0 = y_0 - a_1 x_0$.

For the general case, the function to be minimized is

$$\mathrm{e}(a_0, a_1, \ldots, a_m) = \sum_{i=1}^{n} \left[y_i - \sum_{j=0}^{m} a_j \varphi_j(x_i) \right]^2. \tag{6.27}$$

If $m = 2$ and $k = 1$ in (6.26b), so involving only a_0 and a_1 and one constraint, the full derivatives of f and θ with respect to, say, a_0, at a relative extremum must involve the chain rule to account for the dependency of the coefficients, so

$$\frac{\mathrm{de}}{\mathrm{d}a_0} = \frac{\partial \mathrm{e}}{\partial a_0} + \frac{\partial \mathrm{e}}{\partial a_1}\frac{\mathrm{d}a_1}{\mathrm{d}a_0} = 0 \tag{6.28a}$$

$$\frac{\mathrm{d}\theta}{\mathrm{d}a_0} = \frac{\partial \theta}{\partial a_0} + \frac{\partial \theta}{\partial a_1}\frac{\mathrm{d}a_1}{\mathrm{d}a_0} = 0. \tag{6.28b}$$

Equation (6.28b) is true because the constraint is a constant, so $\mathrm{d}\theta/\mathrm{d}a_j = 0$.

Combining (6.28a) and (6.28b) and solving for $\mathrm{d}a_1/\mathrm{d}a_0$ shows

$$\frac{\mathrm{d}a_1}{\mathrm{d}a_0} = \frac{-\partial \mathrm{e}/\partial a_0}{\partial \mathrm{e}/\partial a_1} = \frac{-\partial \theta/\partial a_0}{\partial \theta/\partial a_1}. \tag{6.29a}$$

With this, the $\mathrm{d}a_1/\mathrm{d}a_0$ terms can be eliminated in (6.28a,b), eliminating this additional unknown, and producing a solvable system of two equations in two unknowns (the a_0, a_1). That is, this is exactly the same as the simple case of a constrained straight-line fit, where we were able to reduce the number of degrees of freedom by using one of the constraints to solve for one of the unknowns and reduce the original equation to a solvable system.

For convenience, rearrange (6.29a) by multiplying through by $(\partial \mathrm{e}/\partial a_1)/(\partial \theta/\partial a_0)$ to get

$$\frac{-\partial \mathrm{e}/\partial a_0}{\partial \theta/\partial a_0} = \frac{-\partial \mathrm{e}/\partial a_1}{\partial \theta/\partial a_1} = \lambda. \tag{6.29b}$$

Then set both equal to the same constant, λ, referred to as a **Lagrange multiplier**. This allows rewriting (6.29b) as two separate equations by rearranging each ratio $= \lambda$ in (6.29b) as

$$\frac{\partial e}{\partial a_0} + \lambda \frac{d\theta}{da_0} = 0 \qquad (6.30a)$$

$$\frac{\partial e}{\partial a_1} + \lambda \frac{d\theta}{da_1} = 0. \qquad (6.30b)$$

Alternatively, simply define one augmented equation, the **Lagrangian**, as

$$L(a_0, a_1) = e(a_0, a_1) + \lambda[\theta(a_0, a_1)], \qquad (6.31)$$

from which the two original ones are recovered as

$$\frac{dL}{da_0} = \frac{\partial e}{\partial a_0} + \lambda \frac{d\theta}{da_0} = 0 \qquad (6.32a)$$

$$\frac{dL}{da_1} = \frac{\partial e}{\partial a_1} + \lambda \frac{d\theta}{da_1} = 0. \qquad (6.32b)$$

More generally, for k constraints, the derivatives of the Lagrangian are given as

$$\frac{dL}{da_j} = \frac{\partial e}{\partial a_j} + \sum_{l=1}^{k} \lambda_\ell \frac{d\theta_l}{da_j} = 0. \qquad (6.32c)$$

The Lagrangian gives $m + 1$ equations for $k + m + 1$ unknown coefficients ($m + 1$ unknown a_j and k unknown λ_l).[5] With the original k constraint equation(s), together (6.32c) and (6.26b) provide the required $k + m + 1$ equations, closing the system.

General Solution for Linear Constrained Fit

The *constraint equations* are given in matrix form as

$$\mathbf{F}\mathrm{x} = \mathbf{y}. \qquad (6.33)$$

For example, the constraints that a third-degree polynomial must pass through the pairs (x_0,y_0) and (x_1,y_1) are given in the form of (6.33) as

$$\begin{bmatrix} 1 & x_0 & x_0^2 & x_0^3 \\ 1 & x_1 & x_1^2 & x_1^3 \end{bmatrix} \begin{bmatrix} a_0 \\ a_1 \\ a_2 \\ a_3 \end{bmatrix} = \begin{bmatrix} y_0 \\ y_1 \end{bmatrix}. \qquad (6.34)$$

(This system is underdetermined – four unknowns in only two equations.) Matrix **F** has a form similar to that of the standard matrix **A**, in **Ax** = **b**, except that in this case, the equality is exact, so when rewritten as **Ax** – **b** = **e**, we get

$$\theta = \mathbf{FX} - \mathbf{y} = \mathbf{0}, \qquad (6.35a)$$

the *standard form* for the constraint equations.

[5] It is $m + 1$ because the coefficients (and corresponding basis constituent terms) are indexed as $j = 0,1,\ldots,m$.

Given the constraints, the curve-fitting problem requires minimization, with respect to the a_j, of the original error function for the curve, as

$$e = \mathbf{e}^T\mathbf{e} = (\mathbf{A}x - \mathbf{b})^T(\mathbf{A}x - \mathbf{b}), \quad (6.35b)$$

giving the Lagrangian:

$$L(a_j, \lambda_k) = (\mathbf{A}x - \mathbf{b})^T(\mathbf{A}x - \mathbf{b}) + \lambda^T(\mathbf{F}x - \mathbf{y}), \quad (6.35c)$$

where vector $\boldsymbol{\lambda}$ contains the k Lagrange multipliers, λ_k – one Lagrange multiplier for each constraint equation contained in **F**.

Then, in this matrix form,

$$\frac{\partial L(a_j)}{\partial a_j} = \frac{\partial}{\partial x}[(\mathbf{A}x - \mathbf{b})^T(\mathbf{A}x - \mathbf{b}) + \lambda^T(\mathbf{F}x - \mathbf{y})] = 0, \quad (6.35d)$$

which is solved (see boxed solution details that follow) to give

$$x = (\mathbf{A}^T\mathbf{A})^{-1}\mathbf{A}^T\mathbf{b} - (\mathbf{A}^T\mathbf{A})^{-1}\mathbf{F}^T\lambda, \quad (6.36)$$

which in turn is now combined with the constraint equations (6.35a) and solved to give

$$\lambda = [\mathbf{F}(\mathbf{A}^T\mathbf{A})^{-1}\mathbf{F}^T]^{-1}[\mathbf{F}(\mathbf{A}^T\mathbf{A})^{-1}\mathbf{A}^T\mathbf{b} - \mathbf{y}]. \quad (6.37)$$

The elements (Lagrange multipliers) of vector $\boldsymbol{\lambda}$ are solved from known quantities in (6.37). These values of the λ_k are then used in (6.36) to determine the values of the a_j in vector **x**. Note that this substitution is easy, since all of the required sub-products in (6.36) are available after computing (6.37), so the final solution of **x** in (6.36) is obtained simply by performing the matrix multiplication of these sub-products. Those a_j (i.e., the elements of **x**) minimize the squared error while satisfying the imposed equality constraints.

Box D6.1 Note Regarding the Data Matrix and Constraint Equation

There is sometimes confusion as to whether the constraint equation (e.g., $y_a = a_1 + a_2 x_a$) should also be included in the main system of equations: $\mathbf{Ax} = \mathbf{b}$. For example, in a straight-line fit, if we are fitting the curve to the set of data that includes the point (4,5) so that this pair represents one row in the main equation $\mathbf{Ax} = \mathbf{b}$. If a constraint demands that the line pass exactly through that point, (4,5), it might seem that this point is included in two equations: the main one and the constraint equation. However, by manipulating the equations, it is clear that the Lagrangian gives the slope as $\sum_{i=I}^{n}[(y_i - 5) + (x_i - 4)]$, and thus the constraint equation falls out as 0 in the derivatives of the Lagrangian, leaving it in the constraint equation only. Therefore, one can either include the constraint in the original set of data or drop it – in either case, the system handles it properly and makes use of the relevant information only once.

Box D6.1 (Cont.)

D6.1 Solution of (6.35)

The matrix differentiation of (6.35d) is given as

$$\begin{aligned}
0 &= \frac{\partial}{\partial \mathrm{x}}\left[(\mathbf{A}\mathrm{x}-\mathbf{b})^{\mathrm{T}}(\mathbf{A}\mathrm{x}-\mathbf{b})+\lambda^{\mathrm{T}}(\mathbf{F}\mathrm{x}-\mathbf{y})\right] \\
&= \frac{\partial}{\partial \mathrm{x}}\left\{[(\mathbf{A}\mathrm{x})^{\mathrm{T}}-\mathbf{b}^{\mathrm{T}}](\mathbf{A}\mathrm{x}-\mathbf{b})+\lambda^{\mathrm{T}}\mathbf{F}\mathrm{x}-\lambda^{\mathrm{T}}\mathbf{y}\right\} \\
&= \frac{\partial}{\partial \mathrm{x}}\left\{[(\mathrm{x}^{\mathrm{T}}\mathbf{A}^{\mathrm{T}}-\mathbf{b}^{\mathrm{T}}](\mathbf{A}\mathrm{x}-\mathbf{b})+\lambda^{\mathrm{T}}\mathbf{F}\mathrm{x}-\lambda^{\mathrm{T}}\mathbf{y}\right\} \\
&= \frac{\partial}{\partial \mathrm{x}}\left(\mathrm{x}^{\mathrm{T}}\mathbf{A}^{\mathrm{T}}\mathbf{A}\mathrm{x}-\mathbf{b}^{\mathrm{T}}\mathbf{A}\mathrm{x}-\mathrm{x}^{\mathrm{T}}\mathbf{A}^{\mathrm{T}}\mathbf{b}+\mathbf{b}^{\mathrm{T}}\mathbf{b}+\lambda^{\mathrm{T}}\mathbf{F}\mathrm{x}-\lambda^{\mathrm{T}}\mathbf{y}\right) \\
&= \frac{\partial}{\partial \mathrm{x}}\mathrm{x}^{\mathrm{T}}\mathbf{A}^{\mathrm{T}}\mathbf{A}\mathrm{x}-\frac{\partial}{\partial \mathrm{x}}\mathbf{b}^{\mathrm{T}}\mathbf{A}\mathrm{x}-\frac{\partial}{\partial \mathrm{x}}\mathrm{x}^{\mathrm{T}}\mathbf{A}^{\mathrm{T}}\mathbf{b}+\frac{\partial}{\partial \mathrm{x}}\mathbf{b}^{\mathrm{T}}\mathbf{b}+\frac{\partial}{\partial \mathrm{x}}\lambda^{\mathrm{T}}\mathbf{F}\mathrm{x}-\frac{\partial}{\partial \mathrm{x}}\lambda^{\mathrm{T}}\mathbf{y} \\
&= \mathbf{A}^{\mathrm{T}}\mathbf{A}\mathrm{x}+\mathbf{A}^{\mathrm{T}}\mathbf{A}\mathrm{x}-\mathbf{A}^{\mathrm{T}}\mathbf{b}-\mathbf{A}^{\mathrm{T}}\mathbf{b}+\mathbf{F}^{\mathrm{T}}\lambda \\
&= 2\mathbf{A}^{\mathrm{T}}\mathbf{A}\mathrm{x}-2\mathbf{A}^{\mathrm{T}}\mathbf{b}+\mathbf{F}^{\mathrm{T}}\lambda \\
&= \mathbf{A}^{\mathrm{T}}\mathbf{A}\mathrm{x}-\mathbf{A}^{\mathrm{T}}\mathbf{b}+\mathbf{F}^{\mathrm{T}}\lambda/2.
\end{aligned}$$

This is manipulated to give

$$\mathbf{x} = (\mathbf{A}^{\mathrm{T}}\mathbf{A})^{-1}(\mathbf{A}^{\mathrm{T}}\mathbf{b}-\mathbf{F}^{\mathrm{T}}\boldsymbol{\lambda}/2).$$

Then define the Lagrange multipliers as $2\lambda_k$ (for convenience) so that the factor $\boldsymbol{\lambda}/2$ reduces to $\boldsymbol{\lambda}$, in which case,

$$\mathbf{x} = (\mathbf{A}^{\mathrm{T}}\mathbf{A})^{-1}(\mathbf{A}^{\mathrm{T}}\mathbf{b}-\mathbf{F}^{\mathrm{T}}\lambda). \tag{D6.1.1}$$

Now consider the constraint equations so that we can solve for $\boldsymbol{\lambda}$. First, in order to allow easy substitution of (6.35a) into (D6.1.1) above, pre-multiply both sides of (D6.1.1) by $\mathbf{F}$:

$$\mathbf{F}\mathrm{x} = \mathbf{F}\,(\mathbf{A}^{\mathrm{T}}\mathbf{A})^{-1}(\mathbf{A}^{\mathrm{T}}\mathbf{b}-\mathbf{F}^{\mathrm{T}}\lambda). \tag{D6.1.2a}$$

Now, replace $\mathbf{Fx}$ with $\mathbf{y}$ from (6.35a), giving

$$\mathbf{y} = \mathbf{F}\,(\mathbf{A}^{\mathrm{T}}\mathbf{A})^{-1}(\mathbf{A}^{\mathrm{T}}\mathbf{b}-\mathbf{F}^{\mathrm{T}}\lambda).$$

This is rearranged to solve for $\boldsymbol{\lambda}$:

$$\begin{aligned}
\mathbf{y} &= \mathbf{F}\,(\mathbf{A}^{\mathrm{T}}\mathbf{A})^{-1}\mathbf{A}^{\mathrm{T}}\mathbf{b}-\mathbf{F}\,(\mathbf{A}^{\mathrm{T}}\mathbf{A})^{-1}\mathbf{F}^{\mathrm{T}}\lambda \\
\mathbf{y}-\mathbf{F}\,(\mathbf{A}^{\mathrm{T}}\mathbf{A})^{-1}\mathbf{A}^{\mathrm{T}}\mathbf{b} &= -\mathbf{F}\,(\mathbf{A}^{\mathrm{T}}\mathbf{A})^{-1}\mathbf{F}^{\mathrm{T}}\lambda \\
\mathbf{F}\,(\mathbf{A}^{\mathrm{T}}\mathbf{A})^{-1}\mathbf{A}^{\mathrm{T}}\mathbf{b}-\mathbf{y} &= \mathbf{F}\,(\mathbf{A}^{\mathrm{T}}\mathbf{A})^{-1}\mathbf{F}^{\mathrm{T}}\lambda \\
\lambda &= \left[\mathbf{F}(\mathbf{A}^{\mathrm{T}}\mathbf{A})^{-1}\mathbf{F}^{\mathrm{T}}\right]^{-1}\left[\mathbf{F}(\mathbf{A}^{\mathrm{T}}\mathbf{A})^{-1}\mathbf{A}^{\mathrm{T}}\mathbf{b}-\mathbf{y}\right].
\end{aligned} \tag{D6.1.2b}$$

6.4 Robust Curve Fits

We will not explicitly cover techniques of robust curve fitting. Often they involve "brute force" computational techniques that require minimizing either the L_1 norm, or, alternatively, they find that curve such that 50 percent of the squared errors lie above the curve and 50 percent below it (this is known as the **least median of squares**, or **LMS**). For details on how to perform a LMS curve fit, see the book by Rousseeuw and Leroy, 1987 – this technique can give a remarkably robust fit (i.e., the curve will be highly insensitive to asymmetrical scatter and the presence of outliers in the data).

The use of robust curve fitting requires careful forethought. Draper and Smith (1981, second edition of their book) offered the opinion that use of robust curve-fitting techniques was inadvisable, at least at the time of their writing.[6] This is because rules had not been formulated to decide which technique, given which circumstance, was most appropriate. They suggested that, if the model is wrong, change the model, not the estimation technique. For example, consider the case where there is a strong clustering of points about a straight line, but a fair number of points lie well above this "obvious" line. One might be tempted to use a robust curve-fitting technique to produce a line that effectively ignores the influence of the points well above the line. This may be valid (which is effectively stating that the points above the line are outliers and therefore not representative of the data distribution), but before doing this one must be certain that the data warrant such an approach. If the points are *not* outliers, then one appropriate action would be to attempt to better understand the distribution of the data (clearly not a Gaussian distribution, so least squares would not be appropriate). The principle of maximum likelihood can then be used with the appropriate distribution to determine the optimal manner with which to estimate the coefficient values. Remember, you can always try multiple approaches and see how sensitive the result is to the approach used. If it is sensitive, you need to understand why, and then decide which approach or approaches are most justified.[7]

6.5 Regression/Calibration

The techniques described thus far and in Chapter 5 often are presented under the title of **regression**. Regression, like all of the smooth curve-fitting techniques, seeks to quantify the relationship between a dependent variable and one or more independent variables. However, for regression, the focus is such that, given knowledge of the independent variables (for regression called **predictor variables**, or previously, x values, the deterministic variables), we can predict the value of the dependent value, now called the

[6] In their third edition, they recommend that robust curve fitting may be advisable for heavy-tailed distributions, but note that the application involves more computational work and more assumptions about the procedure to be employed.

[7] I once published a paper that presented three different approaches, discussed the merits of each and "justified" the result used.

response variable (previously our Y values, the random variable). Thus we wish to compute the response variable for measured or specified predictor variable values.

Regression uses the same mathematical and statistical tools as presented in this and the previous chapter. However, in smooth curve fitting we sometimes are not interested in predicting one variable, given knowledge of the other. For example, we may simply wish to orthogonally fit a curve to a set of data to produce a contour plot. We may be interested in the uncertainty in the fit to give an idea how much "slop" there is in the fit at any particular location, but we are not interested in predicting the Y value given some value of X, other than what is necessary to locate the position of the contour.

Like smooth curve fitting, regression will involve fitting the variable with the error, the response variable, to that which contains little or no error, the predictor variable. Indeed, when this describes the relationship of interest, that is, *when we actually wish to predict the variable containing the uncertainty, given knowledge of the variable with no uncertainty, regression is a direct application of the techniques already discussed.*

However, it often occurs that we may actually wish to predict the value with little uncertainty, given knowledge of the variable containing the uncertainty. In other words, we may wish to predict X given knowledge of Y, even though Y contains the error and must be regressed on X. This situation is known as the **calibration** problem (it also goes by the name of **inverse regression** if one actually reverses the X and Y, even though the error is still in Y).

For example, consider the situation where we wish to measure some quantity in nature, such as the temperature of seawater, using an affordable and transportable instrument. Prior to making measurements in the field, we must first calibrate our instrument. We do this by using the instrument to make measurements under controlled conditions where we know with great precision and accuracy the actual values of seawater temperature, and we calibrate the instrument by fitting a curve that best describes the relationship between measured metal expansion/contraction (thought to be proportional to true temperature) and true temperature; the former being the response variable, Y, (containing the uncertainty) and the latter being the predictor variable, X, (containing little to no uncertainty). From this we obtain a relationship such as

$$Y = a_1 + a_2\mathrm{X} \tag{6.39}$$

We now go into the field and make measurements, Y (metal expansion/contraction), from which we wish to estimate the true temperature, X. So, we wish to invert the previous equation:

$$\hat{\mathrm{X}} = \frac{Y - a_1}{a_2}. \tag{6.40}$$

A natural inclination might be to regress X on Y instead of Y on X so that the regression is set up to reflect the true desired response and predictor variables, but that would be incorrect, since the error being minimized is contained in Y. Consequently, the calibration problem involves the inverted relationship of (6.40), whereupon the original regression was of the form of (6.39).

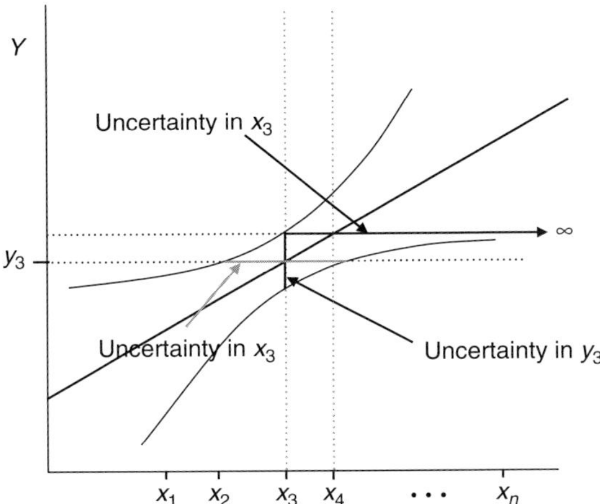

Figure 6.3 For calibration problems, this particular case shows how, in some extreme situations, the original regression that must be inverted to give the relationship needed can result in an infinite error at some predicted points; unusual, but possible.

For these problems, some special considerations are required, though nothing that doesn't involve the same basic tools already employed. Typically, we will perform the regression (Y on x) and then compute the uncertainties in this regression in the usual manner. For the actual application where we will be measuring Y and wish to estimate x, we will invert the regression equation to give x(Y) using the best-fit regression coefficients from regressing Y(x). The primary difficulty comes in determining the uncertainties in the predicted values $\hat{x}$, given measured Y. Draper and Smith (1998) have an excellent discussion of this (their §3.2 on inverse regression) – specifically, if the slope of the original regression was small or poorly determined (i.e., had a relatively large variance).[8] In this case, Figure 6.3 shows the original regression, and uncertainty limits (for a particular α) indicate how one would invert the graph for predicting a value $\hat{x}_0$ for a given measured Y_0 (and the error envelope is also predicted where the horizontal dotted line intersects the envelope); one difficulty comes in inverting the uncertainties at a slightly higher value of Y at Y_1 (the horizontal line does not intersect the lower uncertainty envelope in the figure).

6.6 Correlation Coefficient

When we regress one variable against another, we are implicitly stating that some of the variability in one of the variables can be described, given knowledge of the other variable. In the situation where this relationship between the two variables is described by a straight line, we can describe the degree to which this assumption is true through use of the **correlation coefficient, *r*.**

[8] Either of these suggests that you are probably not using the best instrument for the variable you wish to determine.

The sample correlation coefficient, r, between two variables X and Y sampled as ordered pairs (X, Y) is given by

$$r = \frac{s_{xy}}{s_x s_y} = \frac{\frac{1}{n-1}\sum_{i=1}^{n}(x_i - \bar{x})(y_i - \bar{y})}{\left[\frac{1}{n-1}\sum_{i=1}^{n}(x_i - \bar{x})^2\right]^{1/2}\left[\frac{1}{n-1}\sum_{i=1}^{n}(y_i - \bar{y})^2\right]^{1/2}}. \tag{6.41}$$

As seen, this parameter is normalized covariance (see equation (3.9)) – it is the covariance between X and Y normalized by the standard deviations of X and Y. From this perspective, ignoring for the moment the implications arising from the regression of Y on X, the advantage of this normalization over straight covariance is that the absolute value of r is meaningful, which is not the case for covariance. Specifically, *with covariance, a large number can either reflect the fact that there is considerable covariation between X and Y (recall the meaning of covariance in terms of a joint probability distribution), or that the actual numerical values of X and Y are large though they "covary" little. This is not the case with the correlation coefficient.*

The correlation coefficient has the desirable property that its value is restricted,

$$-1 \leq r \leq 1,$$

where

1 represents a perfect correlation where the values of X and Y covary exactly so that Y varies about its mean in an identical manner to that in which X varies about *its* mean. Plotted as a scatter plot, the values would lie precisely on a straight line. A change in Y of 1 unit corresponds to a proportional change in X, so Y is perfectly predicted, given knowledge of X.

−1 indicates a perfect negative correlation where the values of Y vary about their mean in a manner exactly the opposite of that in which X varies about *its* mean. Plotted as a scatter plot, the values would lie exactly on a straight line with a negative slope, so again, Y is perfectly predicted, given knowledge of X;

0 indicates no *linear* correlation between X and Y at all. Change in X is unrelated to any change in Y (though they may show a very strong nonlinear relationship, which is not measured by r). The two variables are *linearly* independent.

Values in between the limiting extremes indicate the relative degree with which two random variables covary.

6.6.1 Interpreting a Correlation Coefficient

The correlation coefficient is used in one of two ways:

1) assessing whether there is a "significant" degree of linear dependence between two (or more) variables (at some α)

2) assessing how much of the variance in one variable can be estimated, on average, relative to the other, given by r^2

Significance

The significance of r is done via classic hypothesis, where the null hypothesis (H_0: $r = 0$) is what you hope is rejected at some level of significance (α). In the case of r, it requires a transformation (the Fisher transformation) so that r is approximately normally distributed, so that the Z-statistic (normalized Gaussian) can be used to define the critical regions of the test for any desired level of significance. Unfortunately, the distribution of r is dependent upon how close it is to 0, and thus several different statistical tests are potentially involved to test the significance. Good books like Edwards (1976), Draper and Smith (1981) and Dougherty (1990) have extensive and clear discussions worth reading for your own particular case.

However, for most (nonsequential) data, a t-test (small sample size) or normal test (big sample size) can be used to test the significance of r. For a sample size of $10 < n < 50$ the t-test is used; examine value of t for your choice of α for a two-tailed test when you don't care about the sign of r (i.e., you only want to know if the variables are related). The test is performed with

$$\mathrm{t} = r\left(\frac{n-2}{1-r^2}\right)^{1/2}. \tag{6.42}$$

You will then compare the value of t for your sample, to the value of t for a specific level of significance you feel is sufficient to reject the null hypothesis (H_0: $r = 0$) or, as stated before, you will find α of the t you get for your r-value, allowing you to reject H_0 and then decide (and justify) if that is sufficient to claim that the two variables likely are related.

For example, for two variables, x and Y, and a sample size of $n = 25$, you get $r = 0.4$; then apply (6.42), for $\alpha = 0.1$, so for two-sided test $|t_{\alpha/2}| = 1.71$ (reject H_0 if $|t| \geq 1.71$), your t value is $\mathrm{t} = 0.4\left(\frac{25-2}{1-0.4^2}\right)^{1/2} \approx 2.1$. This t is bigger than 1.71, so you reject the null hypothesis (H_0: $r = 0$) and accept that the two variables are related, with a chance of 1 in 10 that you are wrong. In fact, the value 2.1 for $n = 25$ is close to being significant at $\alpha = 0.05$ (you will be wrong rejecting the null hypothesis (that is, stating the two variables are not unrelated) 5 times out of 100).

Significance of r is more complicated if you are correlating sequential data (Chapter 7). In that case, the value you use for n is reduced from that actual number of data points in the data because sequential data are related to their neighboring points, meaning the neighbors are not providing raw independent information, which reduces the number of independent data points (n; called the **effective degrees of freedom, EDOF**) used in the formula. We will resort to use of the bootstrap (see Appendix 2) to deal with that added complexity.

Explained (Shared) Variance (Coefficient of Determination, r^2)

Consider interpreting r in the context of a linear straight-line regression. Recall that the fitted line, y_{pi}, is given as

$$y_{pi} = a_0 + a_1 x_i, \tag{6.43a}$$

and since $a_0 = \bar{y} - a_1\bar{x}$,

$$\begin{aligned} &= \bar{y} - a_1\bar{x} + a_1 x_i \\ &= \bar{y} + a_1(x_i - \bar{x}). \end{aligned} \tag{6.43b}$$

From this, it is seen that when $x_i = \bar{x}, y_{pi} = \bar{y}$. Thus, the straight-line regression line will always pass through $\bar{x}$ and $\bar{y}$ (Figure 5.10).

From equation (5.11b), we know that $a_1 = s_{xy}/s_x^2$, and by definition (6.41) $r = s_{xy}/(s_x s_y)$, so

$$a_1 = r\frac{s_y}{s_x} \tag{6.44}$$

(when regressing X on Y, $a_1 = rs_x/s_y$). Also, combining (5.51) and (6.41), it is seen that

$$s_{rms} \approx s_y(1 - r^2)^{1/2}. \tag{6.45}$$

This relationship is approximate because the $n-2$ in the denominator of equation (5.51) as described in the second paragraph below (5.51) has been replaced with $n-1$ to give (6.45), so the value given here is incorrect by $(n-2)/(n-1)$. For large n, this ratio is trivial and the approximation is accurate for most purposes except very small sample sizes.

If there is no relationship (no covariance) between X and Y, $r=0$, and the best-fit linear line between Y and X is with $a_1 = 0$ (because $r = 0$, and $a_1 \propto$ r). Therefore, the best prediction of y is for $a_1 = 0$ (no slope, so that y = a constant mean)

$$y_{pi} = \bar{y} \tag{6.46}$$

and

$$s_{rms} = s_y. \tag{6.47}$$

That is, for the case of no correlation, a constant mean value of y is the best prediction of y, given any value of x, since there is no dependence of Y on X at all. Also, the rms of the best-fit line is equal to the sample standard deviation (i.e., the scatter of the y values about $\bar{y}$).

On the other hand, if $r = 1$, then $a_1 = s_y/s_x$ and $s_{rms} = 0$ – that is, there is no scatter about the regression line at all. With some additional manipulation, it is fairly easy to show that the variance of Y can be decomposed into two parts, one part being that which is described by the linear relationship between X and Y, given by r^2; the other being that part of the variance that is independent of the relationship with X, given by $1 - r^2$.

Consequently, the value of r^2 is a measure as to just how well X and Y are linearly related; specifically, r^2 *describes how much of the variance in Y can be described (linearly) on average, given knowledge of the variance in* X. For example, if $r^2 = .80$, then, given knowledge of one variable, you can describe (on average) 80 percent of the variance observed in the other. Therefore, r^2 is a measure of the degree of linear relationship or association between the two variables.

If the variables, X and Y have been standardized (i.e., the means removed and divided through by the standard deviation), then $s_y = s_x = 1$ so $a_1 = r$ and the slope of the regression curve therefore gives a direct measure of how the two variables covary – the farther from a perfect slope of 1-to-1, the poorer the correlation or covariation.

Also, while the correlation coefficient describes the degree of linear association between the two variables, it does not give an indication of the magnitude of change in X versus a change in Y – it simply indicates to what degree a change in X will be accompanied by a corresponding (predictable) change in Y. The slope, a_1, of the regression equation, on the other hand, gives exactly this information. That is, if $a_1 = 2$, then a change in X of 1 unit will be accompanied, to some unknown extent (the extent given by r), by a change of 2 units in Y. So if $r = .32$ and $a_1 = 2$, then the 2 units of change in Y accompanying the 1 unit of change in X will only describe about 10 percent (r^2) of the total change in Y (in terms of variance). In this respect, r and a_1 are complementary.

Unlike the regression coefficients, r is not dependent upon whether Y is being regressed on X or X is being regressed on Y. In both cases, the shared variance is the same, even though the actual regression slopes will differ, as shown by

$$a_{1y} = rs_y/s_x \tag{6.48}$$

$$a_{1x} = rs_x/s_y. \tag{6.49}$$

Multiplying (6.48) by (6.49) gives

$$r^2 = a_{1y}a_{1x}. \tag{6.50}$$

It is important to bear in mind that a strong correlation does not say anything about cause and effect or a causal relationship between the variables. It simply indicates that there is a strong degree of covariation. But this card of "no causality" is being overplayed, in my opinion, to the point of implying that a good correlation says nothing about a causal relationship – it is *consistent* with and supportive of a causal relationship if you have hypothesized that two variables have a linear causal relationship, and the r-value is significant at some acceptable level of significance.

Box 6.1 Caution Regarding Regression

Often, we have relationships of the form $y = ax$, in which it may be tempting to recast the equation as $a = y/x$. You might then imagine that you can estimate the "best" value of the "a" coefficient by averaging all of the y_i/x_i. This, however, does *not* give the "best" fit slope. Instead, this gives the average slope, and the associated scatter is representative of the rms scatter in the observations. On the other hand, the best-fit slope is given by the optimization formulas derived by the Principle of Maximum Likelihood developed earlier, and the associated scatter in these is usually significantly smaller than that that associated with the rms scatter, obtained by averaging the y/x (in other words, the uncertainty in the slope that can be expected while still yielding the best fit slope is typically much smaller than the scatter about the best-fit curve). Therefore, it is important to remember that these are two different quantities indicative of two different aspects of the observations.

Also, it is important to not "over-fit" your data. Typically, it is better to start with a simple model (low-order curve, with relatively few degrees of freedom or unknown parameters); then, if the residuals show systematic variability as discussed in §5.7, go to a higher-order model. Otherwise, you may be introducing extreme variability in your fitted curve as it attempts to describe more and more of the actual noise in the data. Recall that an nth-order polynomial will fit n data points exactly (thus, there will be no rms scatter, giving a seemingly "perfect" fit). However, in order to achieve this perfect fit, the polynomial will oscillate wildly, most likely in an unphysically justified manner.

6.7 Take-Home Points

1. For a weighted least-squares fit, multiply every value in each row of the **A** matrix and **b** vector by the weight, typically $1/\sigma_i$, where i is the variable in the row and σ is its precision. With this form, the better (smaller) the precision, the bigger the weight, so that particular row will be weighted more and given a bigger role in the fit. In matrix form, you can do this by forming a diagonal matrix (**w**) of the weights, and pre-multiplying **A** and **b** by it, so $\mathbf{wA}\mathrm{x} = \mathbf{wb}$, with the solution as before: $\mathrm{x} = [\mathbf{A}^T\mathbf{w}^T\mathbf{wA}]^{-1}\mathbf{A}^T\mathbf{w}^T\mathbf{w}$.
2. For a fit that must obey a certain equality constraint (e.g., the curve must pass through a certain point or points), write the constraints as $\theta_k = \mathrm{f}_k(a_0, a_1, \ldots, a_m) = c_k$ or in matrix form: $\mathbf{F}\mathrm{x} = \mathbf{y}$, then $\theta = \mathbf{F}\mathrm{x} - \mathbf{y}$. Then combine the error constraint equations with the standard error equations as $(\mathbf{A}\mathrm{x} - \mathbf{b})^T(\mathbf{A}\mathrm{x} - \mathbf{b}) + \lambda^T(\mathbf{F}\mathrm{x} - \mathbf{y})$, where $\boldsymbol{\lambda}$ is a diagonal matrix with Lagrange multipliers (unknown constants) on each row. The solution is $\mathrm{x} = (\mathbf{A}^T\mathbf{A})^{-1}\mathbf{A}^T\mathbf{b} - (\mathbf{A}^T\mathbf{A})^{-1}\mathbf{F}^T\lambda$ and $\lambda = -[\mathbf{F}(\mathbf{A}^T\mathbf{A})^{-1}\mathbf{F}^T]^{-1}[\mathbf{F}(\mathbf{A}^T\mathbf{A})^{-1}\mathbf{A}^T\mathbf{b} - \mathbf{y}]$.
3. For a "robust curve fit," one in which we do not wish to have suspect data points (e.g., exceptionally large values that may be artifacts of the sampling) to impose too much affect on the fit, we appeal to robust fits. Such fits are not as elegant and may require brute

force, but options include the minimax approach (minimize the largest error), minimize the L_1 norm or minimize the least *median* of squares (i.e., 50 percent of the squared error will be above the curve and 50 percent below). It is sometimes recommended that, instead of using robust fits, you simply fit the correct model, with the result that you will not have to worry about this (i.e., if the model is wrong, change the model, not the fit procedure).

4. Calibration is the situation when you typically regress a variable against a parameter, but in fact you desire to be able to predict that parameter via knowledge of the variable (e.g., you regress an isotope ratio against well-known dated material so that eventually you can look at such a ratio and predict its age). For this, you will regress $Y = f(X)$, then invert the regression, $X = f(Y)$. This is not a problem, though in some pathological cases the uncertainty in this inverted relationship for a predicted value of x may be infinite.
5. Correlation is a normalized covariance giving a value, $r = s_{xy}/(s_x s_y)$, such that the actual value of r has meaning, whereas a large covariance may reflect true covariability between two variables, or the fact that the variables are large values. $-1 \leq r \leq 1$, where a value of $|r| = 1$, is a perfect correlation (a one-for-one correspondence between the two variables) and $r = 0$ is no relationship.
6. If r is significant (at some level of significance, α, you feel is appropriate), that means the odds of the relationship being by chance is unlikely (accepting an 100α percent chance that you are wrong), and r^2 indicates the amount of variance in one variable you can predict on average, given knowledge of the other. So the value of r indicates whether it is likely the relationship between the two variables is by chance, and r^2 indicates how much of one variable can be described by the relationship. So, if r is highly significant but small, it is unlikely to occur by chance, but it doesn't describe much of the variance (i.e., other processes must be active to account for more of the variance).

6.8 Questions

Pencil and Paper Questions

1. Use the principle of maximum likelihood for showing how you would determine the best estimate of the coefficients (β_0, β_1) for a linearly varying mean of Y, $\mu(t) = \beta_0 + \beta_1 t$, given 200 independent observations, $y(t)$, each with a Gaussian distribution and same σ^2. A normal distribution for a variable X is given by PDF, $f(x) = \frac{1}{\sigma\sqrt{2\pi}} exp\left[-\frac{1}{2}\left(\frac{x_i - \mu}{\sigma}\right)^2\right]$. Show the steps of your work for one of the two coefficients, and simply describe how you would get the other one.
2. Define the following with a brief discussion:

   a. Optimal curve fitting (the meaning) and the method of minimizing the square errors (narrate conceptually and give the matrix expressions used)
   b. Over-determined system of equations, the normal equations, and why we can't solve such problems as $\mathbf{x} = \mathbf{A}^{-1}\mathbf{b}$, but can solve interpolation problems that way
   c. Constrained regression and the Lagrangian

d. Weighted regression (definition)
e. Uncertainty in optimal parameter values (why it exists and how it is defined)

3. For x, Yobservations (2, 1), (3, 2), (4, 4), (5, 4), (6, 5), compute the slope and intercept of the best-fit straight line that passes through (5, 4).
4. Consider a constrained least squares fit of a fifth-order, fourth-degree polynomial to a set of x, Y observations (regress Y on x), which is constrained to pass through the points (x_a, y_a) and (x_b, y_b), using the method of Lagrange multipliers.

   a. Give the initial (error) function to be minimized and the constraint equations.
   b. Give the Lagrangian.
   c. Show the matrix form of the function being minimized, the error equation and the constraint equations; describe the contents of each matrix or vector.
   d. What is the system of equations that can now be solved to determine the optimal coefficients of the constrained regression line?

5. Consider a weighted least-squares fit of a straight line to a set of x, Y observations (regress Y on x), with each data pair showing (x_i,y_i,σ_i); σ_i being the standard error for y_i.

   a. Write out the weight matrix.
   b. Give the initial (error) function to be minimized including the weight matrix.

6. You have correlated two time series and find that they have a strongly significant (α = 0.02) value of r = 0.08 for a relationship between x and y, where theory states we expect y to be controlled by x. Explain how to interpret this result in context of the theory it is testing.

7. a. For regressing Y on X (where n = 1000), describe the situation(s) in which you would perform

      i. standard regression
      ii. constrained regression
      iii. orthogonal regression

   b. Give two implications if your best-fit slope coefficient is 0.2 and the estimate of its standard deviation is 0.1.
   c. Why is it always best to use a level of significance of 5 percent (for a 95 percent confidence interval)?

Computer-Based Questions

8. Least-squares regression to parameterize radiative fluxes

Introduction

The energy balance at the Earth's surface can be written as

$$Q_{net} = L + S + LE + SE,$$

where L is the net longwave radiation, S the net shortwave radiation, LE the net latent heat flux and SE the net sensible heat flux. LE and SE are turbulent fluxes and L and S are radiative fluxes. Regarding the radiative fluxes, the shortwave radiation S is energy from the sun (visible, near-ultraviolet and near-infrared) and is primarily downward, though there is an upward component for radiation that is reflected rather than absorbed. Once absorbed, Earth's surface re-emits the energy at longer wavelengths as longwave radiation, L. Some of this longwave radiation emitted upward by the Earth doesn't make it to space (it is absorbed by clouds and greenhouse gases) and gets re-emitted downward.

In this problem, you will parameterize upward and downward longwave radiative fluxes (from NCEPNCAR CDAS-1) along the equatorial Pacific as functions of sea surface temperature (SST; Reynolds/Smith OI) and specific humidity (NCEP-NCAR CDAS-1).

Load and Prepare the Data

There are six vectors in www.cambridge.org/martinson the upward longwave flux (lwflxup), the downward longwave flux (lwflxdown), the SST (sst), the specific humidity (qa), and time vectors for sst (tsst) and everything else (tflx). The SST data are reported weekly while everything else is reported daily, so you'll need to interpolate the SST into a new vector ssti onto the daily grid. You can do this using an interpolant of your choice (but explain the rationale for the interpolant you do choose). Note after interpolation that there are a few days at the beginning and end of the series where the data are not defined (because extrapolation would have been necessary) and they are likely stored as NaN.

Make a figure with four subplots and plot the interpolated SST along with the original SST in the top subplot, the specific humidity below that, the upward flux below that, and finally the downward flux at the bottom. Label all axes and save this figure.

Perform the Regressions

You'll perform a regression for both upward and downward longwave radiation.

Upward Flux

A perfect blackbody of temperature T will emit radiation according to $Q = \sigma T^4$, where $\sigma = 5.67 \times 10^{-8}$ Wm^{-2} K^{-4} (the Stefan–Boltzmann constant). Assuming the sea surface to be a blackbody, we can approximate the upward longwave flux to follow this function of SST. Now that we have a theoretical model for our fit, we will fit a single function of the polynomial basis to our data: the fourth-degree term, with no constant (since we want the intercept to be zero). Store the system of equations as $\mathbf{Ax} = \mathbf{b}$ and solve. Report the single model parameter and its units. How does it compare to σ? Make a new figure, and in a subplot plot the original upward flux as a function of T along with your predicted values.

Downward Flux

Much of the longwave emitted from the surface doesn't make it to space: it is absorbed by greenhouse gases in the atmosphere and then re-emitted, partially downward. Water is one of the strongest greenhouse gases, so we could hypothesize some relationship between atmospheric water content and downward longwave flux. We'll use the specific humidity, which is the mass ratio of atmospheric water to dry air. In a new figure, make a scatter plot of the downward longwave flux as a function of specific humidity.

Does there appear to be a relationship? In the central range of the flux values, the relationship looks linear, but there appears to be some tapering at each of the ends. Let's try a cubic (third-degree) polynomial fit. So this time, we are retaining the first four terms of the polynomial basis. Set up the matrix equation $\mathbf{Ax} = \mathbf{b}$ and solve it. Store the four model parameters in a table, predict the downward longwave flux, and plot the prediction in the same subplot.

Appropriateness of the Fit

In a two subplot figures, as before, plot the residuals about each fit. If there is a systematic pattern to the residuals, it suggests that either we have not included enough constituent terms of our basis or we have chosen the wrong basis. Are these fits appropriate? Discuss.

Goodness of the Fit

Much like the variance measures spread about the mean, the root mean square error quantifies spread about our regression curve. Compute the root mean square error for each of the two fits and report the values (with units). To give them context, also report them as signal-to-noise ratios: (range of data)/(rms error). Explain. Note that when you compute the rms error, the degrees of freedom are N minus the number of estimated parameters.

Uncertainty of the Fit

The regression parameters we solved for are themselves random variables, so they have uncertainty. Use the matrix equations developed in §D5.7 to find the uncertainty in the parameter of the fit of SST to upward longwave radiation and in the four parameters of the fit of specific humidity to downward longwave radiation. You will obtain a matrix, and the variances of the coefficients are the diagonal elements. Take the square root to get values in the original units, and report the values.

Part III

Sequential Data Fundamentals

Simon Shama notes, in the first sentence of the Preface to his book on the French Revolution,[1] the response of Chinese Premier Zhou Enlai (Premier from 1949–1976) to the question, "What was the importance of the French Revolution?" Zhou is quoted as saying, "*It is too soon to tell.*" (I think the Premier was trying to say, "*the specifics of our question dictates the required length of our sample.*")

Sequential data are sequences of data in which the order of occurrence of the observations is important. Such data sequences are usually called **time series**, since it is common to have the data vary as a function of time. *The name "time series," however, is typically applied even if the data vary as a function of space* or another independent variable.[2] Unfortunately, many people still get confused when hearing the phrase "time series," thinking that the data must vary as a function of time – this is not the case! Other names often used for such order-dependent data sequences include "data series," "spatial series," "records," "signals," "traces" and "waveform," among others (many disciplines have their own names for such things, but they are all united in that "order-dependence" deal). Unless the user specifies otherwise, these words can be used interchangeably.

Time series analysis is the field of study that focuses on the analysis of sequential data. It is designed to evaluate the character of the observations and nature of their ordering, to provide information that is useful for classification, extracting signal from noise, identifying consistent relationships between observations and predicting future observations. The beauty of time series analysis is that by using the order dependence of the observations, one can extract a significant amount of additional information from the data relative to the statistical techniques previously considered for non-sequential data.

We begin by developing the technical tools of the trade: **convolution, serial correlation** and **Fourier analysis**. These topics are very important in their own right, but they also provide the foundation from which **spectral analysis, forecast (predictive) models, filter theory, empirical orthogonal function analysis** and other techniques draw.

[1] *Citizens: A Chronicle of the French Revolution*, Simon Schama, 1989, Knopf, pp. 948.

[2] There are instances in which spatial data are treated differently enough from temporal data that the name "spatial series" should be used.

7 Serial[1] Products

7.1 Overview

This chapter introduces the most fundamental operations that are performed on time series, that are used in nearly every time series analysis technique. Most of these involve multiplying numbers together throughout the time series and adding those products together. This operation can show how a value in a time series is related to its neighboring values (even far away neighbors), how two different time series are related (even if one has to be shifted relative to the other to find that relationship), how a time series can be smoothed (via the well known running average, which leads to better methods for smoothing) or otherwise modified linearly (filtering), simulate how many specific physical phenomenon are produced (convolution) and numerous other insights.

Because the order in which the data in the time series adds considerable additional information, this leads to the need for additional statistical information to classify

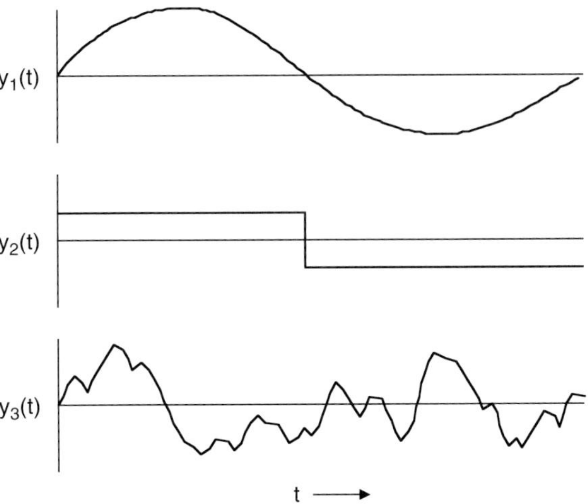

Figure 7.1 Example of three different time series for which the mean and variance are the same, but the series are clearly different, because of the order in which the observations occur, requiring new tools to identify this distinguishing characteristic.

[1] We are not talking about Cheerios™© or Wheaties™©, but rather mathematical series.

sequential data. For example, Figure 7.1 shows three data sets that all have the same mean, variance and standard deviation. However, these time series are distinctly different in character because the nature in which the observations are ordered is different.

Basic Tools

Serial products

Serial products are used for a couple of different operations involving the sum of products of numbers. The two most common such products involve convolution and serial correlation. While very similar in their mathematical structure, the subtle differences between these result in fundamental differences in their interpretation and application.

Convolution represents one of the most important operations of time series analysis while also being one of the most physically meaningful.

Convolution is often considered in terms of **linear filter theory,** which attempts to describe the modification of a series after it passes through a linear filter, where the **filter** is something (say, a black box) that modifies the series as it passes through it.[2] The filter may

1) amplify, attenuate or delay the series, or,
2) modify or eliminate specific "components" of the series.

Convolution is the mathematical operation that performs this modification, given known characteristics of the filter.

Convolution is simply the operation for *linearly* modifying a series by a filter. In Figure 7.2 we convolve the seismic pulse, with the earth acting as a filter, to produce the modified series observed on a seismograph. In this case, convolution simulates a physical process and can be used to model the effects of the process of interest. Of course, whether or not the mathematical procedure of convolution can actually describe the physical process of interest depends upon the process; those criteria that must be formally satisfied are developed as needed below.

In general the process of convolution represents the following:

1) Smoothing data – running means, weighted means, removing specific frequency components, etc., all involve convolution
2) Recording observations by an instrument that (1) responds at a rate slower than the rate at which the observations change; (2) produces a weighted mean over some narrow interval of observation; (3) has lower resolving power than the observations require; etc. In each

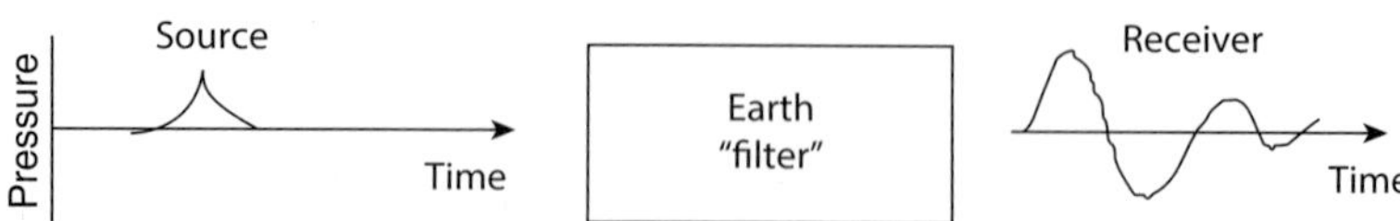

Figure 7.2 Schematic of a seismic pulse traveling through the earth to a receiver located some distance away from the source of the pulse. In this case, the earth acts as a filter to the seismic pulse. The pulse, after traveling through the earth filter, is attenuated and its frequency characteristics are modified.

[2] Linear filter theory is discussed in more detail and more formally later – here, it is introduced at the level required for convolution only.

case, the recording process is performing a convolution on the true process being measured; thus, the data collected are convolved versions of the true phenomenon.
3) Conduction and convection of heat can be represented as a convolution process.

Deconvolution, or **inverse filtering**, is the process of *unscrambling* a convolved signal to determine the shape of the filter or the shape of the input signal. Here are some examples of deconvolution:

1) Since the shape of a seismic pulse and received seismic series are known, we could deconvolve to determine the filtering properties of the layers of the earth through which the pulse passed. In this respect, deconvolution provides knowledge of the geological structure of the earth (or the medical equivalent, using ultrasound, providing knowledge of the detailed structure of your innards).
2) Sediments falling to the seafloor are subject to mixing due to benthic life (worms, etc.). This mixing smears any series being recorded in the sedimentary record. For a pulse of known shape (e.g., a volcanic ash layer, which should approximate a spike) and given the (smeared) recorded version of this pulse, we could deconvolve to determine the characteristics of the mixing filter that acted to smear the entire recorded series. Given the characteristics of the filter, we could then use this information to deconvolve the entire recorded (smeared) series in order to remove the mixing influence and focus directly on the (unsmeared) series.

Therefore, the concepts of convolution and deconvolution, or filtering and inverse filtering, are complementary, though as seen later, the techniques employed are often quite different in practice. Consequently, deconvolution is treated separately in Chapter 13, after more tools have been developed.

Serial correlation provides an indication of the degree of correlation between neighboring points within a time series (known as **autocorrelation**) or the degree of correlation between different time series (known as **cross-correlation**).

7.2 Statistical Considerations

7.2.1 Definitions and Assumptions

A **random process**, or **stochastic process**, denoted as X(t) (continuous) or X_t (discrete), is the *time series equivalent of a random variable – it is an ordered sequence composed of random variables at each time t* (or whatever the independent variable represents). Each individual observation within the time series is a random variable that is described by a probability mass function, PMF. That is, the specific value that is sampled at each point in the time series represents one of an entire range of possible values. The frequency of occurrence of any particular value at any particular time is described by the PMF. The PMF can vary with every data point in the time series, and thus it too can be a function of time. However, unlike the case with nonsequential data, in time series the relationship *between* observations is also described by probability functions, which is where the concept of the autocorrelation or cross-correlation enters the picture.

This means that there must be a joint PMF describing the series, specifically providing the probability of realizing any particular value as a function of time and as a function of the previous value or sets of values in the sequence.

An individual time series represents one realization of a random process, drawn from the entire population of time series possible for the random process. This entire population of time series is called the **ensemble,** and you often need to determine the most representative estimate – that is, most representative time series – of the process or ensemble.[3]

Even if you have a very good idea of the physics or process responsible for producing the observed time series, it will invariably represent a random process because of noise in the real world and in your measuring devices. Thus, in all cases the actual value at each time t can only be estimated in a probabilistic sense and not predicted precisely. So, your data may fundamentally represent a known physical phenomenon predictable by common laws of physics, and thus some mathematical equation (a deterministic component), but the addition of natural noise introduces a degree of uncertainty that foils your ability to predict precisely the actual measured values. If you were to repeat the experiment, your second realization would differ from the first by some amount because of random error in the system. Such time series represent random, or stochastic, processes.

Labeling something as a random process does not imply that there is no order or predictability to the process (though some sources define it otherwise). Often, the precision with which the values of a random process can be predicted depends only on the signal-to-noise ratio (**SN**). If each realization is very similar to the others, then the SN is high and the predictability good, but the time series still represents a random process.[4] In other words, if the variances of the datum at each sampled position are relatively small, the stable underlying nature (signal) of the process will be distinctly obvious. One of the main goals of time series analysis is defining this stable portion of the random process: the signal.

In simple statistics of nonsequential data, you are interested in characterizing a random variable by estimating its distribution (PDF/PMF) or the properties (moments) of its distribution. This distribution describes the population, which includes all of the values of the random variable and, therefore, the probability of drawing any one particular value of the random variable. When drawing a sample (subset) from the population, you assume it is representative of the population and use it to derive the relevant properties concerning the whole population. Similarly, for random processes, the collection of all possible outcomes (time series) that together represent the random process form an ensemble. Thus, the random process is characterized by the properties of the ensemble. A single time series represents only one realization from this ensemble, or one realization of the random process. From this one time series (or, better, from several), you wish to deduce information concerning the entire ensemble, which in turn represents the random process. *The ensemble is typically classified in the time domain by*

[3] Some authors refer to "ensemble" as the collection of realizations making up their sample. This is technically incorrect, as the ensemble is the collection of all possible realizations (i.e., the "population"), but such use is fine as long as its meaning is clearly indicated.

[4] Although the term "time series" can be applied to both deterministic and random processes, in statistical circles, *time series analysis* really applies to the analysis of random processes only, and typically only those that are stationary (defined later).

its autocorrelation function (acf) or by its frequency domain equivalent, the power spectral density (PSD).

Finally, it is important to realize that the ensemble must be defined in context with the particular study being conducted: if the ensemble being sampled is not representative of the ensemble desired (the **target ensemble**), then the collected time series represent a biased sample and subsequent analyses will produce biased results. For example, one set of time series collected in the Red Sea measuring the change in the depth of the surface mixed layer of the ocean may be representative of the ensemble consisting of all possible mixed layers in that particular sea. However, this will differ from the ensemble consisting of all possible mixed layer depths throughout all of the world's oceans. Similarly, that ensemble representing all mixed layer depths possible in the winter months may differ considerably from that representing all mixed layer depths in the summer months.

Characterizing a Random Process

Time series analysis is oriented toward obtaining a relevant description of a random process. Such can be characterized by the joint probability function, $f_{X(t1)X(t2)\ldots X(tn)}(x_1,x_2,\ldots,x_n)$, that describes the probability of getting the values of x_1 *and* x_2 *and* ... *and* x_n over some finite range of the random process. For example, what is the probability of getting some value of $x = x_1$ at time 1, and some different value of $x = x_2$, at time 2, etc., out until time n? This is a *joint* PMF, since it is taking into account the probability of getting a set of values, one particular value for each time. That probability will be limited to the possible values for each time, so the joint PMF is limited in the values that can be obtained, hence constraining the nature of the realizations that can be obtained (i.e., can't just get anything). Previously, for multivariate but nonsequential data, we used a joint PMF to estimate the probability of getting a set of particular values in one experiment (for example, at one time), but here the set is spread over n time steps – one value for each time, t.

In general, determining this joint probability function is impractical (if not impossible), and as with simple random variables, the *random process is typically described by the lower-order moments of its distribution*. Alternatively, we might describe the time series by a model based on a few parameters derivable from the series itself (the topic of Chapter 14, Linear Parametric Modeling).

The first- and second-order moments, or mean, μ_t, and variance, σ_t^2, of the random process, which vary as a function of time, are given in the usual manner through the expectance operator. From this,

$$\mu_t = E[X_t] \tag{7.1}$$

$$\sigma_t^2 = E[(X_t - \mu_t)^2]. \tag{7.2}$$

For example, in Figure 7.3 notice that the PDFs for several data points are shown, and each has a different mean and variance according to their position in time.

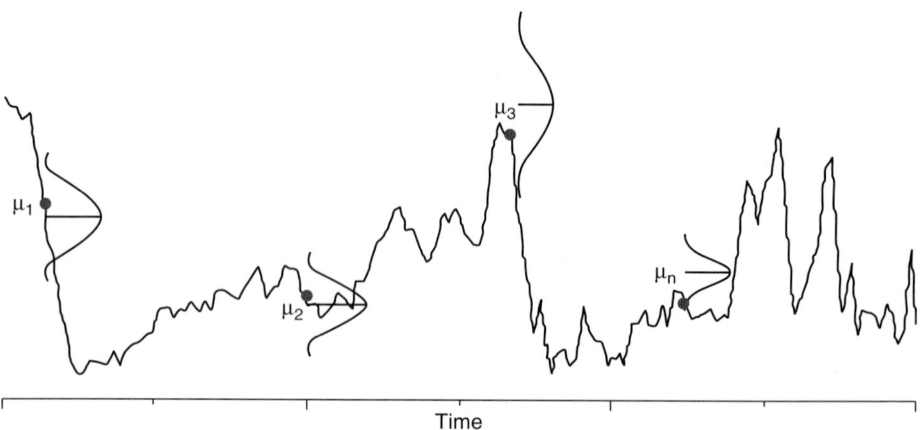

Figure 7.3 300,000 years of Earth's climate as measured by an isotope of oxygen as described for the example of Figure 3.2. As a random process, each datum in the sequence has a range of values it might occupy, specified by the PDF shown for four points as an example.

Figure 7.3 shows the last ~300 kyrs of Earth's sea level (equivalent to climate, as defined by the amount of continental ice). In this case however, an average of all time series ever recorded (usually by deep sea or lake sediments and glacial records) will not give the *ensemble* for sea level rise, since Earth has only run this experiment once – there is only one realization. But it would be the ensemble of interest, being all possible time series recording of that one experiment of interest (and what actually matters). An average of all records would be the best estimate of that one experiment. For this, the variability at each time in the average (called a "stack" in paleoclimate parlance) represents the noise present in the recorded version at each location where a record was obtained (as well as other sources of uncertainty, such as imprecision in dating the values needed to make sure that values recorded at the same time were being averaged).

In addition to the simple univariate moments, for a random process, we define **bivariate moments**; the moments of the joint probabilities between any *two* points in the time series – the simplest of the joint probability moments.

The first bivariate moment is the **autocovariance function (acvf)**,

$$\gamma_{XX}(t_i, t_k) = E[[X(t_i) - \mu(t_i)][X(t_k) - \mu(t_k)]], \tag{7.3a}$$

where $\gamma_{XX}(t_i,t_k)$ is the *true* ensemble acvf. This function provides the first bivariate moment for every pair (x_i,x_k) in the ensemble. So, even though this is only the first joint moment, it can still be a huge function and amount of data. The information is easily presented in a matrix form, as shown later in this section.

For the true ensemble, this requires averaging over all realizations possible in order to determine how the value at the ith point covaries with that at the kth point, as depicted in Figure 7.4. That is, equation (7.3) describes the average of all pairs of products of $(X_{t_i} - \mu_{t_i})$ times $(X_{t_k} - \mu_{t_k})$, giving the autocovariance $\gamma_{XX}(t_i,t_k)$ for any combination of time series points, i and k. So, this is essentially a sum of the form (as defined in equation

(3.9)), summing over the true N realizations of the ensemble (if this were the *sample* acvf, we would have to estimate the mean at each time, t, and lose a degree of freedom in the process, requiring that we divide by N − 1):

$$\gamma_{xx}(t_i, t_k) = \frac{1}{N}\sum_{i=1}^{N}[[X_j(t_i) - \mu_j(t_i)][X_j(t_k) - \mu_j(t_k)]], \tag{7.3b}$$

where $X_j(t_i)$ is the x value at time t_i in realization j; likewise for the t_k points. So we are computing the covariability between the points in the N realizations at t_i and t_k. This is then a function of the covariability between all pairs of points in the random process.

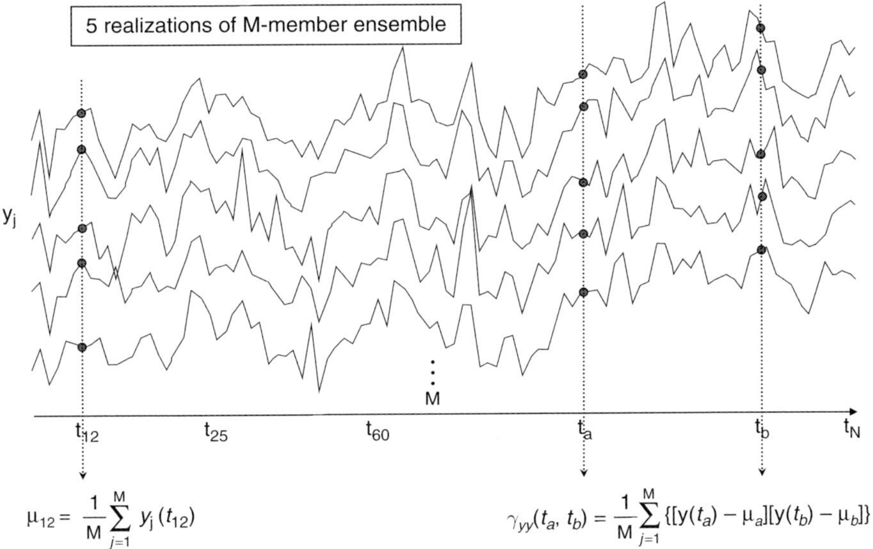

Figure 7.4 Example of computation of the mean and autocovariance of time series, in this case showing five members of an M-member ensemble. In the example, the five realizations shown are exceptionally similar; they do not have to look this similar to share the mean and acvf.

As with standard covariance, the ensemble acvf as defined in (7.3) can be normalized to produce the *true* ensemble **autocorrelation function (acf)**,

$$\rho_{xx}(t_i, t_k) = \frac{\gamma_{xx}(t_i, t_k)}{\sigma_{t_i}\sigma_{t_k}}, \tag{7.4a}$$

or, in summation form,

$$\frac{1}{N}\sum_{j=1}^{N}\left\{\frac{[X_j(t_i) - \mu_j(t_i)][X_j(t_k) - \mu_j(t_k)]}{\sigma_j(t_i)\sigma_j(t_k)}\right\}. \tag{7.4b}$$

Typically, knowledge of the mean, variance and autocorrelation function provides the most fundamental statistical moments used to characterize a random process. The autocovariance function actually provides the variance at any time position, since

$\sigma_t^2 = \gamma_{XX}(t_i, t_k)$ *when i = k. Thus, knowledge of the mean and acvf is all that is needed to provide the important lower-order univariate and bivariate moments. When these moments are estimated and used to characterize a random process, the overall approach is typically referred to as estimation in the time domain.*[5]

7.2.2 Estimation

You need to estimate the population characteristics through examination of your sample, consisting of multiple, independent realizations drawn from the ensemble. Each realization is an entire time series. So, to estimate the mean, $\overline{x}_t$, you must average the values of the time series at the time, t, across all realizations (as shown for the true mean in Figure 7.4). You must do the same to estimate the variance at each time, t.

Likewise, the sample acvf, $C_{XX}(t_j,t_k)$, you estimate from your sample the covariance between different points in time. The true acvf, $\gamma_{XX}(t_j,t_k)$, quantifies the relationship between points at two distinct times (the relationship between these two specific times t = *j* and *k*), as recorded in *all* of the different possible realizations from the ensemble.

Typically, you will not have as many realizations (time series) available as you would have nonsequential data realizations in a sample. Consequently, your estimated ensemble statistical parameters will suffer from a dearth of realizations (e.g., consider estimating the mean of a population from a sample of size 1 or 2). We will get around this restriction by taking advantage of stationarity when possible.

Stationarity

Much of the theoretical development of classical time series analysis is based on the *assumption* that the stochastic process represents a **stationary** process. Stationarity implies the process is in an equilibrium state, not in a transient state. Therefore, while a stationary process may still fluctuate in time, its statistical properties are no longer dependent upon time – its mean state, as well as higher-order moments, no longer changes with time. *The PMF of a stationary process is not a function of time, nor are the joint PMF and statistical moments.*

This definition is *extremely* restrictive and has strong implications for analysis and interpretation (mostly for uncertainty analysis), as is demonstrated below and seen in subsequent chapters.

Specifically, stationarity implies for a random process X_t that

$$E[X_t] = \mu \tag{7.5a}$$

$$E[(X_t - \mu_t)^2] = \sigma^2 \tag{7.5b}$$

for the first two moments.

[5] Later, we will estimate the equivalent of these moments within the *frequency domain* using the methods of spectral analysis. This latter approach is referred to as **estimation in the frequency domain**.

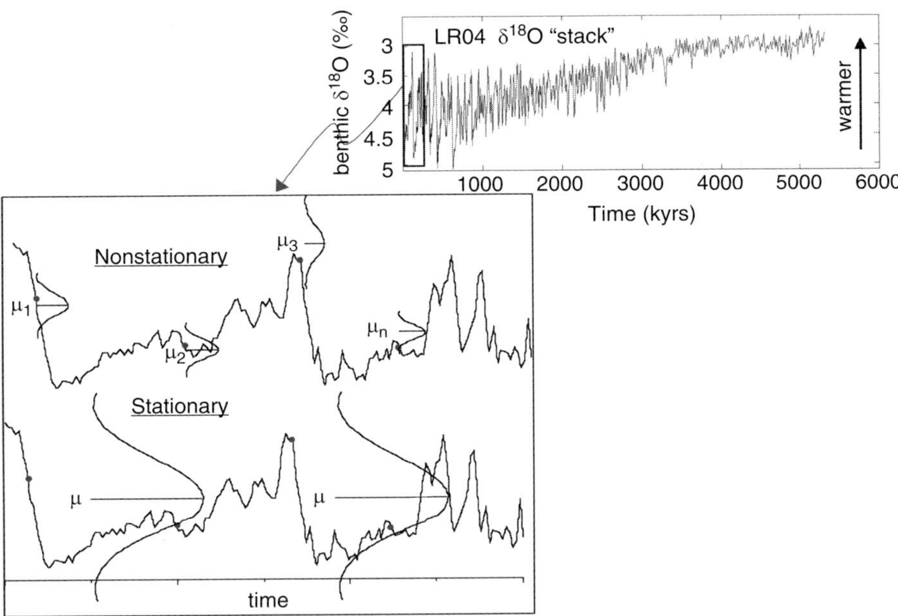

Figure 7.5 Clearly this LR04 series for climate is not stationary, as it is obvious that the mean and variance decrease farther back in time (to the right; noting that the abscissa gets smaller as it goes upward). If it was stationary, it would be difficult to explain the systematically smaller mean and variance over the period ≥3000 kyrs ago.

Now consider the implications of (7.5) for the series in Figure 7.5 that cannot hold: a single mean and PMF is not valid for the entire >5 million years of the series. Clearly, there is a tremendous change in the mean and variance over time – the series is not stationary. So it seems we have been defeated in our ability to apply stationarity solving the problem of determining statistical moments of the time series with few realizations for this example – but we haven't, as will be seen later when we discuss evolutionary time series analysis.

For the first bivariate moment, stationarity implies

$$\begin{aligned} \mathrm{E}[(X_{t_j} - \mu)(X_{t_k} - \mu)] &= \mathrm{E}[(X_{t_j} - \mu)(X_{t_{j+\tau}} - \mu)] \\ &= \mathrm{E}[X_i, X_{i+\tau}] - \mu^2 \\ &= \mathrm{Cov}[X_i, X_{i+\tau}]. \end{aligned} \tag{7.5c}$$

Relationship (7.5c) indicates that for stationary processes, the acvf is no longer dependent upon the absolute positions of t_i and t_k, but *only on the size of the lag, τ* (time separation), between the two values. That is, since the statistical moments are no longer a function of time, the covariance between values separated by τ units of time is independent of any translation of the origin. Consequently, the absolute times, t_j and t_k, at which the X are being compared is no longer important; only the distance, τ, between the points matters (Figure 7.6).

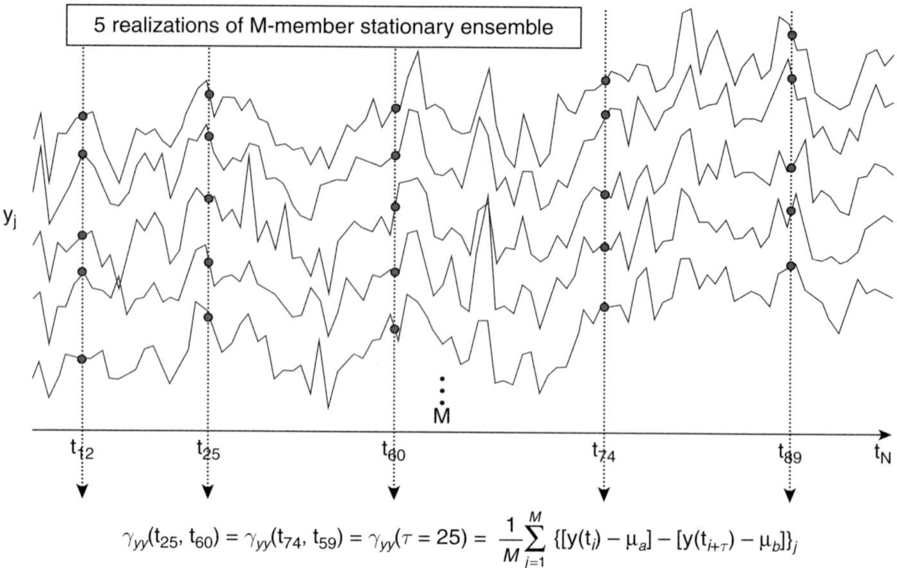

Figure 7.6 Implications of stationarity for computing acvf. In this case, only the time separation (τ) between data points in the series matters, not the specific time points as shown in Figure 7.4.

Implications of Stationarity to the acvf and acf

Consider the true acvf and acf for a random process, Y, in terms of matrices. First, we place y' (values of y residuals *about their mean* at time i) in a vector Y':[6]

$$Y' = \begin{bmatrix} y_1' \\ y_2' \\ \vdots \\ y_n' \end{bmatrix} \tag{7.6}$$

Any single realization (time series) of the random process Y can be stored as such.

If we examine $\mathrm{E}[\mathrm{Y}'\mathrm{Y}'^{\mathrm{T}}]$, we see that it is equivalent to

$$\begin{matrix} & & [\; y_1' & y_2' & \cdots & y_n' \;] \\ \mathrm{E}[\mathbf{Y}'\mathbf{Y}'^{\mathrm{T}}] = & \begin{bmatrix} y_1' \\ y_2' \\ y_3' \\ \\ y_n' \end{bmatrix} & \begin{bmatrix} \mathrm{E}[y_1'y_1'] & \mathrm{E}[y_1'y_2'] & \cdots & \mathrm{E}[y_1'y_n'] \\ \mathrm{E}[y_2'y_1'] & \mathrm{E}[y_2'y_2'] & & \mathrm{E}[y_2'y_n'] \\ \mathrm{E}[y_3'y_1'] & \mathrm{E}[y_3'y_2'] & & \\ & & & \\ \mathrm{E}[y_n'y_1'] & \mathrm{E}[y_n'y_2'] & & \mathrm{E}[y_n'y_n'] \end{bmatrix} \end{matrix} \tag{7.7a}$$

or, more succinctly (noting that, $\mathrm{Var}[\mathrm{Y}] = (\mathrm{Y} - \mathrm{E}[\mathrm{Y}])(\mathrm{Y} - \mathrm{E}[\mathrm{Y}])^{\mathrm{T}} = \mathrm{E}[\mathrm{Y}'\mathrm{Y}'^{\mathrm{T}}] = \Sigma_{\mathrm{Y}}$ the **covariance matrix** of **Y**),

[6] The mean at time i is computed as shown schematically in Figure 7.5, computing the mean at time $t = 12$.

$$\Sigma_Y = \begin{bmatrix} \sigma_1^2 & \gamma_{12} & & \cdots & \gamma_{1n} \\ \gamma_{21} & \sigma_2^2 & \gamma_{23} & & \gamma_{2n} \\ \gamma_{31} & \gamma_{32} & & & \gamma_{3n} \\ \vdots & & & \ddots & \gamma_{n(n-1)} \\ \gamma_{n1} & \gamma_{n2} & & \gamma_{n(n-1)} & \sigma_n^2 \end{bmatrix}. \tag{7.7b}$$

Here, the principal diagonal elements contain the variance of the y_i at each time i, while the cross diagonal points contain the covariance between pairs of data points at two different times.[7] For example, the element at row 2, column 1 contains the covariance between the values of the random process occurring at time 1 and those occurring at time 2. If the variance doesn't change with time (i.e., if the process is stationary), then the diagonal immediately next to the main diagonal of the covariance matrix contains the covariance between all points in the series separated by 1 point in time, whereas one diagonal away from that is the covariance for all pairs separated by two points in time, etc. In this case, the complete acvf of (7.3) is contained in $\mathbf{\Sigma_Y}$, the covariance matrix of the random process Y.

The covariance matrix of a *stationary* process becomes

Nonstationary

$$\Sigma_Y = \begin{bmatrix} \sigma_1^2 & \gamma_{12} & & \cdots & \gamma_{1n} \\ \gamma_{21} & \sigma_2^2 & \gamma_{23} & & \gamma_{2n} \\ \gamma_{31} & \gamma_{32} & & & \gamma_{3n} \\ \vdots & & & \ddots & \gamma_{(n-1)n} \\ \gamma_{n1} & \gamma_{n2} & & \gamma_{nn-1} & \sigma_n^2 \end{bmatrix} \quad \gamma_{\tau=1} \quad \gamma_{\tau=1}$$

Stationary

$$\Sigma_Y = \begin{bmatrix} \sigma^2 & \gamma_1 & \gamma_2 & \cdots & \gamma_{n-1} \\ \gamma_1 & \sigma^2 & \gamma_1 & \cdots & \gamma_{n-2} \\ \gamma_2 & \gamma_1 & & \cdots & \gamma_{n-3} \\ & & & & \vdots \\ \gamma_{n-1} & \gamma_{n-2} & & \cdots & \sigma^2 \end{bmatrix}$$

If we multiplied $\mathbf{\Sigma_Y}$ by another matrix as follows,

$$\begin{bmatrix} 1/\sigma_1^2 & 1/\sigma_1\sigma_2 & \cdots & 1/\sigma_1\sigma_n \\ 1/\sigma_2\sigma_1 & 1/\sigma_2^2 & \cdots & 1/\sigma_2\sigma_n \\ 1/\sigma_3\sigma_1 & 1/\sigma_3\sigma_2 & \cdots & 1/\sigma_3\sigma_n \\ & & \vdots & \\ 1/\sigma_n\sigma_1 & 1/\sigma_n\sigma_2 & \cdots & 1/\sigma_n^2 \end{bmatrix} \begin{bmatrix} \sigma^2 & \gamma_{12} & \gamma_{12} & \cdots & \gamma_{1n} \\ \gamma_{21} & \sigma^2 & \gamma_{22} & \cdots & \gamma_{2n} \\ \gamma_{31} & \gamma_{32} & & \cdots & \gamma_{3n} \\ & & & \vdots & \\ \gamma_{n1} & \gamma_{n2} & & \cdots & \sigma^2 \end{bmatrix} = \rho_Y, \tag{7.8a}$$

we generate the **correlation matrix ($\boldsymbol{\rho}_\mathbf{Y}$)**

$$\rho_\mathbf{Y} = \begin{bmatrix} 1 & \rho_{12} & \cdots & \rho_{1n} \\ \rho_{21} & 1 & \cdots & \rho_{2n} \\ \rho_{31} & \rho_{32} & \cdots & \rho_{3n} \\ & & \vdots & \\ \rho_{n1} & \rho_{n2} & \cdots & 1 \end{bmatrix}, \tag{7.8b}$$

[7] This is why this matrix is often call the **variance-covariance matrix**, but I prefer the less-bulky "covariance matrix," leaving it to the reader to realize that variances lie along the principal diagonal.

where the diagonal elements are now the correlation of the series with itself at time i and the off-diagonal terms are the correlations between the points at different times, i and j. Thus, the correlation matrix contains the complete acf of (7.4).

More generally, stationarity requires that

$$\mathrm{f}_{X_{t_1}X_{t_2}\cdots X_{t_n}}(x_1,x_2,\ldots,x_n) = \mathrm{f}_{X_{t_{1+\tau}}X_{t_{2+\tau}}\cdots X_{t_{n+\tau}}}(x_1,x_2,\ldots,x_n), \tag{7.9}$$

where this relationship states that the joint probability function of X(t) is unchanged with any translation of the origin by an amount τ. That is, all of the joint moments (i.e., those describing the relationships between two or more points in the series) do not change with the actual times or points being compared, but only change with how far apart the points or times being compared are (as specified by the lag, τ).

Since the variance is constant in a stationary process, the acf can be simplified to

$$\begin{aligned} \rho_{XX}(\tau) &= \frac{\gamma_{XX}(\tau)}{\sigma^2} \\ &= \frac{\gamma_{XX}(\tau)}{\gamma_{XX}(0)}, \end{aligned} \tag{7.10}$$

which is now a function of lag, τ, only (before, the variance was a function of time, so the denominator was $\sigma_j\sigma_k$, but here, $\sigma_j = \sigma_k$, so $\sigma_j\sigma_k = \sigma^2$). Both the correlation and covariance matrices are symmetrical, since $\sigma_{ij} = \sigma_{ji}$ and $\rho_{ij} = \rho_{ji}$.

Degree of Stationarity

Suppose now that the mean does *not change* over time (i.e., the series always fluctuates, on average, about some constant mean value), but the variance *does change* with time. Such a series is **stationary of order one**. If both the mean and variance do not vary over time and (7.10) is satisfied, but the third moment (skewness) does vary, then the process is **stationary of order 2**, etc. *For most purposes, if a process is stationary of order 2, it is sufficient to satisfy requirements needed to make statistical deductions about the data in the frequency domain* (as discussed in following chapters on Fourier and spectral analysis). Consequently, any process is called *stationary* if it is stationary of at least order 2 and **nonstationary** if not. More specifically, a process that is stationary of order 2 is called **weakly stationary** or **wide-sense stationary (WSS)**. If it is stationary to higher order it is called **strongly stationary**. If it is stationary with respect to all moments and joint moments, the process is **completely stationary.**[8]

In general, stationarity is a strong restriction (even in its most basic form of weak stationarity), since it disallows predictable trends, shapes, etc. In science, we are often analyzing a series that occurs as a particular shape (varying mean) for which stationarity cannot be assumed. However, weak stationarity can often be satisfied by simply removing any systematic trend (for example, take the derivative of the series), deterministic

[8] If a process is stationary of order 2 but the process has a Gaussian joint PDF, then it is completely stationary, since all higher-order moments are a function of the first two for that distribution.

component or estimate of a variable mean of a time series.[9] This can be done by fitting the time series with a curve. The residuals about that best-fit curve (i.e., subtract the curve from the series) are treated as a stationary process (at least of order 1). Other means of treating nonstationary components (such as pure sinusoidal components) are discussed later.

Because removal of a trend or other systematic variability in the mean is fairly easy, violation of a constant mean is usually a simpler problem than violation of a time-invariant higher order moment, such as changes in variance with time. In this latter case, correcting for stationarity can only be accomplished by examining segments of the time series over which the process does seem to satisfy the stationarity conditions. This treatment of time series – breaking them into shorter, contiguous or overlapping "windows" in which stationarity is approximately satisfied (e.g., Figure 7.7, though note that while the mean value is approximately stationary in each window, variance is questionable) and evaluating each window separately – is one form of analyzing **nonstationary** time series (also called **transient**, or **evolutionary** time series). This procedure is often referred to as *evolutionary time series analysis*. In its simplest form, evolutionary time series analysis involves examining the nature in which any statistical moment changes as a function of time. For example, one might plot the variance within each window against the central time of the window.

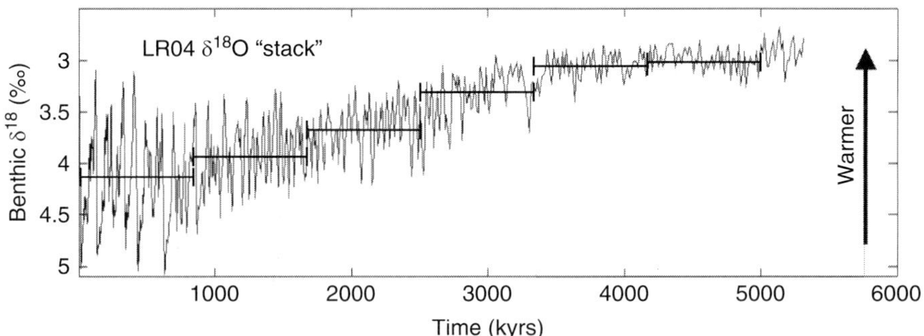

Figure 7.7 As shown in Figure 7.5, the LR04 time series is nonstationary. Here it is clear how that series might be broken into contiguous windows that appear to be stationary in the mean and variance.

A consequence of stationarity is that, if a time series continues without a change in its statistical properties, it must be infinite in extent. If the time series went to zero after (or before) some time, then its variance would also go to zero, which would differ from any nonzero variance before that point. This causes technical problems (later) in Fourier analysis that must (and will) be overcome.

[9] A stationary data series cannot contain a deterministic component (unless it is a constant only), since that would imply a known value (i.e., a change in the mean) that varied with time according to the deterministic law. On the other hand, the statistical evaluation does not include a deterministic component anyway, since it is a component without random uncertainty, and thus we expect that any such component is identified and removed before performing the statistical analysis. Actually, this can be a difficult imposition, leading to some "mixed" (deterministic and random) component analyses in practice.

Ergodicity ('ergō •dis •ity)

Another statistical concept is often drawn upon that, if applicable, facilitates the analysis of a single realization of a random process. As discussed, (7.1) through (7.4) show that the "true" statistical moments of a random process involve averaging over all realizations in the ensemble for the time interval of interest. That is, the mean, μ_{t1}, is obtained for the time t_1 by averaging every possible value of X that can occur at time t_1 (taking every possible time series that could occur from the ensemble and averaging the values that occur at t_1 in each of those realizations). Likewise, the true acf involved averaging all products of $X(t_j)X(t_k)$ – that is, it involved all possible values that can occur at the time t_j and the time t_k. **Ergodicity** is the property that allows the use of time averaging instead of the desired ensemble averaging.

Specifically, *a random process is ergodic if all of its statistical moments can be determined from a single realization using time averaging*. That is, if the process is ergodic, then the mean, for example, can be determined by averaging all of the values occurring over the time spanned by the single realization instead of averaging all of the values in the ensemble as shown in Figure 7.8.

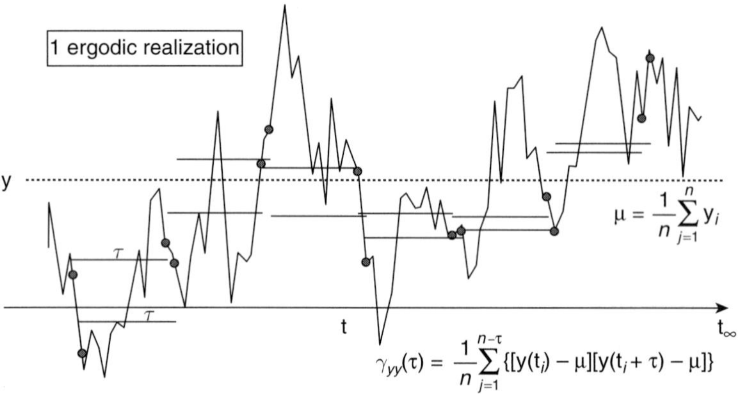

Figure 7.8 Computation of the mean and acvf for an ergodic process, showing that the moments are computed by using all values in the time series for the mean, and every pair of points separated by some τ for the acvf (i.e., every point in the series compared to its partner τ units away).

The implications of this are tremendous. Instead of estimating $\bar{x}_t = \sum_{j=1}^{n} x_j(t)$ where the $j = 1, \ldots n$ are the average of all values in each of N time series *realizations* at time t_j, we now simply estimate a single mean: $\bar{x}_t = \frac{1}{n}\sum_{i=1}^{n} X_i$, where the average is over the n data points in a single realization (Figure 7.8).

Similarly, for the autocovariance function, ergodicity allows that the acvf can be estimated by comparing all values that are offset by a lag, τ, in a *single* realization.[10] This

[10] Note that previously we stated that if a process is stationary, we can compute its acvf by examining the correlation between any pair of points separated by some time τ, regardless of the absolute time at which the points occur, as shown in Figure 7.6. This is not the same as ergodicity, which goes one step farther and states that the acvf itself can be completely determined by evaluating a single realization over its full time extent. That is, one complete realization provides a complete sample of the total variability displayed in the

allows us to define the acvf in terms of a *serial product* (= a summed series of products), which is how you will usually estimate it in practice.

As with stationarity, you can define the order of ergodicity. For example, if the mean computed by time averaging converges in the limit to the ensemble average, then the process is called **ergodic in the mean**. If the time-averaged acvf converges in the limit to the ensemble-averaged acf, the process is said to be **autocovariance ergodic**, and so on.

If a process is both ergodic in the mean and autocorrelation ergodic, and also stationary of order 2 (though order 4 is actually required in general for ergodicity to rigorously hold), then it satisfies most of the requirements for drawing statistical conclusions from an analysis of the data using standard techniques. Note that if a process has zero mean and is Gaussian, then ergodicity automatically holds. Also, for non-Gaussian processes, ergodicity does typically hold if the process is stationary (see Marple (1987) for a nice discussion regarding this).

Hereafter, it is assumed, unless stated otherwise, that all processes to be analyzed are stationary of order 2 and ergodic (or that we have already identified and removed any nonstationary component). From a practical standpoint, this assumption allows us to break a single time series into several segments and treat each segment as if it were a different realization of the random process. In this manner, *we are able to employ standard averaging techniques (and the Central Limit Theorem) to single time series.*

In *summation* form, the sample acvf, $C_{YY}(t_j,t_k)$, is now the acvf as a *serial product*:[11]

$$C_{YY}(t_j t_k) = \frac{1}{n-1}\sum_{t=1}^{n}[(y_j-\bar{y})(y_k-\bar{y})]_t. \tag{7.11}$$

Random Process Noise

Now consider how noise manifests itself in time series. To a data analyst, noise is often a relative term describing any component(s) of a data set that is present in addition to the “signal” of interest. For example, if you are interested in the overall trend of global climate change, then local seasonal cycles superimposed on the longer-term global trend would essentially represent noise.[12] Such noise serves to disguise, blur or otherwise make more difficult the direct examination of the signal of interest. This leads to the common statement that one person’s noise is another’s signal. Note, however, that this definition of noise may not always coincide with the statistician’s general definition, in which noise is the uncertainty in the observations prescribed by the PMF(t).

full ensemble of realizations. Stationarity never stated that we only needed a single realization, it only stated that we no longer had to be concerned with the correlation as a function of the actual time interval over which it occurred, but one would still require an analysis of the complete ensemble of realizations to compute the true acvf.

[11] Some people refer to the serial product (especially when there is no normalization) as the **acvsp**, or **acsp** (for autocorrelation).

[12] If you do remove a “component” of the series, treating it as noise relative to the signal of interest, it is important that the component removed is not related (correlated) to the signal of interest; otherwise you may actually be removing some of the signal, or information related to the signal, that could be important to the analysis or interpretation.

White noise (which has a very specific definition) is probably more representative of what a person in general would consider to be straight "noise" – unlikely to be anyone's signal (except for those people studying "noise").[13] The key feature of white noise is that *each individual datum in a white noise series is uncorrelated to all other points (neighboring and distant) in the series*. As such, it essentially violates the key feature of a time series: that the order in which the values occur is important. This lends itself to quantitative identification.

There is often confusion over the *probability distribution* of white noise. The reason for this is twofold. First, the definition of white noise is *independent* of the distribution of the data points. A random variable and stationary random process may have the same PMF – only the acf distinguishes the two, with the acf indicating the nature of the ordering of random variables in the random process. Therefore, the distribution is unimportant – the only criterion is that the individual points in the white noise series are uncorrelated. Typically, though not always, white noise is considered to have a rectangular distribution (equal probability of obtaining any value in the series). When one is dealing with noise that shows a Gaussian distribution, it is often referred to as Gaussian white noise. Second, the actual existence of recorded white noise is unachievable, since it represents a band-*unlimited* process (discussed in Chapter 11). If it could exist, it can be shown that it could not have a Gaussian distribution. However, since we can only approximate white noise in practice, as long as the values are uncorrelated, the distribution is not restricted.

We can also define a variety of "colored" noises (discussed later in terms of power spectral density), whereby the random series has a particular form of covariability between observations (though the series is still random).

7.3 Convolution

Convolution Basics

Convolution, our most basic tool, is easily visualized by examining the convolution of discrete functions.

First, consider the representation of a discrete time series in terms of a sequence of discrete **impulses**. Each impulse is loosely considered an instantaneous pulse in time of a given amplitude.[14] Thus a discrete time series or sequence of impulses, say g(t), as they are about to pass through a filter, would appear graphically as shown in Figure 7.9.

In the example of Figure 7.9, g(t), contains the following ordered sequence: g(t) = {5, 4, −2, −3, 2, −1, −4, 3}. This discrete sequence can be thought of as individual instantaneous impulses, each separated in time by 1 unit, or it can represent discrete

[13] The most famous case of great things coming from noise was the Nobel Prize-winning work of Penzias and Wilson in 1978 for discovering that their noise (in the form of static as with an untuned radio or TV) was actually the Cosmic Microwave background of the Big Bang.

[14] Impulses are more rigorously defined later. The use of impulses anticipates conveniences resulting from the linearity of convolution.

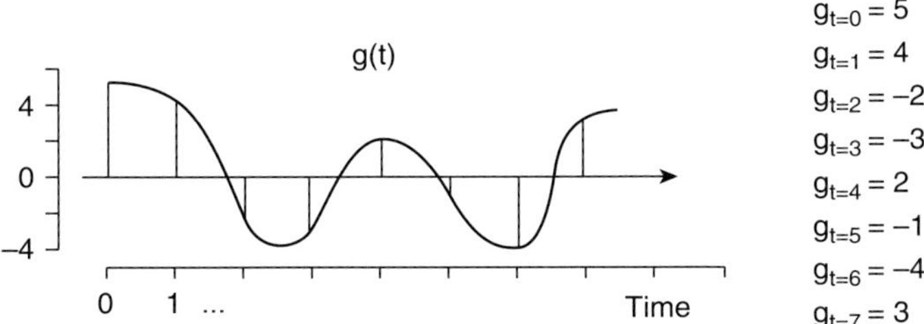

Figure 7.9 Example of an *ordered sequence of impulses*, with the values of each impulse.

samples of a continuously varying time series, which, if continuously sampled, would vary continuously from the top of each impulse as estimated by the continuous line in Figure 7.9.

As shown, using standard notation, time increases to the right in **unit delays**. That is, each increment of t to the right corresponds to an impulse that occurs 1 unit *after* the impulse immediately to the left.[15] Therefore, the impulse at time $t = 7$ (say, 7 seconds), occurred 6 seconds after the impulse at $t = 1$. The impulse at time $t = 0$ corresponds to the *initial* event in g(t).

A **noncausal** or **nonrealizable** system or sequence is one that contains one or more impulses at negative times. For example, an impulse at a time $t = -1$ indicates that the impulse occurred 1 unit *before* the initial time of $t = 0$, or day zero in g(t) above. *This is unacceptable in physical systems, since something cannot occur before it has actually started.*

If g(t) represents the volume of ash that settles each day for a week after a volcanic eruption, then $g_{t=0}$ corresponds to the first day's ash volume, $g_{t=1}$ corresponds to the second day's ash volume, etc., until $g_{t=7}$ corresponds to the last day's ash volume; $g_{t=-1}$ would be the nonrealizable amount of ash that fell the day before the ash started falling. If the ash layer were settling on the seafloor, subject to bioturbation (a filtering process that can be represented as a convolution), then obviously the ash that fell on $t = 0$ will be the first ash exposed to the bioturbation, and thus be the first to enter the filter. The ash falling on $t = 1$ will enter the filter second, etc. Finally, the ash at $t = n$ will be that entering the filter on the last day of the eruption.

So, with respect to passing a sequence through a filter, pictorially, if g(t) is drawn as it is in Figure 7.9, with the unit delays increasing to the right (as is standard on time plots, January would be on the left, then February to the right of that, etc.), you would have to reverse the figure before feeding it through a filter, so that the leftmost impulse would be passed through first and the rightmost impulse passed through last.

Impulse Response Function

Now, consider a single impulse passing through a filter on its way to a recording device, as shown in Figure 7.10.

[15] This is opposite of studies of the ice age climate (e.g., LR04 in our previous examples).

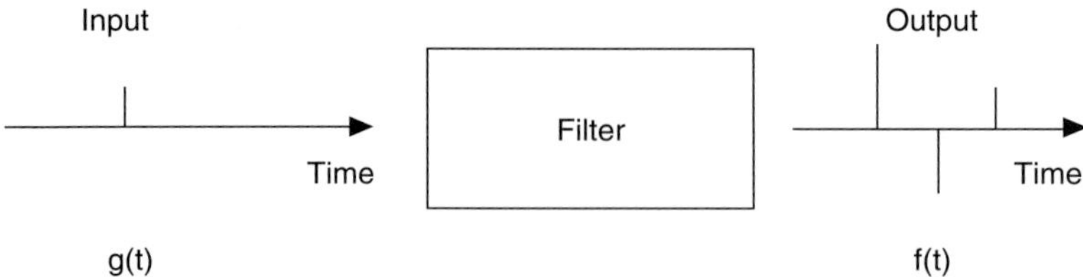

Figure 7.10 Schematic of *impulse response function*, as the response of passing a unit impulse through a filter.

In discrete form, if g(t) and f(t) are evenly spaced sequences of samples (assume unit spacing) in time and the amplitude of the impulse is unity, then

$$\begin{array}{cccccccc} g(t) = & \ldots & 0, & 0, & 1, & 0, & 0, & \ldots \\ & & \uparrow & \uparrow & \uparrow & \uparrow & \uparrow & \\ & & t=-2 & t=-1 & t=0 & t=1 & t=2 & \end{array} \tag{7.12}$$

is the input series. This sequence g(t) represents a **unit impulse**. That is, g(t) is zero at all times other than t = 0, at which it takes on the value of unity, though a unit impulse needn't be centered at time zero.

The unit impulse is input to the filter. The sequence of unit-spaced output values (as shown in the above example),

$$\begin{array}{cccccccc} f(t) = & \ldots & 0, & 0, & 5, & -4, & 2, & \ldots \\ & & \uparrow & \uparrow & \uparrow & \uparrow & \uparrow & , \\ & & t=-2 & t=-1 & t=0 & t=1 & t=2 & \end{array} \tag{7.13}$$

is output from the filter, and it represents the modified form of g(t) after passing through the filter. In a physical sense, the *first* value emerging from the filter does so at t = 0.

Impulse response function is the name given to the output sequence in this example, since it represents the response of the filter to an impulse. In fact, *a filter is represented by its impulse response function, and consequently it represents a fundamental property of the filter*. So, for us to deal with a filter, we must first be able to characterize it, which requires knowledge of its impulse response function.

Before passing the impulse through the filter, the filter had only been a black box whose contents were unknown. Conceptually, this simply indicates that if we effectively multiply the contents of a mysterious black box by 1, the result (output from the black box) provides us with a non-altered indication of what is contained within the black box.

Discrete Convolution

Now consider a more complicated input to the filter, as illustrated in Figure 7.11. Because the series have *constant spacing* (unit delays), they can be represented as simple sequences of numbers:

$$g(t) = 3,\ 2,\ 1,\ 2 \tag{7.14}$$

$$h(t) = 4,\ 3,\ -1, -2,\ 1, \tag{7.15}$$

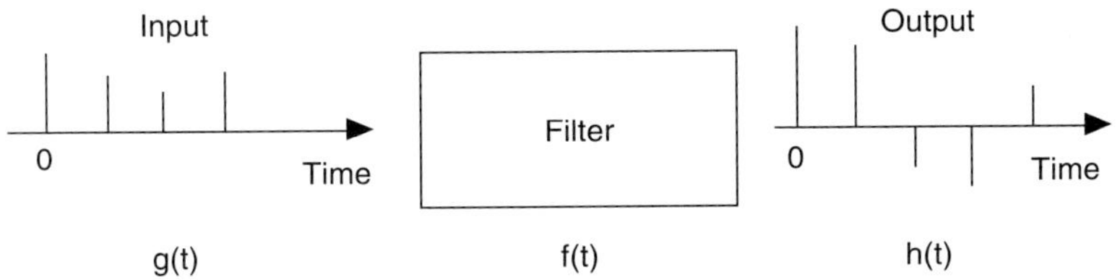

Figure 7.11 Example of a more complicated series being passed through a filter.

where we now ignore the leading and trailing zeros (the values needn't be integers, which are used here only for convenience).

The output series, h(t), is given by the *convolution* of g(t) with f(t), where f(t) is the filter. This **discrete convolution** is given as the following serial product:

$$h_k = \sum_{i=0}^{N-1} f_i g_{k-i} \tag{7.16a}$$

$$= \sum_{i=0}^{N-1} g_i f_{k-i}, \tag{7.16b}$$

where:

$N = \max(N_g, N_f)$, where N_g is the number of values in g(t) and N_f is the number of values in f(t); $\max(a, b)$ indicates the larger of a, b

k = **lag**, which represents the number of unit delays in the output sequence, h(t). There are at most $N_k = N_g + N_f - 1$ lags. k goes from 0 to $N_k - 1$, where $k = 0$ represents h(t) at t = 0.

i = number of unit delays in the input sequence and impulse response function of the filter. So, for the numbers given for g(t) and h(t) in (7.14) and (7.15), $g_{i=0} = 3$, $g_{i=1} = 2$, etc.

This operation of convolution is indicated by the symbol * (regardless of whether the convolution is discrete or continuous), so

$$h(t) = f(t)^* g(t), \tag{7.17a}$$

or simply

$$h = f^* g = g^* f. \tag{7.17b}$$

Visualizing Convolution

The discrete convolution and concept of a lag are best seen by looking at the convolution operation in a different way. Consider the two sequences on separate strips of paper:

g_i : | g_0 g_1 g_2 g_3 |

f_i : | f_0 f_1 f_2 . |

These strips show the two sequences increasing in time to the right. As explained in the previous discussion, the input sequence must be reversed before being passed through the filter, giving

g_3 g_2 g_1 g_0 .

Now consider the sequence that is generated by (7.16) for each lag k.

Lag 0: The 0th lag of the output sequence, h, (i.e., at $h_{k=0}$) is generated by aligning the paper as shown below and multiplying across the overlapping segments:

$$\left.\begin{array}{lllll} \boxed{g_3 \;\; g_2 \;\; g_1 \;\; g_0} & & \\ \quad\quad\quad\;\; \times \;\; \times & & \\ \quad\quad\quad\quad\quad \boxed{f_0 \;\; f_1 \;\; f_2} \end{array}\right\} h_0 = g_0\, f_0$$

So, the 0th lag of the output series (i.e., the first value output from the filter) is given as the product of the first term in the input series (g_0) with the first term of the filter (f_0).

This is seen in the summation of equation (7.16a) for $k = 0$. In this example, where the summation limit is given by N = 4, the summation of (7.16) is expanded to give

$$h_0 = f_0 g_0 + f_1 g_{-1} + f_2 g_{-2} + \ldots, \tag{7.18}$$

but, as seen, there are no terms in g that correspond to negative index values (we have defined g_i so that it begins at time 0). Therefore, all terms in the sum are 0 except for the first term ($f_0 g_0$), so $h_0 = f_0 g_0$.

Lag 1: The first lag of h is given by sliding the paper with the input sequence (g) one increment to the right, consistent with the series moving one more unit of delay into the filter:

$$\left.\begin{array}{lllll} \boxed{g_3 \;\; g_2 \;\; g_1 \;\; g_0} & & \\ \quad\quad\quad\quad\quad\; \times \;\; \times & & \\ \quad\quad\quad\quad\; \boxed{f_0 \;\; f_1 \;\; f_2} \end{array}\right\} h_1 = g_1\, f_0 + g_0\, f_1.$$

The convolution is the sum of the products of overlapping elements. As with lag zero, this product comes directly from (7.16a) by setting $k = 1$.

This simple process of advancing the reversed input series one more increment into the filter and then summing the product of overlapping elements is repeated. The sum gives h_k, where the lag k indicates the number of units of time that have passed since the series first entered the filter. So, $k = 0$ gives an indication of what the filtered sequence looks like that is first passed from the filter, $k = 1$ represents the component emerging from the filter one unit of time later, etc.

The graphical representation of the discrete convolution given above is actually seen to be analogous to the physical process performed by real filters. It is also a good graphical representation of the summation product given by equation (7.16).

Box 7.1 Conceptualization of Convolution

Consider a volcanic ash layer that enters a bioturbation zone[16] on the seafloor, and later we see the mixed signal in a sediment core. The first value output from the filter must have passed through the entire filter before emerging from the bottom of it in order to be preserved unaltered in the non-mixed sediment. So, does this mean that we can never actually observe h_0, and that the first point observed is really h_n, where n is the index equivalent to the entire width of the filter plus one (i.e., this would be the first lag emerging from the filter)? No; the way to consider this is that the filter's impulse response function is actually a measure of what comes out of the filter at each lag, so by its construction the first term from the filter, h_0, is really an indication of what comes out first after passing through the filter. So, when the ash first lands it is mixed throughout the mixing zone, and sometime later some fraction of it emerges from the base of the mixing layer. Then, after the ash has fallen for a while, the main pulse has more time to move through the mixing layer, and stronger and stronger concentrations emerge from the base. Eventually, the entire mass is passed through (it all sums back to the original mass of ash that entered the filter, meaning that the filter coefficients have to sum to 1, or else the filter is nonconservative). This gives the mixed-ash output sequence, but its sequence is not actually a time picture of the sequential operation that is happening within the filter; rather, it is an indication of the cumulative effects that grow as more and more ash enters the filter and is allowed to be mixed together.

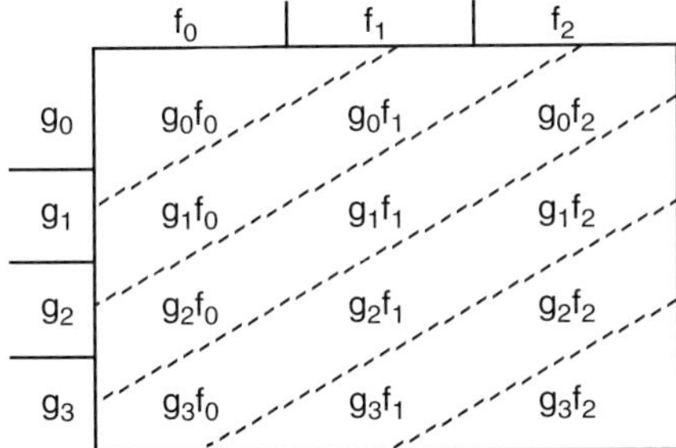

Figure 7.12 So, h_0 is the upper-left-most diagonal, h_1 is the sum along the next diagonal down, etc. The fact that we add the products along the reverse diagonals reflects the fact that the input sequence must be reversed; otherwise we would simply add along the standard diagonals.

As seen from this approach, the total number of lags in the output sequence, N_k, is given by the number of elements in the larger sequence, N_g in this case, plus one less than the number of elements in the smaller sequence, N_f. So, $N_k = N_g + N_f - 1$.

[16] A zone that is subject to mixing by critters living on the seafloor or immediately below it (e.g., worms).

Alternatively, we can obtain the same results by forming a matrix and adding the elements along "reverse" diagonals, as shown in Figure 7.12.

Linear Superposition

Another convenient and intuitive method for considering convolution involves the concept of linear superposition. With this, any particular series, g_t, can be considered as a **linear superposition** of single impulses of amplitude g_i, and zero elsewhere. Then $g_t = g^{(0)} + g^{(1)} + \ldots + g^{(n)}$, where $g^{(0)} = (3,0,\ldots)$, $g^{(1)} = (0,2,0,\ldots)$, etc., for the previous example, where g(t) = 3, 2, 1, 2.

Convolution is a linear operation, so, $f*(g+h) = f*g + f*h$. Therefore, we can consider the convolution of each individual (scaled) impulse with the filter, and simply add the results. The convolution of a single impulse of amplitude a_1 with a filter f_t gives the impulse response function, scaled by an amplitude of a_1. So, by passing each impulse of g_t through the filter, each impulse making up the total sequence in g_t replicates and scales the impulse response function about the position of the impulse. The convolved series is then simply the linear superposition of the scaled, offset impulse response functions.

For the previous example, $h = f*[g^{(0)}+g^{(1)}+\ldots+g^{(n)}] = f*g^{(0)} + f*g^{(1)} + \ldots + f*g^{(n)}$. So at any particular lag k, the filter output is the linear sum of the scaled impulse response functions. At lag $k = 2$, impulses g_0 will have passed through the filter, giving g_0*f; g_1 will have passed through the first two elements of f, giving $g_1*(f_0, f_1)$, and g_2 will have just entered, giving g_2*f_0, for a sum of h_2 = the complete filter scaled by g_0, plus (offset by one), the first two terms of the filter scaled by g_1 plus (offset by 2) the first filter term scaled by g_2.

By linear superposition: $h_{k-2} = h^{0}_{(2)} + h^{1}_{(2)} + h^{2}_{(2)} = 3f_2+2f_1+1f_0$. This is identical to the result directly produced from (7.16), but in this case we are considering the result in terms of three scaled impulse functions that have entered the filter at different times and independent of one another, though their smearing within the filter linearly combines them such that the net result is the linear superposition of each smeared series.

Running Average

Now, consider the convolution of the two functions as shown in Figure 7.13, where each value of f(t) has the same value, 1/5.

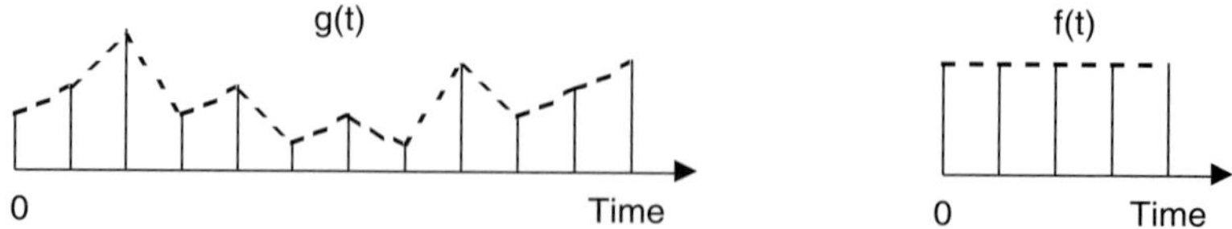

Figure 7.13 So, h_0 is the upper-left-most diagonal, h_1 is the sum along the next diagonal down, etc. The fact that we add the products along the reverse diagonals reflects the fact that the input sequence must be reversed; otherwise we would simply add along the standard diagonals.

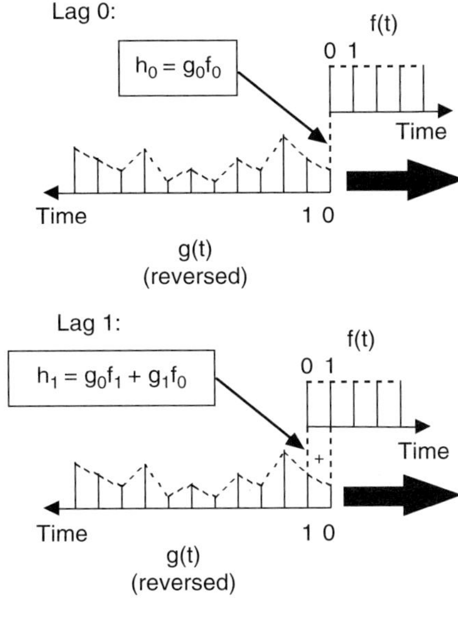

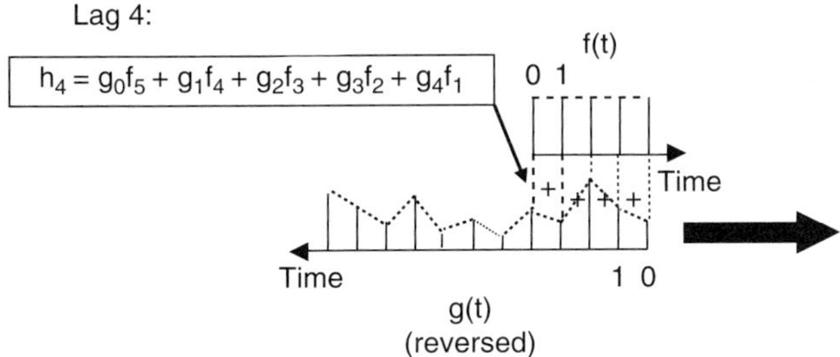

Figure 7.14 Series of impulses being passed through a filter (i.e., being convolved) with a filter impulse response function of five constant height impulses, which is a five-point running average (if all impulses of f(t) sum to 1 (i.e., each has a height of 1/5)).

If we look at this convolution with the moving-strips-of-paper approach, we get the sequence in Figure 7.14.

Or, in summation form,

$$\begin{aligned} h_0 &= g_0/5 \\ h_1 &= (g_0 + g_1)/5 \\ &\vdots \\ h_4 &= (g_0 + g_1 + g_2 + g_3 + g_4)/5 \; h_0 = g_0/5\,. \\ h_5 &= (g_1 + g_2 + g_3 + g_4 + g_5)/5 \\ &\vdots \\ h_{16} &= g_{11}/5 \end{aligned} \tag{7.19}$$

This is simply a five-point **running average** (or **moving average**) of g(t), which produces h(t), a smoothed version of g(t), as in Figure 7.14.

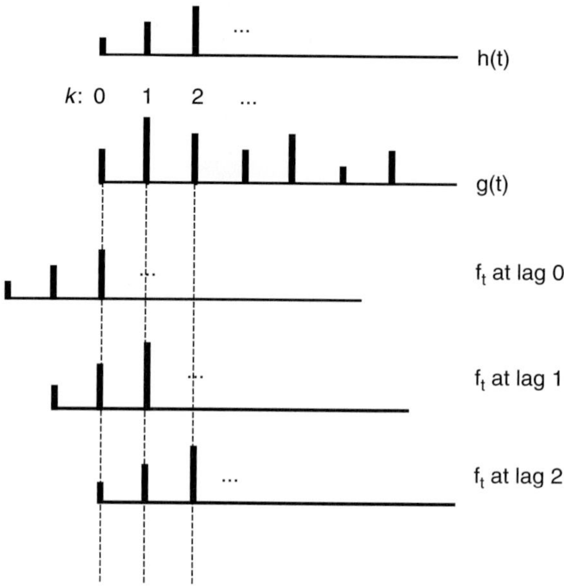

Figure 7.15 Another way of viewing convolution or passage of a series though a filter is with linear superposition. Here, the original series is multiplied by the value of f_t at the appropriate lag, offset by the lag amount, and then all of the overlapping values are added to give the final answer.

An n-point moving average would result if f(t) consisted of n points, each with a value of $1/n$, though you can employ different "weights" (values of f_t) as long as they sum to one.[17]

From a practical standpoint, n is usually an odd number of points and the filter is designed so that the first nonzero element starts at lag $(n - 1)/2$. To understand this offset, consider a three-point running average, as shown in Figure 7.15.

$$f_i = \frac{1}{3}, \frac{1}{3}, \frac{1}{3} \tag{7.20}$$

Then, the superimposed components appear as in Figure 7.15.

As seen here, the first lag ($k = 1$) in h(t), the smoothed series, should correspond to the average of the three points in g(t) centered about the second point, g_1. That is, h_1 should be equal to $(g_0 + g_1 + g_2)/3$. However, here it is seen that the desired average of the three points is actually equal to h_2. Therefore, when computing running averages, the output series should be shifted by one lag higher.

Also, the first and last $(n - 1)/2$ lags of the convolution output are ignored (or else you would end up with more data points than in the original, unsmoothed series) and the first and last $(n - 1)/2$ points of the retained output requires special consideration, since they do not represent the average of n g(t) points because half of the points in f(t) still do not overlap any points in g(t). Typically, you either ignore these points too (so the smoothed sequence is actually shorter than the original series) or you make some assumptions

[17] Different strategies for smoothing time series are discussed in more detail later, in Chapter 13.

about the values of g(t) beyond the ends. Some common treatments include a *reflective end condition*, where the points before the first point in g(t) and after the last point in g(t) are simple reflections of the points in g(t) (e.g., so, $g_{-1} = g_1$; $g_{-2} = g_2$; $g_{n+1} = g_{n-1}$; etc.). Alternatively, the g(t) sequence can be assumed to be periodic, so the next point following the last point in g(t) is equal to the first point in g(t); this is the case when smoothing in the frequency domain, as discussed in Chapter 13.

Z-Transform

The **Z-transform** is a tool that provides another alternative means of computing a convolution, though it will prove useful in a variety of other operations as well.[18] The Z-transform is accomplished by representing ordered sequences (discrete time series) as polynomials in a variable Z. For example, if g(t) = 3, 2, 1, 2, and f(t) = 5, −4, 2, then, in terms of their Z-transforms,

$$G(Z) = 3Z^0 + 2Z^1 + 1Z^2 + 2Z^3 \tag{7.21}$$

$$F(Z) = 5Z^0 - 4Z^1 + 2Z^2, \tag{7.22}$$

where each coefficient represents the value of g(t) and f(t) at successive time points (unit delays). In this respect, the variable Z represents the **unit delay operator**. So,

$$g_0 Z^0 = 3Z^0 = 3 = g(t=0) \tag{7.23a}$$

$$g_1 Z^1 = 2Z^1 = g(t=1), \tag{7.23b}$$

where

$$g_n Z^n \text{ represents } g(t=n). \tag{7.23c}$$

Multiplication by Z gives

$$ZG(Z) = 3Z^1 + 2Z^2 + 1Z^3 + 2Z^4 \tag{7.24}$$

so the product is G(Z) delayed by one unit of time, that is, a simple shift of g(t) by one unit, hence the phrase *unit delay operator*.

So, the Z-transform is the representation of *evenly spaced* sequences as polynomials in Z.

Convolution in Z-Transforms

The convolution of g(t) and f(t) in terms of Z-transforms is given as the *product* of the transformed sequences. So,

$$g*f = G(Z)F(Z). \tag{7.25}$$

This is the representation of a serial product in Z-transforms.

Consider an example of this product for the two sequences above:

[18] Do not confuse this Z-transform with statistical standardization (when normalizing data to a 0 mean and standard deviation of 1).

$$\begin{aligned} G(Z)F(Z) &= (g_0Z^0 + g_1Z^1 + g_2Z^2 + g_3Z^3)(f_0Z^0 + f_1Z^1 + f_2Z^2) \\ &= g_0f_0Z^0 + g_1f_0Z^1 + g_2f_0Z^2 + g_3f_0Z^3 + g_0f_1Z^1 + \\ &\quad g_1f_1Z^2 + g_2f_1Z^3 + g_3f_1Z^4 + g_0f_2Z^2 + g_1f_2Z^3 + \\ &\quad g_2f_2Z^4 + g_3f_2Z^5, \end{aligned} \tag{7.26a}$$

combining like terms:

$$\begin{aligned} G(Z)F(Z) = g_0f_0Z^0 + (g_1f_0 + g_0f_1)Z^1 + (g_2f_0 + g_1f_1 + g_0f_2)Z^2 + (g_3f_0 + g_2f_1 \\ + g_1f_2)Z^3 + (g_2f_2 + g_3f_1)Z^4 + g_3f_2Z^5. \end{aligned} \tag{7.26b}$$

Inserting the values for G(Z) and F(Z) from (7.14) and (7.15),

$$H(Z) = 15Z^0 - 2Z^1 + 3Z^2 + 10Z^3 - 6Z^4 + 4Z^5, \tag{7.27a}$$

so

$$h(t) = 15, -2, 3, 10, -6, 4. \tag{7.27b}$$

Thus, the output of the series g(t), after passing through a filter with impulse response f(t), yields the above-computed series h(t). Z-transforms have numerous beneficial properties especially useful for filter theory and linear parametric modeling.

Special Cases of Convolution

Semi-Infinite Series

The convolution of semi-infinite series can be computed if both series are infinite in the same direction. That is,

$$\{g_0\ g_1\ g_2 \ldots\} * \{f_0\ f_1\ f_2 \ldots\} = \{h_0\ h_1\ h_2 \ldots\}. \tag{7.28}$$

If the series are infinite in opposite directions, then the convolution does not exist,

$$h(t) = \{\ldots g_{-2}\ g_{-1}\ g_0\} * \{f_0\ f_1\ f_2 \ldots\}, \tag{7.29}$$

since the summation can clearly not be constructed – the 0th lag of h(t) would involve the product of $g_{-\infty}f_0$.

Two-Sided Functions

Up until this point, we have only considered one-sided functions – that is, functions for which the negative lags = 0. **Two-sided sequences** (of finite extent) are represented as

$$\begin{matrix} \ldots & g_{-2}Z^{-2} & g_{-1}Z^{-1} & g_0Z^0 & g_1Z^1 & g_2Z^2 & \ldots \\ & \uparrow & \uparrow & \uparrow & \uparrow & \uparrow & . \\ & t=-2 & t=-1 & t=0 & t=1 & t=2 & \end{matrix} \tag{7.30}$$

The significance of this scheme is realized only when dealing with filters.

Causal or realizable filters are those in which the first term of the impulse response function is Z^j, where $j \geq 0$. That is, all values such as $f_{-2}Z^{-2}$ $f_{-1}Z^{-1}$ have the coefficient $f_{-i} = 0$. In the context of a physical system, *a causal filter is one that does not respond before it is forced*. So, if

$$\begin{aligned} g(t) &= \text{input series which begins at some time } t = 0 \text{ and progresses forward in time, so} \\ &= g_0Z^0 + g_1Z^1 + \ldots, \end{aligned}$$

then, if f(t) has a nonzero coefficient for any Z^j where $j < 0$ (i.e., if the filter is *not* causal), $g(t) * f(t)$ will have at least one value for that Z^j term. In other words, the output series will show a response *prior* to the onset of the forcing. Obviously, for physical systems, all filters should be causal (realizable). This, of course, seems obvious, but care must be taken to keep this fact in mind, since *many numerical filters with desirable properties are actually noncausal* (as will be seen later). Actually, careless application of a running filter often leads to noncausal results. For example, consider the smoothing by running average of a step function (e.g., the ash fall deposited on the seafloor), as shown in Figure 7.16.

First difference of a series can be computed by convolving the series with

$$f(t) = \{1 - 1\}. \tag{7.31}$$

Convolution Properties

Convolution rules are easily determined by manipulation of the summation in equation (7.16) for the convolution operation. In particular, some of the more useful rules obeyed by convolution include (for functions f(t), g(t) and h(t))

$$f * g = g * f \tag{7.32}$$

$$h * (f * g) = (h * f) * g \tag{7.33}$$

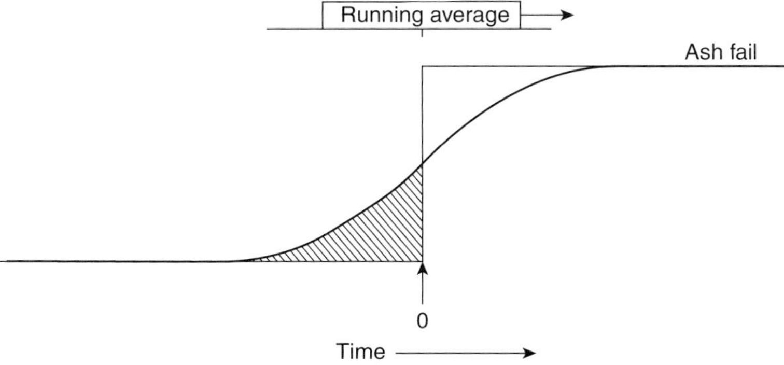

Figure 7.16 Example of smoothing a step function (e.g., ash layer in a deep sea sediment core), whereby centering the running average filter places the origin such that half of the filter has coefficients with negative lags. As a consequence, some of the ash has been mixed to times before the ash has fallen (hatched area), so the filter is noncausal.

$$h * (f + g) = h * f + h * g, \tag{7.34}$$

where this last rule represents the linear property.

The linear rule applies to all **linear filters**, so if

$$g_1(t) * f(t) = h_1(t) \tag{7.35a}$$

$$g_2(t) * f(t) = h_2(t) \tag{7.35b}$$

then

$$[g_1(t)+g_2(t)] * f(t) = h_1(t)+h_2(t) \tag{7.35c}$$

Factoring of a filter is very important in linear filter theory and inverse filtering. Any polynomial (such as that resulting from a Z-transform) can be factored into terms with the form

$$(a_0 + a_1 Z^1). \tag{7.36}$$

Thus, the impulse response function of any filter whose Z-transform is linear in Z can be expressed as a cascade of simple two-term filters (called **dipoles**). We will make use of this when *deconvolving* (unmixing) time series.

7.4 Serial Correlation

Serial correlation takes on one of two specific forms, **cross-correlation** or **autocorrelation**, depending on what is being correlated. In both cases, the mathematical operation is a slight variant on the definition of convolution.

Serial Cross-Covariance and Cross-Correlation

Convolution, as shown above, involves the reversal of one of the series being convolved. Consider the summation of (7.16) when neither series is reversed. In that case, the summation is given as

$$h_k = \sum_{i=0}^{N-1} y_i x_{k+i} \tag{7.37a}$$

The above sum is sometimes indicated using the star symbol, $\star$, so

$$h = y \star x. \tag{7.37b}$$

When $k = 0$, (7.37a) gives the value of h at the 0th lag,

$$h_0 = \sum_{i=0}^{N-1} x_i x_i, \tag{7.38a}$$

or, normalized so as to represent a covariance at lag 0,

$$S_{fg} = \frac{1}{n-1}\sum_{i=1}^{n}[(y_i - \bar{y})(x_i - \bar{x})]. \tag{7.38b}$$

This summation is nothing more than $\mathbf{y}^T\mathbf{x}$, where $\mathbf{y}$ and $\mathbf{x}$ are column vectors. With the removal of the means and $(1/n - 1)$ normalization of (7.38b), this serial product is the *sample covariance* between them, s_{yx} = Cov[y, x]. The correlation, ρ_{yx}, is this normalized sum divided by estimates of $\sigma_y\sigma_x$ (i.e., $\rho = \sigma_{yx}/\sigma_y\sigma_x$). Therefore, this summation is proportional to the covariance and correlation.[19]

Now consider the sum at lag 1:

$$h_1 = \sum_{i=0}^{N-1} y_i x_{i+1}. \tag{7.39}$$

This is seen to give the correlation between the time series y(t) and x(t) (assume that the time series are standardized), when x(t) has been displaced relative to y(t) by one sampling unit. Therefore, we are no longer comparing the pairs (y_0,x_0), (y_1,x_1), ... but rather, (y_0,x_1), (y_1,x_2), ...

So, each lag of h_k is simply a measure of the degree of correlation between y(t) and x(t) at various offsets between the two time series. This sum is therefore called **serial cross-correlation** or **cross-covariance, sequence, products** or **function** (depending upon the normalization used). Specifically, if the sum is divided by $(n - 1)s_y s_x$, then the sum is the sample cross-correlation; otherwise, if it is only divided by $(n - 1)$, the sum is the sample cross-covariance. If the series are standardized already, the sum divided by $(n - 1)$ is the sample cross-correlation. However, in nonstatistical applications, the sum (7.37a) is often referred to as the cross-correlation function (though more rigorous notation limits use of the term "function" to the cross-correlation or cross-covariance of continuous functions, as opposed to discrete sequences, as are presented here).[20]

The cross-correlation function between functions y(t) and x(t) is often abbreviated as **ccf,** and the cross-covariance function as **ccvf**, particularly when they are representing the sum of (7.37a), even if ignoring the form of the normalization and statistical implications (discussed below). The statistical notation is such that R_{yx} or r_{yx} is often used to represent the sample ccf – that is, our best estimate of the true ccf, ρ_{yx}. The sample ccvf is often represented by C_{yx} or c_{yx}, which is the sample-based estimate of the true ccvf, γ_{yx}. From a statistical standpoint, if when estimating the ccf or acf via a serial product such as (7.38b) (i.e., assuming ergodicity), some people will indicate them as ccsp and acsp, to reveal the estimation via working on a single realization through time instead of across multiple realizations.

[19] Actually, the form of normalization can be slightly more complicated than just dividing by $n - 1$ or $(n - 1)\sigma_y \sigma_x$, as discussed later in this section.

[20] The continuous forms of these sequences are given later, though they are essentially the same as the definition here replacing the summation operator with an integral.

The statistical considerations of these functions, as well as that of the acvf and acf discussed immediately below, are presented at the end of this chapter. For now, we simply want to become acquainted with the mathematical operations, general nomenclature and fundamental implications.

Serial Autocovariance and Autocorrelation

Serial Autocorrelation or **autocovariance** (again depending on which normalization is used and ignoring, for now, statistical formalizations), is similar to serial cross-correlation except that in this case, instead of comparing two different time series in the sum, the comparison is made between one series and that same series itself shifted in time. That is, for n overlapped points: $n = \mathrm{N} - |\mathrm{k}|$,

$$\textbf{acvf (or acvsp)} : \mathrm{h}_k = \mathrm{c}_{\mathrm{ff}}(k) = \frac{1}{n-1}\sum_{i=0}^{n-1} \mathrm{g}_i \mathrm{g}_{k+i}. \tag{7.40a}$$

The values (7.40a) at the different lags provide an indication of the covariance between neighboring values in a single time series. Note that when $k = 0$, so that the series g is being compared directly to itself without any time shift whatsoever, then $\mathrm{h}_0 = s_{\mathrm{g}}^2$ – that is, the covariance with itself (which is the *variance* of g). The autocovariance function is often abbreviated as **acvf**, and the sample acvf given as $\mathrm{C(k)}_{\mathrm{ff}}$ or c_{ff}, with the true acvf, γ_{ff}:

$$\textbf{acf (or acsp)} : \mathrm{h}_k = \mathrm{r}_{\mathrm{ff}}(k) = \frac{1}{[(n-1)s_{\mathrm{g}}^2]}\sum_{i=0}^{n-1} \mathrm{g}_i \mathrm{g}_{k+i}. \tag{7.40b}$$

The autocorrelation function of a function f(t) is often abbreviated as **acf**. The sample acf is given as R_{ff} or r_{f}, and the true acf as ρ_{ff}.

The concept of the acvf is extremely important, as it provides a fundamental *joint*-statistical moment for time series, as is discussed later. *This statistic is as important to sequential data as variance is to univariate (nonsequential) random variables, and as covariance is to multivariate (nonsequential) random variables*. Autocovariance and autocorrelation are used extensively throughout the remainder of the text.

Note that (7.40) is seen to contain the serial autocovariance product at lag k (i.e., h_k) in each element as

$$= \begin{bmatrix} \mathrm{h}_0 & \mathrm{h}_1 & \mathrm{h}_2 & \cdots & \mathrm{h}_n \\ \mathrm{h}_1 & \mathrm{h}_0 & \mathrm{h}_3 & \cdots & \\ \mathrm{h}_2 & \mathrm{h}_3 & \mathrm{h}_0 & \cdots & \\ & & \vdots & & \\ \mathrm{h}_{n-2} & \mathrm{h}_{n-1} & & & \\ \mathrm{h}_{n-1} & & & & \end{bmatrix}. \tag{7.41}$$

Interpretation of Correlation Functions

Lags of the Correlation Sequence

Notice that in the summations given by (7.37a) and (7.40), the lags from 0 to $N-1$ give the overlap between the two series when one is shifted to the right relative to the other.

Lag 0: That is, at $k=0$, the two series being correlated, considering the strip of paper approach, are aligned as follows.

For cross correlation,

$$\begin{array}{cccc} g_0 & g_1 & g_2 & g_3 \\ \times & \times & \times & \\ f_0 & f_1 & f_2 & \end{array},$$

or for autocorrelation,

$$\begin{array}{cccc} g_0 & g_1 & g_2 & g_3 \\ \times & \times & \times & \times \\ g_0 & g_1 & g_2 & g_3 \end{array}.$$

Note that neither series g(t) nor f(t) has been reversed. *If the means have been removed*, this sum of products, divided by $(n-1)$ or $(n-1)s_f s_g$, gives the degree of covariance or correlation, respectively, between the overlapping components of f(t) and g(t). If the series have been standardized (mean removed and divided by the standard deviation), then the sum divided by $(n-1)$ gives the degree of correlation directly.

Regardless of the normalization, the lag 0 is defined when both sequences are aligned in time so that each overlapping pair represents the same time.

Lag 1: At lag $k=1$, we get, as seen by examining the products from (7.37a) and (7.40) for $k=1$, the following:
for cross-correlation,

$$\begin{array}{ccccc} g_0 & g_1 & g_2 & g_3 & \\ & \times & \times & \times & \\ & f_0 & f_1 & f_2 & \end{array},$$

or for autocorrelation,

$$\begin{array}{ccccc} g_0 & g_1 & g_2 & g_3 & \\ & \times & \times & \times & \\ & g_0 & g_1 & g_2 & g_3 \end{array}.$$

Here, we get the covariance or correlation between the overlapping points, which are shifted one point to the left (assume that the upper series is the one shifted).

This continues for N^+ lags, where $N^+ = \max(N_f, N_g)$ as before, after which there are no more overlapping points if the longer sequence is the one being shifted; otherwise

$N^+ = \min(N_g, N_f)$, where min(a,b) returns the smaller of either a or b. This does not give any of the values in which the upper series is shifted to the *right*. Recall that with convolution, we actually start with the upper series shifted all the way to the left and eventually end up with it shifted all the way to the right at the final $N_g + N_f - 1$ lag. Here, however, because neither series is reversed, we start with the series completely aligned (centered), and only shift to the left with the positive lags. Therefore, the lags representing the upper series shifted to the right are given as *negative lags*.

Lag −1: Consider, $k = -1$.

For cross-correlation,

$$\begin{array}{cccccc} & \boxed{g_0 & g_1 & g_2 & g_3} \\ & \times & \times & & \\ \boxed{f_0 & f_1 & f_2} & & & \end{array},$$

or for autocorrelation,

$$\begin{array}{ccccc} & g_0 & g_1 & g_2 & g_3 \\ & \times & \times & \times & \\ g_0 & g_1 & g_2 & g_3 & \end{array}.$$

Notice that for the autocorrelation, because of symmetry, this lag is actually identical to the overlap for $k = 1$. Consequently, *for autocovariance and autocorrelation, the negative lags are identical to the positive lag values and the autocorrelation function is symmetrical about the 0 lag.*

Box 7.2 Cautionary Note for Computing Serial Correlations

Since serial correlation, and convolution as well, involves the summation of products of numbers, when computing the correlation or convolution of exceptionally long series it is important to pay attention to the precision of the numerical representation on the computer (e.g., single or double precision, given whatever bit representation of the numbers). The summed products can exceed the number of significant digits represented by the computer, after which all additional products added to the sum are lost. Therefore, it is a good idea to estimate the magnitude of the serial products and make sure that the computational precision is set large enough to handle the necessary number of digits. Often, a triangular-shaped acf or ccf as a function of lag manifests when the numerical precision is inadequate to accommodate the magnitude of the numbers encountered in the serial products, though triangular shapes can occur naturally as well – e.g., when a sequence's variability is small relative to its mean or you convolve two rectangular functions.

As seen by inspection of the above figures, the negative lags can continue for $N^- = \min(N_g, N_f) - 1$ if the longer sequence is being shifted; otherwise, $N^- = \max(N_f, N_g) - 1$. In the end, regardless of whether the longer or shorter sequence is shifted, we obtain a total of $N^+ + N^- = N_g + N_f - 1$ lags, just as we did in convolution. Of these, N^- are negative lags and N^+ (which includes the 0th lag) are positive.

With the ccf and acf, the series f(t) does not designate a filter (we are no longer convolving a series by passing it through a filter), as both f(t) and g(t) are simply time series being correlated at different offsets.

Shape of the Correlation Sequence

An acf typically has a general form that looks similar to that of Figure 7.17 – always largest at the zero lag, where the correlation of the series with itself is a perfect 1. The degree of correlation dies off with increasing lags after the maximum value of 1 at $k = 0$. As the correlation decreases from 1 to 0 near the small lags, the nonzero correlation reflects a system "memory." In other words, the value at any one point has some memory of the value k sample points away, since there is a correlation between all pairs of points separated by an amount k.

Consider an unlikely situation in which the acf has a value of 1.0 at lag 5. This means that the correlation coefficient, describing the degree of linear association between the time series and itself (when shifted 5 values over), is equal to 1. That is, we can consider this as the correlation of two entirely different time series involving comparison of the pairs (g_1, g_5), (g_2, g_6), . . ., (g_{n-5}, g_n). When plotting these pairs on a graph with g(t) along the abscissa and g(t + 5) along the ordinate, the pairs of points fall perfectly along a straight line (the regression line), and the correlation coefficient of $r = 1$ indicates that 100 percent of the variance in g(t + 5) can be predicted by knowledge of g(t). In other words, the value of $r = 1$ at lag 5 indicates that the time series varies in a manner such that, given knowledge of any point in the series, you could predict (with 100 percent accuracy) the value of the time series five points into the future (or into the past).

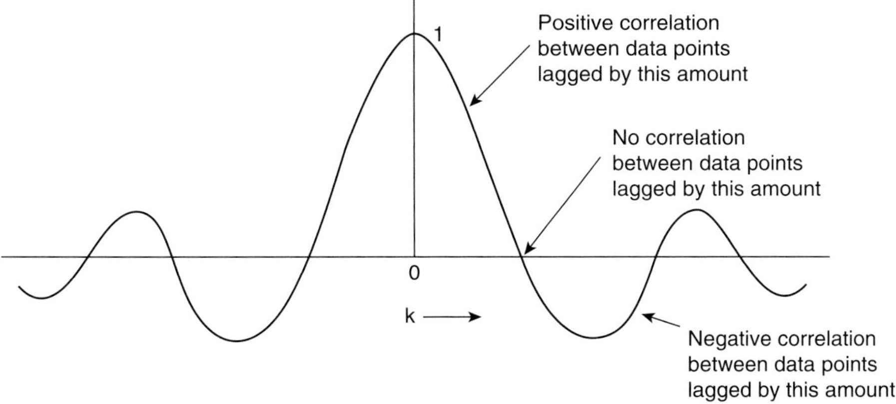

Figure 7.17 General form of a typical autocorrelation function (acf), with sections labeled regarding its interpretation.

If the acf had a value of .9 for lag 5, then this would indicate that, given knowledge of any point in the time series, you could predict ~81% (r^2) of the variance about the mean, of the value of the time series five points away. However, given the natural variability in the data, your prediction would still not account for ~19 percent of the variance at each point five points away from the present position.

This same sort of reasoning and interpretation applies to cross-correlation functions. The main difference with the ccf is that the function needn't be symmetrical, since the negative lags will not be the same as the positive lags, and the maximum value of the ccf is not necessarily at the 0th lag, as it always is in the acf (at the 0th lag of the acf, the correlation will always be perfect, and cannot be better). Also, the ccf is of course comparing the relationship between two different series, not just one series. For the ccf, the lag at which the maximum ccf value occurs indicates the amount of linear shift required in one series relative to the other in order to achieve the highest degree of linear correlation between the two series. Also, as with the acf, the system memory is indicated by how long (i.e., over how many lags) the ccf is significantly different from zero when moving away from the lag with the maximum ccf value.

For example, if the maximum value of the ccf occurs at lag 8, giving an r value of, say, 0.8, then this implies that, given knowledge of one time series, ~64 percent (r^2) of the variance about the mean of the value of the second time series 8 points into the future can be predicted.

A typical ccf will look like that in Figure 7.18 (only not nearly as smooth as that schematic). The longer an acf (or ccf) stays significantly different from zero (positive or negative r values) as you move away from the 0th lag, the more system memory it has. That is, high acf values continuing out to say, 20 lags, would indicate that the system "memory" is such that given any value in the time series, you could predict with some degree of confidence (given by r^2 and its error) the value within the time series to be

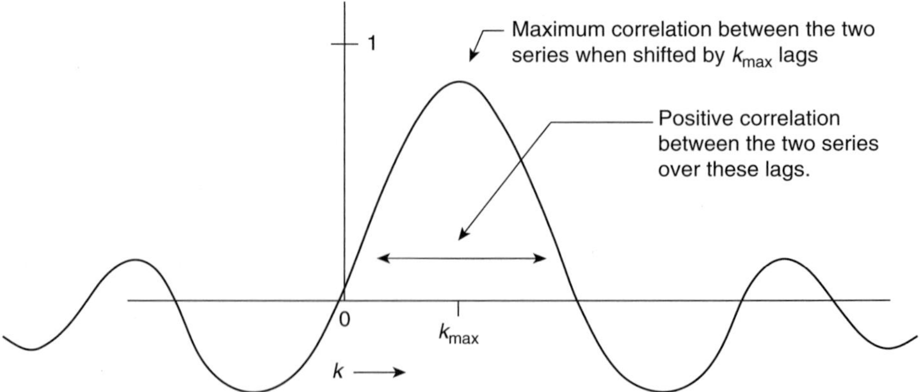

Figure 7.18 General form of a typical crosscorrelation function (acf), with sections labeled regarding its interpretation.

expected 20 points away. Obviously, this information is highly desirable if you wish to make predictions concerning future behavior (such as for the stock market), given knowledge of the present and past behavior of the time series. In fact, for time series, the nature of the acf represents a fundamental characteristic of the time series, since it provides important information regarding the nature of the ordering of the observations and their relationships.

Correlation/Decorrelation Length

The system memory is often given in terms of a **correlation length** or **decorrelation length**. These terms, despite having a completely opposite implication, are synonymous – equivalent to noting that a glass is half full or half empty. Formally, the length provides a measure, in lags, of the length or time (in units of lags) that the correlation remains statistically different from zero. Loosely, it is the number of lags from the maximum correlation to the lag at which the correlation is no longer statistically distinguishable (*at all higher lags*) from 0 for a given level of significance. If the acf passes zero but continues strongly (significantly) into negative r values, then there is still a correlative relationship, and the decorrelation length occurs once the r values tend to zero for keeps. It can be defined for a cross-correlation function or an autocorrelation. Formally, von Storch and Zwiers (2002) show it to be

$$\tau_{\mathrm{d}} = 1 + 2\sum_{k=0}^{\infty} \rho(k). \tag{7.42}$$

This gives τ_{d} in terms of lags. It can be put in units of time by multiplying it by Δt (time increment between lags). In addition to being the distance to the lag at which the correlation is not different from 0 (statistically), in some cases this length may be defined as the number of lags until the correlation drops off to (1) 1/e of its maximum value (1 in an acf, ≤ 1 in a ccf), (2) 1/2 the maximum value, (3) a specified correlation value, i.e., a value above which you can still predict a given percentage of the variance or (4) some other explicitly stated criterion.

Its interpretation is similar to that described previously for interpreting the length of the system memory – in fact, this is just a formal definition of system memory. For example, if it is stated that the spatial surface temperature structure of some region is such that it has a correlation length of 10 km, this implies that you can still predict the temperature 10 km away from any particular measurement within some specified confidence interval. With this information, a sampling program would obviously not need to sample the temperature every fraction of a km, but would just need to sample it better than every 10 km (actually, we would like several realizations of less than 10 km, to provide some statistical stability in the measurements).

White Noise Series

Recall that *each individual point in a white noise series is uncorrelated to all other points (neighboring and distant) in the series*. Thus, it is a purely random series that has *no* system memory (correlation length is zero). Given this definition, the autocorrelation function of white noise is a unit impulse at lag 0 (representing the correlation of the white noise series to itself, with no offset in time). At all other lags, the autocorrelation function is effectively zero, since by definition there is no correlation between neighboring points.[21]

The autocovariance function of white noise is similar to the autocorrelation function, except that at the origin (0th lag), the value is not 1 but instead is the variance, σ_ε^2, of the white noise process.

Colored Noise

In addition to white noise, which has a distinct acf of 1 at lag 0 and of 0 elsewhere, there are other types of noise that are also characterized by the nature with which neighboring points are related in the series. Most notable of these is red noise, which is nothing more than the integration of a white noise series. As a consequence, red noise, often called a "random walk," has unlimited variance, since the variance can grow without bound as one integrates over longer and longer time scales.

What distinguishes red noise and other "colors" of noise is their characterization in the frequency domain, as is discussed in a later chapter.

Notation and Normalization Considerations

Notation

Some cautionary remarks are warranted to avoid confusion in the literature and when making computations. First, it is extremely common (in fact it is nearly ubiquitous in nonstatistical books, papers, etc.) to refer to the summations in (7.37a) and (7.40) as cross-correlation and autocorrelation – *given no normalization*. That is, the simple sum of products, unnormalized, is frequently called cross-correlation or autocorrelation. The words cross-covariance and autocovariance are infrequently used outside statistical literature. Therefore, it is important to always be aware of the true meaning of the summations that you encounter for proper interpretation, and to make it clear how you are defining all terms.

Often, you may only be interested in the shape of the ccf or acf so that you can get the correlation length or other such information. In that case, the normalization is relatively unimportant, though the means must be removed. However, if you are actually interested in the correlation or covariance values, then the means must be removed and the normalizations applied. To avoid confusion, unless otherwise mentioned, I will follow

[21] The values of the acf at all lags other than zero is still not exactly zero, due to the fact that random chance occasionally introduces some correlation between neighboring points.

the precise definitions stated previously (i.e., covariance or correlation imply application of the appropriate normalization).

Normalization

When estimating the cross-covariance or autocovariance or correlation functions, it is seen that we divide by $(n - 1)$ or $(n - 1)s_f\ s_g$ for covariance or correlation, respectively. However, the sample size and standard errors (i.e., the n, s_f and s_g) are changing constantly as a function of lag as we continuously lose some of the compared series in the overlap at the ends. For the ccf of the previous example, at $k = -1$, we have

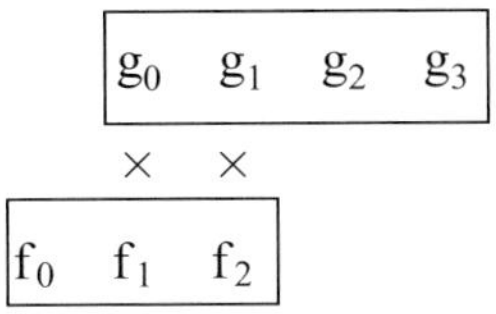

$$\begin{array}{cccccc} & & g_0 & g_1 & g_2 & g_3 \\ & & \times & \times & & \\ f_0 & f_1 & f_2 & & & \end{array}$$

where the comparison of the series involves only two points in each time series. That is, our correlation is lagged to the point that we are now only investigating the relationship between the two series over the final two points in the series. If we were computing the *sample correlation coefficient, r*, between these overlapping segments, we would estimate the mean and standard deviations from the overlapped points, subtract off those means, then sum $g_0f_1 + g_1f_2$, and divide by $(n - 1) = 1$ and the estimates of σ_f and σ_g (s_f and s_g).

For the ccf or acf, re-computing the mean and standard deviation as a function of overlapped points is neither practical nor justified. That is, to justify the calculation of the acf or ccf by examining the relationship between neighboring points as a serial product in a single time series or pair of series (as opposed to ensemble averaging across the ensemble), it is implicitly assumed that the time series are stationary. This means that the statistical moments *do not vary* with time. Consequently, the means and standard deviations do not change with a smaller and smaller amount of overlap (i.e., with shorter segments of the original time series). Therefore, *the best estimate of the means and standard deviations – those that are estimated from the entire time series – are always used in the normalization, regardless of lag.*

In this respect, the actual values of the ccf or acf at higher lag values, where the amount of overlapped points is small, is different from standard sample correlations determined if we treated the overlapping segments as individual time series and computed the correlation coefficient using the standard formula. This is because the estimated values of the means and standard deviations of the short segments will deviate from those values computed for the longer time series. Recognize that this is simply a matter of statistical estimation and not a fundamental difference in the definition of correlation versus autocorrelation at any lag k. That is, the true correlation or autocorrelation at a lag k would be given by the deviations in f and g about the true means, μ_f and μ_g, and then normalized by the true standard deviations, σ_f and σ_g. Here we are stating that the estimation of these moments should be made based on as much information as is available. For autocorrelation, that involves the full time series, not just the number of values involved in the computation at any lag. Consequently, we have better estimates of the moments, and thus a greater confidence that our estimated correlation at any lag is closer to the true correlation for that lag (the Central Limit Theory says that the error in the estimate of the mean decreases inversely with the number of points used to make the estimate).

Normalizations and Statistical Properties of the Sample acvf

Consider the form of the **sample acvf** designated $C_{gg}(k)$ and formulated to estimate $\gamma_g(k)$, the **true acvf**. Assume a mean of 0 for g(t), or g_t, and modify the equations to include the subtraction of the true mean from g_t (allowing us to ignore the loss of a degree of freedom in our normalization because of the estimate of the mean). So, the sample acvf, written as a function of lag k, is

$$\hat{C}_{gg}(k) = \frac{1}{n-|k|}\sum_{i=1}^{n-|k|} g_i g_{i+k}, \tag{7.43}$$

where the form of (7.43) shows the explicit number of terms in the sum for each lag as $n - |k|$, k being the present lag being computed, n being the number of points in g_t.

The expected value of the acvf, $C_{gg}(k)$, at any lag k is determined with the expectance operator in the same manner as it is used to compute the mean of any function. That is,

$$E[C_{gg}(k)] = E\left[\frac{1}{n-|k|}\sum_{i=1}^{n-|k|} g_i g_{i+k}\right]. \tag{7.44a}$$

Recall that $E[aX + bY] = aE[X] + bE[Y]$, so the above sum can be rearranged as

$$= \frac{1}{n-|k|}\sum_{i=1}^{n-|k|} E[g_i g_{i+k}]. \tag{7.44b}$$

By definition, the expectance of $E[g_i g_{i+k}]$, where g_i have zero mean, is the true auto-covariance function, $\gamma_g(k)$. That is, $E[g_i g_{i+k}]$ is the true covariance between g_i and g_{i+k}, or between g_i and itself at each lag k. Therefore,

$$E[C_{gg}(k)] = \frac{1}{n-|k|}\sum_{i=1}^{n-|k|} \gamma_g(k), \tag{7.44c}$$

and the $n - |k|$ additions of the constant $\gamma_g(k)$ for any particular lag k equals $(n - |k|)\gamma_g(k)$, so

$$\mathrm{E}[\mathrm{C_{gg}}(k)] = \gamma_{\mathrm{g}}(k). \tag{7.44d}$$

Therefore, the sample acvf as defined in (7.43) is an *unbiased* estimate of the true acvf, given knowledge of the true mean, μ, of the series. If we estimate the mean and subtract that from g by a similar, though more tedious process, it can be shown that the denominator in the normalization constant in (7.43) must be $n - |k| - 1$, as previously derived, to yield a zero bias.

Preferred Normalization

A better definition of the acvf, and the one we will use for the remainder of the text, uses 1/n as the normalization constant (i.e., for all lags divide by the total number of data points, n, in the time series). In this case, the sample acvf, $\mathrm{C_{gg}}(k)$, is defined as

$$\mathrm{C_{gg}}(k) = \frac{1}{n}\sum_{i=1}^{n-|k|} \mathrm{g}_i \mathrm{g}_{i+k}. \tag{7.45}$$

For this definition, the expectance is

$$\mathrm{E}\left[\mathrm{C_{gg}}(k)\right] = \frac{n - |k|}{n}\gamma_{\mathrm{g}}(k). \tag{7.46}$$

The formulation of the acvf given by (7.45) is thus a *biased* estimate of $\gamma_g(k)$, since the expectance is not the true acvf but rather $(n - |k|)/n$ times the true acvf. For $k \ll n$ (i.e., for the lower order lags), the bias $\gamma_g(k) - \mathrm{E}[\mathrm{C_{gg}}(k)] = [1 - (n - |k|)/n]\gamma_g(k)$, or $-(|k|/n)\gamma_g(k)$, is quite small, so if only dealing with low-order lags, either form of the acvf, (7.43) or (7.45), is acceptable. When considering the variance of these two forms of normalization, however, the picture becomes clearer regarding which form is "better" when going to larger lags.

Note that the approach presented here for computing the mean of the acvf is directly applicable for computing the mean of the ccvf, acf and ccf (the latter two involve some additional constants). For either definition (7.43) or (7.45), the sample autocorrelation function is the sample autocovariance at each lag divided by the 0th lag of the acvf (= sample variance of g_t) – this definition facilitates computation of the acf or ccf statistical moments.

Variance of acvf and ccvf

We typically think of the "best" estimator of a population or ensemble parameter as the one with the smallest variance (the smallest uncertainty). However, in the case of an unbiased estimator such as $\mathrm{C_{gg}}(k)$ in (7.43), such a criterion is not necessarily the right one. A better measure of the best statistic might instead be taken as the one that gives the **minimum mean squared error (MMSE)** – that is, the one that gives the smallest sum of squared errors between the estimated value and true value. Conceptually, this sounds like something that an estimator with the smallest variance would satisfy, but that is not the case. The variance represents the spread about the mean. If the mean estimate is biased, then the variance accounts for the random mismatch between the estimate and estimated mean value, while any bias accounts for an additional mismatch between the estimate and true value.

For example, in an extreme (though unlikely) scenario, if an estimate is extremely biased but has almost zero variance (i.e., it is known with extremely good precision but terrible accuracy), then the small variance is a misleading indicator of the quality of the estimate, since one's first reaction would be that the estimate is extremely good. Rather, the estimate is extremely precise, but quite inaccurate regarding how well it is estimating the value of interest. The MMSE provides a more balanced perspective of the overall quality of an estimate, since it is given by the variance of the estimate *plus* the bias squared. Together, these give the total mismatch between the estimate and true values. If the bias is 0, then the best estimate is given by the estimator with the smallest variance, as always, which is equivalent to the least-squared error estimate.

In the case of $C_{gg}(k)$ where the bias is not zero, this biased estimate of the acvf usually has a smaller MMSE than the unbiased estimate $C_{gg}(k)$ for larger lags – we call the unbiased estimate $C_{ugg}(k)$ to differentiate it from the biased estimate. For the latter, the magnitude of the bias, $|k/n|\gamma_g(k)$, gets larger with increasing lag k, for $C_{gg}(k)$, but the *variance of the unbiased estimate*, $C_{ugg}(k)$, gets even larger with lags, since the estimates at these lags are based upon very few data points. For that reason, *the biased form of the acvf, $C_{gg}(k)$* (7.45), *with* 1/n *as the normalization, gives a smaller minimum mean-squared error at larger lags and is the more common form to use*.

The sample acvf offers a good (as defined by its expected value) estimate of the true acvf, now consider an estimate of its variance. As with all random variables and processes, because of noise in the data, the sample acvf differs from the true acvf and we must estimate the magnitude of scatter by which the sample acvf may deviate from the true acvf. This is most easily done by considering the covariance.

The covariance between $C_{gg}(k)$ and $C_{gg}(k+j)$ is given by

$$\begin{aligned}\mathrm{Cov}[C_{gg}(k), C_{gg}(k+j)] &= \mathrm{E}\Big[\Big\{C_{gg}(k) - \mathrm{E}[C_{gg}(k)]\Big\}\Big\{C_{gg}(k+j) - \mathrm{E}[C_{gg}(k)]\Big\}\Big] \\ &= \mathrm{E}\Big[\Big\{C_{gg}(k) - b\gamma_g(k)\Big\}\Big\{C_{gg}(k+j) - b\gamma_g(k+j)\Big\}\Big],\end{aligned} \tag{7.47}$$

where $b\gamma_g(k) = \mathrm{E}[C_{gg}(k)]$ is the mean of $C_{gg}(k)$. For an unbiased estimate $b = 1$, but for the biased estimate $b = (n - |k|)/n$, according to (7.44). Note that the variance of $C_{gg}(k)$ is given by the above expression when $j = 0$ (no lag). This is the standard form of the covariance (in this case an autocovariance) of a function with a nonzero mean, since $C_{gg}(k)$ will not necessarily have a zero mean even though g_t does.

That is, for two series x_t and y_t, with nonzero means, the covariance is given as $\mathrm{Cov}(x_t, y_t) = \mathrm{E}[(x_t - \bar{x})(y_t - \bar{y})]$. Similarly, the acvf of a single series x_t represents the *covariance* of a series with itself offset by k lags (data points), or y_t is nothing more than x_t offset by k lags in the above covariance. Therefore, the acvf of x_t is given by $C_{xx}(k)$, which estimates $\mathrm{Cov}(x_t, x_{t+k}) = \mathrm{E}[(x_i - \bar{x}_i)(x_{i+k} - \bar{x}_{i+k})]$.

Equation (7.47) is reduced by expanding the expression inside the expectance operator and separating terms,

$$
\begin{aligned}
&= \mathrm{E}[\mathrm{C_{gg}}(k)\mathrm{C_{gg}}(k+j)] \;-\; \mathrm{E}[\mathrm{b}\gamma_{\mathrm{g}}(k)\mathrm{C_{gg}}(k+j)] \\
&\quad - \mathrm{E}\Big[\mathrm{b}\gamma_{\mathrm{g}}(k+j)\mathrm{C_{gg}}(k)\Big] + \mathrm{E}[\mathrm{b}\gamma_{\mathrm{g}}(k)\mathrm{b}\gamma_{\mathrm{g}}(k+j)] \\
&= \mathrm{E}[\mathrm{C_{gg}}(k)\mathrm{C_{gg}}(k+j)] \;-\; \mathrm{b}\gamma_{\mathrm{g}}(k)\mathrm{E}[\mathrm{C_{gg}}(k+j)] \\
&\quad - \mathrm{b}\gamma_{\mathrm{g}}(k+j)\mathrm{E}[\mathrm{C_{gg}}(k)] + \mathrm{b}^2\gamma_{\mathrm{g}}(k)\gamma_{\mathrm{g}}(k+j)\;,
\end{aligned} \tag{7.48a}
$$

where the means, product of means and b, being constants, have been moved outside of the expectance operator, recalling, $\mathrm{E}[\mathrm{aX}] = \mathrm{aE}[\mathrm{X}]$, where a is a constant.

$\mathrm{E}[\mathrm{C_{gg}}(k+j)] = \mathrm{b}\gamma_{\mathrm{g}}(k+j)$ and $\mathrm{E}[\mathrm{C_{gg}}(k)] = \mathrm{b}\gamma_{\mathrm{g}}(k)$, so (7.48a) is reduced to

$$
\begin{aligned}
&= \mathrm{E}[\mathrm{C_{gg}}(k)\mathrm{C_{gg}}(k+j)] \;-\; \mathrm{b}^2\gamma_{\mathrm{g}}(k)\gamma_{\mathrm{g}}(k+j) \\
&\quad - \mathrm{b}^2\gamma_{\mathrm{g}}(k)\gamma_{\mathrm{g}}(k+j) + \mathrm{b}^2\gamma_{\mathrm{g}}(k)\gamma_{\mathrm{g}}(k+j) \\
&= \mathrm{E}[\mathrm{C_{gg}}(k)\mathrm{C_{gg}}(k+j)] \;-\; \mathrm{b}^2\gamma_{\mathrm{g}}(k)\gamma_{\mathrm{g}}(k+j).
\end{aligned} \tag{7.48b}
$$

Substituting (7.47) (i.e., using the biased form of the acvf) into (7.48b) for $\mathrm{C_{gg}}(k)$ and $\mathrm{C_{gg}}(\mathrm{k+j})$, assuming that the $\mathrm{g_t}$ are normally distributed and making some simplifications appropriate for large n, effectively eliminating b, (7.48b) is reduced to the following simplified form (see Priestley (1981), pages 324–330 for details; a more complicated form results when $\mathrm{g_t}$ is not normally distributed):

$$
\mathrm{Cov}[\mathrm{C_{gg}}(k), \mathrm{C_{gg}}(k+j)] \approx \frac{1}{2n}\sum_{i=-\infty}^{\infty}[\gamma_{\mathrm{g}}(i)\gamma_{\mathrm{g}}(i+j) + \gamma_{\mathrm{g}}(i+k+j)\gamma_{\mathrm{g}}(i-k)] \tag{7.49}
$$

where $\gamma_{\mathrm{g}}(i)$ represents the true acvf of $\mathrm{g_t}$ and we are estimating the covariance between each lag of $\mathrm{C_{gg}}(\mathrm{k})$ against those values offset by j lags, $\mathrm{C_{gg}}(k{+}j)$. For simplicity, the sum is given here from $-\infty$ to ∞, to make sure that all of the positive and negative lags of the acvf are included. Obviously, in practice this sum only includes that range for which lags exist. This is from $-N_{\mathrm{g}}-1$ to $+N_{\mathrm{g}}+1$, where N_{g} is the number of points in series $\mathrm{g_t}$.

Equation (7.49) effectively states that the *covariation between various lags of the acvf is given by the convolution of the true acvf with lagged versions of itself.* That is, if $\mathrm{Cov}[\mathrm{C_{gg}}(k), \mathrm{C_{gg}}(k+j)] = \mathrm{h}(k)_j$, then the covariance between any lag value, k, in the acvf, and the value located j lags away from k, is given as

$$
\mathrm{h}(k)_j \approx \gamma_{\mathrm{g}} * \gamma_{\mathrm{g}} + \gamma_{\mathrm{g}}(k) * \gamma_{\mathrm{g}}(k), \tag{7.50}
$$

where the symmetry of the acvf allows the sums in (7.49) to be written as either correlation sums or convolution sums (as expressed in (7.50)).

The net effect of this is that while the true acvf may decay to zero by some lag, because of the convolution (smearing) of the sample acvf as shown in (7.50) *the estimated (sample) acvf will not decay to zero as rapidly as the true acvf (i.e., variance from the nonzero lags is "leaked" into the zero lag region in the sample acvf).*

The variance of $\mathrm{C_{gg}}(k)$ – that is, the measure of the mean squared scatter in $\mathrm{C_{gg}}(k)$ – is given by (7.49) when $j = 0$:

$$\begin{aligned}\mathrm{Cov}[\mathrm{C}_{ZZ}(k), \mathrm{C}_{ZZ}(k)] &= \mathrm{Var}[\mathrm{C}_{ZZ}(k)] \\ &\approx \frac{1}{2\mathrm{n}} \sum_{i=-\infty}^{\infty} [\gamma_{\mathrm{g}}^{2}(i) + \gamma_{\mathrm{g}}(i+k)\gamma_{\mathrm{g}}(i-k).]\end{aligned} \tag{7.51}$$

A similar type of expression is obtained for the covariance and variance of the **sample acf** denoted by $\mathrm{r}_{\mathrm{gg}}(k)$ (versus the **true acf** given by $\rho_{\mathrm{g}}(k)$) by substituting ρ_{g} for γ_{g} in the above expressions. Similarly, the variance of the ccvf and ccf follow the exact same path, only the second series, f_{t}, is substituted for g_{t} at the lagged positions.

Recall that $\gamma_{\mathrm{g}}(k)$ is the true acvf at lag k. To actually use (7.49) to estimate the variance of the acvf, we must use our estimates of $\gamma_{\mathrm{g}}(k)$, $\mathrm{C}_{\mathrm{gg}}(k)$.

Variance of acf for White Noise Series

Consider now the variance of the sample autocorrelation function, $\mathrm{r}_{\mathrm{zz}}(k)$, for a Gaussian white noise series. Determination of this allows us to assess whether values of the acvf or acf are significantly different from values expected for simple white noise.

This variance is derived directly from examination of (7.51), simplified by assuming g_{t} to be white noise. In that case, we know that $\rho_{\mathrm{g}}(\mathrm{k})$, as opposed to $\gamma_{\mathrm{g}}(k)$ for the acvf, is 0 for all lags $\neq 0$, and is 1 for lag 0, so the acf equivalent to (7.51) reduces to

$$\mathrm{Var}[\rho_{\mathrm{gg}}(k)] \approx \frac{1}{n}. \tag{7.52}$$

That is, the only nonzero term in the sum of (7.49) occurs once when $i = 0$ in (7.51), substituting ρ, the acf, for γ, the acvf. At that index, the first term in the sum, $\rho_{\mathrm{g}}^{2}(0) = 1$, and the second term, $\rho_{\mathrm{g}}(0+k)\rho_{\mathrm{g}}(0-k) = \rho_{\mathrm{g}}^{2}(0) = 1$ (because of symmetry of the acf). For all other values of i, the true acf of white noise is zero. Therefore, the sum reduces to 2 and the acf equivalent to (7.51) reduces to $1/n$. So, from (7.52), the standard deviation of $\rho_{\mathrm{gg}}(k)$ is $\sim n^{-1/2}$.

The covariance of $\rho_{\mathrm{gg}}(k)$ given by the acf equivalent to equation (7.51) is

$$\mathrm{Cov}[\rho_{\mathrm{gg}}(k),\ \rho_{\mathrm{gg}}(k+j)] \approx 0 \tag{7.53}$$

for all lags $j \neq 0$, for arguments similar to those provided in computing (7.52).

Therefore, for large n (an assumption imposed to get the simplified forms of (7.51) and (7.53)), the acf of a white noise series has an expected mean value of 0 and a standard deviation of $n^{-1/2}$. The 95 percent confidence limits are given by $\pm 1.96\ n^{-1/2}$. You can thus place these confidence limits on any plot of an acf, and if the acf at lags immediately greater than 0 lie above or below these limits, we reject the hypothesis that the series is white noise, accepting a 5 percent chance of being wrong. However, recall that with noise, 5 out of 100 series will show acf values greater than these confidence limits (thus the potential for being wrong). We will discuss the interpretation of these limits in more detail when we discuss forecast modeling.

For small n, because of the bias in the estimates of the acf associated with using the sample mean of the g_{t} instead of the true μ, the expected value is not 0 but instead $-1/n$. In this case, the variance is distributed about $-1/n$ instead of 0. Obviously, as n goes to infinity, the expected value goes to 0.

7.5 Take-Home Points

1. Sequential data are sequences of data in which the order of occurrence of the observations is important.
2. For sequential data, the "population" of time series is known as the *ensemble*.
3. Convolution is one of the most useful operations applied to time series. This is a linear operation for modifying a series by means of passing it through a filter. The filter can serve to smooth and eliminate certain frequency components, and is used in solving the equation for the conduction and convection of heat. The filter is defined by its *impulse response function*, the response of a filter to the passing of a single pulse (i.e., multiplying the black box by 1). An n-point running mean is an example of convolution where the filter is n points, each of amplitude n^{-1}.
4. A causal (or realizable) filter is one that does not respond before it is forced. A running mean is not realizable.
5. Deconvolution (inverse filtering) is used to unscramble a convolved series. E.g., if a seismic pulse is passed through layers of the earth (a filter), the recording seismograph picks up a convolved version of the pulse after it has passed through the Earth filter. Deconvolution is used to unscramble the recorded signal in order to determine the nature of the filter (layers of the Earth) that the pulse traveled through (or, similarly, for the medical ultrasound scan of a body). Deconvolution and inverse filtering are described in detail in Chapter 13.
6. Sequential data require additional statistical parameters to describe them in order to capture the nature of their sequential relationships. Most important are the lowest-order bivariate moments: autocovariance and autocorrelation functions that provide the degree of covariance or correlation between neighboring points, separated by a distance of $n\Delta t$ (n is the number of lags).
7. Stationarity is the property of statistical moments of the time series, that they do not change with time. It comes in various degrees, according to how many moments are stationary. Stationarity of order 1 has the mean unchanging; order 2 has mean and variance unchanging; etc. Stationarity of order 2 (also known as Wide-Sense Stationary, WSS) is usually good enough for applying many of the methodologies we will use.
8. Ergodicity is the property whereby all moments of an ensemble can be attained from a single time series (of bi-infinite length). Assuming this, we can compute the various bivariate moments from a single series.
9. Autocorrelaton sequence or function (acf) is the correlation of a series with itself at various offsets. That is, at "lag" 0, it is the correlation of the time series with itself (yielding a perfect 1); then at lag 1, the time series is shifted by one time point (lag 1) and correlated with its unlagged self, so we are effectively seeing how points are correlated with their closest neighbor. This continues by lagging the series by 2 lags, etc. It produces a function of correlation as a function of lag. This is the most fundamental bivariate statistical moment.
10. Autocovariance function (acvf) is similar to the acf, but in this case, correlation is replaced by covariance as a function of lag.
11. Cross-correlation function (ccf) is the same as acf, but in this case it is the lagged relationship between two different time series.
12. Cross-covariance function (ccvf) is the same as acvf, but with two different time series.

13. White noise is a series in which there is not relationship between neighboring points (so acf is 1 at the 0 lag, and effectively 0 at all other lags).

7.6 Questions

Pencil and Paper Questions

1. Regarding the following time series:

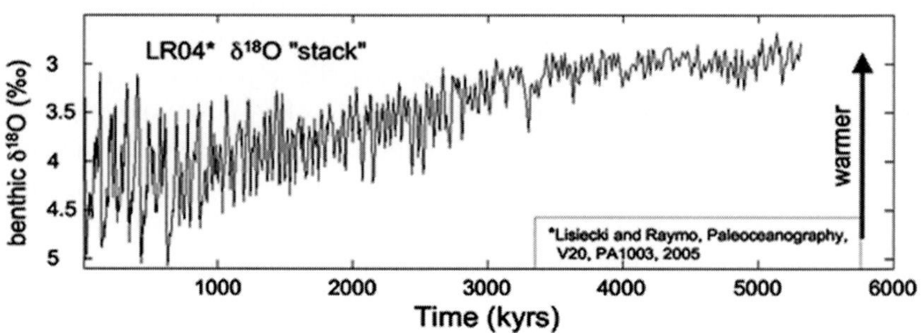

   a. Is it stationary, and if so, to what order?
   b. What is the best way to compute its mean and variance?
   c. What is the value of the 0th lag of its autocorrelation function?

2. a. What is the difference between the acvf and acf?
   b. I have a white noise series. What is the value of the 0th lag of its acf and of its acvf?
   c. I computed the acf by computing across 100 realizations (not using a serial product). Is my time series stationary? (Explain your answer.)
   d. Are the mean and PDF for the series drawn below correct? (Explain your answer.)
   e. Draw the mean and PDF for the opposite answer to d (i.e., if d was correct, draw the incorrect mean and PDF; if it was incorrect, draw the correct ones).

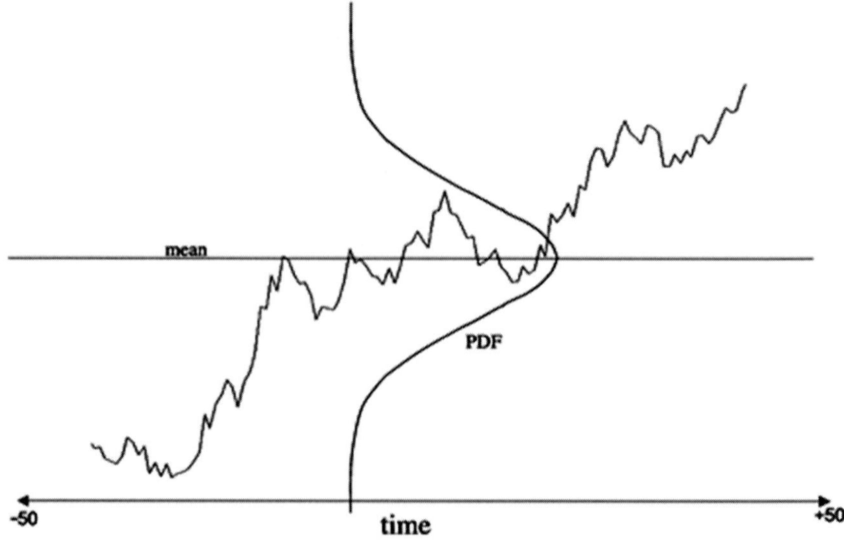

3. For random processes *X* and *Y* with zero means, and stored in *vectors* **X** and **Y**:

   a. What is $E[\mathbf{XY}^T]$?
   b. What is $E[\mathbf{XX}^T]$?
   c. What is Var[**X**]?
   d. If *X* does not have a zero mean, show how to create the same product as that represented by question 3.c, using the expectance operator.

Computer-Based Questions

4. The autocorrelation function (ACF) is a fundamental joint statistical moment for time series, as important to sequential data as the variance is to univariate data. It reveals how different data points in the time series correlate with one another. One way to define the sample autocorrelation function is

$$R_{xx}(k) = \frac{1}{n\sigma_x^2}\sum_{i=1}^{n-|k|} x(i)x(i+k).$$

This form introduces a bias, although it is largely negligible for examining values at small lags (and when *n* is large). Here you will examine the nature of the ACF of two very different time series: first, a geophysical series with strong periodic components, and second, white noise.

Geophysical Series. First analyze the Lisiecki and Raymo (2005; data set of compiled benthic $\delta^{18}O$ measurements www.cambridge.org/martinson). These data serve as a proxy for global climate conditions over the past 5.3 million years. Firstly, the data are not evenly spaced in time. Use a linear interpolant to interpolate these data onto evenly spaced data with $\Delta t = 1000$ years. Secondly, computation of the ACF assumes the data are stationary. Try fitting a cubic polynomial to the data and then subtracting that. Plot the original data and then the residual. Does the series now appear stationary? For this problem, look only at the most recent one million years.

You're now ready to compute the acf. Plot the acf as a function of lag, and label your axes. Identify four maxima in the acf and record their lag. Given the definition of the acf and what these data are, do the lag-values make sense? Explain briefly. Now make some estimates of the decorrelation time scale of the dataset. There are many ways of doing this – you don't have to do anything fancy. How many independent observations does this imply that you have?

White Noise. Generate a Gaussian white noise series. Is this data set truly white noise? Be quantitative: examine, for example, the values at the first 100 lags. Now generate 10,000 realizations of the white-noise process. Compute the acf of each of these and average them for an estimate of the population acf. Plot this estimate on top of the single-sample acf. In a separate plot, plot the variance of the 10,000 realizations of the acf as a function of lag. Focus only on small lags (again, maybe the first 100 lags) and explain these findings.

8 Fourier Series

8.1 Overview

Fourier or **harmonic analysis** involves the "decomposition" of a time series into a sum of sinusoidal components (sines and cosines). The decomposition is nothing more than fitting the n values of a time series with n sinusoids of specific ("Fourier") frequencies. Thus, Fourier analysis is simply a specific type of *interpolation*, but one that only holds for time series sampled at even intervals of the independent variable, giving special properties to the sinusoidal interpolant. The fitted function, given this particular basis, is known as a **Fourier series**, and because of the special properties of the sinusoids, their fit provides unique insights to the underlying structure and composition of the data series being analyzed. Sinusoids are nature's natural cycles, being nothing more than the motion of a spot on the side of rotating circle as it rolls along at a steady speed.[1]

When the data points of the time series are separated by even intervals of the independent variable (i.e., the data points are "***evenly spaced***"), the properties of the sinusoids (they become "orthogonal") allow the interpolation solution procedure to be greatly simplified. The simplified procedure is called the **discrete Fourier transform**, and it provides an extremely efficient method (applicable to any orthogonal basis, not just sinusoids) with which to perform the interpolation and determine the coefficient values multiplying the sinusoids, providing the interpolated fit. Because sinusoids are **periodic** functions, their actual characteristics needed to interpolate the time series are often of interest, and their coefficients are usually presented in the form of a **spectrum**.

As a problem in interpolation, Fourier analysis is a means for analyzing deterministic data; namely, the exact fit to the data, not taking into account the uncertainties and problems associated with noise in the data. When applying the tools and concepts of Fourier analysis to stochastic process data – that is recognizing that the data contain noise – additional considerations are adopted, as was the case in curve fitting: a best fit must be determined and the best-fit coefficients must be evaluated, as must the uncertainty in the fit of the sines and cosines. This involves **spectral analysis**. In other words

[1] You can draw a sinusoid by attaching a pencil to a wheel, pointing outwards: as the wheel is moved at a steady pace by an axle in its center, the movement of the wheel will create a perfect sinusoid. Or jam an indelible marker, drawing end out, into the outer edge of a tire on a car and drive at a constant speed along a wall (then please send photos: one of the sinusoid, and another of the car's owner).

(as used here), Fourier analysis represents the interpolation problem useful for deterministic series, whereas spectral analysis represents the curve-fitting problem useful for series containing noise.

Fourier sines and cosines are the natural eigen functions for complex natural systems,[2] which makes them ideal for decomposing such systems. They follow a number of rules that make analysis and interpretation of your time series exceptionally insightful. Fourier series are used in the solution of certain ordinary differential equations and have been studied and used in innumerable ways, making them well understood and a cornerstone of analysis.

In practice, Fourier and spectral analysis represent some of the most powerful and often used methods in data analysis in all fields of science, engineering, and mathematics. Also, the Fourier transform itself, like convolution and many of the other methodological tools forming the base of our analysis techniques, is invaluable in analytic problem-solving as well as data analysis.

8.2 Introduction

Fourier series involve a set of particular sines and cosines that combine to interpolate a data set or time series. We have already devoted time to solving the general interpolation problem, so you might question the need for spending additional time in developing this special case of interpolation. There are a couple of reasons: this special case is later shown to be applicable to interpolation (and least squares problems) using *any* orthogonal basis function for fitting the data – not just for particular sines and cosines. Therefore, the development of the Fourier series is in some respects a natural extension of the more general discussion of interpolation provided in Chapter 4. However, because of the considerable attention, significant insights, and intuitive comprehension provided by the sine and cosine functions, it makes sense to initiate this general development through the particular case of Fourier series and the Fourier transform.

Before developing the special form of solution appropriate for this interpolant using the Fourier transform, an overview of basic properties and definitions of periodic functions is required. Most fundamental is the concept of a periodic function.

8.3 Periodic Functions

8.3.1 Definitions and Concepts

A **periodic function** is one that repeats in time or length (i.e., its independent variable), so

$$f(t + T) = f(T + nT) = f(t) \tag{8.1}$$

where: T = the **period** of the function if it repeats in *time* (or it is the **wavelength** if it repeats in *space*) and

[2] This is explained in the chapter on Empirical Orthogonal Function analysis (Chapter 15).

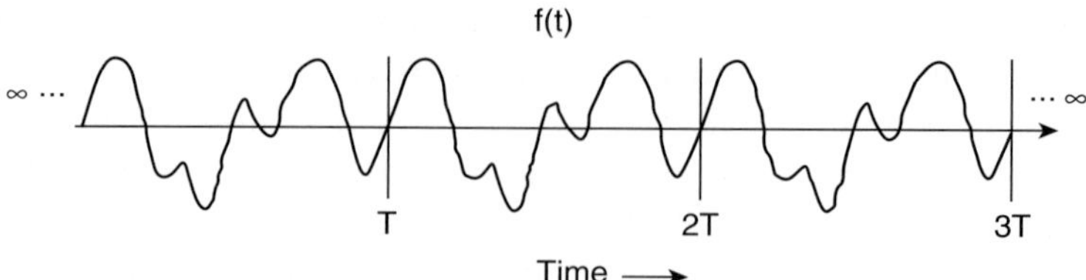

Figure 8.1 Example of a periodic time series that repeats every T units of time.

n = an integer, 0,1,2, . . . The period, T, represents the length of the independent variable after which the time series is repeated. So, for example, Figure 8.1 shows a periodic time series that repeats every T units of time.

The first point in this series is repeated every T units of time. This is true for every point in the time series, so it satisfies the relationship given by (8.1) – an important consequence being: *in order to satisfy* (8.1) *exactly, the time series, f(t), must be infinitely long*. While it may seem esoteric, this generates an unfortunate, troublesome issue that we must later address.

As a general rule, when working with periodic time series, the domain of the independent variable is limited as

$$-\frac{T}{2} \leq t < \frac{T}{2}. \tag{8.2}$$

In other words, the time series is typically presented so that the last point in the series is the point just before the first point repeats in the series, due to the periodicity over T. In this manner, each unique data point is represented just once within the domain represented by (8.2). Any point outside the domain, including the point at T/2, is already represented within the domain at some integral multiple of T (so the value at T/2 is already represented by the point at T/2 − T = −T/2).

Another way to look at periodicity, so that this restriction makes conceptual sense, is by considering that the data lie on a *closed loop*, where the circumference of the loop represents one complete period of the series. In other words, imagine a time series in a circular form, such as is shown in Figure 8.2.

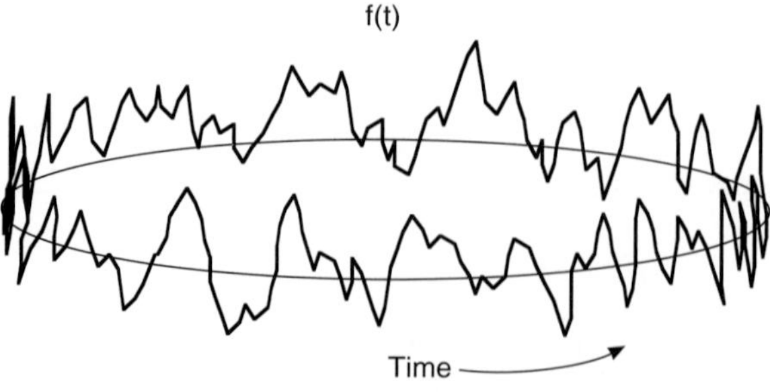

Figure 8.2 One complete period of a periodic time series, wrapped into a closed loop, so that the time series is infinitely long in both directions.

In this respect, it is clear why one would not want the data series to include points at both −T/2 and at +T/2, since this would represent plotting 2 points at the same position on the loop. In other words, the last point in the series is always followed by the first point in the series, regardless of where you make the first point in the series. ***This cyclic representation should always be kept in mind when dealing with periodic data or with analyses that assume that the data are periodic (as discrete Fourier analysis does).*** To put this in standard linear form, you simply disconnect the loop at any point and straighten it out. In that form it would represent just one period of the series. This one period, however, completely represents the unique information of the entire (infinitely long) periodic time series.

For periodic time series, the integral over one full period is independent of where in the series the period begins. That is,

$$\int_{-T/2}^{T/2} f(t)dt = \int_{-T/2+a}^{T/2+a} f(t)dt,$$

where a is a constant. This is shown graphically in Figure 8.3. The shaded region and hatched regions represent the same area because they both encompass one full period of the data.

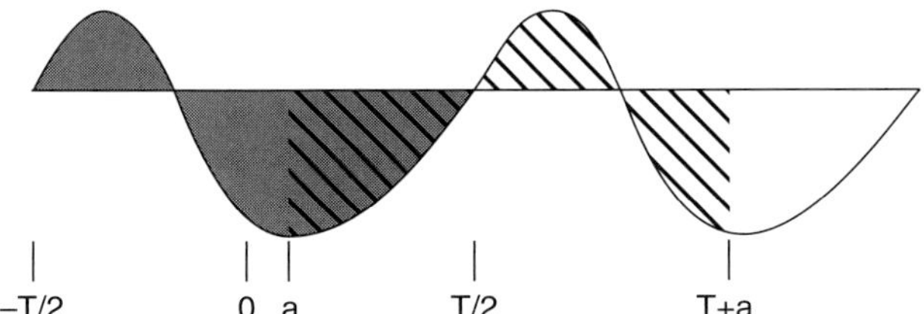

Figure 8.3 Integration over one period of a periodic time series. Both solid-filled and diagonal-line-filled regions give the same integral, since both are over a full period, though they start and stop at different points in the series.

Also,

$$\int_{a}^{b} f(t)dt = \int_{a+T}^{b+T} f(t)dt.$$

Sines and Cosines

The simplest periodic functions are the sine and cosine functions (Figure 8.4). These trigonometric functions can be written in a variety of forms. For example,

$$A\sin 2\pi ft \text{ or } A\sin(2\pi ft + \varphi) \text{ or } A\sin[2\pi f(t + t_\varphi)] \quad (8.4a)$$

$$A\cos 2\pi ft \text{ or } A\cos(2\pi ft + \varphi) \text{ or } A\cos[2\pi f(t + t_\varphi)] \quad (8.4b)$$

where: A = **amplitude** (height from 0) of the sine or cosine;

f = **frequency** (1/T; T = period or wavelength).

Sometimes, the phrase **rotational frequency** is used for this term to avoid confusion with *angular frequency*:

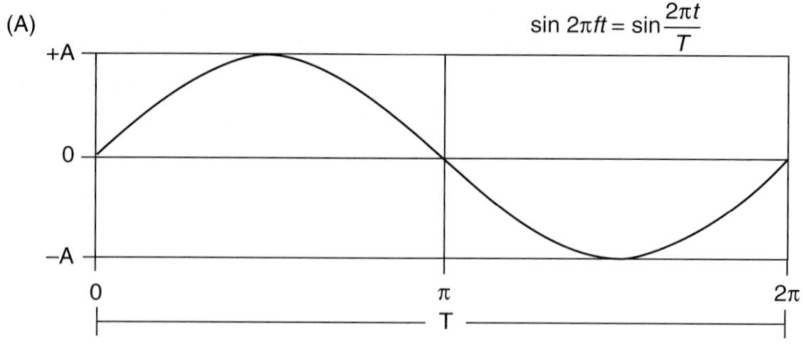

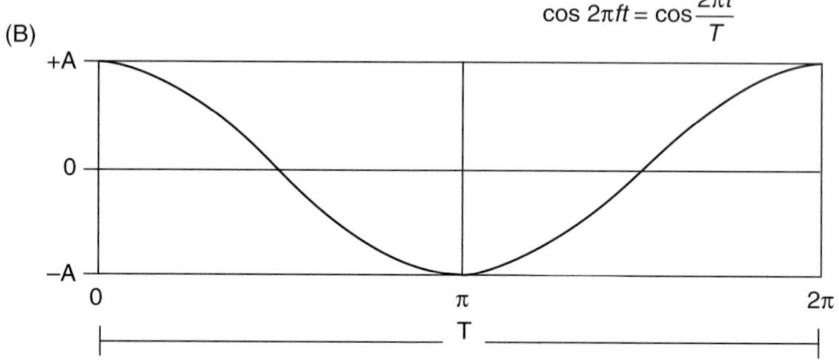

Figure 8.4 **(A)** A generic sine, with zero phase (or a cosine with phase angle $-\pi/2$), has amplitude, A = 1, and period, T. **(B)** A generic cosine, with zero phase and amplitude A = 1.

$\varphi = 2\pi f t_\varphi$ = **phase angle** or phase displacement;
t_φ = translation of the origin (constant displacement in t).
In addition, two additional convenient forms include

$$2\pi f = \begin{cases} \omega \ (= \textbf{angular frequency} \text{ if T is a period in } \textit{time}) \\ k \ (= \textbf{wavenumber} \text{ if T is a wavelength in } \textit{space}) \end{cases}$$

So, (8.4a) can be rewritten in terms of angular frequency or wavenumber as

$$A\sin\omega t \text{ or } A\sin(\omega t + \varphi) \text{ or } A\sin[\omega(t + t_\varphi)] \tag{8.4c}$$

$$A\sin kt \text{ or } A\sin(kt + \varphi) \text{ or } A\sin[k(t + t_\varphi)] \tag{8.4d}$$

In either case, ω or *k* is in radians, though it can be given in degrees or grads.

Note that the relationship between angular and rotational frequency, $\omega = 2\pi f$, is similar to that between natural log (ln) and log base 10 (log) where ln = 2.303log. In both cases (frequency and logs), the two forms are related by a constant and thus represent two slightly different means for presenting the same quantity. Which form is used is often simply a matter of convenience and comfort.

The specific position (time or length; ωt or *k*t) in the waveform is the **phase** (most common when dealing with propagating waves). Though related, do not confuse *phase* with *phase angle*, φ. The latter represents a displacement of the origin, while the former represents the location in the waveform (e.g., at its peak or trough). The two are related, since the phase angle is the phase at time 0. Some people do not differentiate between phase and phase angle, so make sure you are clear on how the word is being used. Here, both words are used interchangeably to mean phase angle.

Since the amplitude and phase determine the form of a sinusoid of any particular frequency, waveforms are often represented in **polar form**, in terms of amplitude (A) and phase (φ), as seen in Figure 8.5. Here, the radius of the circle is equal to the amplitude of the sinusoid and the phase is given by the arc length along the circumference of the circle. In this form, the drawn radius rotates around the circle at a rate of ω (or k). Therefore, a sinusoid is a convenient representation of constant rotational motion, where the rotational rate is given by the angular frequency or wavenumber. The polar form readily conveys this interpretation.

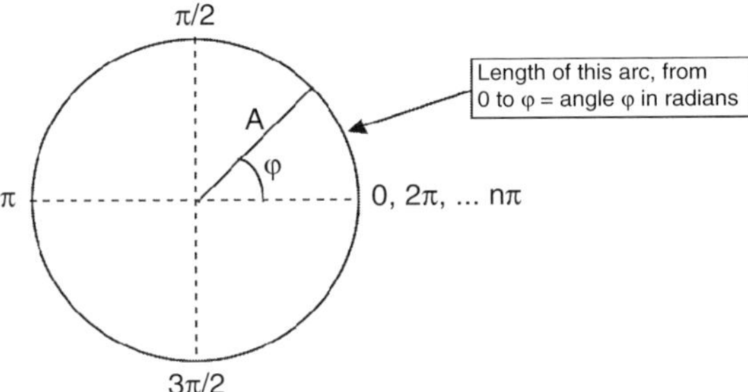

Figure 8.5 Polar form of a sinusoid, giving primary characteristics in terms of amplitude (= radius of circle), and phase angle as length of arc.

Regardless of how they are presented, sines and cosines represent fundamental physical processes and mathematical functions, and, as such, they naturally arise in a variety of situations.

Even and Odd Functions

Cosines and sines represent simple examples of even and odd functions, respectively. An even function (e.g., cosine) is one that is symmetrical about the ordinate, while an odd function (e.g., sine) is antisymmetrical about the ordinate. Therefore,

$$f_e(-x) = f_e(x) \tag{8.5a}$$

$$f_o(-x) = -f_o(x), \tag{8.5b}$$

where $f_e(x)$ is an even function and $f_o(x)$ is an odd function.

Knowledge of the symmetry properties of a function often proves helpful when manipulating the function. Another useful property involves the integration of even and odd functions.[3] Specifically,

[3] For discrete time series, the integration properties also hold for summation ***only*** if the series is sampled at even increments of the independent variable.

$$\int_{-a}^{a} f_e(x)dx = 2\int_{0}^{a} f_e(x)dx \tag{8.6a}$$

$$\int_{-a}^{a} f_o(x)dx = 0. \tag{8.6b}$$

These properties are apparent upon simple inspection of cosine and sine functions drawn from $\pm 2\pi$.

In general, any function f(t), whether it displays any symmetry or not, can be decomposed into a sum of an even, $f_e(t)$, and odd, $f_o(t)$, function as follows:

$$\begin{aligned} f(t) &= \frac{1}{2}[f(t) + f(t) + f(-t) - f(-t)] \\ &= \underbrace{\frac{1}{2}[f(t) + f(-t)]}_{\text{even function}} + \underbrace{\frac{1}{2}[f(t) - f(-t)]}_{\text{odd function}} \\ &= \quad f_e(t) \quad + \quad f_o(t). \end{aligned} \tag{8.7}$$

It is easy to prove that $(1/2)[f(t) + f(-t)]$ is an even function and $(1/2)[f(t) - f(-t)]$ is an odd function: substitute $-t$ for t in each. For the even function, the result is identical whether we use t or $-t$, since $[f(t) + f(-t)] = [f(-t) + f(t)]$; for the odd function $[f(t) - f(-t)] = [f(-t) - f(t)] = -[f(t) - f(-t)]$. Thus, $f_e(t) = f_e(-t)$ and $f_o(t) = -f_o(-t)$.

Basic Trigonometric Identities

The following identities are very useful for Fourier analysis and for understanding the nature of periodic time series:

$$\sin(x + y) = \cos(y)\sin(x) + \sin(y)\cos(x) \tag{8.8a}$$

$$\cos(x + y) = \cos(y)\cos(x) - \sin(y)\sin(x) \tag{8.8b}$$

$$\cos(x)\cos(y) = \frac{1}{2}[\cos(x + y) + \cos(x - y)] \tag{8.8c}$$

$$\cos(x)\sin(y) = \frac{1}{2}[\sin(x + y) - \sin(x - y)] \tag{8.8d}$$

$$\sin(x)\sin(y) = \frac{1}{2}[\cos(x - y) - \cos(x + y)] \tag{8.8e}$$

Because the cosine is an even function and the sine an odd function, the expansions of identities (8.8a) and (8.8b), for $-y$, as opposed to $+y$, are easily given as

$$\begin{aligned} \sin(x - y) &= \cos(-y)\sin(x) + \sin(-y)\cos(x) \\ &= \cos(y)\sin(x) - \sin(y)\cos(x) \end{aligned} \tag{8.9a}$$

and

$$\begin{aligned} \cos(x - y) &= \cos(-y)\cos(x) - \sin(-y)\sin(x) \\ &= \cos(y)\cos(x) + \sin(y)\sin(x) \end{aligned} \tag{8.9b}$$

Consider the standard cosine with angular frequency ω and phase angle, φ. From (8.8b),

$$A\cos(\omega t + \varphi) = A\cos(\varphi)\cos(\omega t) - A\sin(\varphi)\sin(\omega t), \tag{8.10}$$

but since φ is a fixed angle (the initial phase of the sinusoid at time 0), Acos(φ) is a constant, as is Asin(φ), so (8.8a) can be rewritten as

$$A\cos(\omega t + \varphi) = a\cos(\omega t) - b\sin(\omega t), \tag{8.11a}$$

where:

$$a = A\cos(\varphi) \tag{8.11b}$$

$$b = A\sin(\varphi) \tag{8.11c}$$

and, by inverting (8.11b,c),

$$\varphi = \tan^{-1}\frac{b}{a} \tag{8.11d}$$

$$A = (a^2 + b^2)^{1/2} \tag{8.11e}$$

Equations (8.11d,e) are easily obtained by combining (8.11b,c). Specifically, solve for the phase angle (8.11d) by dividing (8.11c) by (8.11b):

$$\frac{\sin(\varphi)}{\cos(\varphi)} = \tan(\varphi) = \frac{b}{a} \tag{8.12a}$$

so

$$\varphi = \tan^{-1}\frac{b}{a}.^{4} \tag{8.12b}$$

Similarly, we can square (8.11b) and (8.11c) and add them to solve for the A,

$$a^2 + b^2 = A^2[\cos^2(\varphi) + \sin^2(\varphi)], \tag{8.13a}$$

and, recalling that $\sin^2 x + \cos^2 x = 1$, yields

$$A = (a^2 + b^2)^{1/2}. \tag{8.13b}$$

Therefore, you can easily switch between a single sinusoid with phase φ and amplitude A, on the one hand, and the sum of a pure sine and a pure cosine (i.e., sines and cosines without a phase shift) on the other.[5] This means that any single sinusoid can be decomposed as in Figure 8.6.

[4] The computation of the phase angle from the arc-tangent requires care on a computer, since the sign of a and b determines which quadrant the phase angle lies within, and some forms of the arc-tangent function do not differentiate this. For example, $\tan^{-1}(-b/a)$ is the same as $\tan^{-1}(b/-a)$, yet these two pairs of a and b represent uniquely different phase angles.

[5] Here *sinusoid* is used generically – that is, it represents either a sine or a cosine function.

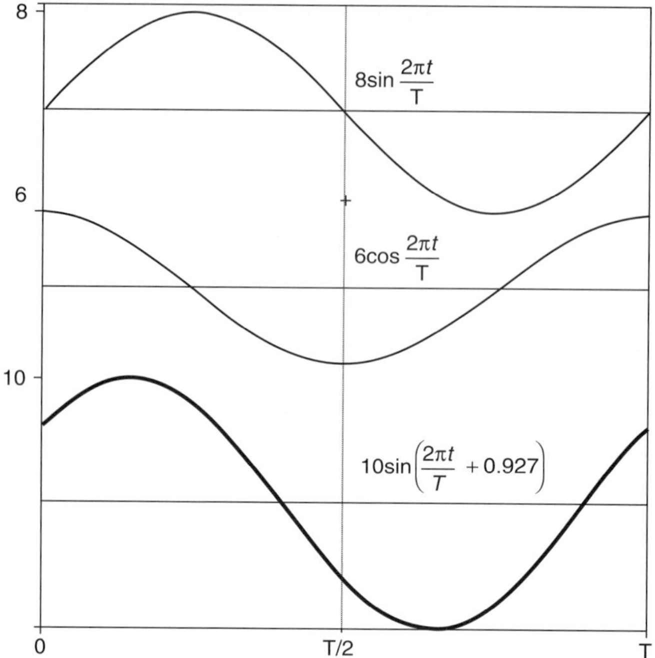

Figure 8.6 Adding the pure sine (amplitude 8) and pure cosine (amplitude 6), both having the same frequency and no phase angle, sum to give the lower sinusoid (thick line) of amplitude 10 and phase angle 0.927. This follows directly from equation (8.11a).

In Figure 8.6 the bottom (bold-line) sinusoid $10\sin[(2\pi t/T) + 0.927]$ is synthesized (from 8.8a) as the sum of a pure cosine of amplitude $a = 10\sin(0.927) = 6$, and a sine of amplitude $b = 10\cos(0.927) = 8$. So, from (8.11a),

$$10\sin\left(\frac{2\pi T}{T} + 0.927\right) = \underbrace{10\sin(0.927)}_{a}\cos\left(\frac{2\pi T}{T}\right) + \underbrace{10\cos(0.927)}_{b}\sin\left(\frac{2\pi T}{T}\right)$$
$$= 6\cos\left(\frac{2\pi t}{T}\right) + 8\sin\left(\frac{2\pi t}{T}\right). \tag{8.14}$$

Now consider adding two cosines of the same frequency but different amplitudes and phases. In that case, the trigonometric identity (8.8b) can be used in combination with (8.11) as

$$\begin{aligned} A_1\cos(\omega t + \varphi_1) + A_2\cos(\omega t + \varphi_2) &= A_1\cos(\varphi_1)\cos(\omega t) - A_1\sin(\varphi_1)\sin(\omega t) \\ &\quad + A_2\cos(\varphi_2)\cos(\omega t) - A_2\sin(\varphi_2)\sin(\omega t) \\ &= (A_{1c} + A_{2c})\cos(\omega t) - (A_{1s} + A_{2s})\sin(\omega t) \\ &= a\cos(\omega t) - b\sin(\omega t) \\ &= A\cos(\omega t + \varphi) \end{aligned} \tag{8.15}$$

where:
$a = A_{1c}+A_{2c}, b = A_{1s}+A_{2s}, A_{1c} = A_1\cos(\varphi_1), A_{2c} = A_2\cos(\varphi_2), A_{1s} = A_1\sin(\varphi_1)$, $A_{2s} = A_2\sin(\varphi_2)$, and the amplitude and initial phase angle of the resulting cosine is $A = \left[(A_{1c}+A_{2c})^2 + (A_{1s}+A_{2s})^2\right]^{1/2}$, and $\varphi = \tan^{-1}[(A_{1s}+A_{2s})/(A_{1c}+A_{2c})]$.

Therefore, the sum of the two cosines of the same frequency produces a new single cosine of that same frequency, ω, but of a different amplitude and phase relative to the two cosines that went into the sum. Similarly, from (8.11), a sine and cosine of the same frequency sum to produce a new single cosine of the same frequency but with a nonzero phase angle.

Pure Sinusoid as Stochastic Process

The assumption of stationarity has considerable implications in spectral analysis, which have led to the analysis focusing on the amplitude or power spectrum, with little emphasis (or attention at all) to the phase of the series. Specifically, consider a process that generates a time series with a perfect or near-perfect sinusoidal component in it. Your initial impression may be that the process is not stationary, since the mean of the series clearly changes with time in a predictable manner, save for some pesky noise. However, while it is true that the single realization of the process contains a predictable sinusoidal component, the process is still stationary, with little ability to predict the value of the time series at a particular time in another realization of the process.

For example, consider a rotating tank of water (rotation makes it behave a bit like the real world oceans on a rotating Earth) where a sensor reports the height of the fluid at a particular location each second of the experiment, which starts after we drop a large stone into the tank next to the tank wall. The signal is dominated by a wavelike pattern as the primary motion. For any particular experiment, the time series shows a rather regular and predictable pattern, giving the impression that the mean is changing with time according to the dominant waveform. It is true that for the one realization, you can gain an excellent understanding and characterization of the realization and make predictions of future times for it.[6]

However, when considering the ensemble average of all realizations that characterize the process itself, the mean height at the location does not change in time, but instead is zero for all times. Likewise, we do not gain an ability to predict the height at any particular time for future experiments in the tank, even if each experiment is dominated by the same general waveform. The reason is that in each experiment the phase of the dominant waveform is different, depending on subtle differences in the initial conditions (e.g., exactly where the object was dropped into the water relative to the wall), spin-up characteristics (time required to spin to a stable state), how synchronous the start time is with the impact of the object, etc. Consequently, if we average together all of the different waveforms from the different experiments, we find that the changing phase leads to cancellation of the different peaks and troughs leading to a zero mean – that is, the *process* is stationary and *stochastic*.

[6] In general, spectra provide a very poor means for prediction, except when the time series is dominated by perfect sinusoidal components (with stable phase through time), in which case prediction can be quite good.

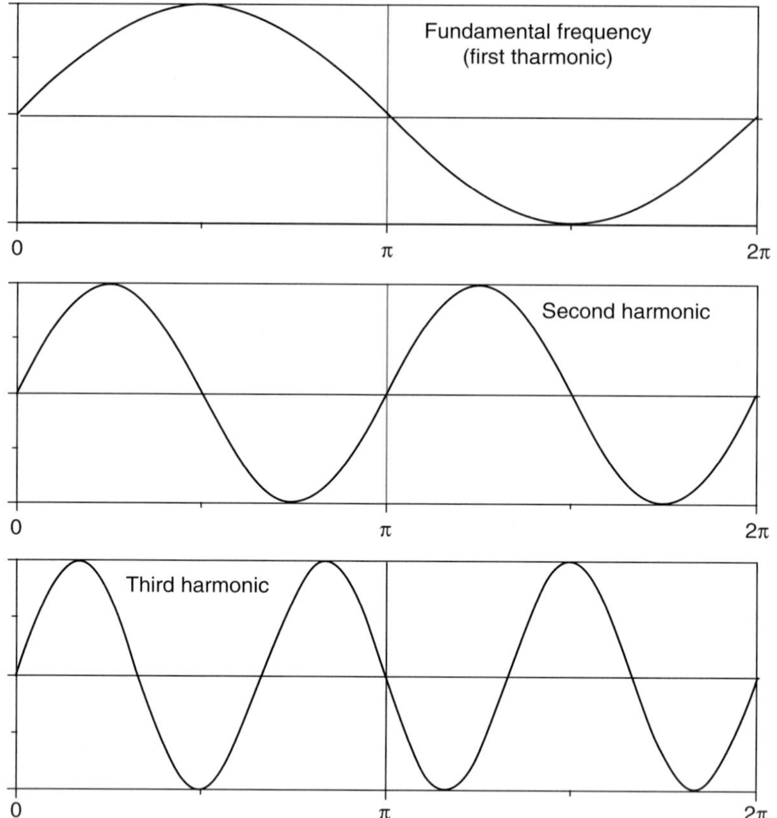

Figure 8.7 Example of first three harmonics for a series of the same length. Independent of initial phase, the harmonics make one, two, and three full cycles over the length of the series.

Harmonics

Harmonics or **overtones** are sines and cosines with frequencies that are integer multiples of a fundamental frequency.

The **fundamental frequency** (or **first harmonic**) is given by $f_1 = 1/T$. This waveform completes one full cycle over the period T, and since this is independent of phase, it is irrelevant whether it is a sine or cosine. In Figure 8.7 this is demonstrated by a sine (we could have an initial phase shift and the relationship would still be the same).

The **second harmonic**, f_2, is the sinusoid that makes two complete cycles or oscillations over period T. Its frequency is twice that of the first harmonic, since it makes twice as many complete oscillations in the same period, so $f_2 = 2f_1$.

The **third harmonic**, f_3, makes three complete oscillations over the same period T, so $f_3 = 3f_1$.

Continuing in this same manner (I'm hoping you see the pattern here), the ***n*th harmonic**, f_n, makes n complete oscillations over the period T, and $f_n = nf_1$.

Therefore,

$$
\begin{aligned}
f_1 &= f_F = \frac{1}{T};\ \omega_1 = 2\pi f_1 \\
f_2 &= 2f_1 = \frac{2}{T};\ \omega_2 = 4\pi f_1 \\
f_3 &= 3f_1 = \frac{1}{T};\ \omega_3 = 6\pi f_1 \\
&\vdots \\
f_n &= nf_1 = \frac{n}{T};\ \omega_n = 2n\pi f_1.\text{[7]}
\end{aligned}
\tag{8.16}
$$

The linear superposition of harmonics (i.e., waveforms that are an integer multiple of the fundamental frequency 1/T) always produces a function that is periodic over T.

Modulation

When two sinusoids (sines or cosines) of different frequencies are linearly combined, they produce a composite signal that has a periodic pinching and swelling of the entire time series, known as a **beat**. If the frequencies of the sinusoids being combined are similar, this "beating" can be considerable.

Specifically, consider

$$y(t) = \cos\omega_1 t + \cos\omega_2 t, \tag{8.17}$$

and rewrite the two angular frequencies as

$$\omega_1 = \overline{\omega} + \delta\omega \tag{8.18a}$$

$$\omega_2 = \overline{\omega} - \delta\omega \tag{8.18b}$$

Then

$$\delta\omega = \frac{\omega_1 - \omega_2}{2} \tag{8.19a}$$

$$\overline{\omega} = \frac{\omega_1 + \omega_2}{2} \tag{8.19b}$$

Substituting (8.18) into (8.17) gives

$$y(t) = \cos(\overline{\omega}t + \delta\omega t) + \cos(\overline{\omega}t - \delta\omega t), \tag{8.20}$$

and applying (8.8c) gives

$$y(t) = 2\cos(\delta\omega t)\cos(\overline{\omega}t). \tag{8.21}$$

[7] Note that some people (acousticians, and Courant, for example) consider that the harmonics are the higher frequency forms of the fundamental frequency. This is consistent with the physics of harmonics in instruments. Given this definition, the fundamental frequency is not considered to be a harmonic at all, and consequently they call $2f_F$ the first harmonic; $3f_F$ the second harmonic, etc. This notation is thus offset from the present (mathematically more common) notation by one. Therefore, you must be careful to clearly indicate how you are defining the terms "first harmonic," etc., and be equally careful in making sure of how others define it (this can get very confusing sometimes, when people start talking about the first harmonic without differentiating whether they are referring to the fundamental frequency or $2f_F$). Unfortunately, both forms have a certain intuitive appeal. I will consistently use the notation as defined here in (8.16).

Examination of this equation shows that the second cosine term, $\cos(\overline{\omega}t)$, with period $T = 2\pi/\overline{\omega}$ has a *time-varying amplitude* given by $\cos(\delta\omega t)$, which itself has a period of $T = 2\pi/\delta\omega$.[8] When one function multiplies another, it is said to **modulate** the function it multiplies. That is, in the example just given, the function $2\cos(\delta\omega t)$ modulates the function $\cos(\overline{\omega}t)$. Therefore, $2\cos(\delta\omega t)$ is the **modulation function**, and it essentially forms an **envelope** within which the modulated function is contained. In the case of sinusoids, since the modulation function is periodic it is called a beat, and its frequency of modulation is called the **beat frequency**.

For example, consider two cosines with frequencies given by $\omega_1 = 11\pi/5$ and $\omega_2 = 9\pi/5$. The addition of these two cosines is shown in Figure 8.8. Alternatively, according to (8.21), this is equivalent to multiplying two cosines of the form $2\cos(\delta\omega t)\cos(\overline{\omega}t)$, where $\delta\omega = (\omega_1 - \omega_2)/2$ and $\overline{\omega} = (\omega_1 + \omega_2)/2$. That product is shown in Figure 8.9.

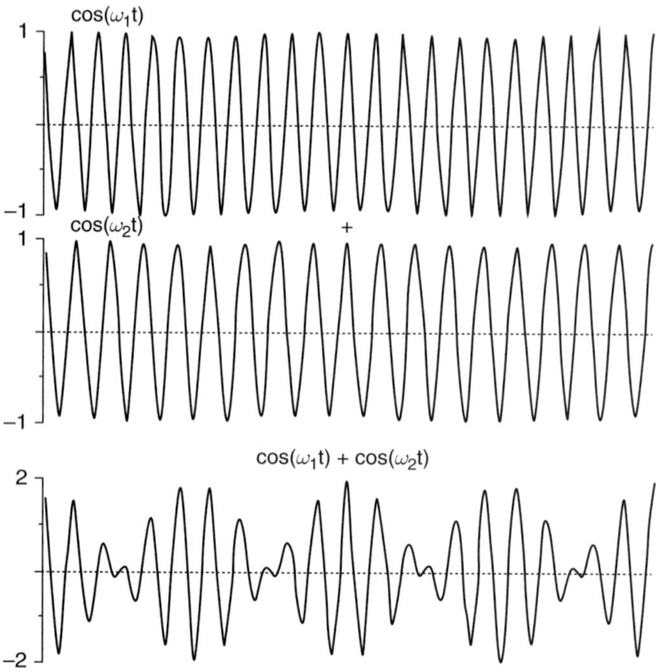

Figure 8.8 The addition of two cosines of similar frequency, giving a modulated cosine with the average frequency.

In those figures, the concept of modulation is clearly seen. The product of the cosines "modulates," as shown by the bold lines forming the envelope of the modulated function, $\cos(\overline{\omega}t)$, in Figure 8.9. The period of oscillation (or modulation) of the envelope, or beat frequency, is given as $T_{beat} = \pi/\delta\omega$, while the period of the modulated series is $T = 2\pi/\overline{\omega}$. The beat frequency is one-half the frequency of the modulation function. That is, the $\cos(\overline{\omega}t)$ is modulated by a function $2\cos(\delta\omega t)$, which has a frequency of $\delta\omega$. The beat frequency is one-half of this modulation frequency, or $\delta\omega/2$. This reflects that fact that the modulation makes complete cycles of the modulated

[8] The period is determined by rewriting $2\cos(\delta\omega t)$ in the standard angular frequency form as $2\cos(2\pi t/T)$, so $2\pi t/T = \delta\omega t$ and $T = 2\pi/\delta\omega$.

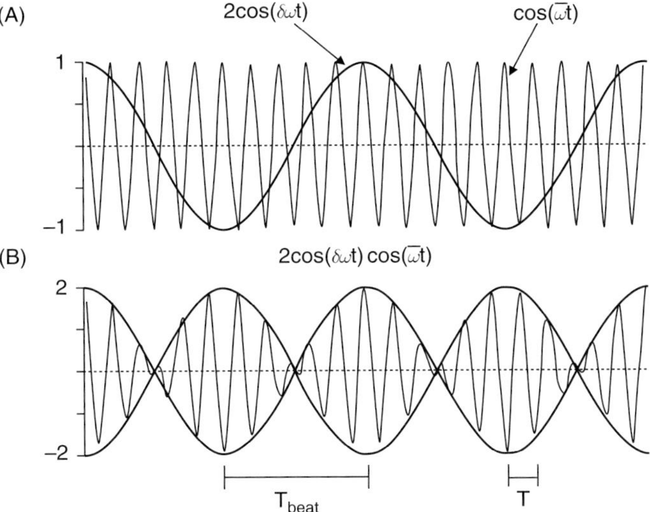

Figure 8.9 **(A)** Average frequency cosine, $\cos(\overline{\omega}t)$, and difference frequencies (the modulating function, $2\cos(\delta\omega t)$, bold line superimposed on the average frequency series) of Figure 8.8 are multiplied (as in 8.21). This multiplication gives the same modulated series as in Figure 8.8, where the two original cosines are added together. **(B)** The envelope is clearly outlined here as the solid bold line in the modulated series.

waveform over each half cycle, because it is symmetrical about the x-axis (abscissa), as seen in the figure.

Also, the relative spacing between peaks in the modulated packets is not regular. Careful examination of the spacing reveals a switch in phase at the period of the beat. Specifically, the peaks are periodic between the zero nodes of the envelope, with the period of the modulated series $T = 2\pi/\overline{\omega}$. But when crossing the zero nodes of the envelope, there is a 180° phase shift. Therefore, each "packet" of the modulated waveform is essentially an inverted form of the preceding (and following) packets. This is due to the x-axis symmetry of the modulating function (i.e., the modulating cosine oscillates symmetrically about zero, flipping the function it modulates when it goes negative).

As $\delta\omega$ gets smaller (i.e., as the two frequencies get closer together), the period of the beat curve gets longer and the phenomena becomes more noticeable. In acoustics, often two frequencies ω_1 and ω_2 are too high to hear, but the beat is within an audible range.

8.4 Fourier Series

8.4.1 Interpolation with Fourier Sines and Cosines

A **Fourier series** is that sequence of sines and cosines that interpolates the specific time series.

Consider fitting a series of n data points, y_i, with a set of sines and cosines that are periodic over the length of the time series, that length given by T. This is a standard continuous (global) interpolation problem, as we've previously examined, and it is given as

$$a_0 \cos 0\,\omega t_i + b_0 \sin 0\,\omega t_i + a_1 \cos 1\,\omega t_i + b_1 \sin 1\,\omega t_i + a_2 \cos 2\,\omega t_i + b_2 \sin 2\,\omega t_i + \ldots = y_i, \quad (8.22a)$$

[9]

where $\omega = 2\pi/T$ and n terms are included in the sum, so there are as many terms of the basis as there are data points.

The first two terms reduce to a_0 (since $\cos 0 = 1$) and 0 (since $\sin 0 = 0$), so b_0 drops out completely, requiring that an additional term be added at the end of the series in order to have n terms for n data points.[10] Thus,

$$a_0 + a_1 \cos 1\omega t_i + b_1 \sin 1\omega t_i + a_2 \cos 2\omega t_i + b_2 \sin 2\omega t_i + \ldots = y_i. \quad (8.22b)$$

Written as a system,

$$\begin{aligned} a_0 + a_1 \cos 1\omega t_1 + b_1 \sin 1\omega t_1 + a_2 \cos 2\omega t_1 + b_2 \sin 2\omega t_1 + \ldots &= y_1 \\ a_0 + a_1 \cos 1\omega t_2 + b_1 \sin 1\omega t_2 + a_2 \cos 2\omega t_2 + b_2 \sin 2\omega t_2 + \ldots &= y_2 \\ &\vdots \\ a_0 + a_1 \cos 1\omega t_n + b_1 \sin 1\omega t_n + a_2 \cos 2\omega t_n + b_2 \sin 2\omega t_n + \ldots &= y_n \end{aligned} \quad (8.23a)$$

or

$$a_0 + \sum_{j=1}^{\le n/2} [a_j \cos(j\omega t_i) + b_j \sin(j\omega t_i)] = y_i. \quad (8.23b)$$

This sum of sines and cosines of increasing harmonic terms is called a **Fourier series.**[11] The a_j and b_j coefficients are called the **Fourier coefficients**. Once we solve for the Fourier coefficients, the sines and cosines just discussed superimpose to interpolate the time series in the standard way.

The Fourier series gives n equations in n unknowns and can be written in standard matrix form:

$$\begin{bmatrix} 1 & \cos\omega t_1 & \sin\omega t_1 & \cos 2\omega t_1 & \sin 2\omega t_1 & \cos 3\omega t_1 & \sin 3\omega t_1 & \ldots \\ 1 & \cos\omega t_2 & \sin\omega t_2 & \cos 2\omega t_2 & \sin 2\omega t_2 & \cos 3\omega t_2 & \sin 3\omega t_2 & \\ & & & \vdots & & & & \\ 1 & \cos\omega t_n & \sin\omega t_n & \cos 2\omega t_n & \sin 2\omega t_n & \cos 3\omega t_n & \sin 3\omega t_n & \end{bmatrix} \begin{bmatrix} a_0 \\ a_1 \\ b_1 \\ a_2 \\ b_2 \\ a_3 \\ b_3 \\ \vdots \end{bmatrix} = \begin{bmatrix} y_1 \\ y_2 \\ \vdots \\ y_n \end{bmatrix}$$

[9] This series can be written as

$$A_0 \cos(0\omega t_i + \varphi) + A_1 \cos(1\omega t_i + \varphi) + A_2 \cos(2\omega t_i + \varphi) + \ldots = y_i \quad (F8.1)$$

However, recall that for the interpolation problem, we wish to establish a system of the form $\mathbf{Ax} = \mathbf{b}$ – that is, one that is linear in the unknown coefficients. That linear form is easily achieved by employing the trigonometric identity, (8.8b), which allows expansion into the pure sines and cosines given in (8.22).

[10] A more thorough discussion regarding these first two terms appears later.

[11] The limit of the sum, as given in (8.23b), is discussed in more detail and given more precisely in the next section.

This system, $\mathbf{Ax} = \mathbf{b}$, is solved for the n coefficients of this sine and cosine series in the usual (matrix) way for interpolation ($\mathbf{x} = \mathbf{A}^{-1}\mathbf{b}$).

8.4.2 Interpreting the Fourier Series

Assume for the moment that we have solved the system given in (8.24) for the a_i and b_i coefficients that cause the sum of sines and cosines given in (8.22) to interpolate the n data points. We can then construct the pure sines and cosines with these amplitudes. Alternatively, we can combine them into cosines (dropping the sines) with amplitude and phase according to (8.11a), where the amplitude of each cosine harmonic, i, is given as $(a_i + b_i)^{1/2}$ and its phase $\tan^{-1}(b_i/a_i)$.

Since each sine and cosine varies about zero, the series has a zero mean, unless the $a_0 \neq 0$. That coefficient is the mean of the final interpolated series – all of the sines and cosines vary about this value.[12]

The first cosine and sine terms represent the fundamental frequency, or first harmonic. Therefore, they both complete one full cycle over the length of the time series at amplitudes determined by the values of a_1 and b_1 (and with no phase displacement).

For the example of Figure 8.10 where $n = 33$ data points, we have a mean (a_0 coefficient) and 16 harmonic terms, each with an amplitude and phase, the combination of the a_i and b_i coefficients, that together sum to exactly fit the data. The a_0 coefficient, representing the mean of the series is not shown in the figure.

This example shows that the irregularly shaped time series shown at the bottom of the Figure 8.10 (and discretely sampled at 33 positions) is completely decomposed into a series of periodic functions (cosines). If you are interested in examining whether the time series represents a periodic process to some degree, then examination of these components for particularly large amplitude harmonic terms may be an indication of underlying periodic components.

Amplitude Spectrum

In the previous figure, the components of the interpolant (i.e., all of the individual cosines in the fitted curve) were explicitly plotted to show the characteristics of each cosine. While that is a perfectly acceptable manner in which to view the components of the interpolant, that same information is more readily displayed by plotting the amplitude of each harmonic term (i.e., the A_i for each cosine, recalling $A_i = (a_i + b_i)^{1/2}$) as a function of the particular harmonic, i. Such a plot is called the **spectrum,** or, more explicitly, the **amplitude spectrum.**[13]

For this example, the spectrum is shown in Figure 8.11.

[12] This is not true for discrete data that are separated at irregular intervals of time or space.

[13] There are actually many forms with which this graph can be displayed. For example, the amplitude might be presented as the A_i values, as the squared values, or as the squared values divided by the total variance of the time series. The harmonic may be presented as the frequency, period, or angular frequency. Also, the axes of the graph may be linear, log, log-linear, etc. Each of these subtly different forms may go by a different name and emphasize a different aspect of the fitted cosines, a matter that is discussed in context later. However, regardless of the specific form of the display, the general information content is the same – the amplitude of the various cosines that interpolate the time series are displayed as a function of the different harmonics.

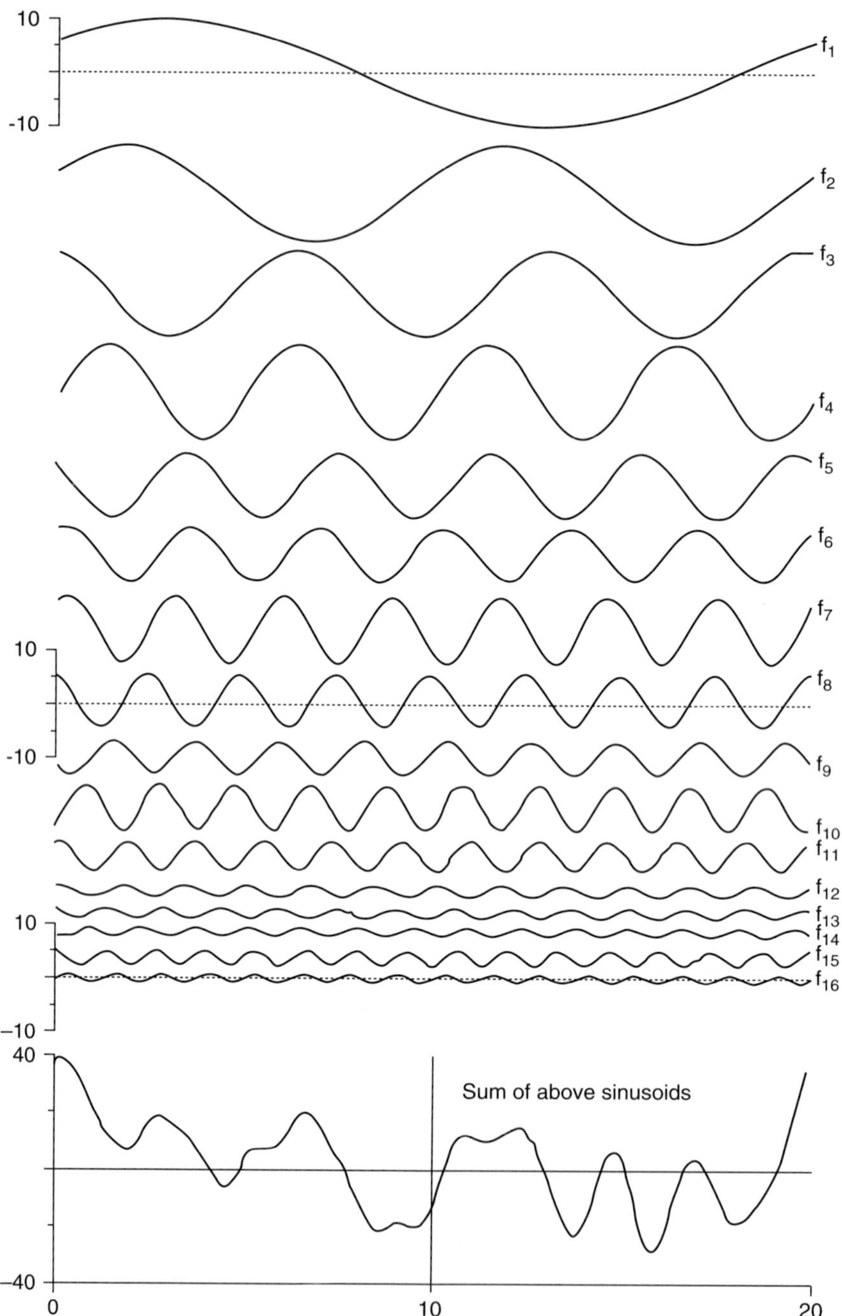

Figure 8.10 Sixteen individual cosines (one for each harmonic), with amplitude and phase, sum to give the more complicated time series at the base of the cosines.

This shows the amplitudes of the various cosines added in Figure 8.10. The harmonic number simply represents the harmonic to which the associated amplitude corresponds. So, for example, harmonic 1 signifies the first (or fundamental) harmonic, represented in

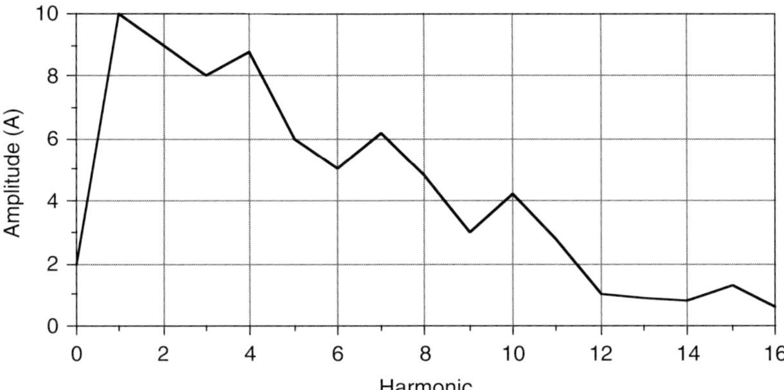

Figure 8.11 Plot of the amplitude of each of the 16 fitted cosines in the example of Figure 8.10.

Figure 8.10 by f_1 (which has an amplitude of 10). The harmonic labeled 0 corresponds to the a_0 coefficient, which represents the mean, as previously discussed.

The higher harmonic terms in this example have relatively small amplitudes, and there is no single harmonic that overwhelmingly dominates. Spectra that show a general dominance of the lower-order harmonics (low-frequency components) with smaller and smaller amplitudes at the higher frequencies are termed **red spectra**.[14] Such spectra are widespread in the physical sciences and represent a particular form of random noise. We define red spectra more rigorously later, as well as numerous variations on this plot, each of which has specific desirable characteristics.

Phase Spectrum

A similar plot can be made to display the phase angle (or displacement) for each of the harmonics. Such a plot is called the **phase spectrum**.[15] The phase spectrum for the example of Figure 8.10, is shown in Figure 8.12.

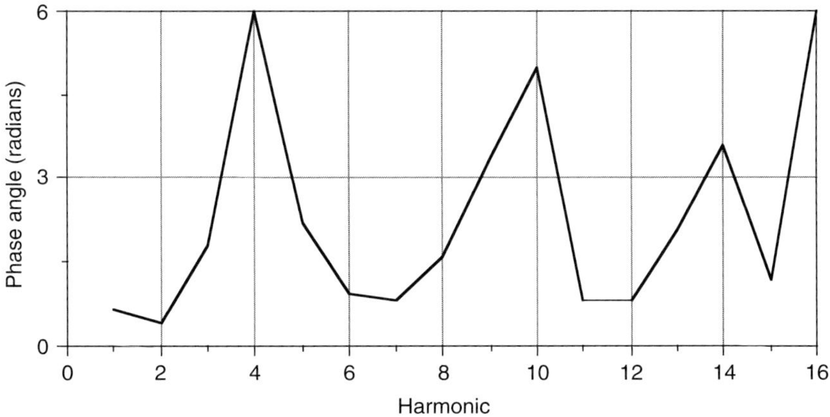

Figure 8.12 Phase angle for each harmonic of the example in Figure 8.10.

[14] Red spectra have a very specific slope in the amplitude spectrum, so there are other "colors" often applied to spectra dominated by lower-frequency components, such as pink, etc., according to their actual slope. More about this later in Chapter 11, Spectral Analysis.

[15] Caution must be used here, since the term *phase spectrum* formally describes a plot of phase differences between two time series in *cross spectral analysis*, a topic that is covered later. As always, considerable care must be taken to avoid confusion in the use of the terms.

In general, phase spectra are more erratic than amplitude spectra, and as discussed later regarding stochastic processes, they are often ignored.

8.5 Take-Home Points

1. Series that are periodic over their length (T) are ones in which the next point that implicitly follows the last point in the series is equal to the first point in the series. Or, $f(t) = f(t + T) = f(t + nT)$. T is the period, but if the series varies in space, T is the wavelength.
2. Sines and cosines are the most natural periodic functions, repeating every $2j\pi$, where j is any integer.
3. Any sinusoid with rotational frequency ($f = 1/T$), amplitude (A) and phase (φ), can be written as $A\cos(2\pi f + \varphi) = + a\cos(2\pi f) + b\sin(2\pi f)$, where $A = (a + b)^{1/2}$, $\varphi = \tan^{-1}(b/a)$. Or, in terms of angular frequency, $\omega = 2\pi/T$.
4. Harmonics are frequencies that are integer multiples of a fundamental frequency. The fundamental frequency is $f_1 = 1/T$, i.e., it completes one full cycle over the length of the series. Second harmonic is $f_2 = 2f_1 = 2/T$, so it makes two full cycles over the length of the series. Generalized, the nth harmonic is $f_n = nf_1 = n/T$, making n full cycles over the length of the series.
5. A Fourier series is a sequence of sines and cosines that interpolates (exactly fits) a specific time series (y_i), written as $a_0 + \sum_{j=1}^{\leq n/2} [a_j \cos(j\omega t_i) + b_j \sin(j\omega t_i)] = y_i$. Fundamental to this series is that the harmonics are specific to the length (T) of the series being fit. Different series with different lengths have different "Fourier" harmonics.
6. Amplitude spectrum is the presentation of the a and b (Fourier) coefficients combined as an amplitude of the interpolated Fourier series as a function of harmonic, or $A(\omega)$, where $A = (a + b)^{1/2}$.

8.6 Questions

Pencil and Paper Questions

1. For a time series y_i of four evenly spaced data points (sampling interval $\Delta t = 1$ s), write out each term of its Fourier series (actually write out each term of the sum). Show the exact solution for the a_i and b_i coefficients. What is the sampling rotational frequency?
2. Write out the Fourier series for a time series containing 100 data points, with spacing at $\Delta t = 1$ s in standard matrix form (**Ax** = **b**), showing the first five and last three rows of the **A** matrix and **x** and **b** vectors, using angular frequency ω_i, including all details explicitly.

9 Fourier Transform

9.1 Overview

This chapter takes advantage of orthogonal functions (in this case an orthogonal basis – the Fourier sine and cosines of Chapter 8) to develop an extremely easy and efficient manner for finding the coefficients of that basis to yield the interpolated fit of the Fourier series. This approach can actually be used to solve for the coefficients of any orthogonal-basis interpolant.

In Chapter 8, the coefficients of the Fourier series were determined by inverting $\mathbf{Ax} = \mathbf{b}$ as a standard interpolation problem. But because the Fourier sinusoids (the harmonics) form an orthogonal basis, we are afforded some elegant reductions determined by Lagrange, who in the 1800s took advantage of the orthogonality relationships between the various harmonics.[1] This allows an efficient analytic solution to the problem of computing the coefficients of the Fourier series.

This convenient and efficient approach is known as the **Fourier transform**. The Fourier transform can also be thought of as a means for converting a time series from the **time domain** (its original domain, which can be time, space or other) into the **frequency domain** (the domain of the component sinusoids). In other words, once we have the Fourier coefficients, we can look at the original time series in terms of the interpolating cosinusoids, at individual frequencies, which also completely describe the time series from a different perspective.

9.2 Discrete Periodic Data

The fundamental restriction with the discrete Fourier transform is that the orthogonality conditions upon which it relies only hold for the case of evenly spaced data. If the data are not sampled at even intervals of the independent variable (e.g., time), Δt, you must either interpolate them to even spacing or resort to a different approach.

To proceed with this analytical approach, we now consider that the data are evenly spaced. So, the times, t_i, can be given in terms of the *even sampling interval*, Δt, as

$$t_i = i\Delta t. \tag{9.1}$$

[1] Lagrange actually used only the sine terms over a range from 0 to π.

as can the *period of the time series*, T,

$$T = n\Delta t. \tag{9.2}$$

Here it is important to differentiate between a truly periodic function in which the known period is T and a data set that is most likely not actually periodic over any length. We develop the theory by *assuming* that the data are periodic over their sampled length so that the next point occurring beyond the last data point sampled replicates the first point in the series (recall the cyclic representation shown previously for a periodic function). So, *we are requiring that the data be evenly spaced and assuming that the data are periodic over the length at which they were sampled.* We will later quantify the implications of the periodic assumption.

Given (9.1) and (9.2) and the above, required assumption, consider the following example. Here, eight data points where $\Delta t = 1$ are shown in Figure 9.1.

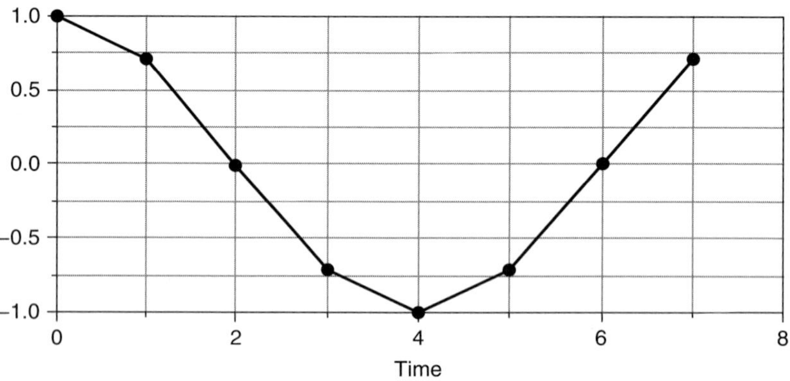

Figure 9.1 Example of an 8-point periodic time series with $\Delta t = 1$.

Equation (9.2) states that this time series is periodic over time, T = 8. Because we assume that the data are periodic over the $n = 8$ values, any additional data points would replicate those we already have. That is, we are assuming that additional samples preceding the first data point and following the last data point are predictable due to the definition that $y(t) = y(t \pm T)$. So, five additional data points, indicated with an **x** give that extended curve as shown in Figure 9.2.

Consistent with the "rule" given in (9.2), the time series is sampled so that it completely represents one full period; the next data point lying beyond the nth point occurs at time = T, so the data value at this point is equivalent to the first point in the time series (this periodic assumption will be relaxed later on). For this example, the point at the origin, $y(1) = y(1 + T) = y(1 + n\Delta t) = y(1 + 8) = y(9)$. Alternatively, you could adopt a numbering scheme where the data lie at positions $i = 0,1,2, \ldots, n - 1$, but I find indexing from 1 to n to be more intuitive, and it affords some intuitive indexing later, so that one will be used here. Either (actually, *any*) index numbering scheme is acceptable, as long as the scheme is used consistently throughout the calculations.

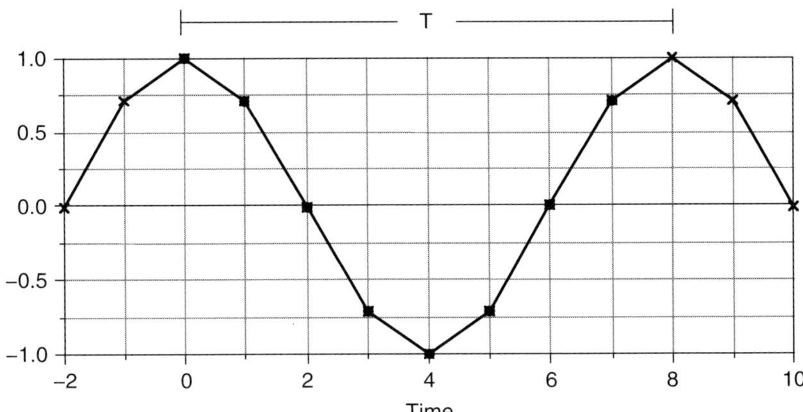

Figure 9.2 Extension (in both directions) of points in Figure 9.1 time series, indicated by $\boldsymbol{x}$.

9.2.1 Fourier Series in Summation Form

Consider rewriting the Fourier series of (8.23b) in terms of the even spacing as

$$y_i = a_0 + \sum_{j=0}^{\leq n/2} \left(a_j \cos \frac{2\pi ji}{n} + b_j \sin \frac{2\pi ji}{n} \right) \tag{9.3a}$$

or

$$y_i = a_0 + \sum_{j=0}^{\leq n/2} (a_j \cos \omega_j \mathrm{t}_i + b_j \sin \omega_j \mathrm{t}_i), \tag{9.3b}$$

where $\omega_j = j2\pi/T = j2\pi/n\Delta t = 2\pi \mathrm{f}_j$, for harmonic j.

The inequality sign "≤" in the summation limit reflects how the number of terms in the sum changes according to whether there is an even or odd number of data points, n, in the time series.

When n Is an Even Number

In this case, the limit of the sum in (9.3) is given by $j = n/2$, and the last terms in the sum are

$$a_{n/2} \cos \frac{2\pi(n/2)\mathrm{t}_i}{n\Delta \mathrm{t}} + b_{n/2} \sin \frac{2\pi(n/2)\mathrm{t}_i}{n\Delta \mathrm{t}}. \tag{9.4}$$

The times, t_i, are at discrete intervals of Δt defined by (9.1), so $\mathrm{t}_i = i\Delta \mathrm{t}$. Therefore, (9.4) can be rewritten as

$$a_{n/2} \cos \frac{2\pi(n/2)\mathrm{t}_i}{n\Delta \mathrm{t}} + b_{n/2} \sin \frac{2\pi(n/2)\mathrm{t}_i}{n\Delta \mathrm{t}} = a_{n/2} \cos (i\pi) + b_{n/2} \sin (i\pi), \tag{9.5}$$

where (9.5) holds for all i, since i is always an integer and $\sin(i\pi) = 0$ for all integer values of i. In this case, the last sine term does not contribute to the Fourier series sum.

So, with the summation index j going from $j = 1$ to $n/2$, for n = even number, we have the Fourier coefficients represented in the vector of unknown coefficients **x**, from (8.24):

$$\mathbf{X}^{\mathrm{T}} = [a_0, a_1, a_2, \ldots a_{n/2}, b_1, \ldots b_{(n/2)-1}], \tag{9.6}$$

where the $b_{n/2}$ term dropped out as just shown.

Thus, the highest sine term present is at $j = [(n/2) - 1]$, or one term less than the $(n/2)$th term. Consequently, the total number of terms in the Fourier summation is

$$1 + n/2 + n/2 - 1 = n. \tag{9.7}$$

So, we have as many terms in the Fourier series as there are data points. This makes the system well posed, and the system can be solved for the unknown coefficients.

For example, if $n = 12$ we have $a_0, a_1, a_2, a_3, a_4, a_5, a_6, b_1, b_2, b_3, b_4, b_5 = 12$ unknowns for 12 equations (data points) and the highest terms are given by $j = n/2 = 6$ (for which only the cosine term survives).

When *n* Is an Odd Number

When the time series contains an odd number of data points so that n is an odd number, the limit of the sum in (9.3) is given by $j < n/2$, or more specifically, $(n - 1)/2$. For example, if $n = 13$, then $n/2 = 6.5$, and the nearest value of the *integer* $j < n/2$ is $6 = (n - 1)/2$. In this case, the last terms in the sum of (9.3) are now given by

$$a_{(n-1)/2} \cos \frac{2\pi[(n-1)/2]\mathrm{t}_i}{n\Delta \mathrm{t}} + b_{(n-1)/2} \sin \frac{2\pi[(n-1)/2]\mathrm{t}_i}{n\Delta \mathrm{t}}. \tag{9.8}$$

In this case, neither of the terms in (9.8) can be eliminated by reduction. Therefore,

$$\mathbf{X}^{\mathrm{T}} = [a_0, a_1, a_2, \ \ldots \ a_{(n-1)/2}, b_1, b_2, \ \ldots \ b_{(n-1)/2}], \tag{9.9}$$

which gives a total number of Fourier terms as

$$1 + (n-1)/2 + (n-1)/2 = n, \tag{9.10}$$

again giving us as many unknowns as equations.

For example, with $n = 13$, the summation limit is again 6, i.e., $(n - 1)/2 = 6$, but in this case the last sine term does not drop out of the summation, so we retain the b_6 term giving a total of 13 unknown coefficients: $a_0, a_1, a_2, a_3, a_4, a_5, a_6, b_1, b_2, b_3, b_4, b_5, b_6$.

9.2.2 A Most Excellent Form of the Discrete Fourier Series

A succinct form of the Fourier series takes consideration of the terms corresponding to $j = 0$. In this case, the cosine and sine terms are given as

$$a_0 \cos(0) + b_0 \sin(0) = a_0. \tag{9.11}$$

Given this, the a_0 term outside of the summation in all of the previous forms of the Fourier series can be taken inside the summation by starting the sum at $j = 0$. Now the Fourier series is written in a succinct form as

$$y_i = \sum_{j=0}^{\leq n/2} (a_j \cos \omega_j t_i + b_j \sin \omega_j t_i) \tag{9.12}$$

or, since the $\omega_j t_i = 2\pi j t_i / n\Delta t$ and $t_i = i\Delta t$, any of these forms can be rewritten explicitly in terms of the harmonics, j, and data points, i, as

$$y_i = \sum_{j=0}^{\leq n/2} \left(a_j \cos \frac{2\pi ji}{n} + b_j \sin \frac{2\pi ji}{n} \right), \tag{9.13}$$

a succinct and convenient form used throughout the remainder of this chapter.
Finally, in anticipation of a later result, (9.13) is often written as

$$y_i = \frac{a_0}{2} + \sum_{j=1}^{\leq n/2} (a_j \cos \omega_j t_i + b_j \sin \omega_j t_i). \tag{9.14}$$

This particular form is advantageous and popular, since it compensates for a factor of 2 that arises in the Fourier transform.

9.2.3 Fourier Frequencies

In the various forms of the Fourier series given above, the frequencies of the sines and cosines are always *harmonics* of the length ($n\Delta t$) of the time series, or T, and are given as

$$\omega_j = \frac{2\pi j}{n\Delta t} = \frac{2\pi j}{T}, \tag{9.15}$$

where

$$0 \leq j \leq \frac{n}{2} \tag{9.16}$$

or

$$j = 0, 1, \ldots \frac{n-1}{2}, \underset{\text{if } n \text{ is even}}{\underset{\uparrow}{\frac{n}{2}}} \tag{9.17}$$

Therefore,

$$0 \leq \omega_j \leq \frac{\pi}{\Delta t}, \tag{9.18a}$$

or

$$\omega_j = 0, \frac{2\pi}{n\Delta t}, \frac{4\pi}{n\Delta t}, \ldots \frac{(n-1)\pi}{n\Delta t}, \underset{\substack{\uparrow \\ \text{if } n \text{ is even}}}{\frac{\pi}{\Delta t}} \tag{9.18b}$$

The jth frequency, or harmonic, ω_j, is referred to as the jth **Fourier frequency.**[2] As such, *Fourier frequencies refer to a very specific set of frequencies, particular to a given time series, of length T.* These frequencies always force any sine and cosine functions to oscillate an integer number of times over the length of the time series. For time series of different lengths, the Fourier frequencies are different.

Nyquist Frequency

The Fourier frequency for harmonic $j = n/2$ is

$$\begin{aligned} \omega_N &= \pi/\Delta t \\ &= 2\pi/2\Delta t \qquad \text{(angular frequency)} \end{aligned} \tag{9.19a}$$

or, in terms of rotational frequency, $\omega_N = 2\pi f_N$ and

$$f_N = 1/2\Delta t \qquad \text{(rotational frequency)}. \tag{9.19b}$$

This particular frequency is referred to as the **Nyquist frequency.**[3] As discussed above, *this frequency is only included in the Fourier series when there is an even number of data points* – otherwise, the highest frequency component in the Fourier series occurs for harmonic $(n - 1)/2$.

The Nyquist frequency represents the highest harmonic that *may be* included in the Fourier series, and is the highest frequency component that can be identified in the sampled form of the time series (the implication of the qualifier "sampled form" is discussed in more detail later). This harmonic has a period of one full cycle every 2 data points. That is, the angular frequency is $2\pi/T$, where in the case of this particular harmonic, $T = 2\Delta t$. Thus, a complete oscillation occurs every other data point. The sinusoid term at this frequency is always zero, as seen from (9.5). Therefore, *the Nyquist frequency represents a cosine only with no phase, and as a consequence it does not give a full representation of the highest frequency present*, which may have a nonzero phase (requiring the presence of a sine term in the real time series).

The Nyquist frequency has special significance in Fourier analysis, as will be seen.

[2] Note that we usually do not differentiate between Fourier angular frequency ($= 2\pi j/T$) and simple Fourier rotational frequency ($= j/T$). As with many of the terms we use, care must be taken to clearly indicate the specific use of the term, because in some cases (e.g., integration) the form can lead to additional terms.

[3] The latter (rotational frequency) form, $1/2\Delta t$, is the more common form when discussing this Nyquist frequency.

9.2.4 Summation Properties and Orthogonality Conditions

Lagrange took advantage of several characteristics of harmonic components for time series sampled at evenly spaced intervals. For convenience, we still assume that $\Delta t = 1$, so the period, $T = n\Delta t = n$. This assumption simplifies some of the relationships that would otherwise be scaled by a constant if $\Delta t \neq 1$. Also, for all relationships, the frequencies, ω_j, are harmonics of the period T, so $\omega_j = 2\pi j/T$ or $2\pi j/n\Delta t$.

These relationships include the summation property of sines and cosines over the period T – the continuous (integral) form of these relationships is also given for later reference:

Summation Property for Cosines

The **summation property for cosines** is

$$\sum_{i=1}^{n} \cos(\omega_j t_i) = \begin{cases} 0 & \text{if } j \neq 0 \quad (9.20a) \\ n & \text{if } j = 0 \quad (9.20b) \end{cases}$$

or, in integral form,

$$\int_{-T/2}^{T/2} \cos(\omega_j t_i) dt = \begin{cases} 0 & \text{if } j \neq 0 \quad (9.20c) \\ T & \text{if } j = 0 \quad (9.20d) \end{cases}$$

That is, these relationships describe the result when *summing or integrating over one full period of a cosine.*

Summation Property for Sines

The **summation property for sines** is

$$\sum_{i=1}^{n} \sin(\omega_j t_i) = 0 \qquad \text{all } j \tag{9.21a}$$

or, in integral form,

$$\int_{-T/2}^{T/2} \sin(\omega_j t) dt = 0 \qquad \text{all } j. \tag{9.21b}$$

These summation properties indicate that the sum over one full period equals 0. This is easily seen by inspecting a sine or cosine over one period. Consider the case of a cosine, sampled *evenly* (solid squares) in Figure 9.3. This example makes it clear why, for the discrete case, the data must be evenly sampled and cannot include both points y_0 and y_T. Essentially, the values added together over the first half period exactly cancel the values added over the second half because of the symmetry of the harmonic. However, this symmetry is not preserved if the spacing is uneven and therefore, the values needn't cancel. That is, the unevenly spaced samples (open circles) do not sum to zero. Similarly, inclusion of the point (cross) at f(t=8) in the sum of the evenly spaced samples (squares)

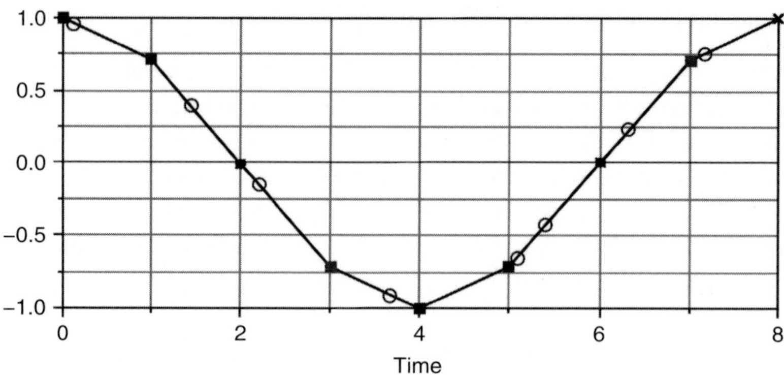

Figure 9.3 Example showing why data points have to be evenly spaced in time for summation property (and orthogonality) to hold. When summing over one period, evenly spaced points (squares) exactly cancel; this is not the case for the uneven spacing (e.g., the open circles). It is that simple! Orthogonality, the foundation of the Fourier transform, takes advantage of this simple cancellation. Otherwise, no can do.

causes the sum to exceed – that is, we are now one point beyond the period and are including the first point in the sum twice.

Orthogonality Conditions (Major Importance)

The three **orthogonality conditions** are a direct consequence of the summation properties, with help from the trigonometric identities of (8.8). Recall that we are still assuming $\Delta t = 1$ (for convenience) and evenly spaced data (by necessity) as before.

First Orthogonality Condition

The **first orthogonality condition** is

$$\sum_{i=1}^{n} [\cos(\omega_j t_i)\cos(\omega_k t_i)] = \begin{cases} n/2 & j = k \neq 0 \neq n/2 \\ n & j = k = 0 = n/2 \\ 0 & j \neq k \end{cases} \quad \begin{matrix} (9.22a) \\ (9.22b) \\ (9.22c) \end{matrix}$$

or, integral form,

$$\int_{-T/2}^{T/2} [\cos(\omega_j t)\cos(\omega_k t)]dt = \left\{ \begin{matrix} T/2 & j = k \neq 0 \\ T & j = k = 0 \\ 0 & j \neq k \end{matrix} \right\} \quad \begin{matrix} (9.22d) \\ (9.22e) \\ (9.22f) \end{matrix}$$

Second Orthogonality Condition

The **second orthogonality condition** is

$$\sum_{i=1}^{n} [\cos(\omega_j t_i)\sin(\omega_k t_i)] = 0 \quad (\text{all } j \text{ and } k) \qquad (9.23a)$$

Box D9.1 Derivation of Orthogonality Conditions

Proof of the orthogonality relationships follows directly from proof of the summation properties and trigonometric identities of (9.8).

First, consider the proof of the summation property for cosines in (9.20). This proof is most easily shown using the integral relationship. For $j \neq 0$, the proof is shown by substituting $u = \omega_j t$; $du = \omega_j dt$, so $dt = (1/\omega_j)du$, which gives

$$\int_{-T/2}^{T/2} \cos(\omega_j t)dt = \frac{1}{\omega_j}\int_{-T/2}^{T/2} \cos(u)du$$

$$u = \omega_j t = \frac{2\pi jt}{T}$$

$$\frac{du}{dt} = \omega_j;\ dt = \frac{du}{\omega_j}$$

$$= \frac{1}{\omega_j}\sin(u)\bigg|_{-T/2}^{T/2}$$

$$= \frac{T}{2\pi j}\sin\left(\frac{2\pi j}{T}\frac{T}{2}\right) + \frac{T}{2\pi j}\sin\left(\frac{2\pi j}{T}\frac{T}{2}\right)$$

$$= 0, \tag{D9.1.1}$$

recalling that sine is an odd function, so $\sin(-x) = -\sin(x)$ and $\sin(\pi j) = 0$ for all j.

For the discrete case, one proof is obtained by a successive expansion of the cosine terms using (9.8b) and breaking up the $\omega_j t = 2\pi ji/n$ into a sum of $\pi + (2ji - n)\pi/n$. An easier proof can be made using the theory of distributions, which we will cover later.

In any case, the discrete and continuous form is relatively easy to conceptualize when we realize that over one period, a sinusoid (sine or cosine) is anti-symmetrical about the x axis: half of the waveform is positive while the other half is negative. Thus, over a full period, the anti-symmetrical halves negate one another when summing or integrating, to give zero.

For $j = 0$, then,

$$\int_{-T/2}^{T/2} \cos(0)dt = \int_{-T/2}^{T/2} dt = T. \tag{D9.1.2}$$

For the discrete case,

$$\sum_{i=1}^{n} \cos(0) = \sum_{i=1}^{n} 1 = n. \tag{D9.1.3}$$

Here it is seen that the discrete case will differ from the integral relationship by the factor Δt, since $T = n\Delta t$. Only when $\Delta t = 1$ are the two relationships, (D9.1.2) and (D9.1.3), identical.

Box D9.1 (Cont.)

The proof of the summation property for sines, (9.21), follows the same logic as is used above. For $j \neq 0$, substituting $u = \omega_j t$ gives

$$\int_{-T/2}^{T/2} \sin(\omega_j t)dt = \frac{1}{\omega_j}\int_{-T/2}^{T/2} \sin(u)du$$
$$= \frac{-1}{\omega_j}\cos(u)\Big|_{-T/2}^{T/2}$$
$$= \frac{-T}{2\pi j}\left[\cos\left(\frac{2\pi j}{T}\frac{T}{2}\right) - \cos\left(\frac{-2\pi j}{T}\frac{T}{2}\right)\right]$$
$$= 0, \tag{D9.1.4}$$

recalling that a cosine is an even function so cos(−x) = cos(x).

For $j = 0$, since sin(0) = 0, obviously, the sum and integral over all i is still 0, so

$$\sum_{i=1}^{n} \sin(0) = 0. \tag{D9.1.5}$$

The first orthogonality relationship is shown by employing the trigonometric identity for the product of two cosines given by (9.8c), so

$$\cos(\omega_j t)\cos(\omega_k t) = \frac{1}{2}\{\cos[(\omega_j + \omega_k)t] + \cos[(\omega_j - \omega_k)t]\}. \tag{D9.1.6}$$

For $j = k \neq 0$, then, using (D9.1.6),

$$\int_{-T/2}^{T/2} \cos(\omega_j t)\cos(\omega_k t)dt = \frac{1}{2}\int_{-T/2}^{T/2} \cos(2\omega_j t)dt + \frac{1}{2}\int_{-T/2}^{T/2} dt. \tag{D9.1.7}$$

We have already shown for (D9.1.1) that the first integral on the right-hand side is 0. Therefore, only the second integral remains, which, as seen from (D9.1.2), reduces to

$$= T/2.$$

For $j = k = 0$, both the integral and discrete case are easily shown:

$$\int_{-T/2}^{T/2} \cos(0)\cos(0)dt = \int_{-T/2}^{T/2} dt = T, \tag{D9.1.9}$$

Box D9.1 (Cont.)

or, for the discrete case,

$$\sum_{i=1}^{n}[\cos(0)\cos(0)] = n. \tag{D9.1.10}$$

Similarly, for $j = k = n/2$ (for the discrete case),

$$\sum_{i=1}^{n}\cos(\omega_j t_i)\cos(\omega_k t_i) = \frac{1}{2}\sum_{i=1}^{n}\cos(2\omega_j t_i) + \frac{1}{2}\sum_{i=1}^{n}\cos(0), \tag{D9.1.11}$$

where $2\omega_{n/2}t_i = 2[2\pi(n/2)i\Delta t/n\Delta t] = 2\pi i$, so (D9.1.11) reduces to

$$\frac{1}{2}\sum_{i=1}^{n}\cos(2\pi i) + \frac{n}{2}. \tag{D9.1.12}$$

Since i is always an integer, $\cos(2\pi i) = 1$ for all i, so (D9.1.12) reduces to

$$\frac{n}{2} + \frac{n}{2} = n. \tag{D9.1.13}$$

Finally, when $j \neq k$, the integral

$$\int_{-T/2}^{T/2}\cos(\omega_j t)\cos(\omega_k t)dt \tag{D9.1.14a}$$

is simplified by application of (D9.1.6) for the product of two cosines, so (D9.1.14a) becomes

$$= \frac{1}{2}\int_{-T/2}^{T/2}[\cos(\omega_j + \omega_k)]dt + \frac{1}{2}\int_{-T/2}^{T/2}[\cos(\omega_j - \omega_k)]dt \tag{D9.1.14b}$$

$$= \frac{1}{2}\int_{-T/2}^{T/2}\cos(\phi_{j+k}t)dt + \frac{1}{2}\int_{-T/2}^{T/2}\cos(\phi_{j-k}t)dt. \tag{D9.1.14c}$$

Both integrals in (D9.1.14) have already been shown by summation property for cosines (9.20) to be equal to zero, so (D9.1.14) reduces to 0 itself.

The discrete case for (9.22c) is proven in a manner analogous to the integral proof here, and then relying on the relationship of (9.20) to reduce the sums to 0. The proofs for the second two orthogonality relationships closely track the proofs given above for the first orthogonality condition.

or, integral form,

$$\int_{-T/2}^{T/2} [\cos(\omega_j t_i)\sin(\omega_k t_i)]dt = 0 \quad (\text{all } j \text{ and } k). \tag{9.23b}$$

Third Orthogonality Condition
The **third orthogonality condition** is

$$\sum_{i=1}^{n} [\sin(\omega_j t_i)\ \sin(\omega_k t_i)] = \begin{cases} n/2 & j = k \neq 0 \neq n/2 \quad (9.24a) \\ 0 & \text{otherwise} \quad (9.24b) \end{cases}$$

or, in integral form,

$$\int_{-T/2}^{T/2} [\sin(\omega_j t)\sin(\omega_k t)]dt = \begin{cases} n/2 & j = k \neq 0 \quad (9.24c) \\ 0 & \text{otherwise} \quad (9.24d) \end{cases}$$

Notice that the orthogonality conditions describe the *covariance* between sines and cosines of various harmonics. When $j = k$, the conditions describe the variance of the cosine or sine at that particular harmonic. The true variance of a sinusoid is given by $E[\cos^2 x] - E[\cos x]^2$, but this second term is zero since the cosine has a zero mean. Therefore, the mean does not need to be estimated when computing the sample variance, so the latter is given by $(1/n)\Sigma\cos^2 x = 1/2$ in the above relationships (except when $k = j = 0$ or $n/2$). If the mean did have to be estimated, then the variance would be given by $1/(n - 1)$. Actually, since the sinusoids are deterministic functions, this variance is simply a measure of the distribution about the mean with time; it is not a statistical moment, since there is no uncertainty in the values at any time t.

9.3 Discrete Sine and Cosine Transforms

Now we start down the beautiful road to the famous "Fourier transform" (the initial development of which is often attributed to people other than Fourier). The relationships given by (9.20) through (9.24) are used to find an analytical solution for the coefficients of the Fourier series. We start by working with the Fourier series in the form of (9.12), repeated here:

$$y_i = \sum_{j=0}^{\leq n/2} (a_j \cos\omega_j t_i + b_j \sin\omega_j t_i). \tag{9.12}$$

As always, $\omega_j = 2\pi j/n\Delta t$ and $t_i = i\Delta t$.

Now multiply both sides of (9.12) by a cosine harmonic, $\cos(\omega_k t_i)$, where $\omega_k = 2\pi k/n\Delta t$:

$$y_i \cos \omega_k t_i = \sum_{j=0}^{\leq n/2} (a_j \cos \omega_j t_i \cos \omega_k t_i + b_j \sin \omega_j t_i \cos \omega_k t_i). \tag{9.25}$$

If we sum over the n data points which represent one period (i.e., $T = n$; recall, $\Delta t = 1$ for convenience) and $i = 1,2, \ldots, n$, we can employ the orthogonality conditions. So,

$$\sum_{i=1}^{n} y_i \cos \omega_k t_i = \sum_{i=1}^{n} \left[\sum_{j=0}^{\leq n/2} (a_j \cos \omega_j t_i \cos \omega_k t_i + b_j \sin \omega_j t_i \cos \omega_k t_i) \right]. \tag{9.26}$$

Interchanging the order of the sum and expanding gives

$$= \sum_{j=0}^{\leq n/2} \left(a_j \sum_{i=1}^{n} (\cos \omega_j t_i \cos \omega_k t_i) + b_j \sum_{i=1}^{n} (\sin \omega_j t_i \cos \omega_k t_i) \right). \tag{9.27}$$

Equation (9.27) can be solved immediately for all frequencies, ω_k, using the three orthogonality conditions directly. However, to see these solutions more clearly, consider the nature of the summation for various values of k.

When k = 0, then $\omega_k = \omega_0 = 0$, so $\cos(\omega_0 t_i) = \cos(0) = 1$ and (9.27) reduces to

$$\sum_{i=1}^{n} y_i = \sum_{j=0}^{\leq n/2} \left(a_j \sum_{i=1}^{n} (\cos \omega_j t_i) + b_j \sum_{i=1}^{n} (\sin \omega_j t_i) \right). \tag{9.28a}$$

Expanded with respect to the sum over j,

$$\sum_{i=1}^{n} y_i = a_0 \sum_{i=1}^{n} (\cos \omega_0 t_i) + b_0 \sum_{i=1}^{n} (\sin \omega_0 t_i) + a_1 \sum_{i=1}^{n} (\cos \omega_1 t_i) + b_1 \sum_{i=1}^{n} (\sin \omega_1 t_i) + \ldots. \tag{9.28b}$$

We know from the summation property for cosines, (9.16), that except for ω_0 ($j = 0$), the sum of a cosine over its period is equal to zero. Similarly, from the summation property for sines, (9.21), this sum is equal to zero for all frequencies (including ω_0). Therefore, in the sum of (9.28b) the only term that does not reduce to zero corresponds to the a_0 coefficient term. That is

$$\begin{aligned} \sum_{i=1}^{n} y_i &= a_0 \sum_{i=1}^{n} (\cos 0) \\ &= a_0 n. \end{aligned} \tag{9.29}$$

Rearranging (9.29) to solve for a_0 gives

$$a_0 = \frac{1}{n} \sum_{i=1}^{n} y_i = \bar{y}. \tag{9.30}$$

Therefore, by multiplying the Fourier series through by a cosine of frequency zero and summing over all data points, we obtain a solution for the a_0 coefficient that says that the a_0 coefficient is the mean of the time series. (Was that cool, or what?)

When k = 1, then $\omega_k = \omega_1$, (9.27) becomes

$$\sum_{i=1}^{n} y_i \cos\omega_1 t_i = \sum_{j=0}^{\leq n/2}\left(a_j \sum_{i=1}^{n}(\cos\omega_j t_i \cos\omega_1 t_i) + b_j \sum_{i=1}^{n}(\sin\omega_j t_i \cos\omega_1 t_i)\right). \quad (9.31a)$$

Expanded,

$$\begin{aligned}\sum_{i=1}^{n} y_i \cos\omega_1 t_i &= a_0 \sum_{i=1}^{n}(\cos\omega_0 t_i \cos\omega_1 t_i) + b_0 \sum_{i=1}^{n}(\sin\omega_0 t_i \cos\omega_1 t_i)\\ &\quad a_1 \sum_{i=1}^{n}(\cos\omega_1 t_i \cos\omega_1 t_i) + b_1 \sum_{i=1}^{n}(\sin\omega_1 t_i \cos\omega_1 t_i) + \ldots\ .\end{aligned} \quad (9.31b)$$

For this case, the orthogonality conditions, (9.22) and (9.23), can be applied: all of the sums involving the products of sines with cosines equal 0. The only sum involving the products of cosines with cosines that is not 0 is the a_1 term, where $j = k$ and the sum equals $n/2$. Therefore, (9.31) reduces to

$$\begin{aligned}\sum_{i=1}^{n} y_i \cos\omega_1 t_i &= a_1 \sum_{i=1}^{n}(\cos\omega_1 t_i \cos\omega_1 t_i)\\ &= a_1 n/2\end{aligned} \quad (9.32)$$

so

$$a_1 = \frac{2}{n}\sum_{i=1}^{n} y_i \cos\omega_1 t_i. \quad (9.33)$$

For $k \neq 0$ and $\neq n/2$, since the orthogonality relationships are the same as above (for $k = 1$) for all other values of $k \neq 0 \neq n/2$, extension of the above case, for $k = 1,2, \ldots, (n-1)/2$, in general gives

$$a_k = \frac{2}{n}\sum_{i=1}^{n} y_i \cos\omega_k t_i. \quad (9.34)$$

For k = $n/2$, $\omega_k = \omega_{n/2}$, (9.27), in expanded form, becomes

$$\begin{aligned}\sum_{i=1}^{n} y_i \cos\omega_{n/2} t_i &= a_0 \sum_{i=1}^{n}(\cos\omega_0 t_i \cos\omega_{n/2} t_i) + b_0 \sum_{i=1}^{n}(\sin\omega_0 t_i \cos\omega_{n/2} t_i) + \ldots\\ &\quad + a_{n/2} \sum_{i=1}^{n}(\cos\omega_{n/2} t_i \cos\omega_{n/2} t_i) + b_{n/2} \sum_{i=1}^{n}(\sin\omega_{n/2} t_i \cos\omega_{n/2} t_i).\end{aligned} \quad (9.35)$$

Again, using the orthogonality conditions (9.22) and (9.23), all sums in this series equal 0 except for the term corresponding to $\omega_{n/2}$ which, from (9.22b), equals n. So, (9.35) reduces to

$$\sum_{i=1}^{n} y_i \cos \omega_{n/2} t_i = a_0 \sum_{i=1}^{n} (\cos \omega_{n/2} t_i \cos \omega_{n/2} t_i)$$
$$= a_{n/2} n, \tag{9.36}$$

so

$$a_{n/2} = \frac{1}{2} \sum_{i=1}^{n} y_i \cos \omega_{n/2} t_i. \tag{9.37}$$

Equations (9.30), (9.34) and (9.37) have provided summations of a similar form for the a_k coefficients. For convenience, we can arbitrarily define the constants a_0 and $a_{n/2}$ as $a_0/2$ and $a_{n/2}/2$ (that is, since a_0 is a constant, then we can just define a new constant, $a_0{}' = a_0/2$, which is simply twice the mean of the time series). Given, this definition of a_0 and $a_{n/2}$ (i.e., that they are divided by 2), all a_k coefficients take on a similar form, given by

$$a_k = \frac{2}{n} \sum_{i=1}^{n} y_i \cos \omega_k t_i, \tag{9.38}$$

for all k, that is, $k = 0,1, \ldots,(n - 1)/2, n/2$ for n an even number; for odd n, the highest $k = (n - 1)/2$. This equation is known as the **discrete cosine transform**. It states that all of the values of the a_k coefficients (the cosine terms) of the Fourier series can be determined simply by summing the product of each data point times a cosine of a given harmonic evaluated at the same position of t_i as the data point it multiplies.

Similarly, if we multiply the Fourier series by $\sin(\omega_k t_i)$ and sum over i, applying the orthogonality conditions as above, we obtain

$$b_k = \frac{2}{n} \sum_{i=1}^{n} y_i \sin \omega_k t_i \tag{9.39}$$

for $k = 1, \ldots,(n - 1)/2$. Recall that for $k = 0$ and $k = n/2$, $b_k = 0$ always, so b_0 and $b_{n/2} = 0$ always, and thus they do not require special attention as the a_0 and $a_{n/2}$ terms did. This equation, (9.39), providing a solution for all of the b_k coefficients (the sine terms) of the Fourier series, is the **discrete sine transform**.

9.3.1 Interpretation of the Discrete Cosine and Sine Transforms

The coefficients to the Fourier series are easily determined using the simple equations (cosine and sine transforms) given by (9.38) *and* (9.39). Conceptually, the orthogonality of harmonic sines and cosines allows that each of the Fourier coefficients (the a_k and b_k) can be solved for using (9.38) and (9.39) separately – that is, without any dependence on any of the other Fourier coefficients. So, *the orthogonality means that the sines and cosines are completely independent of one another, to the point that any change in one Fourier coefficient has no influence on the values of any of the other coefficients,* and it is this characteristic that allows us to solve for the a_k and b_k in such a simple manner.

Another insight to these transforms is gained by considering their form in (9.38) and (9.39). These transform equations are directly proportional to the *covariance* between the y_i values and the sine or cosine waveforms (of frequency f_k). That is, the sample covariance is given for two series as

$$\hat{\sigma}_{xy}^2 = \frac{1}{n-1}\sum_{i=1}^{n}(y_i - \bar{y})(x_i - \bar{x}). \tag{9.40a}$$

For $x_i = \cos(\omega_k t_i)$ or $\sin(\omega_k t_i)$, both have zero mean, so

$$\hat{\sigma}_{xy}^2 = \frac{1}{n-1}\sum_{i=1}^{n}(y_i - \bar{y})x_i = \frac{1}{n-1}\sum_{i=1}^{n}y_i x_i - \frac{\bar{y}}{n-1}\sum_{i=1}^{n}x_i. \tag{9.40b}$$

Recalling that both the cosine and sine harmonics sum to zero over the length of y (the fundamental period), we see that the second term on the right-hand side (containing Σx_i) is eliminated, leaving the original form of the cosine or sine transform. That is, $\mathrm{Cov}[XY] = \mathrm{E}[XY] - \mathrm{E}[X]\mathrm{E}[Y]$ and $\mathrm{E}[X] = 0$, so $\mathrm{Cov}[XY] = \mathrm{E}[XY]$. While the sample covariance is normalized by $(n-1)$ instead of n (as is the case in the discrete sine and cosine transforms), this normalization factor can be effectively ignored, since $(n-1)/n$ is ~1 for large n. From this, it is seen that the a_k and b_k coefficients from (9.38) and (9.39) are nothing more than twice the covariance between the data (y_i) and the sines and cosines used to fit the data.

Alternatively, dividing the covariance by the standard deviations of the y_i and of the sines and cosines (the latter being $2^{-1/2}$) gives a correlation coefficient, the square of which describes the degree of shared variance between the data and sines and cosines. In that respect, it is seen that the a_k and b_k are proportional to these correlation coefficients. That is, for the kth harmonic,

$$r_k = \frac{\frac{1}{n-1}\sum_{i=1}^{n}y_i x_i}{s_y \sigma_x} = \frac{\frac{1}{n-1}\sum_{i=1}^{n}y_i x_i}{s_y 2^{-1/2}}, \tag{9.41}$$

which, for the cosine (r_{ck}) harmonics,

$$r_{ck} = \frac{na_k}{2(n-1)s_y 2^{-1/2}}$$

and, since $n/(n-1) \approx 1$ for large n, and $2 \times 2^{-1/2} = 2^{1/2}$,

$$\approx \frac{a_k}{s_y 2^{1/2}}$$

$$a_k \approx \sqrt{2}(s_y r_{ck}). \tag{9.42a}$$

Likewise, the b_k are proportional to the correlation between the sine harmonics and the time series, y_i, as

$$b_k \approx \sqrt{2}(s_y r_{sk}), \tag{9.42b}$$

where r_{ck} and r_{sk} are the correlation coefficients when replacing, in (9.41), x_i by either $\cos\omega_k t_i$ (for r_{ck}), or by $\sin\omega_k t_i$ (for r_{sk}). So both the a and b coefficients are approximately (within $(n-1)/n$) equal to $\sqrt{2}s_y r$.

Therefore, *the cosine and sine transforms can essentially be interpreted as describing the degree of linear correlation between the data and each individual cosine and sine wave (at the Fourier frequencies)*. The better the relationship – that is, the closer the data are to varying precisely with the individual sines or cosines – the higher the correlation coefficient and the larger the a_k and b_k coefficient value. This implies that the interpolant constructed to fit the data using the sines and cosines is dominated by those sines and cosines that most strongly resemble the actual periodic behavior of the data. Sines and cosines that show little covariation with the data contribute little to the final interpolant curve. If there is no relationship whatsoever between the time series and a particular harmonic, the coefficient value for that harmonic would be zero.

The above relationships can be quantified in terms of the described variance of each harmonic term. The orthogonality conditions reveal that the various harmonic terms do not covary (a sine and cosine do not covary, nor do cosinusoids of different frequencies, i.e., $k \neq j$). This implies that the fraction of the variance contributed by each harmonic to the overall series variance, given by s_y^2, is unique, and it is described by r^2. For the cosine harmonic terms.

$$r_{ck}^2 \approx \frac{a_k^2}{2s_y^2}. \tag{9.43a}$$

For the sine harmonic terms,

$$r_{sk}^2 \approx \frac{b_k^2}{2s_y^2}. \tag{9.43b}$$

So,

$$a_k^2 \approx 2s_y^2 r_{ck}^2 \tag{9.44a}$$

$$b_k^2 \approx 2s_y^2 r_{sk}^2. \tag{9.44b}$$

The Fourier coefficients are proportional to the correlation of the time series with the cosine and sine harmonics, as is the variance described by r^2 (shown in Chapter 6).

9.4 Continuous Sine and Cosine Transforms

If y is a continuous function of t over the range T (representing an infinite number of data points), then

$$y(t_i) = \sum_{i=1}^{\infty} (a_j \cos \omega_j t_i + b_j \sin \omega_j t_i). \tag{9.45}$$

In this case, in order to determine the a and b coefficients, we multiply $y(t)$ by $\cos(\omega_k t)$ and $\sin(\omega_k t)$ as before, but now we *integrate* over T (from −T/2 to T/2) to obtain a set of integral relationships to which the integral forms of the orthogonality conditions, (9.22) through (9.24), and integral relationships, (9.20) and (9.21), can be applied. This gives the **continuous cosine and sine transforms**:

$$a_k = \frac{2}{T} \int_{-T/2}^{T/2} [y(t) \cos(\omega_k t)] dt \tag{9.46}$$

$$b_k = \frac{2}{T} \int_{-T/2}^{T/2} [y(t) \sin(\omega_k t)] dt, \tag{9.47}$$

respectively, where $k = 0, 1, \ldots, \infty$.

Notice in this continuous case, when $j = k = T/2$ (corresponding to $n/2$ in the discrete case), that the continuous orthogonality condition (9.22d) shows that for $a_{T/2}$ or any integer multiple, m, of it,

$$a_{mT/2} = \frac{2}{T} \int_{-T/2}^{T/2} [y(t) \cos(\omega_k t)] dt. \tag{9.48}$$

Thus, no special attention is required to make the $a_{mT/2}$ coefficients show the same relationship as given in the general form of the continuous cosine transform of (9.47). That is, we needn't define $a_{mT/2} = a_{mT/2}/2$ as was done in the discrete case, though the a_0 still requires the special treatment, as before.

9.5 The Fourier Transform

9.5.1 Complex Numbers

The use of complex notation simplifies much of the algebra associated with Fourier analysis and the previous transformations; it is therefore mathematically convenient to use. First, we review the basic rules and properties of complex numbers required for this particular application.

Euler Relation

The Euler relation is one of the more fundamental properties of a complex number. It is given as

$$e^{i\omega t} = \underbrace{\cos(\omega t)}_{\text{real part}} + i\underbrace{\sin(\omega t)}_{\text{imaginary part}}, \tag{9.49}$$

where the complex number, i, is[4]

$$i = \sqrt{-1}. \tag{9.50}$$

Alternatively,[5]

$$e^{-i\omega t} = \cos(\omega t) - i\sin(\omega t). \tag{9.51}$$

Any real-valued function (i.e., real data) is represented as a complex-valued function with a zero imaginary part.

Inverse Euler Relationships

These relationships, sometimes referred to as the **de Moivre relationships**, are given by combining (9.49) and (9.51):

$$\cos(\omega t) = \frac{1}{2}(e^{i\omega t} + e^{-i\omega t}) \tag{9.52a}$$

$$\sin(\omega t) = \frac{1}{2}(e^{i\omega t} - e^{-i\omega t}). \tag{9.52b}$$

Complex Conjugate

The complex conjugate (denoted by *) of a complex function f(x) is given by $f^*(x)$ where

$$f(x) = R(x) + iI(x) \tag{9.53}$$

$$f^*(x) = R(x) - iI(x), \tag{9.54}$$

where R(x) = real part of f(x) and I(x) = imaginary part (the part multiplied by i) of f(x).

[4] In some literature, such as in engineering, the complex number is represented by the letter j instead of i.

[5] Relationship (9.51) follows directly from (9.49), since $e^{-i\omega t} = e^{i(-\omega)t} = \cos(-\omega t) + i\sin(-\omega t) = \cos(\omega t) - i\sin(\omega t)$, because $\cos(-x) = \cos(x)$ and $\sin(-x) = -\sin(x)$.

Magnitude

The magnitude of a complex number is given by

$$|f(x)| = [f(x)f^*(x)]^{1/2} \tag{9.55a}$$

$$= [R^2(x) + iI^2(x)]^{1/2}. \tag{9.55b}$$

Polar Representation

The polar form of a complex number, $R(\omega) + iI(\omega)$, is given as

$$|A(\omega)|e^{-i\varphi(\omega)}, \tag{9.56}$$

where:[6] $|A| = [R^2(\omega) + I^2(\omega)]^{1/2}$ is the magnitude (as above), $R(\omega)$ and $I(\omega)$ are the real and imaginary components of the complex number, and

$$\varphi(\omega) = \tan^{-1}[I(\omega)/R(\omega)].$$

Using the Euler relationship, (9.56) can be rewritten as

$$A(\omega)[\cos(\omega) - i\sin(\omega)]. \tag{9.57}$$

9.5.2 Orthogonality Conditions for Complex Form of Sines and Cosines

From the Euler relationship (9.49), it is clear that the sine and cosine terms that appear in the Fourier series (9.14) can be combined into a single complex exponential term. There is only one orthogonality relationship governing the summed or integrated product of complex numbers (it just keeps getting better). This is given by

$$\sum_{i=1}^{n}(e^{i\omega_j t_i})(e^{-i\omega_j t_i}) = \begin{cases} n & \text{if } j = k \quad (9.58a) \\ 0 & \text{otherwise} \quad (9.58b) \end{cases}$$

or, in integral form for the continuous case,

$$\int_{-T/2}^{T/2}(e^{i\omega_j t_i})(e^{-i\omega_j t_i})dt = \begin{cases} T & \text{if } j = k \quad (9.58c) \\ 0 & \text{otherwise} \quad (9.58d) \end{cases}$$

Note that there is no case in which the summed (or integrated) products equals $n/2$ (or $T/2$), so in complex form there will be no need to define any of the coefficients in any special manner (e.g., $a_0 = a_0/2$). As before, the summed product in (9.58) is only valid for data that are sampled at even intervals of t.

[6] Note that these forms of amplitude and phase are consistent with the forms derived previously for describing the amplitude and phase of a single sinusoid based on the sum of a pure (without phase) sine and cosine term of the same frequency. That is, we had $A\cos(\omega t + \varphi) = a\cos(\omega t) - b\sin(\omega t)$, and $A = (a^2 + b^2)^{1/2}$, $\varphi = \tan^{-1}(b/a)$.

Box D9.2 Derivation of the Complex Orthogonality Condition

Proof of the orthogonality condition (9.58) is straightforward.

For $j = k$,

$$\begin{aligned}\sum_{i=1}^{n}(e^{i\omega_j t_i})(e^{-i\omega_j t_i}) &= \sum_{i=1}^{n} e^{i(\omega_j-\omega_j)t_i}\\ &= \sum_{i=1}^{n} e^{0}\\ &= n\end{aligned} \qquad (D9.2.1)$$

or, in the continuous case,

$$\int_{-T/2}^{T/2}(e^{i\omega_j t_i}e^{-i\omega_j t_i})dt = \int_{-T/2}^{T/2} e^{0}dt = T. \qquad (D9.2.2)$$

For $j \neq k$, the solution is found by expansion of the $e^{\pm i\omega t_i}$ terms in (9.58) using the Euler relationships (9.49) and (9.51). This transforms the sums and integrals to products of sines and cosines as before, to which the standard orthogonality conditions are then directly applied.

For example,

$$\sum_{i=1}^{n}(e^{i\omega_j t_i}e^{-i\omega_k t_i}) = \sum_{i=1}^{n}\{[\cos(\omega_j t) + i\sin(\omega_j t)][\cos(\omega_k t) - i\sin(\omega_k t)]\}. \qquad (D9.2.3a)$$

Expansion of the product on the right-hand side gives

$$\begin{aligned}&= \sum_{i=1}^{n}\cos(\omega_j t_i)\cos(\omega_k t_i) + \sum_{i=1}^{n}\sin(\omega_j t_i)\sin(\omega_k t_i)\\ &\quad i\sum_{i=1}^{n}[\cos(\omega_k t_i)\sin(\omega_j t_i)] - i\sum_{i=1}^{n}[\cos(\omega_j t_i)\sin(\omega_k t_i)].\end{aligned} \qquad (D9.2.3b)$$

Application of the standard orthogonality relationships, (9.22) through (9.24), to the individual summed terms then reduces this equation to the orthogonality conditions given in (9.58) (all sums are zero except for the first two sums, when $j = k$). The identical procedure is used to prove the conditions for the continuous case.

9.5.3 Complex Discrete Fourier Transform

Complex Form of the Fourier Series

Now use the inverse Euler relations of (9.52) to rewrite the standard Fourier series in complex form. Starting with the Fourier series as

$$y_i = \sum_{j=0}^{\leq n/2} [a_j \cos(\omega_j t_i) + b_j \sin(\omega_j t_i)]. \tag{9.59}$$

Applying the inverse Euler relationships,

$$= \sum_{j=0}^{\leq n/2} \left[\frac{a_j}{2} (e^{i\omega t} + e^{-i\omega t}) + \frac{b_j}{2} (e^{i\omega t} - e^{-i\omega t}) \right]. \tag{9.60}$$

Rearranging gives

$$= \frac{1}{2} \sum_{j=0}^{\leq n/2} \left(a_j + \frac{b_j}{i} \right) e^{i\omega_j t} + \left(a_j - \frac{b_j}{i} \right) e^{-i\omega_j t}. \tag{9.61}$$

Noting that $(b/i)(i/i) = -ib$, equation (9.61) is further reduced by multiplying through by (i/i), giving

$$= \frac{1}{2} \sum_{j=0}^{\leq n/2} (a_j - ib_j) e^{i\omega_j t} + (a_j + ib_j) e^{-i\omega_j t}. \tag{9.62}$$

Negative Frequencies

Since the second term contains $e^{-i\omega_j t_i}$, consider the effects of negative frequencies (i.e., the $-\omega_j$) to simplify this expression further (see derivation (D9.3)),

$$(a_{-j} - ib_{-j}) e^{i\omega_{-j} t} = (a_j + ib_j) e^{-i\omega_j t}. \tag{9.63}$$

The second term of (9.62) can be dropped if we write the sum over negative frequencies as well as positive ones. That is, the **complex form of the Fourier series** can be written as

$$y_i = \frac{1}{2} \sum_{j>-n/2}^{\leq n/2} (a_j - ib_j) e^{i\omega_j t}, \tag{9.64}$$

where $-n/2 < j \leq n/2$.[7] So, while (9.62) gives two terms for each positive index of j, this form, (9.64), gives the $a_j + ib_j$ terms for the negative values of the index j (i.e., half of the previous terms) and gives the $a_j - ib_j$ terms (the other half of the terms) with the positive index values of j. In this respect, we still end up with the same number of terms. So, for any one time, i, we still get all of the frequency terms by assigning half of them as negative and the other half as positive frequencies.

[7] The sum over the j frequencies would be more symmetrical as $-n/2 \leq j \leq n/2$ instead of $-n/2 < j \leq n/2$. However, the latter form is used (as explained in detail later) to take advantage of the fact that, in the previous case of the discrete cosine transform, the a_0 and $a_{n/2}$ terms required special treatment (they were defined to be twice the true value). If we sum over the j terms as defined in (9.64), this special treatment is not required when using this range of j frequencies for the complex form of the Fourier series.

Box D9.3 Negative Frequencies in the Complex Fourier Series

The relationship in (9.63) is shown by considering the individual coefficients as defined by the discrete cosine and sine transforms at negative frequencies. For the cosine transform (that gives the functional form of the a_j coefficient values),

$$\begin{aligned} a_{-j} &= \frac{2}{n}\sum_{i=1}^{n} y_i \cos(\omega_{-j}t_i) \\ &= \frac{2}{n}\sum_{i=1}^{n} y_i \cos\left(\frac{-2\pi j t_i}{T}\right), \end{aligned} \tag{D9.3.1a}$$

but since the cosine is an even function, i.e., cos(−x) = cos(x), then

$$= \frac{2}{n}\sum_{i=1}^{n} y_i \cos\left(\frac{2\pi j t_i}{T}\right), \tag{D9.3.1b}$$

so

$$a_{-j} = a_j. \tag{D9.3.2}$$

Similarly, for the discrete sine transform at negative frequencies,

$$\begin{aligned} b_{-j} &= \frac{2}{n}\sum_{i=1}^{n} y_j \sin(\omega_{-j}t_i) \\ &= \frac{-2}{n}\sum_{i=1}^{n} y_j \sin(\omega_j t_i), \end{aligned} \tag{D9.3.3}$$

since sine is an odd function, i.e., sin(−x) = −sin(x). So,

$$b_{-j} = -b_j. \tag{D9.3.4}$$

Finally, $\omega_j = 2\pi j/T$, so $\omega_{-j} = -2\pi j/T = -\omega_j$. Thus, in the Euler relationships,

$$e^{i\omega_{-j}t_i} = e^{-i\omega_j t_i}. \tag{D9.3.5}$$

So, for negative j,

$$(a_{-j} - b_{-j})e^{-i\omega_{-j}t_i} = (a_j + ib_j)e^{-i\omega_j t_i}. \tag{D9.3.6}$$

Defining a complex amplitude,

$$J_j = \frac{1}{2}(a_j - ib_j) \tag{9.65}$$

equation (9.64) can be rewritten in polar form as

$$y_i = \sum_{j>-n/2}^{\leq n/2} (J_j e^{i\omega_j t_i}) \tag{9.66}$$

where, from (9.54), (9.65), (D9.3.2) and (D9.3.4),

$$J_{-j} = J_j^*. \tag{9.67}$$

This equation (9.66) represents the **general complex form of the Fourier series**.

Discrete Fourier Transform
Multiply (9.66) by $e^{-i\omega_k t_i}$ and sum over t_i,

$$\sum_{i=1}^{n} y_i e^{-i\omega_k t_i} = \sum_{i=1}^{n} \sum_{j>-n/2}^{\leq n/2} \left[(J_j e^{i\omega_j t_i}) e^{-i\omega_k t_i} \right]. \tag{9.68}$$

Rearranging the order of the summations, complex orthogonality can be applied directly to the inner sum:

$$= \sum_{j>-n/2}^{\leq n/2} J_j \sum_{i=1}^{n} (e^{i\omega_j t_i} e^{-i\omega_k t_i}). \tag{9.69}$$

The orthogonality conditions of (9.58) indicate that the inner sum is only nonzero when $j = k$, in which case the sum equals n. Following the logic used previously for the discrete cosine and sine transforms,

$$J_j = \frac{1}{n} \sum_{i=1}^{n} y_i e^{-i\omega_j t_i}, \tag{9.70}$$

where $-n/2 < j \leq n/2$.

This equation (9.70) is the ***discrete Fourier transform*** of the series y_i, and equation (9.66) is the ***discrete inverse Fourier transform***. So, given the J_j values, you can perform the inverse transform to reconstruct the original data. This simply re-states the interpolation condition originally specified when defining the Fourier series previously. In other words, *the inverse Fourier transform performs the interpolation of the time series, given knowledge of the Fourier coefficients – while the discrete Fourier transform gives those coefficient values*. Alternatively, *the Fourier transform converts your time series from the time domain to the frequency domain and the inverse transform from the frequency domain to the time domain (two different perspectives for viewing the same data)*.

When we solved for the a_j and b_j coefficients using the discrete cosine and sine transforms of (9.38) and (9.39), we had to define the a_0 and $a_{n/2}$ coefficients specially, as one-half the value computed by the cosine transform of (9.38). For the complex form of (9.70), this is not the case. That is, the J_0 and $J_{n/2}$ coefficients do not require special consideration. This is because of the form of the sum over the negative to positive frequencies. Note that $J_{-n/2} = J_{n/2}$, because at $j = n/2$ the sine term drops out and the cosine term, $a_{-n/2} = a_{n/2}$, since the cosine is an even function. If we computed the term twice, at $-n/2$ and $n/2$, we would have $2a_{n/2}$, as was the case for the discrete cosine transform. Here, by only computing the term at $j = n/2$ and not again at $-n/2$, there is no need to make this special treatment of the $a_{n/2}$ term. This is why the frequency range is given above as $-n/2 < j \leq n/2$. Likewise, the $j = 0$ term only occurs once in the sum in (9.66). Therefore, $J_0 = \frac{1}{n}\Sigma y_i = \bar{y}$, not $2\bar{y}$ as was the case using the discrete cosine transform of (9.38). Thus there is no need to define the a_0 coefficient as $a_0/2$, either.

9.5.4 Real-Valued Time Series

For a real-valued series, y_i,

$$y_i = R_i + iI_i \tag{9.71a}$$

$$= R_i. \tag{9.71b}$$

since the imaginary part is zero. Consequently,

$$y_i = y_i^*. \tag{9.72}$$

Now consider the transform of this real series by substituting (9.71a) into the discrete Fourier transform (9.70), giving

$$J_j = \frac{1}{n}\sum_{i=1}^{n}(R_i + iI_i)\mathrm{e}^{-i\omega_j \mathrm{t}_i}, \tag{9.73a}$$

which, from (9.71b) reduces to

$$\begin{aligned} &= \frac{1}{n}\sum_{i=1}^{n} R_i \mathrm{e}^{-i\omega_j \mathrm{t}_i} \\ &= \frac{1}{n}\sum_{i=1}^{n} R_i[\cos(\omega_j \mathrm{t}_i) - i\sin(\omega_j \mathrm{t}_i)]. \end{aligned} \tag{9.73b}$$

Inspection of (9.73b) at negative frequencies shows that

$$J_{-j} = J_j^* \tag{9.74}$$

because the cosine terms are even functions, cos(−x) = cos(x), while the sin(−x) = −sin(x). Thus, *for real data, the transform (J_j) is completely determined by the positive values of j. That is, knowledge of the values of the transform for the positive frequencies (positive j values) gives, by the relationship of* (9.74), *the values of the transform at the negative frequency values*. Therefore, consistent with the earlier development, the real-valued time series, written in complex form, are still completely determined by the n (non-negative) values of a_j and b_j,

$$J_j = \frac{1}{2}(a_j - ib_j), \tag{9.75}$$

where $j = 0, 1, \ldots, \leq n/2$. It is this symmetry that is expressed by the factor of 2 that appears in the discrete cosine and sine transforms of (9.38) and (9.39).

9.5.5 Complex Continuous Fourier Transform

For y Continuous. Then in a manner similar to that used above, for a continuous series sampled over a range T,

$$y(\mathrm{t}) = \sum_{j=-\infty}^{\infty} \left(J_j \mathrm{e}^{-i\omega_j \mathrm{t}}\right) \tag{9.76}$$

is the **complex continuous inverse Fourier transform** and

$$J_j = \frac{1}{\mathrm{T}} \int_{-\mathrm{T}/2}^{\mathrm{T}/2} y(\mathrm{t})\mathrm{e}^{-i\omega_j \mathrm{t}} \mathrm{dt} \tag{9.77}$$

is the **complex continuous Fourier transform** where $-\infty \leq j \leq \infty$.

These two equations represent special cases of the general Fourier transform and inverse, discussed below.

9.6 Fourier Transform of Non-Periodic Data

So far, we have developed the Fourier series and Fourier transform by assuming that the time series are periodic over their sampled length. That is, if the data contain n evenly spaced data points, then we are fitting these data with harmonic sines and cosines of a fundamental period, $n\Delta\mathrm{t}$, giving fundamental frequency $\mathrm{f_F} = 1/\mathrm{T} = 1/n\Delta\mathrm{t}$. Since these sines and cosines are exactly repeated every period, infinitely, in both directions ($\pm$t), the fitted curve (the interpolant) will also repeat every T units.

More likely than not, your time series is not exactly periodic over its sampled length ($n\Delta$t). Consequently, we need to determine the implication of violating the assumption of periodicity over length $n\Delta$t, for interpreting fitting perfectly periodic functions (the Fourier transform results) to nonperiodic time series. Specifically, *we must be able to evaluate how frequencies other than those that are harmonics of the time series length* (f_F) *will appear in our discrete Fourier transform*.

For this, we need the continuous Fourier transform and inverse transform. With those, we can develop the classic sampling theorem that builds the framework from which we can evaluate the discrete transform results in terms of any frequency components.

9.6.1 General Continuous Fourier Transform

Extending the development of the discrete Fourier transform to the case of continuous time series (using the continuous orthogonality relationships) leads to the **general form of the Fourier transform**, given by

$$F(f) = \int_{-\infty}^{\infty} f(t)e^{-i2\pi ft}dt, \tag{9.78a}$$

where f = rotational frequency (in inverse time as cycles per unit time, or space as cycles per distance). *This general (continuous) form differs from the previous form(s) in that it is not strictly dependent upon Fourier frequencies and is not restricted to a period of T or wavelength of λ.*

The function F(f), the Fourier transform, is essentially the *continuous function* of interpolating sines and cosines at any frequency f (not just the harmonic ones). F(f) is thus the continuous analog to the discrete Fourier transform, and has a similar interpretation.

The corresponding **general form of the inverse Fourier transform** is given by

$$f(t) = \int_{-\infty}^{\infty} F(f)e^{i2\pi ft}df. \tag{9.78b}$$

Unlike the continuous Fourier series of (9.76) for a continuous time series restricted to a finite interval of time, T, we now allow (actually, demand) the time series f(t) to continue indefinitely in both directions with no assumption about periodicity.

The form of (9.78) is one of several possible forms. The other common forms differ only by a scaling factor. For example, the nice, clean form involving angular frequency, ω,

$$F(\omega) = \int_{-\infty}^{\infty} f(t)e^{i\omega t}dt, \tag{9.79a}$$

requires a $(2\pi)^{-1/2}$ scaling factor in the inverse transform:

$$f(t) = \frac{1}{2\pi}\int_{-\infty}^{\infty} F(\omega)e^{i\omega t}d\omega. \tag{9.79b}$$

Therefore, to preserve symmetry in the form of the relationship (which also makes some of the Fourier properties discussed below more symmetrical in form), the form given in (9.78) is used throughout the remainder of the text.

Here, **Fourier transform pairs** are designated by

$$f(t) \Leftrightarrow F(f), \tag{9.80}$$

where F(f) is often referred to as the **frequency spectrum** of the time series, f(t). Following this convention, the transform of a function f(t), which varies as a function of time (or space), is represented in the frequency domain by its spectrum F(f), which varies as a function of frequency. So we can *examine the time series in the time domain or the frequency domain – same data, different perspectives.*

Box 9.1 Example of Analytic Fourier Transform

Consider the Fourier transform of the function

$$f(t) = ae^{-\lambda t} \qquad t \geq 0,$$

where a and λ are constants. Then, the Fourier transform is given by

$$\begin{aligned} F(f) &= a\int_0^\infty e^{-\lambda t} e^{-i2\pi f t} dt \\ &= a\int_0^\infty e^{-(i2\pi f + \lambda)t} dt. \end{aligned}$$

This integral is easily solved by substituting

$$u = -(i2\pi f + \lambda)t$$
$$du = -(i2\pi f + \lambda)dt,$$

giving

$$\begin{aligned} &= \frac{-a}{i2\pi f + \lambda}\int_0^{-\infty} e^u du \\ &= \frac{-a}{i2\pi f + \lambda} e^u \Big|_0^{-\infty} \\ &= \frac{a}{i2\pi f + \lambda}. \end{aligned}$$

This result is put into standard complex form by multiplying through by $\frac{-i2\pi f + \lambda}{-i2\pi f + \lambda}$, giving

$$\begin{aligned} &= \frac{-ia2\pi f + a\lambda}{(2\pi f)^2 + \lambda^2} \\ &= \underbrace{\frac{a\lambda}{(2\pi f)^2 + \lambda^2}}_{\text{real part}} - \underbrace{\frac{a2\pi f}{(2\pi f)^2 + \lambda^2}}_{\text{imaginary part}}. \end{aligned}$$

Rewriting this in polar form in terms of amplitude and phase then gives

$$\begin{aligned} F(f) &= A(f)e^{-i\varphi(f)} \\ &= [R^2(f) + iI^2(f)]e^{-\tan^{-1}[I(f)/R(f)]}, \end{aligned}$$

where R(f) and I(f) are the real and imaginary components, respectively.

Box 9.1 (Cont.)

The amplitude spectrum of the function, $f(t) = ae^{-\lambda t}$ for $t \geq 0$, is thus given by

$$\begin{aligned} A(f) &= [R^2(f) + iI^2(f)]^{1/2} \\ &= \left\{ \frac{a^2\lambda^2 + a^2(2\pi f)^2}{[(2\pi f)^2 + \lambda^2]^2} \right\}^{1/2} \\ &= \left\{ \frac{a^2[\lambda^2 + (2\pi f)^2]}{[(2\pi f)^2 + \lambda^2]^2} \right\}^{1/2} \\ &= \frac{a}{[(2\pi f)^2 + \lambda^2]^{1/2}}, \end{aligned}$$

and the phase is given by

$$\begin{aligned} \varphi(f) &= \tan^{-1}[I(f)/R(f)] \\ &= \tan^{-1}[-2\pi f/\lambda]. \end{aligned}$$

For the actual transform of a discrete time series, the discrete Fourier transform is always used, involving Fourier frequencies only. The general integral form, (9.78), on the other hand, is used for analytical transforms (e.g., solving differential equations).

9.6.2 Existence of the Fourier Integral (A Formality)

The existence of the Fourier integral is dependent upon conditions usually satisfied by most functions of interest and definitely satisfied by all data sets. Specifically, if

$$\int_{-\infty}^{\infty} |f(t)| dt < \infty, \tag{9.81}$$

the Fourier integral will exist.

Some additional functions that do not satisfy the above condition, (9.81), can be handled by satisfying

$$f(t) = a(t)\sin(2\pi ft + \varphi), \tag{9.82}$$

where $a(t + \gamma) < a(t)\gamma$, where γ is a large value in time. Then, if

$$\int_{-\infty}^{\infty} \left| \frac{f(t)}{t} \right| dt < \infty, \tag{9.83}$$

the Fourier transform, exists.

Theory of Distributions

Additional functions such as the cosine, which do not satisfy the above conditions, can be handled if one extends the Fourier theory to include the impulse function or **Dirac delta** function, $\delta(t)$.

This function requires an introduction to the **Theory of Distributions.**

Impulse Function

The impulse function is defined by the following property:

$$\int_{-\infty}^{\infty} \delta(t - t_0)x(t)dt = x(t_0), \tag{9.84a}$$

or, writing $t = t + t_0$, giving

$$\int_{-\infty}^{\infty} \delta(t)x(t + t_0)dt = x(t_0). \tag{9.84b}$$

$\delta(t\text{-}t_0)$ is a spike with infinite amplitude and infinitesimal width located at $t = t_0$. This integral samples $x(t)$ at t_0, thus its name: **sifting integral**. Unlike the discrete unit impulse introduced previously, this function is defined in terms of continuous functions. Specifically, it is defined in terms of a sinusoid,[8] as

$$\delta(t) = \lim_{a\to\infty} \frac{\sin(at)}{\pi t}. \tag{9.85}$$

Given this latter description and (9.84), consider the Fourier transform of the impulse function of amplitude c. That is, if $h(t) = c\delta(t)$, then the Fourier transform of $h(t)$ is given by

$$\begin{aligned} H(f) &= \int_{-\infty}^{\infty} c\delta(t)e^{-i2\pi ft}dt \\ &= c\int_{-\infty}^{\infty} \delta(t)e^{-i2\pi ft}dt \\ &= ce^0 \\ &= c. \end{aligned} \tag{9.86}$$

Two fundamental identities for $\delta(\tau)$ critical for sampling theory are

$$\delta(\omega) = \frac{1}{2\pi}\int_{-\infty}^{\infty} e^{i\omega t}dt \tag{9.87}$$

and

$$\int_{0}^{\infty} \cos(\omega t)dt = \frac{1}{\pi}\delta(\omega). \tag{9.88}$$

[8] To understand a sinusoid form like sin(at) in terms of the more normal form, which involves $2\pi ft$, simply set $a = 2\pi f = 2\pi/T$, so the period of the sine is $T = 2\pi/a$. So, as a goes to infinity, the period goes to zero and the frequency goes to infinitely fast.

Introduction of the Frequency Shifting property (§9.8.5) allows generalization of this to allow the Fourier transform of the cosine at other harmonics.

Given this latter description and (9.84), consider the Fourier transform of the impulse function of amplitude c. That is, if $h(t) = c\delta(t)$, then the Fourier transform of h(t) is given by (9.86). That is, the integral of the impulse times x(t) yields $x(t_0)$. Here, $t_0 = 0$ and $x(t) = e^{-i2\pi ft}$. So, x(0) in this case is e^0. Therefore, *an impulse scaled by amplitude c, located at the origin, in the time domain, is equal to a constant amplitude of c for all frequencies, f, in the frequency domain.*

9.7 Fourier Transform Properties

We are interested in transforming a data set from the time domain to the frequency domain, where it can be examined for periodic components, while recognizing that it is unlikely in most situations that the period of sampling is exactly the fundamental period of all harmonics present in the time series. Therefore, we wish to examine the consequences of transforming a time series into a sequence of sines and cosines at frequencies other than the true frequencies that dominate the periodic components in the data. In order to evaluate this situation, we must develop a set of Fourier transform properties that aid in this process.

The following Fourier transform properties are the most basic and important (their proofs are given below).

9.7.1 Symmetry Property

If

$$f(t) \Leftrightarrow F(f), \tag{9.89a}$$

then

$$F(t) \Leftrightarrow f(-f). \tag{9.89b}$$

In other words, if you know the transform of a given function's form, you effectively know the transform of a function with the transform's form. That is, it doesn't really matter what side of the transform the function is defined on, except that a sign change is required for asymmetrical or odd symmetrical functions. So, if the Fourier transform of a function yields some other functional form, then by simply replacing the f with t in the F (f) function, you know that the Fourier transform of a function F(t) is the same function as f(t), but with the t replaced by −f. If the f(−f) is an even function, then f(−f) = f(f); if it is an odd function, then f(−f) = −f(f).

Another important transform pair arises from the Theory of Distributions above; the transform of an impulse of amplitude c is a constant c (i.e., the value of c at all frequencies), and symmetry shows that the transform of a constant c (i.e., value of c at

Box 9.2 Example of Gate Function Transform Pair

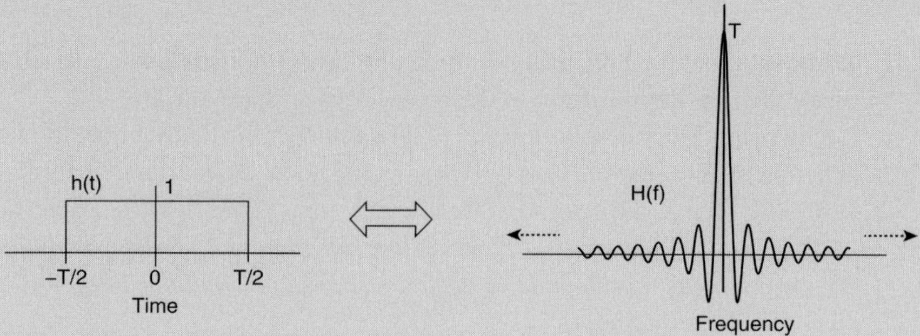

Figure 9.4 Transform pair of rectangular function with sinc function.

Consider the transform of a **rectangular function**, or more colorfully, a "**boxcar**" or "**gate**" function, shown as Figure 9.4 and defined as

$$f(t) = \begin{cases} k & \text{constant if } |t| \leq \ T/2 \\ 0 & \text{elsewhere,} \end{cases} \tag{9.90a}$$

so

$$\begin{aligned} F(f) &= \int_{-T/2}^{T/2} k e^{-i2\pi f} dt \\ &= k \int_{-T/2}^{T/2} (\cos 2\pi ft - i \sin 2\pi ft) dt \end{aligned}$$

defining $u = 2\pi ft$, so $du = 2\pi f dt$, gives

$$\begin{aligned} &= \frac{k}{2\pi f} \sin 2\pi ft \Big|_{-T/2}^{T/2} + \frac{ik}{2\pi f} \cos 2\pi ft \Big|_{-T/2}^{T/2} \\ &= \frac{k}{2\pi f} [\sin \pi fT + \sin \pi fT] + \frac{ik}{2\pi f} [\cos \pi fT - \cos \pi fT] \\ &= k \frac{\sin(\pi ft)}{\pi f}. \end{aligned} \tag{9.90b}$$

The function $k \sin(\pi fT)/\pi f$ is known as the **sine cardinal function**, or simply the **sinc function**. It plays a fundamental role in Fourier analysis and is discussed in more detail later. Here, however, it can be seen that the function is a simple sinusoid that is modulated by a function of the form $k/\pi f$.

The transform of a constant and of the sinc function are related through the symmetry relation, so the transform of a boxcar is the sinc function,

$$f(t) = \begin{cases} k & \text{constant if } |t| \leq \ T/2 \\ 0 & \text{elsewhere} \end{cases} \Leftrightarrow F(f) = k \frac{\sin(\pi fT)}{\pi f}, \tag{9.91a}$$

and by symmetry, the transform of the sinc function is a boxcar:

$$f(t) = k \frac{\sin(\pi tT)}{\pi t} \Leftrightarrow F(f) = \begin{cases} k & \text{constant if } |f| \leq T/2 \\ 0 & \text{elsewhere} \end{cases}. \tag{9.91b}$$

all times) is an impulse of amplitude c. So *a "spike" transforms to a flat line, and vice versa* (Figure 9.4).

$$f(t) = c\delta(t) \Leftrightarrow F(f) = c \tag{9.92}$$

9.7.2 Linearity

If

$$f(t) = g(t) + h(t) \tag{9.93}$$

and

$$g(t) \Leftrightarrow G(f) \tag{9.94a}$$

$$h(t) \Leftrightarrow H(f), \tag{9.94a}$$

then

$$f(t) \Leftrightarrow F(f) = G(f) + H(f) \tag{9.95a}$$

or

$$g(t) + h(t) \Leftrightarrow G(f) + H(f). \tag{9.95b}$$

Therefore, the Fourier transform is a linear operator (useful for linear systems analysis) and the transform of a composite (additive) system is the sum of the transforms of the components.

Box 9.3 Example of Linear Trend

For example, consider a function g(t) modified by a linear trend, h(t) = (mt + b), so

$$f(t) = g(t) + h(t) \quad t < T/2. \tag{9.96}$$

Then

$$\begin{aligned} F(f) &= \int_{-\infty}^{\infty} g(t)e^{-i2\pi ft}dt + \int_{-\infty}^{\infty} h(t)e^{-i2\pi ft}dt, \\ &= G(f) + L(f), \end{aligned} \tag{9.97}$$

where L(f) is the Fourier transform of a linear trend (between ±T/2). Therefore, you could examine the transform of the unmodified g(t) by

$$G(f) = F(f) - L(f). \tag{9.98}$$

9.7.3 Scaling

If

$$f(t) \Leftrightarrow F(f), \tag{9.99}$$

then

$$f(at) = \frac{1}{|a|}F(f/a). \tag{9.100}$$

Therefore, if a time scale is expanded ($a > 1$) by a constant amount (e.g., T is now aT), so too is the period expanded and the frequency contracted (cycles per now-longer period) to lower frequencies, since $f_1 = 1/aT$, while the amplitude is decreased by $1/|a|$. The contraction of the frequency axis is expected, since time is inversely proportional to frequency. Decreasing the amplitude maintains a constant area relative to the unscaled function, preserving the energy of the function.

At first glance this might seem a bit esoteric, but the scaling property can be very helpful. This property shows how the spectrum will change when you change the sampling interval. For example, if $\Delta t = 1$, you will get one result, but if $\Delta t = 2$, the above property comes into play with $a = 2$, the frequency axis will be contracted with $f_1 = 1/2T$ (first full cycle over twice as long time) and the amplitudes reduced relative to the initial spectrum.

The result of *scaling the frequency axis* instead of the time axis, as shown above, is just an application of the symmetry property to the time scaling transform pair (because the scaling constant can be positive or negative, the negative sign in the symmetry relationship is ignored). Therefore,

$$\frac{1}{|a|}F(f/a) \Leftrightarrow f(at).$$

Box 9.4 Example of Scaling Law

For example, consider a function g(t), which has a questionable time scale (e.g., one based on the estimated age to a paleomagnetic reversal).[9] If the date of the reversal is later found to be 10 percent older, then the time axis of g(t) is stretched by the constant $a = 1.1$. From the scaling property, we needn't recalculate the spectrum of g(t). Instead, we can simply modify it by

$$g(t_{new}) = g(1.1t), \tag{9.101}$$

so

$$f(1.1t) \Leftrightarrow \frac{1}{|1.1|}F(f/1.1). \tag{9.102}$$

Therefore, the spectrum of the function, given the new age estimate, diminishes all amplitudes by a uniform factor of 1/1.1 (~0.9), and all of the frequencies are lower (= slower; periods extended = longer) by ~10 percent.

[9] For those new to this concept, the Earth's magnetic field reverses its polarity once in a rare while (on average, about once every half-million years) – i.e., the south pole and north pole reverse, so your compass many millions of years ago may have showed N actually pointing to the south.

9.7.4 Time Shifting

If

$$f(t) \Leftrightarrow F(f), \tag{9.103}$$

then

$$f(t-t_0) \Leftrightarrow F(f)e^{-i2\pi ft_0}. \tag{9.104}$$

Rewriting F(f) in (9.104) in polar form (9.56) gives

$$\begin{aligned} F(f)e^{-i2\pi ft_0} &= A(f)e^{i\varphi(f)}e^{-i2\pi ft_0} \\ &= A(f)e^{i[\varphi(f)-cf]}. \end{aligned} \tag{9.105}$$

Therefore, a shift in the time axis (translation) does not affect the amplitude spectrum at all. It only affects the phase spectrum by linearly shifting the phase by cf, where $c = 2\pi t_0$. Thus, the slope of the phase shift (i.e., it increases linearly with frequency) is directly proportional to the magnitude of the time shift t_0.

Box 9.5 Example of Time Shifting

For example, consider a function g(t) that has passed through a linear filter, producing an output h(t). If the linear filter does nothing more than introduce a "lag" into the input function – that is, if the output is simply equal to the input shifted by a time, t_0 – then the input and output have identical spectra except that the phases of the two functions are shifted by a linearly changing amount. The slope of the phase shift is equal to the amount of time delay added by the filter. Consider an input series that is a pure sine with zero phase (i.e., the series starts at y(0) = 0 and period T = 100 yrs). If the filter delays the output by 25 years, then the output will be the same sinusoid, but in that case the output series starts one-quarter of the way through the sinusoid, equivalent to a 90° phase shift, so y(25) = 1. If there is another pure sine in the series of T = 70 (i.e., shorter wavelength so higher frequency), then the 25-year shift is one-third the period, and y(25) for that period is $\sin(2\pi/3)$, because we are now one-third of the way through the sine. With higher frequencies, this 25-year shift is a larger fraction of the period T, thus the phase shift is relatively larger for each higher frequency – nicely summarized as a linear change in phase as a function of wavelength. So, a constant time shift will introduce a small shift in phase to long period cycles and a relatively larger shift to shorter period cycles in the form of a linear trend.

9.7.5 Frequency Shifting

Frequency shifting is obtained from the symmetry rule, applied to the time shifting property. So if

$$f(t) \Leftrightarrow F(f), \tag{9.106}$$

then a shift in the spectrum of f_0 at each frequency gives

$$f(t)\, e^{i2\pi f_0 t} \Leftrightarrow F(f - f_0). \tag{9.107a}$$

From this, we can now compute the Fourier transform for the different harmonics of sinusoids. So if

$$\int_0^{\infty} \cos(\omega t)dt = \frac{1}{\pi}\delta(\omega), \tag{9.107b}$$

then invoking the frequency shifting to get the integral of $\cos(\omega_j t)$ gives

$$\int_0^{\infty} \cos(\omega_j t)dt = \frac{1}{\pi}\delta(\omega - \omega_j), \tag{9.108a}$$

and for the Fourier transform of the complex exponential (transform of a sine and cosine)

$$\int_0^{\infty} e^{i\omega_j t} e^{-i\omega_j t}dt = 2\pi\delta(\omega - \omega_j). \tag{9.108b}$$

Given the above, the existence of the Fourier integral for the cosine function is now possible. This transform is

$$\begin{aligned} \int_{-\infty}^{\infty} \cos(\omega_j t)e^{-i\omega_j t}dt &= \int_{-\infty}^{\infty} \frac{1}{2}(e^{i\omega_j t} + e^{-i\omega_j t})e^{-i\omega_j t}dt \\ &= \int_{-\infty}^{\infty} \frac{1}{2}\left(e^{i(\omega-\omega_j)t} + e^{-i(\omega+\omega_j)t}\right)dt \\ &= \pi\delta(\omega - \omega_j) + \pi\delta(\omega + \omega_j). \end{aligned} \tag{9.108c}$$

9.7.6 Symmetrical Function Transforms

Consider a real function f(t), which, as described previously, can be decomposed into a sum of an even, $f_e(t)$, and odd, $f_o(t)$, function:

$$f(t) = f_e(t) + f_o(t). \tag{9.109a}$$

From the linearity property, we know that the transform of such a function is given as

$$F(f) = F_e(f) + F_o(f). \tag{9.109b}$$

Even Functions

The transform of an even function is given as

$$f_e(t) \Leftrightarrow R_e(f), \tag{9.110}$$

where $R_e(f)$ is the real component of the transform, itself an even function given by

$$R_e(f) = \int_{-\infty}^{\infty} f(t)\cos(2\pi ft)dt. \tag{9.111}$$

Therefore, since the sine terms are eliminated, the transform does not have any b_j coefficients: *all phases are zero* (i.e., *there is no phase in an even function, so* $\varphi = 0$ *for all frequencies*). This reflects the fact that only perfect cosines are evenly distributed about the origin, and any phase shift of the origin (or introduction of a sine component) would destroy the symmetry.

Odd Functions

The transform of a real odd function is given as

$$f_o(t) \Leftrightarrow iI_o(f), \tag{9.112}$$

where $I_o(f)$ is the imaginary component of the transform, itself an odd function given by

$$I_o(f) = \int_{-\infty}^{\infty} f(t)\sin(2\pi ft)dt. \tag{9.113}$$

For complex waveforms or series, the real and imaginary parts are each decomposed into even and odd components and then transformed as above for the real functions.

General Transform Pairs

Given the above relationships, the following transform pair types can be derived.

f(t) type	F(f) type
real	real part even imaginary part odd
real even	real even
real odd	imaginary odd
imaginary	real part odd imaginary part even
real even, imaginary odd	real
real odd, imaginary even	imaginary
imaginary even	imaginary even
imaginary odd	real odd
complex even	complex even
complex odd	complex odd.

Knowledge of these transform pairs is helpful for reducing, simplifying or checking transforms. Also, while you will never have imaginary data,[10] it is often useful to take 2-dimensional data (e.g., spatial vectors such as velocity) and treat them as a 1-dimensional complex quantity, in which case knowledge of the above transform relationships can help to reduce the transform problem and deduce important relationships.

Box D9.4 Derivation of Fourier Transform Properties

Symmetry. Symmetry relationship is easily shown by consideration of the inverse Fourier transform,

$$f(t) = \int_{-\infty}^{\infty} F(f) e^{i2\pi ft} df, \tag{D9.4.1}$$

and

$$f(-t) = \int_{-\infty}^{\infty} F(f) e^{-i2\pi ft} df, \tag{D9.4.2}$$

giving the inverse transform. Now, rename symbols by swapping t and f in (D9.4.2) to give

$$f(-f) = \int_{-\infty}^{\infty} F(t) e^{-i2\pi ft} dt. \tag{D9.4.3}$$

Linearity. Proof of the linearity property is directly shown by examination of the Fourier transform:

$$\begin{aligned} F(f) &= \int_{-\infty}^{\infty} [g(t) + h(t)] e^{-i2\pi ft} dt \\ &= \int_{-\infty}^{\infty} g(t) e^{-i2\pi ft} dt + \int_{-\infty}^{\infty} h(t) e^{-i2\pi ft} dt \\ &= G(f) + H(f). \end{aligned} \tag{D9.4.4}$$

Scaling. Proof of the scaling of the time axis property is shown directly by transforming the scaled axis. First consider the case for $a \geq 0$:

$$F(f) = \int_{-\infty}^{\infty} f(at) e^{-i2\pi ft} dt. \tag{D9.4.5}$$

[10] If you do, then consult with your imaginary friend to have them converted to real data.

Box D9.4 (Cont.)

Substitute $u = \text{at}$, so $\text{d}u = \text{adt}$, giving

$$= \frac{1}{a}\int_{-\infty}^{\infty} f(u)e^{-i2\pi(f/a)u}du$$
$$= \frac{1}{a}\int_{-\infty}^{\infty} f(u)e^{-i2\pi(f')u}du$$

where $f' = f/a$, so

$$\begin{aligned} &= \frac{1}{|a|}F(f') \\ &= \frac{1}{|a|}F(f/a), \end{aligned} \tag{D9.4.6}$$

where the absolute value sign simply reflects the restriction that the scaling constant $a \geq 0$.

If $a < 0$, again substitute $u = \text{at}$, but notice that in this case the limits of integration are reversed,

$$= \frac{1}{a}\int_{\infty}^{-\infty} f(u)e^{-i2\pi(f/a)u}du,$$

which is equivalent to reversing the integration limits and multiplying by -1, giving

$$\begin{aligned} &= \frac{1}{-a}\int_{\infty}^{-\infty} f(u)e^{-i2\pi(f/a)u}du \\ &= \frac{1}{|a|}F(f/a), \end{aligned} \tag{D9.4.7}$$

where the absolute values sign again indicates $a > 0$ (since the negative value of a is multiplied by -1). Therefore, for all values of a,

$$F(f) = \frac{1}{|a|}F(f/a). \tag{D9.4.8}$$

<u>**Time Shifting**</u>. Proof of the time shifting property follows an approach similar to that used for the scaling proof:

$$F(t - t_0) = \int_{-\infty}^{\infty} f(t - t_0)e^{-i2\pi ft}dt.$$

Box D9.4 (Cont.)

Substitute $u = t - t_0$, so $du = dt$, giving

$$\begin{aligned} &= \int_{-\infty}^{\infty} f(u) e^{-i2\pi(fu+t_0)} du \\ &= \int_{-\infty}^{\infty} f(u) e^{-i2\pi fu} e^{-i2\pi ft_0} du \\ &= e^{-i2\pi ft_0} \int_{-\infty}^{\infty} f(u) e^{-i2\pi ftu} du \\ &= e^{-i2\pi ft_0} F(f). \end{aligned} \tag{D9.4.9}$$

Even Functions. The transform of an even function is shown directly as

$$\begin{aligned} F_e(f) &= \int_{-\infty}^{\infty} f_e(t) e^{-i2\pi ft} dt \\ &= \int_{-\infty}^{\infty} f_e(t)[\cos(2\pi ft) - i\sin(2\pi ft)] dt \\ &= \int_{-\infty}^{\infty} f_e(t)\cos(2\pi ft) dt - i\int_{-\infty}^{\infty} f_e(t)\sin(2\pi ft) dt. \end{aligned} \tag{D9.4.10}$$

However, the product of an even function times an odd function is an odd function, and the integral of an odd function over a symmetrical interval about the origin is 0. Therefore, the integral of the imaginary term is equal to 0, leaving

$$= \int_{-\infty}^{\infty} f_e(t)\cos(2\pi ft) dt. \tag{D9.4.11}$$

Odd Functions. The transform of an odd function is shown directly as above:

$$\begin{aligned} F_o(f) &= \int_{-\infty}^{\infty} f_o(t) e^{-i2\pi ft} dt \\ &= \int_{-\infty}^{\infty} f_o(t)[\cos(2\pi ft) - i\sin(2\pi ft)] dt \\ &= \int_{-\infty}^{\infty} f_o(t)\cos(2\pi ft) dt - i\int_{-\infty}^{\infty} f_o(t)\sin(2\pi ft) dt. \end{aligned} \tag{D9.4.12}$$

In this case, the real term is the product of an odd and even function, and thus the integral is equal to 0, leaving

$$= -i\int_{-\infty}^{\infty} f_o(t)\sin(2\pi ft) dt. \tag{D9.4.13}$$

Box D9.4 (Cont.)

Note that this integral is the product of two odd functions that produce an even function. However, $F_o(f)$ is an odd function, since the independent variable, f, occurs only in the sine term of the integrand.

9.8 Fourier Transform Theorems

Finally, in addition to the preceding Fourier transform *properties*, there are a few Fourier transform *theorems* involving differentiation, integration and convolution that we require in order to interpret the implications of a discrete Fourier transform of real data.

9.8.1 Differentiation Theorem

If

$$f(t) \Leftrightarrow F(f), \tag{9.114}$$

then, if $f(t) \rightarrow 0$ as $t \rightarrow \pm\infty$,

$$f'(t) \Leftrightarrow -i2\pi f F(f), \tag{9.115}$$

where the prime indicates the derivative with respect to t.

So the transform of the derivative of a function results in a linear scaling of the frequency function and reversal of real and imaginary parts, since F(f) is multiplied by *i*. This reversal of the real and imaginary parts only affects the phase of the transform, not the amplitude.

9.8.2 Integration Theorem

If

$$f(t) \Leftrightarrow F(f) \tag{9.116a}$$

and f(t) has a zero mean, so

$$\int_{-\infty}^{\infty} f(t)dt = 0, \tag{9.116b}$$

then

$$\int_{-\infty}^{t} f(t)dt \Leftrightarrow \frac{1}{i2\pi f} F(f). \tag{9.117}$$

Therefore the integral of a function up to a time t has a transform equal to that of the f(t) itself, scaled by $1/i2\pi f$. This integral is similar to that of a cumulative distribution function and is sometimes useful when F(f) is known.

9.8.3 Convolution Theorem (Major Importance)

First, we define the **convolution integral** (analogous to the discrete convolution sum),

$$f * g = h(\tau) = \int_{-\infty}^{\infty} f(t)g(\tau - t)dt, \tag{9.118}$$

where τ is the lag (as lag k is in the discrete case), but unlike k does not have to be an integer.

At $\tau = 0$,

$$h(0) = \int_{-\infty}^{\infty} f(t)g(-t)dt, \tag{9.119}$$

so, after reversing g(t), the functions are multiplied together and the product integrated as in the discrete case.

At $\tau = 1$

$$h(1) = \int_{-\infty}^{\infty} f(t)g(1 - t)dt, \tag{9.120}$$

so, f(t) is multiplied by g(t) after it has been reversed and shifted by one t unit, and the product is then integrated. Note that τ is not restricted to integer numbers as in the discrete case, but instead takes on an infinite number of values, reflecting the continuous nature of this integral.

Therefore, (9.118) describes a reversing of g(t) and passing it over f(t), just like the discrete operation.

Several properties of the convolution integral of relevance here include

$$f*g = g*f \tag{9.121a}$$

$$f*(g*h) = (f*g)*h \tag{9.121b}$$

$$f*(g+h) = f*g+f*h. \tag{9.121c}$$

The relationship between convolution and the Fourier transform leads to one of the most useful theorems in analysis as stated by the **convolution theorem**. This theorem states that *convolution in the time domain is equal to multiplication in the frequency domain, and vice versa*. So,

$$f(t)*g(t) \Leftrightarrow F(f)G(f), \tag{9.122a}$$

and in polar coordinates,

$$f(t) * g(t) \Leftrightarrow |F(f)|e^{i\theta_F}|G(f)|e^{i\theta_G} = |F(f)||G(f)|e^{i(\theta_F+\theta_G)}. \tag{9.122b}$$

This result provides the basic tool from which (among other things) the continuous Fourier transform is related to the discrete Fourier transform.

Application of the symmetry relationship to the convolution theorem result of (9.122) shows that

$$f(t)g(t) \Leftrightarrow F(f)*G(f). \tag{9.123}$$

Note that the symmetry relationship actually states that the convolution in the frequency domain should be with the negative frequencies, or F(−f)*G(−f). However, because of symmetry relationships (as shown in the derivation section that follows), the change in sign is not required.

This theorem is of fundamental importance: *convolution in one domain is multiplication in the other*. Recollection of this theorem will help your conceptual understanding of many operations that we will apply to the time series (e.g., smoothing via running averages, a convolution) and to the spectrum.

9.8.4 Autocovariance Theorem (Wiener–Khinchin Relationship)

Analogous to the convolution theorem is the autocovariance theorem. First, consider the continuous form of the autocovariance function:

$$C_{ff}(\tau) = \int_{-\infty}^{\infty} f(t)f(t+\tau)dt. \tag{9.124}$$

As with the continuous convolution integral, this form of the acvf is analogous to the serial product used to compute the discrete acvf.

$$C_{ff}(\tau) = \frac{1}{n}\sum_{i=1}^{n-|k|} f_i f_{i+\tau}, \tag{9.125}$$

assuming here that the function f(t) has zero mean, or the mean has been previously removed.

The autocovariance theorem states that the Fourier transform of the acvf as given by (9.124) is directly equal to the **power spectrum** of f(t), where the power spectrum is $|F(f)|^2$, or the **power spectral density function (PSD)**, so

$$C_{ff}(\tau) \Leftrightarrow |F(f)|^2. \tag{9.126}$$

This naming convention acknowledges the relationship between a PDF for non-sequential random variables and the comparable PSD for random processes (sequential data) – the PSD gives the distribution of power as a function of frequency bands. The PSD is sometimes called the **spectral density function (SDF)**.

9.8.5 Cross-Covariance Theorem

The **cross-covariance theorem** is analogous to the autocovariance theorem. First, consider the continuous form of the cross covariance function:

$$C_{\mathrm{fg}}(\tau) = \int_{-\infty}^{\infty} \mathrm{f}(\mathrm{t})\mathrm{g}(\mathrm{t}+\tau)\mathrm{dt}. \tag{9.127}$$

As with the continuous autocovariance function, this form of (9.127) is analogous to the sampled cross-covariance function,

$$C_{\mathrm{fg}}(\tau) = \frac{1}{n}\sum_{i=1}^{n-|k|} \mathrm{f}_i \mathrm{g}_{i+\tau} \tag{9.128}$$

The cross-covariance theorem states that the transform of the cross-correlation function (or cross-covariance) yields the **Cross-Spectral Density function (CSD)**, the covariance between the two time series as a function of frequency bands,

$$C_{\mathrm{fg}}(\tau) \Leftrightarrow \mathrm{F}^*(\mathrm{f})\mathrm{G}(\mathrm{f}). \tag{9.129}$$

This transform pair effectively allows you to consider the correlation between two time series as a function of frequency. This is discussed in more detail in Chapter 12.

9.8.6 Parseval's Theorem

Parseval's Theorem states that the energy of the signal f(t) is equal to the energy of its transform, where *energy* is defined as $\int_{-\infty}^{\infty} \mathrm{f}^2(\mathrm{t})\mathrm{dt}$, and *power* (energy per time) is equal to $\frac{1}{\mathrm{T}}\int_{-\infty}^{\infty} \mathrm{f}^2(\mathrm{t})\mathrm{dt}$. Therefore,

$$\int_{-\infty}^{\infty} |\mathrm{f}(\mathrm{t})|^2\mathrm{dt} = \int_{-\infty}^{\infty} |\mathrm{F}(\mathrm{f})|^2\mathrm{dt}. \tag{9.130}$$

This theorem is used in a variety of situations, which are discussed later in context.

Distribution of Variance in Fourier Series

An extension of Parseval's theorem comes in the form of conservation of variance between the time series and its transform. Consider the sample variance of the time series being fit with the sines and cosines:

$$s_y^2 = \frac{1}{n-1}\sum_{i=1}^{n} (y_i - \bar{y})^2. \tag{9.131}$$

Box D9.5 Derivation of Fourier Transform Theorems

Differentiation

Proof of the differentiation theorem requires expansion of the Fourier transform using integration by parts and a couple of assumptions (yeah, there are always assumptions). First, transform the derivative:

$$F(f) = \int_{-\infty}^{\infty} \frac{\partial f(t)}{\partial t} e^{-i2\pi ft} dt. \tag{D9.5.1}$$

Consider integration by parts in which

$$\int_a^b u dv = uv|_a^b - \int_a^b v du. \tag{D9.5.2}$$

For (D9.5.1), define

$$u = e^{-i2\pi ft} \qquad du = -i2\pi f e^{-i2\pi ft} dt \tag{D9.5.3a}$$

$$dv = \frac{\partial f(t)}{\partial t} dt \qquad v = f(t). \tag{D9.5.3b}$$

Substituting (D9.5.3) into (D9.5.2) with limits $a = -\infty$ and $b = \infty$ gives

$$F(f) = f(t)e^{-i2\pi ft}\Big|_{-\infty}^{\infty} + i2\pi f \int_{-\infty}^{\infty} f(t)e^{-i2\pi ft} dt. \tag{D9.5.4}$$

If f(t) → 0 as t → ±∞, then the first term on the right-hand side is equal to zero. Alternatively, we need f(t) → 0 as t → −∞ and f(∞) = finite, since $lim_{t\to\infty} \frac{f(t)}{e^t} = 0$. In either case, given these restrictions on the function f(t), (D9.5.4) reduces to

$$F(f) = i2\pi f F(f). \tag{D9.5.5}$$

Convolution

Proof of the convolution theorem is by direct application of the Fourier transform to the convolution integral. If h(t) = f(t)*g(t), then

$$H(f) = \int_{-\infty}^{\infty} h(t)e^{-i2\pi ft} dt. \tag{D9.5.6}$$

Applying the definition of the convolution integral, (9.118), to h(t) in (D9.5.6) gives

$$= \int_{-\infty}^{\infty} \left[\int_{-\infty}^{\infty} f(\tau)g(t-\tau)d\tau \right] e^{-i2\pi ft} dt. \tag{D9.5.7}$$

Box D9.5 (Cont.)

Notice that the lag has been defined here as t, not τ (which is representing time). Interchanging the order of integration, which assumes continuity of the integrand over ±∞, gives

$$= \int_{-\infty}^{\infty} f(\tau)\left[\int_{-\infty}^{\infty} g(t-\tau)dt\right] e^{-i2\pi ft}d\tau. \quad \text{(D9.5.8)}$$

Defining $u = t - \tau$, so $du = dt$, then gives

$$\begin{aligned} &= \int_{-\infty}^{\infty} f(\tau)\left[\int_{-\infty}^{\infty} g(u)e^{-i2\pi f(u+\tau)}du\right] d\tau \\ &= \int_{-\infty}^{\infty} f(\tau)\left[\int_{-\infty}^{\infty} g(u)e^{-i2\pi fu}e^{-i2\pi f\tau}du\right] d\tau \\ &= \int_{-\infty}^{\infty} f(\tau)e^{-i2\pi f\tau}\left[\int_{-\infty}^{\infty} g(u)e^{-i2\pi fu}du\right] d\tau \\ &= \int_{-\infty}^{\infty} f(\tau)e^{-i2\pi f\tau}G(f)d\tau \\ &= G(f)\int_{-\infty}^{\infty} f(\tau)e^{-i2\pi f\tau}d\tau \\ &= G(f)F(f). \end{aligned} \quad \text{(D9.5.9)}$$

The inverse relationship of the convolution theorem – that is, that multiplication in the time domain is equal to convolution in the frequency domain – is shown by application of the symmetry relationship to (9.122). This states that $f(t)g(t) \leftrightarrow F(-f)*G(-f)$. However, consider the convolution integral of (9.118) in which $-f$ has been substituted for t, so that $df = -dt$. This gives

$$F^{*}G = -\int_{\infty}^{-\infty} F(-f)G(\tau + f)df. \quad \text{(D9.5.10)}$$

Now substitute $u = \tau + f$, $du = df$, giving

$$= -\int_{\infty}^{-\infty} F(\tau - u)G(u)du. \quad \text{(D9.5.11)}$$

Reversing the order of integration gives

$$\begin{aligned} &= \int_{-\infty}^{\infty} F(\tau - u)G(u)du \\ &= F(f) * G(f). \end{aligned} \quad \text{(D9.5.12)}$$

Box D9.5 (Cont.)

Therefore, the convolution of F(−f)*G(−f) is equivalent to convolving F(f)*G(f), that is, without reversing the sign.

Autocovariance Theorem

Proof of this theorem comes by direct transformation of the autocovariance function (assume a lag of t) and closely follows the proof of the convolution theorem:

$$\begin{aligned} C_{\mathrm{ff}}(\mathrm{f}) &= \int_{-\infty}^{\infty} C_{\mathrm{ff}}(\tau)\mathrm{e}^{-i2\pi \mathrm{ft}}\mathrm{dt} \\ &= \int_{-\infty}^{\infty}\left[\int_{-\infty}^{\infty} \mathrm{f}(\tau)\mathrm{f}(\tau+\mathrm{t})\mathrm{d}\tau\right]\mathrm{e}^{-i2\pi \mathrm{ft}}\mathrm{dt} \\ &= \int_{-\infty}^{\infty} \mathrm{f}(\tau)\left[\int_{-\infty}^{\infty} \mathrm{f}(\mathrm{t}+\tau)\mathrm{e}^{-i2\pi \mathrm{ft}}\mathrm{dt}\right]\mathrm{d}\tau. \end{aligned} \tag{D9.5.13}$$

The time-shifting theorem can be directly applied to the inner integral to give

$$\begin{aligned} &= \int_{-\infty}^{\infty} \mathrm{f}(\tau)\mathrm{e}^{i2\pi \mathrm{f}\tau}\mathrm{F}(\mathrm{f})\mathrm{d}\tau \\ &= \mathrm{F}(\mathrm{f})\int_{-\infty}^{\infty} \mathrm{f}(\tau)\mathrm{e}^{i2\pi \mathrm{f}\tau}\mathrm{d}\tau \\ &= \mathrm{F}(\mathrm{f})\mathrm{F}^{*}(\mathrm{f}) \\ &= |\mathrm{F}(\mathrm{f})|^{2}. \end{aligned} \tag{D9.5.14}$$

Note that if the proper normalized form of the acf, in (9.125), were used here instead of the autocovariance form, then the result here would be changed only by a scaling factor, since the normalization factor is a constant and therefore comes out of the integral.

Cross-Covariance Theorem

Proof of this theorem is identical to that for the autocovariance theorem above, but in this case, g(τ + t) is substituted for the f(τ + t) in (D9.5.13). This leads directly to the stated result.

Parseval's Theorem

Parseval's theorem is proven by direct application of the convolution theorem. Alternatively, consider

$$\int_{-\infty}^{\infty} |\mathrm{f}(\mathrm{t})|^{2}\mathrm{dt} = \int_{-\infty}^{\infty} \mathrm{f}(\mathrm{t})\mathrm{f}(\mathrm{t})\mathrm{dt}. \tag{D9.5.15}$$

Box D9.5 (Cont.)

Rewriting one f(t) in terms of its inverse Fourier transform gives

$$\begin{aligned}\int_{-\infty}^{\infty} |f(t)|^2 dt &= \int_{-\infty}^{\infty} f(t)\left[\int_{-\infty}^{\infty} (f)e^{i2\pi ft} df\right] dt \\ &= \int_{-\infty}^{\infty} F(f)\left[\int_{-\infty}^{\infty} f(t)e^{i2\pi ft} dt\right] df \\ &= \int_{-\infty}^{\infty} F(f)F^*(f) df \\ &= \int_{-\infty}^{\infty} |F(f)|^2 df.\end{aligned} \tag{D9.5.16}$$

Substituting the Fourier series for y_i gives

$$s_y^2 = \frac{1}{n-1}\sum_{i=1}^{n}\left[a_0 + \sum_{i=1}^{\leq n/2} a_j \cos(\omega_j\, t_i) + b_j \sin(\omega_j\, t_i) - \bar{y}\right]^2, \tag{9.132}$$

where $\omega_j = j\omega_1 = j2\pi/T$. Since $a_0 = \bar{y}$, the a_0 and $\bar{y}$ terms in (9.132) cancel. Completing the square, rearranging and reducing through use of the orthogonality conditions allows explicit reduction of (9.132) for evenly spaced data, as

$$s_y^2 \approx \frac{1}{2}\sum_{i=1}^{\leq n/2} (a_j^2 + b_j^2) \tag{9.133a}$$

$$\approx \frac{1}{2}\sum_{i=1}^{\leq n/2} A_j^2. \tag{9.133b}$$

So, (9.133) indicates that the variance of the time series is equal to the sum of the square of the amplitudes of the various sinusoids summed in the composite curve fitting the data.[11] Consequently, A_j^2/s_y^2 *gives the fraction of variance described by any one particular*

[11] The approximately equal sign, ≈, in (9.134) reflects the dropping of a factor $n/(n-1)$, which for large n is approximately 1.

sinusoidal component. This allows you to determine the distribution of variance as a function of periodic components and determine if any particular harmonic or group of harmonics explains a disproportionately large percentage of the variance in the time series. This information allows you to quantify the degree of periodicity represented by the data and determine how that periodicity is distributed among the various harmonics (i.e., the PSD being a PDF for the time series in the frequency domain).

9.9 Fast Fourier Transform

The **Fast Fourier Transform (FFT)** is a computationally fast and efficient *algorithm* for computing the discrete Fourier transform. An FFT for a power of two data points reduces the number of operations from n^2 to $n\log_2 n$ (n = number of data points), so an FFT will transform a 2048-sample time series in less than 1/1000 of the operations required for an efficient standard discrete Fourier transform algorithm. There are numerous versions of FFTs, though they all essentially use the same principle of breaking the data into smaller and smaller segments so that the transform of each segment becomes very quick and efficient.[12] The most common form of the FFT requires that the number of data points (n) in the discrete time series is a factor of 2. That is, $n = 2^N$. This is typically accomplished by padding the time series (after removing the mean) with zeros in order to achieve the requisite power-of-two number of data points. If you don't have a zero-mean time series, remove the mean before padding.

Kanasewich (1981; pages 122–125; especially his figure 9.8) shows that the addition of zeros to the original n data points to make a new n, n_2, which is a factor of 2, can lead to improved features in, say, the transform of the Daniell window (discussed later) if the ratio of n to n_2 is ~0.7. This very specific case offers an example of how the addition of zeros to the data *can* affect the result (despite the popular conception to the contrary), but for the most part it has no impact other than speeding up the computation for very large time series.

Other differences between the FFT and standard transform:

1) The addition of zeros changes the fundamental period of harmonics, which must be accounted for when labeling the frequency axis of the PSD and any other spectral display.
2) Padding with zeros also improves the frequency resolution (because longer T leads to lower (smaller) $f_1 = 1/T = \Delta f$, giving the appearance of finer frequency resolution, attained by sampling more closely in the frequency domain).
3) Addition of zeros essentially violates the requirement of statistical stationarity: i.e., the variance changes from the true value to zero once the zone of zeros is entered. This should not be too bad a problem when adding just a few zeros, but when adding enough to almost double the time series, the statistical analysis to determine the uncertainty in the Fourier coefficients must be done very carefully. It can also change the variance, so in some contexts it is necessary to weight the PSD to compensate for this – again discussed in Kanasewich (pg. 126).

[12] See Brigham (1974), chapter 6, for details, or Kanasewich (1981) for a summary.

These implications are discussed in more detail as they arise in context in later chapters.

9.10 Take-Home Points

1. Fourier frequencies are those that are harmonics of the period (T) of the series being fit, so $f_j = j/T$, where j is an integer and represents the jth Fourier frequency.
2. Nyquist frequency is the highest frequency that can be resolved in the Fourier fit: $f_N = 1/2\Delta t$. Any higher frequency in the series will be resolved falsely as a lower-frequency harmonic, known as an *alias*.
3. The Fourier series: $y_i = \sum_{i=0}^{\leq n/2} a_j \cos(\omega_j t_i) + b_j \sin(\omega_j t_i)$ contains $n/2$ harmonics if n, the number of points in the time series, is an even number; otherwise, for an odd-number n, it contains $(n-1)/2$. In the latter case, the spectrum does not contain the Nyquist frequency.
4. The length (T) of a time series dictates the fundamental harmonic ($f = 1/T$) that is the lowest frequency resolved in the spectrum, and it also serves as the frequency spacing ($f_1, f_2, \ldots f_n$), since $f_1 = 1/T = \Delta f$, $f_2 = 2\Delta f$, $f_n = n\Delta f$. The time step, Δt, of the time series dictates the highest-frequency of the spectrum (Nyquist frequency).
5. Sines and cosines sum to 0 over a cycle when sampled at a constant Δt. If multiplied and summed over an integer number of cycles, they are orthogonal *when sampled at a constant* Δt. This allows a direct and simple solution to the Fourier series, known as the *Fourier Transform*: $J_j = \frac{1}{n}\sum_{i=1}^{n} y_i e^{-i\omega_j t_i}$, where $J_j = \frac{1}{2}(a_j - ib_j)$ (a complex amplitude). The inverse transform, converting back to the time domain, is $y_i = \frac{1}{2}\sum_{j>-n/2}^{\leq n/2}(a_j - ib_j)e^{i\omega_j t_i}$.
6. The convolution theorem, one of the most important in Fourier analysis, is that multiplication in the time domain is equivalent to convolution in the frequency domain, and vice-versa.
7. The autocovariance theorem states that the Fourier transform of the acf or acvf gives power spectral density (PSD), the way the power of the time series is distributed as a function of frequency – the frequency domain analog to a probability density function (PDF) for describing non-sequential data. The PSD is the fundamental characterization of a time series in an ensemble.
8. The cross-covariance theorem states that the Fourier transform of the ccf or ccvf gives the cross-spectral density function, the covariance between two time series as a function of frequency.
9. Parseval's theorem states that the energy of the time series is equivalent to that of the PSD. Alternatively, the variance of the time series is equal to sum of its spectrum.
10. Fast Fourier transform (FFT) is an *algorithm* for a very fast (and efficient) Fourier transform.

9.11 Questions

Pencil and Paper Questions

1. For a time series y_i of five evenly spaced data points (sampling interval $\Delta t = 2$ s), write out each term of its Fourier series (actually, write out each term of the sum). Show the exact solution for the a_i and b_i coefficients. What is the sampling angular frequency? What is the Nyquist frequency? Is it included in the series?
2. Write out the Fourier series for a time series containing 101 data points, with spacing at $\Delta t = 1$ s, in standard matrix form ($\mathbf{Ax} = \mathbf{b}$), showing the first five and last three rows of the $\mathbf{A}$ matrix and $\mathbf{x}$ and $\mathbf{b}$ vectors, using rotational frequency f_i, including all details explicitly.
3. a. Define, for a time series, y_i, with n evenly spaced data points ($\Delta t = 5$)

    1. fundamental Fourier frequency, f_F
    2. period of the fundamental Fourier frequency, T
    3. Nyquist frequency
    4. Fourier frequencies, as angular frequencies, ω
    5. the jth harmonic

   b. Sketch a time series. Then, immediately below it, sketch to a comparable scale

    1. the corresponding cosine for the zero Fourier frequency
    2. the corresponding cosine for the fundamental Fourier frequency
    3. the corresponding cosine for the fourth harmonic

   c. Write the formula (be explicit in all terms shown, e.g., f, summation limits, etc.) for

    1. discrete cosine transform
    2. discrete Fourier transform

Computer Questions

For the convolution theory, compute a 25-point running average on the LR04 data www.cambridge.org/martinson in the time domain, plotting the two series that are being convolved. Then repeat in the frequency domain, showing the two amplitude spectra superimposed being multiplied. Superimpose the two different results to show that they are identical.

10 Fourier Sampling Theory

10.1 Overview

Typically, you will be analyzing a discretely sampled finite segment of a continuous process in an attempt to understand it, determine its underlying structure, composition, or whatever – essentially, looking for some signal in it. It is important to know how well that sampled version captures the true continuous process. This chapter shows how to determine that. It is done via sampling theory, which follows a clear set of operations on the time series (in the *time domain*), with the equivalent operation on the Fourier series (in the *frequency domain*), relating each operation in its respective domain through the convolution theorem (multiplication in one domain is convolution in the other).

Sampling theory starts by performing the continuous Fourier transform on an infinitely long continuous time series, giving the true continuous Fourier spectrum (that which we hope to recover from the sampled time series); then it discretely samples the continuous series, showing the impact of this on the continuous true spectrum. It continues in this manner until showing the analysis of the discrete series and the overall impact on the true continuous spectrum. This side-by-side comparison clearly shows exactly what we are losing and why, allowing us to develop rules to avoid or minimize any lost or distorted information.

The beauty is that this will show that if our discrete sampling interval, Δt, is small enough (fast enough in time, or close enough in space), we can capture the true continuous series. Otherwise, if the sampling is too slow, we suffer from something known as **aliasing**, high-frequency information disguised as lower-frequency power. Likewise, when we apply Fourier analysis we are implicitly assuming that the segment sampled exactly repeats itself over this length, justifying the fitting with harmonics of this sampled length. To the extent that this is not true, the fit will introduce bias in the form of **leakage**. Aliasing can often be avoided; leakage, while its affects can be minimized with clever truncation schemes, can only be avoided if your discrete series is exactly cyclic over the sampled length – something absolutely clear from sampling theory, the focus of this chapter.

10.2 Sampling Theorem

The Sampling Theorem tells us that if a continuous time series, y(t), is band-limited – that is, if its Fourier transform is zero for all frequencies greater (higher) than some frequency, f_c (with period T_c) – then y(t) can be uniquely determined from knowledge of its sampled values, given a sampling interval, $\Delta t < 1/2f_c$. In other words, if we sample y(t) at discrete points that have a spacing in time, $\Delta t < 1/2f_c$ (or $< T_c/2$, so the sampling interval is less than half of the cutoff period, giving you two data points per highest-frequency wavelength), the discrete sampled time series, f_t, contains enough information to allow a *complete* description of the continuous time series y(t) – we lose no information by only considering a discretized version of y(t), given a close enough sampling interval, Δt. Wow – we can describe the band-limited continuous time series *completely*, even if we only sample a discretized version, as long as we sample at short enough time intervals!

If we do not satisfy this sampling requirement, our sample spectrum will suffer – whereby higher frequencies not resolved by the sampling interval will appear in the sample spectrum disguised as lower-frequency information, distorting the true lower-frequency power, an infliction known as aliasing, whose name will become apparent below.

10.2.1 A Sampling Theorem Derivation

We start by sampling an infinitely long continuous time series, y(t). Distribution theory indicates that this can be accomplished by multiplying y(t) by an infinite series of Dirac delta functions, δ(t), each spaced Δt apart: a **Dirac comb**. The Dirac comb is denoted by Δ(t) (this is a *function*, as indicated by the parentheses around the t; it is not the sampling interval, Δt); likewise for its transform (frequency equivalent), Δ(f).

In order to understand how the sampling of the Dirac comb in time affects the true continuous spectrum, we need to know the form of the Dirac comb in the time and frequency domains. These forms are, in time,

$$\Delta(\mathrm{t}) = \sum_{j=-\infty}^{\infty} \delta(\mathrm{t} - j\Delta\mathrm{t}), \tag{10.1}$$

with a transform into frequency of

$$\Delta(\mathrm{f}) = \frac{1}{\Delta\mathrm{t}} \sum_{j=-\infty}^{\infty} \delta(\mathrm{f} - j\Delta\mathrm{t}^{-1}). \tag{10.2}$$

Graphically, Δ(t) is shown in time in Figure 10.1, and its transform, Δ(f), is shown in frequency in Figure 10.2.

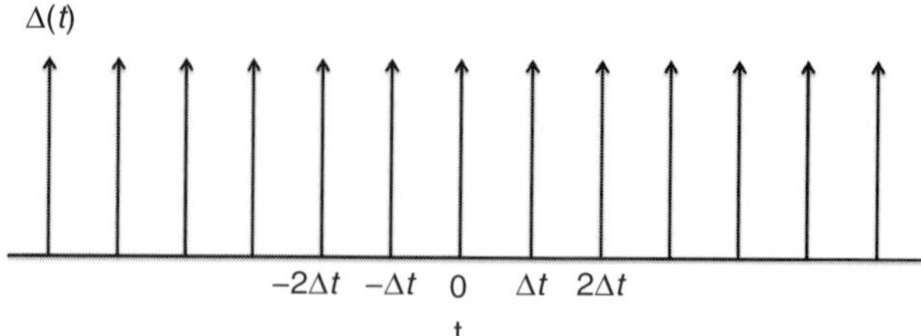

Figure 10.1 Dirac comb in time domain with impulse spacing of Δt.

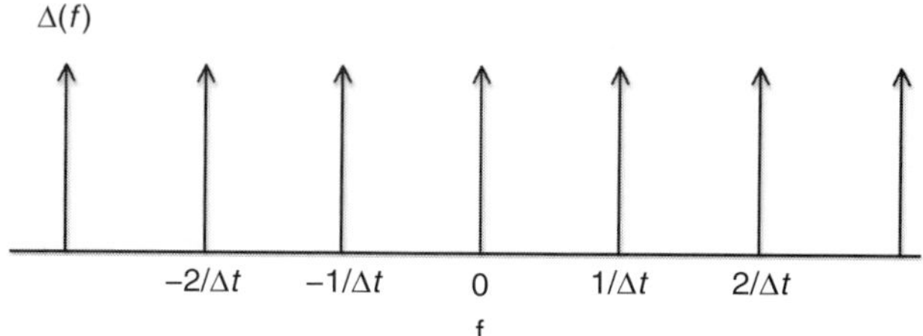

Figure 10.2 Fourier transform of Dirac comb, $\Delta(t)$, of Figure 10.1, giving Dirac comb in frequency domain, $\Delta(f)$, with impulses spaced at $1/\Delta t$, i.e., centered at harmonics with fundamental frequencies of period $T = \Delta t$.

Box D10.1 Building a Dirac Comb

In Time

We build the Dirac comb by adding an infinite number of cosines of specific harmonic frequencies. This is clearly seen by examining this sum, starting with just a single cosine, then increasing the number of cosines of harmonics,

$$g_1(t) = 1 + 2\cos(2\pi f_1 t), \tag{D10.1.1}$$

where we define the fundamental harmonic in the usual fashion, $f_1 = 1/T = 1/\Delta t = 1$, and add the one giving a nonzero mean.

Then we go to three cosines of harmonics of the fundamental ($1/T = 1/\Delta t$),

$$g_3(t) = 1 + 2\sum_{j=1}^{3} \cos(2\pi j f_1 t), \tag{D10.1.2}$$

and 100 harmonics,

$$g_{100}(t) = 1 + 2\sum_{j=1}^{100} \cos(2\pi j f_1 t). \tag{D10.1.3}$$

As suggested by the progression with additional cosines, in the limit, an infinite number of harmonic cosines sum to form a *train of infinite amplitude impulses at Δt units apart*. This sequence is illustrated in Figure D10.1.1.

Box D10.1 (Cont.)

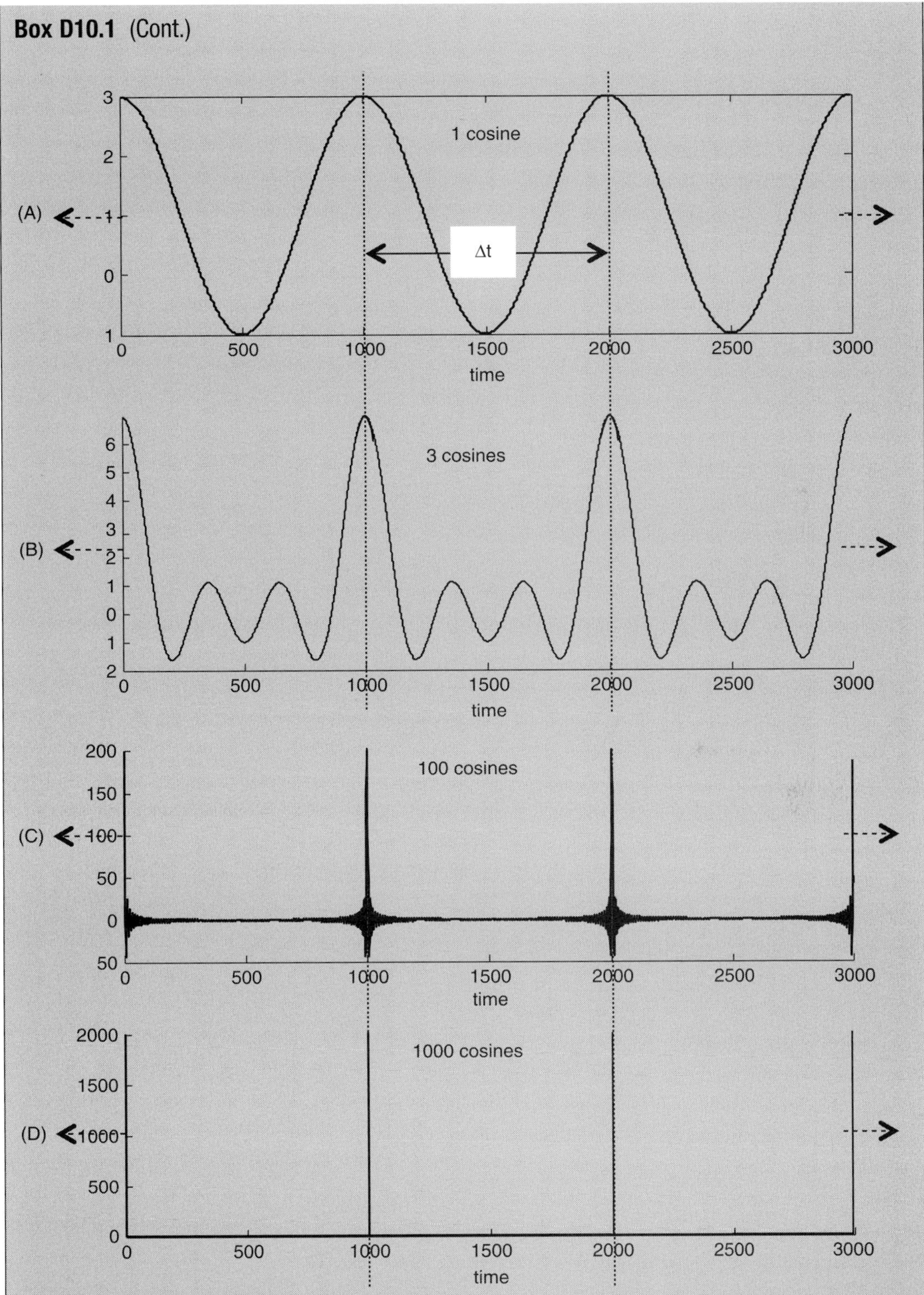

Figure D10.1.1 Example of building a Dirac comb by adding an infinite number of harmonic cosines of the same fundamental frequency ($1/\Delta t$); showing here only the first four cycles of infinitely long cosines. **(A)** 1 cosine; **(B)** 3 cosines; **(C)** 100 cosines; **(D)** 1000 cosines, clearly approaching a series of pure impulses.

Box D10.1 (Cont.)

In this limit of adding infinite cosine terms of frequency $f_1 = \Delta t^{-1} = 1$, the sum converges to the Dirac comb:

$$\sum_{j=-\infty}^{\infty} \cos(2\pi j f_1 t) = \sum_{j=-\infty}^{\infty} \delta(t - j\Delta t). \tag{D10.1.4}$$

Here we have dropped the factor of 2 in front of the sum and the mean value of 1, because the cosine is an even function and the negative values of j are equivalent to the positive values, i.e., $\cos(-j) = \cos(j)$. So, as written above, when $j = -3$, it produces a cosine term equivalent to that produced when $j = +3$, and thus we get $2\cos(2\pi j f_1 t)$ terms, as was the case using the form (D10.1.2). Similarly, we get two identical terms for each of the $\pm j \neq 0$ index values. When $j = 0$, we get $\cos(0) = 1$, hence providing the value 1 to the series as before. Therefore, (D10.1.4) is identical in form to (D10.1.3).

In Frequency

The Fourier transform of the infinite comb, $g_\infty(t)$, from direct application of (9.108c), is

$$\int_{-\infty}^{\infty} \left(\sum_{j=-\infty}^{\infty} \delta(t - j\Delta t) \right) e^{-i2\pi f t} dt = \sum_{j=-\infty}^{\infty} \delta(f - j f_0) \tag{D10.1.5}$$

where $f_0 = \Delta t^{-1}$. Thus, we get a Dirac comb in frequency, with the impulses separated by $1/\Delta t$.

For this development of the Dirac comb, the addition of an infinite number of cosines of harmonic frequencies so they all reinforce at the fundamental period ($T = \Delta t = 1$, in this case) leads to an infinite sequence in time of infinite-amplitude impulses spaced the fundamental period apart ($\Delta t = 1$). The Fourier transform of this comb leads to an infinite sequence of unit amplitude impulses spaced f_0 units (Δt^{-1}) apart. For Dirac combs with spacing other than unity ($\Delta t = 1$), the scaling property of the Fourier transform is invoked. Recall that if $y(t)$ has a transform $F(f)$, then $f(at)$ has a transform $(1/|a|)F(f/a)$. If we scale the time axis of the Dirac comb so that $\Delta t \neq 1$, then the transform of the comb is scaled by Δt^{-1}, as are the amplitudes of the impulses in the frequency domain. Then,

$$\Delta(t) = \sum_{j=-\infty}^{\infty} \delta(t - j\Delta t), \tag{D10.1.6}$$

Box D10.1 (Cont.)

with a transform of

$$\Delta(\mathrm{f}) = \frac{1}{\Delta \mathrm{t}} \sum_{j=-\infty}^{\infty} \delta(\mathrm{f} - j\mathrm{f}_0). \tag{D10.1.7}$$

This is the more general form of the Dirac comb and its transform for any given sampling interval, Δt.

Therefore, *a sequence of impulses in time separated by Δt and extending from ±∞ has a Fourier transform that is also a sequence of impulses of amplitude 1/Δt and separated by Δf = 1/Δt from ±∞.*

Given the transform pair, $\Delta(\mathrm{t}) \Leftrightarrow \Delta(\mathrm{f})$, the transform of the sampled time series (the transform of the continuous time series multiplied by the Dirac comb, as shown with the sifting integral of (9.84) in Chapter 9) can also be given, via the convolution theorem, as the convolution of the true spectrum F(f) with the frequency version of the Dirac Comb, so $y(\mathrm{t})\Delta(\mathrm{t}) \Leftrightarrow \mathrm{F(f)}{*}\Delta(\mathrm{f})$. This operation holds for continuous or discrete series via integration or summation, respectively.

For convenience, we use the discrete form of these operations, assuming the Dirac comb in time to be a sequence of unit amplitude impulses, so we sample by multiplication, $y(\mathrm{t})\Delta(\mathrm{t}) = \mathrm{f_t}$, the discrete y(t) sampled every Δt. This operation in the time domain is equivalent to convolving the true spectrum, F(f) with Δ(f).

First, consider the convolution of a function F(f) with a single unit impulse; this operation replicates the function F(f) (recall the definition of the impulse response function in Chapter 7). This is shown in Figure 10.3.

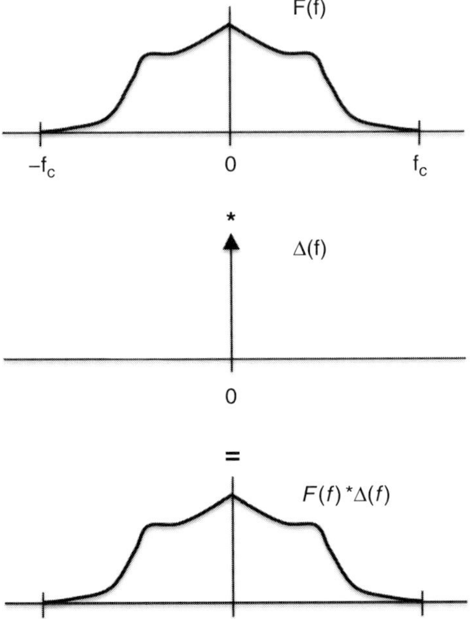

Figure 10.3 Convolution of impulse with spectrum replicates spectrum, centered about the location of the impulse.

Next, consider the convolution of the function F(f) with impulses located at f = 0 and $\pm f_1$, where $f_1 = 1/\Delta t$. Again, the spectrum is replicated about each impulse – if the impulses obey the sampling theorem ($\Delta t < 1/2f_1$), the replicated spectra *will not* overlap (Figure 10.4A); if they do not obey ($\Delta t < 1/2f_1$), the replicated spectra *will* overlap (Figure 10.4B).

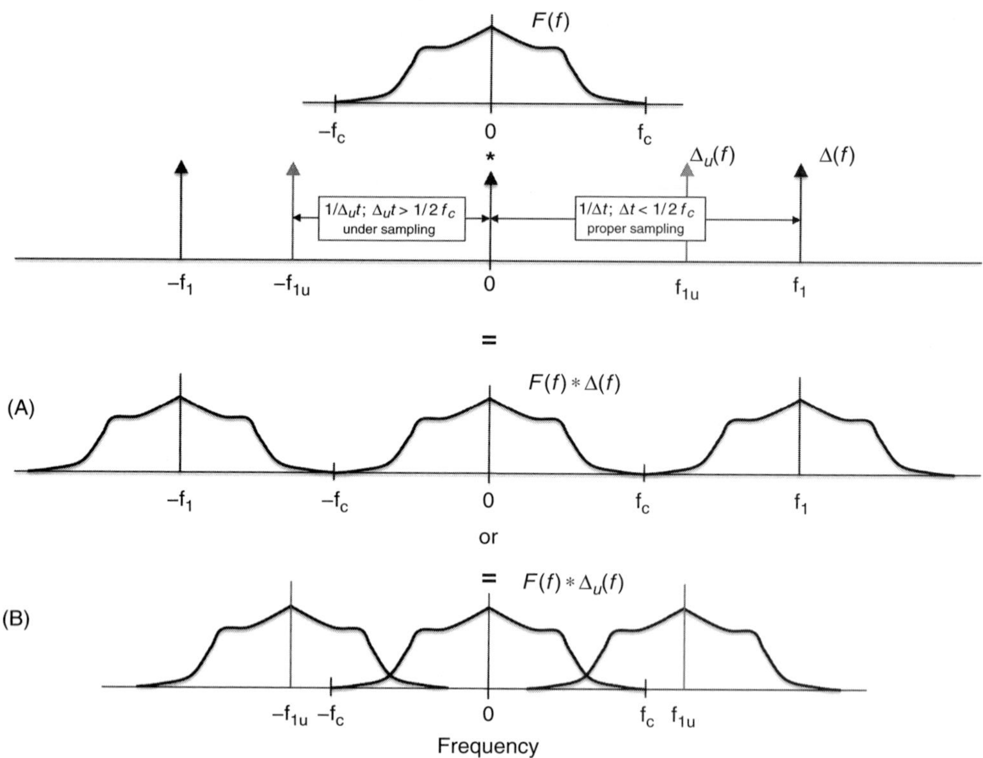

Figure 10.4 Convolution of Δ(f) (in this case, two sets of three impulses) with true spectrum, yields replication of spectrum at spacing of impulses ($f_1 = 1/\Delta t$). Subscript u represents undersampled case. **(A)** Δt satisfies the requirement that $\Delta t < 1/2f_c$; **(B)** $\Delta_u t$ (light grey impulses) violates that requirement: $\Delta_u t > 1/2f_c$, so the series is undersampled, resulting in frequency impulses that are too close together, $f_{1u} > (2\Delta_u t)^{-1}$, together causing overlap of the replicated spectrum, which must be spaced at least $2f_c$ apart to avoid overlap.

Regardless of whether they obey the sampling theorem or not, the convolution F(f)*Δ(f) is equivalent to the multiplication in the time domain of $y(t)\Delta(t) = f_t$. When:

$\Delta t < 1/2f_1$ Δt is small enough (fast enough in time or close enough in space), this has a transform (shown in Figure 10.4A) that is the same as the true transform F(f), over the frequency range $\pm f_c$, except that the scaling law (of §9.8.3) has scaled the amplitude of F(f) by Δt^{-1}.

$\Delta t > 1/2f_1$ Δt is too big (slower in time or wider in space), the impulses in the frequency domain are closer together (given $f_1 = 1/\Delta t$) and the spectra overlap in $F(f)*\Delta(f)$ (as shown in Figure 10.4B).

Now for the grand prize: recovery of the true spectrum. For the case of proper sample size, we isolate the true spectrum of Figure 10.4A and return it to its original amplitude by multiplying the spectrum of f_t by a rectangular function, H(f), of width $2f_c$ and amplitude Δt (Figure 10.5). So, in the frequency domain,

$$F(f) = \begin{cases} H(f)[F(f) * \Delta(f)] \\ H(f)F_f \end{cases}, \tag{10.3a}$$

or, the equivalent operations in time domain,

$$f(t) = \begin{cases} h(t) * [f(t)\Delta(t)] \\ h(t) * f_t \end{cases}, \tag{10.3b}$$

where F(f) is the true spectrum and f(t) the continuous time series: F_f and f_t, their respective discrete forms.

The inverse Fourier transform of the frequency domain rectangular function, F(f), is the sinc function given by (9.90), here for a width $2f_c$, amplitude Δt (*k* in (9.90)):[1]

$$h(t) = 2\Delta t f_c \frac{\sin 2\pi f_c t}{2\pi f_c}. \tag{10.4}$$

So, the inverse transform of this rectangular function is equal to the sinc function scaled by $2\Delta t f_c$, with the side lobes of the sinc having period $f_c^{-1} = 2\Delta t$.

Then, the convolution in (10.3b) is given by the sum

$$\begin{aligned} f(t) &= \sum_{j=-\infty}^{\infty} f_j h_{t-j} \\ &= \Delta t \sum_{j=-\infty}^{\infty} f(j\Delta t) \frac{\sin 2\pi f_c (t - j\Delta t)}{\pi (t - j\Delta t)}, \end{aligned} \tag{10.5}$$

where the continuous times, t, are the "lags" in this convolution over the discrete points at $j\Delta t$.

For a sampling interval $\Delta t < 1/2f_c$ and a band-limited time series, where $F(f) = 0$ for $|f| > f_c$, the continuous time series is completely recovered by knowledge of the sampled series only. That is, there is no loss of information associated with the sampling process – the continuous time series is obtained by convolving the discrete time series with the appropriate sinc function, or by simply truncating (Figure 10.5) the infinitely repeated true spectrum to isolate a single original continuous spectrum, as shown below.

[1] Equation (9.90) is transforming from time to frequency, whereas here we are going from frequency to time.

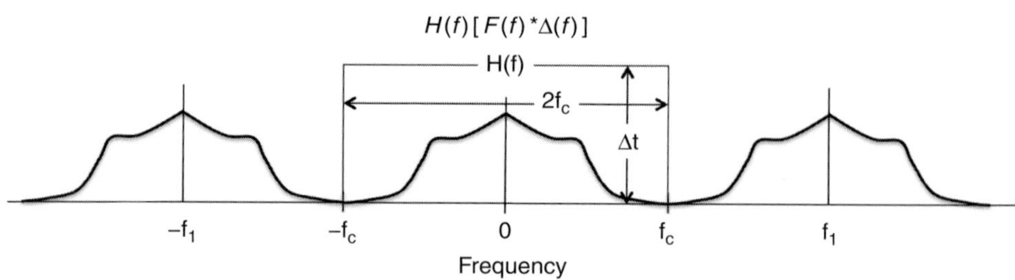

Figure 10.5 Replicated true spectrum is the spectrum of the sampled time series from Figure 10.4A. Multiplying this by the rectangular function of width $2f_c$ and height Δt isolates the original true spectrum and returns it to its original amplitude. This operation is equivalent to the convolution of discrete time series with the transform of rectangular function (= sinc function).

10.2.2 Aliasing

Sampling theorem tells us that the *sampled* (discrete) time series, f_t, provides a complete representation of its continuous "parent" time series, $y(t)$, and its true spectrum F(f), when $y(t)$ obeys the following restrictions:

1) band-limited [$F(f) = 0$ for $|f| > f_c$]; $y(t)$ contains no information at frequencies higher than f_c
2) sampled at an interval $\Delta t < 1/2f_c$; at least two data points per fastest cycle

Aliasing is the consequence of sampling the continuous time series at a sampling interval of $\Delta t \geq 1/2f_c$ in violation of the second sampling theorem restriction above. If the first restriction is violated, the second can never be satisfied.

The precise nature of aliasing can be viewed from several perspectives. First, the highest-frequency component resolved by a discrete Fourier transform for an even number of data points, *n*, is the **Nyquist frequency**, $f_{n/2} = f_N = 1/T_N = 1/2\Delta t$, so $2\Delta t = T_N = 1/f_N$, showing that there are two sampled data points over the period of the Nyquist frequency (T_N). So if the Nyquist frequency is for a cycle with a 1-year period, Δt will require more than two data points per year, or one data point in under 6 months. The cutoff period $T_c = 1/f_c$, so if $\Delta t > 1/2f_c$ then $T_c < 2\Delta t$, thus making $T_c < T_N$. The highest frequency (shortest period) that can be resolved is the Nyquist, but in this case f_c has a shorter period (higher frequency signal). This by itself doesn't sound too bad, as it simply means that the discrete Fourier transform does not resolve all of the high-frequency components present (i.e., those between f_N and f_c). Unfortunately, the unresolved high-frequency components, f_a (where $f_N < f_a \leq f_c$), will appear in the discrete spectrum disguised as lower-frequency components.

For example, Figure 10.6 shows a continuous time series, $y(t) = \cos(2\pi f_1 t)$ – a single cosine with period T = 12 hours, so its rotational frequency is $f_1 = 1/12 \approx 0.083$ (angular frequency, $\omega_1 = 2\pi/12 \approx 0.52$). This time series is band-limited, since there are no frequency components higher than (or lower than) f_1 present. So the highest-frequency component present in this simple series is $f_c = f_1$. The sampling theorem tells us that $y(t)$ will be resolved by a discretized version if the sampling interval, $\Delta t < 1/2f_c$, or faster than

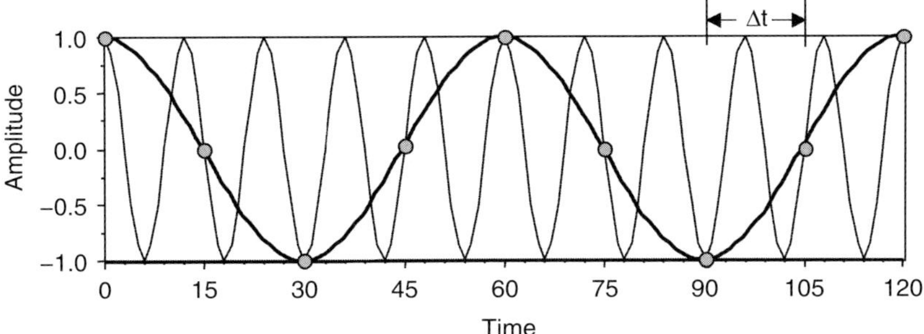

Figure 10.6 To properly resolve the plotted sinusoid, we need more than two data points per period of oscillation. Here, the period of the sinusoid is 12 hours, and we are sampling every 15 hours (light grey dots). As seen, those samples appear as a much longer period oscillation, repeating every 60 hours. Therefore, it looks like a lower-frequency sinusoid (of T = 60), adding to any power already existing at that frequency.

once every 6 hours. The 6-hour sampling (two points per cycle) will provide the Nyquist frequency. In Figure 10.6, $y(t)$ is sampled with $\Delta t = 15$, and as a consequence of this **undersampling** the discretized version, f_t, *appears* as if it were a cosine with period $T_a = 60$ (bold line). Only when $\Delta t \leq 1/2f_c$ (i.e., when $\Delta t \leq 6$), is the sampling interval sufficient to prevent this frequency from appearing as a lower frequency. Note that this effect of the higher-frequency component looking like a lower-frequency component is a direct consequence of the even sample interval Δt. If $y(t)$ had been sampled at *uneven* intervals of t, this relationship would not hold, but then the Fourier harmonics would not be orthogonal so you could not utilize the Fourier transform, and the harmonics would not even be a basis for the time series. As suggested graphically above, frequency components that are unresolved manifest as lower-frequency components (hence the word "alias"). In fact, every frequency, f_a, not in the range resolved by the sampling interval

$$0 \leq |f| \leq |f_N| \qquad (f_N = 1/2\Delta t) \tag{10.6}$$

has an alias that does lie in that range. This alias is known as the ***principal alias*** *of the unresolved frequency component.*

The principal alias is computed by considering how aliasing is actually introduced in the sampling process. For this, recall that the sampling process involves multiplying the continuous time series, $y(t)$, by the Dirac comb, which consists of a series of unit impulses spaced Δt apart. The convolution theorem tells us that this multiplication in the time domain is equivalent to convolution in the frequency domain, or $F(f)*\Delta(f)$. As shown previously when $\Delta t \leq 1/2f_c$, the replicated $F(f)$ do not overlap in the convolution of $F(f)*\Delta(t)$, as shown in Figure 10.4A.

Now consider the case where $f_c > f_N$ (the cutoff frequency is higher than the Nyquist frequency), shown in Figure 10.4B and highlighted in Figure 10.7. In this case, the sampling interval is too large, so the impulses in $\Delta(f)$ that are spaced $1/\Delta t$ apart become too close together and the replicated $F(f)$ now overlap. In the region of overlap, the convolution is the sum of the overlapped parts (recall the linear superposition rule of

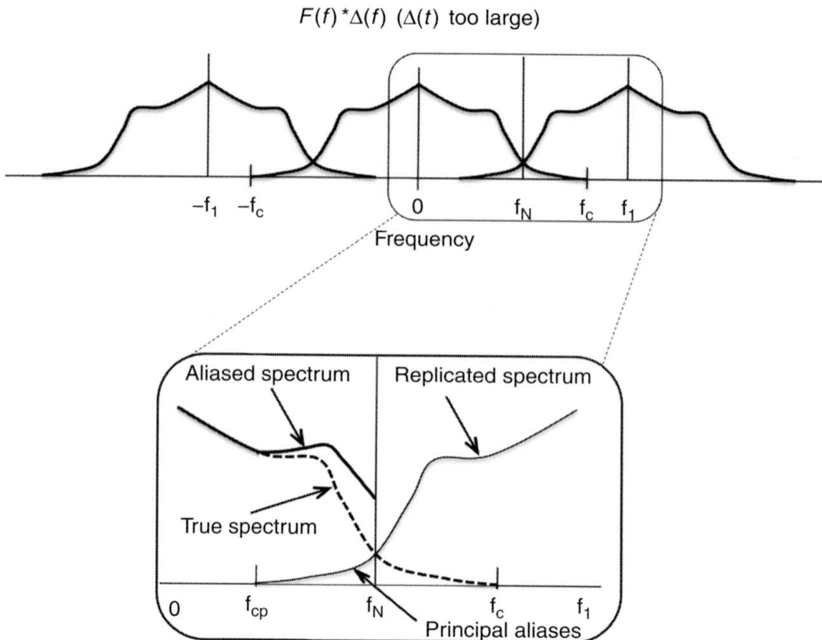

Figure 10.7 For undersampling (Δt is too big, such that $\Delta t > 1/2f_c$), the replicated true spectrum overlaps itself as shown in Figure 10.4B and repeated here in the top panel. The lower boxed inset shows the implication of this overlap – a portion of the replicated spectrum overlaps the true spectrum within resolved frequency range $-f_N$ to f_N and adds to true spectrum, causing aliased portion of spectrum.

convolution). The principal aliases are here seen to be the overlapped portion of the true spectrum from the replicated F(f) centered about f_1. These principal aliases add to the true spectrum in the resolved frequency range ($0 \leq |f| \leq f_N$), resulting in a distorted (i.e., aliased) spectrum – the sum of the true spectrum and principal aliases. In this example, the aliasing obscures the true spectrum over all frequencies higher than f_{cp}, the principal alias of f_c.

Given this source of aliasing, it is seen (see Figure 10.8) that the overlapped region or principal alias is equivalent to "folding" the true spectrum about f_N back into the resolved frequency range ($0 - f_N$).

So, if a frequency component in y(t), f_a, lies outside the resolved range given by (10.6) (i.e., $0 \leq |f| \leq |f_N|$), it will have a principal alias, f_{ap}, given by

$$f_{ap} = f_N - |f_N - f_a|. \tag{10.7}$$

This describes the folding operation depicted above. Because of this nature of aliasing, the Nyquist frequency is often called the **folding frequency**.

In the case of extreme aliasing (Figure 10.9), where $f_N - |f_N - f_a| < f_0$, the principal alias of the high cutoff frequency, f_{cp}, is less than zero and will fold back into the resolved range by folding against f_0, as shown in Figure 10.9. In this case, the entire spectrum within the resolved range is aliased (distorted by the addition of principal aliases to the true spectrum).

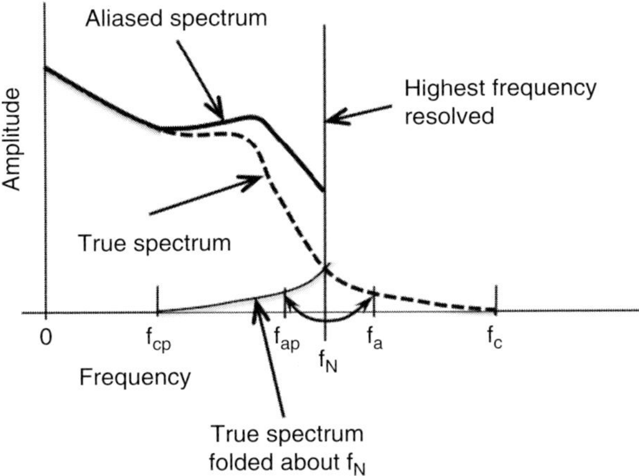

Figure 10.8 Detail of the overlapped region from Figure 10.7, showing that the aliased portion of spectrum is equivalent to taking true spectrum and folding it back about f_N and adding that folded-back portion to the existing power already in the folded-back region to give the aliased spectrum.

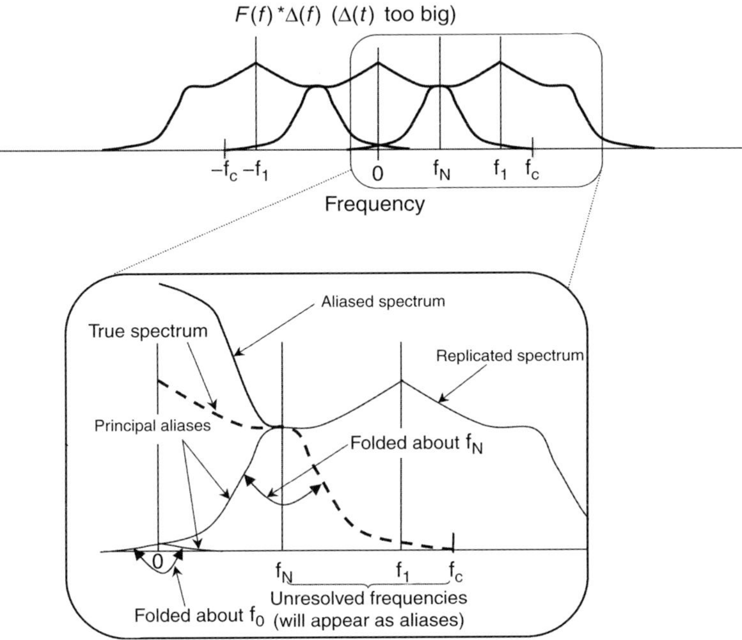

Figure 10.9 Severe undersampling leads to the principal aliases being folded about the Nyquist frequency (f_N), and that fold overlapping the zero frequency (f_0) and folding back into the resolved frequency range about f_0.

Here, in order to compute the principal aliases, the folded part of the spectrum must be folded about f_N, and again about 0.

In either of the above aliased cases, if we now multiply $F(f)*\Delta(f)$ by a rectangular function of width $2f_c$, as required to isolate the transform of F(f) (shown previously in

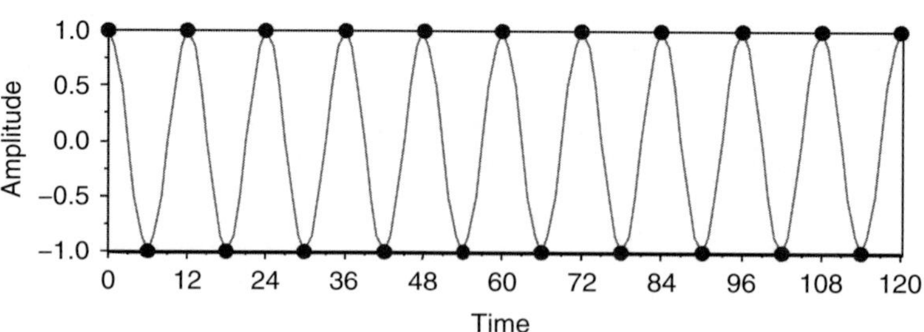

Figure 10.10 Pure cosine with zero phase sampled at $1/2f_c$; i.e., at the Nyquist sampling rate of two data points per cycle. In this case with no phase, the cosine is represented exactly.

Figure 10.5), we get a distorted spectrum that is not representative of F(f) due to the overlapped portions. Therefore, the continuous function cannot be uniquely recovered from the sampled series because the sampling interval Δt was too large and introduced aliasing. The latter distorts the true spectrum, preventing recovery of the continuous time series from the aliased, discrete one.

Sampling at the Nyquist Frequency

The sampling theorem tells us that aliasing is avoided if $\Delta t \leq 1/2f_c$. Now consider the case of sampling exactly at $\Delta t = 1/2f_c$, the highest frequency component in the continuous time series.

If the data is a pure cosine (zero phase, so no sinusoid component), sampling at $\Delta t = 1/2f_c$ gives sample points as shown in Figure 10.10. Recall that the Nyquist frequency, only present in a Fourier transform if the time series has an even number of data points, is $f_{n/2}$. At this frequency, the $b_{n/2}$ Fourier coefficient is zero. The $a_{n/2}$ term represents the $\cos(n\pi/T) = (-1)^{(n)}$, so $a_{n/2} = (-1)^{(n)}$. Therefore, at the Nyquist frequency, the cosine term fitting this highest frequency component in the data oscillates between the maximum and minimum values of the cosine. Since there is no sine term ($b_{n/2} = 0$), there is no phase to the fitted cosine. In this respect, the true cosine is exactly reproduced. Now consider if the highest-frequency component in the data is a cosine with 45° phase. Then sample at $\Delta t = 1/2f_c$. In this case (Figure 10.11), this frequency component is sampled at the same times as previously (Figure 10.10), but in this case those times are at points with amplitude of ±0.707. These points are fit with a pure cosine (i.e., one without the actual 45° phase) with an amplitude of 0.707. Therefore, the fitted cosine does not reproduce the actual phase ($\varphi = \tan^{-1}(b/a)$) of the true cosine, resulting in a smaller amplitude ($A = (a^2 + b^2)^{1/2}$) than is present in the original phenomenon being observed (though not sampled), because, as shown in §9.3.3, the sine coefficient is $b = 0$ at the Nyquist frequency.

Finally, consider if the highest frequency component in the data is a cosine with a 90° phase (i.e., a pure sinusoid). Then, sampling at $\Delta t = 1/2f_c$, again at the same times as previously, we see in Figure 10.12 that we sample exactly at the zero crossing points, and therefore this highest frequency component is not represented at all in the sampled

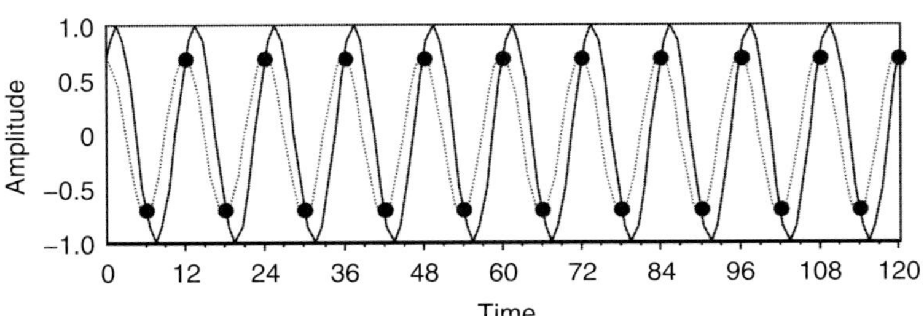

Figure 10.11 Again, sampling at the Nyquist sampling rate, but in this case the cosine has a 45° phase, and since Nyquist has no sine term (*b* coefficient), phase is not captured, and the dashed line fitting discrete points shows the proper frequency cosine but the wrong amplitude.

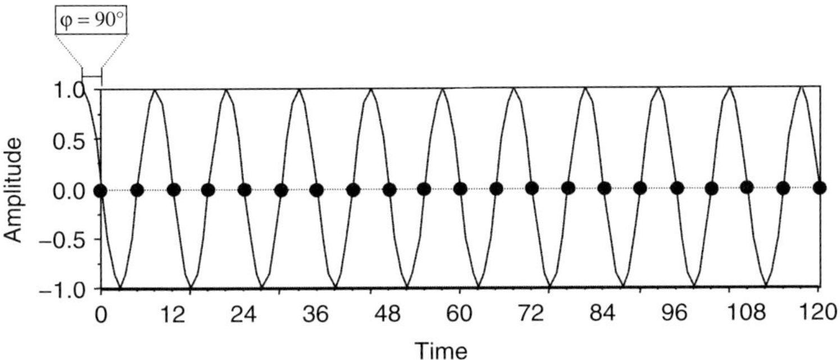

Figure 10.12 Again, sampling at the Nyquist sampling rate, but in this case the highest frequency is a pure sine (cosine with 90° phase), and since Nyquist has no sine term (*b* coefficient), no waveform is captured.

waveform. This is consistent with the $b_{n/2}$ coefficient being zero – in fact this is *why* the $b_{n/2}$ coefficient is zero. This component is fit with a cosine of zero amplitude (i.e., the $a_{n/2}$ coefficient has a value of 0).

Therefore, while using a sampling interval of $\Delta t = 1/2f_c$ avoids aliasing, as shown above, the sampling interval must actually be smaller than this, i.e., $\Delta t < 1/2f_c$, in order to fully resolve the highest frequency component in the original (band-limited) continuous process being sampled. *That is, you should sample the waveform at a sampling interval that is smaller than* $1/2f_c$.

How to Avoid Aliasing

1) Sample at an interval of $\Delta t < 1/2f_c$, where f_c is the highest frequency component present in the continuous-time series being sampled.
2) Filter the time series *before* sampling (to force the time series to a band-limited range, with a known highest-frequency component f_c, which can be resolved by the desired sampling interval); this avoids aliasing, but completely eliminates the unsampled frequencies.
3) Focus only on that portion of the spectrum that is not aliased.

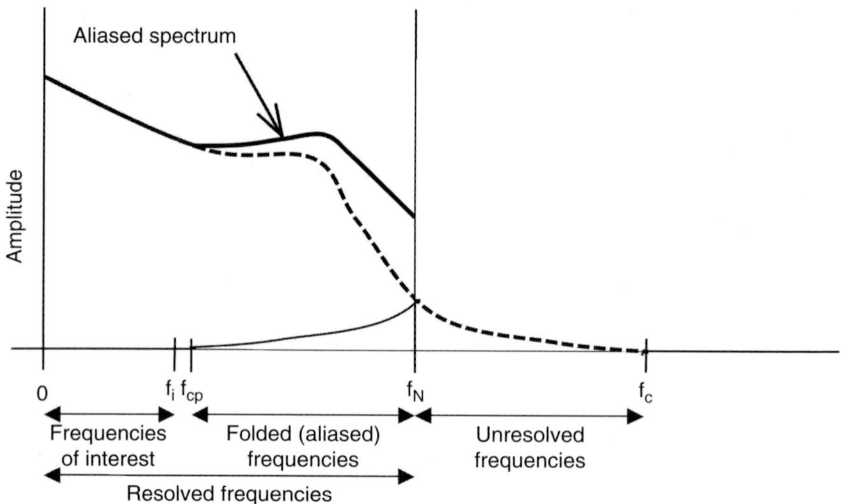

Figure 10.13 One means of avoiding aliasing is to pick a sampling rate (defining the folding frequency, $f_N = 1/2\Delta t$) that will leave the frequency range of interest untouched by the folded frequencies, as shown here.

4) Sample the process simultaneously at more than one Δt so that each series has a different f_N, which allows for the separation of the true spectrum and principal aliases.[2] This is reasonable in concept, though noise in the system may limit the success of this approach.

In general, one must be careful either to choose Δt so that the amplitudes of frequency components higher than f_N and smaller than f_c are either small or nonexistent or that f_N is high enough so that the aliased part of the spectrum is at frequencies higher than those of interest (as shown in Figure 10.13).

Therefore, you want to choose Δt so that f_i, the highest frequency of interest, is lower than the lowest aliased (folded) frequency, f_{cp}, or

$$f_i < f_N - (f_c - f_N) = 2f_N - f_c, \tag{10.8a}$$

or, since $f_N = 1/2\Delta t$,

$$f_i < 1/\Delta t - f_c. \tag{10.8b}$$

So, in words, the sampling theorem indicates

1) You can recover $y(t)$ completely from f_t if you can recover F(f) completely from F(f) $*\Delta(f)$. In other words, sampling of $y(t)$ results in replicating F(f), centered about the impulses in $\Delta(f)$.

[2] With a different Nyquist frequency for each sampled version of the series, only the unaliased portion of the spectrum will be consistent in each of the aliased spectra, allowing one to uniquely solve (or in a least-squares sense) for the true and aliased components.

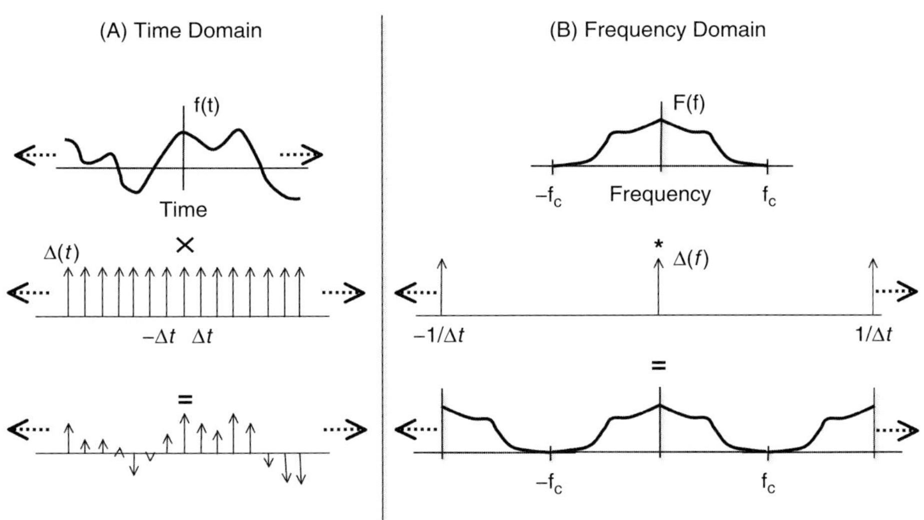

Figure 10.14 Summarizing the sequence of steps for sampling a continuous time series. **(A)** shows mathematical operations in the time domain for discretizing continuous time series (you perform this in the real world by sampling the process at discrete time intervals) giving f_t; **(B)** comparable operations in frequency domain, recalling that multiplication (x) in one domain is convolution (*) in the other. Shown here is $\Delta t < 1/2f_c$ giving a replication of the true continuous spectrum that can be isolated by multiplying a single true spectrum from the final step.

2) If Δt (the sampling interval) is $< 1/2f_c$, then F(f) *will not* overlap itself in $F(f)*\Delta(f)$. F(f) is then completely recovered by "cutting out" F(f) by multiplication with a rectangular function (Figure 10.4A).
3) If $\Delta t > 1/2f_c$, F(f) *will* overlap itself in $F(f)*\Delta(f)$ and we cannot isolate F(f) exactly. Therefore, we cannot recover $y(t)$ from the *undersampled* f_t (Figure 10.4B).

Often a continuous phenomenon might be recorded as a continuous signal that is then sampled at discrete intervals. In this case, since all instruments have a finite response time, the continuous recording of the signal will be band-limited if the instrument averages over the sampling interval. Therefore, the sampled version is also band-limited (this is effectively filtering the time series during sampling). Furthermore, as discussed below, all time series except for those representing pure white noise are band-limited as required. However, it is possible that the limiting band is higher than can reasonably be sampled. In such cases, the above considerations must be taken into account. Also, nature sometimes serves to filter a process or recorded series (e.g., sediment mixing) so that the data are effectively filtered before sampling, which tends to create a band-limiting, or at least reduce f_c further.

10.2.3 Conceptual Understanding of Sampling Theorem

Our stroll through the sampling theorem shows why the theorem works. Conceptually, the theorem can be explained by considering the nature of band-limited time series. In particular, in band-limited time series there is always some *correlation* between nearby values, so consequently we do not need to know all of the values in the series to uniquely define it. That is, since there is a correlation between neighboring values, neighboring values are related to one another, so you don't need to sample all of the values in order to uniquely define the time series.

Consideration of an autocorrelation function, $R_{ff}(\tau)$, makes it most clear why white noise is the only series that is *not* band-limited. For all time series, at the origin (lag = 0), there is a perfect correlation of the time series with itself, so $R_{ff}(0) = 1$. At higher lags, $R_{ff}(\tau) \neq 0$, reflecting a correlation between values in the time series offset by τ sample points. Only in the case of white noise, where $R_{ff}(\tau) \approx 0$ for all lags away from the origin, is this not true. In this latter case, $|F(f)|^2 = \sigma_\varepsilon^2$ for all frequencies (i.e., power is constant for all frequencies and thus never goes to zero giving f_c as required here); therefore, a white noise time series is not band-limited and cannot be represented by a discretely sampled version of it.[3]

Restating the above: *all sampled time series are band-limited except for those representing white noise* (but for white noise, we don't have to worry about aliasing anyway, because there are no frequency peaks or valleys to alias).

10.3 Relationship between Discrete and Continuous Transform

We have just seen that if properly done, discrete sampling of a continuous time series does not limit our ability to estimate the true spectrum of the continuous series. However, we must still consider the impact of truncating the infinite-length time series to a finite length. Knowledge of the sampling theorem and the convolution theorem allows us to fully evaluate the relationship between the discrete Fourier transform and continuous Fourier transform. In other words, we can now quantify how the truncation to finite length and discrete sampling impacts our ability to estimate the true spectrum of a random process.[4]

10.3.1 Sampling the Time Series

The *sampling* of an infinite length, nonperiodic continuous time series $y(t)$ with transform F(f) and its influence on the spectrum of the sampled series (as previously shown for the sampling theorem) is summarized in Figure 10.14. While the temporal sampling interval is small enough to prevent aliasing in the example, it is still for the case of an *infinite length* time series. Therefore, f_t must be truncated to a finite length.

[3] This result is directly from the autocorrelation theorem (§9.9.4), and the transform of an impulse (equation (9.86)).

[4] Brigham (1974) provides an excellent pictorial guide to this relationship, from which some of the present discussion is fashioned.

10.3.2 Truncating the Time Series

So far, we are encouraged because with proper sampling we can avoid aliasing, but the other aspect of sampling the continuous time series is that we are only sampling a finite length, which is equivalent to truncating the infinite series, leading to convolution of the true spectrum by the transform of the truncation function – uh oh, this will distort the true spectrum that we worked so hard to preserve with proper selection of Δt. The amount of distortion and severity depends upon the truncation function and nature of the spectrum.

The simplest case is to multiply f_t by a gate (or boxcar) function,[5] h(t), of width T centered about t = 0, where

$$h(t) = \begin{cases} 1 & |t| \leq (T/2 - \Delta t/2) \\ 0 & \text{elsewhere} \end{cases}. \tag{10.9}$$

The h(t) and its transform H(f), the sinc function, are shown in Figure 10.15.[6]

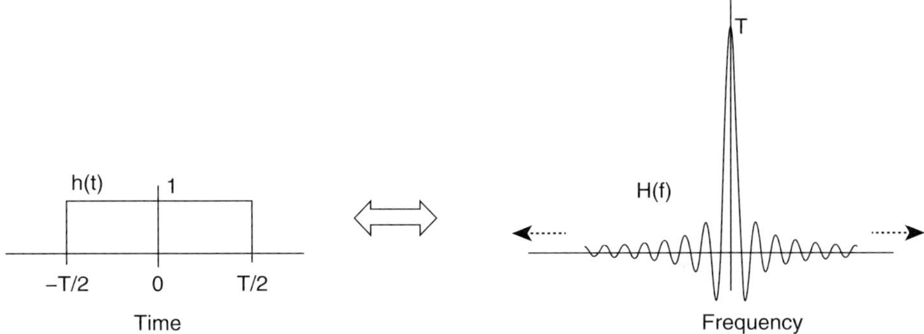

Figure 10.15 Rectangular (gate) function that multiplies continuous infinite time series and its sinc function transform that convolves the true spectrum.

Multiplying the infinitely long continuous time series by the rectangular function is equivalent to convolving the true spectrum by the gate's transform, the sinc function (Figure 10.16).

Unfortunately, even in the absence of aliasing, the true spectrum is now distorted by this convolution, which tends to smear the spectrum as well as introduce ripples (often called **ringing**). This influence due to the truncation process is often referred to as **leakage** because it "leaks" power from one set of frequencies to another (recalling that each convolved point is a weighted average of all of the points in the true spectrum, weights defined by the sinc function). It manifests in the sampled spectrum as ***bias***. Only when $y(t)$ is perfectly periodic over the period of truncation, T, will leakage be avoided – that is, will not introduce distortion into the spectrum.

[5] Note that the rectangular function can be defined in a variety of ways. For example, it needn't be centered about t = 0 and its limits can be written in any manner, so long as its width captures one period – its definition in (10.9) is convenient, though the transform of h(t) is most easily computed by translating the origin so that h(t) = 1 for $|t| \leq \pm T/2$.

[6] This transform pair was established in example of Box 9.5 in Chapter 9.

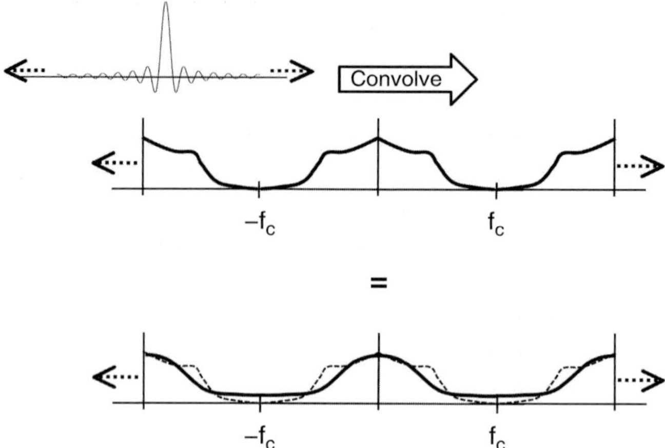

Figure 10.16 Multiplying time series by the truncation function (gate as in Figure 10.15) is equivalent to convolving the true repeated spectrum by transform of the gate, which is the sinc function. This convolution smears features in the true spectrum, some of which can be classified as "leakage" from high-power regions to low-power regions (dashed line in lower figure is the true spectrum; solid line is the convolved spectrum).

If the sinc function were an impulse, there would be no change in the convolved spectrum, since each point in the true spectrum would be replicated. But an impulse in the time domain could only be obtained if the truncation function were a constant (i.e., a straight line that imparts no truncation).[7] You can construct truncation windows that truncate more gently or with larger T, making them more like a straight line, which would yield sinc-like functions with tighter central lobes but, typically, larger side oscillations. There has been quite an effort at realizing the "optimum" truncation function such that the sinc-like function has minimal oscillations and a tight central lobe. Priestley (1981) discusses a number of these, but in doing so notes that the "optimal" function is dependent upon one's definition of "optimal." Acceptable truncation schemes are discussed in more detail in the next chapter when considering the uncertainties associated with estimates of the power spectrum.

While the sampling steps described here have produced a discrete, finite-length time series, the spectrum is still infinite in length (but $F(f) = 0$ for $|f| > |f_c|$) and continuous in frequency, so it too must be sampled and truncated.

10.3.3 Sampling the Spectrum

Mathematically, we obtain the **sample spectrum** by discretizing the continuous spectrum, which is comparable to discretizing the continuous time series. In this case, you multiply the true spectrum by a frequency Dirac comb, $\Delta_f(f)$, where

[7] As shown in Chapter 9, when defining the impulse function in §9.7.2.

$$\Delta_f(f) = \sum_{j=-\infty}^{\infty} \delta(f - j\Delta f), \tag{10.10}$$

consistent with the definition of a Dirac comb in the time domain. Here, the impulses are spaced Δf apart, where $\Delta f = 1/T$, with T being the width of the rectangular function used to truncate the infinite time series in (10.9) (T = length of your sampled time series as always) – so the impulses in (10.10) are placed at the Fourier frequencies for the truncated time series. That is, the continuous time series has been truncated so that it has a sampling period of T. Multiplication of the *transform* of f_t with $\Delta_f(f)$ is equivalent to convolving f_t with the *inverse transform* of $\Delta_f(f)$, given as $\Delta_f(t)$.[8] As developed previously for the time domain case, this inverse transform (recall the symmetry property of Fourier transform pairs) is given by

$$\Delta_f(t) = \frac{1}{\Delta f} \sum_{j=-\infty}^{\infty} \delta(t - j\Delta f^{-1}), \tag{10.11}$$

shown here in Figure 10.17.

This convolution of f_t with $\Delta_f(t)$ in the time domain results in the replication of the finite-length (sampled length) time series, centered about each impulse located at T in the time domain. As shown in Figure 10.17, because the time Dirac comb impulses are separated by a distance of T, the replicated time series will not overlap, which would have caused a problem comparable to aliasing in the frequency domain. This is assured because we sample the spectrum at Fourier frequencies, which are harmonics of 1/T, so the temporal impulses are spaced at $(1/T)^{-1}$. The complete process is shown in Figure 10.18.

This convolution shows why the spectrum must be sampled at the Fourier frequencies (or closer intervals). If $f_l > 1/T$ (i.e., T_l is too short), then the impulses in $\Delta_f(t)$ will be spaced closer together than T, which is the width of f_t. In that case, the replicated f_t will overlap, resulting in an aliased time series (analogous to the aliased spectrum, *but in this case meaning that the harmonics do not interpolate the time series properly, because the improperly defined harmonics did not form a proper basis for the series*). Therefore, in order to avoid this, $\Delta f = f_l \leq 1/T$. This means that we cannot use a fundamental frequency for the Fourier series that assumes a fundamental period for the data that is *shorter* than

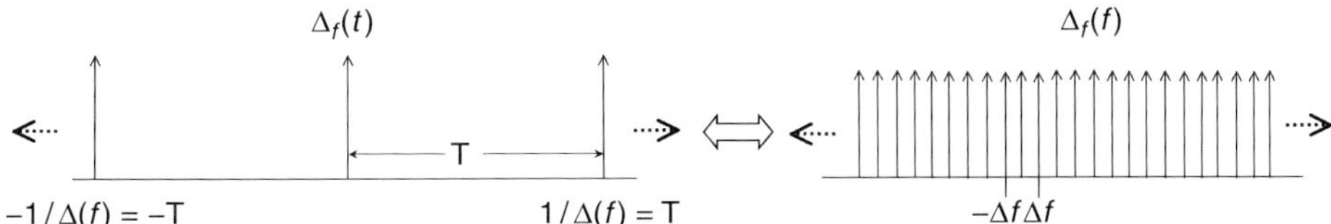

Figure 10.17 For sampling the continuous spectrum, we *multiply* it by the Dirac comb in the frequency domain (**right**), where the impulses are separated by a distance of the fundamental frequency (1/T), giving the Fourier frequencies. This operation is equivalent to *convolving* the discretely sampled time series, f_t, by the transform of the frequency comb, giving the Dirac comb in the time domain (**left**), impulses separated by T.

[8] This notation indicates the frequency comb, Δ_f transformed to the time domain, and thus a function of t, (t); so $\Delta_f(t)$.

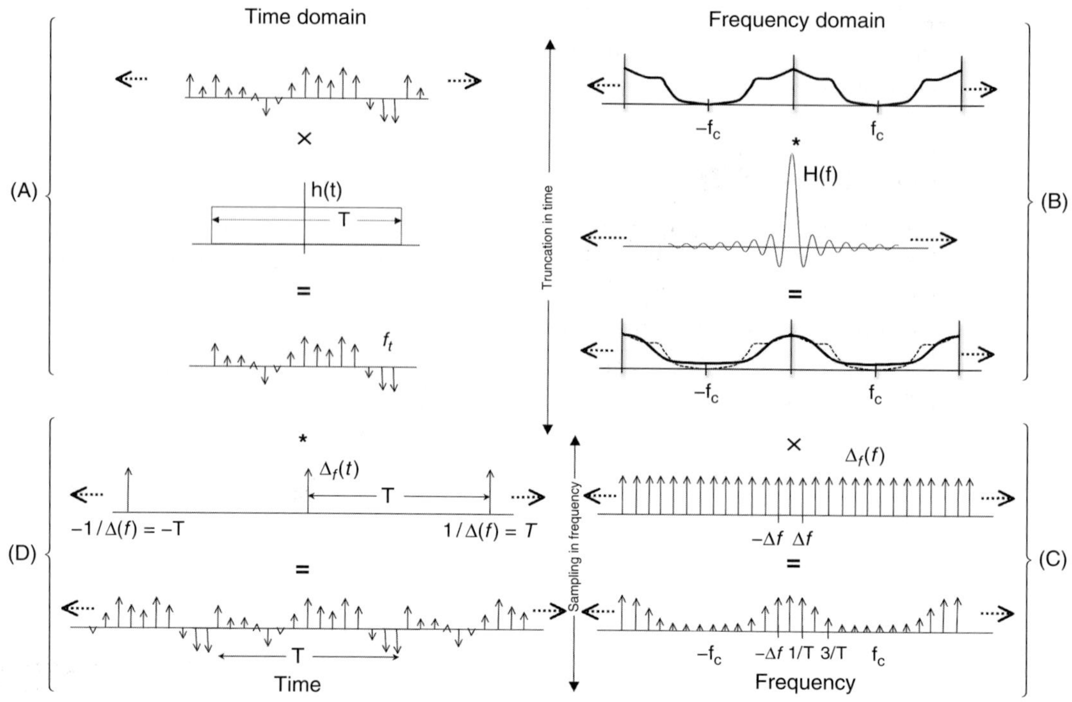

Figure 10.18 This follows the steps accomplished in Figure 10.14. **A and B**: truncation of the time series to a finite length (**A**) and the consequential smearing of the true spectrum via convolution with the sinc function (**B**). **C and D**: smeared spectrum, [F(f)*Δ(f)]*H(f), sampled discretely by Dirac comb, $\Delta_f(f)$, with impulses separated by the fundamental frequency, $f_1 = \Delta f = 1/T$, to get sample spectrum (**C**); corresponding convolution in the time domain (**D**) does not alias time series since the impulses are spaced at T, so no overlap of the finite length time series, f_t.

the length of the sampled time series. However, we can use a fundamental period that is *longer* than the sampled length of the time series – e.g., you can add zeros to the end of your discrete time series to extend it, making a larger extended T_e, resulting in a finer frequency sampling, $\Delta f = 1/T_e$, useful to resolve spectral peaks that are closer together than the Fourier Δf at 1/T.

10.3.4 Resulting Discrete Spectrum and Time Series

Because both the time and frequency domain functions have been convolved with a series of impulses – f_t with $\Delta_f(t)$ in time and F(f) with Δ(f) in frequency – both functions are *periodic* over the spacing of the impulses in the convolutions, or in *n* discrete values. The *n* discrete spectral values, spaced at Δf = 1/T, lying between $-f_N$ and $+f_N$ and spanning a frequency range of 1/Δt ($2f_N$ going from the negative to positive Nyquist, hence 2/2Δt), are representative of the entire periodic spectrum. Similarly, the *n* discrete data points, spaced at $\Delta t = 1/2f_c$, lying between −T/2 and +T/2, and spanning a time range of 1/Δf, are representative of the entire periodic time series. So, the *n* time samples and *n* frequency samples represent one period of the time series and spectrum.

Given the operations of sampling and truncation in time and of sampling in frequency, the *transform* of a discrete, finite-length time series is related to the true spectrum of the continuous time series from which it was sampled by

Box 10.1 Periodicity in the Spectrum

The periodicity in the spectrum indicates the presence of higher and higher frequencies in the data. Consider any one frequency component, $\cos(2\pi j/n\Delta t)$, in the spectrum. As shown above, the periodicity in the spectrum suggests that the amplitude and phase of this jth harmonic component is repeated every $1/\Delta t$ frequency units. However, at the discrete sample times, $t_i = i\Delta t$, $i = 0,1,2,\ldots$, the periodicity of $1/\Delta t$ in frequency is shown by

$$\begin{aligned}\cos(2\pi f_j t_i) &= \cos\left[2\pi\left(f_j + \frac{1}{\Delta t}\right)t_i\right] \\ &= \cos\left[2\pi\left(\frac{j}{n\Delta t} + \frac{1}{\Delta t}\right)i\Delta t\right] \\ &= \cos\left[2\pi\left(\frac{ji}{n} + i\right)\right] \\ &= \cos\left[\left(\frac{ji2\pi}{n} + i2\pi\right)\right] \\ &= \cos(i2\pi)\cos\left(\frac{ji2\pi}{n}\right) - \sin(i2\pi)\sin\left(\frac{ji2\pi}{n}\right) \\ &= \cos\left(\frac{ji2\pi}{n}\right).\end{aligned}$$

Multiplying numerator and denominator by Δt then gives

$$= \cos(2\pi f_j t_i).$$

In other words, at the discrete sampled (integer) positions, $t_i = i\Delta t$, in time, the cosine of frequency f_j takes on values identical to those it does for the much higher frequency, $f_j + 1/\Delta t$. Therefore, the periodicity in frequency simply indicates that at the sampled positions in time, these higher-frequency cosines can also describe the sampled data set, as can any other cosines at frequencies $f_j \pm i/\Delta t$. That is, as long as we examine *any* one of the replicated sample spectrums, we will capture f_t.

This is equally true of the time series. That is, we can just as well sample the time series at positions $y(t + iT)$ to get the values located at $y(t)$. Due to the periodicity (assumed) in the data, these values are indistinguishable. In practice, we simply consider that the sampled spectrum represents one complete period of frequencies closest to the frequency origin (i.e., for the lowest values of f).

In both cases, we avoid this periodicity by truncating with a rectangular function in the respective domain, to recover the original finite-length series.

$$F_{f_t}(f) = \{[F(f) * \Delta(f)] * H(f)\}\Delta_f(f). \tag{10.12}$$

If the operations in (10.12) are written out explicitly, it is seen that this is the exact form of the discrete Fourier transform of f_t.

Equation (10.12) and the preceding discussion also show that a discrete Fourier transform can differ from the continuous one by two effects:

1) aliasing (from time domain sampling)
2) leakage (from time domain truncation)

It shows why the discrete time series is treated as if it were periodic over its sampled period, T, regardless of whether it is or isn't. The degree to which this is not true (i.e., that the time series is not periodic over T) influences the degree of distortion (bias) due to leakage.

10.3.5 Leakage

Only for time series that are *exactly periodic* over the sampled period, T, is leakage avoided. The reason for this is seen by examining the sinc function, which is convolved with the spectrum when truncating the time series.

Leakage is introduced when f_t, the discrete time series of infinite length, is truncated by multiplication with the rectangular function, h(t) of width T. This corresponds to a convolution of the spectrum, F(f)*Δ(f) (F(f) replicated at $1/\Delta t$ intervals), with the transform of the rectangular function, H(f), a sinc function. The sinc function has the form (derived previously)

$$H(f) = T\frac{\sin \pi f T}{\pi f}. \tag{10.13}$$

If the data are truly periodic over T, then the true spectrum, F(f) consists of a series of impulses at the frequencies $f = jf_1$, where $f_1 = 1/T$, the fundamental frequency, and $j = 0,1,2,\ldots,\infty$, the harmonics of this fundamental frequency. This spectrum would then contain one impulse for each pure (harmonic) cosine contained in the time series. The sampled spectrum is [F(f)*Δ(f)]*H(f). The order of convolution is unimportant, as stated in equation (9.36), so instead of first considering the convolution of the Dirac comb, Δ(f), on the true spectrum, we can consider it on the sinc function. In that case, the convolution Δ(f)*H(f) replicates H(f) at each impulse in Δ(f) centered $f_1 = 1/T$ apart. So, at $f = jf_1 = j/T$, $H(f) = T\sin(j\pi)/j\pi = 0$ at all values of j except for $j = 0$, where $H(f) = T$.

In other words, at the Fourier frequencies, jf_1, the overlap in the replicated sinc functions corresponds to the zeros of H(f), so there is no distortion introduced at the Fourier frequencies due to this convolution. Furthermore, for a periodic series, the non-Fourier frequency components must also be zero, since they are the ones that are not periodic over T. Consequently, *for truly periodic data, truncation in the time domain (over the true period) does not introduce any leakage, so the discrete spectrum is identical to the continuous spectrum (assuming no aliasing has been introduced).*

Minimizing the Effects of Leakage

Statistically, leakage manifests as a bias (addressed in Chapter 11). Conceptually, *some* of the influence of leakage can be considered in context of the extent to which periodicity of your time series is nonperiodic. If it is perfectly periodic, convolution with the truncation window does not cause any distortion, as just shown. Conversely, if your series is truncated at some location that is highly nonperiodic, the leakage can be severe. Consider a pure cosine (or two), represented as lines (impulses) in the true spectrum. When convolving the spectrum with the sinc function, these impulses effectively capture (replicate) the sinc function centered at their frequencies and scaled to their large amplitude, with side lobes appearing in high frequencies (Figure 10.19).

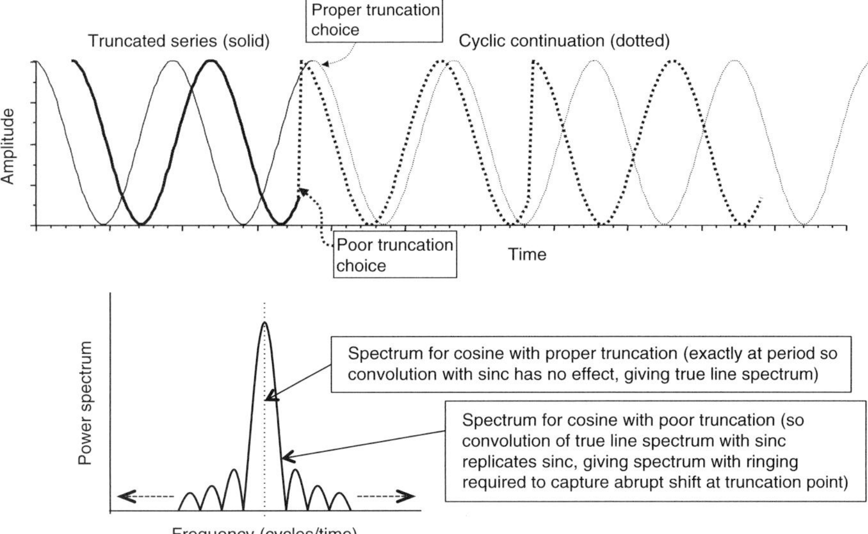

Figure 10.19 Example of leakage resulting from poor truncation of a pure cosine. **Thin** time series is truncated at the actual period, resulting in a line spectrum at the proper frequency (thin impulse) of the pure cosine. **Thick** time series shows truncation in the middle of a cycle; its spectrum is essentially the sinc function amplitude squared, having been replicated by the true spectrum impulse when the sinc function was convolved with the true spectrum during the truncation of the time series.

The side lobes in poorly truncated (thick line of Figure 10.19) series contribute the higher frequency components (ringing) required to capture the abrupt discontinuity at the end of the truncated series, keeping in mind that the Fourier series is an exact fit to your data whether your data represent a subsection of an infinitely long time series or are the entire finite-length time series without truncation; the problem is that this spectrum producing the exact fit is not representative of the true spectrum for the process. Steps can be taken to minimize distortion introduced by this leakage. Specifically, it is clear that the truncation interval should be "adjusted" (i.e., collect more data or throw out some of the existing data) in order to minimize the discontinuity between the first and last data points, at t_1 and t_n, or to most closely truncate at the period of a dominant sinusoid in the time series. It can also be treated by employing different truncation windows, known as "tapers." In practice, this simple precaution can keep this particular influence of leakage to a fairly small distortion. This is not unexpected, since ~90 percent of the power of the sinc function lies within its main lobe, so the power introduced by the side lobes is fairly small even in the worst cases, though the smearing by the main lobe can be appreciable.

Unfortunately, even when the ends are well behaved (have minimal discontinuity), leakage still influences the spectrum for nonperiodic data. This is because convolution with a sinc function is smearing the spectrum with a weighted running mean. If the spectrum has fairly steep slopes in it, when the sinc function is convolved over the slopes, the running mean tends to flatten the slope. Therefore, power in the spectrum is "leaked" from the frequencies with the higher power to those with the lower power and vice versa. The spectrum of the truncated data series is therefore flatter than that of the true spectrum. Along these same lines, power from the frequency components not represented by the Fourier frequencies is leaked into those components that are resolved (thus preserving the total variance of the series).

Data Tapers

One can minimize the side lobes of the sinc function convolving the spectrum (which introduces the distortion during the convolution). This is accomplished by truncating the time series with a more gently sloping rectangular function (called a **taper, fader, window**, and other names). That is, use a smoother truncation function, leading to a sinc-like function with better convolution characteristics.

The most common taper forms look like those schematically represented in Figure 10.20. Others include Parzen (similar to Hanning, but decays quicker and more steeply), Hamming (like Hanning) and Bartlett–Priestly (quadratic with "best" properties satisfying specific criteria).

All of these tapers have a transform that is less oscillatory than the sinc function, but broader. Therefore, multiplication of the time series with one of these will result in convolution with a function in the frequency domain that smears peaks more than the sinc function but reduces the rippling effect far away from spectral peaks. In general, the difference between the various tapers in practice is fairly small and often imperceptible in casual use, though in special situations one may need to use tapers of specific shapes to minimize certain undesirable influences.

Multiplying by a taper essentially eliminates the discontinuity across the ends of the truncation interval. However, dampening the f_t near the ends does not eliminate the effects of the main lobe on smearing the spectrum, so leakage is not eliminated. Also, the transform will represent the time series after its having been tapered, which is different from the untapered time series. That is, if the Fourier series is reconstructed (i.e., the inverse transform computed), the time series that has been fit by the sines and cosines has the amplitudes reduced near the ends, so this is not the same as the straight truncated series. This usually is only important when comparing series of different lengths to one another.

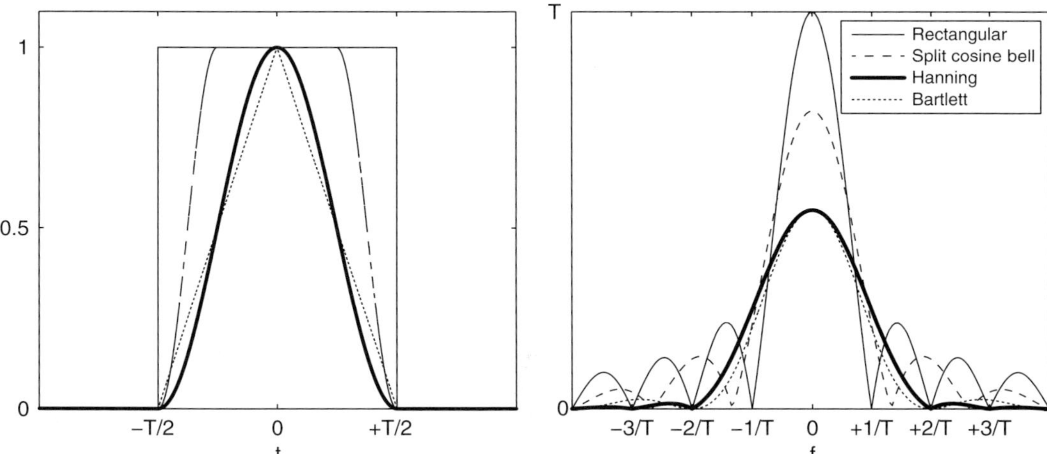

Figure 10.20 Schematic of different common tapers and their transforms. Those tapers with more gentle slopes to end have smaller side lobes and a lower but broader central lobe. The latter tends to cause worse smearing in close proximity to the portion of spectrum being convolved, but leaks less power from distant frequencies from the smaller side lobes.

Leakage introduces bias in the sample spectrum, and attempts to reduce this are discussed with statistical implications of the Fourier analysis in Chapter 11.

10.4 Other Sampling Considerations

Based on the preceding discussions, several points can now be summarized regarding the transformation from the time domain to the frequency domain. In particular, the *length* of the discrete time series defines (actually, limits) the period of the fundamental Fourier frequency, f_1 – the *lowest* frequency that can be resolved in the frequency domain. Since all other Fourier frequencies are integer multiples of this fundamental frequency, $f_j = jf_1$, the discrete frequency interval, Δf, in the frequency domain is equal to $f_1 = 1/T$. Therefore, the *length of the time series also dictates the frequency resolution in the sample spectrum*. Note however, that the period T is the period of the transformed time series. Thus, if you have padded zeros to the end of the time series in order to achieve some power-of-2 sample values (in order to use a fast Fourier transform), then T is the length of the time series including the padded zeros.

Similarly, the *sample spacing* in the time domain, Δt, dictates the *highest* frequency that can be resolved in the frequency domain. That is, this highest frequency is the Nyquist frequency (for an even number of sample points), $f_N = 1/2\Delta t$. Therefore, if you pad your Fourier transform with zeros before inverse-transforming back to the time domain, you will decrease Δt in the inverse-transformed time series. This is exactly analogous to padding the time series with zeros before transforming to decrease the Δf frequency sampling interval. In this respect, padding the frequency domain series with zeros allows you to interpolate the data (using a sine and cosine interpolant) to a finer sampling interval in time, since $\Delta t = 1/2f_N$.

So, ***length of time series (T) sets lowest frequency and frequency resolution (1/T) in the frequency domain; sampling interval sets highest frequency resolved (1/2Δt), controlling degree of aliasing***.

10.5 Take-Home Points

1. Aliasing and leakage are fundamental problems associated with sampling of a continuous time series at discrete sampling intervals Δt. Aliasing occurs if Δt is too big (according to the Nyquist frequency it resolves; $f_N = 1/2\Delta t$) to resolve the highest frequency. In that case, the unresolved frequency components will manifest as lower frequency components in the spectrum. This can be avoided with proper sampling.
2. Leakage is the consequence of multiplying an infinitely long time series by a truncating function to a finite length (i.e., your sampled time series). According to the convolution theorem of Chapter 9, this multiplication in time is the same as convolution in the frequency domain. So the true spectrum is convolved with the Fourier transform of the truncation function, which tends to smear and pull in power from distant frequencies, moving (leaking) power from high points to low, low to high, etc., causing leakage. This can be minimized by making sure the time series is truncated such that the first point in the series is consistent with being the next point after the last point (do this by removing some additional points if necessary, or applying a taper that accomplishes this by diminishing the end points to zero).

10.6 Questions

Pencil and Paper Questions

1. a. In two parallel columns (one in the time domain, the corresponding steps in the frequency domain in the second), show through schematic diagrams the sequence of events from sampling the infinitely long continuous random process with true spectrum, through to truncation of the infinite length to your finite-length sample. Make figures big enough to show that you understand each step. Within the sequence, show where the following are introduced:

      1. aliasing
      2. leakage

   b. How do you eliminate or reduce the above two effects (be explicit)?

Computer-Based Questions

2. State conceptually why unevenly spaced data cannot produce aliased spectra. And, if that is the case, why should we not just perform a simple **Ax** = **b** solution for the a_i and the b_i? So, generate a time series at a super high temporal resolution (Δt is very, very small). Include in the series several cosines lower than the Nyquist frequency (that will be well resolved) and include a high amplitude cosine that has a harmonic higher than the Nyquist frequency (i.e., $f_a < 1/2\Delta t$) that should be aliased. Then subsample that series at an irregular time interval and solve for the sine and cosine coefficients via $\mathbf{x} = \mathbf{A}^{-1}\mathbf{b}$, and produce the amplitude or power spectrum. Then subsample the original series at a Δt comparable to the average Δt in the unevenly sampled series, so they will give the same harmonics, and solve again using **Ax** = **b**, *and* via the Fourier transform. Are the three spectra the same? Which ones are the same and which is not the same? Explain the answer (*hint*: work out the error covariance matrix in the **Ax** = **b** cases and examine the off-diagonal elements).

11 Spectral Analysis

11.1 Overview

So far, we have considered the technical aspects and consequences of fitting a series of harmonic sines and cosines to a set of discrete data – an interpolation problem known as Fourier analysis. In cases where you are simply interested in examining a time series in the frequency domain (i.e., decomposing it into frequency components) or performing interpolation, Fourier analysis provides all of the procedural information required to accomplish that goal. The spectrum of sine and cosine amplitudes produced in a Fourier analysis will exactly interpolate the time series being analyzed. However, there are times you need to best estimate the ensemble from which the time series was drawn (i.e., estimate what the most likely realization will look like when created by the generating process). For nonsequential data, we did this by forming the probability density function (PDF, or its discrete analog, the PMF) – that is, how the random variable is distributed as a function of actual value. The time series equivalent to this is the **power spectral density function, PSD (the Fourier transform of the autocovariance function)** – that is, how the time series is distributed as a function of frequency. Here too, we will make the claim that our realizations (the time series we have collected) are representative of the true ensemble, but in this case, after considering the uncertainties that enter a raw Fourier interpolation, we must apply a means of making a better, more reliable estimate of the most representative time series realization and true spectrum from which it was drawn. **Spectral analysis** is mainly concerned with methods for estimating the PSD.

Here I adopt a nomenclature such that Fourier analysis is the mathematical interpolation process, while spectral analysis concerns itself with the statistical aspects of the Fourier analysis (which is not necessary if the time series is deterministic, since there is no uncertainty in a deterministic series).[1] Time series are data, they are subject to uncertainty, and therefore the general statistical analysis of time series, which is dominated by spectral analysis, is referred to as "time series analysis." Time series analysis is often compartmentalized into (1) analysis in the time domain (e.g., via serial products and parametric modeling, as discussed later) or (2) analysis in the frequency domain (spectral analysis).

[1] This definition is not universally used, though I have never seen the reverse definition used.

This chapter presents the classic methods of spectral analysis, sometimes referred to as Blackman–Tukey spectral analysis, since their book (Blackman and Tukey, 1958) initially presented much of this material. A brief overview of some of the newer variants of spectral analysis is also presented.

Ergodicity plays more than one role in spectral analysis: (1) it allows us to get away with a complete description of a typical realization by merely describing the mean, variance and acf, and (2) it plays a major role in allowing a method for reducing the variance (uncertainty) in the sample spectrum. Without ergodicity, we are forced to analyze subsections of the time series over periods of time for which stationarity might be reasonably approximated (= evolutionary time series analysis), or via a relatively new methodology known as wavelet analysis (not covered within).

11.2 Noise in the Spectrum

Without noise in your time series, there would be no need for much of this chapter, but since a total lack of noise is highly unlikely, we start by examining how noise manifests in the spectrum.

A *stationary* random process is typically characterized by its mean and acvf (which gives its variance as well). If the random process is a Gaussian process (i.e., the data points in the process are distributed jointly with a multivariate Gaussian PDF), then these moments completely describe the PDF.[2] For non-Gaussian processes, they represent the most fundamental moments used to describe the process. The acvf is what distinguishes a random process time series from a simple random variable, since both may have the same PDF, but the acvf describes the fundamental information of how the time series data points covary in time for the random process.

The autocovariance theorem states that the Fourier transform of the acvf gives the PSD function directly.[3] Therefore, a random process can be described directly from the PSD (which includes the mean of the series – the 0th frequency – and its variance, by Parseval's theorem). However, since the acvf represents a function of a random variable, it too is a random variable, as is its Fourier transform (for the same reason). Therefore, estimating the true PSD becomes a problem in statistical estimation. That is, as with all other statistical moments, we make estimates of the true distribution parameters. The estimates are based on estimators designed to give the best approximation of the true value from the observed (sampled) values. In the case of spectral analysis, we wish to estimate the true PSD using one or more realizations drawn from the ensemble.

First consider the "raw" power spectrum computed from the discrete Fourier transform of perfect sinusoids of the same fundamental frequency (1/T). Graphically, the results appear as a **line spectrum** (Figure 11.1) – a graph in which the power (amplitudes

[2] PMF for each point is Gaussian, and covariance between any two points in series is Gaussian (the probability of getting a specific value at time t_j given the value at time t_i follows a Gaussian distribution).

[3] Both the acf and acvf transform to the PSD, but I prefer the acvf and will use this for ensuing discussions of the PSD.

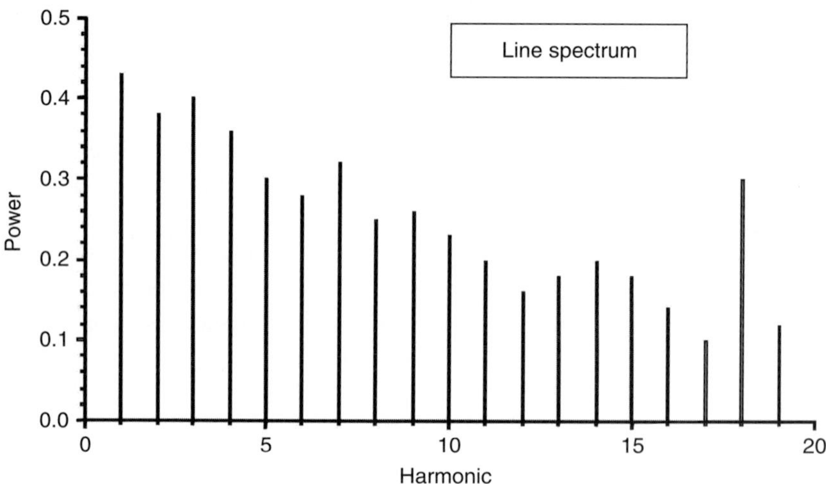

Figure 11.1 Line spectrum (impulses) result for pure sinusoids in the time series truncated exactly at the period of the sinusoids. Such perfectly periodic signals give rise to "line spectra" (most of us will never see such spectra from data). This is most typical of deterministic components in your time series.

squared) or amplitudes of each of the fitted cosines are drawn as vertical lines whose height is proportional to the power.

Added together, the cosines at their amplitude and phase exactly interpolate the discrete time series, as previously shown. Therefore, what we have called aliasing and leakage represents "erroneous frequency information in the spectrum," which *only* applies when trying to recover the true, continuous, aperiodic function from which the discrete, finite data have been sampled. *This Fourier transform does what it is mathematically formulated to do: it defines a Fourier series that interpolates the discrete data, passing through each point exactly, regardless of whether it has aliased or leaked the "true" spectrum.*

Now consider a related situation: a series of m data points that deviate from a straight line by small irregularities (noise). Were it not for the presence of a small amount of noise in the data, they would in fact lie on a straight line. If we fit the data with an nth order polynomial where $n = m$ (analogous to the Fourier problem, only here we use the polynomial basis instead of the cosine basis), we will get an exact fit to the data, so the rms error is 0, but the fitted curve will be highly oscillatory and unlikely to resemble the true character of the data (a straight line).

If, on the other hand, we were to fit the points with a polynomial of order 2 (a straight line), we would get a least-squares fit that likely captures the true essence of the dominant information contained in the data (we might then look at the shape of the residuals for any additional, but less dominating, information contained in what we believe to be the noise).

For example, consider the example of Figure 11.2: six noisy data points fit by a sixth-order polynomial interpolant (wiggly line) and a second-order polynomial (solid straight

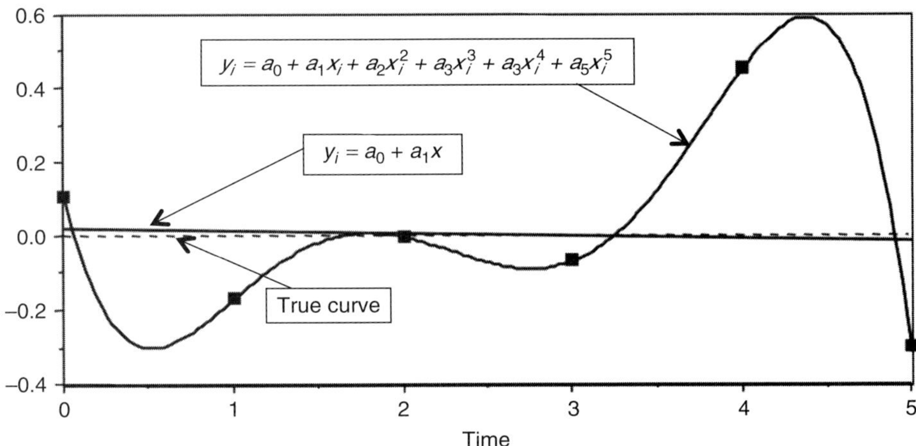

Figure 11.2 Six data points from a straight-line noisy process yielding the six solid squares. Noise is not accounted for in the sixth-order polynomial fit that exactly fits each point and is a bad representation of the *true process* shown by the horizontal dashed line at 0 amplitude. Least-squares fit of a second-order polynomial (bold straight line) results in the solid line that is slightly skewed relative to a true flat line, so the least-squares fit gives good approximation to the true straight line.

line) least-squares smoothed fit. The true curve is the constant value of zero (dashed line).

This highlights the difference between interpolation and least-squares fitting, as previously discussed. For a statistical interpretation of the *stable* portion of the data, we need the least-squares fit; for an exact fit of the data, regardless of the distortion imparted by such a fit owing to the presence of noise, we interpolate. The dominant problem with the least-squares approach is that we need to know, or have a good estimate of, the true number of coefficients representing the embedded signal, or its actual functional form.

The sensitivity of an exact fit to noise is further highlighted by considering the consequence of fitting the *n*th-order polynomial to multiple realizations of the same process. For example, consider the 10 realizations (m = six data points each) of Figure 11.3. Each represents the same process: a straight line with value 0.5. The noise present in the samples is Gaussian, so the discrete observations vary randomly about the true curve by a small amount dictated by the variance of the distribution. Each realization has been fit exactly with a sixth-order polynomial of the form

$$\begin{aligned} y_i &= a_0 + a_1\mathrm{x}_i + a_2\mathrm{x}_i^2 + a_3\mathrm{x}_i^3 + a_4\mathrm{x}_i^4 + a_5\mathrm{x}_i^5 \\ &= \sum_{j=0}^{n} a_j\mathrm{x}_i^j, \end{aligned} \tag{11.1}$$

where $j = 0,1,2,3,4,5$, for $y(\mathrm{x}_i)$. For every realization, the value of each coefficient, a_j, in the fitted polynomial is presented in a plot of amplitude (value) of the coefficient versus

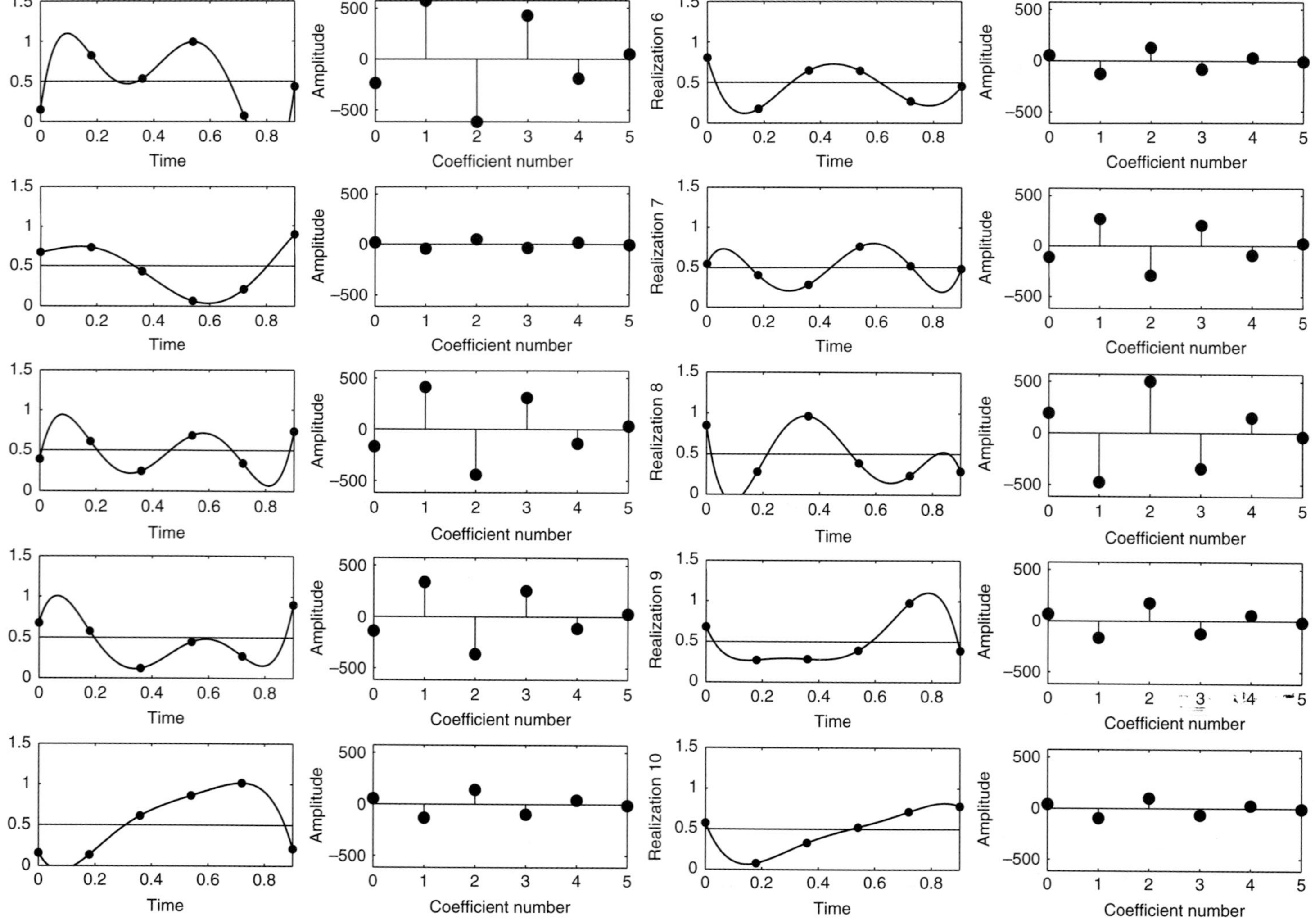

Figure 11.3 Ten realizations of the same straight-line process with small Gaussian noise randomly added to each of six points defining the random process. Each process fit with an exact-fit sixth-order polynomial (value of fit coefficients given in the right column). The true answer is a straight line with value 0.5 in time series; and with polynomial coefficients, all zero except for $a_0 = 0.5$. Ordinate of polynomial coefficients is different for each realization.

j, the coefficient number. This is presented as analogous to a line spectrum for a Fourier interpolant. In this manner, the sensitivity to noise is clearly revealed by the differences in the various interpolant curves shown in the time domain for each realization as well as the difference in the coefficients of the interpolant. The true solution has $a_0 = .5$ and $a_1 = a_2 = a_3 = a_4 = a_5 = 0$.

The fitted polynomial curves show dramatically different shapes for each realization, and the "spectra" show that the coefficients of the fitted polynomials are equally diverse and different from the true values.

These differences between the different realizations are analogous to the case of non-time-series data in which each observation drawn from a population by itself can show a considerable deviation from the most representative value of the population – the mean, for example. In that case, we estimated the most representative value by averaging the various observations. We can employ that same strategy here – average (or ***stack***) the 10 time series in an effort to estimate a single, most representative version as well as to provide information regarding the "spread" amongst the various realizations (i.e., employ the Central Limit theorem to reduce noise).

Figure 11.4 shows the results of this averaging: the average of the 10 points at each time position (the large open circles) and the polynomial fitted to this average realization (the bold line) comes significantly closer to matching the true curve (the flat line of 0.5) than any single realization.

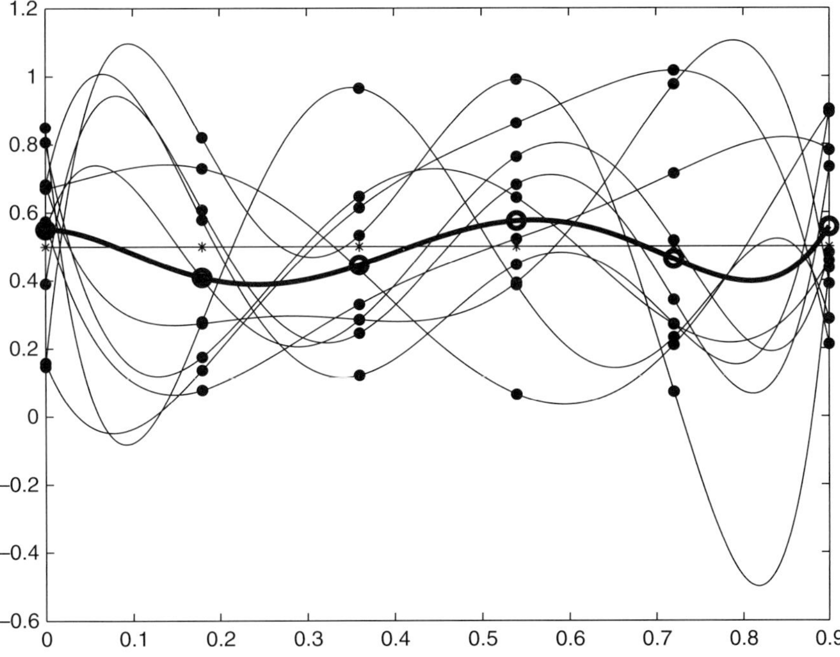

Figure 11.4 Summary plot with each of the 10 realizations superimposed from the example of Figure 11.3. Bold line is sixth-order polynomial fit to average of data points (black dots). Bold line also results from polynomial fit using average coefficients (shown in Figure 11.5) of the 10 realizations.

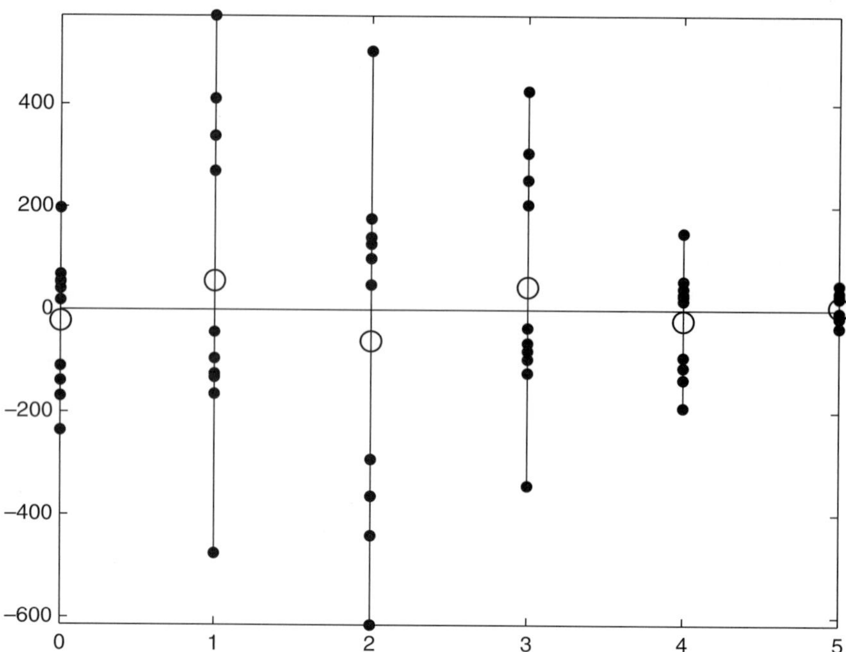

Figure 11.5 Polynomial coefficient values (black dots) for each of the 10 realizations of Figure 11.3 and their average (open circles). Average coefficients do a good job of approximating true coefficient values of 0 for all but the 0th coefficient, which is well approximated at the true value of 0.5. Construction of a polynomial using the averaged coefficients gives the bold curve in Figure 11.4.

Because the polynomial curves fitted to each realization interpolate the discrete points, the averaging operation just described could have been carried out just as effectively by averaging the actual polynomial functions for each realization. Since the polynomials are linear in the coefficients, this is identical to averaging the corresponding j coefficient values for each of the 10 realizations (as done in Figure 11.5).

The open circles in the Figure 11.5 "spectrum" of polynomial coefficient values represent the averages for each individual coefficient. These averaged values provide a good approximation to the true "spectrum" of coefficient values (0.5 at $j = 0$; 0.0 for all other j). Also, constructing the sixth-order polynomial from these six averaged coefficient values produces the same curve depicted by the average (bold line) polynomial in the previous "time domain" figure (Figure 11.4).

Therefore, in the case of **multiple realizations**, *averaging of the spectra coefficients* determined for each realization provides a much better estimate of the true "spectrum" of coefficient values and a most "representative" time series for the ensemble – consistent with estimating the mean from a sample of size n. As n approaches infinity (or the actual size of the ensemble), the estimate of the mean approaches the true mean of the population. Similarly, with time series, as the number of realizations approaches infinity or the true size of the ensemble, the average of the spectrum coefficients should approach the most representative time series of the ensemble. *But*, you cannot achieve this result for stochastic processes by *averaging the time series*, since each realization can have different phases of the same fundamental cosines, leading to a cancellation of the sines and cosines in the average (transform of the acvf gives the PSD without phase).

11.3 More Stable Estimates of the Fourier Coefficients

With spectral analysis we have the same sensitivity to noise only in this case, instead of using polynomials of order n, we are using n sinusoids. We are therefore fitting the data exactly and the *fit is sensitive to the distribution of the noise in the data*. If the noise present in the data was slightly different, the Fourier series fit might change dramatically as in the polynomial example above, and thus the amplitude and phase spectra would be different. If your time series only requires Fourier interpolation, is deterministic, or is for a single event of interest – e.g., an earthquake where the ensemble is of no interest – you don't need to consider what is the most representative time series. But if your time series is of a process you wish to characterize – e.g., seismograms for the San Andreas earthquakes, ice age cycles or tides – then you need to consider how to make the spectrum more representative of the statistically stable portion of the data (i.e., of the ensemble).

Here we address the same methods that were used for the polynomial case above, but starting with the most practical first: (1) averaging the spectral coefficients from several realizations, (2) segmenting a single realization providing multiple, but shorter, realizations for averaging and (3) least squares fit.

11.3.1 Smoothing the Spectrum

Ensemble Averaging

For the most general case, when the signal is not truly periodic, consider averaging the spectral coefficients from the spectra of multiple realizations. That is, take the spectra of numerous realizations of the process and average those spectra to produce a single best estimate of the underlying, statistically stable signal contained in the realizations. If you had an infinite number of realizations, then your averaged spectra would represent the spectrum of the ensemble. Though *ensemble* is to time series data what *population* is to non-time series data, it is sometimes convenient to use the word ensemble to represent the suite of realizations available, analogous to a *sample* – in this respect, this procedure (averaging results from the available realizations) is often referred to as **ensemble averaging**.

While this analogy to the polynomial case is conceptually obvious, several points must be considered before we blindly apply that intuitive and heuristically demonstrated approach. In the polynomial example, we averaged the coefficients based on analogy with computation of a mean and the linearity of the polynomial in the coefficients. However, in the case of the more complicated functions involved in PSD estimates (spectral power is not linear in the two coefficients, being the sum of those coefficients squared), we must first determine the distribution of the coefficients, and from that, the most appropriate method of combining the information from several realizations to make the best approximation of the true spectral values. This will also provide the information necessary for determining spectrum uncertainty.

Chi-Squared Distribution of the PSD Estimates

Since estimates $\hat{a}_j$ and $\hat{b}_j$ of the true Fourier coefficients are linear combinations of the y_i values, they are normally distributed random variables (by the Central Limit Theorem). Also, the a_j and b_j coefficients are independent (as shown by the orthogonality relationships). Therefore, **the power spectrum, or PSD, $A_j^2 = a_j^2 + b_j^2$, is the sum of two independent squared normally distributed random variables.**

A sum of n squared values of a standardized normally distributed random variable, Z_i, where $Z_i = N(0,1)$

$$Y = \sum_{i=1}^{n} z_i^2 \tag{11.2}$$

produces a variable Y that has a *chi-squared distribution*, χ_v^2, with $v = n$ degrees of freedom. This should not be too surprising given the similarity of the form of (11.2) to the chi-squared goodness of fit statistic (3.46). What should be surprising is that we get a chi-squared distribution instead of a Gaussian – has the Central Limit Theorem failed us? No; for highly asymmetrically distributed random variables (such as squared values) it takes a larger sum before converging to a Gaussian, which is the case here. As n gets larger, the chi-squared distribution approaches a symmetrical Gaussian, as expected.

A χ_v^2 distribution with $v = n$ degrees of freedom (in this case, the sum of n independent squared normal deviates) has a mean of n and a variance of $2n$ for the case of $Z_i = N(0,1)$. The mean is easily shown by examining the expectance of the sum in (11.2). By definition,

$$E[Z] = 0 \tag{11.3}$$

and

$$E[Y] = E\left[\sum_{i=1}^{n} z_i^2\right] = \sum_{i=1}^{n} E[z_i^2],$$

recalling, $E[Z^2] = Var[Z] + \mu_Z^2$, so

$$= \sum_{i=1}^{n} E[\sigma_Z^2 + \mu_Z^2].$$

Since $Z_i = N(0,1)$, $\mu_Z^2 = 0$ and $\sigma_Z^2 = 1$, so

$$= \sum_{i=1}^{n} 1 = n. \tag{11.4a}$$

By a similar, though more tedious approach,[4]

$$\mathrm{Var}[Y] = 2n. \tag{11.4b}$$

For the case in which Z does not have zero mean and unit variance, we must compensate as always for the loss in degrees of freedom associated with estimating the mean and variance in order to standardize Z, in which case the mean and variance of Y are generalized by $\mathrm{E}[Y] = \nu$ and $\mathrm{Var}[Y] = 2\nu$, where ν is the degrees of freedom ($n - 2$ if both the mean and variance were estimated for Z).[5]

Further, **when the normally distributed random variables, z_i, are not standardized and have a distribution $\mathbf{N(0,\sigma^2)}$**, then the mean and variance are

$$\mathrm{E}[Y] = \mathrm{E}\left[\sum_{i=1}^{n} z_i^2\right] = \sum_{i=1}^{n} \mathrm{E}[\sigma_Z^2 + \mu_Z^2]$$

$\mu_Z = 0$, but $\sigma_Z \neq 1$, so,

$$= \sum_{i=1}^{n} \sigma_z^2 = n\sigma_z^2 = \nu\sigma_z^2 \tag{11.5a}$$

and

$$\mathrm{Var}[Y] = 2\nu\sigma^4. \tag{11.5b}$$

So, for the PSD where $A_j^2 = a_j^2 + b_j^2$, each estimate of power is the sum of two squared normal distributions, $\mathrm{N}(0,\sigma^2)$, indicating that the power spectrum ordinate values (the powers) have a chi-squared distribution with 2 degrees of freedom ($= \chi_2^2$).

Combining Realizations

When dealing with aperiodic continuous time series, our finite-length time series represent a truncated version of the continuous series. Truncation in time corresponds to convolution in the frequency domain, which serves to "smooth" the true spectrum (i.e., leakage). In this case, the power estimate for each harmonic consists of a *weighted sum* of χ_ν^2 distributed variables, so the estimates may not be χ_ν^2 distributed.

Consequently, in order to accommodate a variety of truncation forms (and smoothing functions), we must consider the *ratio* of the estimated power spectrum to the true

[4] Best done by actually inserting the mathematical function for the chi-squared distribution (which is what this sum of squared normal deviates produces) into the expectance operator and solving analytically. This is often done for the more general case of the gamma distribution, of which the chi-squared distribution is a special case.

[5] Note that if the z_i are not independent (as in all time series except white noise); their squared sum is still a chi-squared distribution, but the degrees of freedom are reduced to reflect the shared information contained by the correlation between the variables being summed.

spectrum, a ratio that is approximately χ_v^2 distributed.[6] So, if we call $|F_j|^2$ our power estimate and p_j the true power spectral estimate for the jth harmonic, then $A_j^2/p_j \propto \chi_v^2$, or $A_j^2/ap_j = \chi_v^2$ distributed where a is the constant of proportionality. Because we want a χ_v^2 distribution that has the same mean and variance as the above ratio in the limit, and recalling that the mean and variance of a χ_v^2 distribution are v and 2v, you can solve for the value of $a = 1/\text{v}$.

So, $\text{v}A_j^2/p_j$ is approximately χ_v^2 distributed: the number of degrees of freedom is related to the sum of the squared truncation window weights (discussed later, when we deal with various truncation windows; See Priestley (1981), chapter 6, for some of the derivations, as well as Jenkins and Watts (1968), section 6.4.) In general, the proportionality factor, v/p_j, can be derived in a variety of ways and involves general asymptotic, or Taylor-series, expansions. For a Gaussian white noise process, the specific form is easily shown since the actual values of p_j are a known constant equal to the variance of the noise. Though this ratio is approximate and stems from asymptotic limits, it has been shown to be fairly decent for relatively short time series.

For the simple case where the continuous time series is truncated by a rectangular function, then, as already stated, v = 2. This represents a special case of the χ_v^2 distribution known as the **exponential distribution** (both the chi-squared and exponential distributions are special cases of the **gamma distribution**).

The functional form of the general χ_v^2 distribution for a random variable x is

$$f_{\chi_v^2}(x) = \frac{x^{[(v/2)-1]}}{2^{v/2}\Gamma(v/2)} e^{(-x/2)}, \tag{11.6}$$

where $\Gamma(\text{v}/2)$ is the **gamma function** given by

$$\Gamma(v/2) = \int_0^{\infty} e^{-t} t^{[(v/2)-1]} dt. \tag{11.7}$$

For v = 2 and $\Gamma(1) = 1$

$$f_{\chi_v^2} = \frac{1}{2} e^{-x/2}. \tag{11.8}$$

Hence, the χ_v^2 distribution does take on an exponential distribution shape for v = 2.

Since the actual quantity that has the χ_v^2 distribution is $\text{v}A_j^2/p_j$, equation (11.6) is rewritten explicitly as

$$f_{\chi_v^2}\left(vA_j^2/p_j\right) = \frac{v(A_j^2)^{[(v/2)-1]}}{p_j 2^{v/2}\Gamma(v/2)} e^{(-vA_j^2/2p_j)} \tag{11.9}$$

[6] If your data represent a *general linear process model*, this approximate χ_v^2 distribution will hold for data *not* normally distributed, because each data point represents an infinite sum of X_i. By the Central Limit Theorem, the distribution of the X_i will be Normal. For the rectangular function, the distribution is still χ_v^2.

or, for $\nu = 2$, equation (11.9) reduces to the exponential distribution, f_e, as

$$f_e\left(2A_j^2/p_j\right) = \frac{1}{p_j} e^{(-A_j^2/p_j)}. \quad (11.10)$$

The term $x^{[(\nu/2)-1]}$ in the numerator of equation (11.6) refers to the random variable, which is A_j^2 for this case.[7] The *constants* ν/p_j multiply that random variable (that is, they are not also raised to the power). Since χ_ν^2 is a sum of squared variables, this distribution is only for positive values of the random variable x.

So, the *ratio* of the sample power (A^2) to true power (p_j) has an exponential distribution (χ_ν^2 with $\nu = 2$), as shown in Figure 11.6. This distribution has a long tail, and thus what would be considered outlier values in a normal distribution can be expected to occur more often in this exponential distribution. That is, you should expect large differences in the ratio (sample to true power), signifying small power or large sample estimate.

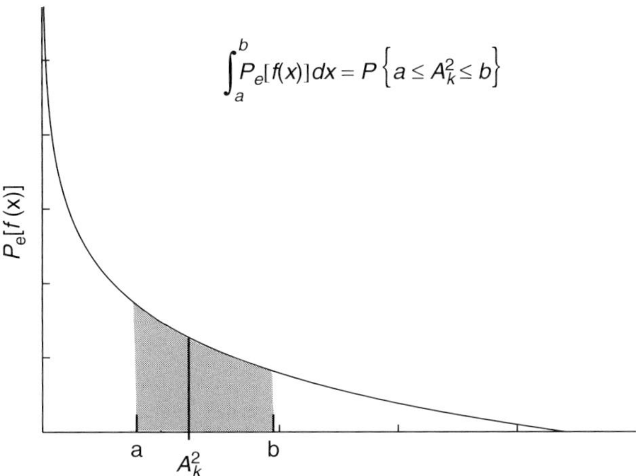

Figure 11.6 PDF for exponential distribution (P_e) for each estimate of spectral power. Asymmetrical bounds forming a probability of a true signal can be defined for any level of significance. The probability of obtaining a value within those limits is given by the integral of the PDF within those bounds (= shaded area under the curve).

Best Way to Combine Multiple Realizations

Suppose you have n realizations and corresponding estimates of their power spectrum. Use the *principle of maximum likelihood* (Chapter 3, §3.6) to determine how to best estimate the p_j using the n discrete estimates (one estimate for each realization). In other words, how can you estimate p_j such that the likelihood of obtaining the estimates that you *did* obtain would have the maximum probability of occurring. For convenience,

[7] The variable that shows a chi-squared distribution is $\nu A_j^2/p_j$, but only the A_j^2 is a random variable, since the ν and p_j are constants.

define $A_k^2 = I_k (= a_k^2 + b_k^2)$, $k = 1, 2, \ldots n$. Then, form the likelihood by taking the product of the distribution of the estimates

$$\begin{aligned} \mathrm{L}(I) &= \prod_{k=1}^{n} \left[\frac{1}{\mathrm{p}} \mathrm{e}^{(-I_k/\mathrm{p})} \right] \\ &= \mathrm{p}^{-n} \exp\left[\sum_{k=1}^{n} (-I_k/\mathrm{p}) \right]. \end{aligned} \tag{11.11}$$

Taking the log,

$$\ln[\mathrm{L}(I)] = -n[\ln(\mathrm{p})] - \frac{1}{\mathrm{p}} \sum_{k=1}^{n} I_k. \tag{11.12}$$

As usual, find the maximum likelihood by computing the derivative with respect to p and setting this equal to 0:

$$\frac{\mathrm{d}}{\mathrm{dp}} \ln[\mathrm{L}(I)] = -\frac{n}{\mathrm{p}} + \frac{1}{\mathrm{p}^2} \sum_{k=1}^{n} I_k = 0. \tag{11.13}$$

Solving for p,

$$\mathrm{p} = \frac{1}{n} \sum_{k=1}^{n} I_k. \tag{11.14}$$

Therefore, *the best estimate of the true power spectral estimate at a given frequency is given by averaging the n estimates you have for each spectrum*. We suspected that simple averaging of the spectra should give a better estimate, and the above shows this to be the case, even though the I_k are sums of squared quantities, and not simply linear combinations of the original a_k and b_k.

Uncertainty in Individual Estimates

After averaging the spectra, you must consider the statistical significance of the individual power spectral estimates that make up your **ensemble average**. This requires knowledge of the statistical moments of the p_j.

Mean

The true power spectrum, p(f), is given by the Fourier transform of the true autocovariance function, $\gamma(\tau)$:

$$\mathrm{p(f)} = \sum_{\tau=-\infty}^{\infty} \gamma(\tau) e^{-i2\pi \mathrm{f}\tau}. \tag{11.15}$$

Since $\gamma(\tau)$ is an even function the above transform (as shown in §9.7.6) reduces to just the cosine transform (where the 0 lag term comes out as the variance of the time series, so it is not counted twice),

$$\mathrm{p(f)} = \sigma_y^2 + 2\sum_{\tau=1}^{\infty} \gamma(\tau)\cos(2\pi \mathrm{f}\tau)\ ; \tag{11.16}$$

likewise for the *sample* PSD,

$$I(\mathrm{f}) = \sum_{\tau \geq -(n-1)}^{n-1} \hat{R}(\tau)e^{-i2\pi \mathrm{f}\tau}\ , \tag{11.17}$$

where I(f) and $\hat{R}(\tau)$ are the sample-based estimates of p(f) and $\gamma(\tau)$, respectively. The expected value of the I_j estimate (i.e., I(f) at the jth harmonic), is given by

$$\begin{aligned} \mathrm{E}[I(\mathrm{f})] &= \mathrm{E}\left[\sum_{\tau \geq -(n-1)}^{(n-1)} \hat{R}(\tau)e^{-i2\pi \mathrm{f}\tau}\right] \\ &= \sum_{\tau \geq -(n-1)}^{n-1} \mathrm{E}[\hat{R}(\tau)e^{-i2\pi \mathrm{f}\tau}] \\ &= \sum_{\tau \geq -(n-1)}^{(n-1)} \mathrm{E}[\hat{R}(\tau)]e^{-i2\pi \mathrm{f}\tau}, \end{aligned} \tag{11.18}$$

where this last reduction occurs because the cos(2πfτ) (or exponential) term is a constant for each value of τ (that is, it is deterministic).

The expectance of $\hat{R}(\tau)$ in (11.18) is given by

$$\begin{aligned} \mathrm{E}[\hat{R}(\tau)] &= \mathrm{E}\left[\frac{1}{n}\sum_{i=1}^{n-|\tau|} y_i y_{i+\tau}\right] \\ &= \frac{1}{n}\sum_{i=1}^{n-|\tau|} \mathrm{E}[y_i y_{i+\tau}] \\ &= \frac{1}{n}\sum_{i=1}^{n-|\tau|} \gamma(\tau) \\ &= \frac{n-|\tau|}{n}\gamma(\tau). \end{aligned} \tag{11.19}$$

$\mathrm{E}[y_i y_{i+\tau}]$ *defines* covariance (assuming y has 0 mean) as a function of lag, τ. That is, we previously defined $\mathrm{E}[(X - \mu)^2] = \sigma^2$, which is what we have here as a function of lag τ. This in turn is the definition of $\gamma(\tau)$ (the true autocovariance function).

So the above shows that the expected value of the estimated (sample) autocovariance function is *proportional* to the true autocovariance (i.e., it is biased, except at lag zero, and in the limit as $n \to \infty$).[8]

Returning to our determination of E[I(f)], (11.19) can now be substituted into (11.18) for $\hat{R}(\tau)$, giving

$$\begin{aligned} \mathrm{E}[I(\mathrm{f})] &= \sum_{\tau=-\infty}^{\infty} \frac{n-|\tau|}{n} \gamma(\tau) \mathrm{e}^{-i2\pi \mathrm{f}\tau} \\ &= \sum_{\tau=-\infty}^{\infty} \varphi(\tau)\gamma(\tau)\mathrm{e}^{-i2\pi \mathrm{f}\tau}. \end{aligned} \tag{11.20}$$

Equation (11.20) shows that the expected value of I(f) is the Fourier transform of the *product* of $\varphi(\tau)$ $\gamma(\tau)$, that product being the *convolution* of the Fourier transforms of $\varphi(\tau)$ and $\gamma(\tau)$. Though we already proved this, we are approaching it from a different direction here. Specifically, if we replace the true acvf with the inverse of its Fourier transform (i.e., the PSD), we get

$$\begin{aligned} &= \sum_{\tau=-\infty}^{\infty} \varphi(\tau) \left[\sum_{k=-\infty}^{\infty} \mathrm{p}_k \mathrm{e}^{i2\pi \mathrm{f}_k t} \right] \mathrm{e}^{-i2\pi \mathrm{f}_k \tau} \\ &= \sum_{\tau=-\infty}^{\infty} \varphi(\tau) \left[\sum_{k=-\infty}^{\infty} \mathrm{p}_k \mathrm{e}^{-i2\pi \mathrm{f}_k \tau} \right] \\ &= \sum_{k=-\infty}^{\infty} \mathrm{p}_k \sum_{\tau=-\infty}^{\infty} \varphi(\tau) \mathrm{e}^{-i2\pi \mathrm{f}_k \tau} \end{aligned}$$

$$\mathrm{E}[I(\mathrm{f})] = \sum_{k=-\infty}^{\infty} \mathrm{p}_k \mathrm{W}\mathrm{f}_k. \tag{11.21}$$

The lagged sum in (11.21) is a convolution, so

$$\mathrm{E}[I(\mathrm{f})] = \mathrm{W}(\mathrm{f}) * \mathrm{p}(\mathrm{f}). \tag{11.22}$$

The function $\varphi(\tau)$, is a triangular tapering window – it is at its maximum (= 1) at zero lag and it linearly decreases in amplitude with bigger lags, going to 0 at lag n. This triangular taper is known as a **Bartlett** window or taper. Its Fourier transform is the **Fejer kernel** or **truncated periodogram spectral window**, W(f). You get a triangle function by convolving two rectangular functions. That convolution in time is the multiplication of the

[8] Since it is not equal to the true covariance, this makes our estimate of the autocovariance a **biased** estimate – we could make it an **unbiased** estimate by normalizing the autocovariance function by $1/(n - |\tau|)$, though this usually is not done, based on arguments given previously (in Chapter 7, §7.4.5) that indicate the biased estimate actually gives a smaller overall rms error (see equations (7.47) and (7.48) and accompanying discussion).

Fourier transform in frequency: the product of two sinc functions. Hence, the Fejer kernel is essentially the sinc function squared:[9]

$$W(f) = \frac{k^2 \sin^2(\pi f T)}{(\pi f)^2}. \tag{11.23}$$

That is, *the expected value of the estimated power spectrum is the convolution of the transform of the triangular tapering window (the Fejer kernel) with the true spectrum.*

Examination of the inverse transform of the Fejer kernel reveals that the weights of the spectral window sum to one, and it is thus conservative (i.e., this convolution preserves the original power of the PSD):

$$\varphi(\tau) = \int_{-f_N}^{f_N} W(f) e^{i2\pi f\tau} df \tag{11.24}$$

at lag 0,

$$\varphi(0) = \int_{-f_N}^{f_N} W(f) df = 1. \tag{11.25}$$

In other words, the values of W(f) sum to 1 and the expected value of I(f) is a weighted average of the true power spectrum.

Examination of (11.23) shows that W(f) approaches a delta function as $T \to \infty$ (as always, convolution with a delta function in (11.22) simply replicates the true spectrum). So, for large T,

$$E[I_j] \approx p_j. \tag{11.26}$$

This shows that the power spectral estimates are **asymptotically unbiased** (since they approach the true power spectrum in the limit as T goes to infinity), though for finite lengths of the time series it is a **biased** estimator. For finite-length time series, the bias (difference between the true spectrum and expected spectrum) is shown to be of order $\log(n)/n$.[10] It also indicates that with smaller T, the expected value is a smoothed version of the true power spectrum.

Variance

The variance is derived in a manner similar to that described above, only it is now easier, since much of the groundwork has already been done (the covariance term of I^2(f) drops out due to the orthogonality of the different harmonics, f), leading to

$$\begin{aligned} \mathrm{Var}[I_j] &= E[I_j^2] + E[I_j]^2 \\ &= E[I_j]^2 \end{aligned}$$

[9] The Fejer kernel appears in multiple forms. Even this form can have a different scaling factor.
[10] See Priestley (1981), page 418, for this.

from (11.26):

$$\approx \mathrm{p}_j^2. \tag{11.27}$$

As seen from (11.27), the variance of each estimate is large. In fact, *the standard deviation is equal to the magnitude of the value being estimated* – ouch, that is a big uncertainty (e.g., the power is $10^4 \pm 10^4$)!

Even worse, the value of *the variance is independent of the number of data points in the realization*, so as we increase the number of data points, we do not decrease our variance of each estimate, so the power spectrum is **not** a *consistent* estimator; its variance doesn't tend to zero in the limit as n goes to infinity. So we can't reduce the variance by adding data points to our time series realization. This is not surprising, since the increase in the number of data points in a realization simply increases the number of spectral estimates; the extra points do not go toward better statistical properties of the original number of spectral estimates.

Covariance

Finally, the estimates of I_j are uncorrelated and independent. This too is shown with the expectance operator in which the orthogonality relationships of the sum over a period of the various products of cosines, sines and cross-products results in the covariance being 0 for all lags (i.e., between all neighboring power spectral estimates). Of course, the only terms that survive through the sum of the products are for the covariance between a single point and itself, which is the variance. This result impacts how we are justified in manipulating the power spectral estimates to achieve statistical stability.

Power Spectrum Moments Summarized

After averaging the spectra, we must consider the statistical significance of the individual power spectral estimates that make up our **ensemble average**. Each individual estimate at harmonic j, $I_j = a_j^2 + b_j^2$ is the sum of two squared independent normally distributed random variables (for a rectangular truncation function of your infinitely long time series), and as a consequence,

$$I_j \propto \chi_v^2 \tag{11.28}$$

$$\mathrm{E}[I_j] \approx \mathrm{p}_j \tag{11.29}$$

$$\mathrm{Var}[I_j] \approx \mathrm{p}_j^2 \tag{11.30}$$

$$\mathrm{Cov}[\mathrm{p}_j, \mathrm{p}_{j+\tau}] = 0. \tag{11.31}$$

The uncertainty in the raw estimate is as large as the estimate itself, and increasing the number of data points in the time series does nothing more than increase the number of harmonics resolved, not decrease the uncertainty. The raw estimate of the power spectrum is referred to as a *periodogram*, though there are numerous accounts of the exact meaning of the name, all more or less captured by this most general use here. Also, the estimate is badly biased unless the number of data points in the time series approaches infinity (which is rather difficult to achieve and process). Statistically, we see that the periodogram could easily be the poster child of bad statistics. But there are a number of approaches available to overcome some of these defects.

11.3.2 Confidence Intervals for Averaged Estimates

We showed via the principle of maximum likelihood that the best way to combine spectra is via averaging. Here we examine the reduction in the uncertainty that can be obtained by averaging n spectra. If you average together n, I_j estimates together at each frequency j, the averaged values, $\overline{I}_j$, are random variables with a chi-squared distribution with $\nu = 2n$ degrees of freedom. That is, each I_j has a chi-squared distribution, the sum of two squared, normally distributed random variables and the sum of n chi-squared, distributed variables, each with two degrees of freedom, is the same as a single sum of two n squared, normally distributed variables. To appreciate the effect of this increase in the number of degrees of freedom, consider the actual form of the variable, which is distributed as a χ^2_ν distribution,

$$\chi^2_\nu = \nu\overline{I}_j/\mathrm{p}_j\ , \tag{11.32}$$

where the $\overline{I}_j$ is the n–sample averaged power spectral estimate at each frequency or frequency band centered at j, p_j is the true power spectral value and ν is the "**equivalent" (or "effective") degrees of freedom** (the meaning of the qualifier, "equivalent," will become more apparent later when we consider other truncating window shapes in which case the value of ν must be modified to accommodate the influence of the truncating function).

Given (11.32), we compute confidence intervals (**CI**) for our power spectral estimates and examine how the averaging of values has helped by increasing the number of degrees of freedom. Specifically (shown in Figure 11.7),

$$\mathrm{P}\left\{\chi^2_\nu(\alpha/2) \le \nu\overline{I}_j/\mathrm{p}_j \le \chi^2_\nu(1-\alpha/2)\right\} = 100(1-\alpha)\%. \tag{11.33}$$

In words, the probability that the quantity $\nu\,\overline{I}_j/\mathrm{p}_j$ lies between the values of $\chi^2_\nu(\alpha/2)$ (the smaller error limit to the left in Figure 11.7) and $\chi^2_\nu(1-\alpha/2)$ (the larger limit to the right in the figure) is $100(1-\alpha)$ percent, where α is the level of significance. This is a two-sided test, so we compute and interpret the left (lower) and right (upper) limits as follows, arranging so that the confidence intervals can be placed on the sample spectrum or on the true spectrum (p_j). Example of these CI for different α is shown in Figure 11.8.

To place the CI on the sample spectrum ($\overline{I}_j$), rearrange so that the p_j is bracketed by the limits. Start by inverting the ratio (which requires reversal of the limits of (11.33)):

$$\mathrm{P}\left\{\left[\chi^2_\nu(1-\alpha/2)\right]^{-1} \le \mathrm{p}_j/\nu\overline{I}_j \le \left[\chi^2_\nu(\alpha/2)\right]^{-1}\right\} = 100(1-\alpha)\% \tag{11.34a}$$

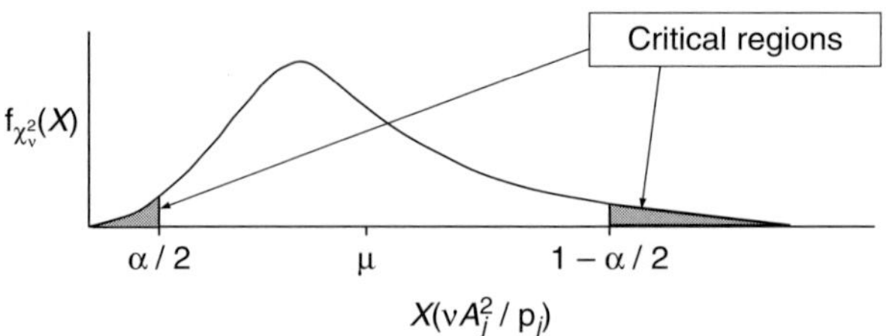

Figure 11.7 Chi-squared distribution, showing critical regions defining values of the random variable X that would not likely occur in such a distributed variable for a level of significance α. I.e., for $\alpha = 0.1$, each shaded zone contains 5 percent of the values expected, so 90 percent of the values we expect will fall between these limits – this being stated by equation (11.33).

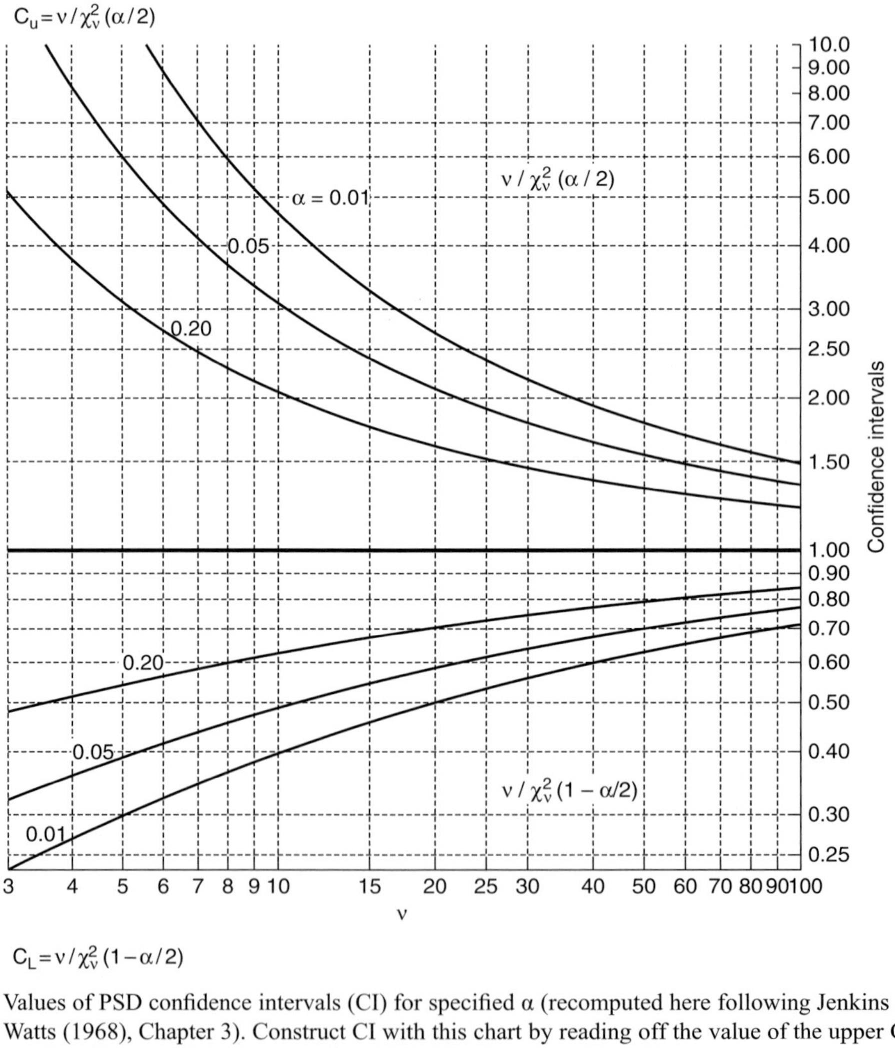

$C_u = \nu / \chi^2_\nu (\alpha/2)$

$C_L = \nu / \chi^2_\nu (1 - \alpha/2)$

Figure 11.8 Values of PSD confidence intervals (CI) for specified α (recomputed here following Jenkins and Watts (1968), Chapter 3). Construct CI with this chart by reading off the value of the upper CI contours (C_U in equation (11.34)) and lower ones for C_L for a specific α value and degrees of freedom (ν), then multiply the averaged I_j value by those values, plotting log power, use equation (11.35). Note how rapidly CI tightens with increasing ν, while also becoming more symmetrical as the chi-squared distribution approaches a Gaussian. The CI from this graph are for placement about the sample PSD, as in Figure 11.9A.

so

$$\mathrm{P}\left\{\nu\overline{I}_j/\left[\chi^2_\nu(1-\alpha/2)\right] \le \mathrm{p}_j \le \nu\overline{I}_j/\left[\chi^2_\nu(\alpha/2)\right]\right\} = 100(1-\alpha)\%\,, \tag{11.34b}$$

or

$$\mathrm{P}\left\{\mathrm{C_L}\overline{I}_j \le \mathrm{p}_j \le \mathrm{C_U}\overline{I}_j\right\} = 100(1-\alpha)\%\,, \tag{11.34c}$$

where $\mathrm{C_L} = \nu/\left[\chi^2_\nu(1-\alpha/2)\right]$ and $\mathrm{C_U} = \nu/\left[\chi^2_\nu(\alpha/2)\right]$. So, the probability that the true spectral value lies within the limits defined by the left and right terms in (11.34c) is $100(1-\alpha)$ percent. If the *estimated* true spectrum lies outside of the CI, then the sample power is not consistent with the true spectrum at that frequency (given a 100α percent chance of being wrong in that conclusion). In other words, at those frequencies that the sample spectrum lies outside of the CI, it is unlikely that the sample spectrum is equal to p_j at those frequencies (i.e., you reject the null hypothesis with an α percent chance of being wrong).

For placing the CI on the true spectrum, do not invert the ratio as done above,

$$\mathrm{P}\left\{\chi^2_\nu(\alpha/2) \le \nu\overline{I}_j/\mathrm{p}_j \le \chi^2_\nu(1-\alpha/2)\right\} = 100(1-\alpha)\%\,, \tag{11.34d}$$

so multiply out the ν and p_j, giving

$$\mathrm{P}\left\{\mathrm{p}_j\chi^2_\nu(\alpha/2)/\nu \le \overline{I}_j \le \mathrm{p}_j\chi^2_\nu(1-\alpha/2)/\nu\right\} = 100(1-\alpha)\% \tag{11.34e}$$

or

$$\mathrm{P}\left\{\mathrm{p}_j\mathrm{C}^*_\mathrm{L} \le \overline{I}_j \le \mathrm{p}_j\mathrm{C}^*_\mathrm{U}\right\} = 100(1-\alpha)\%. \tag{11.34f}$$

For this case, since we did not invert the ratio relative to the original form of (11.33), the multiplying constants are opposite of that case as well ($\mathrm{C}^*_\mathrm{L} = \mathrm{C}_\mathrm{U}^{-1}$ and $\mathrm{C}^*_\mathrm{U} = \mathrm{C}_\mathrm{L}^{-1}$) as are the CI. Examples of these two forms of CI placement are shown in Figure 11.9.

Examination of the form of the confidence limits and the graph of $\mathrm{C_L}$ and $\mathrm{C_U}$ versus ν in Figure 11.8 (adapted from Jenkins and Watts (1968)) reveals several interesting relationships.

First, notice that for a fixed α, as one increases the number of degrees of freedom, the width of the error bars (confidence interval) decreases. Therefore, by averaging the $\overline{I}_j$ of several realizations, we increase our degrees of freedom from 2 to $2n$ and quickly decrease the variance (uncertainty) in our estimated values.

Second, because the chi-squared distribution is not symmetric, we get an asymmetrical distribution of the error bars about each power spectral estimate, though they become increasingly symmetric as ν is increased (as $\nu \to \infty$, the chi-squared distribution goes to a normal distribution; in fact, by $\nu \approx 100$, the normal distribution makes an excellent approximation to the chi-squared distribution – as per the Central Limit Theorem).

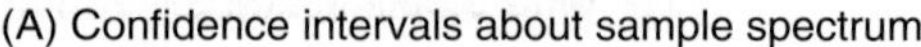
(A) Confidence intervals about sample spectrum

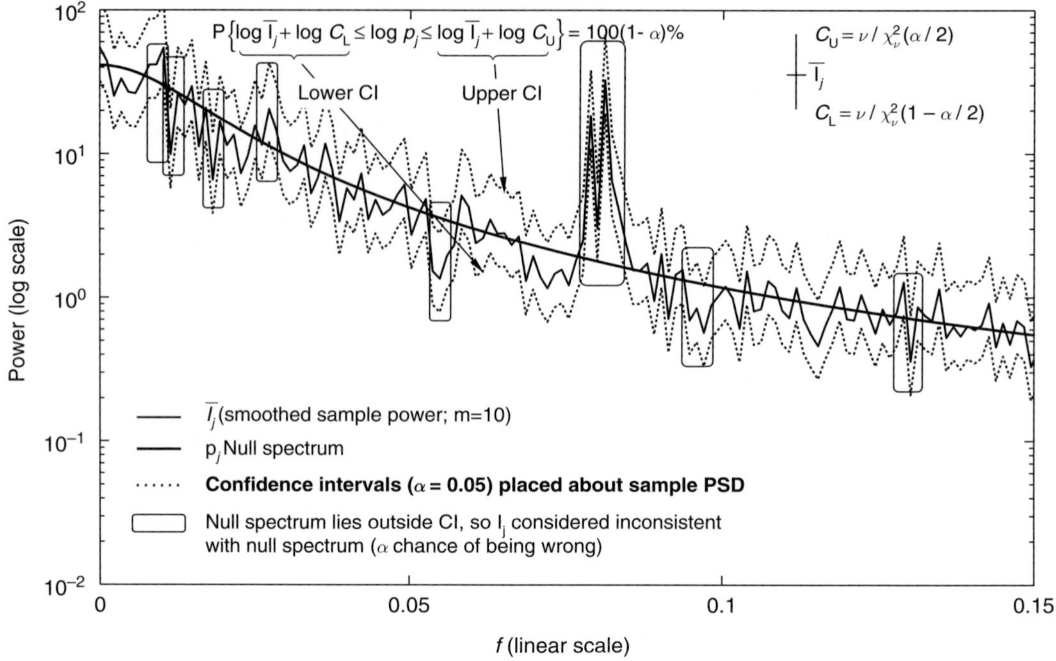

(B) Confidence intervals about null spectrum

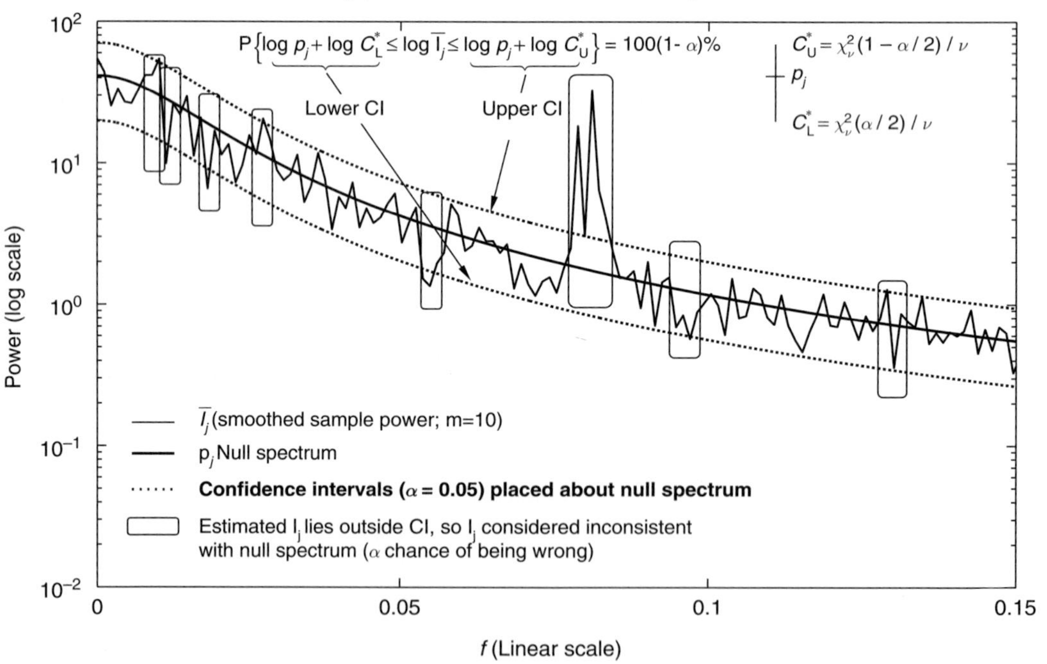

Figure 11.9 In all cases, null spectrum is either a noise spectrum or some known spectrum thought to be the ensemble PSD; ordinate (power, the $\bar{I}_j$) is $\log_{10}$ scale; peaks or valleys exceeding the CI, for which the null hypothesis is rejected, are shown in boxes. **(A)** Schematic showing confidence interval about the sample PSD. **(B)** CI presented as an envelope about the null spectrum. **(C)** Examining how to assess significance on significant valley, by placing the single error bar on the valley, showing that it lies outside of the null spectrum. You have $(1-\alpha)100$ percent probability (95 percent, for this example) that the boxed features are not from the null spectrum (null hypothesis

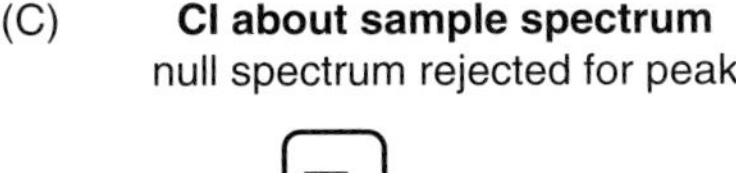

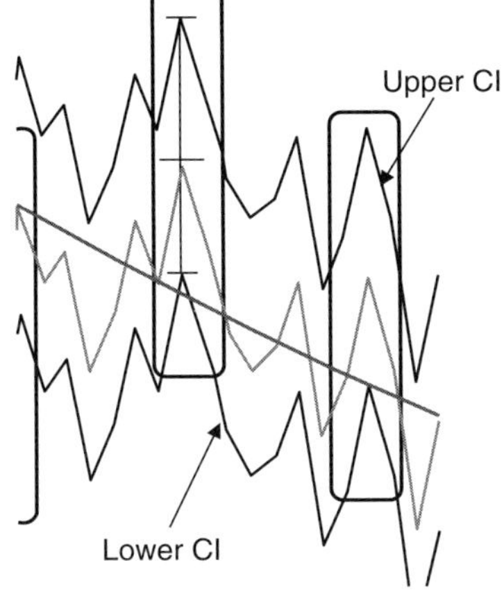

CI about null spectrum
null spectrum rejected for peak

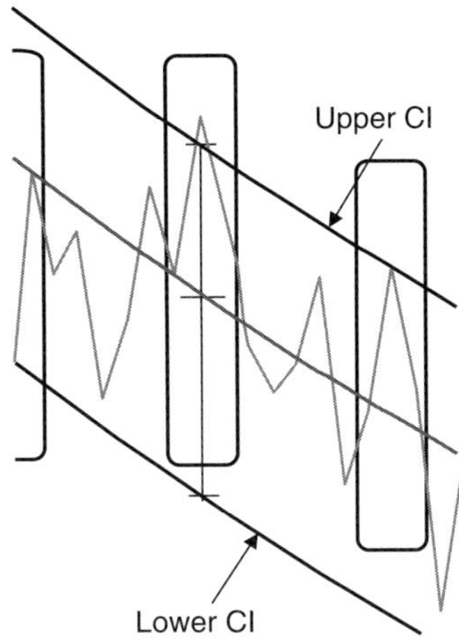

Peak to far right satisfies the equality part of the ≤ in the expression:
$P\{\log\overline{I}_j + \log C_L \leq \log p_j \leq \log\overline{I}_j + \log C_U\} = 100(1-\alpha)\%$

Figure 11.9 (cont.)

Third, the values of C_L and C_U are scaled by $\overline{I}_j$. So, the larger the power spectral estimate at any one frequency, the larger the error bar for that estimate.

Log Power Spectrum

That dependence of the error bar on the value of $\overline{I}_j$ is overcome by looking at $\log \overline{I}_j$:

$$P\left\{\log C_L + \log \overline{I}_j \leq \log p_j \leq \log C_U + \log \overline{I}_j\right\} = 100(1-\alpha)\%. \qquad (11.35)$$

With this form, the limits distributed about the sample PSD values are now fixed constants, $\log C_L$ and $\log C_U$.

Caption for Figure 11.9 (cont.)

rejected). You accept the possibility that you have about a 5 percent chance of being wrong when rejecting the null hypothesis. And, by definition, you'd expect that 5 out of 100 spectral estimates will yield values outside of the CI by chance – in this case, we have 120 estimates, with 12 estimates lying outside or equal to the CI. So, by chance we expect as many as 6 values lying outside the CI, yet we show 12. Which are "real," and which are simply by chance? They could all be real, 6 do not have to be by chance. But, like all analyses, you have to put *all* of the evidence together (e.g., is there some theory predicting certain peaks and valleys? Do you have other realizations showing same or different results?) and come up with the overall most internally consistent result; are there other data suggesting certain frequencies, etc.?

Spectral Interpretation: Significance of Spectral Peaks and Valleys

A typical power spectrum plot with 95 percent confidence interval using a log ordinate scale gives a single CI for the entire frequency range, as shown in Figure 11.9 (which shows both the single error bar as well as the CI as an envelope. Plotting either or both is acceptable – it is cleaner to just use the error bar, but the CI envelope is sometimes easier to interpret, unless the spectrum has too much variability (in that case, just show the error bar).

Null Spectrum and Testing Significance of Spectral Power

As done previously, we wish to use the confidence intervals (**CI**) to test the null hypothesis (I_j is your sample power estimate at harmonic j):

$$\begin{aligned} \mathrm{H}_0 &: \quad I_j = \text{noise} \\ \mathrm{H}_1 &: \quad I_j \neq \text{noise}. \end{aligned} \tag{11.36}$$

You can only assign a probability (level of significance) of being *wrong* for *rejecting* the null hypothesis when it is in fact true (saying nothing about the significance of being correct if you accept the null hypothesis).[11] Hence, we choose the **null spectrum** (that spectrum representative of the null hypothesis) that you want to reject – typically a noise spectrum, often estimated as a smooth fit through the center of the sample spectrum, or by fitting an autoregressive order 1 process, AR(1), to the sample spectrum (see Equation 11.37 below).

For those values of power at specific frequencies lying outside or equal to ($\leq$) the confidence limits, you reject the null hypothesis. These spectral power values are not from random noise, recognizing that you have a $\alpha 100$ percent chance of being wrong about that conclusion. If you want to compare your PSD to a known spectrum, use the known PSD as the null spectrum, though you will be rejecting those values that are not consistent with the known spectrum, so you will state that those particular frequencies are unlikely to be from your known PSD at the significance level α.

Spectral Gaps

While testing for the significance of peaks is common (almost natural), you need to also check for missing power: valleys that lie below the confidence interval. In many phenomena there is a continuum of power, and a gap in that continuum is referred to as a **spectral gap**. Such a gap suggests that there may be two different processes forming the phenomenon you have sampled (one for the continuum of power above the gap and another for that below the gap) or that some other process may be at work to remove energy from that particular frequency band. From that perspective, gaps are as interesting as peaks – maybe even more so.

Null Spectrum for Noise

There are a variety of ways of determining the null spectrum for noise. First there are the standard classes of *colored noise*.[12] The colors are based on analogies to light spectra, the

[11] See Chapter 3, §3.8 for review.

[12] Schroeder (1990) discusses noise in his chapter 5. He uses a slightly different naming convention and approaches the subject from a different perspective than that given here (his initial comments are somewhat amusing).

primary ones being (1) white noise (flat spectrum with equal power at all frequencies, as for white light), (2) red noise (power concentrated in lower frequencies; it has a slope of −2 on a log power versus log frequency plot; produced in nature by a sluggish system that tends to integrate a white noise forcing – also known as a random or drunkard's walk or Brownian motion, which leads some to call this "brown" noise instead of red) and (3) blue noise (power concentrated in higher frequencies). **Fractal noise** is not uncommon, and it takes on a linear form in PSD plots at a slope related to the fractal dimension. **Flicker noise** (1/f) is common in electrical circuits. You may also have a very specific noise consistent with your phenomenon that dictates the noise spectrum. Specific types and colors of noise can be generated through parametric models (e.g., low-order AR models of Chapter 14). The choices are unlimited. *In practice, unless you have a specific noise you are questioning, the most appropriate noise shape for your data is a smoothed curve running through your sample PSD – something you would get from extreme smoothing or fitting a smooth curve (e.g., low-order polynomial) or parametric model through your spectrum.*

Probably the best estimate of the null spectrum is using an autoregressive order 1 model (AR(1); AR models discussed in Chapter 14). The following formula provides an estimate of the AR(1) null spectrum:

$$\mathrm{p}_k = \frac{1 - r_1^2}{1 + r_1^2 - 2r_1 \cos(2\pi k/n)}, \tag{11.37}$$

where r_1 is the lag 1 value of the sample autocorrelation function for your time series (if $r_1 = 0$, null spectrum will be for a white noise process), k is harmonic. You will need to scale this to your spectrum by, for example, looking at the ratio of the total variance (sum of power) in your estimated and AR(1) model then use that ratio to scale the null spectrum to fit the estimated one.

Once you place the confidence intervals on the null spectrum, you are looking for those peaks and valleys in your sample spectrum that lie outside. But, always remember, *a 90 percent confidence limit by definition implies that, on average, 10 out of every 100 power spectral estimates will exceed the confidence limits just by chance (hence the 10 percent uncertainty you have for rejecting the null hypothesis when it is true).* Furthermore, because of the long-tailed distribution, *some of these deviations are expected to be quite large and to therefore look convincingly "real."* Also, you may find peaks at frequencies you expect, but they may not be significant at your chosen level of significance. You are the analyst. Find the level at which they are significant, and decide if that is good enough for you to continue your investigation (making this choice clear in any presentation).

We are attempting to "improve" our sample PSD (by averaging, etc.) in order to get the best estimate of the ensemble that the realization was drawn from. The uncertainty (variance) we computed above for our sample spectrum is relative to that ensemble; it's not saying anything about whether we believe certain features are or are not present in the time series analyzed. So, how can you tell that a peak standing above the CI is "real" (i.e., a characteristic of the ensemble) or not? As in many statistical analyses, *your confidence about the significance of certain spectral peaks is bolstered when these same peaks show up with significant power in multiple realizations of the same process – and those particular cycles*

Box 11.1 Interpreting Cycles That Are not "Real"

The fact that we expect several peaks to lie outside of the confidence intervals by chance does not mean that the *individual* realization being analyzed does not actually contain those harmonics. We have already shown that a sample spectrum will exactly fit the time-series data points as an interpolant must. If you analyze a noise time series and get a spectrum containing two large-amplitude harmonics that lie within the CI, you will interpret them as noise, or not "real," at the chosen significance level. But for this particular realization, they *are* real in the sense that they do show a strong cyclic presence in that single realization. When you tell someone those cycles are not real, and they look at you like *you're* not real, you must explain that they do indeed appear in this realization but they are unlikely to appear in other realizations from the same ensemble. That is, this particular realization dominated by these two harmonics is unlikely to be representative of the *ensemble* for the process being sampled. We are averaging (or smoothing) the spectrum to reduce the huge variance that allows such large-amplitude harmonics from noise in order to find the most accurate representation of the process you are sampling. On the other hand, if you are only interested in that particular realization (e.g., a single earthquake seismogram or some particular seafloor bathymetry) and don't care what the ensemble's most representative series may look like, then you don't need to do any smoothing, since the sample PSD you get describes that phenomenon perfectly.

are consistent with other information you have about the process you are investigating (the latter being the most important factor in all analyses – all or most of the evidence points to a single consistent answer; in that respect, the confidence intervals constitute nothing more than one more piece of evidence; typically, they are not so powerful as to trump other pieces of information, though in some cases they certainly are, and those cases are usually obvious).

Box 11.2 Why White Noise Cannot Exist

As seen in Chapter 9, §9.8.7, the variance of the time series is equal to the sum of the power within the PSD-resolved frequencies. So pure (i.e., continuous) white noise would have infinite variance and generate a constant power equal to σ_ε^2 to $\pm\infty$ frequencies (hence, with infinite integrated power). This alone is unfeasible, as is the notion that anything can be recorded continuously, meaning there will always be some Δt between data points, making the white noise band limited, as defined above – this also preventing infinite constant power, since it will drop to 0 after f_c. Whether

Box 11.2 (Cont.)

a process can even exist in nature that generates a time series continuously is debatable.

When making a finite-length discrete band-limited white noise series, its PSD will be a constant of average power, σ_ε^2. But if you want to satisfy Parseval's theorem (§9.8.6), making the integral (sum of discrete) PSD frequencies, you will have to divide that by $1/N$ in order to satisfy the rule that the energy of the time series is equal to that of the PSD (N being the number of frequencies).

White Noise

White noise (Chapter 7) is noise in which each individual point is *uncorrelated* to all other points (neighboring and distant) in the time series. Thus, the *autocorrelation* function (acf) of white noise is a unit impulse at lag 0, and essentially zero at all other lags by definition, since there is no correlation between neighboring points, but the noise does lead to random scatter at the nonzero lags with variance of $1/n$ in the acf, as shown in equation (7.54). The Fourier transform of an impulse (Chapter 9, equation (9.86)) approximates a flat line – *equal amplitude or power at all frequencies*.

The *autocovariance* function (acvf) of white noise is similar to the autocorrelation function, except that at the origin (0th lag), the value is not 1 but instead is the variance, σ_ε^2, of the white noise process. Therefore, the Fourier transform of this, a unit impulse scaled by σ_ε^2, gives the same constant (flat) spectrum, but in this case the constant value of the PSD is σ_ε^2 (c in equation (9.86)).

Band-limited white noise is discrete white noise where each datum in the series is independent of the others, but each datum is separated by some time, Δt. Hence, there is no frequency higher than $f_c = 1/2\Delta t$, so $|f| \le f_c$.

11.3.3 Single-Realization Treatment

Segmenting the Data

While the preceding ensemble averaging represents an ideal manner by which to improve the statistical properties of the power spectral estimates, reality often rears its ugly head and leaves you stuck with just a single realization of the stochastic process (and one that was amazingly difficult to obtain). In that case, we might attempt to make use of ergodicity (i.e., "averaging over time is equivalent to averaging over the ensemble at one time") to approximate the above procedure. You do this by breaking up a long time series into a number – say, m – of shorter segments, and treat each segment as if it were a separate realization. Then compute the PSD for each segment and average them as before (average the PSD, *not* the time series segments).[13] The underlying assumption

[13] This is because the PSD does not contain phase information, so how the various cycles are phased relative to one another can vary, changing the shape of the time series. But the PSD will not change, so average the PSD, not the time series.

being used here is that the shorter segments in time are representative of any one segment anywhere else in time.

Consider the example of a time series of total length 3T in Figure 11.10. You break this series into three segments ($m = 3$). Each segment now has length T and, as a consequence, the frequency sampling interval Δf of each segment is 1/T as opposed to 1/3T, as it is for the full-length series.

The averaged spectrum, constructed by averaging the three segment spectra shows an improvement (reduction) in the size of the confidence limits, since we now have $2m = 6$ degrees of freedom versus just 2 otherwise. However, as a tradeoff, we *lose* **spectral resolution**. That is, because of the shorter segments, our Δf (frequency sampling interval) is reduced by a factor of three relative to that of the longer series.

Comparison of the resolved harmonics is shown in Figure 11.11. The bold spectral estimates are at the harmonics, which are preserved after the segmenting (i.e., $\Delta f = 1/T$).

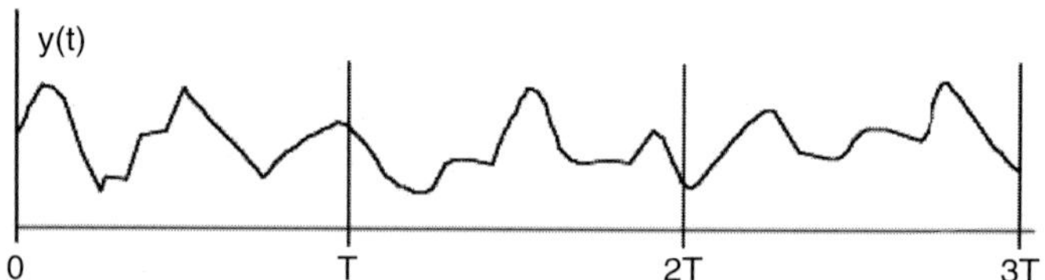

Figure 11.10 This single time series is segmented into three equal-length segments each for separate analysis and then for their spectra to be averaged, simulating the ensemble averaging approach discussed above.

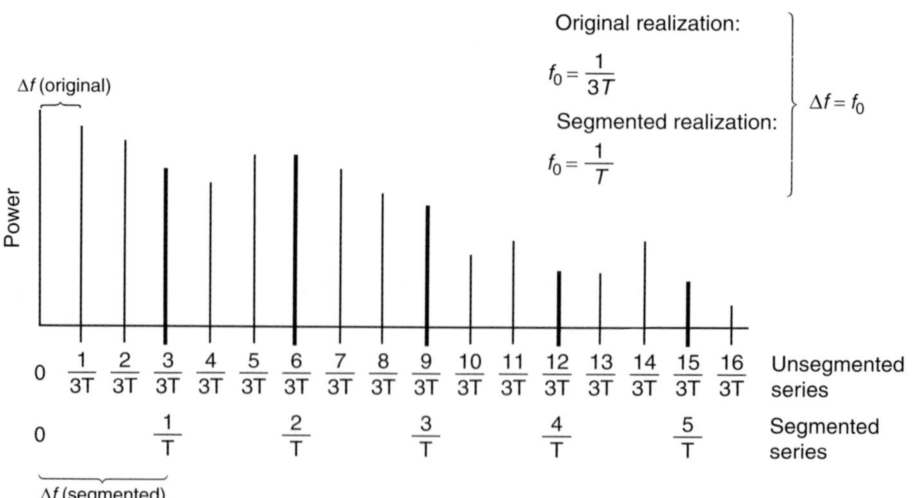

Figure 11.11 Schematic PSD showing frequency resolution for time series of Figure 11.10 before segmenting and after segmenting, showing the reduction in spectral resolution by $(m - 1)/m$.

Estimates at these harmonics have smaller variance than the estimates at any of the harmonics in the unsegmented spectrum, but of course we lose the ability to resolve independent frequency information at the more closely spaced Δf interval (= 1/3T) in the longer series.

The tradeoff between **resolution** and **variance** reduction is the main consequence of all techniques geared toward improving spectral estimates with a single realization. We see later that a third player also enters into the tradeoff: we will be increasing the **bias** of our estimates as we reduce the variance.

Smoothing the Spectrum

The consequence of our recent segmenting of the data was a loss of frequency resolution, suggesting that the center frequency of the new resolved band (bold harmonics in Figure 11.11) is essentially the sole representative of the neighboring frequencies that are no longer explicitly resolved. This suggests another approach to improving the statistical properties of our single spectrum. We have shown that the power spectral estimates at each frequency are independent and uncorrelated to the spectral estimates at neighboring frequencies. Therefore, just as smoothing a time series (averaging neighboring values) can improve its statistical properties by reducing high-frequency noise, we might consider the effects of averaging m neighboring spectral estimates (those just lost by the segmenting approach),

$$\overline{I_j} = \frac{1}{m} \sum_{i=-(m-1)/2}^{(m-1)/2} I_{j+i} , \tag{11.38}$$

where m is an odd integer (and directly analogous to m used in the segmented approach).

This operation is shown graphically in Figure 11.12. In the figure, the spectrum appears at first glance to have maintained all of the frequency information in the averaged spectrum, but just as in the segmented approach, we have actually lost resolution, since only the estimates separated by more than $m - 1$ harmonics are uncorrelated and independent. In other words, the averaging procedure has correlated neighboring estimates over the width of the smoothing window, and therefore the information between every $m - 1$ harmonic is no longer contributing additional information. We could simply drop the non-bold frequencies in the figure and still retain all of the independent information in the smoothed spectrum. Therefore, this technique is accomplishing essentially the same effect as the segmenting approach, and the net result is an increase to $\nu = 2m$, as before, and a resolution loss of $(m - 1)/m$ frequency bands.

Using this approach, it is somewhat easier to see the effects of the segmenting approach (or any of the methods). Predominantly, you can see that the averaging tends to smear spectral peaks (this is the simple manifestation of losing resolution). That is, if we average neighboring spectral estimates, the center frequency of the averaged band represents the mean power of all frequency components lying within the band, so the resolution is now limited to the width of the averaging band (the **bandwidth**). You can also see that this effect is fine if the true spectrum is of a smooth nature itself, in which

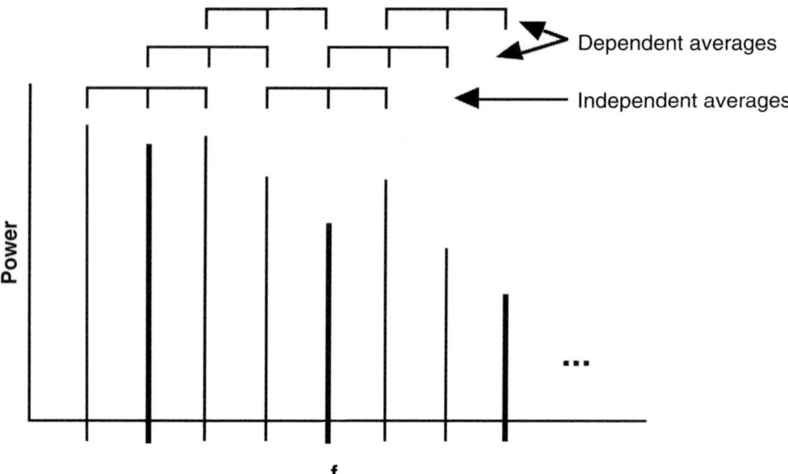

Figure 11.12 Schematic of PSD for spectral smoothing. Bold lines show independent frequency estimates (so, again, for averaging three neighboring frequencies, as with segmenting the data into three segments, we lose frequency resolution).

case the decreased resolution is obviously worth the improved statistical properties. However, if the true spectrum is characterized by sharp peaks, then the improved stability is at a cost of obscuring the true information. We will consider the smoothing (averaging) across neighboring frequencies in more detail later, when considering effects of tapering the data before transforming (which is another smoothing operation).

Truncating the Autocovariance Function

A third operation exists which produces the same results as the above two cases. This involves the truncation of some of the higher lags in the autocovariance function (acvf) before transforming. Recall that the transformation of the autocovariance function directly yields the power spectrum. When $n \gg m$, segmenting a time series with m segments is approximately equal to truncating the acvf at N/m, N being the number of lags in the acvf.

When the acvf is truncated prior to transformation at N/m lags (where N is the total number of lags; m, as in the above cases), the net effect is essentially equivalent to the preceding cases. That is, the frequency resolution is decreased by $(m - 1)/m$ and the degrees of freedom increased to $2m$. Intuitively we might expect some improvement in the improved precision (decreased variance) of our spectral estimates from this approach because the lags being eliminated represent correlations of the data at very large lags, using very few points to compute those correlations. Therefore, truncation of these large lags is in effect eliminating the poorest quality covariance estimates, leaving only those estimates made from a majority of data points.

This method of stabilizing the power spectral estimates is the oldest and has the largest body of theoretical background. The other methods were developed after the development of the fast Fourier transform (FFT), and represent the more common approaches now used – though you should use the method you are most comfortable with. The actual implementation of these techniques requires consideration of tapering windows and other practical aspects considered next.

11.3.4 Practical Considerations

Smoothing as a Spectral Window

As noted, proper interpretation of the spectrum in terms of its statistically stable properties (i.e., in estimating the true PSD of the ensemble) requires some form of smoothing to decrease the variance of the individual estimates. Specifically, if our smoothed power spectral estimates are given by averaging neighboring points as

$$\overline{I_j} = \frac{1}{m} \sum_{i=-(m-1)/2}^{(m-1)/2} I_{j+i} , \tag{11.39}$$

then, *if the true power spectrum is approximately constant over the interval* m, the expectance of the smoothed estimates is as before (for the not-averaged estimates), while the variance of the estimates is now reduced by $1/m$ (recall that in this case, $\nu = 2m$), so

$$\mathrm{E}[\overline{I_j}] = \mathrm{p}_j \tag{11.40}$$

and

$$\mathrm{Var}[\overline{I_j}] = \mathrm{p}_j^2/m. \tag{11.41}$$

These results follow conceptually if we consider that, since the individual estimates of I_j are uncorrelated and independent, the variance reduction reflects the fact that we are effectively averaging m independent estimates and benefiting from the Central Limit Theorem (the original variance is divided by m). The beauty of this result is that $\overline{I_j}$ represent *consistent* estimates of the true p_j. The raw I_j estimates are inconsistent, but the smoothed estimates are consistent because the variance goes to 0 as m goes to infinity.

The expectance, on the other hand, only remains unbiased as long as the averaged frequency estimates have approximately the same mean as the central value. Consider the consequence when this is not the case (most likely to be true). Then, the averaging of neighboring peaks cannot converge to the true estimate at the frequency j as we increase m, because the values being averaged have different mean values themselves (except for white noise).

Consider the expectance of $\overline{I_j}$ directly when computed by smoothing over m neighboring frequency bands:

$$\begin{aligned}
\overline{I_j} &= \frac{1}{m} \sum_{i=-(m-1)/2}^{(m-1)/2} I_{j+i} \\
\mathrm{E}[\overline{I_j}] &= \mathrm{E}\left[\frac{1}{m} \sum_{i=-(m-1)/2}^{(m-1)/2} I_{j+i}\right] \\
&= \frac{1}{m} \sum_{i=-(m-1)/2}^{(m-1)/2} \mathrm{E}[I_{j+i}].
\end{aligned} \tag{11.42}$$

We showed previously (equation 11.22) that $E[I_j] = W_k * p_k$, where the W_k are the weights of the transform of $\varphi(\tau)$ (the triangular Bartlett window), the truncation function multiplying our autocovariance function before transformation to the power spectrum. Recall that for the Bartlett taper, these weights describe the Fejer kernel:[14]

$$W_k = \left[k \frac{\sin(2\pi\tau)}{2\pi\tau}\right]^2 \tag{11.42a}$$

and

$$W_k * p_k = \sum_{j=-(n-1)}^{(n-1)} W_{k-j} p_j \, , \tag{11.42b}$$

where the lags of the true power spectrum have been written as j.

From this,

$$E[I_{j+i}] = \sum_{j=-(n-1)}^{n-1} W_{k-j-i} p_j . \tag{11.43a}$$

We are computing the mean of the averaged spectra in (11.42), which now with (11.43a) can be rewritten as

$$= \frac{1}{m} \sum_{i=-(m-1)/2}^{(m-1)/2} \sum_{j=-(n-1)}^{n-1} W_{k-j-i} p_j . \tag{11.43b}$$

Rearranging the order of summation,

$$\begin{aligned} &= \frac{1}{m} \sum_{j=-(n-1)}^{n-1} p_j \sum_{i=-(m-1)/2}^{(m-1)/2} W_{k-j-i} \\ &= \frac{1}{m} \sum_{j=-(n-1)}^{n-1} p_j W_{k-j}^{SP} \\ &= p_k * W_k^{SP} \, , \end{aligned} \tag{11.43c}$$

where

$$W_k^{SP} = \frac{1}{m} \sum_{i=-(m-1)/2}^{(m-1)/2} W_{k+i} . \tag{11.44}$$

That is, the expected value of each smoothed power spectral estimate, $\overline{I}_j$, is the convolution of the true power spectral values, p_j, with a spectral window that consists of the

[14] A consequence of normalizing the acvf with $1/n$, as shown in (11.20) through (11.23).

sum of several Fejer kernels, each offset by a single lag and summed over the width of the averaging window (and that sum divided by the number of added kernels). Graphically, this is shown in Figure 11.13.

This window has a very nearly rectangular form (comparable to a running average) resulting from the averaging of the m offset Fejer kernels. The result is called the **smoothed periodogram spectral window**. As $m \to \infty$, this smoothing goes to a pure weighted average of all spectral estimates.

Essentially, this shows that the expected value of the smoothed I_j estimates is nothing more than the running average smoothing of the true power spectrum. Of course, if the true spectrum is a constant, then the expected values of $\overline{I_j}$ are simply equal to the true values. However, as p_j deviates from a constant, we can see that the expected values also deviate, and thus we introduce a bias into our smoothing procedure.

Smoothing as a Lag Window

When you smooth the spectrum by truncating the autocovariance function and transforming, that operation is a multiplication in the time (lag) domain, so the transform (requiring the cosine term only, since the autocovariance is an even function) is

$$\overline{I_j} = \sum_{i=-M}^{M} \lambda_i R_i \cos 2\pi f_j \tau_i \ , \tag{11.45}$$

where λ_i is a set of weights describing the truncation function. The simplest case is where $\lambda_i = 1$, a rectangular or gate function, for all $|i| \leq M$; $M = N/m$ where N is the total number of lags and m is as defined previously, the number of frequencies averaged in a smoothing approach. The weights λ_i define a **lag window**. The value of M is the **truncation point**. In the frequency domain, this operation is a convolution of a form similar to that of (11.43). In fact, the smoothing of the periodogram in general can be expressed either by (11.43) or by (11.45) for all of the smoothing methods discussed previously.

As an example, consider one of the most popular shapes of lag window, the cosine form of the **Tukey–Hanning window**, given by

$$\lambda_i = \frac{1}{2}\left[1 + \cos\left(\frac{2\pi i}{2M}\right)\right], \tag{11.46}$$

where $i = 0,1,2, \ldots,M$.

This corresponds to transformation of the autocovariance that has been truncated by a *lag window*, sketched in Figure 11.14.

Alternatively, this is the same as *convolving* the power spectral estimates by the transform of the Tukey–Hanning window. The transform of a lag window forms a **spectral window**, which in this case (Figure 11.15) has a form similar to a sinc function but has a similar form but smaller side lobes. While it does have smaller side lobes, the central lobe in this window is broader than that of the sinc function. Therefore, this

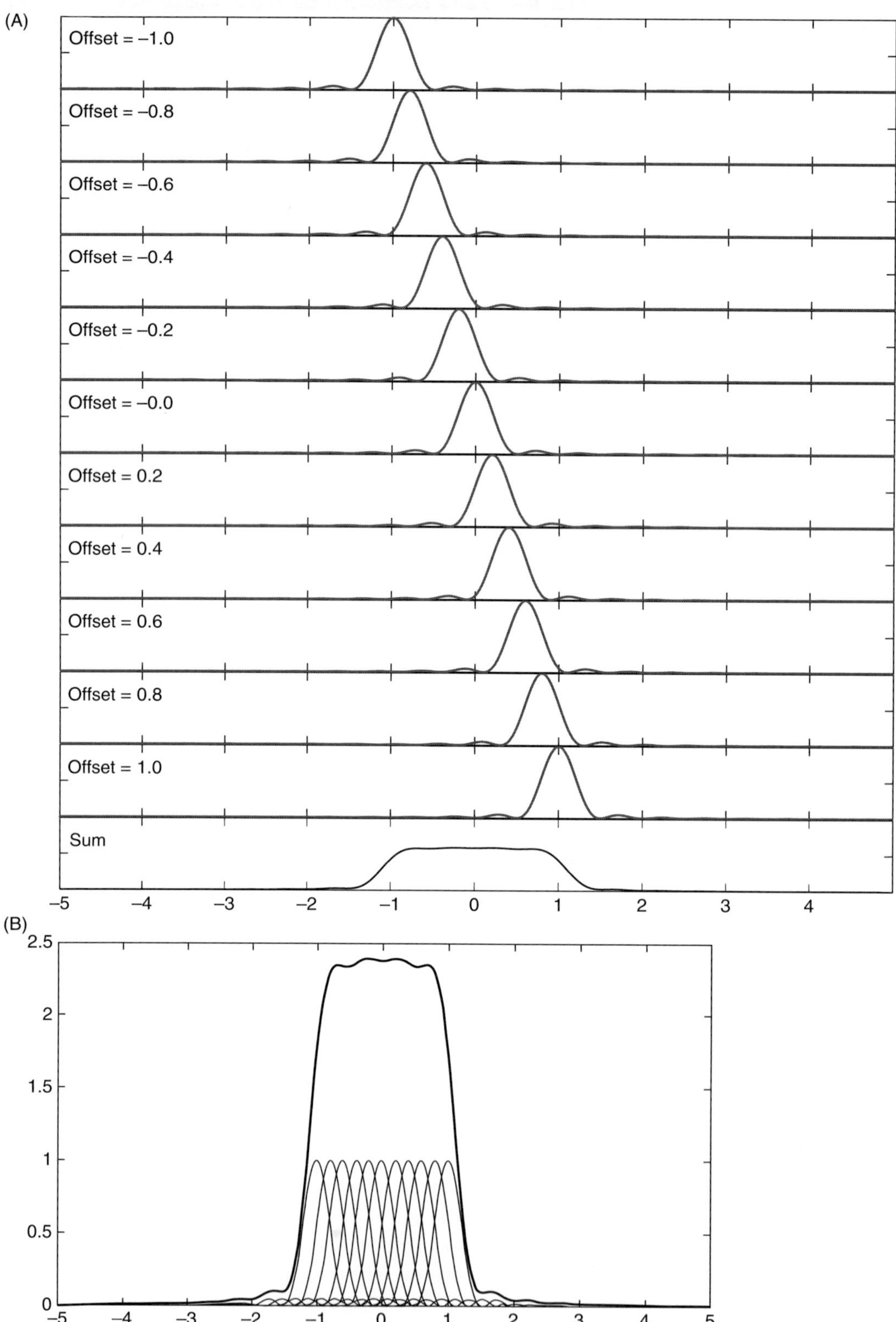

Figure 11.13 **(A)** Example of the sum of 11 offset Fejer kernels producing a nearly rectangular window, comparable to a running average filter. **(B)** Expanded view of all offset Fejer kernels and their sum superimposed.

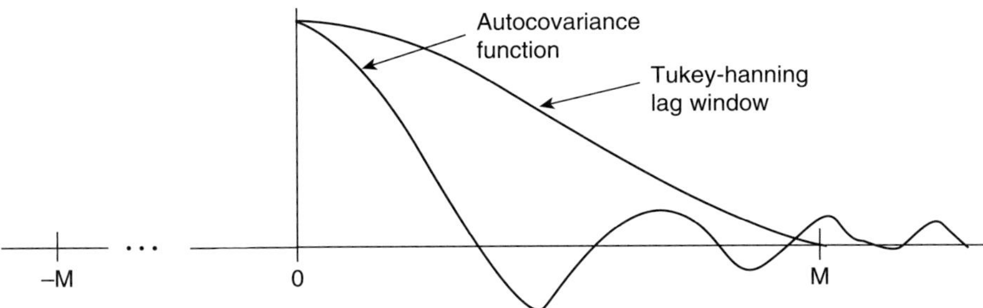

Figure 11.14 Schematic of the Tukey–Hanning lag window multiplying acvf.

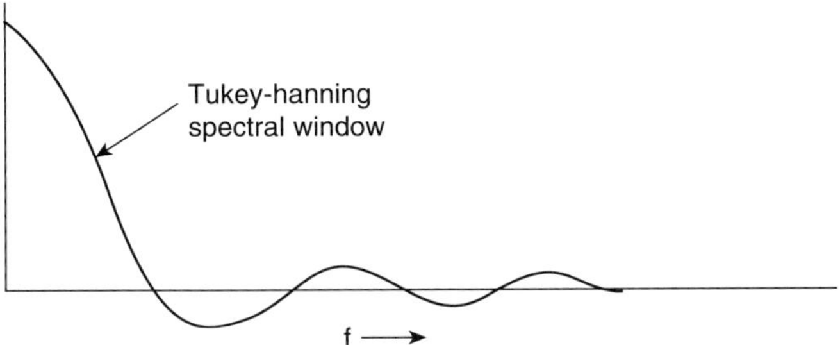

Figure 11.15 Schematic of the Tukey–Hanning spectral window.

window tends to average by giving more weight to nearby neighboring spectral estimates (relative to the sinc function) and less weight to distant spectral estimates.

The Tukey–Hanning window has $8n/3M$ equivalent degrees of freedom (i.e., $\nu = 8n/3M$), so if you transform the autocovariance function after multiplying it by a Tukey–Hanning lag window (11.46), the variance is reduced according to χ^2_ν that decreases with increasing ν.

Examination of the Tukey–Hanning spectral window indicates that this window will introduce some bias into the estimate, and this amount is given by $\pi^2 p''(f)/4M^2$, where $p''(f)$ represents the second derivative of the true power spectrum at f. Therefore, the bias is proportional to the curvature of the spectrum, which means that in the presence of peaks or troughs the bias will be largest. One can compute the bias for each of the various spectral windows commonly used. If you wish to experiment, the bias of several popular windows is given in table 6.1 (pg. 463) of Priestley (1981); degrees of freedom for each window are given in his table 6.2 (pg. 467).

Bandwidth

Consider now the concept of **spectral bandwidth, B_s**. The bandwidth of the true spectrum is given (most commonly) by the distance across the narrowest peak at the half-power points, as shown in Figure 11.16.

This quantity is a measure of the width of the narrowest feature of the true power spectrum, and thus an indication of the resolution (Δf) required of the estimated power

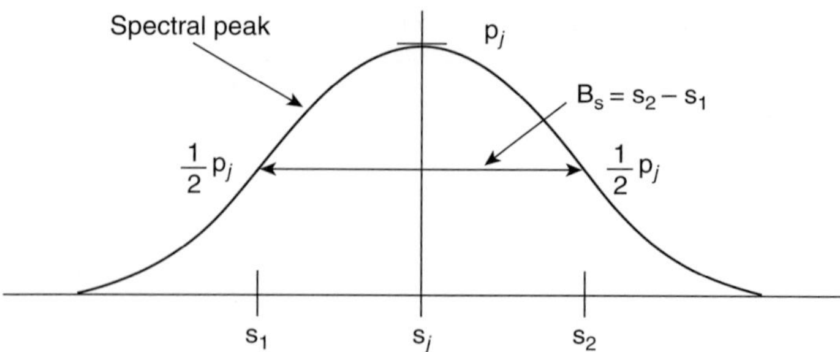

Figure 11.16 Schematic showing definition of spectral bandwidth (width of the half-power points of the narrowest spectral feature).

spectrum. As you average neighboring frequency estimates together, you lose spectral resolution (Figure 11.12). So, in the straightforward case where we simply average m neighboring estimates (all of which are given equal weight), we lose $m - 1$ frequency estimates between each m estimates. In other words, we have a window bandwidth, B_w, of m, which results in our losing our ability to resolve information about the nature of the spectrum over the width of this window.

While the measure of window bandwidth is straightforward in Figure 11.12 (the window is rectangular in shape = **Daniell window**), we require some measure of the window bandwidth for a window like that found when using the Tukey–Hanning window, shown above. Several different methods for estimating this value have been put forth. Most differ only in subtle theoretical properties. Bandwidths are included in Priestley's (1981) table 7.1 (pg. 527).

Window Bandwidth versus Truncation Point

From examination of the lag window versus the spectral window and our general knowledge of transform properties, we can see some relationship between window bandwidth and lag window truncation point. In general, as the truncation point gets larger (more lags included in the transform), the transform of the lag window gets "tighter" (approaches a delta function) or the window bandwidth gets smaller. So, while we want to make the truncation point as small as possible to get the best estimate of the power spectrum (that with the smallest variance), we do so at the expense of resolution, since the window bandwidth is a measure of the frequency interval, or band, over which neighboring frequencies have been averaged, and are therefore no longer independent and uncorrelated. Priestley (1981) gives the measure of window bandwidth in his table 7.1 (pg. 527) as a function of truncation point. Obviously, if one has an idea of the *true* spectral bandwidth of the random process, then the truncation point should be such so as to produce a $B_w \leq B_s$ (i.e., resolving the spectral bandwidth required for your data).

In general, *decreasing the truncation point (or making* m *larger for the other methods) serves to (1) decrease variance (more stable estimates), (2) increase bandwidth (less resolution) and (3) increase bias (converge farther from true spectrum)*. That is, a smaller truncation point corresponds to multiplying the autocovariance function by a *narrower* taper, and thus convolving the true spectrum with the transform of this narrower taper. The more narrow a taper, the broader the central lobe of its transformed function (typically of sinc-like form), thus convolving the true spectrum at each frequency with a wider band. In the limit, a taper equivalent to an impulse function, the sinc function, reduces to an infinite-length straight line, as shown in (9.86). This results in equally averaging spectral estimates over the entire frequency band. Consequently, you achieve more smoothing (reduced variance) but less resolution, and since there is more likelihood that the true spectrum is not linear over a broader band you are more likely to smooth toward a value that is different from the true unsmoothed value, thus increasing the bias.

11.3.5 Least-Squares Spectral Estimates

Finally, there is an obvious approach for obtaining most representative estimates of the Fourier coefficients: a least-squares (LS) fit of the n data points using fewer than n sinusoids. For this, we can set up the linear (in the coefficients) system and solve it through one of the matrix techniques described previously. But for this case, because of the orthogonality of sines and cosines over an interval T for evenly spaced samples, the system can conveniently be solved analytically. Because the LS employs a model, this method of estimation is a different animal than the more generic methods described above. *This LS approach is useful only in those special cases where you have an actual model describing your data, with sinusoids exactly periodic over the length of the time series (rarely true).* This method has its own uncertainty analysis and testing for significance because of the model.

The model assumes that your data consist of m (deterministic) harmonic terms plus some zero-mean random noise, ε_t,

$$y_i = \sum_{j=0}^{m} [a_j \cos(2\pi f_j t_i) + b_j \sin(2\pi f_j t_i)] + \varepsilon_i, \tag{11.47}$$

where $m < n/2$ if n is even and $m < (n-1)/2$ if n is odd. The term ε_i represents the noise at each discrete time point, i.

You determine the optimal values of the m coefficients (a_j and b_j) – those that minimize the sum of the squared error, e – measured in the usual manner:

$$\begin{aligned} e &= \sum_{i=1}^{n} \varepsilon_i^2 \\ &= \sum_{i=1}^{n} (y_{\text{observed}} - y_{\text{computed}})_i^2 \\ &= \sum_{i=1}^{n} \left(y_i - \sum_{j=0}^{m} [a_j \cos(2\pi f_j t_i) + b_j \sin(2\pi f_j t_i)] \right)^2. \end{aligned} \tag{11.48}$$

Taking the derivative of (11.48) with respect to the coefficients and applying orthogonality conditions yields (see derivation (D11.1) for details) the discrete cosine and discrete sine transforms(!),

$$a_j = \frac{2}{n}\sum_{i=1}^{n} y_i \cos 2\pi f_j t_i \tag{11.49a}$$

$$b_j = \frac{2}{n}\sum_{i=1}^{n} y_i \sin 2\pi f_j t_i . \tag{11.49b}$$

These solutions for the Fourier coefficients of the least-squares problem are identical to those obtained from the Fourier transform for the time series (the interpolation problem). We can accomplish a least-squares fit by simply dropping an arbitrary number of the coefficients from the standard Fourier transform result.

This outcome provides some conceptual insights into the nature of orthogonal functions such as the sines and cosines used here. Orthogonal functions – whether we use sines and cosines or any other orthogonal function – are *completely* independent of one another, so the fit of any one harmonic term does not depend in any way on the fit of the other harmonics. Therefore, you should not expect that the solution for one harmonic will change simply because you are using fewer harmonic functions when doing the least-squares fit. The fit of the jth harmonic term is not dependent on the fit of the other harmonic terms, and thus its solution does not change, whether it is the only harmonic being fit or simply one of n harmonics being fit. This feature of orthogonal functions is fully apparent in the form of the solutions themselves, which, for any coefficient a_j, involves only the sum of the data values multiplied by a sine or cosine of the jth harmonic (the covariance between the time series and the function being fit). Consequently, each coefficient is determined from a single equation, independent of the other coefficient values and harmonics. This relationship is responsible for the ease of solution and practicality of the entire Fourier technique.

Choosing Harmonics for the Least-Squares Estimate

You are now left with determining how many (and which) harmonic terms to keep to produce the most representative fit. For this case, *we are estimating terms of a best-fit Fourier (cyclic) model* given by (11.47), so the sines and cosines being LS fit are for a *deterministic* process whose mean varies cyclically – this is different than the approach we use for finding the frequency content and assessing its uncertainty for a stochastic process (discussed above).

You can estimate the optimal number of terms to retain via a test of the null hypothesis for the individual harmonics. That is, you estimate for each frequency whether the power of the particular harmonic term is significantly different from 0. If it is, you assume that the frequency represents a legitimate harmonic component in the data and retain it in the model. If the power is statistically indistinguishable from 0 at some level of significance, you drop that coefficient from the model (remember that you have just computed the standard deviation of the coefficient, equation (11.27), allowing you to make that determination as done in Chapter 3 hypothesis testing). In this manner, you hope to identify true harmonic components in the data from which you can construct the model. The residual from the model then represents the random noise, ε_t.

Identification of Statistically Significant Peaks for the LS model
If the least-squares model of (11.47) actually represents your data, *pure sinusoids, perfectly periodic over the length of the time series embedded in white noise – a condition rarely realized naturally* – you can employ a test to determine which

Box D11.1 Derivation of Least-Squares Fourier Fit

The minimum of e with respect to the coefficients a_j and b_j is obtained by taking the derivative of e with respect to each coefficient and setting it to zero (forming the system of normal equations):

$$\frac{\partial e}{\partial a_k} = 2\sum_{i=1}^{n}\left[\left\{y_i - \sum_{j=0}^{m}[a_j \cos(2\pi f_j t_i) + b_j \sin(2\pi f_j t_i)]\right\}\cos(2\pi f_j t_i)\right] = 0 \tag{D11.1.1a}$$

and

$$\frac{\partial e}{\partial b_k} = 2\sum_{i=1}^{n}\left[\left\{y_i - \sum_{j=0}^{m}[a_j \cos(2\pi f_j t_i) + b_j \sin(2\pi f_j t_i)]\right\}\sin(2\pi f_j t_i)\right] = 0 \tag{D11.1.1b}$$

for $k = 0,1,2, \ldots,m$.

Interchanging the order of summation and rearranging gives the standard form of the normal equations (one equation for each a_k coefficient and one for each b_k coefficient):[15]

$$\sum_{i=1}^{n} y_i \cos(2\pi f_k t_i) = \sum_{j=0}^{m}\left[\left\{\sum_{i=1}^{n}[a_j \cos(2\pi f_j t_i) + b_j \sin(2\pi f_j t_i)]\right\}\cos(2\pi f_k t_i)\right] \tag{D11.1.2a}$$

and

$$\sum_{i=1}^{n} y_i \sin(2\pi f_k t_i) = \sum_{j=0}^{m}\left[\left\{\sum_{i=1}^{n}[a_j \cos(2\pi f_j t_i) + b_j \sin(2\pi f_j t_i)]\right\}\sin(2\pi f_k t_i)\right] \tag{D11.1.2b}$$

[15] Equation (D11.1.2a) is the same form as (9.26) in the derivation of the discrete cosine transform of Chapter 9, so go there to see this solved more explicitly if you find the treatment to be too brief here.

Box D11.1 (Cont.)

We know from the orthogonality relationships that the sum over n of cosine times sine terms always equals zero, so these cross-products drop out, leaving

$$\sum_{i=1}^{n} y_i \cos(2\pi f_k t_i) = \sum_{j=0}^{m} a_j \left\{ \sum_{i=1}^{n} [\cos(2\pi f_j t_i)] \cos(2\pi f_k t_i) \right\} \quad \text{(D11.1.3a)}$$

and

$$\sum_{i=1}^{n} y_i \sin(2\pi f_k t_i) = \sum_{j=0}^{m} b_j \left\{ \sum_{i=1}^{n} [\sin(2\pi f_j t_i)] \sin(2\pi f_k t_i) \right\} \quad \text{(D11.1.3b)}$$

The orthogonality conditions can be directly applied to the products on the right-hand side of (D11.1.3): the sum over a period of the product of two cosines, $\Sigma\cos(\omega_j t_i)\cos(\omega_k t_i) = 0$ for all $j \neq k$; $= n/2$ for all $j = k \neq 0 \neq n/2$; and $= n$ for $j = k = 0 = n/2$. Similarly, for $\Sigma\sin(\omega_j t_i)\sin(\omega_k t_i) = n/2$ for all $j = k \neq 0 \neq n/2$. So, defining the a_0 coefficient as $a_0/2$ and $a_{n/2}$ as $a_{n/2}/2$ yields the solutions for the a_k and b_k coefficients as

$$a_k = \frac{2}{n} \sum_{i=1}^{n} y_i \cos(2\pi f_k t_i) \quad \text{(D11.1.4a)}$$

and

$$b_k = \frac{2}{n} \sum_{i=1}^{n} y_i \sin(2\pi f_k t_i). \quad \text{(D11.1.4b)}$$

coefficients are significantly different from those values expected for the white null spectrum (the ε_t). That is, you can apply a test to select those harmonics that are so large that they are unlikely to have been produced by the noise. This requires what was just established in the previous two sections: knowledge of the distribution of the null spectral values and their lower-order moments (expected values and their variance).

For this particular type of purely cyclic time series, the peaks in the PSD will be lines, forming "line spectra," because they are perfectly periodic over the length of the time series by definition (thus, not smeared by leakage). The null hypothesis (H_0) states that the Fourier coefficient values are 0; the alternate hypothesis, H_1, states that Fourier coefficient values are not 0 (i.e., the peak being tested is significantly different from zero, and thus representative of a true harmonic component). So,

$$H_0 : a_j, b_j = 0 \tag{11.50a}$$

$$H_1 : a_j, b_j \neq 0. \tag{11.50b}$$

The test statistic used here is the **Fischer *g* statistic,**[16]

$$g = \frac{\max_{1 \leq k \leq m} |F_k|^2}{m^{-1} \sum_{j=1}^{m} |F_j|^2}, \tag{11.51}$$

where m is the number of harmonics (j) in the sample spectrum. This g statistic is a slight variant on the χ_2^2 distribution, and it must be used instead of directly using the χ_2^2 distribution since you do not actually know the variance of noise in the data series, and you must estimate it as the variance of the residual noise about the deterministic sine and cosine components within the series. You estimate this variance using a subtle variant of Parseval's theorem, stating that the variance of the spectrum is equivalent to the variance of the series. In (11.51), the denominator is a large-sample estimate of the variance of the noise, since the sum in the denominator is over all frequencies except those of any frequency components that have been identified as being significantly different from zero (as discussed later).

The g distribution is based on the χ_ν^2 distribution for $\nu = 2$. This is a special case of the chi-squared distribution that is equivalent to an *exponential distribution* whose cumulative distribution function (CDF) is approximately given by

$$P[g \leq g_\alpha] \sim [1 - \exp(-g_\alpha/2)]^m, \tag{11.52a}$$

and for convenience, the complement of this CDF – that is, the probability that the test statistic g – is greater than some critical value, g_α, is

$$P[g > g_\alpha] \sim 1 - [1 - \exp(-g_\alpha/2)]^m. \tag{11.52b}$$

As with all statistical tests, you choose a critical value of g (g_α) such that the probability that g is greater than g_α, $P[g > g_\alpha] = \alpha$, where α is the level of significance selected (Figure 11.17).

Having chosen the level of significance that is acceptable for considering a peak as nonzero, you compute the critical value of g via the approximation

[16] See Priestley (1981) for a more thorough description of this test and other tests to identify signal (sines and cosines) in noise. The discipline of finding deterministic signal in noise is often referred to as "signal processing."

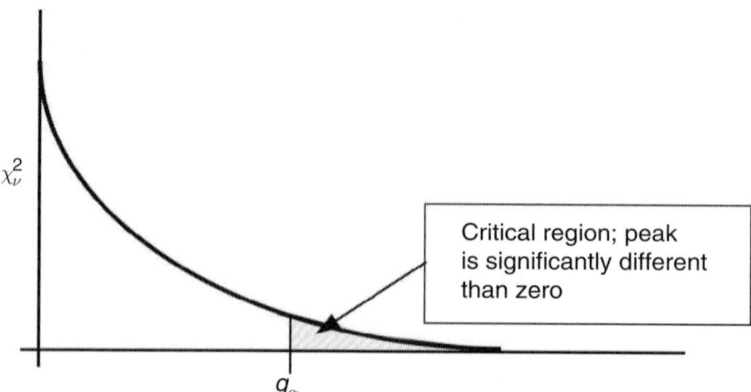

Figure 11.17 Chi-squared distribution for $\nu = 2$, appropriate for the g statistic of the power spectral estimates. At a specified level of significance (α), the critical value g_α determines values failing the null hypothesis (i.e., spectral peak is likely not noise).

$$g_\alpha \approx 1 - (\alpha/m)^{[1/(m-1)]}. \tag{11.53}$$

Then compute the power spectrum of the data series and find the largest peak in it (i.e., the largest $|F_j|^2$). Plug this value into the numerator of (11.51) and the sum of the power spectral values divided by m into the denominator of (11.51). This provides a value of g. If the value of g exceeds that of g_α, then that power spectral line is significantly different from zero at the 100α percent level, the null hypothesis is rejected, and you conclude that the time series y_i contains that periodic component.[17] If so, eliminate this peak from all the remaining peaks and repeat this operation (i.e., m is reduced by 1), now looking for the second-largest peak (so, the largest peak is eliminated). Repeat this operation until you arrive at the first peak that no longer exceeds the value of g_α. Ultimately, all of those peaks that exceed the g_α are considered significantly different from the null spectrum (at the α significance level), and no other deterministic sinusoids are present. The remaining ones just make up the white noise.

This seemingly complicated iterative procedure was developed to account for the fact that the actual level of noise, σ_ε^2, in the data is unknown, so this procedure takes the uncertainty of this estimate into account. We have not assumed stationarity of the data in this case, but we *have* assumed that the noise present is white noise and that the periodic components present are periodic over the length of the series (and thus are Fourier frequencies of this length series). There are several other test statistics available for this case, each of which has specific desirable properties – see Priestley (1981) (§6.1.4) or Percival and Walden (1993) (§10.9) for the details of these. If you will be using this test, it is worth reviewing the above references to consider the other, possibly better, alternative tests, often involving subtle variants of what has been presented here. It may be worthwhile to use several of the alternatives

[17] Specifically, you conclude that that cycle was not likely generated by the noise, recognizing that you have a $\alpha 100$ percent chance of being wrong in that conclusion (i.e., it was unusually large, but possible for noise).

and see if they make a meaningful difference in which coefficients are accepted as being genuine signals. If you do get different results, make sure you have a thorough understanding of the tests to determine why you get different answers and which test is most appropriate for your data.

The significant peaks thus make up the periodic components of the time series in the least-squares model of equation (11.47). Note that sometimes it may be desirable to compute the power spectrum at finer frequency intervals than are required so that you will be more likely to capture the actual frequency components of the signal. This is simply done by padding zeros to the time series before transforming, which effectively lengthens the series and thus gives a longer T, so the frequency interval (1/T) is smaller. Padding zeros does increase spectral resolution, it simply interpolates the PSD to higher Δf, providing a smoother-looking spectrum.

This procedure works best if the true periodic components making up the time series are at the Fourier frequencies. If this is not the case, then this test becomes decreasingly reliable as the true frequencies deviate more and more from the Fourier frequencies (though a high signal-to-noise ratio in the data will help to extend the usefulness of this test in such cases). Realize also that *this test is based on the <u>unlikely assumption</u> that* (11.47) *is a reasonable model of your data set (i.e., that the data really are composed of some periodic components and some amount of white noise)*. If this is not the case, then you need to use one of the other tests available in the references mentioned (Priestley (1981) and Percival and Walden (1993)), or, more likely, use the confidence for standard PSD estimates as discussed previously.

Comparable tests exist for models using different types of noise (e.g., for colored noise, see Percival and Walden (1993), §10.11). These tests are a bit more complicated but are typically more likely to be representative of real data series that have truly periodic components over the length of the sampled series.

11.4 Spectral Estimation in Practice

You perform spectral analysis when you suspect that there are periodic (cyclic) components in your time series and you wish to identify them quantitatively or when you suspect your data show spectral coloring of noise (useful for identifying some forms of physical processes). For this you will follow these steps:

1) Ensure that time series are evenly spaced for application of the Fourier Transform (Chapter 10).
2) Apply the Fourier transform to bring the time series into the frequency domain (Chapter 10).
3) Estimate the null spectrum (i.e., the noise spectrum stating there are no cyclic components present other than those expected, given the random process for which your data were drawn – described by your null spectrum) (§11.3.2).

4) Perform some form of spectral smoothing to give reduced uncertainty in power estimates (at the expense of reduced frequency resolution and increased bias in estimates) (§11.3.3).
5) Generate confidence interval for desired level of significance (α) to assess which, if any, frequency bands contain power above or below that expected from the null spectrum (§11.3.2).

Given the considerations of this chapter, we can now present the actual procedure followed to produce a good spectral estimation, given the limitations imposed by the data set at hand.

11.4.1 Sampling

First, you must think about the *sampling* of the physical process itself: determine (1) the length of the series, T, and frequency resolution ($\Delta f = 1/T$) required and (2) the best value of Δt to provide the desired range of frequencies and highest frequency ($f_N = 1/2\Delta t$).

Length of Time Series

The time series, T, must be long enough to provide information over the lowest frequency of interest. For example, if one desires to study the seasonal temperature variability of Paris, that requires a time series that covers at the very least one year (try to get as many full cycles of the lowest frequency component of interest). The fundamental harmonic will be at 1/T, and all other harmonics will be of higher-frequency components (at integral multiples of 1/T).

For statistical stability, you will need to choose a truncation point of less than T at say, M, after which your lowest frequency will be at $1/M$. Therefore, in order to resolve the lowest frequency of interest with some statistical stability, the total time series must be long enough so that the truncation point is still large enough to resolve the period of interest – hence, the comment about obtaining many full cycles. For example, if you need to resolve 1 year, obtaining a 10-year record allows you to break it into three segments, each containing 3 full years, and the variance is improved by 6v. *Longer is better.*

Sampling Interval

The sampling interval, Δt, must be small enough to resolve the highest frequency of interest and prevent aliasing (assuming one has some information about the highest frequency (f_N) in the process or its natural cutoff frequency, f_c). So, $\Delta t \leq 1/2f_c$ or $\Delta t < 1/2f_N$. Often in the natural sciences, the true spectrum is red in color – that is, most of the variance of the signal is contained in the lower frequencies, and the variance at higher frequencies drops off toward zero or a noise "floor" (recall that the transform of a signal plus noise is the transform of the signal plus the transform of the noise, which is a constant (σ_ε^2, the variance of the noise). In such a case, one expects to see low values in the high frequency range, which tends to support the notion that any aliasing encountered is either negligible or nonexistent.

11.4.2 Smoothing

Next we must consider the best spectral or lag window shape and width to use for smoothing purposes.

Spectral/Lag Window Shape

You need to average the raw sample spectrum to reduce the variance of the estimates. You smooth the spectrum by convolving it with the spectral window. This is the same as multiplying the time series by the inverse transform of the spectral window, which is the lag window. Therefore, instead of smoothing the PSD, we can directly produce a smoothed PSD by multiplying the time series first with the lag window.

First you must determine which lag window you will use. The actual window shape chosen should reflect the nature of the true spectrum and any particular requirements of the specific investigation. In general, though, most of the windows (except the Bartlett) that are frequently presented are quite acceptable for general usage. There is a tremendous body of literature about the properties and characteristics of certain windows. For example, Priestley (probably the most complete source on window designs and characteristics) shows that a window he calls the Bartlett–Priestley window shows "optimal" properties. That is, (1) this window is non-negative everywhere (since the true spectrum must be everywhere positive, windows with negative values in places are sometimes considered inconsistent with the general theory) and (2) the window offers the largest decrease in variance at the least introduction of bias. If, however, one were to accept negative windows, then the Tukey–Hanning window is optimal. In either case, the criteria used to define "optimal" serve more of a theoretical basis than any real practical purpose. Probably the most popular windows are the Tukey–Hanning and Parzen (all books on spectral analysis give most of these windows; see Priestley (1981) for nice tables summarizing their properties).

Truncation Point

The truncation point chosen will determine the window bandwidth of the sample spectrum, because it defines T, and the frequency interval (controlling bandwidth), $\Delta f = 1/T$. This step is quite important and actually plays a dominant role in the nature of the estimated spectrum. Obviously the ideal way to choose the truncation point is by having first-hand knowledge of the true spectral bandwidth (B_s). In that case, simply choose the truncation point so as to produce a window width $\leq B_s$, or ideally, $B_w \approx (1/2)B_s$ if possible. Lacking knowledge of B_s, three other techniques are commonly employed for choosing the truncation point.

ACVF Inspection

Conceptually, we can see a relationship between the rate at which the autocovariance function decays to zero and the bandwidth of the time series by considering the limiting cases. First, consider the case of pure noise. With this, we know that the acvf should have a value of σ^2 at $\tau = 0$ and be zero for all lags other than zero. The transform of this function is then a constant that has a $B_s = \infty$. Alternatively, consider the acvf of a pure sinusoid. It is periodic (a sinusoid itself) and never decays to zero. The transform of this function is a single spike or a delta function that has an effective bandwidth of 0. Thus it is seen that the more rapidly the acvf decays to zero, the wider the bandwidth.

Traditionally, the truncation point is chosen from the acvf at the point where the acvf goes to zero, since after this point there is very little contribution to the spectrum. The problem with this method of selecting the truncation point is that it is only good for getting an idea of the general shape of the power spectrum. If there is a single, narrow peak actually contained in the data, then this technique tends to gloss over it.

Window Closing

"Window closing" acknowledges the fact that we don't really know B_s and therefore can only attempt to gain insights about it by using several truncation points, and then examining and comparing the estimates from each. In particular, you generally choose three values of M: one for a large B_w, one for a medium B_w and one for a narrow B_w. For example, we might select the medium B_w as the value of M where the acf goes to zero. The other two M are chosen such that the ratio of smallest/largest $M = 4$ (as a general rule).

In all, the window closing method (or some variation thereof) probably offers the most information, since the degree of change in the estimates as a function of M provides additional information concerning the stability of the estimates in general. That is, if for all three bandwidths, the spectral shape is essentially preserved, then you can feel quite comfortable drawing conclusions concerning the distribution of the variance as a function of frequency. On the other hand, if the spectral shapes changes wildly from one bandwidth to the next, you should narrow your range of M values until you find a range that is more stable and provides a reasonable estimate of the bandwidth—then make valid conclusions. In such a case (actually, in *all* cases), additional realizations of the process will be very helpful for drawing valid conclusions.

Specified Variance Reduction

Alternatively, you can choose M to achieve some desired reduction of the variance (via an incease in the degrees of freedom), or variance/bias product, or to achieve the desired resolution, etc. This approach, however, ignores any special considerations required by different true spectral shapes.

Welch's (or Weighted) Overlapped Segment Averaging (WOSA)

One of the most popular overlapped segmented data approaches for variance reduction in the PSD was developed by Welch (1967). He introduced the idea of breaking a time series of length n into K overlapped contiguous segments of length N_s:

$$\left.\begin{aligned} X_{1,j} &= X_j \\ X_{2,j} &= X_{j+D} \\ &\vdots \\ X_{K,j}(j) &= X_{j+(K-1)D} \end{aligned}\right\} j = 0,\ 1, \ldots, N_S - 1. \tag{11.54}$$

For this, each segment (length of segments, N_S, long enough to resolve the lowest frequency desired, as described above) is tapered, and then overlapped to recover some of the information lost in the tapering. This produces a better variance reduction than non-overlapping segments. For this, you make the overlap approximately one-half the block size ($D = N_S/2$), then taper, and compute the PSD.

Taper each segment k with a sequence $\{h_t\}$ and compute the sample PSD ($jf = j/N_S$ is the rotational frequency of the jth harmonic),

$$\hat{S}_k(f) = \left| \sum_{j=0}^{N_S-1} h_j X_{k,j} e^{-i2\pi jf} \right|^2, \tag{11.55}$$

then average the K segment-PSD to give the smoothed WOSA PSD estimate:

$$\hat{S}^{WOSA}(f) = \frac{1}{K} \sum_{k=1}^{K} \hat{S}_k(f). \tag{11.56}$$

Welch and Percival and Walden use the Hanning taper:

$$h_t = \left(\frac{2}{3(N_S + 1)} \right)^{1/2} \left[1 - \cos\left(\frac{2\pi t}{N_S + 1} \right) \right]. \tag{11.57a}$$

With this method, each frequency estimate now has ν degrees of freedom,

$$\nu \approx 2K \left\{ 1 + 2 \sum_{j=1}^{K-1} \left[\left(1 - \frac{j}{K} \right) \left| \sum_{t=1}^{K} h_t h_{t+jD} \right|^2 \right] \right\}^{-1}, \tag{11.57b}$$

where D is the fraction of N_S that shifts each segment (e.g., for a 50 percent overlap $D = N_S/2$).

Multitaper Method of Spectral Analysis (MTM)

An even better method of using multiple tapers is a method developed by Thomson (1982) known as the Multitaper Method (MTM). This uses a series of orthogonal tapers that utilize all of the data instead of segments. Here we will present a simple overview (for more detail of this method, see the excellent presentation in Percival and Walden (1993)).

MTM uses multiple (K) tapers. Each taper is applied to the full-length raw time series (N values; fundamental harmonic, $1/N\Delta t$, and Nyquist frequency, $f_N = 1/2\Delta t$) and the sample PSD estimated. Averaging these sample PSDs (each using a different orthogonal taper) gives a smoothed estimate of the PSD with minimum bias and loss of spectral resolution for maximum variance reduction. For each tapered time series, estimate the PSD:

$$\hat{S}_k^{MT} = \left| \sum_{t=0}^{N-1} h_{t,k} X_t e^{-2\pi ft} \right|^2. \qquad (11.58)$$

Thomson calls these K spectra "eigenspectra," and as with WOSA, you then average the K eigenspectra to form the estimated MTM PSD:

$$\hat{S}^{MTM}(f) = \frac{1}{K} \sum_{k=0}^{K-1} \hat{S}_k^{MT}(f). \qquad (11.59)$$

If you do not taper, you are truncating your time series with a gate function, which is equivalent to convolving the true PSD with the squared sinc function (spectral window of the gate function). This amounts to performing a running weighted average of the true PSD with the squared sinc function (squared because you are working with power = amplitude squared). That smears local power by averaging neighboring power over the central lobe of the sinc function and then pulling in power from distant frequencies due to the side lobes of the sinc function. This introduces bias into the sample spectrum.

Ideally, you'd like your taper to have a spectral window most closely approximating a delta function for which convolution with the true PSD simply replicates that PSD with no ill effects. The problem here is that, to get a delta spectral window, your taper must be an infinitely long constant, and for that you must neither truncate nor taper your infinitely long time series.

Many tapers have been developed to concentrate power in the central lobe and reduce it in the side lobes. This is the concentration problem[18] for which we wish to find that sequence that concentrates the most power in a band-limited central lobe while it minimizes the power outside of the central lobe (minimizing the leakage of spectral power from the most distant parts of the PSD). As an optimization problem, Slepian (1978) found the best set of tapers with ideal properties to minimize leakage. He did this by rewriting the problem into an eigenfunction problem for which the eigenvalues (λ_i) that are closest to 1 in value do the best job of concentrating the power in the central lobe ($1-\lambda_i$ is the fraction of power associated with leakage). The 0th-order eigenfunction is the optimal solution, and the higher-order eigenvalues are very close to 1, providing excellent anti-leakage properties; the remainder drop off rapidly. The tapers are known as discrete prolate spheroidal sequences (DPSS), or simply, Slepian functions. The DPSS are orthogonal, and when applied to the full length of the time series they produce tapered time series that are independent. If the $\hat{S}^{MT}(f)$ has terms that are pairwise uncorrelated and have a common variance, then each (according to the Central Limit Theorem) will have variance reduced by $1/K$, or, equivalently, the averaged spectrum $\hat{S}^{MT}$ will have $2K$ effective degrees of freedom (reducing

[18] Not the same as that problem older people experience on a regular basis.

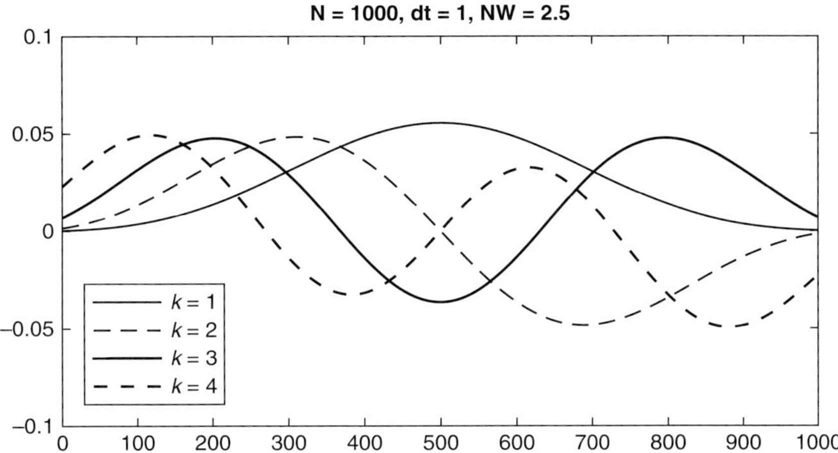

Figure 11.18 First four DPSS, which are multiplied as tapers in the time domain modifying the entire length of the time series (length 1000 in this case). The tapers are orthogonal and as a consequence produce tapered time series that are independent.

variance by the same amount). Furthermore, since MTM uses the entire length of the time series (thus, not changing the frequency range still resolves lowest $1/(N\Delta t)$ and highest $1/(2\Delta t)$ frequencies) and has resolution $2NW\Delta t$ built into the DPSS. Figure 11.18 shows the first four DPSS.

$2NW$ is the bandwidth when $\Delta t = 1$, $W = j/N$, so the bandwidth is given as $2j$ (the number of neighboring harmonics that are averaged over in the convolution of the PSD with the spectral window of the tapes, consistent with the smoothing of PSD by averaging neighboring peaks, m in (11.39), only in this case using an optimal smoother). When $\Delta t \neq 1$, $W = j/N\Delta t$, and the bandwidth becomes $2NW\Delta t$. In either case, $2NW\Delta t = 2j$ the number of neighboring frequencies over which the spectrum is smoothed.

Software performing the MTM (e.g., Matlab) usually requires that you choose the bandwidth you desire – how many neighboring j harmonics you want the central lobe of the spectral window for the tapers to smooth over – and from that compute W to inform the MTM software about how many tapers (DPSS) you desire ($2NW$ is often recommended, whereas other sources suggest $2NW-1$). The tradeoff is that the wider the bandwidth $2NW$, the less resolution you end up with and the greater the potential loss of fine-scale information, but the better the reduction in variance, given the use of more tapers (but if you exceed $2NW$ tapers, you may begin to use tapers that will introduce leakage). So you need to choose a reasonable bandwidth and variance reduction. This is not unlike the spectral smoothing approach of §11.3.4, and like that in practice, you may wish to find that balance using window closing. This is an exceptionally good method for spectral analysis; it uses all of the data and minimizes leakage while giving an optimal tradeoff in resolution and variance reduction.

11.4.3 Technique Used

Finally, you must decide the specific manner in which you intend to actually carry out the spectral estimation and smoothing process. That is, are you going to (1) truncate the acvf via multiplication with one of the various lag windows and then transform, (2) segment the data, transform each segment and average the PSD estimates or (3) compute the power spectrum first and then smooth it directly by convolution with a spectral window?

The approach used will as likely depend on what software is available as it will on which one you feel most comfortable with. Historically, the truncated acvf dominated the approaches (mainly because of the tremendous body of literature addressing this approach). However, the other approaches are now in frequent use as well, given the advent of the FFT. Because the transform of short segments is very efficient both computationally and in terms of storage. Press et al. (1986) claim[19] that for fixed data (i.e., those in which the Δt and T are imposed upon us and not easily changed), the near-optimum method is to make overlapping segments, taper each segment, and then average the spectra of the segments. They provide the equivalent degrees of freedom for this operation, thus allowing usage of the standard confidence interval that was given previously.

Three Final Considerations

1) While it often doesn't matter which spectral window you use in practice, some of the windows were designed to be applied as spectral windows and others as lag windows. The simplest example of this is the Daniell (rectangular) spectral window mentioned earlier. In the frequency domain, this is simply $= 1/m$ for m frequencies and $= 0$ elsewhere (a straight m-point moving average). Computationally speaking, this is extremely easy to implement in the frequency domain to average m neighboring values. However, in the time domain, the corresponding lag window is the transform of the rectangular window, the sinc function. This function is *infinite* in extent. Therefore, unless you and your computer have a way to deal with infinitely long functions, use of this window as a *lag window* becomes difficult. There are approximations you can use: truncate the sinc function, making it an approximation to the real lag window. This is a fairly simple concept, but is frequently overlooked until one becomes frustrated at the programming level.

2) We originally discussed leakage associated with a mismatch in the beginning and end points of the time series (which introduces high-frequency components), but have failed to bring it up in the spectral smoothing context (smoothing increases bias by leaking power from neighboring, even-distant spectral estimates). We addressed the former through application of a taper, but have not discussed how that interacts with the convolution associated with the spectral smoothing done to improve the variance of each spectral estimate, though for the MTM it is addressed explicitly.

[19] This claim precedes the introduction of Thomson's multitaper method (which I suspect would now be chosen as the best).

If you segment the data and then average the power spectra of the segments, it is sometimes necessary to taper each segment before transforming (though you should pick segment boundaries that are least susceptible to this problem). Therefore, in that case, you are convolving the F_j by the transform of the taper and effectively convolving again by averaging the spectra of the segments. The net effect is that you are doing one single smoothing operation using a modified spectral window created by convolving the two spectral windows together. That is, (d*t)*s = d*(t*s) = d*ms, where d = data, t = taper, s = smoother and ms = modified smoother (all in the frequency domain).

The effect of most tapers is to actually eliminate some data information, so it tends to increase the uncertainty (variance) of the spectral estimates ($\leq 10 - 20$ percent), though the taper is designed to reduce the bias (via leakage) of the estimates. So, the taper increases the variance (bad) while decreasing the bias (good); MTM avoids this problem by using all of the data with orthogonal tapers. The smoothing then decreases the variance (good) while increasing the bias (bad), and the net effect is a compromise in which we simply could have done everything with one window, using some intermediate characteristics.

If you use a specific taper and smoother you could convolve them to make a modified smoother (reducing the need to taper the data), or if you wish to work out the statistical consequences of the modified smoother you might do so. It seems that many people don't want to work out the new statistical properties of the modified window, in which case working out the taper and smoother separately makes perfect sense and probably doesn't hurt anything unless you want to automate some large processing job.

3) Finally, in cases where the true spectrum shows a wide dynamic range (e.g., when the power at one end of the spectrum exceeds that at the other end by a large amount, say by orders of magnitude), then it is *conceptually* advantageous to flatten the spectrum prior to estimating the true spectrum. This procedure is known as **pre-whitening** (i.e., we wish to manipulate the data so that the estimated spectrum appears nearly white). The reason for this is obvious when one considers that the bias in the estimate of the spectrum is proportional to the curvature of the true spectrum and bandwidth.

The expectance of the smoothed power spectral estimates is the smoothed true spectrum. If the spectrum is averaged over the wide dynamic range about f_a, the smoothed estimate will be strongly biased (i.e., it will be much larger or smaller than it should be). This bias reflects leakage resulting from the fact that the smoothed estimate will contain power from frequency bands large distances away from the band for which the estimate is being made and the dynamic range suggests that the smoothed power at f_a will be dramatically different power than originally.

One way to handle this is to convolve the time series with some function so that the convolved series has a spectrum that is essentially flat (white). The spectral estimates of this flat spectrum can then be made in the usual manner, knowing that the bias will be minimal. After that, the spectrum is corrected by removing the effects of the convolving function by division of the transform of that function in the frequency domain. Actually, for this to work, the pre-whitening has to do more than simply remove the slope. Rather, it must also remove peaks and valleys, and curvature in general, which is very difficult to accomplish in practice.

While conceptually meritorious, this technique has been criticized due to a variety of problems. Essentially, the most serious problem is that one needs to know the shape of the true spectrum in order to properly pre-whiten. Also, to properly preserve information about (true) sharp peaks, the convolving function must have an impulse response function with a very long tail, and therefore we require an extremely long time series, which may not be an option available to you.

When the spectral dynamic range is attributed to trends in the data, one can effectively pre-whiten by removing the trend from the data prior to transforming. This can be accomplished by fitting an appropriate function (e.g., a low-order polynomial) to the data and subtracting it out (the residuals will then be transformed and should have an approximately white noise distribution). Alternatively, one might remove a polynomial trend by performing a first difference in the time series, which is another effective way of removing trends. Note that some tapers have also been designed to effectively address this problem, though *Priestley*, who criticizes pre-whitening, also criticizes these tapers and claims that it would just be easier to pre-whiten than to go to the trouble of defining some complicated, optimal taper.

11.4.4 Practical Consideration for Transforming acvf

You know that the transform of the acvf directly yields the PSD. But how can this be, when there are many more lags in the acvf than data points in the time series? The fundamental period $T = 1/n\Delta t$, where $n = n_t$ is the number of data points in time series, while $n_a = 2n_t + 1$ for the acvf (i.e., a considerably larger T, giving different Fourier harmonics: $f_i = i/T$). When n is large, the acvf is effectively $2n$, so it will have a considerably more dense (2x) frequency resolution ($\Delta f = T - 1 = 2n\Delta t$). Fortunately, the lags are spaced the same as Δt, so the Nyquist frequency is the same for both ($1/2\Delta t$), and Δf is an integer multiple. We are interpolating the same PSD at a different resolution (according to the size of n).

11.5 Bootstrap Testing with Time Series

11.5.1 Generating Colored Noise Time Series

Testing the significance of results involving time series can be a difficult task, given standard methods (if for no other reason than it is difficult to properly estimate the effective degrees of freedom in a time series, since the data points are autocorrelated and thus not independent, despite a seemingly endless list of formulas for doing so). But, as with nonsequential data, the bootstrap method can be conveniently used to repeatedly perform the same statistical test using artificial time series displaying colored noise, including noise that preserves the important characteristics of your sampled time series.

Preserving the mean and variance is rather easy, but here we must also preserve the lowest-order bivariate moment, the acvf of the time series (dictating to what degree neighboring points are dependent upon one another, reducing the total degrees of freedom).

Generation of noise time series involves the following: (1) define your noise spectrum, (2) compute the power spectrum of that noise spectrum (directly in the frequency domain, or via transform of a series for which you wish to duplicate the noise color), (3) generate a random phase for the amplitude spectrum (preserving any symmetry relationships required by working in polar coordinates) and (4) invert the amplitude and random phase spectrum back into the time domain, producing a time series with the required noise coloring.

The above considerations are presented here in context of an example. The example consists of a time series of surface precipitation at a single location on the Earth, **p**(t). We wish to see how this precipitation correlates with climate elsewhere on the globe (in hopes of linking it with a more coherent large-scale climate pattern, such as El Niño, that might provide some predictive capabilities). We start by correlating **p**(t) with time series of surface air temperature **T**(t) at various locations around the world (e.g., in a global grid). Ideally, we plot the r (or r^2) value at each grid location. We now need to consider the significance of such correlations, which are subject to a variety of problems: (1) for every 100 correlations we make, we expect 5 on average to show significance at the 95 percent confidence level (Katz and Brown (1991) refer to this problem as being one of **multiplicity);** (2) because **T**(t) in some locations is highly correlated with **T**(t) at other locations, we might expect an even higher than usual number of "highly significant" correlations, since a high correlation with one of these intercorrelated regions might naturally lead to stronger correlations at the covarying locations; and (3) if the time series have n points in them, unless the **p**(t) is white noise, the significance of each correlation is dependent upon the **effective degrees of freedom** (**EDOF**), which is less than n because of autocorrelation. There are techniques that have been developed to address each of these particular problems, though it is somewhat easier and more intuitive to employ the bootstrap method to assess significance in a manner that addresses each of these issues.

Choosing the Spectral Coloring

You will assume that your time series was noise drawn from an ensemble defined by a noise PSD (**PSD$_N$**) – and you must define the PSD$_N$. This is essentially the null spectrum you choose for your PSD, as discussed in §11.3.2. Alternatively, you might want to test your results against simple colored noise; if so, you have to decide what the **spectral coloring** is. Obviously it makes sense to simulate coloring consistent with that in the sample PSD (**PSD$_S$**) for your time series, **p**(t). In that case you have several approaches available for this: (1) fit a low-order polynomial to the PSD$_S$, (2) fit the PSD$_S$ with a low-order autoregressive process (e.g., using the formula in equation (11.37) for an AR(1) fit) or (3) smooth the PSD$_S$ until a smooth curve results.

Recall that white noise (a time series in which every value is independent of all others) displays a flat spectrum, whereas a red spectrum (integrated white noise, very common in natural systems) displays a slope of −2 (on a log power axis) so that there is more variance with lower-frequency components of the data. Fractal processes often lead to other spectral slopes that can be simulated, if appropriate.

Alternatively, you may choose to adopt the *most conservative approach* – preserve the exact PSD_S (suggesting that it, even the peaks and valleys, is representative of the noise process). All artificial time series from your PSD_N will have a generally similar look, but because of the random phasing, this test will show just how representative your actual time series is if it was drawn from PSD_N (the noise process).

Now Construct the Artificial Time Series

Once a representative noise power spectrum (PSD_N) is decided upon, it must be constructed as described above (obviously, if you are using the original PSD_S for this, you have already created what you need). This gives $\mathbf{A}^2(\mathrm{f})$. Now generate a random series for the phase, φ_f. Then inverse transform as

$$y_\mathrm{t}^+ = \sum_{\mathrm{f}=0}^{\mathrm{f_N}} A_\mathrm{f} e^{i\varphi_\mathrm{f} \mathrm{t}} \qquad \text{for } 1 \leq \mathrm{t} \leq n \qquad (11.60)$$

and repeat as many times as desired (the more, the better[20]), saving each y^+ as a realization of your process if from noise.

Now Repeat Your Original Statistical Test

For this example, you have correlated the time series of precipitation from one location on Earth to thousands of surface air temperature (T) time series at stations covering the Earth. You need to assess where those correlations are most likely not generated from noise, but more likely consistent with some hypothesis involving global climate connectivity. So, now repeat the correlations using your new artificial precipitation time series. After correlating all of them, construct a histogram (**$\mathbf{PMF_N}$**) of r values for all noise (bootstrap) correlations for each grid station i (**$\mathbf{PMF_{Ni}}$**) and for all stations combined (**$\mathbf{PMF_{Nt}}$**). Construct a similar PMF for the combination of all sample correlation r values (**$\mathbf{PMF_{St}}$**). Comparison of the means of the two "total station" PMFs reveals whether the observed series consistently differs from that expected from noise (you can evaluate that difference with a t-test). Also, determine if there are more highly significant correlations (a longer tail) in the sample PMF than in the noise. Because the bootstrap noise time series have the same acvf as the sample time series, they all have the same number of EDOFs, and both are exposed to the same degree of multiplicity, so the bootstrap accommodates those problems.

[20] It's just a loop in the computer code, so I usually do 10^4 or more, and increase until my bootstrap PMF looks like a PDF.

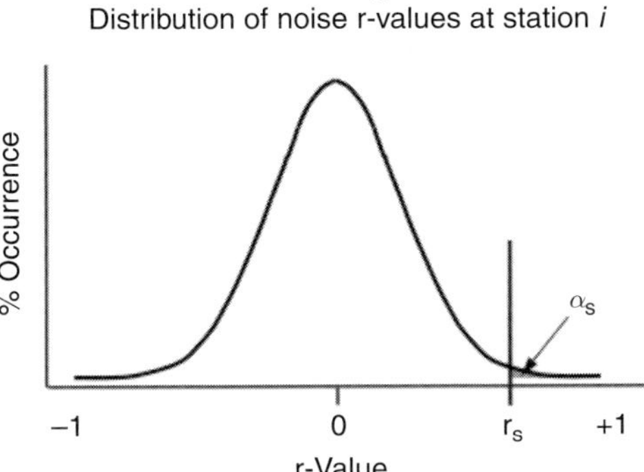

Figure 11.19 Example of a bootstrap PMF_{Ni} with actual sample r-value for this particular station, i. Likelihood of getting an r-value this high from the noise is only α_s. This α_s is found by adding all PMF_{Ni} values from r_s to 1, or all values from −1 to r_s and subtracting that from 1. In this example, it is clear that obtaining an r value as high as r_s is very slim if generated by the noise process.

To clearly identify which stations show r values that are unlikely to occur from the noise, use the bootstrap PMF_{Ni} to determine the likelihood of achieving the r value actually achieved for that station (Figure 11.19). From this, you can contour the likelihood (the α or p value), the lowest α most likely not from the noise (you choose the contour level most meaningful from your understanding of the problem).

11.6 Take-Home Points

1. The sum of ν squared Gaussian random variables form a chi-square distribution (χ^2_ν) with ν degrees of freedom, so a single-power spectral estimate ($a^2 + b^2$) should have a chi-squared distribution of 2 degrees of freedom. But because your time series is truncated, each PSD estimate is a weighted average of neighboring estimates (by convolution with the Fourier transform of the truncating function), so the estimate is not a simple chi-squared estimate. Fortunately, the ratio of the estimated amplitude to true spectrum value is proportional to a chi-squared distribution, so $A_j^2/\mathrm{p}_j \propto \chi^2_\nu$ or, $A_j^2/\mathrm{ap}_j = \chi^2_\nu$. By satisfying Parseval's theorem, the constant of proportionality is a = 1/ν, so $\nu A_j^2/\mathrm{p}_j$ is approximately χ^2_ν distributed.
2. The uncertainty (variance) of power in a power spectral density function (PSD) estimate is $4\sigma^4$, hence a single standard deviation is 200 percent of the value – this is ridiculously large! In order to get a more stable estimate, you have to average spectra (or some variant of this) to reduce it to a reasonable precision.
3. Because the spectral estimates are chi-square, confidence intervals are not symmetrical.
4. If you have multiple realizations of your data (multiple time series), you should average their spectra to reduce the uncertainty (dramatically).

5. Without multiple spectra, you have several options for reducing the uncertainty (all of these give the same reduction of variance):

   a. Invoke ergodicity and segment time series into several (m) segments, then average spectra of each segment.
   b. Average m neighboring spectra values.
   c. Truncate the N-length acf as N/m before transforming.

   For the above, you need, if possible, to make the segments long enough to allow you to resolve the lowest frequency you desire (1/T). Reducing the variance will lead to a broader spectral resolution ($\Delta f = 1/T$).

11.7 Questions

Pencil and Paper Questions

1. a. What distribution (PDF) does each estimate of power follow (e.g., Normal, etc.)?
   b. What do you need to do in order to reduce the uncertainty of the power estimates?
   c. Name three ways to improve the uncertainty of a PSD, and show how to use each to give same reduction in uncertainty.
   d. What is the null spectrum and how can you estimate it?
2. a. Your time series is 100 points long. How long does it have to be to decrease the uncertainty of each PSD estimate by a factor of 2?
   b. Your time series spans 100 years of data, with $\Delta t = 1$ yr. You need to resample this 100 years at what Δt in order to be able to resolve a rotational frequency of .005?
   c. Your PSD for your time series of $n = 500$ shows 6 peaks, all of which lie above the 95 percent confidence interval. Are they real? And how do you know your answer?
   d. If one of the peaks in example 2.c is not real, does that mean it doesn't really exist in the time series? And if it does exist, how can it not be real?

Computer-Based Questions

3. For the fist million years of LR04:

   a. Plot the PSD.
   b. Plot the null spectrum on a smoothed PSD plot (draw a horizontal line whose length indicates the spectral resolution (the smoothing width)).
   c. Draw the 96 percent confidence interval about the null spectrum; compute the CI from equation (11.6).
   d. Change the amount of smoothing and replot the new confidence errors.

e. Give the frequencies (and their periods) of each peak exceeding the confidence interval.

4. Generate a bootstrap version of the first million years of LR04 and plot its PSD on top of that in question 3.a. (They should be identical. If they are not, you likely did not do the random phase in the polar form.) Now plot the bootstrap version in the time domain immediately below a plot of the original series.

12 Cross-Spectral Analysis

12.1 Overview

So far, we have focused on the analysis of a single random process. Here, we consider the *relationship between pairs of random processes* (X_1, X_2), such as identifying the degree to which two series contain the same distribution of variance as a function of time or of frequency. Simple relationships include (1) one series is thought to represent the response, after passing through a linear filter, of the other, and (2) we may have several series that are all thought to represent different responses to the same forcing. Cross-spectral analysis is the frequency-domain approach designed for these types of problems. In some respects, these techniques are analogous to those of linear correlation and regression, except that they are carried out in the frequency domain, and as a consequence they provide a different perspective affording additional insights regarding the nature of the relationship.

Most of the procedures used in this analysis are direct and straightforward extensions of spectral analysis techniques used for the analysis of single processes discussed in the previous chapter. As was the case for univariate analysis, here we wish to simplify the analysis by describing any pair of processes through their lowest-order statistical moments. That is, the mean, μ_i, $i = 1, 2$, and the lowest-order bivariate moments – in this case, the autocovariance function (acvf) that describes how any two points, separated by a lag τ, *within* a single process are related (i.e., given the points at one time, t, how well you can predict the values at another time, $t + \tau$), and the cross-covariance function (ccvf) that describes how any two points, separated by a lag τ, *between* the pair of processes are related.

12.2 Joint PDF Moments in the Time Domain

We start with the *time domain*, beginning with estimation of the following moments:[1] mean,

[1] Please excuse the dry presentation of the various functions that must be introduced here for use in the remainder of the chapter.

$$E[X_1] = \mu_1 \tag{12.1a}$$

$$E[X_2] = \mu_2, \tag{12.1b}$$

autocovariance (acvf),

$$\gamma_{X_1,X_1}(\tau) = \gamma_{11}(\tau) = E[(X_1(t) - \mu_1)(X_1(t+\tau) - \mu_1)] \tag{12.2a}$$

$$\gamma_{X_2,X_2}(\tau) = \gamma_{22}(\tau) = E[(X_2(t) - \mu_2)(X_2(t+\tau) - \mu_2)], \tag{12.2b}$$

and cross-covariance (ccvf),

$$\gamma_{X_1,X_2}(\tau) = \gamma_{12}(\tau) = E[(X_1(t) - \mu_1)(X_2(t+\tau) - \mu_2)] \tag{12.3a}$$

$$\gamma_{X_2,X_1}(\tau) = \gamma_{21}(\tau) = E[(X_2(t) - \mu_2)(X_1(t+\tau) - \mu_1)]. \tag{12.3b}$$

Since $\gamma_{12}(\tau) = \gamma_{21}(-\tau)$ for real processes, either of the two forms in (12.3) is sufficient for real data. Also, while the acvf is an even function, the ccvf is not, so knowledge of both positive and negative lags is required for the ccvf.

As before, the corresponding autocorrelation functions (acf) are given by

$$\rho_{11}(\tau) = \frac{\gamma_{11}(\tau)}{\gamma_{11}(0)} = \frac{\gamma_{11}(\tau)}{\sigma_1^2} \tag{12.4a}$$

$$\rho_{22}(\tau) = \frac{\gamma_{22}(\tau)}{\gamma_{22}(0)} = \frac{\gamma_{22}(\tau)}{\sigma_2^2}, \tag{12.4b}$$

and corresponding cross-correlation function (ccf),

$$\rho_{12}(\tau) = \frac{\gamma_{12}(\tau)}{\sqrt{\gamma_{11}(0)\gamma_{22}(0)}} = \frac{\gamma_{12}(\tau)}{\sigma_1\sigma_2} = \rho_{21}(-\tau). \tag{12.5}$$

Alternatively, (12.4) and (12.5) can be generalized as

$$\rho_{ij}(\tau) = \frac{\gamma_{ij}(\tau)}{\sqrt{\gamma_{ii}(0)\gamma_{jj}(0)}}, \tag{12.6}$$

with $i = j$ (for the acf), and $i \neq j$ (for the ccf); likewise for the acvf and ccvf.

If two random processes are **uncorrelated**, then the *true* ccf between them is *everywhere* zero.[2] However, because the variance (uncertainty) in the sample ccf estimates are functions of the acf of the time series (equation (7.55)), you can get very large artificial

[2] The *sample* ccf will have random fluctuations about zero because of noise.

values of the ccf, which could be erroneously interpreted as significant. The only way this can be avoided is to "pre-whiten" the time series by filtering them first, though this approach suffers to some degree from the same problems as it does in the spectral analysis case. Here, the overall goal is to attempt to produce an acf for each series that looks similar to a white noise acf. This then reduces the degree of autocorrelation, which in turn reduces the variance in the sample ccf and thus the likelihood of getting artificially large ccf values. For white noise series, the variance in the sample ccf is as it was for the acf, $\mathrm{Var}[\hat{R}_{12}(\tau)] \approx 1/N$, so values in the range of $\pm 1/\sqrt{N}$ are not significantly different at one standard deviation from pure noise.

The methods for estimating the acvf (or its frequency domain equivalent, the PSD) for each of the pairs of processes is unchanged, and thus follows the exact procedures already laid out for estimation of these quantities (§7.4.2). Here, we need only consider how to best estimate the ccf, or its frequency-domain equivalent, the **cross-spectral density (CSD) function** (comparable to the PSD, but here showing the *covariance* as a function of frequency).

12.2.1 Linear Causal Relationship

Before estimating any form of covariability, it helps to consider how processes may actually be related in a manner addressed via standard linear regression (or other linear methods). The simplest case relating two random processes is through a simple linear filter system. That is, X_1 is the input to a linear system that gives X_2 as the output plus some random noise, ε_t (assume X_1 and ε_t have zero means). This system is good for describing the relationship between related time series frequently examined in science.

Formally, given a linear filter, X_1 and X_2 are related by

$$X_{2k} = \sum_{\mathrm{t}=1}^{N} \mathrm{f_t} X_{1(k-\mathrm{t})} + \varepsilon_\mathrm{t}, \qquad (12.7)$$

or, recalling the sliding strips of paper,

$$\left.\begin{array}{ll} & \boxed{\mathrm{f}_0 \quad \mathrm{f}_1 \quad \mathrm{f}_2} \\ X_1 & \boxed{\ldots \quad x_2 \ x_1 \ x_0} \rightarrow \end{array}\right\} X_2 \quad x_{2(0)} = \mathrm{f}_0 x_{1(0)}.$$

Simple, One-Term Filter

Now consider the special case in which the impulse response function of the filter, f_t, contains only one term, f_0. Then (12.7) reduces to

$$X_{2\mathrm{t}} = \mathrm{f}_0 X_{1\mathrm{t}} + \varepsilon_\mathrm{t} \qquad (12.8)$$

and the ccvf between X_1 and X_2 is given as

$$\begin{aligned}\gamma_{12}(\tau) &= \mathrm{E}[X_{1t}X_{2(t+\tau)}] \\ &= \mathrm{E}[X_{1t}(\mathrm{f}_0 X_{1(t+\tau)} + \varepsilon_{t+\tau})] \\ &= \mathrm{f}_0\mathrm{E}[X_{1t}X_{1(t+\tau)}] + \mathrm{E}[X_{1t}\varepsilon_{t+\tau}].\end{aligned} \tag{12.9a}$$

X_1 is not correlated with the noise, so this reduces to

$$= \mathrm{f}_0\gamma_{11}(\tau). \tag{12.9b}$$

In words, (12.9) shows that the ccvf between two time series are related by a simple scaling factor. Therefore, $\gamma_{12}(\tau)/\gamma_{11}(\tau) = \mathrm{f}_0$, though this division might not be stable, given that $\gamma_{11}(\tau)$ can pass through zero in multiple locations.

The cross-correlation function for this above system is

$$\rho_{12}(\tau) = \frac{\gamma_{12}(\tau)}{\sqrt{\gamma_{11}(0)\gamma_{22}(0)}}, \tag{12.10}$$

where

$$\gamma_{11}(0) = \sigma_1^2$$

and, for a zero-mean input series (so $\mathrm{E}[X]^2 = 0$),

$$\begin{aligned}\gamma_{22}(0) &= \sigma_2^2 = \mathrm{Var}[X_2] = \mathrm{E}[X_2^2] \\ &= \mathrm{E}[(\mathrm{f}_0 X_1 + \varepsilon_t)^2] \\ &= \mathrm{f}_0^2\mathrm{E}[X_1^2] + 2\mathrm{f}_0\mathrm{E}[X_1\varepsilon_t] + \mathrm{E}[\varepsilon_t^2].\end{aligned}$$

Since both X_1 and ε_t have zero means, then $\mathrm{Var}[X_1] = \mathrm{E}[X_1^2]$ and $\mathrm{Var}[\varepsilon_t] = \mathrm{E}[\varepsilon_t^2]$,

$$= \mathrm{f}_0^2\sigma_1^2 + \sigma_\varepsilon^2. \tag{12.11}$$

Thus, the variance of the output signal is the variance of the input signal scaled by the squared factor in the filter, and further increased by the variance of the noise. The cross-product term dropped out, since we have already stated that the input signal is not correlated with the system noise.

So, the true ccf, (12.10), can be rewritten as

$$\begin{aligned}\rho_{12}(\tau) &= \frac{\mathrm{f}_0\gamma_{11}(\tau)}{\sqrt{\sigma_1^2(\mathrm{f}_0^2\sigma_1^2 + \sigma_\varepsilon^2)}} \\ &= \frac{\mathrm{f}_0\gamma_{11}(\tau)}{\sigma_1^2\sqrt{(\mathrm{f}_0^2 + \sigma_\varepsilon^2/\sigma_1^2)}}\end{aligned}$$

where the acvf divided by the standard deviation, $\gamma_{11}(\tau)/\sigma_1^2$, is the acf, $\rho_{11}(\tau)$, giving

$$= \frac{f_0 \rho_{11}(\tau)}{\sqrt{(f_0^2 + \sigma_\varepsilon^2/\sigma_1^2)}}$$

or

$$= \phi \rho_{11}(\tau), \tag{12.12}$$

where $\phi = f_0/(f_0^2 + \sigma_\varepsilon^2/\sigma_1^2)^{1/2}$; $\sigma_\varepsilon^2/\sigma_1^2$ is the inverse of a **signal-to-noise ratio (SN)** of the input signal. Therefore, the ccf is proportional to the acf of the input signal.

By (12.9) and (12.12), given this model of (12.8), you can solve (in a least-squares sense, using the sample acvf, ccvf, or acf and ccf) for the filter coefficient (scaling factor) and the variance of the noise introduced in the system.

White Noise Input

If X_{1t}, the input to the filter, is white noise (uncorrelated with the system noise, ε_t), then the ccvf in (12.9) is given as

$$\gamma_{12}(\tau) = \begin{cases} f_0 \sigma_1^2 & \tau = 0 \\ 0 & \tau \neq 0 \end{cases} \tag{12.13}$$

and the ccf in (12.12) is given by

$$\rho_{12}(\tau) = \begin{cases} \dfrac{f_0}{\sqrt{f_0^2 + \sigma_\varepsilon^2/\sigma_1^2}} & \tau = 0 \\ 0 & \tau \neq 0. \end{cases} \tag{12.14}$$

This latter form shows the influence of the system noise, ε_t, with variance σ_ε^2. At the 0th lag, the ccf is not equal to 1 (showing perfect correlation between X_1 and X_2), except for the case where there is no noise, so $SN^{-1} = 0$. Otherwise, the more noise leading to a smaller SN (larger SN^{-1}), the larger the additive constant to f_0^2 in the denominator, and thus the larger the value of $(f_0^2 + \sigma_\varepsilon^2/\sigma_1^2)^{1/2}$ relative to f_0, driving $\rho_{12}(0)$ smaller (relative to 1).

Box 12.1 Simple Two-Term Filter

Here is an examination of the addition of another term in the filter, before proceeding to the general N-term filter case. This establishes the pattern for the general case more clearly.

Now, consider the case in which the impulse response function contains two terms: f_0 and f_1, reducing (12.7) to

$$X_{2t} = f_0 X_{1t} + f_1 X_{1(t-1)} + \varepsilon_t; \tag{12.15}$$

or, shown for integer times, the convolution is

$$\begin{aligned} X_{2t=1} &= f_0 X_{1t=1} + \varepsilon_{t=1} \\ X_{2t=2} &= f_0 X_{1t=2} + f_1 X_{1t=1} + \varepsilon_{t=2} \\ X_{2t=3} &= f_0 X_{1t=3} + f_1 X_{1t=2} + \varepsilon_{t=3} \\ &\vdots \\ X_{2t=n} &= f_0 X_{1t=n} + f_1 X_{1t=n-1} + \varepsilon_{t=n}. \end{aligned}$$

The ccvf between X_1 and X_2 in this case is given by

$$\begin{aligned} \gamma_{12}(\tau) &= E[X_{1t} X_{2(t+\tau)}] \\ &= E[X_{1t}(f_0 X_{1(t+\tau)} + f_1 X_{1(t+\tau-1)} + \varepsilon_{t+\tau})] \\ &= f_0 E[X_{1t} X_{1(t+\tau)}] + f_1 E[X_{1t} X_{1(t+\tau-1)}] + E[X_{1t}\varepsilon_{t+\tau}] \\ &= f_0 \gamma_{11}(\tau) + f_1 \gamma_{11}(\tau - 1). \end{aligned} \tag{12.16}$$

The ccvf takes the form of a standard convolution, as seen by comparison to (12.15). Hence, the ccvf is equal to the convolution of the acvf of the input signal with the two-term filter. In other words, the linearity of the convolution operator reveals that the convolution of the input series is carried through the convolution of the (linear) serial products.

In terms of the cross-correlation function, the variance of X_1 is still given as σ_1^2, whereas for X_2 (with zero mean, as with the single-term filter case),

$$\begin{aligned} \mathrm{Var}[X_2] &= E[X_2^2] = E[(f_0 X_{1(t+\tau)} + f_1 X_{1(t+\tau-1)} + \varepsilon_t)^2] \\ &= f_0^2 E[X_{1(t+\tau)}^2] + f_1^2 E[X_{1(t-1)}^2] + E[\varepsilon_t^2] + 2 f_0 f_1 E[X_{1t} X_{1(t+\tau-1)}] + \\ &\quad E[2 f_0 \varepsilon_t X_{1t}] + E[2 f_1 \varepsilon_t X_{1(t-1)}]. \end{aligned} \tag{12.17}$$

Because of stationarity, $E[X_{1t}^2] = E[X_{1(t-1)}^2] = \sigma_1^2$, and as already stated, the noise is not correlated to the input series, so (12.17) reduces to

$$= (f_0^2 + f_1^2)\sigma_1^2 + \sigma_\varepsilon^2 + 2 f_0 f_1 \gamma_{11}(1). \tag{12.18}$$

Box 12.1 (Cont.)

Therefore, the ccf, given by (12.10), is written for this two-term filter case as

$$
\begin{aligned}
\rho_{12}(\tau) &= \frac{f_0\gamma_{11}(\tau) + f_1\gamma_{11}(\tau-1)}{\sqrt{\sigma_1^2[(f_0^2+f_1^2)\sigma_1^2 + \sigma_\varepsilon^2 + 2f_0f_1\gamma_{11}(1)]}} \\
&= \frac{f_0\gamma_{11}(\tau) + f_1\gamma_{11}(\tau-1)}{\sigma_1^2\sqrt{(f_0^2+f_1^2) + \sigma_\varepsilon^2/\sigma_1^2 + 2f_0f_1\gamma_{11}(1)/\sigma_1^2}} \\
&= \frac{f_0\rho_{11}(\tau) + f_1\rho_{11}(\tau-1)}{\sqrt{(f_0^2+f_1^2) + \sigma_\varepsilon^2/\sigma_1^2 + 2f_0f_1\rho_{11}(1)}}
\end{aligned}
\tag{12.19}
$$

or

$$\phi_0\rho_{11}(\tau) + \phi_1\rho_{11}(\tau-1), \tag{12.20}$$

where $\phi_i = f_i/\sqrt{f_0^2 + f_1^2 + \sigma_\varepsilon^2/\sigma_1^2 + 2f_0f_1\rho_{11}(1)}$. So again, the ccf is a convolution of the input signal's acf, but the two terms of the filter, the ϕ_i, are scaled by $[f_0^2 + f_1^2 + \sigma_\varepsilon^2/\sigma_1^2 + 2f_0f_1\rho_{11}(1)]^{-1/2}$.

White Noise Input

If X_{1t}, the input to the two-term filter, is white noise (uncorrelated with the system noise, ε_t), then the ccvf in (12.16) is given as

$$\gamma_{12}(\tau) = \begin{cases} f_0\sigma_1^2 & \tau = 0 \\ f_1\sigma_1^2 & \tau = 1 \\ 0 & \tau \neq 0,1 \end{cases} \tag{12.21}$$

and the ccf in (12.20) is given by

$$\rho_{12}(\tau) = \begin{cases} \dfrac{f_0}{\sqrt{f_0^2 + f_1^2 + \sigma_\varepsilon^2/\sigma_1^2}} & \tau = 0 \\ \dfrac{f_1}{\sqrt{f_0^2 + f_1^2 + \sigma_\varepsilon^2/\sigma_1^2}} & \tau = 1 \\ 0 & \tau \neq 0,1. \end{cases} \tag{12.22}$$

So, for all lags, except for $\tau = 0$, the acvf and acf of the *input* signal is 0. But because there are two terms in the filter that is convolving the acf (when constructing the ccf), two terms in the ccf survive the convolution process, leaving two terms nonzero in the ccf ($\tau = 0,1$).

General Filter

The above examples show how the acvf of the input signal is related to the ccvf between the input signal and output signal for cases when the filter consists of only one or two terms. These examples are easily extended to provide the general filter case, when the filter contains an arbitrary number of independent terms, say N, as described by the original system of (12.7).

First, consider expanding the sum in (12.7) for each output value of t:

$$X_{2\mathrm{t}} = \mathrm{f}_0 X_{1\mathrm{t}} + \mathrm{f}_1 X_{1(\mathrm{t}-1)} + \mathrm{f}_2 X_{1(\mathrm{t}-2)} + \ldots + \mathrm{f}_N X_{1(\mathrm{t}-N)} + \varepsilon_\mathrm{t}. \tag{12.23}$$

The ccvf between X_1 and X_2 in this case is given by

$$\begin{aligned}
\gamma_{12}(\tau) &= \mathrm{E}[X_{1\mathrm{t}} X_{2(\mathrm{t}+\tau)}] \\
&= \mathrm{E}[X_{1\mathrm{t}}(\mathrm{f}_0 X_{1\mathrm{t}} + \mathrm{f}_1 X_{1(\mathrm{t}+\tau-1)} + \mathrm{f}_2 X_{1(\mathrm{t}+\tau-2)} + \ldots + \mathrm{f}_N X_{1(\mathrm{t}+\tau-N)} + \varepsilon_{\mathrm{t}+\tau})] \\
&= \mathrm{E}\left[X_{1\mathrm{t}}\left(\sum_{k=0}^{N} \mathrm{f}_k X_{1(\mathrm{t}+\tau-k)} + \varepsilon_\mathrm{t}\right)\right] \\
&= \sum_{k=0}^{N} \mathrm{f}_k \mathrm{E}[X_{1t} X_{1(t+\tau-k)}] + \mathrm{E}[X_{1t}\varepsilon_t] \\
&= \sum_{k=0}^{N} \mathrm{f}_k \gamma_{11}(\tau - k).
\end{aligned} \tag{12.24}$$

Again, the ccvf is a convolution of the acvf of the input signal with the filter – a generalization of the expressions explicitly derived for the one-term (N = 1) and two-term (N = 2) filter cases given in (12.9) and (12.16).

For conversion to the ccf, the variance of X_2 is computed in a manner identical to that done for the previous simple cases to give

$$\mathrm{E}[X_2^2] = \sigma_1^2 \sum_{i=0}^{N}\sum_{j=0}^{N} \mathrm{f}_i \mathrm{f}_j \gamma_{11}(j-i) + \sigma_\varepsilon^2. \tag{12.25}$$

The ccf is given as

$$\rho_{12}(\tau) = \sum_{k=0}^{N} \phi_k \rho_{11}(\tau - k), \tag{12.26}$$

where

$$\phi_k = \frac{\mathrm{f}_k}{\left[\left(\sum_{i=0}^{N}\sum_{j=0}^{N} \mathrm{f}_i \mathrm{f}_j \rho_{11}(j-i)\right) + \sigma_\varepsilon^2/\sigma_1^2\right]^{1/2}}.$$

So, as with the simpler cases, the ccf is the convolution of the acvf of the input signal with a filter that has been uniformly scaled by the term in the denominator of (12.26).

White Noise Input

As before, if X_{1t}, the input to the N-term filter is white noise (uncorrelated with the system noise, ε_t), then the ccvf in (12.16) is given as, for all N lags τ,

$$\gamma_{12}(\tau) = \begin{cases} f_\tau \sigma_1^2 & \tau \leq N \\ 0 & \tau > N \end{cases} \tag{12.27}$$

and the ccf is given by

$$\rho_{12}(\tau) = \begin{cases} \phi_\tau & \tau \leq N \\ 0 & \tau > N \end{cases} \tag{12.28}$$

where

$$\phi_\tau = \frac{f_\tau}{\left[\left(\sum_{i=0}^{N} f_i^2\right) + \sigma_\varepsilon^2 / \sigma_1^2\right]^{1/2}}.$$

With more terms in the filter, the effect of the signal-to-noise ratio on the ccf is diminished in the correlation at any one lag. More importantly, while the overall correlation is decreased at the lowest-order lags, the influence of more filter coefficients tends to extend the relationship between the input and output series over more and more lags. That is, the filter serves to relate more distant points, since more and more values, at longer and longer times, are being involved in creating the output series.

As is clear from this general case, in the time domain, examination of the ccf or ccvf can give some indication of the nature of the linear filter that relates the two series being compared.

12.3 Frequency Domain Estimation of the ccf

12.3.1 Definitions and Interpretation

Cross Spectrum

Estimation of the ccvf in the time domain involves use of the sample ccvf, as discussed in §7.4.1. The sample ccvf has the same problems as the acvf – that is, the neighboring lags are highly correlated and the sample based estimate is biased by $(n - |k|)/n$, where k is the lag (§7.4.5). Now we estimate the frequency domain equivalent, the **cross-spectral density function (CSD)**, or simply **cross**

spectrum. Its estimate is the **sample cross spectrum**. Alternatively, the Fourier transform of the ccf also gives the CSD, though the results have different units.

First, here are some fundamentals of the CSD. The *true* cross spectrum, $p_{12}(f)$, is defined as the transform of the true ccvf, $\gamma_{12}(\tau)$, or ccf, $\rho_{12}(\tau)$, so

$$p_{12}(f) = \begin{cases} \sum_{\tau \geq -(n-1)}^{n-1} \gamma_{12}(\tau)e^{-i2\pi f\tau} & \text{(cross-covariance)} \quad (12.29a) \\ \text{or} & \\ \sum_{\tau \geq -(n-1)}^{n-1} \rho_{12}(\tau)e^{-i2\pi f\tau} & \text{(cross-correlation),} \quad (12.29b) \end{cases}$$

and the *sample* cross spectrum, using the sample ccvf, $C_{12}(\tau)$, or ccf, $R_{12}(\tau)$, is given by

$$I_{12}(f) = \begin{cases} \sum_{\tau \geq -(n-1)}^{n-1} C_{12}(\tau)e^{-i2\pi f\tau} & \text{(cross-covariance)} \quad (12.30a) \\ \text{or} & \\ \sum_{\tau \geq -(n-1)}^{n-1} R_{12}(\tau)e^{-i2\pi f\tau} & \text{(cross-correlation).} \quad (12.30b) \end{cases}$$

From the cross-covariance theorem (§9.8.5), we know that we can also write this transform pair as

$$C_{12}(\tau) \Leftrightarrow X_1^*(f)X_2(f), \tag{12.31}$$

where the asterisk denotes complex conjugate, so the Fourier transform is $R(f) + iI(f)$ instead of the usual $R(f) - iI(f)$.

Writing the Fourier transforms in polar form as

$$X_1^*(f) = A_1(f)e^{i\varphi_1(f)} \tag{12.32a}$$

$$X_2(f) = A_2(f)e^{-i\varphi_2(f)} \tag{12.32b}$$

shows that the cross spectrum, as defined in (12.31), is equal to

$$I_{12}(f) = X_1^*(f)X_2(f) = A_1(f)A_2(f)e^{i[\varphi_1(f)-\varphi_2(f)]} \tag{12.33a}$$

or, in standard form,

$$= A_{12}(f)e^{-i\varphi_{12}(f)}, \tag{12.33b}$$

where $A_{12}(f) = A_1(f)A_2(f)$, and $\varphi_{12}(f) = \varphi_2(f) - \varphi_1(f)$.

Therefore, the **cross amplitude spectrum**, $A_{12}(f)$, gives an indication as to whether a large-amplitude frequency component in one time series is associated with a large-amplitude component (at the same frequency) in the other time series. The **cross-phase spectrum**, $\varphi_{12}(f)$ (sometimes referred to as the **phase difference**, or simply **phase**

spectrum) shows the difference in phase (for the same frequency component) in the two time series.[3]

One major difference between the CSD and the PSD is that, since the ccvf is not an even function, both the cosine and sine (real and imaginary) components are preserved in the spectrum. As a consequence, the CSD (i.e., this joint-density distribution as a function of frequency) provides phase information as well as amplitude information.

Co- and Quadrature Spectra

The cross amplitude spectrum, A_{12}(f), is an even function, while the cross phase spectrum, φ_{12}(f), is an odd function of frequency. For this reason, it is sometimes convenient to rewrite the cross spectrum in terms of its real and imaginary components as

$$I_{12}(\mathrm{f}) = Co_{12}(\mathrm{f}) - iQ_{12}(\mathrm{f}), \tag{12.34}$$

where Co_{12}(f) is called the **cospectrum** and Q_{12}(f) is called the **quadrature spectrum**.

Consider the composition of these components by rewriting (12.33b) using the Euler relationship:

$$I_{12}(\mathrm{f}) = A_{12}(\mathrm{f})[\cos\varphi_{12}(\mathrm{f}) - i\ \sin\varphi_{12}(\mathrm{f})] \tag{12.35}$$

so

$$Co_{12}(\mathrm{f}) = A_{12}(\mathrm{f})\cos\varphi_{12}(\mathrm{f}) \tag{12.36a}$$

$$Q_{12}(\mathrm{f}) = A_{12}(\mathrm{f})\sin\varphi_{12}(\mathrm{f}). \tag{12.36b}$$

In this form, the amplitude of I_{12}(f), A_{12}, is given in terms of the cospectrum and quadrature spectrum as

$$A_{12}(\mathrm{f}) = [Co_{12}^2(\mathrm{f}) + Q_{12}^2(\mathrm{f})]^{1/2}, \tag{12.37a}$$

and the phase is given as

$$\varphi_{12}(\mathrm{f}) = \tan^{-1}[Q_{12}(\mathrm{f})/Co_{12}(\mathrm{f})]. \tag{12.37b}$$

Consider that both time series, X_k(t), $k = 1, 2$, can be written in its frequency domain representation as

$$X_k(\mathrm{f}) = A_k(\mathrm{f})\cos[2\pi\mathrm{ft} + \varphi_k(\mathrm{f})], \tag{12.38a}$$

[3] To avoid confusion with the phase spectrum constructed for univariate time series spectral analysis, I will break with the standard and only use the term *phase spectrum* for univariate spectra and not for cross spectra, preferring *cross phase spectrum* for that.

then expanding this using the trigonometric identity (8.8b), implicitly recognizing the dependency of the A_k, a_k, b_k and φ_k on frequency,

$$= a_k(\mathrm{f})\cos 2\pi \mathrm{ft} - b_k \sin 2\pi \mathrm{ft}, \tag{12.38b}$$

where $a_k = A_k \cos\varphi_k$ and $b_k = A_k \sin\varphi_k$. Now consider the cospectrum in terms of the a_k and b_k,

$$Co_{12}(\mathrm{f}) = A_1 A_2 \cos[\varphi_1(\mathrm{f}) - \varphi_2(\mathrm{f})], \tag{12.39a}$$

and expanding using the trigonometric identity (8.9b),

$$\begin{aligned} &= A_1 A_2[\cos\varphi_1(\mathrm{f})\cos\varphi_2(\mathrm{f}) + \sin\varphi_1(\mathrm{f})\sin\varphi_2(\mathrm{f})] \\ &= A_1 \cos\varphi_1(\mathrm{f}) A_2 \cos\varphi_2(\mathrm{f}) + A_1 \sin\varphi_1(\mathrm{f}) A_2 \sin\varphi_2(\mathrm{f})] \end{aligned} \tag{12.39b}$$

or

$$= a_1(\mathrm{f})a_2(\mathrm{f}) + b_1(\mathrm{f})b_2(\mathrm{f}). \tag{12.39c}$$

Essentially, (12.39) shows that the cospectrum provides the covariance between the cosine components and the covariance between the sine components of the two time series, that is, of the *in-phase* components.

Similarly, for the quadrature spectrum,

$$\begin{aligned} Q_{12}(\mathrm{f}) &= A_1 A_2 \sin[\varphi_1(\mathrm{f}) - \varphi_2(\mathrm{f})] \\ &= A_1 A_2[\sin\varphi_1(\mathrm{f})\cos\varphi_2(\mathrm{f}) - \cos\varphi_1(\mathrm{f})\sin\varphi_2(\mathrm{f})] \\ &= A_1 \sin\varphi_1(\mathrm{f}) A_2 \cos\varphi_2(\mathrm{f}) - A_1 \cos\varphi_1(\mathrm{f}) A_2 \sin\varphi_2(\mathrm{f})], \end{aligned} \tag{12.40a}$$

or

$$= a_2(\mathrm{f})b_1(\mathrm{f}) - a_1(\mathrm{f})b_2(\mathrm{f}). \tag{12.40b}$$

This is giving the covariance between the amplitudes of the components that are out of phase – that is, between the sine and cosine terms at comparable frequencies.

This interpretation is useful in a variety of studies where, for example, you are interested in integrating over several periods of time. Specifically, integrating over all frequencies, the cospectrum integrates to the covariance between the original time series,

$$\begin{aligned} \int_{-\infty}^{\infty} I_{12}(\mathrm{f})\mathrm{df} &= \int_{-\infty}^{\infty} [Co_{12}(\mathrm{f}) - iQ_{12}(\mathrm{f})]\mathrm{df} \\ &= \int_{-\infty}^{\infty} A_1 A_2 \cos[\varphi_{12}(\mathrm{f})]\mathrm{df} - i\int_{-\infty}^{\infty} A_1 A_2 \sin[\varphi_{12}(\mathrm{f})]\mathrm{df} && (12.41\mathrm{a}) \\ &= \int_{-\infty}^{\infty} A_1 A_2 \cos[\varphi_{12}(\mathrm{f})]\mathrm{df} && (12.41\mathrm{b}) \\ &= \sigma_{12}^2. && (12.41\mathrm{c}) \end{aligned}$$

For real data, the A_k are even functions of frequency (§9.7.6) and the sine is an odd function. Also, the product of two even functions is an even function and the product of an even and odd function is an odd function. Therefore, the second integral on the right-hand side of (12.41a) is of an odd function that integrates to 0. Following similar lines, the first integral is not zero, but in fact integrates to the covariance between the two series.

In other words, because of orthogonality, we know that integrating over frequencies, the quadrature component must drop out while the cospectrum gives an indication of the degree of covariation between the terms that will not drop out as a function of frequency.[4] For this reason, there are numerous measurements, for example those related to the flux of some properties, where the correlated components are those responsible for driving the fluxes, and thus the cospectrum gives an indication of the frequency bands over which the fluxes are being predominantly driven in.

Coherency Spectrum

Finally, there is another form in which information in the cross spectrum is conveniently conveyed. This is **coherency**, often given as **complex coherency**, as a function of frequency f,

$$\kappa_{12}(\mathrm{f}) = \frac{\mathrm{p}_{12}(\mathrm{f})}{[\mathrm{p}_1(\mathrm{f})\mathrm{p}_2(\mathrm{f})]^{1/2}}, \tag{12.42a}$$

where $\mathrm{p}_{12}(\mathrm{f})$ is the true cross spectrum, and $\mathrm{p}_1(\mathrm{f})$ and $\mathrm{p}_2(\mathrm{f})$ are the power spectra (squared amplitude spectra) of the two time series. Alternatively, the **squared coherency spectrum** is defined as[5]

$$\kappa_{12}^2(\mathrm{f}) = |\kappa_{12}(\mathrm{f})|^2 = \frac{|\mathrm{p}_{12}(\mathrm{f})|^2}{\mathrm{p}_1(\mathrm{f})\mathrm{p}_2(\mathrm{f})}. \tag{12.42b}$$

The estimate of the coherency spectrum (the **sample coherency spectrum**) is made by

$$k_{12}(\mathrm{f}) = \left[\frac{\mathrm{Co}_{12}^2(\mathrm{f}) + \mathrm{Q}_{12}^2(\mathrm{f})}{A_1^2(\mathrm{f})A_2^2(\mathrm{f})}\right]^{1/2} = \frac{A_{12}(\mathrm{f})}{[A_1^2(\mathrm{f})A_2^2(\mathrm{f})]^{1/2}}. \tag{12.43a}$$

This formula is in terms of the amplitude of the power spectrum, but is sometimes given with the numerator as $|C_{12}(\mathrm{f})|^2$ = magnitude of the CSD squared, or $A_{12}^2(\mathrm{f})$. Recall the

[4] "Quadrature" is a natural term for this context based on its astronomical definition, which indicates when two bodies are at a 90° angle relative to a third (the mathematical definition "constructing a square equal in area to a given surface," is clearly less appropriate).

[5] Note that the square root of this quantity is often referred to as the "coherency," while the squared value is the "squared coherency," but some authors do not distinguish between the two and simply use "coherency spectrum" always. Also, some people like to differentiate by calling the squared coherency "coherence." As, usual, you must make sure you know how an individual is using the term, and likewise make it clear how you are using it yourself.

relationship between cross-spectrum amplitude and co-spectra and quadrature spectra in (12.37a).

The estimate of the squared coherency spectrum (the **sample squared coherency spectrum**) is made by

$$k_{12}^2(\mathrm{f}) = \frac{Co_{12}^2(\mathrm{f}) + Q_{12}^2(\mathrm{f})}{A_1^2(\mathrm{f})A_2^2(\mathrm{f})} = \frac{A_{12}^2(\mathrm{f})}{A_1^2(\mathrm{f})A_2^2(\mathrm{f})}. \tag{12.43b}$$

Coherency shows the degree of correlation as a function of frequency. The cross-amplitude spectrum gives a measure of the covariance between the two time series as a function of frequency. But, as is the usual problem with covariance, it is a function of the actual magnitude of the quantities being compared, so there is no simple reference for comparison (i.e., a large value may either reflect an excellent correlation or may simply reflect that the numbers being compared were large). The squared coherency takes on values between 0 and 1 and can be interpreted like a standard (squared) correlation coefficient (Chapter 6). $\kappa_{12}^2(\mathrm{f})$ indicates the fraction of power on average expected in one series, given knowledge of the power in that band in the other series.

Consider the value of squared coherency for the case in which the spectra are "unsmoothed" (i.e., where we have performed no smoothing, as was done to the univariate spectra discussed previously). In that case, we are computing the correlation of two pure sinusoids, independent of phase, and even though they may have different amplitudes we only have one estimate from each series, so there is no opportunity to introduce any scatter in the estimate at all. Therefore, *for coherency to have any practical purpose, it only makes sense to compute it over a frequency band (window).* In other words, we need a smoothed coherency spectrum (recall that the smoothing reduces frequency resolution and the remaining frequencies represent a band over which they are smoothed). This makes coherency estimates sensitive to the window width of the smoothed spectrum: if the window is too narrow, we will still be comparing close to a pure sinusoid and the coherency will be artificially inflated; if the smoothing is overdone, the window (bandwidth) will be too wide and the correlation will be reduced by comparing frequency components over too wide a range of frequencies. Typically, if you use a *window closing* technique and choose a stable-looking window in the standard manner, you should get reasonable coherency estimates.[6]

12.4 Statistical Considerations

12.4.1 Amplitude and Phase Spectrum Uncertainty, Unsmoothed

The cross spectrum suffers from the same problems as the PSD: the variances of the individual estimates are independent of the sample size, and therefore the sample cross spectrum is an inconsistent estimator of the true cross spectrum. This problem is overcome in the same manner as it was for the univariate spectral analysis case: you must

[6] Coherency often shows a bias toward 1/2.

smooth the spectrum in order to reduce the variance to some tolerable level while sacrificing resolution and increasing the bias of your estimates. Fortunately, in this case, through examination of the coherency spectrum, you do get some additional feedback not previously available.

The statistical moments of the sample cross spectrum are computed in the same manner as the moments of the sample PSD were, through use of the expectance operator. The first two moments of the sample co- and quadrature spectra are[7]

$$E[Co_{12}(f)] = 0 \tag{12.44a}$$

$$\mathrm{Var}[Co_{12}(f)] = \frac{\sigma_1^2\sigma_2^2}{2} \tag{12.44b}$$

$$E[Q_{12}(f)] = 0 \tag{12.44c}$$

$$\mathrm{Var}[Q_{12}(f)] = \frac{\sigma_1^2\sigma_2^2}{2}. \tag{12.44d}$$

Also, the covariance between the co- and quadrature spectra is zero, and these spectra are uncorrelated with the PSD of the individual time series being compared.

To combine the results of (12.44) into an uncertainty for the amplitude and phase spectrum of the sample cross spectrum, we must consider a new random variable, $4A_{12}^2/\sigma_1^2\sigma_2^2$, which is equivalent to $[2A_1^2/\sigma_1^2][2A_2^2/\sigma_2^2]$ (recall that $A_{12}^2 = A_1^2A_2^2$, and $\nu|I(f)|/p(f) = \nu A^2/p(f)$ is a chi-squared distributed random variable). This is the product of two chi-squared distributed variables (each with 2 degrees of freedom). From this, the distributions of the co- and quadrature spectra can be determined (in the standard manner), and these can be combined to yield the properties of the cross amplitude and cross-phase spectra for uncorrelated white noise,

$$E[A_{12}^2(f)] = \sigma_1^2\sigma_2^2 \tag{12.45a}$$

and

$$\mathrm{Var}[A_{12}^2(f)] = 3\sigma_1^4\sigma_2^4, \tag{12.45b}$$

where the factor of three reflects the fact that two different chi-square processes are now involved in the estimates.

For the cross phase spectrum, it can be argued that the phase will be uniformly distributed (for uncorrelated white noise) over the range $-\pi/2$ to $\pi/2$.

12.4.2 Smoothing the Cross Spectrum

General

Improvement in the statistical properties of the cross spectrum comes from either smoothing the sample cross spectrum after computing it (by averaging neighboring

[7] Considering the co- and quadrature spectra separately is analogous to considering the a_j and b_j coefficients separately in the univariate case.

frequency values as before) or by truncating the ccf before transforming (also as before). As with univariate spectra, with additional smoothing, the smoothed cross spectrum approximates the true cross spectrum in the limit as n goes to infinity.

One difference between the univariate case (PSD) and the bivariate case (CSD) here, however, is the fact that while the acvf is evenly distributed with respect to the lag axis, the ccvf is not. Therefore, the truncation point in the ccf may not be best chosen in a symmetrical sense either. *In extreme situations, the bias introduced by using a symmetrical truncation point (which equates to the frequency smoothing) can be significant. Therefore, in order to avoid this unnecessary trouble, you should always realign your time series before doing a cross-spectral analysis. The realignment should shift one series relative to the other so that the maximum value in the ccvf is at the 0th lag*. In this manner, the ccvf will more closely approximate the symmetrical properties of the acvf, and the use of a single truncation point or frequency-domain smoother will not introduce as large a bias (in general). See Jenkins and Watts (1968) (pp. 399–404) for examples of how to align the ccvf and examine the cross-phase spectrum to see if the alignment requires further adjustment. Regardless, it is more likely that you will introduce larger bias into your CSD via transform of the truncated ccf than by simply averaging neighboring frequencies.

Once the sample cross spectrum has been smoothed (or transformed with a truncated ccvf), the amplitude and phase spectra can be computed in the standard manner using the smoothed co- and quadrature spectra.

Smoothed Coherency Estimates

Once the cross spectra have been smoothed (indicated by an overbar), then the squared coherency can be estimated using the smoothed components, as

$$\overline{\kappa}_{12} = \frac{\overline{A}_{12}(\mathrm{f})}{\left[\overline{A}_1^2(\mathrm{f})\overline{A}_2^2(\mathrm{f})\right]^{1/2}} \tag{12.46a}$$

$$\overline{\kappa}_{12}^2(\mathrm{f}) = \frac{\overline{\mathrm{Co}}_{12}^2(\mathrm{f}) + \overline{\mathrm{Q}}_{12}^2(\mathrm{f})}{\overline{A}_1^2(\mathrm{f})\overline{A}_2^2(\mathrm{f})} = \frac{\overline{A}_{12}^2(\mathrm{f})}{\overline{A}_1^2(\mathrm{f})\overline{A}_2^2(\mathrm{f})}. \tag{12.46b}$$

Uncertainties in Smoothed Estimates

The variance associated with the smoothed cross amplitude, cross phase and coherency estimates can be approximated by assuming small perturbations and then expanding into Taylor series and taking expectances (this is not as neat and straightforward as was the case for univariate spectra). The variance of these three estimators is given by

$$\mathrm{Var}[\mathrm{A}_{12}] \approx \frac{\alpha_{12}^2}{2n}\left(1 + \frac{1}{\kappa_{12}^2}\right), \tag{12.47a}$$

where the α_{12}^2 is the true cross-amplitude spectrum,

$$\mathrm{Var}[\bar{k}_{12}] \approx \frac{1}{2n}\left(\frac{1}{\kappa_{12}^2} - 1\right) \tag{12.47b}$$

and

$$\mathrm{Var}[\bar{k}_{12}^2] \approx \frac{4\kappa_{12}^2}{2n}\left(1 - \kappa_{12}^2\right)^2. \tag{12.47c}$$

The main thing to note from these results is that the variance is related to the degree of coherence between the time series at each particular frequency. In general, as the coherence decreases, the uncertainty in the cross amplitude and phase spectra grows very large (tending to infinity as coherence goes to zero). Thus, in general, the ability to assess the cross spectral results depends to a large extent on the degree of coherence (and the uncertainty goes to zero when the coherence is perfect).

Confidence intervals for coherency in its various forms (squared or not) involves a rather elaborate transformation and becomes more complicated if the smoothing of the spectra involved is anything other than a simple rectangular function. That is, if you segment your data into m segments and average their spectra, or equivalently, if you average m neighboring frequency values in a single spectrum or truncate your acvf at N/m points, the CI are relatively straightforward.

Following Koopmans (1974), if $m > 20$, and $0.4 < \bar{k}_{12}^2 < 0.95$, and significance level, α, the distribution of the coherence squared is approximately normal (with $\mu_{\alpha/2}$ the standardized value for a Normal distribution). For the $(1 - \alpha)100$ percent confidence interval,

$$\begin{aligned}
C_L &= \tanh\{\operatorname{arctanh}(\bar{k}_{12}) - (\mu_{\alpha/2})(2m-1)^{-1/2} - [(2(m-1)]^{-1}\} \\
C_U &= \tanh\{\operatorname{arctanh}(\bar{k}_{12}) + (\mu_{\alpha/2})(2m-1)^{-1/2} - [(2(m-1)]^{-1}\} \\
P\{C_L &\le \bar{k}_{12}^2 \le C_U\} = (1-\alpha)100\%.
\end{aligned} \tag{12.48}$$

Otherwise,

$$\begin{aligned}
\mathrm{Var}[\bar{k}_{12}^2] &= \frac{C_w}{m} 4|\bar{k}_{12}(f)|^2(1 - |\bar{k}_{12}(f)|^2) \\
C_w &= \sum_{j=-(m/2)}^{m/2} \lambda_j^2(f),
\end{aligned} \tag{12.49}$$

where C_w is the sum of the smoothing lag weights squared (i.e., when averaging m neighboring frequencies, the λ_j are the weights used for this averaging).

For $m \le 20$, or smoothing via a nonrectangular taper (averaging window), follow the excellent discussion in Bloomfield (1976) (§9.5, pg. 224–228), Jenkins and Watts (1968) or Priestley (1981).

The uncertainty in the estimate of the smoothed coherency ($\overline{k}_{12}$) (not squared, as above) is treated the same as for a correlation coefficient. Applying the Fisher z-transformation (which we have not discussed), gives

$$\overline{Y}_{12}(\mathrm{f}) = \mathrm{arctanh}[|\overline{k}_{12}|] = \frac{1}{2}\ln\left(\frac{1+|\overline{k}_{12}|}{1-|\overline{k}_{12}|}\right), \tag{12.50}$$

and the uncertainty in this transformed variable is (not valid when $\kappa_{12}(\mathrm{f}) = 0$ or: small)

$$\mathrm{Var}[\overline{Y}_{12}(\mathrm{f})] \approx \frac{\mathrm{C}_w}{m}. \tag{12.51}$$

Key to this is that the uncertainty is now independent of frequency, and so can be plotted as a single line. Assume $\overline{Y}_{12}$ is approximately normally distributed to get confidence intervals for a chosen alpha. Then, take the hyperbolic tangent of each side of (12.50) to put the CI in the original units.

You can also use the bootstrap to evaluate either form of coherency (squared or not) by generating noise time series for the real ones, and compute k_{12} or k_{12}^2, repeatedly building a PMF for the noise coherence to show the likelihood of getting the real values obtained from the data.

12.5 Take-Home Points

1. Cross spectrum gives covariance as a function of frequency between two different time series, given as the product of their individual spectra, $X_1^*(\mathrm{f})X_2(\mathrm{f}) = A_1(f)A_2(\mathrm{f})e^{i[\varphi_1(\mathrm{f}) - \varphi_2(\mathrm{f})]}$, where the left-hand-side product is individual spectra (complex conjugate of series 1) and the right-hand side shows that cross spectrum amplitude is a product of amplitude spectra, and the cross phase spectrum shows differences in phase between the spectral peaks of the two series.
2. Fourier transform of the cross correlation function gives the cross spectral density function.
3. Cospectrum is $Co_{12}(\mathrm{f}) = A_{12}(\mathrm{f})\cos\varphi_{12}(\mathrm{f}) = a_1(\mathrm{f})a_2(\mathrm{f}) + b_1(\mathrm{f})b_2(\mathrm{f})$, which expresses the "in-phase" components of the two series as a function of frequency.
4. Quadrature spectrum is $Q_{12}(\mathrm{f}) = A_{12}(\mathrm{f})\sin\varphi_{12}(\mathrm{f}) = a_2(\mathrm{f})b_1(\mathrm{f}) - a_1(\mathrm{f})b_2(\mathrm{f})$, which expresses the "out-of-phase" components of the two series as a function of frequency, where the a and b are the coefficients of the cosine and sine transforms (recall that $A(\mathrm{f}) = [a^2(\mathrm{f}) + b^2(\mathrm{f})]^{1/2}$ and $\varphi = \tan^{-1}(b/a)$.
5. Co_{12} and Q_{12} serve as the real and imaginary components of a standard spectrum, so $A_{12}(\mathrm{f}) = (Co_{12}^2 + Q_{12}^2)^{1/2}$ and $\varphi_{12}(\mathrm{f}) = \tan^{-1}[Q_{12}(\mathrm{f})/Co_{12}(\mathrm{f})]$.
6. Coherency, $\kappa_{12}(\mathrm{f}) = \left[\frac{Co_{12}^2(\mathrm{f}) + Q_{12}^2(\mathrm{f})}{A_1^2(\mathrm{f})A_2^2(\mathrm{f})}\right]^{1/2} = \frac{A_{12}(\mathrm{f})}{[A_1^2(\mathrm{f})A_2^2(\mathrm{f})]^{1/2}}$, is effectively correlation between two time series as a function of frequency. Individual spectra must

be smoothed for this because otherwise you will always get a perfect correlation between two sinusoids of the same frequency.

7. Coherency squared, $\kappa_{12}^2(f) = \frac{Co_{12}^2(f) + Q_{12}^2(f)}{A_1^2(f)A_2^2(f)} = \frac{A_{12}^2(f)}{A_1^2(f)A_2^2(f)}$, as with r^2 gives the fraction of power on average expected in one series, given knowledge of the power in that band in the other series.

12.6 Questions

Pencil and Paper Questions

1. a. How can you estimate the uncertainty of a correlation coefficient that will take into account the covariability of neighboring points in the time series, and what is the fundamental assumption you will make to use this method?
   b. Describe the various steps required to estimate whether the r value is consistent with what you might expect by random chance if your time series were nothing more than colored noise.

2. For CSD:
   a. Describe two ways to generate a CSD, and describe the fundamental parts.
   b. Describe a typical phase spectrum of a CSD.
   c. Describe how you would normalize a CSD to give something easier to interpret. Why it is easier to interpret, and what is this normalized function called?

Computer-Based Questions

3. For the first and last million years of LR04 (with their means removed):
   a. Plot the PSD for both segments and their CSD.
   b. Plot the squared coherency spectrum for five different degrees of smoothing, and state which degree of smoothing makes most sense and why it does.
   c. Draw the 96 percent confidence interval on the coherency plot.
   d. Give the frequencies (and their periods) of each peak exceeding the confidence interval.
   e. Which frequencies that were significant in the PSD at the 96 percent confidence level are not significant in the squared coherency spectrum?

13 Filtering and Deconvolution

13.1 Overview

Filtering is the ability to isolate a specific component or components of your time series for examination in isolation, or for removal of undesirable components (as you do when smoothing a time series). For example, ocean currents often display what are known as "inertial currents" as well as tides. We can compute the exact frequency of the inertial currents and then use filtering to isolate them from a spectrum. For example, Figure 13.1A shows current speed in an east-west direction from an underwater current meter deployed off the west coast of the Antarctic Peninsula; Figure 13.1B shows the spectrum with inertial frequency band identified; Figure 13.1C shows the inertial current isolated; and Figure 13.1D shows the original current with the inertial current removed. Before removing the inertial cycle, tidal currents were removed in a similar manner. This example is presented to demonstrate the method, but for a real analysis we would have to assume that there was originally some energy in the inertial frequency band, just not the large peak, so we would remove only that amount of the inertial power above a background spectrum (i.e., we assume there is no spectral gap across this band).

This is a very powerful and useful tool and can involve any number of operations. Most frequently, those methods involve convolution (e.g., smoothing via a running average) or spectral analysis (or other forms of orthogonal fitting, such as empirical orthogonal functions, as will be discussed in Chapter 15).

The word **filter** is often used to describe any operation that involves integration, differentiation, summation, differencing, smoothing, selective frequency removal, and other linear operations, and that results in a different representation of your time series. Here I only address the common usage of the term, which involves smoothing or culling of selective frequency components, for which the Fourier approach is ideal.

Bonnie Raitt, in her cover of John Hiatt's song "A Thing Called Love," sings: "whether your sunglasses are off or on, you only see the world you make."[1,2] I'm not positive, but I'm pretty sure she sings this inspired by having seen filtering misused to

[1] In her multiplatinum Grammy-award winning album "Nick of Time," Capitol, 1989, a particular favorite of mine, and apparently of many others.
[2] In his album, "Bring the Family," A&M Records, 1987.

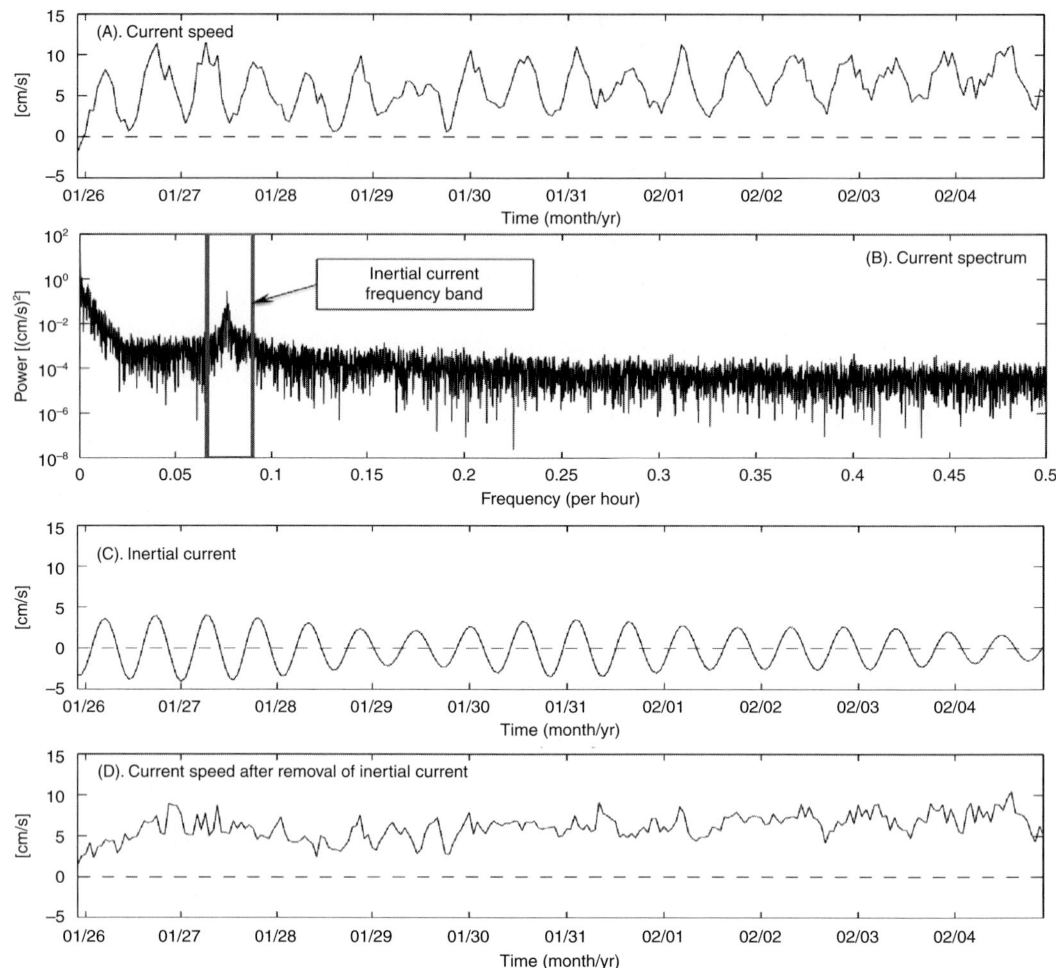

Figure 13.1 Example of filtering panel: **(A)** the original time series, **(B)** the spectrum of current, showing a band whose frequency content will be removed ("inertial band"), **(C)** the time series of removed inertial current, and **(D)** the current after removing the inertial cycle.

the point of analysts forcing the results they want. This can lead to something that applies to all methods: be honest, and don't let your preconceived notions dominate your analysis interpretation (though they should serve as a guide).

The concept of filtering was introduced previously with convolution.[3] There, we defined a filter as a "black box" through which some time-varying input series, x_t, passes, and the emerging output series, y_t, has been modified in some manner (e.g., the signal may have undergone an amplitude modification, phase shift, modification of specific frequency components, etc.). When the black box (filter)

[3] Chapter 7.

represents a **linear, time-invariant system**, filtering depicts a standard convolution operation,

$$y_t = f_t * x_t, \tag{13.1}$$

where f_t is the impulse response of the filter at times t. *The time invariant qualifier of the system simply states that, for the above relationship to hold, the impulse response function, f_t, must not vary with time, so the properties of the filter are constant over time.* We are not locked into the time invariant restriction, though, as time-adaptive approaches also exist.

This system is capable of describing (or closely approximating) a great number of *physical* phenomena. In such cases, the filter must be **causal** or **realizable** (where the output signal is a weighted linear combination of past and present input values only, so $f_t = 0$ for all $t < 0$) and **stable** ($|f_t|$ has a finite sum). However, there is no real need to confine oneself to using causal filters in practice unless one is constrained to do so by the purpose of the analysis (i.e., if one is attempting to describe a physical system). Noncausal filters frequently have superior properties regarding their influence on the phase of the signal being filtered. Clearly, real-time filtering is restricted to causal filters.

13.2 Frequency Domain Representation

13.2.1 Transfer Function

We can rewrite (13.1) in terms of the Fourier transform of the series using the convolution theorem (i.e., convolution in time domain is equal to multiplication in frequency domain),

$$Y_f = F_f X_f \tag{13.2a}$$

$$F_f = Y_f / X_f, \tag{13.2b}$$

so

$$F_f = \left| \frac{Y_f}{X_f} \right| e^{i(\varphi_Y - \varphi_X)}. \tag{13.2c}$$

F_f, the transform of the impulse response of the filter, is called the **transfer function**. Examination of the transfer function (as a function of f) will provide a clear picture of the effect of the filter on the input signal.

Given that property of the transfer function, the frequency domain represents an ideal domain with which to deal with filtering operations. This is contrary to examination of the impulse response function, f_t, because it is *convolved* with the input signal so the output, x_t, at each t, is a weighted combination of products over some range of t. Except for a few impulse response functions, it is seldom clear what the true consequence of the filter is from a time domain representation. One exception to this is the case in which the impulse response is a constant ("boxcar"). If the value of the constant is $1/n$ and the length of the filter is n values (the period, T, of the boxcar), we know that this filter will produce an output signal that is an ***n*-point running average** of the input signal. But in the frequency domain, you see exactly what frequencies are being eliminated to accomplish this smoothing. Recall the frequency response for an n-point boxcar, the sinc function:[4]

$$F_f = \frac{\cos(\pi f n)}{\pi f n}. \tag{13.3a}$$

So, the multiplication of F(f) with X(f) is

$$|F(f)||X(f)|\, e^{i(\varphi_F + \varphi_X)}. \tag{13.3b}$$

Figure 13.2 shows the operation in both domains: (1) convolution with a boxcar of length n for each value $1/n$ and (2) multiplication in frequency domain. From the latter, it is clear that the running average is eliminating high-frequency noise, but it is diminishing the high frequencies with a slow roll-off followed by a damped oscillatory decrease to zero (a seemingly goofy alteration to the unsmoothed input series, X_f).

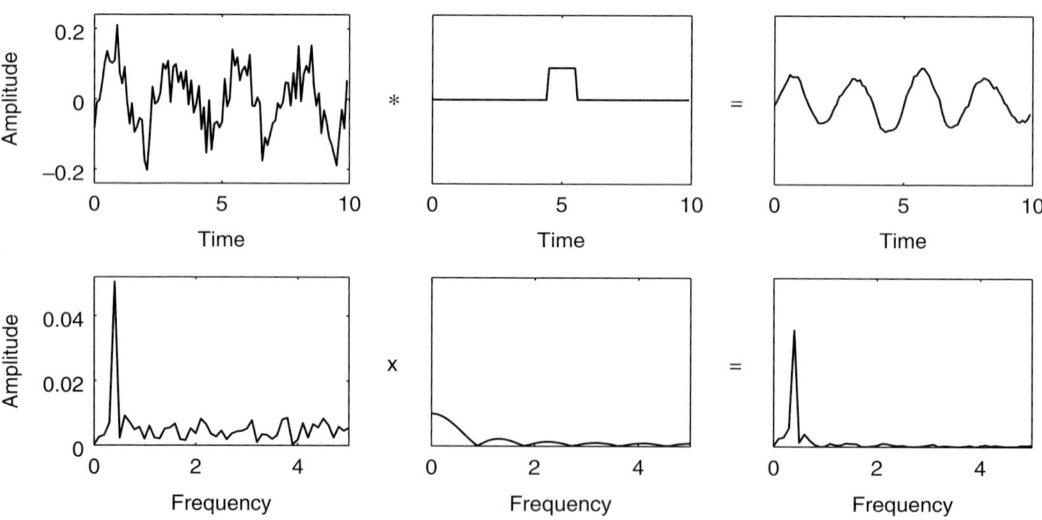

Figure 13.2 Time series and running average filter convolution to make smoothed time series. This is a Fourier transform of Figure 13.1, showing the same operation but with multiplication in the frequency domain. Note that the gain function shows a slow roll-off and decaying oscillations to higher frequencies, showing exactly what the running average is doing.

Gain Function

If we write the transfer function in polar coordinates (as we have done for all of our spectra), the *magnitude* ($(a_j^2 + b_j^2)^{1/2}$) is called the **gain (gain function or gain**

[4] Chapter 9, equation (9.91), where, in this case, $k = 1/n$ and $T = n$.

spectrum) of the transfer function, or gain of the filter. The gain squared is called the **power transfer function**.

Phase Shift

Similarly, the *phase* of the transfer function ($\varphi_j = \tan^{-1}(b_j/a_j)$) is called the **phase shift** of the transfer function or filter. Sometimes this function is simply called the **phase** of the filter, or **phase spectrum**[5] (phase lag function, if working with the negative of the phase shift function).

So, if we were to plot the magnitude and phase spectra of the transfer function (i.e., of the filter), then we would be plotting the gain and phase shift of the filter. The reason for the nomenclature of gain and phase *shift* is shown when we examine the consequence of the product of (13.2),

$$\mathrm{F_f} = |\mathrm{F_f}|\, e^{i\varphi_{\mathrm{F_f}}} \tag{13.4a}$$

$$X_\mathrm{f} = |X_\mathrm{f}|\, e^{i\varphi_{X_\mathrm{f}}}, \tag{13.4b}$$

so

$$Y_\mathrm{f} = \mathrm{F_f} X_\mathrm{f} = |\mathrm{F(f)}||\mathrm{X(f)}|\, e^{i(\varphi_{\mathrm{F_f}} + \varphi_{X_\mathrm{f}})}. \tag{13.5}$$

Examination of (13.5) shows that the magnitude of the output, Y_f, is simply the product of the magnitude of the amplitude spectrum and that of the transfer function. In other words, the magnitude of the transfer function acts to scale the input magnitude, thus controlling any *gain* in magnitude in the output relative to input signal. If the input signal has no variance or power at a particular frequency, then the output of the series will also be void of power at that frequency. The phase of the output signal is the phase of the input signal *shifted* by an amount described by the phase of the transfer function.

Thus, a fundamental property of a linear system is that the output signal can only contain the same frequencies as the input signal, though the magnitudes and phases of the output components may differ from those of the input components.

13.2.2 Phase Shifts and Causality

If a filter is to accomplish its task of modifying an input signal without introducing a phase shift, the transfer function must have zero phase for all frequencies, ($\varphi_\mathrm{f} = 0$ for all f), so the phase-shift spectrum must be flat with a value of zero. Since the phase is given by $\tan^{-1}(b_j/a_j)$, it can only be zero at all f if $b_j = 0$ everywhere. This is only accomplished for real even functions (the transform of a real even function is a *real* even spectrum, so $b_j = 0$;

[5] Recall that for stationarity to apply in classical spectral analysis, the phase of the random process is not given because the process is defined by its PSD, though each individual representation of a time series generated by that process will have a phase associated with it. For this reason, while we looked at the amplitude or power spectrum of a realization, we did not look at the phase in what would obviously be called the "phase spectrum" (as I am using that expression). The phase spectrum is not used in spectral analysis, and this is why the phase shift associated with the transfer function is simply called the phase spectrum here. This is the third different definition of "phase spectrum" you may come across.

§9.7.6). However, if the filter's impulse response function is an even function (in the time domain), this implies that it is symmetrical about the origin. Therefore, the filter will have nonzero coefficients for negative t, which in turn indicates that the filtered output at any particular time t is a weighted, linear combination of past, present and *future* values of the input series. This is a **nonrealizable (noncausal)** filter – such a filter cannot exist in the real world where we only deal with the past and present values of a signal. So, ***any realizable filter introduces a phase shift at some values of* f** (the transfer function will also always be complex, since the imaginary term won't drop out). So, *any filter that does not introduce some phase shift into the output is nonrealizable* (i.e., it cannot be applied in real time).

A simple running average is an example of a noncausal filter. This is most clearly apparent when smoothing, via a running average, some variable that had a step function onset at time 0 (Figure 13.3).

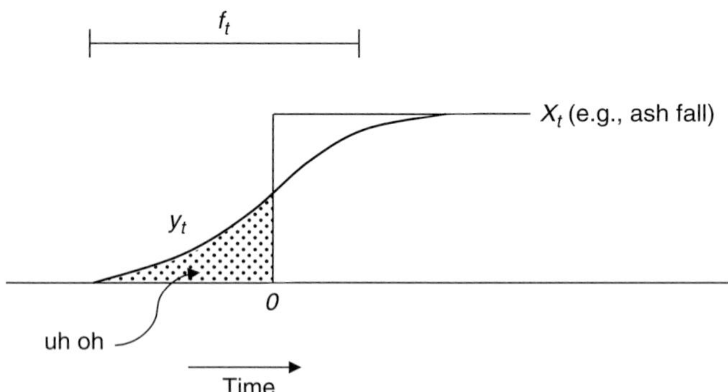

Figure 13.3 Variable showing no concentration until time 0 at onset of some event, such as an ash fall from a volcano. When smoothed by a running average of width f_t (shown), the smoothed result shows a concentration occurring before the start of the event: ash is accumulating before the volcano erupts – interesting, but noncausal (this is okay, in some situations, but not when forced to maintain reality).

13.3 Special Types of Filters

13.3.1 Ideal Filters

We revisit the running average filter for suppressing the rapidly fluctuating portion of a time series (Figure 13.4). As in Figure 13.2 and as highlighted in Figure 13.4, the multiplication of the sinc function with the input signal leaves some high-frequency power in the smoothed series, and even the lowest frequency content that you are trying to preserve is diminished by the slow roll-off in magnitude.

Ideally, we would leave the components at frequencies lower than some frequency (say, f_c) unaltered, while completely eliminating all components higher than that frequency (f_c). Such an ideal filter is accomplished by multiplying the spectrum of the data with a rectangular shaped (boxcar) gain function (shown here, Figure 13.5, for the

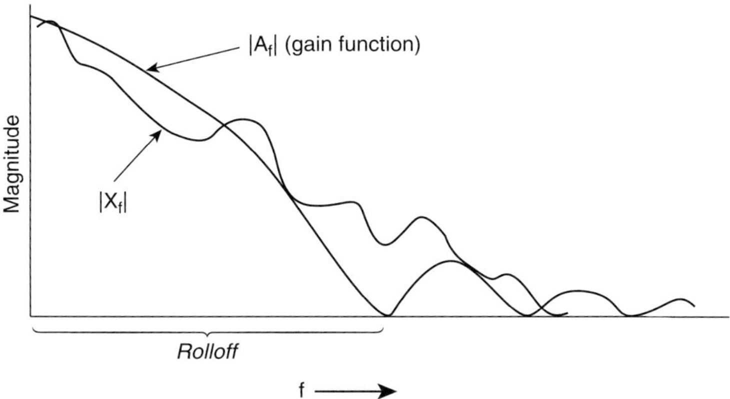

Figure 13.4 Schematic of gain function for running average (gate function) filter. This gain function adjusts the magnitude spectrum, giving the running average smoothing as it appears in the frequency domain, showing exactly what the running average is doing to the frequency content of the time series being smoothed. As expected, it is reducing the high-frequency chatter slowly and eventually oscillatory as a function of frequency. Roll-off is that portion of the transition band that takes the gain to zero, though in this case it then continues in an oscillatory damping to zero.

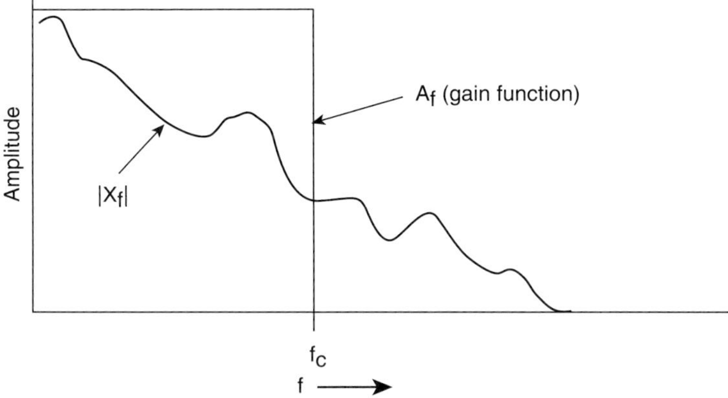

Figure 13.5 Gain function for an ideal low-pass filter for positive frequencies (passes all frequencies lower than f_c while eliminating all higher). The mirror of this is applied to the negative frequencies, zeroing all frequencies higher than $|f_c|$.

positive frequencies, but in practice remember you have negative frequencies also, so the mirror of this gain function must occur for those frequencies).

This is equivalent to convolving in the time domain with an impulse response function the shape of the sinc function (i.e., performing a running average with a weighted averaging window the shape of the sinc function). This sometimes worries people because of the oscillating tails of the sinc function, but that doesn't matter since this operation accomplishes the "smoothing" by removing the high-frequency content you consider to be noise on your signal of interest, and it is trivial to apply in the frequency domain.[6] Also, in the time domain this sinc function is infinite in length and so cannot be synthesized in the real world.

Ideal Low-Pass Filter

The gate function in the above example represents the gain spectrum of the **ideal low-pass filter** because it *passes*, without magnitude modification, the desired low-frequency components while precisely eliminating the undesired high-frequency components. In other words, we define the low-pass filter to have a gain function of

$$|A_f| = \begin{cases} 1 & |f| \leq |f_c| \\ 0 & |f| > |f_c| \end{cases}, \tag{13.6}$$

where f_c represents the cutoff frequency. The frequency band $f \leq |f_c|$ represents the **pass band,** since this is the frequency band (smaller, and therefore lower, than the absolute value of f_c) that will be *passed* through the filter unaltered. The frequency band $f > |f_c|$ is called the **stop band,** since this band *stops* the passage of any components lying within it (in this case, the frequencies higher than the cutoff frequency). This filter is called *ideal* because it cannot be attained (or at least synthesized) in the real world (it requires a time-domain filter with *infinite* elements and is noncausal, as is the running average). It can be closely approximated, though, in which the actual shape approaches that of the rectangle but with slightly rounded corners. In such a case, the frequency band in which the gain function is less than 1 and greater than 0 is called the **transition band** and its rate and shape of descent toward 0 is called the **roll-off**.[7]

Cyclic Convolution

When the impulse response function is infinite in length, as it is for the sinc function when using a boxcar transfer function, the convolution is over infinite time. So, each value in the smoothed (filtered) series involves the weighted linear combination of an infinite number of data points. Since the data are assumed to be periodic, the weighted values introduced from *outside* the sampled interval of the data are assumed equivalent to their periodic component inside the interval.

This infinite convolution is illustrated by showing the periodic y_t unwrapped, as in Figure 13.6. This demonstrates the concept of **cyclic convolution**. Note that this is *always* the manner in which the convolution operation is carried out in practice when done by multiplication in the frequency domain. That is, if two signals are to be convolved by multiplication in the frequency domain, the periodic assumption results in a convolution, as shown in Figure 13.6.

[6] The running average also shows oscillatory behavior in the higher frequency range.

[7] Note that these definitions apply to any type of filter (not just low-pass filters). Also, some people don't differentiate between transition and pass bands. They simply refer to the entire band of frequencies being passed by the filter as the pass band – that is, the band in which F(f) > 0.

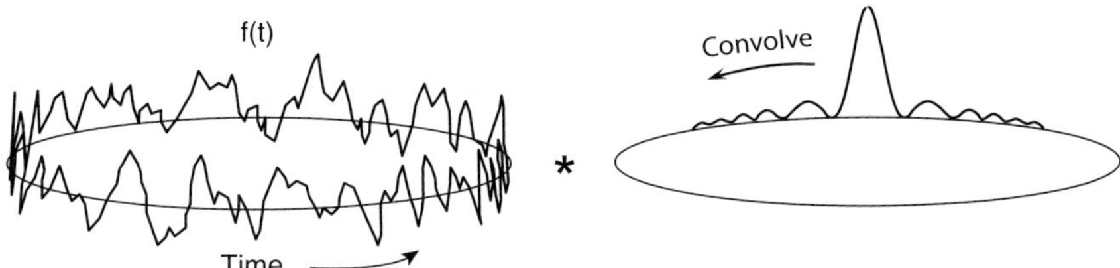

Figure 13.6 Convolution as done by multiplication in frequency domain is cyclic convolution in the time domain.

Other Low-Pass Filter Shapes

While the ideal low-pass filter does an excellent job of eliminating all frequency components higher than some specified cutoff frequency, there are some situations where it is not ideal in practice. In particular, its impulse response function (the sinc function, as defined in Chapter 9) is infinite in length, so it cannot be used to filter (convolve) in real time (i.e., as you collect the data, which is sometimes desired).

The ideal low-pass filter can be *approximated* in the time domain by truncation after some specified number of points. This is equivalent to multiplying the sinc function in the time domain by a gate function, which in turn is equivalent to convolving the transform of the gate function (the ideal filter) in the frequency domain with the transform of the gate you used to truncate the sinc function in the time domain, which will smear the gate and add wiggles. So, in the frequency domain you see the effect of truncating the sinc filter to finite length for use in the time domain.

Given this, the most common compromise for achieving a **f**inite-length **i**mpulse **r**esponse function (known as a **FIR filter**) and non-oscillatory gain function is to construct a low-pass filter that has more gently rounded corners in the frequency domain than the ideal filter.

One functional shape that does an excellent job of accomplishing this compromise is a Gaussian. The transform of a Gaussian is a Gaussian. Therefore, one could create a Gaussian filter designed to die out by a specified cutoff frequency, which would correspond to dying off in the time domain over a given interval; a running average with weights dropping off as a Gaussian, so farther away points symmetrically decrease their contribution to the local average value.

While this filter shape (a Gaussian) is used to a fair extent, its use is not as common as several other shapes whose characteristics for applications such as smoothing spectra are well known. The **split cosine-bell** is one such shape and the **Butterworth filter** is another. These arise from the least-squares approach to filter construction. In general, one can specify an ideal filter shape suited for any specific application, taking into account the considerations discussed above (i.e., a finite-length impulse response function that closely approximates the ideal low-pass filter rectangular shape, or minimizes neighboring points used in averaging, etc.). "Best" shapes (where different shapes are given, according to different definitions of "best") are approximated by both the split cosine-bell and Butterworth shapes. Consequently, these functions are frequently employed to perform low-pass filtering by convolution in the time domain when desired to work in that domain. The split cosine-bell is given by simply fitting a cosine function that decays from a value of 1 to 0 at the desired cutoff frequency. The decay occurs over some small frequency range (the transition band) equal to, say, 10 to 20 percent of the total frequency range. The impulse response function of the split cosine-bell effectively approaches zero at a significantly shorter distance than the sinc function. It is also less oscillatory. The pure cosine function is even quicker to approach zero and less oscillatory, but it has the less-desirable quality of modifying components over all frequencies.

Before you employ a more difficult convolution approach, it is certainly worthwhile to try a quick application of an ideal filter to see if it is acceptable (as it typically will be). It doesn't matter how ugly or long the impulse response function is if the removal of all frequencies within some specified band is accomplished. The result is likely to be precisely what you were hoping for and produces a result that is both easy to interpret and easy to experiment with, using different cutoff frequencies to test its sensitivity.[8]

Ideal High-Pass Filter

The high-pass filter is similar in most construction details to the low-pass filter, except for the obvious difference that this filter will *pass high-frequency components* while eliminating low ones. The practical considerations discussed for low-pass filters hold equally in this case (again, the ideal filter is the rectangular one). The gain function (A_f) looks like that in Figure 13.7:

$$|A_f| = \begin{cases} 1 & |f| \geq |f_c| \\ 0 & |f| < |f_c| \end{cases}. \tag{13.7}$$

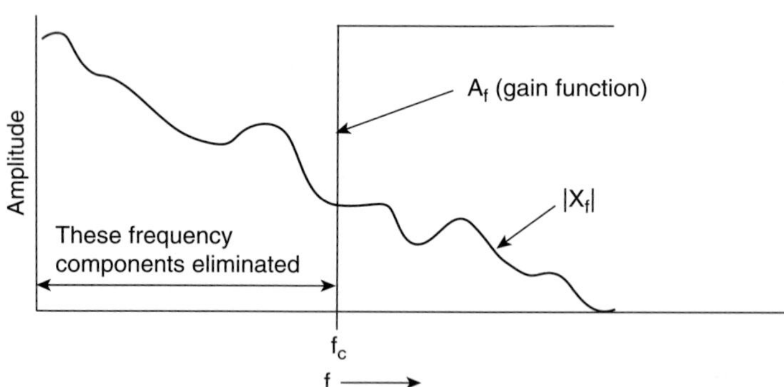

Figure 13.7 Ideal high-pass filter for positive frequencies (passes all frequencies that are higher than f_c, while eliminating all that are lower). The mirror of this is applied to the negative frequencies, zeroing all frequencies lower than $|f_c|$.

[8] Considerable use of the Butterworth (no, this is not a candy bar) and split bell cosine is due to historical reasons. Many filter activities are applied in the time domain during data acquisition, and these filters have been constructed and used for very long times with known results – as good a reason as any for using such. But if you are not in that camp, you will find the application of an ideal filter seemingly trivially easy, which does not make it undesirable. Use what makes the most sense to you and best accomplishes what you are trying to accomplish.

Ideal Band-Pass Filter

The ideal band-pass filter (with a rectangular shape) passes only those frequencies in a specified band. It, too, shares the features and practical considerations discussed with the other two types of ideal filters:

$$|A_{\mathrm{f}}| = \begin{cases} 1 \begin{cases} |\mathrm{f}| \geq |\mathrm{f_L}| \\ |\mathrm{f}| \leq |\mathrm{f_H}| \end{cases} \\ 0 \begin{cases} |\mathrm{f}| < |\mathrm{f_L}| \\ |\mathrm{f}| > |\mathrm{f_H}| \end{cases}. \end{cases} \tag{13.8}$$

The gain function of the ideal band-pass filter looks like that shown in Figure 13.8.

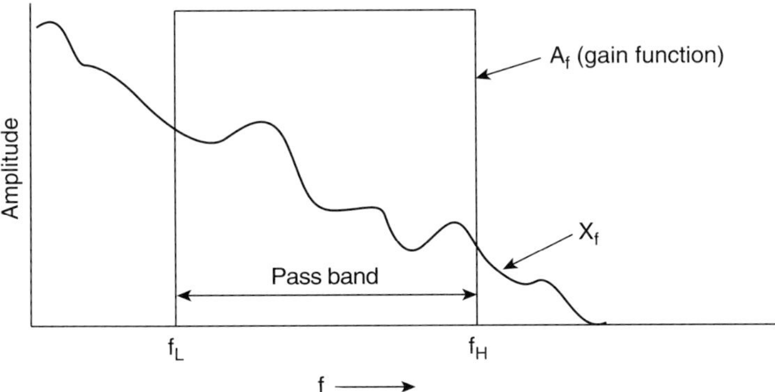

Figure 13.8 Ideal band-pass filter for positive frequencies (passes all frequencies within pass band). The mirror of this is applied to the negative frequencies.

In this case, f_L defines the lowest frequency passed (all lower frequencies are eliminated) and f_H defines the highest frequency passed (all higher frequencies are eliminated).

13.3.2 Cascaded Filters

Cascaded filtering is associated with a linear system in which the output from the first filter acts as the input to the next filter, whose output becomes the input to a next filter, etc. That is, the input is passed sequentially through a series of filters (Figure 13.9).

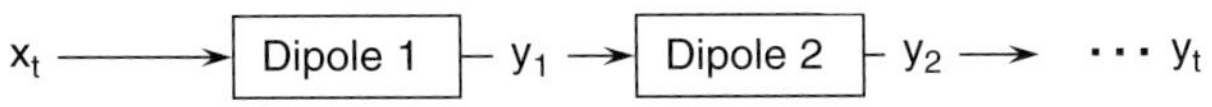

x_t ⟶ Dipole 1 ⟶ y_1 ⟶ Dipole 2 ⟶ y_2 ⟶ ··· y_t

Figure 13.9 Schematic of cascaded filters. In this case, each filter is a dipole (two-term) filter.

This system describes **cascaded filters**. The transfer function describing the net effect is arrived at simply by examination of the system in the frequency domain,

$$Y_1(\mathrm{f}) = \mathrm{F}_1(\mathrm{f})X(\mathrm{f}) \tag{13.9a}$$

$$Y_2(\mathrm{f}) = \mathrm{F}_2(\mathrm{f})Y_1(\mathrm{f}) = \mathrm{F}_2(\mathrm{f})[\mathrm{F}_1(\mathrm{f})X(\mathrm{f})] \tag{13.9b}$$

$$\vdots$$

$$Y_n(\mathrm{f}) = \mathrm{F}_n(\mathrm{f})\mathrm{F}_{n-1}(\mathrm{f})\ldots\mathrm{F}_1(\mathrm{f})X(\mathrm{f}) = \mathrm{F}_i(\mathrm{f})X(\mathrm{f}), \tag{13.9c}$$

where $\mathrm{F}_i(\mathrm{f})$ is the transfer function associated with the ith filter. Therefore, the gain of the net transfer function is given by

$$A_\mathrm{f} = A_1A_2\ldots A_{n-1}A_n = \prod_{i=1}^{n} A_i \tag{13.10}$$

and the phase shift is given by

$$\varphi_\mathrm{f} = \varphi_1 + \varphi_2 + \ldots + \varphi_{n-1} + \varphi_n = \sum_{i=l}^{n} \varphi_i. \tag{13.11}$$

So the gain is the product of the gain functions for each filter and the phase shift is the total phase shift resulting from all of the filters.

In the time domain, this net result is equivalent to the convolution of the input series with the convolution of all of the filters together (in any order).

13.4 Practical Considerations

The actual implementation of the filters discussed thus far (as well as any others) is exceedingly simple in practice. You first compute the Fourier transform of the data series to be filtered. This will provide a set of a_k and b_k coefficient values for both positive and negative frequencies, as well as an a_0 frequency representing the mean of the series. Then, in the case of an ideal filter (low-, high-, or band-pass), you simply set all of those coefficients (the a_k and b_k values) that lie within the stop band equal to zero. That is, the subscript k represents the harmonic number, which translates directly to frequency ($\mathrm{f}_k = k/\mathrm{T}$; $\omega_k = 2\pi k/\mathrm{T}$), so zero those coefficients that represent frequencies within the stop band (for both positive and negative frequencies). The one exception here is the a_0 coefficient. If the series has a nonzero mean, then simply retain this nonzero value.

If the filter has a transition band, which by definition has values between 0 and 1, then multiply the coefficients that lie within this transition band (recall the symmetry of the filter: it applies to both the positive and negative frequencies) by the respective value of the filter at each frequency within this transition band.

For example, if the filter has values of 1 for $k = \pm(1\text{–}6)$ and then a value of 0.75 for $k = \pm 7$, 0.5 for $k = \pm 8$, 0.25 for $k = \pm 9$, and then zero for all higher harmonics, the pass band is for $k = \pm(1\text{–}6)$, the transition band is for $k = \pm(7\text{–}9)$ and the stop band is for |k| ≥ 10, so the a_k and b_k coefficients for the pass band are left unaltered. Within the transition band, the a_k and b_k coefficients are multiplied by 0.75 for $k = \pm 7$, by 0.5 for $k = \pm 8$ and by 0.25 for $k = \pm 9$. For all higher harmonics, $|k| \geq 10$, the a_k and b_k values are set equal to zero.

Then, after the modification of the a_k and b_k values, the inverse Fourier transform is computed (using the modified a_k and b_k coefficients), producing the desired filtered time series.

The above is for filtering a time series decomposed by orthogonal sinusoids. However, filtering can be done with any orthogonal decomposition. For example, in Chapter 15 we will decompose a time series into a set of empirical orthogonal functions. Filtering could just as easily be done by removing some of those functions (modes).

13.5 Inverse Filtering (Deconvolution)

Often, a recorded time series represents some signal that passed through a known filter, though the signal of interest is actually the original one that was input to the filter, not the convolved, recorded one. In such cases, you **deconvolve** to determine the nature of the true input signal.

Deconvolution is appropriate in situations where

1) you wish to remove the effects of a filter that has served to muddle the desired signal (For example, if the time series you have in hand is actually a filtered version of the series of interest, with the filtering being done by the recording device, such as an instrument or natural recorder like deep sea sedimentation or snow accumulation on a glacier, and we would wish to remove the effects of the filter to see the original, or pure, series.)
2) you wish to solve for the filter, given the input and output series (For example, a seismic or sonic pulse of known shape has passed through layers of the Earth or layers of a body (the filter), and an instrument has recorded the distorted pulse after it has passing through this filter, and you wish to estimate the nature of the filter (the Earth or the body) that acted upon (convolved) the initial pulse).

13.6 Exact (Deterministic) Deconvolution

13.6.1 Direct Solution

First, consider deconvolution using the moving-strip-of-paper approach used to describe convolution (Figure 13.10). For this case, given knowledge of the input signal, x_t, and of the filter, f_t, then the output signal, y_t, is easily calculated in the standard manner ($y = f*x$).

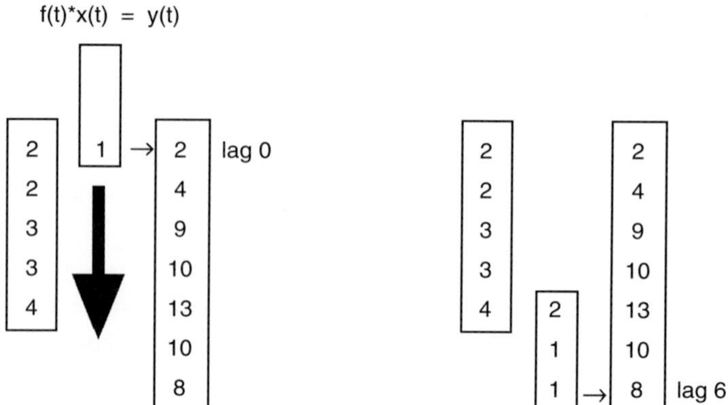

Figure 13.10 Graphic representation of convolution operation: overlapping numbers in f(t) and x(t) are multiplied and their sum is inserted in output y(t). For exact deconvolution, we simply undo this operation.

Solving for the Input Series

Consider the case where the filter, f, and output series, y, are known, and the input, x, is unknown. We know that the same process was followed in the convolution, so from Figure 13.11 it is clear that the first event in y_t is given by

$$y_0 = x_0 \mathrm{f}_0. \tag{13.12}$$

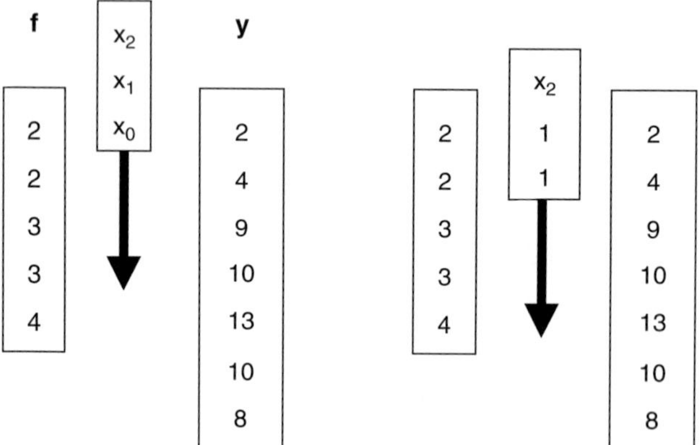

Figure 13.11 Graphic representation of an exact deconvolution operation, here with noise-free data: x values overlapping with f(t) values must be those to yield output value in y(t).

Substituting in the actual values for x_0 and f_0 (recall that the filter, f, and convolved signal, y, are known),

$$x_0 = y_0/\mathrm{f}_0 = 2/2 = 1. \tag{13.13}$$

Similarly, for the next coefficient,

$$y_1 = x_0 \mathrm{f}_1 + x_1 \mathrm{f}_0 \tag{13.14a}$$

$$x_1 = (y_1 - x_0 \mathrm{f}_1)/\mathrm{f}_0 = 1. \tag{13.14b}$$

Following this, it is clear that the general form of the solution for any value of x (say, at x_k) is found from

$$y_k = x_0 \mathrm{f}_k + x_1 \mathrm{f}_{k-1} + \ldots + x_{k-1} \mathrm{f}_1 + x_k \mathrm{f}_0, \tag{13.15}$$

giving

$$\begin{aligned} x_k &= (y_k - x_0 \mathrm{f}_k + x_1 \mathrm{f}_{k-1} + \ldots + x_{k-1} \mathrm{f}_1)/\mathrm{f}_0 \\ &= \frac{1}{\mathrm{f}_0} \left(y_k - \sum_{i=0}^{k-1} x_i \mathrm{f}_{k-i} \right). \end{aligned} \tag{13.16}$$

This will return x_t as in Figure 13.10: $\{x_0, x_1, x_2\} = \{1,1,2\}$.

Now, consider the solution for x_3 (of which there is no x_3):

$$y_3 = x_0 \mathrm{f}_3 + x_1 \mathrm{f}_2 + x_2 \mathrm{f}_1 + x_3 \mathrm{f}_0,$$

so

$$\begin{aligned} x_3 &= (y_3 - x_0 \mathrm{f}_3 - x_1 \mathrm{f}_2 - x_2 \mathrm{f}_1)/\mathrm{f}_0 \\ &= (10 - 1 \times 3 - 1 \times 3 - 2 \times 2)/2 \\ &= 0. \end{aligned} \tag{13.17}$$

This procedure is easily followed to precisely yield all of x_t, even those that are zero. However, as seen below, there is a better means of estimating the values of $\boldsymbol{x}$ for real data in which noise is present.

Solving for the Filter Coefficients

Now consider the case where we wish to determine f_t, given knowledge of y_t and x_t. This is directly found from rearrangement of (13.16). Specifically,

$$y_0 = x_0 \mathrm{f}_0, \tag{13.18}$$

so f_0 is easily determined as

$$\mathrm{f}_0 = y_0/x_0 = 2 \tag{13.19}$$

and

$$y_1 = x_0 \mathrm{f}_1 + x_1 \mathrm{f}_0 \tag{13.20}$$

$$\mathrm{f}_1 = (y_1 - x_1 \mathrm{f}_0)/x_0 = 1. \tag{13.21}$$

In this same manner, we can rearrange (13.16) to exactly solve for any coefficient of the unknown filter, say f_k, by

$$\begin{aligned} \mathrm{f}_k &= (y_k - x_1\mathrm{f}_{k-1} - \ldots - x_{k-1}\mathrm{f}_1 - x_1\mathrm{f}_0)/x_0 \\ &= \frac{1}{x_0}\left(y_k - \sum_{i=0}^{k-1} x_{k-i}\mathrm{f}_i\right) \end{aligned} \tag{13.22}$$

This system is identical in form to that of (13.16), suggesting a generalized expression that we will introduce below in the concept of a "shaping filter."

Direct Solution with Noisy Data

In order for the above procedure to work correctly, whether solving for x_t or $\mathrm{f_t}$, it requires that there be *no noise* in the system. That is, it is clearly seen from its example that the problem is *overdetermined* – we only required the first three coefficients of $\mathrm{f_t}$ and y_t to fully resolve x_t in the above example. All additional information is superfluous, and in fact, if the remaining information is not perfectly noise free, the solution obtained will yield more than just the true three values of x_t (or than the five filter coefficients, if solving for $\mathrm{f_t}$), and even these first three values need not agree with the true values. This is analogous to fitting a straight line to 20 points when in fact we only need 2 points, and any additional points can only cause trouble unless they fall exactly on the line. Otherwise, no single line can be determined.

In the case of deconvolution, the procedure outlined here keeps determining values of unknown values or coefficients until a unique series is found that matches the n input series to the output series. This is analogous to adding additional terms to a polynomial curve being fit to noisy data until eventually you end up with a polynomial of the same order as there are data points being fit.

Consider the original example, only in this case the output series contains noise relative to the true results (Figure 13.12). Repeating the previously shown solution procedure now gives estimated values of the input series as shown in the figure and in Table 13.1. Thus, the solution procedure produces as many nonzero values of the x_t input series as are required to precisely satisfy the known filter and (noisy) output series. As a consequence, the estimated values of the input series, $\hat{x}_t$, are incorrect, and there are too many of them.

In practice, one doesn't expect a perfect fit (i.e., noise-free data), so this method of solution is not practical in general. Therefore, when noise is present (which it nearly always is), we should resort to a method that can more readily accommodate the presence of noise, which, as is often the case, involves a least-squares approach.

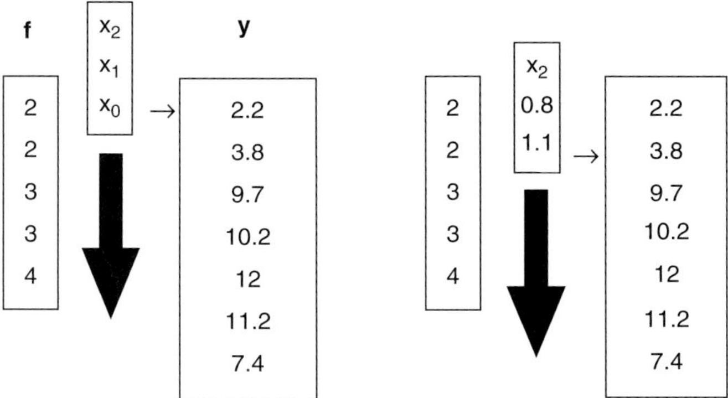

Figure 13.12 Graphic representation of convolution operation for the case of Figure 13.10, though here a small amount of noise has been added to the input series (x_t). Overlapping numbers in f(t) and x(t) are multiplied, with the sum of those products inserted in output y(t). See Table 13.1 for the impact of noise.

Table 13.1 Deconvolution results for noise-free and noisy data.

NOISE-FREE RESULTS		NOISY RESULTS	
Output (**y**)	Estimated input (**x**)	Output (**y**)	Estimated input (**x**)
$y_0 = 2.0$	$\hat{x}_0 = 1$	$y_0 = 2.2$	$\hat{x}_0 = 1.1$
$y_1 = 4.0$	$\hat{x}_1 = 1.0$	$y_1 = 3.8$	$\hat{x}_1 = 0.8$
$y_2 = 9.0$	$\hat{x}_2 = 2.0$	$y_2 = 9.7$	$\hat{x}_2 = 2.4$
$y_3 = 10.0$	$\hat{x}_3 = 0$	$y_3 = 10.2$	$\hat{x}_3 = -0.25$
$y_4 = 13.0$	$\hat{x}_4 = 0$	$y_4 = 12.0$	$\hat{x}_4 = -0.75$
$y_5 = 10.0$	$\hat{x}_5 = 0$	$y_5 = 11.2$	$\hat{x}_5 = 1.1525$
$y_6 = 8.0$	$\hat{x}_6 = 0$	$y_6 = 7.4$	$\hat{x}_6 = -0.7525$

13.6.2 Inverse Filtering

This is how you will accomplish deconvolution with noisy data.

Inverse Series

Before moving on to the least-squares solution, consider the *exact* solution to the deconvolution problem similar to the above example, but now in terms of an inverse filter. Assume that you have the impulse response function of the filter, f_t, and the output series, y_t. This is clearly presented in terms of Z-transforms:

$$Y(\mathrm{Z}) = \mathrm{X}(\mathrm{Z})\mathrm{F}(\mathrm{Z}) \tag{13.23}$$

so

$$\mathrm{X}(\mathrm{Z}) = Y(\mathrm{Z})/\mathrm{F}(\mathrm{Z}) = Y(\mathrm{Z})\mathrm{F}^{-1}(\mathrm{Z}), \tag{13.24}$$

where $F^{-1}(Z)$ is the **inverse filter** = $1/F(Z)$ and is defined by the following (standard) inverse property:

$$F^{-1}(Z)F(Z) = \delta = f_t^{-1} * f_t, \tag{13.25}$$

where δ represents the unit impulse and F(Z) is known, recalling that the unknown in (13.23) is *X*(Z).

If you had knowledge of the input and output series and wished to solve for the filter, f_t, then $F(Z) = Y(Z)/X(Z) = Y(Z)X^{-1}(Z)$, and the exact same procedure is followed, only this time, (13.25) is written as

$$X^{-1}(Z)X(Z) = \delta = x_t^{-1} * x_t. \tag{13.26}$$

Therefore, the mathematics of this section are independent of whether we are actually solving for F(*z*) or *X*(Z).

In both cases, solving (13.23) for the **inverse series**, $F^{-1}(Z)$ for the inverse filter, $X^{-1}(Z)$ for the inverse input series, allows for the solution of the original deconvolution problem: $X(Z) = Y(Z)F^{-1}(Z)$, or $F(Z) = Y(Z)X^{-1}(Z)$. However, the advantage of solving for either inverse series as an intermediate step instead of just directly solving for *X*(Z) (in the case of the inverse filter) or F(Z) (in the case of the inverse input series) is that we often wish to utilize the inverse series itself for deconvolving other series that have passed through the same filter, or that have been subjected to the same input series.

For example, given $F^{-1}(Z)$ for some standard filter such as a sediment mixing zone on some region of the ocean floor, then, any series, *Y*(Z), that has undergone mixing via that same filter, can be immediately deconvolved by multiplying $Y(Z)F^{-1}(Z)$. Or, for a standard input series shape, *X*(Z) (say, a Gaussian shaped pulse for an artificial seismic signal), then the inverse of this shape, $X^{-1}(Z)$, is used to immediately compute the filter responsible for distorting this input signal, given some output signal, i.e., $F(Z) = Y(Z)X^{-1}(Z)$.

In either case, *deconvolution is typically defined as convolution with an inverse series*.

For the case of the inverse filter, consider the actual *polynomial product* given by the Z-transform product in (13.25):[9]

$$1 = (f'_0 + f'_1 Z + f'_2 Z^2 + \ldots)(f_0 + f_1 Z + f_2 Z^2 + \ldots), \tag{13.27}$$

where the primes (′) indicate coefficients to the inverse filter.[10] Rearranging (13.27) gives

$$1 = (f'_0 + f'_1 Z + f'_2 Z^2 + \ldots) = (f_0 + f_1 Z + f_2 Z^2 + \ldots)^{-1}, \tag{13.28a}$$

[9] For convenience, the remainder of this section focuses on the inverse filter instead of the inverse input series, but as just explained, the math would be identical if we were to solve for the latter.

[10] Do not confuse the coefficients of the inverse filter with the *inverse* of the original individual filter coefficients – these are not the same. That is, $f'_0 \neq 1/f_0$; rather, it is equal to the first coefficient of the inverse filter, f_t^{-1}. This is analogous to the fact that $1/(3 + 2) \neq 1/3 + 1/2$.

or, in terms of the Z-transforms,

$$F^{-1}(Z) = \frac{1}{F(Z)}. \tag{13.28b}$$

This represents the exact inverse filter.

Now consider the simple case where the impulse response consists of a dipole: just two terms, $1Z^0$ and f_1Z^1, or 1 and f_1Z. In this case,[11]

$$F^{-1}(Z) = \frac{1}{1 + f_1 Z}. \tag{13.29a}$$

Carrying out the division yields

$$= (-1, \ -f_1, \ f_1^2, \ -f_1^3, \ldots). \tag{13.29b}$$

This inverse filter contains an infinite number of terms. In fact, this is always the case; *inversion of a finite polynomial yields an infinite polynomial.*

For example, if $x_t = 2,2$, $f_t = 1,2$, then the convolution $f * x = y$, where $y_t = 2,6,4$. Given f_t and y_t, and applying the above exact inverse relationship,

$$X(Z) = Y(Z)F^{-1}(Z), \tag{13.30}$$

where

$$F^{-1}(Z) = 1/(1 + 2Z) = (1, \ -2, \ 4, \ -8, \ 16, \ -32, \ldots). \tag{13.31}$$

Now, convolving this infinite series with y_t gives

$$\begin{aligned} Y(Z)F^{-1}(Z) &= (2 + 6Z + 4Z^2)(1 - 2Z + 4Z^2 - 8Z^3 + 16Z^4 + \ldots) \\ &= 2 + 6Z + 4Z^2 - 4Z - 12Z^2 - 8Z^3 + 8Z^2 + 24Z^3 + \\ &\quad 16Z^4 - 16Z^3 - 48Z^4 - 32Z^5 + \ldots \\ &= 2 + 2Z + 0Z^2 + 0Z^3 + \ldots \\ &= 2 + 2Z. \end{aligned} \tag{13.32}$$

[11] Division of 1 by a polynomial 1 + z can be done just like standard long division:

```
             1  -  Z  +  z^2  -  z^3  +  ...
          ____________________________________
  1 + Z  |  1  +  0  +  0   +  0   +  0  +  ...
            1  +  Z
            -------
                 -z  +  0
             -   -z  -  z^2
                 ----------
                        z^2  +  0
                    -   z^2  +  z^3
                        -----------
                               -z^3  +  0
                           -   -z^3  -  z^4
                               ------------
                                        z^4
                                         ...
```

In other words, the original, finite length series of $x_t = 2,2$, is returned by convolution of x_t with the *infinite* series of the inverse filter, f_t^{-1} Conceptually, the growing terms would cancel if we could expand in a finite number of packets as shown above, but the actual infinite series does not converge, since the magnitude of the terms is growing. This will cause a stability problem, as discussed below.

For this case, you also could have solved directly for f_t (not its inverse, which is what was just computed here), and found the exact deconvolution solution for the two coefficients using the approach discussed immediately above.[12]

Now consider the next simplest case, where the filter to be inverted contains three terms instead of the two terms demonstrated above. For example, $F(Z) = 1 + 5Z + 6Z^2$. This can be factored as $(1 + 2Z)(1 + 3Z) = F_1(Z)F_2(Z)$. Equation (13.28) is now rewritten in terms of the product of two simple filter inverses, each being the inverse of a dipole (two-term) filter. That is,

$$\begin{aligned} F^{-1}(Z) &= \frac{1}{1 + 5Z + 6Z^2} \\ &= \frac{1}{(1 + 2Z)(1 + 3Z)} \\ &= \frac{1}{(1 + 2Z)}\frac{1}{(1 + 3Z)} \\ &= (1, \ -2, \ 4, \ -8, \ldots)(1, \ -3, \ 9, \ -27, \ldots), \end{aligned} \tag{13.33}$$

or, more generally, a procedure applicable for any number of coefficients:

$$= (1, \ -f_1, \ f_1^2, \ -f_1^3, \ldots)(1, \ -f_2, \ f_2^2, \ -f_2^3, \ldots). \tag{13.34}$$

Wavelets

The above discussion leads to the concept of a wavelet.[13] A **wavelet** (or **transient**) is a signal of *finite energy* with a *specific start time*. That is, *it is one-sided and has a definite origin or arrival time.*[14] The term "one-sided" is used to imply that the filter has a specific start time and contains no values before that, and, specifically, the start time is at or after time t = 0, so the wavelet is causal, or realizable – that is, it does not respond before it is forced.

[12] If you work this case through, you quickly find that even if you assume that the impulse response function of the filter, f, contains more than two coefficients, the solution shows that all coefficients after the second one are equal to zero. Therefore, in this exact case, the simple exact solution shown initially in §13.6.1 is best. Here, however, we are developing the general properties of an inverse filter that have considerably wider applicability than this simple, exact solution case.

[13] While there is consistency in the nomenclature, wavelets, as discussed here, are not to be confused with the wavelet transform.

[14] The **energy** of a series is the sum of the squares of all values in the series. Therefore, a wavelet with its finite energy is not *stationary,* since a stationary series must maintain constant statistical properties over all time, making its energy infinite (the *power* of a stationary series is finite, since **power** is the energy per unit time). So, the wavelet is a complete package – it is a transient (*nonstationary*) series unlike our typical time series, which is a sampled (finite) portion of an infinite stationary series.

A wavelet that consists of only two terms, e.g., (f_0, f_1), is called a **dipole**, like a two-term filter. *Any wavelet, such as a filter or finite waveform, longer than two terms can be generated by the convolution of a series of dipoles (as in the cascaded filters of* §13.5.2*).* So for the above example with the three-coefficient filter, $F(Z) = 1 + 5Z + 6Z^2$, the filter was written in terms of two dipoles, $f = f_1 * f_2$, the convolution of the individual dipoles, f_1 and f_2, and $y = x*f = x*(f_1*f_2) = (x*f_1)*f_1$.

This follows from the fact that the Z-transform of any wavelet, F(Z), is given as

$$F(Z) = f_0 + f_1 Z + f_2 Z^2 + f_3 Z^3 + \dots. \tag{13.35}$$

F(Z) is a simple polynomial in Z, and all polynomials can be factored. That is, any polynomial can be rewritten in terms of a product of n dipoles (where n is the degree of the polynomial and $n + 1$ is the order of the polynomial being factored),

$$F(Z) = (a_0 + a_1 Z)\,(b_0 + b_1 Z)\,(c_0 + c_1 Z)\dots. \tag{13.36}$$

In each case, the dipole coefficients can be normalized by dividing through by the lead coefficient, giving

$$\begin{aligned} F(Z) &= A[\,(1 + a_1/a_0)Z][\,(1 + b_1/b_0)Z][\,(1 + c_1/c_0)Z]\ \dots \\ &= A(1 + a_n Z)(1 + b_n Z)(1 + c_n Z)\dots, \end{aligned} \tag{13.37}$$

where A is the scaling constant of the normalizations $= a_0 b_0 c_0\ \dots$ The **zeros** of the polynomial occur at the negative inverse of the second coefficient in each dipole. That is, the polynomial is zero where $Z = -a_0/a_1 = -1/a_n$, $Z = -b_0/b_1 = -1/b_n$, etc.). Zeros are important when designing specific filters.

Inverse Filter Stability

Examination of the inverse filter terms in (13.29b) reveals that if the value of $|f_1| > 1$, the convolution with this inverse filter produces a nonconverging series. That is, the convolution sum is infinite and the serial product grows larger with each additional product term, since the magnitude of each term is increasing. This represents an **unstable inverse**. A **stable inverse** is obtained only if $|f_1| < 1$, in which case the convolution sum, while still infinite, converges because the terms are getting smaller and smaller with the larger powers.[15]

For more complicated filters, we can address the question of inverse filter stability by examination of the dipoles. That is, we can rewrite (13.27) in terms of the factored form of the wavelet:

$$\begin{aligned} (f'_0 + f'_1 Z + f'_2 Z^2 + \dots) &= \frac{1}{(1 + a_n Z)(1 + b_n Z)(1 + c_n Z)\dots} \\ &= \frac{1}{(1 + a_n Z)}\frac{1}{(1 + b_n Z)}\frac{1}{(1 + c_n Z)}\dots \\ &= F_0^{-1}(Z)F_1^{-1}(Z)F_2^{-1}(Z)\dots. \end{aligned} \tag{13.38}$$

[15] When $|f_1| = 1$, this is a special case that contains properties of both the stable and unstable inverse. It is treated later.

In this form, we see that the inverse filter is the product of the inverse of a set of simple dipoles.[16] So, if each dipole inversion is stable, then the entire inverse is stable. Conversely, if any of the dipole inverses is unstable, then the wavelet inverse is unstable.

To see this, consider the inverse of one dipole in (13.38):

$$\mathrm{F}_0^{-1}(\mathrm{Z}) = (1, \ -a_n, \ a_n^2, \ -a_n^3, \ldots). \tag{13.39}$$

Whether this inverse is stable or not depends solely on the magnitude of the second coefficient of the normalized dipole. If $|a_n| < 1$, so $|a_1/a_0| < 1$ and $|a_1| < |a_0|$, the inverse *dipole* is stable, and if all of the other dipoles share the same property, then the inverse *wavelet* is stable.

A dipole in which the second coefficient is less than 1 (i.e., the first coefficient is larger than the second, so $|a_n| < 1$ and the inverse is stable) is called a **minimum-delay dipole,** and the corresponding wavelet composed of minimum-delay dipoles is a **minimum-delay wavelet.**[17] If the second coefficient is larger than 1 ($|a_n| > 1$ and the inverse is unstable), the dipole is a **maximum-delay dipole,** and if all the dipoles are maximum-delay in the wavelet, then the wavelet is a **maximum-delay wavelet**.

So, returning to the exact inverse filter problem, we see that if the wavelet is minimum-delay, then it has a stable inverse.[18]

13.7 Best-Fit Deconvolution

A stable inverse filter (of a finite-length filter) has an infinite number of terms; thus, it can provide an exact fit to any series, except that its infinite length does not allow you to implement the deconvolution in practice. Therefore, you must either truncate it to a finite length (which is really only practical for noise-free data) or you must resort to another means for approximating the inverse filter.

13.7.1 Generalizing the Deconvolution Problem

Before developing the general least-squares solution for performing an approximate deconvolution or inverse filtering in the presence of noise (i.e., for real data), first consider how we might generalize the various forms of deconvolution that have been discussed (direct deconvolution, inverse filtering and shaping filters).

In its various guises, we now wish to solve one of the deconvolution problems:

[16] Recall that the *product* of Z-transformed polynomials is equal to the *convolution* of the non-transformed series.

[17] This nomenclature stems from an energy viewpoint – that is, a minimum-delay dipole or wavelet is one in which the first coefficient represents the maximum energy term, and therefore there is a minimum delay in the arrival of the energy.

[18] It will turn out that if the wavelet is maximum delay, a stable inverse can be obtained but is nonrealizable, and in fact is fully *anticipative* in that it requires all future values only for the convolution sum. If it is a mixture of maximum and minimum delay dipoles, then its inverse requires realizable and nonrealizable components – something that is no problem mathematically, and is discussed later.

1) direct deconvolution solving for the input series X(Z), given knowledge of Y(Z) and F(Z):

$$Y(Z)F^{-1}(Z) = X(Z) \tag{13.40a}$$

2) direct deconvolution solving for the filter F(Z), given knowledge of X(Z) and Y(Z):

$$Y(Z)X^{-1}(Z) = F(Z) \tag{13.40b}$$

3) inverse filtering solving for the inverse filter $F^{-1}(Z)$, given knowledge of F(Z):

$$F(Z)F^{-1}(Z) = \delta(Z) \tag{13.40c}$$

4) inverse filtering solving for the inverse input series $X^{-1}(Z)$, given knowledge of X(Z):

$$X(Z)X^{-1}(Z) = \delta(Z) \tag{13.40d}$$

5) shaping filter problem, a generalization of the above, by solving for a shaping filter $F_S(Z)$, as an inverse to *X*(Z), given knowledge of *X*(Z) and $Y_S(Z)$, the latter a desired output shape (the real output or a desired output):

$$X(Z)F_S(Z) = Y_S(Z) \tag{13.40e}$$

The last three problems listed here are essentially identical in form. That is, in each case, one of the series on the left-hand side is known and the series on the right-hand side is known. Of these, (13.40e) might be considered a more general form, since $Y_S(Z)$ can even be defined as the impulse function, in which case $F_S(Z)$ would yield the inverse input series, or an inverse filter if X(Z) is replaced with F(Z). Furthermore, as previously discussed, solving for the inverse series in (13.40c) and (13.40d) solves the inverse problem and allows solution of the deconvolution problems (13.40a) and (13.40b), respectively. Therefore, nothing is gained by considering the forms of (13.40a) and (13.40b) separately, since these direct deconvolution problems can be placed into the general form of (13.40e) if they are rearranged back into their original (non-inverted) form of

$$X(Z)F(Z) = Y(Z). \tag{13.40e}$$

In this form, the series on the right-hand side, Y(Z), is known, as is *one* of the series on the left-hand side, either X(Z) or F(Z), while the other is unknown, and its coefficients must be solved for.

Therefore, (13.40e) represents a general form that can describe any of the specific forms of deconvolution, inverse filtering or shaping filtering problems discussed so far. For this reason, the remainder of the chapter follows the form of (13.40e). In all cases, we seek to determine the values or coefficients of an unknown series (one of the series on the left-hand side), which, when convolved with some observed series, produces some other observed or specified series. For consistency with the previous developments, in the remainder of the chapter we will consider that the filter F(Z) contains the unknown coefficients.

Quantifying the Error

Given the general form of (13.40e), now consider the fact that we are acknowledging the presence of noise in the data so that our unknown filter, F(Z), cannot exactly solve (13.40e). Instead, we are solving

$$X(\mathrm{Z})\mathrm{F}(\mathrm{Z}) = Y(\mathrm{Z}) + \varepsilon(\mathrm{Z}) \tag{13.41a}$$

or

$$X * \hat{\mathrm{f}} = y + \varepsilon = \hat{y}, \tag{13.41b}$$

where ε(Z) represents the error present in each value of Y(Z). That is, given an approximate solution, $\hat{\mathrm{F}}(\mathrm{Z})$, we do not expect that our convolution of $X*\hat{\mathrm{f}}$ will yield an exact fit to y_t, but rather an approximate fit, $\hat{y} = y + \varepsilon$, where the error present in each of the values in y_t is indicated by ε.

We can measure the overall quality of our solution by the total error, e, defined in the standard manner as the sum of the squared errors for each coefficient,

$$\begin{aligned} \mathrm{e} &= \sum_{i=0}^{N} \varepsilon_i^2 \\ &= \sum_{i=0}^{N} (\hat{y}_i - y_i)^2 \\ &= \sum_{i=0}^{N} [(x * \hat{\mathrm{f}})_i - y_i]^2, \end{aligned}$$

and, since the term in parentheses is a convolution,

$$\mathrm{e} = \sum_{i=0}^{N} \left[\left(\sum_{k=0}^{\mathrm{p}} \hat{\mathrm{f}}_k x_{i-k} \right) - y_i \right]^2, \tag{13.42}$$

where y_t is the specified, or true, series that is being approximated, which consists of N+1 values, and $N = m + n$, $i = 0, 1, \ldots, N$; $\hat{y}_i$ is the approximation given by the convolution of $x * \hat{\mathrm{f}}$. The approximation, $\hat{y}_i$, and thus the total error, e, is a function of the m+1 unknown coefficients of $\hat{\mathrm{f}} = (\mathrm{f}_0, \mathrm{f}_1, \ldots, \mathrm{f}_m)$, and the n+1 known values of y_t. The convolution sum in (13.42) involves a sum over p = max(m+1,n + 1), and this for the (m+1) + (n+1) − 1 = N+1 lags, i.[19]

For the different forms of (13.40), the error function, in each case following this exact form, is given as

1) direct deconvolution, solving for the input series directly (without first solving for the inverse input series):

$$\mathrm{e} = \sum_{i=0}^{N} [(\mathrm{f} * \hat{y})_i - x_i]^2 \tag{13.43a}$$

[19] See equation (7.16) and accompanying notation explanation for the limits of the convolution sum.

2) inverse filtering, solving for the inverse filter:

$$\mathrm{e} = \sum_{i=0}^{N}[(\mathrm{f} * \hat{\mathrm{f}}^{-1})_i - \delta_i]^2 \tag{13.43b}$$

3) inverse filtering, solving for the inverse input series:

$$\mathrm{e} = \sum_{i=0}^{N}[(y * \hat{x}^{-1})_i - \delta_i]^2 \tag{13.43c}$$

4) Shaping filter problem, solving for the shaping filter:

$$\mathrm{e} = \sum_{i=0}^{N}[(x * \hat{\mathrm{f}}_S)_i - x_{Si}]^2 \tag{13.43d}$$

13.7.2 Truncated Deconvolution

As stated, we must approximate the infinite inverse filter to finite length. First, we consider the simple case of truncating it. In this case, we can describe the difference between the actual product of F(Z) $\mathrm{F}_\mathrm{T}^{-1}(\mathrm{Z})$, where $\mathrm{F}_\mathrm{T}^{-1}(\mathrm{Z})$ is the truncated approximation to the inverse filter, and δ, the desired product of $\mathrm{F(Z)F}_\mathrm{T}^{-1}(\mathrm{Z})$. As shown below, truncation after n terms results in an error of f_1^{2n} (i.e., truncation after two terms gives an error f_1^4).

Truncated Inverse of a Dipole Filter

From (13.29), the inverse filter of a stable dipole, (1, f_1), has coefficients $1, -\mathrm{f}_1, \mathrm{f}_1^2, -\mathrm{f}_1^3, \ldots$. *Without* truncation,

$$\begin{aligned}\mathrm{F}^{-1}(\mathrm{Z})\mathrm{F}(\mathrm{Z}) &= (1 - \mathrm{f}_1\mathrm{Z}^1 + \mathrm{f}_1^2\mathrm{Z}^2 - \mathrm{f}_1^3\mathrm{Z}^3 \ldots)(1 + \mathrm{f}_1\mathrm{Z}) \\ &= 1 - \mathrm{f}_1\mathrm{Z}^1 + \mathrm{f}_1^2\mathrm{Z}^2 - \mathrm{f}_1^3\mathrm{Z}^3 \ldots + \mathrm{f}_1\mathrm{Z}^1 - \mathrm{f}_1^2\mathrm{Z}^2 + \mathrm{f}_1^3\mathrm{Z}^3 \ldots \\ &= 1,\end{aligned} \tag{13.44}$$

so, for the inverse of a *stable* dipole, (13.43b) has an error given as

$$\mathrm{e} = (1-1)^2 = 0. \tag{13.45}$$

First-Order Truncation

If we truncate $\mathrm{F}^{-1}(\mathrm{Z})$ after the first coefficient, so $\mathrm{F}_\mathrm{T}^{-1}(\mathrm{Z}) = 1$, then

$$\mathrm{F}_\mathrm{T}^{-1}(\mathrm{Z})\mathrm{F}(\mathrm{Z}) = 1(1 + \mathrm{f}_1\mathrm{Z}) = 1 + \mathrm{f}_1\mathrm{Z}, \tag{13.46}$$

so

$$\begin{aligned}\mathrm{e} &= (1-1)^2 + (0 - \mathrm{f}_1)^2 \\ &= \mathrm{f}_1^2.\end{aligned} \tag{13.47}$$

Second-Order Truncation

For truncation after the second term,

$$\begin{aligned} \mathrm{F_T^{-1}(Z)F(Z)} &= (1-\mathrm{f_1Z})(1+\mathrm{f_1Z}) \\ &= 1-\mathrm{f_1Z}+\mathrm{f_1Z}-\mathrm{f_1^2Z^2} \\ &= 1-\mathrm{f_1^2Z^2}, \end{aligned} \tag{13.48}$$

so

$$\begin{aligned} \mathrm{e} &= (1-1)^2 + (0-0)^2 + (0+\mathrm{f_1^2})^2 \\ &= \mathrm{f_1^4}. \end{aligned} \tag{13.49}$$

Continuing in this fashion shows that *the error is always equal to the square of the first truncated (i.e., first ignored) coefficient value.* Since inverse dipole coefficient values are $1, -\mathrm{f_1}, \mathrm{f_1^2}, -\mathrm{f_1^3}, \ldots$, if we truncate after n terms, the error is $\mathrm{f_1^{2n}}$. Furthermore, repeating this procedure for nth-ordered filters and their $n-1$ dipole inverses, one can compute the magnitude of the error for each order of truncation. For example, first-order truncation of a filter consisting of three terms (factored into two dipoles) has $\mathrm{e} = \mathrm{f_1^2} + \mathrm{f_2^2}$.

Unfortunately, this method is only useful when there is no noise in the series you are inverting, so you can get an exact inverse to truncate. Otherwise, the inverse is only useful for deconvolving other series with the exact same distribution of noise (which probably wouldn't be noise if it were replicable to that precision). So, unless this is being done for a theoretical case, truncation is not really practical.

13.7.3 General Least-Squares Deconvolution

While the method of truncation described above is acceptable in many instances, it is easily shown that for the same number of estimated coefficients, the method of least squares produces a smaller error than that resulting from straight truncation. In fact, the method of least squares can be employed to develop an *optimal* unknown filter, one that provides the smallest sum of squared errors, as defined in (13.43), for any specific (finite) number of filter coefficients. Unfortunately, in order to reduce this to the most simple form, we must follow a tortuous path of indexing – apologies, but in the end, worth the reductions.

The error function, (13.43), is a function of the m+1 unknown coefficients of the approximate unknown filter, $\hat{\mathrm{f}}_k$, so $\mathrm{e}(\mathrm{f_0}, \mathrm{f_1}, \ldots, \mathrm{f}_m)$. As shown previously for least-squares problems, we find those coefficients ($\hat{\mathrm{f}}_k$) that provide the minimal error, e, by taking the derivatives of $\mathrm{e}(\mathrm{f_0}, \mathrm{f_1}, \ldots, \mathrm{f}_m)$ with respect to each unknown coefficient and setting these equal to zero. This provides the m+1 *normal equations* for the m+1 unknown coefficients, which is well posed and easily solved via matrix methods.

The derivatives are given as

$$\frac{\partial \mathrm{e}}{\partial \hat{\mathrm{f}}_j} = 2\sum_{i=0}^{N}\left[\left(\sum_{k=0}^{p}\hat{\mathrm{f}}_k x_{i-k}\right) - y_i\right](x_{i-j}) = 0 \tag{13.50}$$

recalling that p = max(m+1,n+1), n+1 being the known values of x_t.

Box 13.1 Operating on Sums

Here is a simple trick for operating on sums: expand the summations (i.e., write out the individual terms in the sum), perform the operation on the expanded terms, then recombine into sums after you see the pattern forming. For (13.50), expand several terms, take the derivative with respect to a couple of coefficients, and then recombine into summations based on the pattern established from the expanded sums.

Rearranging (13.50) yields

$$\sum_{i=0}^{N}\sum_{k=0}^{p}(\hat{\mathrm{f}}_k x_{i-k}x_{i-j}) = \sum_{i=0}^{N} y_i x_{i-j}. \tag{13.51}$$

Changing the order of the summations,

$$\sum_{k=0}^{p}\left[\hat{\mathrm{f}}_k\sum_{i=0}^{N}(x_{i-k}x_{i-j})\right] = \sum_{i=0}^{N} y_i x_{i-j}. \tag{13.52}$$

This describes a system of equations: one equation for each coefficient of the unknown filter, specified by the index j, as seen by examining the left-hand side of (13.50).

Examination of (13.52) reveals that the sums it contains allow reduction of the equation into standard serial products. Specifically, the sum on the right-hand side is the cross-covariance function (**ccvf**) between the input and output series. The ccvf has coefficients $c - j$, but they are defined in an opposite-than-usual sense, i.e., usually the index would be $i + j$, not $i - j$. Therefore, the ccvf is reversed relative to its standard form.[20]

Therefore, (13.52) can be rewritten as

$$\sum_{k=0}^{p}\left[\hat{\mathrm{f}}_k\sum_{i=0}^{N}(x_{i-k}x_{i-j})\right] = c_{-j}. \tag{13.53}$$

Likewise, the inner sum on the left-hand side of (13.53) is like an autocovariance function (**acvf**) of the input series, shifted by j lags relative to its usual form. Thus, the

[20] Actually, the sum on the right-hand side of (13.53) is a *correlation* and not a *convolution,* since neither signal is reversed, in which case there would be an $-i$ index associated with either signal, not $-j$.

Box 13.1 (Cont.)

acvf has coefficients a_{j-k}. This can be understood by starting with the standard acvf and modifying it to be offset by a specified amount. The standard form of the acvf is given as

$$a_L = \sum_{t=0}^{N} x_t x_{t+L}, \tag{13.54}$$

where the index L dictates the lagged (offset) positions of the signals prior to multiplying and summing. If, however, the index $L = j - k$, then

$$a_{j-k} = \sum_{t=0}^{N} x_t x_{t+j-k} \tag{13.55}$$

Setting $i = t + j$, so $t = i - j$, then

$$a_{j-k} = \sum_{i-j=0}^{N} x_{i-j} x_{i-k}. \tag{13.56}$$

This is the form of the inner sum in (13.53). The lower limit of the sum in (13.56) is now at $i - j = 0$, or $i = j$, whereas in (13.53) it is still shown as $i = 0$. This was merely a convenience in (13.53). Because of the offset in lags by j at $i = 0$, there are no overlapping coefficients of the series until $i = j$, so in fact the lower limit could have been as easily written as $i = j$ in (13.53) and (13.56).

The acvf as it is written in (13.53) or (13.56) now has the coefficients at lags $L = j - k$ (i.e., each coefficient a_{j-k} reflects the covariance of y_t with itself, but offset by an amount of $t = j - k$). Conceptually, the writing of L, the acvf lag, in terms of j and k, is simply a means for aligning the terms of the acvf properly with the terms of the ccvf $(c - j)$ and the terms in $\hat{f}_k$ still present in the outer sum of (13.53). In other words, we are dealing with standard serial products, but they must be offset relative to one another so that the sums are sequenced properly.

Substituting (13.56) into (13.53) gives the normal equations to the least-squares problem in a relatively easy to comprehend form:

$$\sum_{k=0}^{p} \hat{f}_k a_{j-k} = c_{-j}. \tag{13.57}$$

Therefore, *the approximate unknown filter coefficients that provide the least-squares fit to the deconvolution or shaping filter problem are those which, when convolved with the acvf of the input series, yield the ccvf of the input signal with the output, or specified shaped, series*. Equation (13.57), which is a simple rearrangement of the normal equations, thus describes a system of equations – one for each coefficient j.

Further Reduction for the Inverse Filtering Problem

Consider the original inverse filter problem where we are trying to determine $F^{-1}(Z)$ in a finite form that minimizes the error between $F(Z)F^{-1}(Z)$ and δ. In the preceding solution procedure using the general form of the problem $X(Z)F(Z) = Y(Z)$, we replace the input series x_t with the known filter, f_t. The unknown filter, f_t, is now the inverse filter, f_t^{-1}, and the output signal (y_t) in the previous solution is replaced with the unit impulse, δ, with coefficients 1,0,0, . . . This gives us $F(Z)F^{-1}(Z) = \delta$.

The ccvf $(c - j)$ in (13.57) is between δ and the original filter f_t, so its coefficients are

$$\sum_{i=0}^{N} \delta_i \hat{f}_{i-j} = c_{-j}. \tag{13.58}$$

For $j = 0$, the summation can be expanded to give

$$\begin{aligned} c_0 &= \delta_0 f_0 + \delta_1 f_0 + \ldots \; + \delta_{m+n} f_{m+n-1} \\ &= f_0, \end{aligned} \tag{13.59}$$

since the only nonzero value of the unit impulse occurs at $\delta_0 = 1$. For $j = 1$, the summation can be expanded to give

$$\begin{aligned} c_{-1} &= \delta_0 f_{-1} + \delta_1 f_0 + \ldots \; + \delta_{m+n} f_{m+n-1} \\ &= 0, \end{aligned} \tag{13.60}$$

since the filter f_t is realizable so that all coefficients for $-j < 0$ are equal to zero (thus canceling out the first term in the above sum). The impulse takes on zero values at all of the other values of j, thus canceling out the remaining terms in the sum as well.

If one continues the above process, it is clear that the ccvf of f_t with δ is equal to f_0 at the 0th lag and zero at all other lags, or $c_0 = f_0$, $c_1 = c_2 = \ldots = c_m = 0$. Therefore, the normal equations given by (13.57) describe the following system:

$$\begin{aligned} &f'_0 a_0 + \; f'_0 a_1 + \; \ldots \; + f'_m a_m = f_0 \\ &f'_0 a_1 + f'_1 a_0 + \ldots + f'_m a_{m-1} = 0 \\ &\vdots \\ &f'_0 a_m + f'_1 a_{m-1} + \ldots + f'_m a_0 = 0 \end{aligned} \tag{13.61}$$

Notice that, since the acvf is an even function, we can define the system in terms of the positive lags of the acvf only (otherwise, in the first equation for example, all of the a_j coefficients would be negative).

This system can be written in matrix form as

$$\begin{bmatrix} a_0 & a_1 & \ldots & a_m \\ a_1 & a_0 & & \\ \vdots & & & \vdots \\ a_m & a_{m-1} & & a_0 \end{bmatrix} \begin{bmatrix} f'_0 \\ f'_1 \\ \vdots \\ f'_m \end{bmatrix} = \begin{bmatrix} f_0 \\ 0 \\ \vdots \\ 0 \end{bmatrix}$$

or

$$\mathbf{A}\mathrm{F}^{-1} = \mathrm{F}. \tag{13.62}$$

Matrix **A** in (13.62) can now be inverted to solve for $\mathbf{F}^{-1}$ to give the coefficients of the inverse filter that best approximates the desired unit impulse response function in a least-squares sense.

The extreme symmetry of the **A** matrix in (13.62) leads to a very efficient solution. This matrix (or any matrix with the same symmetry as that displayed in the **A** matrix) is called a **Toeplitz matrix**. Because of its symmetry, the Toeplitz matrix can be solved in a particularly efficient manner through a recursive method known as **Levinson recursion**. This recursive method is available in most software packages and usually requires that the autocorrelation function be used instead of the auto-covariance function. The solution to the shaping filter (i.e., if we were using some other desired output signal shape instead of the unit impulse function) is very similar to that given above, except that a modified version of the Levinson recursion is required. See Kanasewich (1981) for details (his chapter 14 on deconvolution provides a very good introduction to inverse filtering and includes a nice example of the above technique for a dipole filter).

13.8 Take-Home Points

1. Here a filter is defined as a modification of a time series by convolution with a "filter" defined by its impulse-response function (that is, the output of an impulse passing through the filter; comparable to multiplying a black box by the number 1, and the outcome is whatever was in the box).
2. Because of the convolution theorem ("convolution in one domain is multiplication in the other"), this convolution can also be performed as a multiplication of the Fourier transforms of the time series and the filter. In this form, it is clear precisely what is being done in the frequency domain.
3. Ideal filters are those that perfectly preserve the frequency content over your desired range of frequencies and zero out all other frequency components, so if you desire to smooth a time series by removing the high-frequency components, you do this with an ideal low-pass filter, where you multiply the spectrum by 1 over the desired low-frequency components you wish to preserve, and by zero over all of those frequency components of higher frequency (thus *passing* the low frequencies).
4. The modified output from a filter can be "deconvolved" to determine the input series or the filter. Doing this via least squares presents the most efficient solution to this problem (it allows for noise in the data).

13.9 Questions

Pencil and Paper Questions

1. a. Describe an ideal band-pass filter. Sketch one.
   b. How can you implement the exact same filter in the time domain? Is there any advantage to doing it in one domain over the other?
   c. How do you perform a classic running average in the frequency domain?
   d. What is a causal filter?
   e. Is a running average causal? Sketch an example demonstrating your answer.

2. a. What is deconvolution?
   b. What is an inverse filter? Show formula in terms of Z-transform.
   c. Describe the various steps used to compute a least-squares inverse filter.
   d. What is a shaping filter?

Computer-Based Questions

3. For the first million years of LR04:

   a. Plot the PSD.
   b. Using an ideal filter, isolate the 40 kyr component and plot it in the time domain.
   c. Plot a pure 40 kyr time series on top of the filtered version from LR04.

      (1) Describe the difference and explain it.
      (2) Can choice of a wider or narrower filter window improve on answer c.1?

14 Linear Parametric Modeling

14.1 Overview

Parametric analysis involves the analysis of data in the context of a model – in this case involving some mathematical expression. Given the general form of a model *thought* to describe the process responsible for generating the observations, the analysis is reduced to estimating the values of the model parameters. The time series analyses we have considered so far has been independent of any mathematical model assumptions; we were mainly guided by conceptual models that the data, for example, may contain cycles, or some specific form of a curve. Such approaches are representative of **nonparametric analysis**.

Parametric analysis has the advantage that, given the form of the model, the interpretation of the observations is tailored to most accurately extract the signal. The trade-off is that knowledge of the appropriate form of model is required to utilize parametric approaches.[1] If you assume a certain model that is *inappropriate* for the process, then the results of the parametric analysis can be seriously misleading. Therefore, nonparametric analyses are safer to use in general, though the interpretation of the results are more limited. Conversely, nonparametric models often impose assumptions regarding the distribution of the data (most commonly that they are normally distributed), which may not hold either, in which case the interpretation of the nonparametric results can also be misleading or in error, though most can be modified to work for other statistical distributions.

The most obvious example of parametric modeling comes in the form of spectral analysis. We have focused on classical (nonparametric) methods involving a Fourier transform, where we find the amplitude and phase of specific harmonics. Parametric methods fit a model to the data and the spectrum is determined for any frequency component, by simply applying the model parameters and computing the amplitudes for any specified frequencies. The model can then be used to forecast future values and estimate uncertainties.

[1] Yes, there's always a catch!

14.1.1 Two Fundamental Types of Parametric Model

Deterministic Models

One type of model frequently used for studying physical systems is the **deterministic model**. This type of model assumes that the data represent a pure or strongly deterministic process, though possibly contaminated by noise, for which the appropriate physics driving the system can be adequately represented mathematically. If we wish to use the model to predict future behavior, the model must be **prognostic** – that is, the model must describe the time-dependent behavior of the variables of interest. Usually this means that the system of governing equations will contain a time-derivative term. This is contrary to **diagnostic** models, which do not contain the time dependence and instead simply predict the values of the relevant variables as a function of present values of some variables (often "state" variables). For example, the sensible heat flux across the air/sea interface can be diagnostically determined at any time, given knowledge of the temperature contrast across the interface and the wind velocity at that particular time. Deterministic models are usually computed numerically (using a finite difference or finite element form of the governing equations that lend themselves to numerical computation on a computer) or analytically (i.e., the system can be solved exactly for the variable in question, as a function of time and/or space).

Deterministic models are invaluable for developing an understanding of the physics controlling the process or phenomenon under study. They are also useful for simulating a particular process and for prediction. Unfortunately, the processes under study are sometimes so complex that such models must be significantly simplified, often via empirical relationships, to allow a computationally feasible mathematical description. In general, models of this type must be custom designed for each specific problem under study (though there are a limited number of standard classes of differential equations that can sometimes be drawn from to start).

Stochastic Models

Models of another class have proven useful for describing (or approximating) a broad variety of systems in a more statistical sense. The most common of this class of models is the linear stochastic process model. A **linear stochastic process** (or **linear process**) results from passing white noise through a linear filter. This type of process has been touched upon in previous sections (convolution, filtering and cross spectral analysis). It has been widely studied and is useful in a variety of situations. It is representative of many situations in which the only information we have about the system is its response to white noise (e.g., the passage of random ground motion through some local geological structure). Knowledge of this general response can often reveal a significant amount of information about the filter (physical process) itself – sometimes more than can be revealed by the response to a deterministic input signal, given the previously shown fact that the properties of the white noise input signal can serve to simplify the interpretation of the filter output. In fact, for this reason, it is often (or at least, sometimes) advisable to design an experiment so that the process under study is subjected to a white noise forcing. In this manner, you are assured of studying the nature of the process itself without coloring your interpretation with the properties of the input signal. In other situations, such a model is a useful approximation to a more deterministic situation.

From experience it has been shown that simple linear-process models can describe many physical systems. Therefore, much effort has been invested in estimating which *type* of linear-process model most likely describes the process under study. That is, we often assume that the observations can be described by a linear-process model, but we don't actually know the specific type of model. Given this situation, we must first estimate the model type (and order), and after that we can estimate the optimal model parameters to best fit the model to the data. Given knowledge of the model and its optimal parameter values, one can then use the model to (1) predict the true power spectral density function of the process (**parametric spectral analysis**), (2) predict the values of future observations (**predictive** or **forecast modeling**) and/or (3) derive a strategy to show how the input signal (if under our control) should be structured to minimize certain undesirable features or characteristics of the output (**control modeling**). The last two capabilities are extremely important, particularly in business and economics. They are of less importance in the physical sciences (though they are still very useful in some disciplines – e.g., weather and climate prediction). Note that frequency domain analyses offer a fairly poor predictive capability unless, for example, the data are strongly periodic in a few well-defined frequency bands (i.e., the variance represented by these few frequency bands accounts for a significant amount of variance in the observed time series).

Here, we focus predominantly on the use of linear-process modeling as a tool for estimating the spectrum – that is, for describing the nature of our realization. This is consistent with our focus on describing the observations themselves up to this point.

14.2 Discrete Linear Stochastic Process Models

14.2.1 Definitions

Any **stationary** discrete linear stochastic process can be described by a simple linear system of the form[2]

$$z_t = \varepsilon_t * \Psi_t, \tag{14.1}$$

where z_t represents the observed time series values (random variable) at times (lags) t, ε_t represents white noise (a time series of identically distributed, independent random values with mean $\mu = 0$ and variance $= \sigma_\varepsilon^2$) and Ψ_t is the impulse-response function of the time-invariant causal linear system (filter). This is represented in the standard way as shown in Figure 14.1.

The impulse response, Ψ_t, is generally normalized so that $\Psi_0 = 1$. Then, from (14.1), we write the equation for any particular value of z_t as a convolution:

[2] Our filter notation is altered here to be consistent with the standard forecasting literature (e.g., Box and Jenkins, 1976).

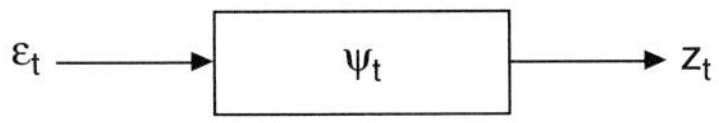

ε_t → ψ_t → z_t

Figure 14.1 Schematic of white noise (ε_t) entering a filter with impulse response function Ψ_t yielding output series z_t.

$$z_t = \varepsilon_t + \Psi_1\varepsilon_{t-1} + \Psi_2\varepsilon_{t-2} + \ldots \tag{14.2a}$$

$$= \sum_{k=0}^{\infty} \Psi_k\varepsilon_{t-k}. \tag{14.2b}$$

Because of the zero-mean noise, z_t has zero mean. Also, the convolution sum starts at $k = 0$, reflecting the causality of the system (required to describe a physical system).

Z-Transform Representation

Equation (14.2) is frequently written in terms of its Z-transform (in the forecasting literature, the unit-delay operator, Z, is often referred to as the **backshift operator** and designated by the letter B instead of Z):

$$z_t = \Psi(Z)\varepsilon_t. \tag{14.3a}$$

This is not in the usual Z-transform form, which is

$$\begin{aligned}
Z(Z) &= \Psi(Z)\varepsilon(Z) \\
&= (\Psi_0 Z^0 + \Psi_1 Z^1 + \Psi_2 Z^2 + \ldots)(\varepsilon_0 Z^0 + \varepsilon_1 Z^1 + \varepsilon_2 Z^2 + \ldots) \\
&= \varepsilon_0\Psi_0 Z^0 + \varepsilon_1\Psi_0 Z^1 + \varepsilon_2\Psi_0 Z^2 + \varepsilon_0\Psi_1 Z^1 + \varepsilon_1\Psi_1 Z^2 + \varepsilon_2\Psi_1 Z^3 + \varepsilon_0\Psi_2 Z^2 + \\
&\quad \varepsilon_1\Psi_2 Z^3 + \varepsilon_2\Psi_2 Z^4 + \varepsilon_3\Psi_2 Z^5 + \ldots \\
&= \varepsilon_0\Psi_0 Z^0 + (\varepsilon_1\Psi_0 + \varepsilon_0\Psi_1)Z^1 + (\varepsilon_2\Psi_0 + \varepsilon_1\Psi_1 + \varepsilon_0\Psi_2)Z^2 + \ldots,
\end{aligned} \tag{14.3b}$$

so, for $t = 0$, corresponding to Z^0, $z_0 = \varepsilon_0\varphi_0 Z^0 = \varepsilon_0$. Recall that we normalized the impulse response above so that $\varphi_0 = 1$ (and $Z^0 = 1$); for $t = 1$, corresponding to Z^1, $z_1 = \varepsilon_1\varphi_0 + \varepsilon_0\varphi_1 = \varphi_0\varepsilon_t + \varphi_1\varepsilon_{t\text{-}1}$; and in general,

$$\begin{aligned}
z_t &= \Psi_0\varepsilon_t + \Psi_1\varepsilon_{t-1} + \Psi_2\varepsilon_{t-2} + \ldots \\
&= \sum_{k=0}^{t} \Psi_k\varepsilon_{t-k}.
\end{aligned} \tag{14.3c}$$

Also, the right-hand side of (14.3a) can be written as

$$\Psi(Z)\varepsilon_t = (1 + \Psi_1 Z + \Psi_2 Z^2 + \ldots)\varepsilon_t, \tag{14.3d}$$

where each product

$$\Psi Z^k \varepsilon_t \; (k = 0, 1, 2, \ldots, t) = \Psi_k\varepsilon_{t-k}, \tag{14.3e}$$

so (14.2), though it now combines elements of Z and straight representation in t, with the implicit rule that the corresponding periods of time, t, are matched to their appropriate unit delays, Z. Note that the sum in (14.2b) is effectively from $k = 0$ to t, because the system is causal so that all other terms from $k = t + 1$ to ∞ (where $t - k < 0$, thus involving the noise prior to $t = 0$) are equal to zero.

In spectral analysis, the Fourier transform of the impulse response function (F_f) is the transfer function; in forecasting models, the *Z-transform* (as opposed to the Fourier transform) of Ψ_t, $\Psi(Z)$, is called the **transfer function**.

The lower moments (statistical properties) of (14.2) can now be computed to reveal any potential limitations or restrictions of the model. These moment calculations also serve as templates for the calculation of the lower moments of the specific models (simplifications of the general form) required later.

14.2.2 Statistical Moments of the General Linear Process

Mean

The expectance of the stochastic process, z_t, in equation (14.2) is given by

$$\begin{aligned} E[z_t] &= E\left[\sum_{k=0}^{\infty} \Psi_k \varepsilon_{t-k}\right] \\ &= \sum_{k=0}^{\infty} \Psi_k E[\varepsilon_{t-k}] \\ &= 0, \end{aligned} \tag{14.4}$$

since the mean of the white noise is zero. If the noise had mean μ_z, then the expectance would have been μ_z. Since the expectance in either case is independent of t, the process z_t is stationary in the mean. Note that this is a direct consequence of the fact that the mean of the white noise (i.e., of the filter input) and the filter coefficients (Ψ_k) do not vary in time either.

Variance

The variance of z_t can be derived directly as above, but since it represents a specific value of the autocovariance function (the 0th lag), a function that we need later, it is more efficient in this case to compute the autocovariance and simply extract the variance from that.

The autocovariance, $\gamma_z(k)$, of z_t is given by (derivation in (D14.1))

$$\gamma_z(k) = \sigma_\varepsilon^2 R_{\Psi'\Psi}(k), \tag{14.5}$$

where $R_{\Psi'\Psi}(k) = \sum_{m=0}^{\infty} \Psi_m \Psi_{p+m} = nR_{\Psi\Psi}(k)$ takes the form of an un-normalized (i.e., not multiplied by $(n-|k|)^{-1}$ or n^{-1}) autocovariance serial product of the filter coefficients and $p = j - i$ and $m = i$. Thus, from (14.5), it is seen that the autocovariance of the general linear process is the (un-normalized) "autocovariance" of the filter scaled by the variance of the white noise being passed through the filter.

Box D14.1 Derivation of Variance of General Linear Process

Defining the true autocovariance function (or series) of any process x_t, as $\gamma_x(k)$, then for z_t (recall it has a zero mean),

$$\begin{aligned}\gamma_z(k) &= \mathrm{E}[z_t z_{t+k}] \\ &= \mathrm{E}\left[\sum_{i=0}^{\infty}\Psi_i\varepsilon_{t-i}\sum_{j=0}^{\infty}\Psi_j\varepsilon_{t-j}\right],\end{aligned} \tag{D14.1.1a}$$

or, rearranging the order of the summations and taking the expectance operator inside the sums,

$$= \sum_{i=0}^{\infty}\sum_{j=0}^{\infty}\Psi_i\Psi_j\mathrm{E}[\varepsilon_{t-i}\varepsilon_{t-j}]. \tag{D14.1.1b}$$

Note that the impulse response function comes outside of the expectance operator, since it consists of constants (non-random variables).

By definition, $\mathrm{E}[\varepsilon_t\varepsilon_{t+k}] = \gamma_\varepsilon$. So, by defining $\tau = t - i$, (D14.1.1b) is rewritten as

$$\begin{aligned}&= \sum_{i=0}^{\infty}\sum_{j=0}^{\infty}\Psi_i\Psi_j\mathrm{E}[\varepsilon_\tau\varepsilon_{\tau+i+k-j}] \\ &= \sum_{i=0}^{\infty}\sum_{j=0}^{\infty}\Psi_i\Psi_j\gamma_\varepsilon(i+k-j).\end{aligned} \tag{D14.1.1c}$$

Setting $p = j - i$ and $m = i$ gives

$$= \sum_{m=0}^{\infty}\sum_{p=-m}^{\infty}\Psi_m\Psi_{p+m}\gamma_\varepsilon(k-p). \tag{D14.1.1d}$$

Rearranging, and noting that the filter is causal so that values of $p < 0$ have zero coefficients, the inner sum can be written, as always, starting at 0. Rearranging the order of summation,

$$= \sum_{p=0}^{\infty}\gamma_\varepsilon(k-p)\sum_{m=0}^{\infty}\Psi_m\Psi_{p+m}. \tag{D14.1.1e}$$

The inner sum, that involving the impulse response function, is of the form of an unnormalized sample autocovariance function, $\mathrm{R}_{\Psi'\Psi}(p)$, allowing

$$= \sum_{p=0}^{\infty}\gamma_\varepsilon(k-p)\mathrm{R}_{\Psi'\Psi}(p). \tag{D14.1.1f}$$

Box D14.1 (Cont.)

Note however, that the autocovariance of the white noise is zero at all lags except for the zero lag, where it is equal to the variance of the noise, σ_ε^2. Thus, (D14.1.1f) is zero for all p except for $p = k$, reducing (D14.1.1f) to

$$\gamma_z(k) = \sigma_\varepsilon^2 R_{\Psi'\Psi}(k). \tag{D14.1.2}$$

That is, when p, the index of the sum in equation (D14.1.1f), is equal to k, then we have the only nonzero value of the autocovariance of the noise (σ_ε^2). The kth lag is thus the only surviving term from the sum. Alternatively, this is another way of stating our earlier assertion (from Chapter 7) that the convolution of an impulse with a function replicates the function scaled by the height of the impulse (σ_ε^2).

The *variance* of the general linear process (again, (D14.1)) is the sum of the squares of the filter coefficients, or

$$\text{Var}[z_t] = \gamma_z(0) = \sigma_z^2 = \sigma_\varepsilon^2 \sum_{p=0}^{\infty} \Psi_p^2. \tag{14.6}$$

Therefore, as found for the mean, the variance of the general linear process (i.e., of z_t) is independent of time, and therefore the process is stationary at least to order 2. This result is not unexpected, since it follows directly from the stationarity of the input noise and time invariance of the filter.

Spectrum

The spectrum of the general linear process is simply the multiplication of the Fourier transform of the transfer function times the transform of the white noise.[3] We know, however, that the noise has a constant transform proportional to its variance. Therefore, the power spectrum (PSD) of z_t is the transform of the impulse response function (the transfer function or gain function) scaled by the variance of the noise.

14.2.3 Moving Average (MA) Process

From (14.6) one can see that if the sum of the filter coefficients is not finite, the variance grows in time and can be unbounded (since the filter impulse response coefficients would not die off to yield a simple constant for the sum). Thus, the restriction (or approximation) of a finite filter is necessary for practical applications, and consistent with physical data.

[3] That is, once again, convolution in the time domain is equal to multiplication in the frequency domain.

Consider the case in which the filter coefficients decay to zero after a given number of terms, q.[4] In this case,

$$z_t = \varepsilon_t + \Psi_1 \varepsilon_{t-1} + \Psi_2 \varepsilon_{t-2} + \ldots + \Psi_q \varepsilon_{t-q}. \tag{14.7}$$

In this form, z_t is called a **moving average process of order *q***, or simply **MA(*q*)**. If we write this equation in the convolution form of (14.1), then the name "moving average" is obvious – the current value of the time series is a weighted average of the preceding q random disturbances plus a current random disturbance (and a mean value, if desired). Therefore the system is fully described by $q + 1$ parameters (or $q + 2$ parameters, if a nonzero mean is included). Since the number of impulse-response function coefficients is finite, their sum will always be finite, and therefore all moving average processes are stationary.

The autocovariance of the MA(q) process is, as before, given by (14.5), only the unnormalized acvf sum is now restricted to the range of nonzero coefficient values:

$$\gamma_z(k) = \sigma_\varepsilon^2 R_{\Psi'\Psi}(k) = \sigma_\varepsilon^2 \sum_{p=0}^{q-k} \Psi_p \Psi_{p+k}. \tag{14.8}$$

From this, it is obvious that the autocovariance function of a moving average process is zero after q lags.

Therefore,

$$\gamma_k = \begin{cases} \sigma_\varepsilon^2 \sum_{p=0}^{q} \Psi_p^2 & k = 0 \\ \sigma_\varepsilon^2 \sum_{p=0}^{q-k} \Psi_p \Psi_{p+k} & 0 < k \le q \\ 0 & k > q \end{cases}. \tag{14.9}$$

Alternatively, dividing the autocovariance through by σ_z^2 (the 0th lag of the autocovariance function, $\gamma_z(0)$) yields the autocorrelation function, $\rho_z(k)$. It has values[5]

$$\rho_z(k) = \begin{cases} 1 & k = 0 \\ \dfrac{\sum_{p=0}^{q-k} \Psi_p \Psi_{p+k}}{\sum_{p=0}^{q} \Psi_p^2} & 0 < k \le q. \\ 0 & k > q \end{cases} \tag{14.10}$$

[4] The letter q is almost always used for this truncation value.

[5] Note that the value of $\rho_z(k)$ is of a similar form to that computed previously for the cross-correlation function in Chapter 12, except that in this case, since we are dealing with an acf rather than a ccf, the normalization in the denominator is less complicated.

Therefore, the autocorrelation function of a moving average process decays to zero by the $(q+1)$th lag. This represents the major diagnostic indicator of a MA(q) process (discussed later) through inspection of a **correlogram** (the plot of an autocorrelation function).

First-Order Moving Average Process

The first-order moving average process is probably the most commonly used MA model.[6] The **first-order moving average process, MA(1)**, is given by

$$z_t = \varepsilon_t - \beta_1 \varepsilon_{t-1}, \tag{14.11}$$

where, following standard notation, we now define the MA process as a subtraction instead of addition of averaged terms and use the symbol β instead of Ψ to designate the impulse response function coefficient (so $\beta = -\Psi$). As before, we could have defined a mean term, μ_z.

This model represents a time series in which each value is a weighted value of the previous random disturbance plus a new disturbance at the present value. The lower moments of the MA(1) process follow directly by application of our previous moment derivations from the general case, given by (14.4) and (14.5), or (14.9) and (14.10), so

$$E[z_t] = 0 \quad (\text{or} = \mu) \tag{14.12}$$

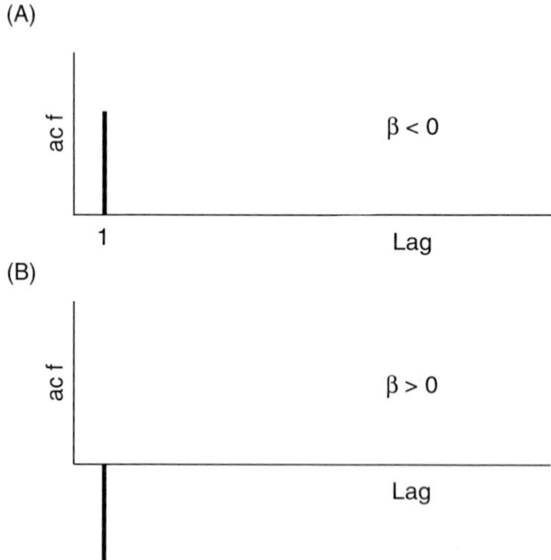

Figure 14.2 Ideal correlograms for MA(1) process where **(A)** the single-filter coefficient is less than zero and **(B)** the single coefficient is greater than zero. The sign of the acf function is simply that of the MA (1) single coefficient.

[6] Compare this to a least-squares fit of a data set. The linear fit is the one often commonly employed – in many cases this is simply due to the simplicity afforded by such lower-order forms.

and

$$\gamma_z(0) = \sigma_\varepsilon^2(1+\beta_1^2) \quad \rho(0) = 1 \tag{14.13a}$$

$$\gamma_z(1) = \sigma_\varepsilon^2(-\beta_1) \quad \rho(1) = -\beta/(1+\beta_1^2) \tag{14.13b}$$

$$\gamma_z(k>1) = 0 \quad \rho(k>1) = 0. \tag{14.13c}$$

These relationships indicate that a MA(1) process has a "memory" that is only 1 sample interval long (Figure 14.2). That is, as seen from the acf functions, any one particular value of z_t is correlated to its neighboring values on each side, but not to any other values in the time series.

The MA(1) process should have a correlogram that is zero at all lags greater than lag 1.

14.2.4 Autoregressive (AR) Process

The moving average process is the natural truncated version of the general linear stochastic process model. However, because the model depends upon knowledge of the preceding random disturbances (the noise that has previously entered the filter), it is not actually used in practice by itself much. Instead, a version of the linear model that is more practical (and easier to use) is one involving a random disturbance and the linear combination of p previous *observations* (i.e., outputs from the filter). That is, one that uses the actual observations, not the unknown random disturbances from which the observations were created. Such a model is called an **autoregressive model of order p**, or **AR(p)**.

This model arises by rearranging (14.2) in terms of ε_t:

$$\varepsilon_t = z_t - \Psi_1\varepsilon_{t-1} - \Psi_2\varepsilon_{t-2} - \ldots. \tag{14.14}$$

This equation is valid for any value of t, so consider it for t − 1:

$$\varepsilon_{t-1} = z_{t-1} - \Psi_1\varepsilon_{t-2} - \Psi_2\varepsilon_{t-3} - \ldots. \tag{14.15}$$

Substituting (14.15) into (14.14) for ε_{t-1} gives

$$\begin{aligned}
\varepsilon_t &= z_t - \Psi_1[z_{t-1} - \Psi_1\varepsilon_{t-2} - \Psi_2\varepsilon_{t-3} - \ldots] - \Psi_2\varepsilon_{t-2} - \ldots \\
\varepsilon_t &= z_t - \Psi_1 z_{t-1} + \Psi_1\Psi_1\varepsilon_{t-2} + \Psi_1\Psi_2\varepsilon_{t-3} + \ldots - \Psi_2\varepsilon_{t-2} - \ldots \\
\varepsilon_t &= z_t - \Psi_1 z_{t-1} + (\Psi_1\Psi_1 - \Psi_2)\varepsilon_{t-2} + (\Psi_1\Psi_2 - \Psi_3)\varepsilon_{t-3} + \ldots.
\end{aligned} \tag{14.16}$$

If we continue in this manner, we can eventually replace all of the ε_{t-j} values on the right-hand side of (14.16) and obtain

$$z_t = \alpha_1 z_{t-1} + \alpha_2 z_{t-2} + \alpha_3 z_{t-3} + \ldots + \varepsilon_t, \tag{14.17}$$

where the new coefficients, α, are linear combinations of the original Ψ_t, as suggested by examination of (14.16).

Truncating (14.17) after p of the α coefficients gives the autoregressive process of order p. The name stems from the observation that this equation is now in the form of a multiple regression equation where *we are regressing z on the p previous observations*, $z_{t-1}, z_{t-2}, \ldots, z_{t-p}$. Hence we're using the previous p *observations* to determine a best fit for the current value (this is seen conceptually when $p = 1$, below).

First-Order Autoregressive Process

As with MA processes, the most common (and actually very useful in practice) auto-regressive process is the **first-order autoregressive process**, or **AR(1)**. The AR(1) process is also referred to as a **Markov process**. It is given by

$$z_t = \varepsilon_t + \alpha_1 z_{t-1}. \tag{14.18}$$

Using this equation, we can reverse the operation we did above by defining z_{t-1} from (14.18),

$$z_{t-1} = \varepsilon_{t-1} + \alpha_1 z_{t-2}, \tag{14.19}$$

and substituting this into (14.18) to give

$$z_t = \varepsilon_t + \alpha_1 (\varepsilon_{t-1} + \alpha_1 z_{t-2}) = \varepsilon_t + \alpha_1 \varepsilon_{t-1} + \alpha_1^2 z_{t-2}. \tag{14.20}$$

Repeating this procedure for z_{t-2} and then z_{t-3}, etc., we see that when we redefine the AR (1) process in terms of the original random disturbances, we get an infinite series that looks like an infinite-order moving average process. That is, the current value, z_t, is a weighted average of the past infinite values of ε_t. Furthermore, had we played a similar substitution game with the MA(1) process, we would have found that *the MA(1) process corresponds to an AR(∞) process just as the AR(1) process corresponds to a MA(∞) process.*

Examination of (14.18) shows why this model is referred to as an auto*regressive* one. Consider the regression equation $y_t = a_0 + a_1 x_t$ that describes a linear relationship between the variables y and x. For regression equations, we know that this equation cannot be solved exactly, given noise in the system, so we allow for noise by rewriting it as $y_t = a_0 + a_1 x_t + \varepsilon_t$. If the intercept is 0 ($a_0 = 0$) and we re-label so that $y_t = z_t$ and $x_t = z_{t-1}$. Then it is seen that (14.18) is exactly this regression equation in which the current value of the output, z_t, is linearly predicted by knowledge of the previous value of z, z_{t-1}. For higher-order AR processes, this same concept applies, only in those cases the regression is a multi-dimensional one (p-dimensions) in which the current value is found linearly by knowledge of the previous p observations.

Further manipulation of (14.18), as indicated in (14.20) and (14.16), shows that the coefficients to the infinite-order moving average form, MA(∞), of the AR(1) process, i.e. the Ψ_t, are given as

$$\Psi_1 = \alpha_1 \tag{14.21a}$$

$$\Psi_2 = \alpha_1^2 \tag{14.21b}$$

$$\Psi_3 = \alpha_1^3 \tag{14.21c}$$

$$\vdots$$

Therefore, in order for the AR(1) process to be stable and stationary, a requirement is that $|\alpha_1| < 1$. That is, recalling that the sum of the Ψ_t must be finite and as defined in (14.21), the only way the infinite Ψ_t can sum to a finite number is if the α_1 magnitude is less than one. This assures a stable sum, stationarity (an infinite sum otherwise grows without bounds beyond the mean) and validity of the model. Note that this requirement is identical to that of a minimum-delay wavelet, which we encountered during inverse filtering (for what ultimately can be shown to be the same reason, involving invertibility).

We can determine the lower-order moments of the AR(1) process directly as before:

$$\mathrm{E}[z_t] = 0 \qquad (\text{or} = \mu_z) \tag{14.22}$$

$$\begin{aligned} \mathrm{Cov}[z_t, z_{t-k}] &= \mathrm{E}[(z_t)(z_{t-k})] \\ &= \mathrm{E}[(\varepsilon_t + \alpha_1 z_{t-1})(z_{t-k})] \\ &= \mathrm{E}[\varepsilon_t z_{t-k} + \alpha_1 z_{t-1} z_{t-k}] \\ &= \mathrm{E}[\varepsilon_t z_{t-k}] + \mathrm{E}[\alpha_1 z_{t-1} z_{t-k}]. \end{aligned} \tag{14.23a}$$

The value of z_{t-k} is uncorrelated to ε_t, since it only involves values of the noise prior to ε_t (see (14.17)). Therefore, z_{t-k} is uncorrelated to ε_t and the expectance of the first term on the right-hand side is 0 (for $k \neq 0$, i.e., for the covariance).[7] So,

$$\begin{aligned} &= \alpha_1 \mathrm{E}[z_{t-1} z_{t-k}] \\ &= \alpha_1 \gamma_z(k-1). \end{aligned} \tag{14.23b}$$

In other words, the autocovariance between k lagged values is equivalent to α_1 times the autocovariance at $k-1$ lagged values. Dividing the results of (14.23) by $\gamma_z(0)\ (= \sigma_\varepsilon^2 + \alpha_1 \gamma_z(-1))$ yields the autocorrelation function. For $k = 0$, $\rho(0) = 1$, whereas for higher-order lags,

$$\rho(1) = \alpha_1 \rho(k-1) \qquad \text{for } k \geq 1. \tag{14.24}$$

So, specifically, for $k = 1$,

$$\begin{aligned} \rho(1) &= \alpha_1 \rho(0) \\ &= \alpha_1. \end{aligned} \tag{14.25a}$$

[7] For the variance, at $k = 0$, the $\mathrm{E}[z_{t-k}\varepsilon_t]$ term becomes $\mathrm{E}[z_t\varepsilon_t]$, and substituting (15.18) for z_t gives $\mathrm{E}[(\varepsilon_t + \alpha_1 z_{t-1})\varepsilon_t] = \mathrm{E}[\varepsilon_t^2] = \sigma_\varepsilon^2$. Therefore, this first term on the right-hand side of (15.23a) does not cancel, but rather goes to the variance of the noise of the input series.

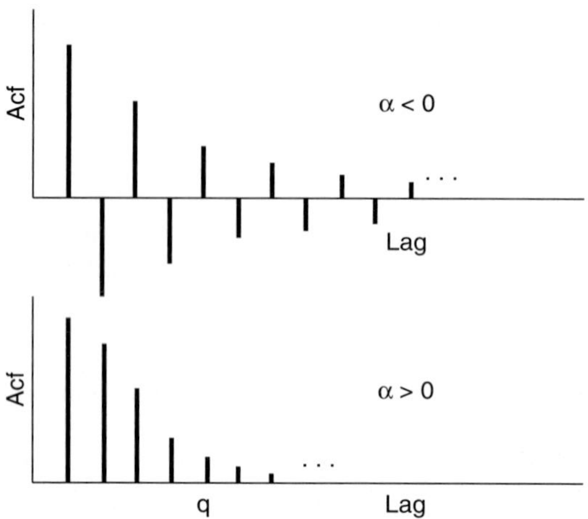

Figure 14.3 Correlograms for an AR(1) process.

For $k = 2$,

$$\rho(2) = \alpha_1 \rho(1) = \alpha_1^2; \tag{14.25b}$$

likewise, for higher k,

$$\rho(3) = \alpha_1^3 \tag{14.25c}$$

$$\rho(4) = \alpha_1^4 \tag{14.25d}$$

$$\vdots$$

or

$$\rho(k) = \alpha_1^k. \tag{14.26}$$

From (14.21), in terms of the original filter coefficients, $\rho(k) = \Psi_k(\text{for } k > 0)$.

Equation (14.26) thus shows that an AR(1) process has an autocorrelation function that decays exponentially with increasing lag. The nature of the decay depends upon the sign of the α_1 coefficient, and will look like Figure 14.3.

14.2.5 Mixed Autoregressive and Moving Average (ARMA) Process

As shown, an AR(1) process is equivalent to an infinite-order MA process and an MA(1) process is equivalent to an infinite-order AR process, but working with first-order processes is significantly easier than working with processes of higher order. For that reason, it is often convenient to use a **mixed autoregressive, moving average model of order *p,q***, or **ARMA(*p,q*)**. That is, when a first-order process is an inadequate description of the observed data (we discuss how to determine this later), it is often easier to use a low-order ARMA model (which can represent higher-order AR or MA models for the reasons discussed above) than to construct a higher-order AR or MA model. As before, the p represents the truncation of the AR process and the q the truncation of the MA process.

The general form of the ARMA model is

$$z_t = \alpha_1 z_{t-1} + \alpha_2 z_{t-2} + \ldots + \varepsilon_t - \beta_1 \varepsilon_{t-1} - \beta_2 \varepsilon_{t-2} - \ldots. \quad (14.27)$$

The same restrictions on the coefficients that apply to the individual models still apply here.

14.2.6 Integrated ARMA (ARIMA) Process

Finally, the restriction imposed on the analysis that the time series must be stationary (in the mean) can be limiting. Therefore, a final general form of the linear stochastic process model can be made through the **integrated autoregressive, moving average (ARIMA) process model**. This model takes into account the fact that a time series that is nonstationary in the mean can often be converted to a stationary one by a first differencing operation. This is intuitive, given the fact that differentiation of a function removes the slope, which in turn represents a form of growth typically expected in non-equilibrium time series. If a first differencing is not adequate to make the time series stationary, then a second differencing may be required (it is rare to have to go beyond a second difference – usually the first difference is adequate).

The ARIMA model therefore requires a differencing operation prior to the analysis, after which the procedures are the same as with stationary data. That is, the observations are differentiated before any analysis is performed. Again, the truncation of the AR process is given by p, and q specifies the MA truncation point. With the ARIMA model, there is a third value, d, which is the order of the differencing required to achieve stationarity in the mean. Therefore, the order of the ARIMA process is given by p,d,q, or **ARIMA(*p,d,q*)**.

Note that this general form can be conveniently used to describe any of the models discussed so far. For example, an AR(1) process is now an ARIMA(1,0,0) process; an ARIMA(1,1,1) process is a first-order ARMA process after having first differenced the data; etc.

This above discussion provides an introduction to the types of general stochastic process models that have proven adequate to describe a great variety of observational data. The next task involves identifying *which* model is the appropriate choice for the data at hand, followed by the job of actually determining the best coefficient values of the chosen model for the particular data set.

14.3 Model Identification and Solution

As just indicated, we can describe any stochastic linear process in terms of the ARIMA (p,d,q) model classification. Therefore, we are left with the task of examining the data in an effort to determine the most appropriate order of the ARIMA model to describe our observations. Following this, we must then determine the optimal values of the model coefficients so that the model most accurately describes the existing data. Accomplishing this, we can then use the model to best forecast (predict) future values of the sampled process to within known confidence limits, and/or estimate the spectrum of the process from the model coefficients.

Given the previous analysis associated with the model descriptions, one might guess that the best tool for model identification is the sample autocovariance function, or sample autocorrelation function. That is, we've determined the general properties of the autocovariance and autocorrelation serial products (acvsp and acsp, respectively) for both MA and AR processes. Consequently, examination of the estimates of either of these functions (e.g., the sample autocorrelation serial product) should help in identifying the appropriate model.

In order to properly extract the required information from the sample autocovariance function, we must consider its statistical properties. This provides the necessary foundation so that we can make the most representative estimate of model type and order where appropriate. Recalling that the sample acvsp is a biased estimator when using the preferred normalization of $1/n$ (see (7.43) in §7.5.5, where $C_{zz}(k) = (1/n)\Sigma z_i z_{i+k}$), its expectance is given by

$$E[C_{gg}(k)] = \frac{n - |k|}{n} \gamma_z(k). \tag{14.28}$$

This bias could be eliminated using the $1/(n - |k|)$ normalization of the acvsp, but in this particular case the biased estimate using $1/n$ normalization yields the minimum least squared error in the acvsp estimates.

However, for $k \ll n$ (i.e., for the lower-order lags), the bias $(n - k)/n$, or $1 - k/n$, is quite small. So, if we assume that the order of the process (q for an MA(q) process or p for an AR(p) process) will generally be $\ll n$, either form of the acvf, normalizing by $1/n$, or $1/(n - |k|)$, should represent a good estimate of the true autocovariance function.

For the uncertainty of the sample acvf, equation (7.51), the main thing to recall is that neighboring values in the acvf are highly correlated, hampering our ability to identify the order of the process by examination of the sample acvf. It may also hamper our ability to distinguish an AR from an MA process. That is, for an AR process we expect that the acf is infinitely long (different from zero over all lags). For the MA process, while the true acf goes to zero after q lags, the sample acf may show smooth ripples (due to the convolution effect) similar to those expected from certain AR processes.

These potential problems are combined with the fact that random variation alone will cause, on average, 5 out of 100 acf estimates to exceed a 95 percent confidence interval (so as to appear to be significantly different from 0). These problems suggest that we look for expected *shapes* in the acf and evaluate the statistical significance in context with expected shapes when attempting to identify a model type from examination of the acf.

Box 14.1 Model Order and Decorrelation Length

All time series other than those of white noise have acfs with nonzero values at lags greater than zero. This implies a degree of correlation between neighboring points, suggesting that the series must obey an AR process of some order. In fact, at first blush it would appear that the order of the AR process must more or less be equal to the decorrelation length, since the definition of the decorrelation length shows how many of the neighboring observations are correlated, and the AR(p) model says that any current observation is correlated to the p preceding observations. However, while this seems conceptually straightforward, the examination of an AR(1) acf clearly shows a decorrelation length much longer than 1 lag. This is because each point is correlated to a neighboring point, so in actuality all points are correlated (e.g., 2nd point is related to 1st, but 3rd is related to 2nd, so 3rd is also related to 1st, etc., and the relationship dies off exponentially if consistent with an AR(1) model). This is consistent with the fact that the AR(1) model is equivalent to an MA(∞) model, and the order of the MA model does coincide with the concept of decorrelation length, where q is equal to the decorrelation length as shown in (14.10).

For example, a first-order AR process should decay approximately exponentially with increasing lags, and an MA(1) process should have the first lag significantly different from zero while the other lags should randomly vary about zero (some of which will exceed our established confidence limits). Remember that *unlike spectral estimates, the acf and acvf are consistent estimators. In other words, as the length of the series increases, the uncertainty in the acf estimates decreases. Thus, it makes no sense to segment the data in these cases in order to achieve improved statistical properties.*

14.3.1 Identifying a Moving Average Process

For the MA(q) process, the acf goes to zero for lags larger than the order of the process (given by the value of q). Therefore, we might hope to determine by examination of the sample acf (1) whether the process can be classified as a moving average process and (2) the order, q, of the process. That is, *if* the time series represents a physical phenomenon that can be modeled or at least classified in terms of a truncated discrete linear stochastic process, then the sample acf constructed from the data should become indistinguishable from zero after q lags, which specifies the order of the process.

A moving average process has a sample acf whose variance is somewhat simplified from the general form. For a process *assumed* to be of order q, then the variance of the sample acf *at lags, k, larger than q* (i.e., where the true acf is identically zero) is given by (derived in the usual manner using the expectance operator)

$$\mathrm{Var}[r_{zz}(k)] \approx \frac{1}{n}\left(1 + 2\sum_{j=1}^{q} \rho_j^2\right). \tag{14.29}$$

Therefore, we expect that the sample acf for a MA(q) process has a mean value of zero at lags $k > q$ and variance given by (14.29) where we use the sample acf, $r_{zz}(j)$, in place of the true ρ_j in the equation. The distribution of the acf for the lags at which the true value is zero (in this case, for lags $> q$) is approximately normal according to Bartlett in an impressive study in 1946.

Given that the value of q is unknown, you substitute the value k for q as the upper limit of the sum and then test the significance of the $(k + 1)$st acf value (i.e., $r_{zz}(k + 1)$). If this lag value exceeds that of, say $\pm 1.96\{\mathrm{Var}[r_{zz}(k)]\}^{1/2}$ (i.e., the value appears to be significantly different from 0 at the 0.05 significance level), then increment q by one in (14.29) and test the next lag value for significance. Repeat this procedure until the test fails, at which point you accept the value of k (not $k + 1$, which is being tested) as being equal to the order of the MA model. Note that you can sometimes get a rough idea of the model order to start this test by comparing the acf against the confidence interval computed for white noise, where its variance is given by (7.52) as $\sim 1/n$.

Remember that we expect that 5 out of 100 lags, on average, will exceed a 95 percent confidence level. However, these values that exceed the confidence interval would be expected to occur randomly throughout the r_k, and not necessarily just near the low-order lags, where the significant ones are expected to occur. Therefore, this variability should have a minimal effect on estimating the order of the model (Figure 14.4).

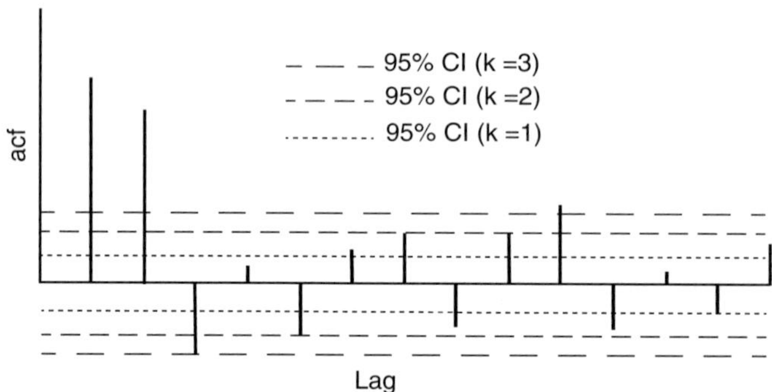

Figure 14.4 Schematic of ideal correlogram for an MA(2) process, using equation (14.29) to determine q. This case is a bit tricky since acf($\tau = 3$) lies on the 0.05 CI. If we chose $\alpha = 0.06$, this would likely be a MA(3) process with acf($\tau = 3$) lying outside the CI. For this case, you have to decide if it is MA (2) or MA(3). I'd probably say MA(3), loosening the level of significance a bit (and explicitly stating the 95 percent problem, leading me to a less stringent level of significance). Ideally you would test both MA(2) and MA(3) to see which gives a better fit.

More quantitative methods for determining the order of the process involve the use of special indices. These essentially require that we compute the MA model for a variety of q values and then compute the "index" for each q. This involves a comparison of the *order* of the model to the amount of variance in the data, explained by the model of the given order.

Ultimately, we choose the value of q that yields the minimum value of the index. This corresponds to finding the best-order model, where "best" reflects the greatest amount of variance described for the lowest-order model. That is, increasing the order of the model introduces a penalty in the index that must be compensated for by a significantly better model fit to the data – if the model only fits the data slightly better, then the index is dominated by the penalty and takes a fairly large value. Otherwise, if the model does a significantly better job of describing the data, then the penalty is overcome and the index takes on a fairly small value. We wish to find the order of the process that gives the minimum index value (equivalent to the most efficient model). See section 5.4.5 in Priestley (1981), page 370, for details concerning the various indices available. In practice, the above use of examining the acf will suffice. Note that indices such as these can be used in general to determine the best order of any model: e.g., we might use such an approach to determine the optimal order of a least-squares polynomial fit to a set of data.

The advantage of the more quantitative methods for order determination lies in the fact that the values of the sample acf (R_{zz}) are highly correlated as shown for the general acf previously. That is, for the MA(q) process, the Cov[$R_{zz}(k)$,$R_{zz}(k+j)$] is approximated at the higher lags by

$$\mathrm{Cov}[R_{zz}(k), R_{zz}(k+1)] \approx \frac{1}{n} \sum_{i=-\infty}^{\infty} [\rho_z(i)\rho_z(i+1)]. \tag{14.30}$$

Consequently, the interpretation of the sample acf can potentially lead to misinterpretation, and obvious care must thus be utilized when estimating the order of the function from the correlogram.

14.3.2 Identifying an Autoregressive Process

While the MA process is known to have a true acf that cuts off after a given number (q) lags, the AR(1) process decays exponentially to zero while higher-order AR processes decay in a more complicated manner (a damped sinusoidal pattern is common). Therefore, the sample acf of an AR process should not simply drop to zero after a given number of lags. This makes estimation of the order of the AR process rather difficult, if based on direct examination of the sample acf. As a consequence, the quantitative indices mentioned above are ideal for determining the order of the AR process.

Alternatively, one can adopt a graphical method for determining the order of the AR process, which is similar to examining the correlogram for the MA process. In this case however, we use something called the **partial autocorrelation function (partial acf)** or another function known as the **inverse autocorrelation function**.

Partial acf

The partial acf can be developed by examining the acf of an AR(p) process. Previously we considered the acvf of an AR(1) process; now consider the autocovariance function of an AR(2) process:

$$z_t = \varepsilon_t + \alpha_1 z_{t-1} + \alpha_2 z_{t-2}. \tag{14.31}$$

For the 0th lag (the variance),

$$\gamma_z(0) = \mathrm{E}[z_t z_t]. \tag{14.32a}$$

Substituting (14.31) for one z_t in (14.32a),

$$\begin{aligned} &= \mathrm{E}[z_t(\varepsilon_t + \alpha_1 z_{t-1} + \alpha_2 z_{t-2})] \\ &= \mathrm{E}[z_t\varepsilon_t] + \alpha_1 \mathrm{E}[z_t z_{t-1}] + \alpha_2 \mathrm{E}[z_t z_{t-2}]. \end{aligned} \tag{14.32b}$$

By definition $\mathrm{E}[z_t z_{t-1}] = \gamma_z(-1)$, and $\gamma_z(-1) = \gamma_z(1) = \gamma_1$ by symmetry. Substituting (14.31) into (14.32b) for z_t in the first term on the right-hand side and applying the above symmetry rule reduces (14.32b) as

$$\begin{aligned} &= \mathrm{E}[(\varepsilon_t + \alpha_1 z_{t-1} + \alpha_2 z_{t-2})\varepsilon_t] + \alpha_1\gamma_1 + \alpha_2\gamma_2 \\ &= \mathrm{E}[\varepsilon_t^2] + \alpha_1 \mathrm{E}[\varepsilon_t z_{t-1}] + \alpha_2 \mathrm{E}[\varepsilon_t z_{t-2}] + \alpha_1\gamma_1 + \alpha_2\gamma_2. \end{aligned} \tag{14.32c}$$

Because $z_{t-1} = \varepsilon_{t-1} + \alpha_1 z_{t-2} + \alpha_2 z_{t-3}$, z_{t-1} is uncorrelated to ε_t, so expectances $\mathrm{E}[\varepsilon_t z_{t-1}] = \mathrm{E}[\varepsilon_t z_{t-2}] = 0$, and therefore (14.32c) reduces to

$$\gamma_z(0) = \sigma_\varepsilon^2 + \alpha_1\gamma_1 + \alpha_2\gamma_2. \tag{14.32e}$$

Repeating this exercise for $\gamma_z(1)$,

$$\begin{aligned} \gamma_z(1) &= \mathrm{E}[z_t z_{t-1}] \\ &= \mathrm{E}[(\varepsilon_t + \alpha_1 z_{t-1} + \alpha_2 z_{t-2}) z_{t-1}] \\ &= \mathrm{E}[\varepsilon_t z_{t-1}] + \alpha_1 \mathrm{E}[z_{t-1}^2] + \alpha_2 \mathrm{E}[z_{t-2} z_{t-1}], \end{aligned} \tag{14.33a}$$

but by stationarity $\mathrm{E}[z_{t-1}^2] = \mathrm{E}[z_t^2] = \gamma_z(0)$ (i.e., the variance is independent of time) and substituting $\tau = t - 2$, $\mathrm{E}[z_{t-2} z_{t-1}] = \mathrm{E}[z_\tau z_{\tau+1}] = \gamma_z(1)$, so[8]

$$\gamma_z(1) = \alpha_1\gamma_0 + \alpha_2\gamma_1. \tag{14.33b}$$

For $\gamma_z(2)$,

$$\begin{aligned} \gamma_z(2) &= \mathrm{E}[z_t z_{t-2}] \\ &= \mathrm{E}[(\varepsilon_t + \alpha_1 z_{t-1} + \alpha_2 z_{t-2}) z_{t-2}] \\ &= \mathrm{E}[\varepsilon_t z_{t-2}] + \alpha_1 \mathrm{E}[z_{t-1} z_{t-2}] + \alpha_2 \mathrm{E}[z_{t-2}^2] \\ &= \alpha_1\gamma_1 + \alpha_2\gamma_0. \end{aligned} \tag{14.34}$$

Continuing this procedure for the higher lags, $j > 2$ gives

[8] Note that had we substituted for z_{t-1} in equation (14.33a) instead of z_t, we would have ended up with $\gamma_z(1)$ as a function of γ_2 and γ_3, which would lead to an unsolvable system since we would always require additional values of $\gamma_z(k)$ in order to solve for any one lag value.

$$\gamma_z(j) = \alpha_1 \gamma_{j-1} + \alpha_2 \gamma_{j-2}. \tag{14.35}$$

Dividing the $\gamma_z(k)$ by $\gamma_z(0)$ gives the acf for the AR(2) process,

$$\rho_z(1) = \alpha_1 + \alpha_2 \rho_1 \tag{14.36a}$$

$$\rho_z(2) = \alpha_1 \rho_1 + \alpha_2, \tag{14.36b}$$

and for lags $j > 2$ (i.e., higher than the order of the process),

$$\rho_z(j) = \alpha_1 \rho_{j-1} + \alpha_2 \rho_{j-2}. \tag{14.36c}$$

Notice that (14.36a) and (14.36b) provide two equations in two unknowns – easily solved for ρ_1 and ρ_2 (given the coefficients, α_1 and α_2). The higher lags are determined iteratively by application of (14.36c). That is, knowledge of ρ_1 and ρ_2 can be used in (14.36c) to determine ρ_3, which can then be used with ρ_2 in (14.36c) to determine ρ_4, etc.

For an AR process of order p, the above procedure is followed, giving the acf of z_t as

$$\begin{aligned} \rho_1 &= \alpha_1 \rho_0 + \alpha_2 \rho_1 + \alpha_3 \rho_2 + \ldots + \alpha_p \rho_{p-1} \\ \rho_2 &= \alpha_1 \rho_1 + \alpha_2 \rho_0 + \alpha_3 \rho_1 + \ldots + \alpha_p \rho_{p-2} \\ &\vdots \\ \rho_p &= \alpha_1 \rho_{p-1} + \alpha_2 \rho_{p-2} + \alpha_3 \rho_{p-3} + \ldots + \alpha_p \rho_0. \end{aligned} \tag{14.37}$$

Again, for lags of order higher than $p, j > p$,

$$\rho_j = \alpha_1 \rho_{j-1} + \alpha_2 \rho_{j-2} + \alpha_3 \rho_{j-3} + \ldots + \alpha_p \rho_{j-p}. \tag{14.38}$$

The well-posed system of equations (14.37) is known as the **Yule–Walker equations**. In matrix form it is given as

$$\begin{bmatrix} 1 & \rho_1 & \cdots & \rho_{p-1} \\ \rho_1 & 1 & & \rho_{p-2} \\ & & & \\ \rho_{p-1} & & & 1 \end{bmatrix} \begin{bmatrix} \alpha_1 \\ \alpha_2 \\ \vdots \\ \alpha_p \end{bmatrix} = \begin{bmatrix} \rho_1 \\ \rho_2 \\ \vdots \\ \rho_p \end{bmatrix}. \tag{14.39}$$

Now consider the following strategy. If we were to assume that the time series could be classified as an AR(1) process, then we could easily solve the Yule–Walker equations for α_1, given an estimate of ρ_1 by the sample acf, $r_{zz}(1)$, so

$$r_{zz}(1) = \hat{\alpha}_1, \tag{14.40}$$

where $\hat{\alpha}_1$ represents the estimate of the true model coefficient, α_1. If the value of $\hat{\alpha}_1$ is significantly greater than zero, we can assume that the order of the process is at least 1, since this estimate of the model coefficient is different from zero (i.e., for an order p AR model, the α_j, with $j > p$, are all equal to zero by definition). We can then repeat this procedure and solve the Yule–Walker equations for $p = 2$ and test whether $\hat{\alpha}_2$ is significantly different from zero. If it is, we then can assume that the model is at least of order 2. This procedure can be followed, comparing each $\hat{\alpha}_j$ against zero until that value of j for which $\hat{\alpha}_j$ is not significantly different from zero. The order of the AR model is then taken to be $p = j - 1$. One might wonder why we bother doing this iteratively, and why we do not just solve the equations for p = large, and then plot all values of α and see which one is insignificant. The reason is that $\gamma(0)$ changes as a function of p (see (14.32)), so the acf normalization also changes with p.

If we were therefore to plot the $\hat{\alpha}_j$ against j, the order of the AR model is given by the last value of $\hat{\alpha}_j$ that is significantly different from zero. This is exactly analogous to examining the acf to determine the order of a MA model, and consequently the $\hat{\alpha}_j$ are called the **partial autocorrelation coefficients**.

The variance of the partial acf for lags greater than p is $\sim 1/n$, so the 95 percent confidence interval on the partial acf is given by $1.96\, n^{-1/2}$. So, where $|\hat{\alpha}_j| \leq 1.96\, n^{-1/2}$, the $\hat{\alpha}_j$ are not significantly different from zero at the 0.05 level of significance, and the order of the AR model is given by the highest value of j for which $|\hat{\alpha}_j| > 1.96\, n^{-1/2}$ (for this level of significance).

14.3.3 Identifying an ARMA Process

Identification of a mixed autoregressive/moving average process is more difficult than identifying either a straight AR or MA process. In general, both the acf and partial acf of an ARMA process show a slow decay to zero. Use of the quantitative indices mentioned previously serves to best determine the order of an ARMA process (see Priestley (1981), section 5.4.5, page 370).

14.4 Parameter Estimation

Once the type of model and its order have been determined, you need to determine the value of the model coefficients (parameters) that best represent the data. Note that we are following a strategy similar to that used for regression problems: first determine the *type* of model most appropriate; then determine the *order* of the model; and, finally, determine the *optimal coefficient values* of the model and their uncertainty. If the graphical methods listed above are used, this is the sequence. If, however, one employs one of the quantitative indices to determine the order of the process, the estimation of the coefficient values occurs simultaneously.

14.4.1 Parameters of an AR(p) Process

Estimation via the Yule–Walker equations

Estimation of the coefficients of the AR(p) process is the simplest of the various models. The most common method used for determining the values of the AR(p) process model are through solution of the Yule–Walker equations given above. That is, by estimating the $\rho_z(k)$ with the sample acf, $r_{zz}(k)$, the matrix equation (14.39) can be inverted to yield the coefficients, α_j. Note that the coefficient matrix of the

Yule–Walker equations is of the Toeplitz type (a Toeplitz matrix is one in which all elements along a single diagonal are the same, though the elements of the different diagonals may be different). Therefore, the inversion of (14.39) can be done using the standard matrix relationships available for solving this form of equation (Levinson recursion is frequently used).

Least-Squares Estimation

Alternatively, the optimal values of the unknown p coefficients of an AR(p) process can be obtained using the principle of maximum likelihood, which leads to the method of least squares for normally distributed noise. This approach is the most intuitive, given that the AR model describes a simple regression of the present value on the previous p values of the data. Therefore, as with standard regression analysis, we are best served by estimating the coefficients that minimize the sum of squared errors between the estimated model values and observed values.

First, consider the form of the AR(p) model (where z_t has zero mean or we have removed the sample mean already):

$$z_t = \alpha_1 z_{t-1} + \alpha_2 z_{t-2} + \ldots + \alpha_p z_{t-p} + \varepsilon_t \tag{14.41}$$

or

$$\varepsilon_t = z_t - (\alpha_1 z_{t-1} + \alpha_2 z_{t-2} + \ldots + \alpha_p z_{t-p}). \tag{14.42}$$

This later form is already in a form suited for the least-squares approach. That is, we wish to minimize the ε_t^2 over all t by choosing those coefficient values, α_j, which minimize the sum,

$$e(\alpha_j) = \sum_{t=p+1}^{n} (z_t - \alpha_1 z_{t-1} - \alpha_2 z_{t-2} - \ldots - \alpha_p z_{t-p})^2. \tag{14.43}$$

Notice that the sum is from $p + 1$ to n. This means that we start with the first p values of z_t so that we needn't require future values, e.g., for $t < 1$.

Taking the derivative of $e(\alpha_j)$ with respect to the unknown α_j and setting this equal to zero gives

$$\begin{aligned} \frac{\partial e(\alpha_j)}{\partial \alpha_j} &= 2 \sum_{t=p+1}^{n} (z_t - \alpha_1 z_{t-1} - \alpha_2 z_{t-2} - \ldots - \alpha_p z_{t-p}) z_{t-j} \\ &= 0. \end{aligned} \tag{14.44}$$

Multiplying through by the z_{t-j} and rearranging gives the normal equations (the well-posed p equations for p unknowns) as

$$\sum_{t=p+1}^{n} z_t z_{t-j} = \alpha_1 \sum_{t=p+1}^{n} z_{t-1} z_{t-j} + \alpha_2 \sum_{t=p+1}^{n} z_{t-2} z_{t-j} + \ldots + \alpha_p \sum_{t=p+1}^{n} z_{t-p} z_{t-j}. \tag{14.45}$$

Assuming that p (order of the model) is small relative to n (the number of sample values), we can approximate the individual sums as

$$\sum_{\mathrm{t}=p+1}^{n} z_{\mathrm{t}-k} z_{\mathrm{t}-j} \approx \sum_{\mathrm{t}=1}^{n} z_{\mathrm{t}-k} z_{\mathrm{t}-j} = nC_{zz}(k-j), \tag{14.46}$$

where the lags $k-j$ are obtained by substituting $\tau = \mathrm{t} - k$ in the sum on the right-hand side. Note that the factor n is introduced because the sum for the sample acvf should be normalized by $1/n$, and since the sum in (14.46) is *not*, then the sum is equal to n times the sample acvf, or $nC_{zz}(k-j)$.

Making this approximation (p is small relative to n, so the contribution of the first $p+1$ values of z_t is negligible relative to the total sum) and dividing through by σ_z^2 reduces the normal equations (14.45) to

$$r_{zz}(-j) = \alpha_1 r_{zz}(1-j) + \alpha_2 r_{zz}(2-j) + \ldots + \alpha_p r_{zz}(p-j). \tag{14.47}$$

In matrix form, these normal equations for the p values of j (i.e., the values of j correspond to the p coefficient values, so $j = 1,2, \ldots, p$) are given as

$$\begin{bmatrix} 1 & r_1 & \ldots & r_{p-1} \\ r_1 & 1 & & r_{p-2} \\ & & & \\ r_{p-1} & & & 1 \end{bmatrix} \begin{bmatrix} \alpha_1 \\ \alpha_2 \\ \vdots \\ \alpha_p \end{bmatrix} = \begin{bmatrix} r_1 \\ r_2 \\ \vdots \\ r_p \end{bmatrix}, \tag{14.48}$$

which are precisely the Yule–Walker equations, rewritten with the sample acf instead of the true acf (remember that the symmetry of the acf allows us to write this matrix in terms of the positive lags only). Therefore, the normal equations of the method of least-squares are simply the Yule–Walker equations that were presented previously.

Solving the above matrix equation thus yields an approximation to the least-squares solution for the coefficients of the AR(p) model. The solution is approximate because of our assumption that $p \ll n$, and therefore that the individual sums of (14.45) could be represented as sample acfs. *If you use a high-order AR model so that* p *is not much less than n, then the above system will not represent a good approximation and the solution to the actual normal equations in* (14.45) *is required (a fact often overlooked).*

14.4.2 Parameters of an MA(q) and ARMA(p,q) Process

Estimating the best values of the unknown coefficients to a moving average process of order q is generally more difficult than estimating the values for the AR(p) process model. No explicit expression for a likelihood function is easily written, and as a regression problem, the system is nonlinear in the coefficients (since they would have to be inverted to be put into a regression format). For these reasons, the parameters (coefficients) of the MA(q) model are usually found via numerical techniques. Box and Jenkins (1976) outline such a technique, and Priestley (1981) briefly outlines a couple of approaches as well (including the Box and Jenkins numerical approach). Marple (1987; see his section 10.3) also presents a method that is relatively straightforward and analogous to that used to solve the AR(p) problem—in fact, it cleverly involves estimation of the MA(q) model in terms of a high-order AR(p) model.

The estimation procedure for the mixed autoregressive/moving average process of order p,q is similar to that used for the pure MA(q) process. That is, in general an assumption concerning the joint distribution of the data is assumed and a numerical solution to the maximum likelihood problem employed.

An ARIMA(p,d,q) model follows the same procedures as above, only the data are differenced d times before the analysis begins. Note that one is often alerted to the nonstationarity of the data by examination of the acf, which will decay very slowly, indicating a nonstationary trend present, *not* an AR process.

14.5 Forecasting

Having determined the model type, order and "best" parameter values, we can now predict future values ($n+k$) of the series by using the standard formula for the appropriate model with the determined coefficient values for values of z_{n+k}. The errors and specific details of this aspect of the linear stochastic process predictive modeling can be found in Box and Jenkins (1976). We have essentially established the foundations of the method here, and its application is fairly straightforward. We skip these details and proceed directly to spectral estimation, given the estimation of the model type, order and parameter values.

14.6 Parametric Spectral Estimation

14.6.1 General

The spectral estimation techniques we considered previously were based on a smoothing of the periodogram. That class of spectral estimation is often referred to as the classic **lag-window estimation technique, Tukey method** or **Blackman–Tukey method**. Since that class of methods is independent of assumptions concerning the particular form of the data, it is a **nonparametric** technique: one that does not require fitting a particular model to the data. In this respect, the lag-window estimation technique represents a general approach to the estimation of the true spectral density function of any sampled process.

Alternatively, if we determine that the data approximate a realization of a stochastic linear process of known type and order, then we can estimate the spectrum directly by determining the model coefficient (parameter) values and using them in the theoretical expression of the spectral density function for the particular model. This represents a **parametric spectral estimation** procedure.

The advantage offered by this latter approach lies in the fact that, if the type of process generating the data is known, then the spectrum can be obtained directly from our knowledge of the theoretical form of the spectrum for that particular process. Recall that in the lag-window technique, we ideally require information on the spectral bandwidth to determine the optimal window bandwidth. Lacking that information, we are left to experiment through such methods as window closing. The parametric model, on the other hand, has an exact theoretical form that is only dependent on the model coefficient values. Of course, the trade-off is that we now have to identify the type of process represented by the data. If we are incorrect, our spectral estimates can be severely misleading. Therefore, *the parametric techniques are strongly model dependent and must be used with caution (if at all) as general spectral estimation methods*. This point is often overlooked, and it is not uncommon to see people use, for example, the maximum entropy method, because it usually produces more spiked spectral peaks, when in fact it is a parametric method only appropriate for AR processes. Fortunately though, use of these techniques requires fitting a model to the data first, and thus, when such techniques are used, at least they should be consistent with the fitted model.

14.6.2 Theoretical Parametric Spectral Representations

Moving Average Process

Recall that the MA process is described by the truncated version of the general linear system equation,

$$z_t = \varepsilon_t^* \beta_t, \tag{14.49}$$

where z_t is the output of the filter (with mean 0) and ε_t is the sequence of white noise (mean 0 and variance σ_ε^2) input to the filter with impulse response β_t. The impulse response is described by $q + 1$ coefficients, $\beta_0, \beta_1, \ldots, \beta_q$ which are normalized so that $\beta_0 = 1$.

Rewritten with respect to a single element of z_t,

$$z_t = \beta_0 \varepsilon_t + \beta_1 \varepsilon_{t-1} + \beta_2 \varepsilon_{t-2} + \ldots + \beta_q \varepsilon_{t-q}. \tag{14.50}$$

Note that the β_0 term has been included here for convenience.

The system described by either (14.49) or (14.50) is referred to as a **causal feedforward system**. This describes any system in which the output is a weighted combination of past and present values of any given input sequence. That is, the standard convolution problem involving the passage of a series through a causal filter to give the output represents a causal feedforward system or filter.

In the frequency domain, this convolution (i.e., the causal feedforward system) is given as the product of the transforms

$$Z(\omega) = B_q(\omega)\varepsilon(\omega). \tag{14.51}$$

Equation (14.51) shows the general form of the Fourier transform of any feedforward linear system (i.e., the input signal to the filter need not be restricted to a white noise process as shown for the MA case). However, in this particular case, since $\varepsilon(\omega)$ is the transform of white noise, we can reduce the equation by writing it in terms of its power spectral representation:

$$|Z(\omega)|^2 = |B_q(\omega)|^2 |\varepsilon(\omega)|^2 = \sigma_\varepsilon^2 \, |B_q(\omega)|^2. \tag{14.52}$$

Box 14.2 Relationship between Z-Transform and Fourier Transform

Written in terms of its z-transform, the feedforward system of (14.49) is given by

$$Z(Z) = B_q(Z)\varepsilon(Z). \tag{14.53}$$

The Z-transform of a series y_t is represented by the polynomial sum in Z as

$$Y(Z) = \sum_{j=1}^{n} y_j Z^j. \tag{14.54}$$

The Z-transform transforms the time sequence y_t into a polynomial in the *complex variable* Z. In polar form, Z can be written as $Z = Ae^{-i\omega}$. When Z has magnitude $A = 1$, a plot of Z over $0 \leq \omega < 2\pi$ describes a circle, called the **unit circle**. The unit circle is a circle about the origin with radius 1. It is formulated as

$$Z = e^{-i\omega}, \tag{14.55}$$

where ω, the angular frequency, simply identifies the position on the unit circle. Each position on the unit circle defines a sine and cosine at phase ω (recall Euler relation: $e^{-i\omega} = \cos\omega - i\sin\omega$).

Now consider the significance of the unit circle. Recall that you can factor any *n*th-order polynomial into *n* dipoles (§13.8.2: Inverse Filtering). For a filter to have a stable inverse, each dipole is required to have the first coefficient value larger than the second. That is, for (g_0, g_1), $g_0 > g_1$, and the normalized dipole is $(1,\alpha)$, where $\alpha = g_1/g_0$ and $\alpha < 1$. This defines a minimum-delay dipole (if all of the dipoles of the wavelet are minimum delay, then the wavelet is minimum delay).

Alternatively, we can now define a minimum delay wavelet as one in which the **zeros** of its polynomial all lie outside the unit circle in the complex plane. The zeros of a polynomial are found by factoring the polynomial and finding those values of Z that zero each factor (i.e., the roots of the polynomial). So, in the above example with dipole $(1,\alpha)$ (or in polynomial form, $1 + \alpha Z$), if $Z = -1/\alpha$, the dipole $= 0$. The value $-1/\alpha$ thus defines a zero of the polynomial. A polynomial has one zero for each factor, so an *n*th-order polynomial has *n* (not necessarily distinct) zeros. *For a wavelet to be minimum delay, each zero must lie outside the unit circle.* That is, if $|\alpha| < 1$, then $|1/\alpha| > 1$. Thus, if each zero lies outside the unit circle, the wavelet is minimum delay and its polynomial converges.

Box 14.2 (Cont.)

The unit circle therefore identifies the region for which a series will converge or diverge (among other things). For this reason, filter theory and filter construction is often approached in terms of the positions of the zeros relative to the unit circle. Also, when *inverting* a filter, the zeros that occur in the denominator represent values of Z that would zero the polynomial and therefore result in dividing by 0. Such zeros are called **poles**.

Now, consider expressing the Z-transform of y_t in terms of its position on the unit circle. We can do so by substituting $e^{-i\omega}$ for Z in the Z-transform. So, rewriting the Z-transform of (14.54) by substituting the relationship in (14.55) for Z gives

$$\begin{aligned} Y(e^{-i\omega}) &= \sum_{j=1}^{n} y_j e^{-i\omega j} \\ &= Y(\omega). \end{aligned} \tag{14.56}$$

So, *expression of the Z-transform in terms of its position on the unit circle describes the discrete Fourier transform*. Conversely, the Fourier transform is the Z-transform evaluated on the unit circle of the complex Z plane. This also shows that the Fourier transform is obtained from the Z-transform by substituting $e^{-i\omega}$ for Z in the polynomial in Z.

This shows that the transform of the acf of the output signal (i.e., the power spectrum of z_t) is equal to the variance of the input noise multiplied by the power spectrum of the impulse response of the filter (i.e., the squared gain function). The variance of the white noise is typically estimated during the solution of the MA(q) parameters.

Consider the implication of this result. The power spectrum of the n values of the z_t time series is obtained directly by the transform of the acf of the impulse response function. However, we know that since the MA process is a truncated version of the general linear stochastic process system, there are only $q + 1$ nonzero terms in the impulse response function. Specifically, (14.10) shows that the acf of the MA(q) process is zero for all lags greater than q, the order of the model. Therefore, this corresponds to taking the transform of a truncated acf, which is precisely the procedure used to obtain a smoothed spectral estimate using the classic lag-window methods. In other words, *the most appropriate manner in which to obtain an estimate of the spectrum of a MA process is the standard lag-window approach*. The acf should be truncated (i.e., the lag window should go to zero) by lag $q + 1$.

Autoregressive Process

Whereas the MA process describes a causal feedforward system, the AR process, defined by

$$z_t = \varepsilon_t + \alpha_1 z_{t-1} + \alpha_2 z_{t-2} + \ldots + \alpha_p z_{t-p} \tag{14.57}$$

describes a **causal feedback system**.

Prediction Operator

The causal feedback system is best understood in terms of a **prediction operator**. First, consider the basic convolution equation,

$$y_t = \sum_{j=1}^{n} z_j \alpha_{t-j}, \tag{14.58}$$

which indicates that the output series, y_t, is a weighted linear combination of present and past values of the input series: z_t, in this case (the reason for using this notation will become apparent shortly). If the filter, α_t, is a prediction operator with **prediction distance** τ, then the output to the series is an estimate of the input series at a future time $t + \tau$. In that case,

$$y_t = \hat{z}_{t+\tau} = \sum_{j=1}^{n} z_j \alpha_{t-j}, \tag{14.59}$$

where the circumflex, as usual, indicates an *estimate* of z at the time $t + \tau$.

The errors between the estimated future values of z and the true values at time $t + \tau$ are given by

$$\begin{aligned} \varepsilon_{t+\tau} &= z_{t+\tau} - \hat{z}_{t+\tau} \\ \varepsilon_{t+\tau} &= z_{t+\tau} - \sum_{i=1}^{n} z_j \alpha_{t-j}. \end{aligned} \tag{14.60}$$

This can be written in terms of its Z-transform to give

$$Z^{-\tau}\varepsilon_t = Z^{-\tau} z_t - z_t A(Z), \tag{14.61}$$

where the multiplication of ε_t and z_t by the unit delay operator (or backshift operator), $Z^{-\tau}$ shifts all values τ units forward (toward the present or into the future). Recall that future values are represented by negative exponents of the unit delay operator. That is, when we deal with causal filters, we only include present and past values of the series that all have Z *exponents* ≥ 0. We are estimating future values here (hence the name prediction operator) that correspond to the negative exponents of Z.

Multiplying both sides of (14.61) by Z_τ gives

$$\begin{aligned} \varepsilon_t &= z_t - Z^\tau z_t A(Z) \\ &= z_t \left[1 - Z^\tau A(Z)\right]. \end{aligned} \tag{14.62}$$

The term $[1 - Z^\tau A(Z)]$ is called the **prediction error operator,** which represents the difference between the zero-delayed unit impulse and the **prediction operator** A(Z) delayed by a **prediction distance**, τ.

Therefore, the error series is the product of the input series, z_t, with the prediction error operator. Conceptually, this states that if we know the prediction error filter, we can determine the predictable part of the future values of the input series to within some random (unpredicted) noise.

This causal feedback system is precisely that described by the AR process of order p, in which case the filter coefficients are already written in terms of a prediction error filter. This is seen by rearranging the AR(p) series in (14.57),[9]

$$\varepsilon_t = z_t - \alpha_1 z_{t-1} - \alpha_2 z_{t-2} - \ldots - \alpha_p z_{t-p}, \tag{14.63a}$$

which can be re-expressed, in terms of a Z-transform and the prediction error operator, $A_p(Z)$, as

$$\varepsilon_t = z_t A_p(Z), \tag{14.63b}$$

where this is the form for a single value of ε at time t, as was the case for (14.3), and

$$\begin{aligned} A_p(Z) &= 1 - \alpha_1 Z - \alpha_2 Z^2 - \ldots - \alpha_p Z^p \\ &= 1 - ZA(Z), \end{aligned} \tag{14.63c}$$

where

$$A(Z) = \alpha_1 + \alpha_2 Z + \alpha_3 Z^2 + \ldots + \alpha_p Z^{p-1}. \tag{14.63d}$$

Therefore,

$$\varepsilon_t = z_t[1 - ZA(Z)], \tag{14.64a}$$

which is in the exact form of the prediction error operator given by (14.62).

Alternatively, while (14.64) is written in terms of a single value of ε at time t, it can be written in terms of the entire series of ε in the standard form as

$$\varepsilon(Z) = z(Z)[1 - ZA(Z)]. \tag{14.64b}$$

So, $A_p(z)$ corresponds to a prediction error filter with prediction distance of one time unit ($Z^\tau = Z^1$), prediction error ε_t and input signal z_t (even though z_t is the output of our feedback system as originally defined, it is equal to the *input* series when the equation is rearranged in terms of a prediction error filter). In this case, we wish to find a filter that converts our known series, z_t, into a sequence of white noise after it passes through the filter (the same as our previous inverse filtering problem).

Box 14.3 Example of Prediction Operator and Standard AR Model

To understand the relationship between (14.64a) and (14.64b) and between the prediction operator and standard AR representation, consider the case where $p = 1$. Then,

[9] Note potential confusion here: if you simply substitute in t = 1, then it immediately looks like the prediction of z requires knowledge of future values of z (at t = 0,−1,−2, . . .), but this simply reflects the fact that we can't make a proper prediction until t = p (i.e., until p values of z have entered the filter), after which point we can start predicting z. Set t = p, and then it is seen that we only require the p previous values of z, not the future ones.

Box 14.3 (Cont.)

$$\varepsilon_t = z_t - \alpha_1 z_{t-1}. \tag{14.65}$$

This is a direct statement that the present value of the observed series, z_t, is predicted by the previous value, z_{t-1}, to within a random noise element, ε_t. Therefore, from this, it is clear that the AR process (of any order) is representative of a prediction operator. It has a prediction length of one time unit, since the present value is only one unit beyond the previous observations.

Now, consider the pseudo-Z-transform of (14.64a):

$$\varepsilon_t = z_t[1 - ZA(Z)], \tag{14.66a}$$

where

$$A(Z) = \alpha_1 Z^0 \tag{14.66b}$$

$$ZA(Z) = \alpha_1 Z^1 \tag{14.66c}$$

$$1 - ZA(Z) = Z^0 - \alpha_1 Z^1. \tag{14.66d}$$

Recall that $z_t Z^k$ represents the *past* point z_{t-k}, which is k time units before the present. So,

$$\varepsilon_t = z_t[Z^0 - \alpha_1 Z^1] = z_t - \alpha_1\ z_{t-1}, \tag{14.67}$$

which shows that for any time, t, the form of (14.64a) gives the exact form of the AR process as written in its standard form (for one particular time).

Now, consider the true Z-transform of (14.64b),

$$\varepsilon_t = z(Z)[1 - ZA(Z)], \tag{14.68a}$$

where

$$\varepsilon(Z) = \varepsilon_0 Z^0 + \varepsilon_1 Z^1 + \varepsilon_2 Z^2 + \ldots \tag{14.68b}$$

$$z(Z) = z_0 Z^0 + z_1 Z^1 + z_2 Z^2 + \ldots. \tag{14.68c}$$

Then, using (14.66d) with (14.68a),

$$\begin{aligned}
\varepsilon(Z) &= z(Z)[Z^0 - \alpha_1 Z^1] \\
&= (z_0 Z^0 + z_1 Z^1 + z_2 Z^2 + \ldots)(Z^0 - \alpha_1 Z^1) \\
&= z_0 Z^0 + z_1 Z^1 + z_2 Z^2 + \ldots - \alpha_1 z_0 Z^1 - \alpha_1 z_1 Z^2 - \alpha_1 z_2 Z^3 - \ldots \\
&= z_0 Z^0 + (z_1 - \alpha_1 z_0) Z^1 + (z_2 - \alpha_1 z_1) Z^2 + (z_3 - \alpha_1 z_2) Z^3 + \ldots,
\end{aligned} \tag{14.69}$$

Box 14.3 (Cont.)

where this error, ε, at any particular time t, as was the case for the expression as formulated in (14.64a), is given by that particular exponent of the Z polynomial. So, for t = 3 (as an example), coinciding with the term $(z_3 - \alpha_1 z_2)Z^3$, the error is

$$\varepsilon_3 = z_3 - \alpha_1 z_2, \tag{14.70a}$$

and thus, for any other time, t, (14.70a) can be generalized by substituting t for 3, giving

$$\varepsilon_t = z_t - \alpha_1 z_{t-1}, \tag{14.70b}$$

which is identical to the form of (14.67).[10]

For any one time, t, consistent with the form of (14.64a), we must examine any one particular time lag given by a particular polynomial value in Z. This is in contrast to the form of (14.64b), in which the entire series is presented and thus all time lags (all polynomials in Z) are given.

Rewriting (14.63b) in terms of z_t gives

$$z(Z) = \varepsilon_t / A_p(Z). \tag{14.71}$$

Comparing (14.71) with the equivalent equation for a MA process (where $z_t = \varepsilon_t B_q(Z)$, see (14.3a)) reveals the basic difference between the feedforward and feedback systems. The difference in form of these two systems leads to another naming convention often employed to distinguish the two types of system. That is, the feedback system of (14.71) has a polynomial in the denominator. Therefore, the zeros of the polynomial represent poles of the system, whereas in the feedforward (MA) model they are zeros of the system, since the polynomial is not a denominator term. In this respect, the feedback system is often called an **all-pole model** and the feedforward system an **all-zero model**.

Writing (14.71) in terms of the power spectrum reveals the form of the spectrum of an AR process:

$$|z(\omega)|^2 = \sigma_\varepsilon^2 / |A_p(\omega)|^2. \tag{14.72}$$

Unlike the MA process, here we see that the finite number of filter coefficients in the denominator is equivalent to an infinitely long acf (recall that the inverse of a finite-length polynomial is an infinite-length one). Therefore, it is clear that *the classic method of estimating the spectrum using a truncated acf is not necessarily the best manner with which to compute the spectrum of an AR process.*

In this case, the best manner with which to compute the spectrum is to first determine the values of the prediction error filter [1 − ZA(Z)], multiply it by z(Z) to get ε(Z) and

[10] Note that this would have worked just as well using t = 2 for the example, instead of t = 3.

then compute the transform of the filter, inverting it and multiplying by σ_ε^2 (the transform of ε_t) to yield the proper estimate of the spectrum.

We have already presented one way to determine the values of A(z) (i.e., the α_i coefficients) by using the Yule–Walker (normal) equations, which represent the least-squares solution to the problem. Note however, that (14.71) presents the AR process in terms of an inverse-filter problem. We addressed such a system previously in the context of deconvolution and inverse filtering. Recall that the solution to the inverse filter problem via the method of least squares gave a matrix equation involving a Toeplitz autocorrelation matrix and a cross-correlation column vector. In fact, examination of the Yule–Walker or least-squares matrix solution presented previously for the AR process is exactly that. In other words, the Yule–Walker equations can also be arrived at by solving (14.71) as an inverse filter problem using a least-squares approach (in essence, that is exactly what we did previously when we computed the least-squares solution to the AR process).

Therefore, given the solution to the Yule–Walker equations, we can simply take the computed coefficient values (the α_i) and compute the power spectrum of them (or take their Z-transform given by $1 - \alpha_1 Z - \alpha_2 Z^2 - \ldots - \alpha_p Z^p$ and substitute $e^{-i\omega}$ for Z and multiply by an equivalent form using Z^{-i} to yield the power spectrum, that is, $|A_p(\omega)|^2 = |A_p(\omega)||A_p{}^*(\omega)|$). We then invert the power spectrum and multiply by σ_ε^2, which is estimated by the rms error between the model (computed from knowledge of the coefficient values) and data to give the best spectral estimate (i.e., the definition of the noise as shown above).

Examination of (14.72) for the AR power spectral representation reveals an advantage of the parametric estimation techniques. That is, since we are computing a model for the data, we need only determine the p unknown coefficient values, after which we obtain a complete (model) description of the data. Given this, there is no limit as to the number of frequencies, or their spacing, at which we can compute the spectral estimates. In other words, we have a *model* of the spectrum based on the p coefficients. Given those values, we can determine the spectrum (for those particular coefficient values) at any frequencies that we want (obviously, the frequencies should be less than the Nyquist frequency, which still specifies the highest-frequency component that we could expect to find in the data, given its sampling interval). In this respect, the parametric spectral estimations are sometimes referred to as **super-resolution methods** (though keep in mind that the validity of the spectrum is strictly dependent upon the relevance of the model).

Maximum Entropy Method (MEM)

While the Yule–Walker normal equations lead to a least squares solution of the AR(p) process coefficients, another method available to solve for the coefficients is known as the **Burg estimation technique,** or **maximum entropy method (MEM)**. The maximum entropy method of spectral estimation is nothing more than a spectral estimation using (14.72), in which case the AR coefficient values are solved using a different approach than the least-squares method.

Both the least-squares (maximum likelihood for Gaussian noise) and MEM methods converge to the same result as the time series (data) involved get longer and longer (in particular, as the number of sample points increases). However, the MEM method is advantageous in instances where the time series is relatively short or consists of a small number of sample points.

This MEM technique is only applicable for data that represent an AR process. Some people know that the MEM is good for short series, and consequently use this method as a general technique whenever their data sets are relatively short. This violates the underlying assumptions of the model upon which the entire method is based. Simple tests by numerous authors have show that the spectral results achieved by MEM on non-AR process series are clearly wrong.

The MEM excels with short time series because it approaches the problem not through the acf, as the least-squares approach does, but by utilizing only the data available. Recall that the variance of the acf of an AR process is inversely proportional to the number of data points. So, for short (i.e., small n) series, the acf that makes up the Toeplitz matrix of the least-squares approach has relatively large errors associated with it. Burg gets around this by *not* minimizing the prediction error in the standard manner, which is minimizing the squared sum of differences between the predicted values and true values. Obviously, we don't know the true values directly, but the filter is a prediction operator and we know that the prediction error arises in the least-squares solution in the form of the cross-correlation between the desired output and actual input (see the least-squares solution in the inverse filtering section). Therefore, the desired output is the input at future times $t + \tau$, so the cross-correlation between desired output and input is simply the acf shifted by the amount of the prediction distance, τ, which we do know.

Burg actually defines the prediction error (P) as the average of the standard prediction error plus **hindsight error**, that is (for the first coefficient),

$$P_1 = \frac{1}{2(n-1)} \sum_{j=1}^{n-1} \left[(z_{j-1} + z_j\alpha_1)^2 + (z_j + z_{j+1}\alpha_1)^2 \right], \qquad (14.73)$$

where the first squared term is the standard prediction error and the second squared term is the corresponding hindsight error. Given an autocorrelation of a series, the result is identical if we convolve forward or backward because the result is an acf, which is an even function. The backward pass corresponds to the hindsight error. In other words, for an AR(1) process, for example, each value in the time series is correlated to the following value, but conversely, each value is also correlated to its preceding value. Therefore, by computing the error in both the forward and backward manners, we can effectively use all of the data points in computing the errors and not limit ourselves to only using those after $p + 1$ time intervals (as was required in the solution to the standard Yule–Walker equations).

The advantages of this definition of the error to be minimized are twofold: (1) the solution for the unknown coefficient utilizes all of the available data points and (2) the definition always leads to a minimum delay wavelet A(z), which is required in order for the filter to be stable and converge.

Having defined the prediction error (P_1), the method proceeds in the usual manner, solving for α_1 by taking the derivative of P_1 with respect to α_1 and setting it equal to zero. This provides the solution for α_1. Following this, the prediction error for an order-2 AR process is defined and the derivative set equal to zero, allowing for solution of α_2. The method proceeds in this recursive manner until all of the unknown coefficients are determined. Following that, (14.72) is used to determine the spectral estimate at the specified frequencies, ω. Note that in practice, this system of equations is written in matrix form, where it is easily manipulated so as to take the standard form of a Toeplitz matrix, etc. In this form, it is then solved in the usual manner using a modified Levinson recursion algorithm (sometimes referred to as Toeplitz recursion).

The method is referred to as the maximum entropy method because it effectively is solving the *constrained* least-squares problem in which it finds the filter coefficients that maximize the disorder (entropy) of the output (that is, we want our input signal to pass through the filter and come out as white noise) while it is constrained so that the power spectrum still satisfies the relationship that it is equal to the transform of the autocorrelation function. This leads to a system of equations constructed with Lagrange multipliers (§6.4.2), which when solved gives (14.72) for the spectral estimate.

ARMA Process

Finally, the ARMA (or ARIMA) process is described by a causal feedback/feedforward system that is given in z-transform as

$$z(Z) = \frac{B_q(Z)}{A_p(Z)} \varepsilon(Z). \tag{14.74}$$

Its spectral form is determined as with the other methods, and is

$$|z(\omega)|^2 = \frac{|B_q(\omega)|^2}{|A_p(\omega)|^2} \sigma_\varepsilon^2. \tag{14.75}$$

This is obviously the most general of the parametric models (it is assumed that if the data are nonstationary, the appropriate differencing is performed prior to any of these analyses and the process is thus an ARIMA(p,d,q) in which the value of d is nonzero). The spectrum is computed by determining the values of the unknown coefficients to both the AR and MA components and then simply computing the power spectrum for each filter and combining them at desired frequencies using (14.75).

The most efficient method of solving for the coefficients (as stated previously) is through one of several numerical schemes designed to determine the optimal order and coefficient values for the ARMA process model, or to use canned software.

14.6.3 Examples

So, while the above parametric models are extremely versatile and powerful, misuse (i.e., forcing an AR model onto an MA system, for example) can lead to misleading results, as shown in Robinson and Treitel (1980) and reproduced here in Figures 14.5 and 14.6.

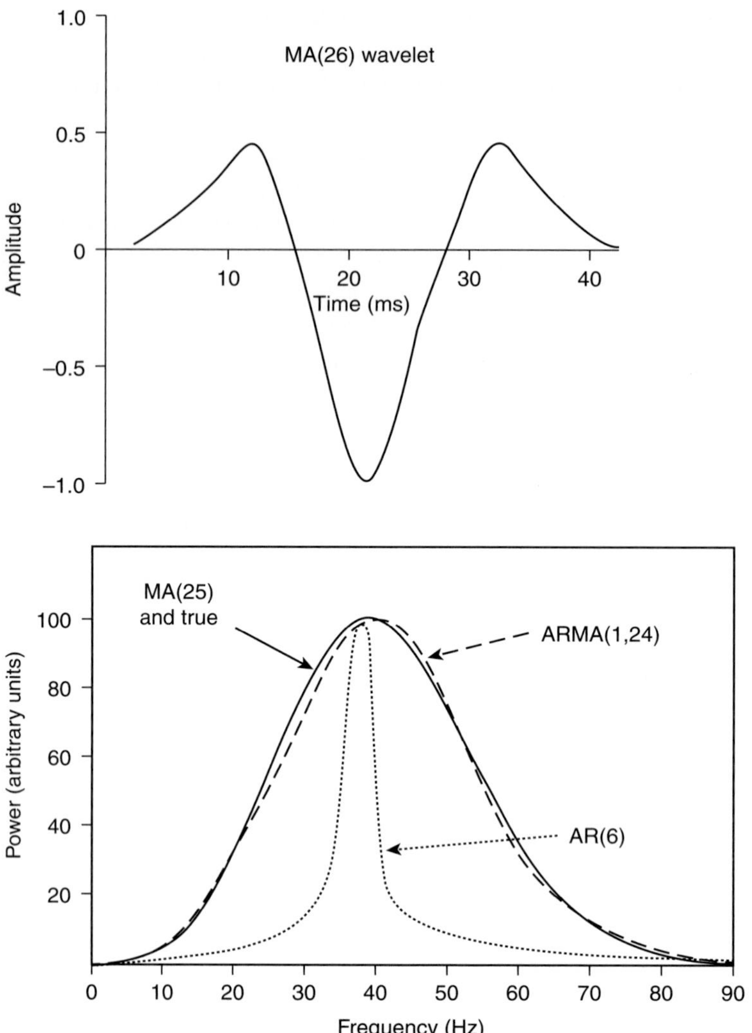

Figure 14.5 A wavelet generated as a MA(26) process and the spectra computed using MA(25) giving true answer, AR(6), which is far too spikey, and ARMA(1,24), which is close to true spectrum (solid line).

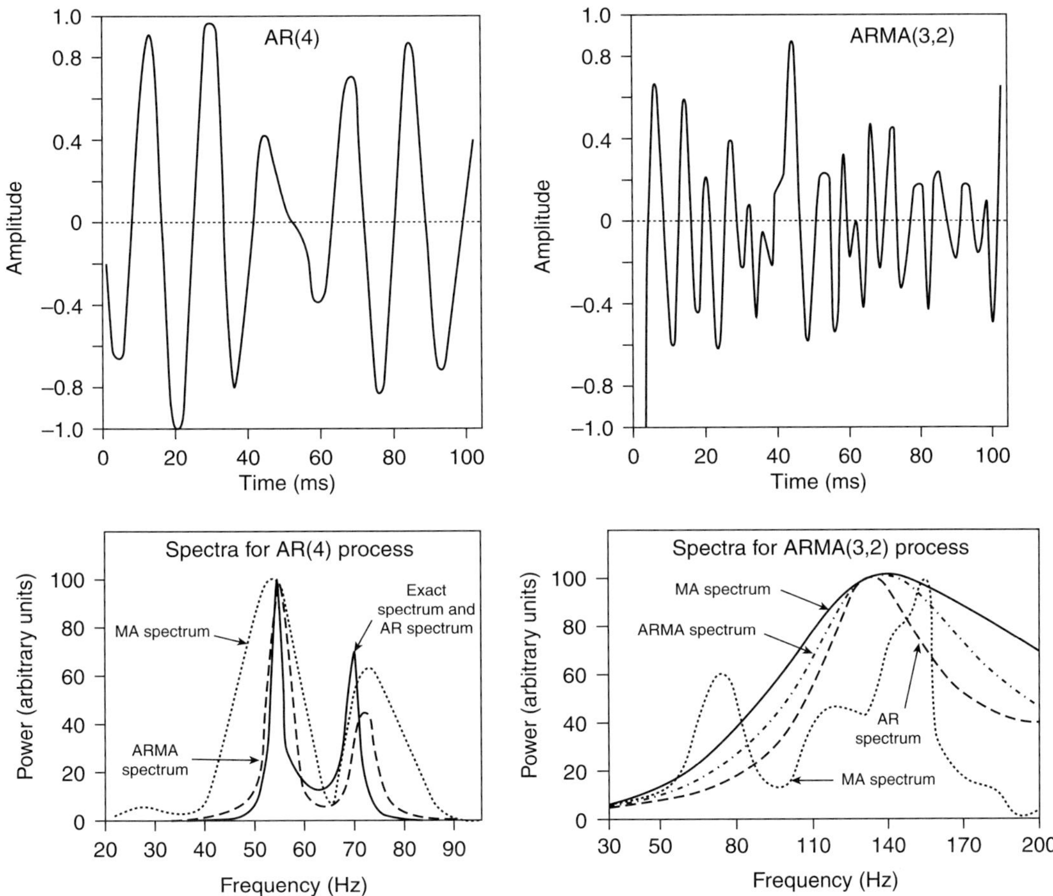

Figure 14.6 Examples of generated processes and the spectra (below the processes) resulting, using different order models. Most alarming is the MA spectrum for the ARMA(3,2) process.

14.7 Take-Home Points

1. Parametric spectral analysis, using a model determined from evaluation of your time series, can be used to identify spectral amplitude at any frequencies, since it is not limited to harmonics.
2. AR(1) models are ideal for estimating the null spectra for confidence intervals on spectra.
3. Prediction error operators allow you to estimate forward in time a specific distance within the amplitude of noise.

14.8 Questions

Pencil and Paper Questions

1. a. For a third-order MA model:
    1. Show the equation describing this model.
    2. Sketch the function for this model that you would use to identify its order.

b. For a third-order AR model:
   1. Show the equation describing this model.
   2. Sketch the function for this model that you would use to identify its order.
c. Define an ARMA.
d. Define an ARIMA.

2. a. What are the pros and cons of using Maximum Entropy Spectra?
   b. Describe some conceptual differences between AR and MA spectra.

Computer-Based Questions

3. For the first million years of LR04:
   a. If you have software for performing linear parametric time series analysis, find the order of the ARMA model for this series.
   b. Use the ARMA model to produce the PSD for the series and plot this on the PSD computed in the usual manner using a Fourier transform.
   c. How do the two PSDs in part 3.b differ, and why?

14.9 Time Series References

Blackman, R. B. and J. W. Tukey, 1958. *The Measurement of Power Spectra*. Dover Publications Inc., New York, NY. 190 pp.

This is the classic text, from the standpoint that this is where the details of classical spectral analysis were first laid out in detail. It is not an easy read, since the authors skip most of the proofs, but as the original reference it is worth looking at. Lots of good tips and insights.

Bloomfield, Peter, 1976. *Fourier Analysis of Time Series: An Introduction*. John Wiley and Sons, New York, NY. 258 pp.

This is a very clear introduction to spectral analysis, with a focus on the statistics. It is presented at a very readable level, not overburdened with proofs, and contains many good practical examples. Definitely one of the better introductory texts. Gives some simple and useful subroutines.

Box, George E. P. and Gwilym M. Jenkins, 1976. *Time Series Analysis Forecasting and Control*. Holden-Day, Inc., Oakland, CA. 575 pp.

This is the classic reference for linear process modeling. The book is very thorough. Considered intimidating by some, but it isn't actually overly difficult. It is comparable to what Draper and Smith (1981) is to regression. A must-have if you're doing much linear process modeling. Otherwise, Nelson is probably more readable and sufficient.

Bracewell, Ronald N., 1965. *The Fourier Transform and Its Applications*. McGraw-Hill, New York, NY. 444 pp.

This has long been the standard reference on Fourier analysis. Normalization constants of some of the statistical things, like the acf and acvf, are not really considered (but this is immaterial for Fourier analysis – just a caution). The development is predominantly with the continuous Fourier transform – no statistics. A new edition has recently been released.

Brigham, E. Oran, 1974. *The Fast Fourier Transform*. Prentice-Hall, Inc., Englewood Cliffs, NJ. 252 pp.

This is an excellent introductory text for learning about the Fourier transform. It is superb for relating the continuous transform to the discrete transform of finite length. Excellent figures throughout. Also good for learning how the FFT works.

Chatfield, C., 1989. *The Analysis of Time Series*. Chapman and Hall Ltd., New York, NY. 241 pp.

This is a very introductory text to time series analysis. It is good from the standpoint that it gives an overview of the discipline, but the details are very brief, so you don't read this one to learn the theory, rather to get a good idea of the general issues. Probably worth having since it's relatively cheap. Highly recommended to beginners.

Elliot, Douglas F., 1987. *Handbook of Digital Signal Processing, Engineering Applications*. Academic Press, New York, NY. 999 pp.

This book is actually a compilation of a number of engineering-oriented papers that have been carefully edited so that notation is consistent and cross-referencing good. Some of the chapters are very interesting and good. Some just give a brief overview of their subject. Predominantly on filtering techniques, with a strong engineering bias (may cause some difficulty due to the terminology used).

Hamming, R. W., 1989. *Digital Filters*. Prentice-Hall, Inc., Englewood Cliffs, NJ. 284 pp.

Hamming is one of early contributors to this field, and his insights are worth seeing. He approaches the subject of Fourier analysis and filtering from somewhat of a different perspective than most other authors. Fairly introductory level.

Jenkins, Gwilym M. and Donald G. Watts, 1968. *Spectral Analysis and Its Applications*. Holden-Day, Inc., San Francisco, CA. 525 pp.

This is a classic. It has an emphasis on the statistics and provides nearly all of the proofs (many texts still simply refer to the proofs found here). It gives excellent overviews of statistics, Fourier analysis and then a thorough treatment of time series analysis in both the time and frequency domains. It is more advanced than, say, Bloomfield (1976), and more thorough and readable than Blackman and Tukey (1958).

Kay, Stephen M., 1988. *Modern Spectral Estimation, Theory and Application*. Prentice-Hall, Inc., Englewood Cliffs, NJ. 543 pp.

Modern text with a good cross-discipline perspective. He comes from an engineering slant, but includes statistical aspects and presents good insights on a number of things. Text and theory are succinct, so it can be difficult in some places. A lot on linear-process models. Comes with a diskette containing programs. This book is somewhat of a companion to Marple (1987) (in fact there is a lot of overlap – this is the more general of the two).

Koopmans, L. H., 1974. *The Spectral Analysis of Time Series*. Academic Press, New York, NY. 366 pp.

This book is fairly advanced and along the lines of Jenkins and Watts (1968), though more complementary than replicative. Has some excellent insights and some other areas that are difficult to follow due to the brevity of the presentation of a difficult subject.

Marple, Jr., S. Lawrence, 1987. *Digital Spectral Analysis with Applications*. Prentice-Hall, Inc., Englewood Cliffs, NJ. 492 pp.

Similar in style to Kay (1988), but this book does less overview and spends considerable effort covering new and specialized methods of spectral analysis (mostly of parametric forms). Includes a diskette of the programs, which is excellent, given the variety of techniques discussed.

Nelson, Charles R., 1973. *Applied Time Series Analysis for Managerial Forecasting*. Holden-Day Inc., San Francisco, CA. 231 pp.

This is an excellent book for studying linear stochastic modeling (predominantly in the time domain). It is presented at an intermediate level in a very clear manner. Presents the relevant statistics and includes forecasting. An easier reader than Box and Jenkins (1976).

Percival, Donald B. and Andrew T. Walden, 1993. *Spectral Analysis for Physical Applications*. Cambridge University Press, Cambridge. 583 pp.

This book contains an excellent discussion on the classic spectral techniques, while providing considerable insights and examples regarding some of the more recent developments in spectral estimation techniques (e.g., nice discussion on bias in estimates). It also provides an excellent description of the method of multitapers. The book is presented at a moderate level.

Press, William H., Brian P. Flannery, Saul A. Teukolsky and William T. Vetterling, 1986. *Numerical Recipes*. Cambridge University Press, Cambridge. 818 pp.

Superb general computational text with relevant theory, practical implementation tips and subroutines for everything.

Priestley, M. B., 1981. *Spectral Analysis and Time Series*. Academic Press, New York, NY. 890+ pp.

This is a very complete and excellent reference for Fourier and spectral analysis, with good reviews of statistics. Covers the classical material in great detail and shows all of the caveats, etc. Many good insights. Fairly advanced, but just reading between the seemingly endless equations can provide good insights (with this and Jenkins and Watts (1968), you would have probably 90 percent of the classical material covered).

Robinson, E. A. and S. Treitel, 1980. *Geophysical Signal Analysis*. Prentice-Hall, Inc., Englewood Cliffs, NJ. 466 pp.

Nice book focusing on geophysical methodologies, often giving discussions and derivations from a different perspective, which can be very useful. Excellent discussion and examples of parametric spectral estimation.

15 Empirical Orthogonal Function (EOF) Analysis

15.1 Overview

Empirical Orthogonal Function (EOF) analysis is designed to find covariability within a data set and create new composite variables that capture that internal dependence, allowing a few composite uncorrelated variables to describe most of the variability (variance) in the data described in the much larger, dependent dataset. This is typically applied to large space-time data (e.g., time series collected at numerous spatial locations), though it can also be used on individual time series, or even nonsequential data – for example, if you have collected air temperature data every day for 10 years at every airport in the United States.[1] Typically, the different time series will show some degree of dependence (e.g., if it is exceptionally hot in one location, it is likely to be hot in others and just average or even cold in others). This means that different time series in different locations are carrying replicate information. EOF analysis finds those relationships in space that share the same time variability and combines them into a single spatial pattern sharing a common time variability. This pattern is found by regression across space so that a precise relationship between neighboring sites is found. Ultimately, the patterns that are found show these covarying locations, producing, for example, a single pattern in space that shows how any one location covaries with all others. So if it is hot in one location, the pattern shows those locations that are also hot, slightly cooler, a lot cooler, average, etc. This single pattern then captures all of the replicate information otherwise stored in the many individual time series. This makes it easier to view how the variable varies in space and time and collapses a potentially huge data set into a minimum number of patterns that capture most of the variance.

15.2 Introduction

Previously, we have fit data with known functions, such as sines, cosines or polynomials, in an effort to determine the fundamental characteristics of the processes they represent

[1] I particularly like the Philadelphia airport (PHL), where there are many excellent restaurants to eat at during weather-related delays at flight destinations.

or to identify some underlying signal buried within noise. But the signal within your data may reflect the linear combination of a set of processes that do not follow some standard (or any) functional form. For example, the distribution of rainfall within a region may represent a linear combination of the temperature, humidity, atmospheric pressure and other physical variables. Each of these variables changes in time in a highly irregular fashion, defying a simple mathematical description. In such a case, how can you decompose the rainfall data into a set of meaningful functions?

Empirical orthogonal function (**EOF**) analysis, or **eigenvector analysis**, is the tool you employ when searching for a set of natural, or *empirical*, orthogonal functions that can be combined linearly to describe the data.[2] If the data are thought to represent the sum of different periodic functions, then we decompose the data set into a set of orthogonal sinusoids (Fourier harmonics). With EOFs, we are decomposing the data into a set of empirical orthogonal functions – functions with no standard mathematical representation, but instead, shapes that, while orthogonal, represent the regularly occurring, temporally coherent major-variance-describing patterns present in the data (just like sines and cosines are regular patterns that have simple mathematical representation, here we will find shapes that regularly occur, regardless of their particular form). This ability to isolate consistent patterns within individual data sets makes EOFs "*data adaptive*."

EOF analysis finds patterns that share the same temporal variability to be consistent throughout the datasets, so that one pattern, orthogonal to the other patterns, can now be constructed, replacing that same pattern hidden in each time series and being repeated multiple times in the data set. Once this first pattern is identified, another pattern is identified, capturing the greatest amount of the remaining temporal variability not already captured by the previous pattern(s).[3] While this is a statistical construct, we hope that the mechanisms driving the phenomenon being examined may have a unique signature that is captured by this methodology – here identified as a pattern that varies in time the same way in each time series in the dataset. Each of the patterns identified has an associated time series showing how the amplitude of the pattern changes in time, known as **principal components** (**PCs**, or, **expansion coefficients**). Each pattern (the **EOF**) and its time series (PC) together are called a **mode**. Individual modes can be isolated and studied as potentially physically meaningful (though they needn't be so), and a subset of modes can be combined to describe the signal (with the remaining, unused set describing noise).

Consider the following example of a spatio-temporal dataset consisting of a measure of upper-ocean freshwater content (called salt deficit, SD_W), measured within a grid in the waters lying along the western (Pacific) margin of the Antarctic Peninsula (Figure 15.1). This grid was sampled every austral summer for 12 years (now over 20

[2] You can decompose a data set into nonorthogonal functions, but the advantage of orthogonal functions is that they are completely independent of one another, thus any part of the signal described by one orthogonal function is no longer available for description by another. Because of this, the different orthogonal functions can be added together linearly to describe the dataset. And it is because of the orthogonality that the solution for the function coefficients is so incredibly simple (i.e., the Fourier transform, an approach that can be applied to any orthogonal function, not just sinusoids). Orthogonality does introduce some limitations to be relaxed when introducing rotated EOFs (not discussed within).

[3] Formally, we identify the common time series describing the pattern, and then other time series that are orthogonal to the other time series, with each having its own different pattern.

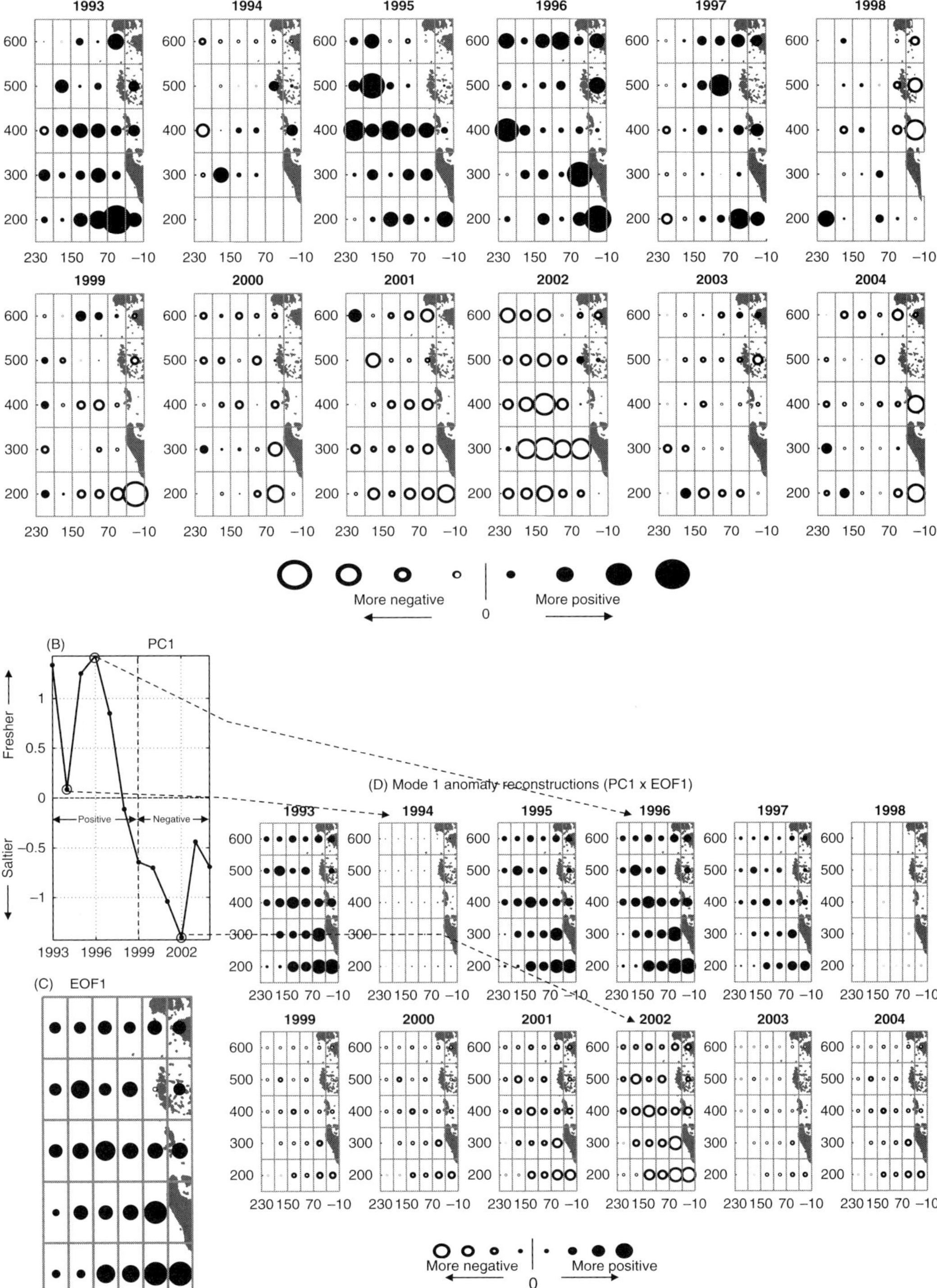

Figure 15.1 (**A**) Anomaly plots (data value minus 12-year average at each grid point) (**B**). Each year, all anomaly maps share a pattern that has an identical time history. The dominant time history is shown by this first PC (PC1), and (**C**) the pattern undergoing this temporal change (the EOF1) in. In this case, this first PC describes ~51 percent of the temporal variance, leaving another 49 percent to be described by other patterns. (**D**) Each year, EOF1 is multiplied by PC1 for

continued on next page

years), giving a *map* of values for each of the 12 years and a 12-year time series in each grid cell. We wish to determine if there is any consistent variability in space and/or time over these 12 years that might help us to better isolate any change leading to an understanding of any physical mechanisms driving such change. For this, we employ empirical orthogonal function analysis.

EOF analysis, as employed within, is a general term that encompasses a family of techniques, such as **principal components analysis** (**PCA**, the statistical name of what I am more generally calling EOF analysis), **canonical correlation analysis** (**CCA**) and **factor analysis**, all of which employ some subtly different form by which the empirical functions are arrived at or treated. Typically, EOF analysis is an exercise in multivariate statistics, and thus it is concerned with decomposing multiple variables into a set of orthogonal functions (equivalent to a multi-dimensional Fourier decomposition). The decomposition is invaluable for, among other things, providing the most amount of information in the most manageable form. Consider a situation where we have measured p different variables. The simplest statistics, i.e., the mean, variance and covariance between variables, requires $p + (p^2 + p)/2$ values to describe them. If the variables were uncorrelated, then the covariance is zero between all pairs, and thus the data description would be accomplished with only two p statistics – a considerable reduction, given a large p! EOF techniques are useful for reducing this massive data set to a smaller number of independent, uncorrelated variables (the covariances are zero), thus reducing the *dimension* or *degrees of freedom* of the data set from p to a smaller size, say, k, that contains most of the variance. With these, the data can be described by two k statistics.[4] The techniques are also useful for identifying coherent, even if functionally irregular, structure in the data.

There is a tremendous body of literature associated with these techniques. EOF analysis can also be used to decompose a single time series into a sum of orthogonal functions,[5] though traditionally it is more typically used to decompose multivariate or multidimensional data.

Caption for Figure 15.1 (cont.)

that year to give the pattern with actual mode 1 values for that year. This primary mode shows a strong decreasing trend: just before 1998, the SD'_W becomes mostly negative (water is saltier), so the sign of this first PC changes from positive to negative to accommodate this (given that all but one cell in the corresponding EOF1 are of the same sign), reaching maximum negative (salty) in 2002 as seen in the anomaly maps, the reconstruction and the PC.

[4] Often we will have removed the mean from the original variables for simplicity, in which case the reduced set has zero mean.

[5] Discussed at the end of this chapter, after the fundamentals of the technique have been derived and examined.

15.3 Eigenvector Analysis

15.3.1 Fundamentals

The heart of EOF analysis centers on the concept of **eigenvectors**, also referred to as **characteristic vectors**. Eigenvectors can be described and derived from a variety of perspectives, but the fundamental point here is a property of a *square* matrix $\mathbf{A}$ that has no counterpart in ordinary algebra.

We begin by asking the question whether it is possible to multiply a vector, $\mathbf{x}$, by matrix $\mathbf{A}$ and produce a new vector, $\mathbf{b}$, that is parallel to the original $\mathbf{x}$. Or, alternatively, for any square matrix $\mathbf{A}$, does there exist one or more non-trivial (i.e., non-zero) vectors $\mathbf{x}$, that when multiplied (transformed) by matrix $\mathbf{A}$, yield the original, though scaled, vector $\mathbf{x}$? In this case,

$$\mathbf{Ax} = \lambda\mathbf{x} \tag{15.1a}$$

or, rearranged,

$$(\mathbf{A} - \lambda\mathbf{I})\mathbf{x} = 0. \tag{15.1b}$$

Here, λ is a scalar such that the new vector $\mathbf{b}$ $(= \lambda\mathbf{x})$ is parallel to $\mathbf{x}$, since parallel vectors can differ only in length (or magnitude), and thus are equal to scalar multiples of one another (see Appendix A1 on Matrix Algebra).

Conceptually, one might imagine that satisfying (15.1) should be a rather difficult task, since the operation of $\mathbf{Ax}$ is equivalent to multiplying every element of $\mathbf{x}$ by an entire column in $\mathbf{A}$, and yet the net result must be the original vector $\mathbf{x}$, altered only by a constant. So,

$$\begin{bmatrix} a_{11} & a_{12} & \ldots & a_{1n} \\ a_{21} & a_{22} & & a_{2n} \\ & \vdots & & \\ a_{n1} & a_{n2} & & a_{nn} \end{bmatrix} \begin{bmatrix} x_1 \\ x_2 \\ \vdots \\ x_n \end{bmatrix} = \lambda \begin{bmatrix} x_1 \\ x_2 \\ \vdots \\ x_n \end{bmatrix} \tag{15.1c}$$

$$\begin{bmatrix} & & & \\ \mathbf{a}_1 & \mathbf{a}_2 & \ldots & \mathbf{a}_n \\ & & & \end{bmatrix} \begin{bmatrix} x_1 \\ x_2 \\ \vdots \\ x_n \end{bmatrix} = \lambda \begin{bmatrix} x_1 \\ x_2 \\ \vdots \\ x_n \end{bmatrix}, \tag{15.1d}$$

or, expanding the multiplication in terms of the $\mathbf{a}_j$ vectors,

$$x_1 \begin{bmatrix} a_{11} \\ a_{21} \\ \vdots \\ a_{n1} \end{bmatrix} + x_2 \begin{bmatrix} a_{12} \\ a_{22} \\ \vdots \\ a_{n2} \end{bmatrix} + \ldots + x_n \begin{bmatrix} a_{1n} \\ a_{2n} \\ \vdots \\ a_{nn} \end{bmatrix} = \lambda \begin{bmatrix} x_1 \\ x_2 \\ \vdots \\ x_n \end{bmatrix}. \tag{15.1e}$$

Therefore, after a rather elaborate multiplication of each element in **x**, the new vector still points in the same original direction – all of the multiplications doing nothing more than ultimately scaling each element of **x** by the same amount.

Alternatively, the difficulty is clearly apparent by inspection of the system that (15.1) represents,

$$\begin{aligned} a_{11}x_1 + a_{12}x_2 + \ldots + a_{1n}x_n &= \lambda x_1 \\ a_{21}x_1 + a_{22}x_2 + \ldots + a_{2n}x_n &= \lambda x_2 \\ &\vdots \\ a_{n1}x_1 + a_{n2}x_2 + \ldots + a_{nn}x_n &= \lambda x_n, \end{aligned} \tag{15.2a}$$

which tends to emphasize the row vector manipulation taking place in (15.1). That is, the elements of **x** are being multiplied by an entire row of **A**, and yet the final product is still just the scaled **x**. In other words, each element x_i is a weighted sum of all other elements of **x**, yet the weighted sum is simply proportional to the original element. We wish to find those elements of **x** that satisfy this criterion for the "weights" contained in any square matrix **A**.

Equation (15.2a) is rearranged to give

$$\begin{aligned} a_{11}x_1 + a_{12}x_2 + \ldots + a_{1n}x_n - \lambda x_1 &= 0 \\ a_{21}x_1 + a_{22}x_2 + \ldots + a_{2n}x_n - \lambda x_2 &= 0 \\ &\vdots \\ a_{n1}x_1 + a_{n2}x_2 + \ldots + a_{nn}x_n - \lambda x_n &= 0 \end{aligned} \tag{15.2b}$$

or

$$\begin{aligned} (a_{11} - \lambda)x_1 + a_{12}x_2 + \ldots + a_{1n}x_n &= 0 \\ a_{21}x_1 + (a_{22} - \lambda)x_2 + \ldots + a_{2n}x_n &= 0 \\ &\vdots \\ a_{n1}x_1 + a_{n2}x_2 + \ldots + (a_{nn} - \lambda)x_n &= 0. \end{aligned} \tag{15.2c}$$

In matrix form,

$$x_1\begin{bmatrix} a_{11}-\lambda \\ a_{21} \\ \vdots \\ a_{n1} \end{bmatrix} + x_2\begin{bmatrix} a_{12} \\ a_{22}-\lambda \\ \vdots \\ a_{n2} \end{bmatrix} + \ldots + x_n\begin{bmatrix} a_{1n} \\ a_{2n} \\ \vdots \\ a_{nn}-\lambda \end{bmatrix} = \begin{bmatrix} 0 \\ 0 \\ \vdots \\ 0 \end{bmatrix} \tag{15.2d}$$

$$\begin{bmatrix} a_{11}-\lambda & a_{12} & \ldots & a_{1n} \\ a_{21} & a_{22}-\lambda & & a_{2n} \\ & \vdots & & \\ a_{n1} & a_{n2} & & a_{nn}-\lambda \end{bmatrix}\begin{bmatrix} x_1 \\ x_2 \\ \vdots \\ x_n \end{bmatrix} = 0. \tag{15.2e}$$

Thus, given carefully chosen elements of **x** and the constant λ, it is possible to sum up the column vectors in the $(\mathbf{A} - \lambda\mathbf{I})$ matrix so that they sum to the **null vector** – that is, they cancel one another out. The vectors of $(\mathbf{A} - \lambda\mathbf{I})$, when scaled by the elements of $\mathbf{x}$, ultimately end up back at the origin.[6]

In this *homogeneous* form (i.e., the equation equals zero), we know that the system (15.2) can only have a nontrivial solution when the determinant is equal to zero:

$$\begin{vmatrix} a_{11}-\lambda & a_{12} & \cdots & a_{1n} \\ a_{21} & a_{22}-\lambda & & a_{2n} \\ & \vdots & & \\ a_{n1} & a_{n2} & & a_{nn}-\lambda \end{vmatrix} = 0. \tag{15.3}$$

That is, the system shown in (15.2e), or represented by the equivalent matrix expression of (15.1b), shows that the weighted sum of the columns in matrix $(\mathbf{A} - \lambda\mathbf{I})$ sum to zero, and this is equivalent to the definition of *dependent* columns. That is, the columns of a matrix are dependent if they can be summed to zero when weighted with the appropriate weights. A matrix with dependent columns is singular (see Appendix 1), and a singular matrix has its determinant equal to zero – thus (15.3) above. $(\mathbf{A} - \lambda\mathbf{I})$ must be singular if a solution to (15.1) or (15.2) exists, other than the trivial solution (i.e., $\mathbf{x} = \mathbf{0}$), which is rather meaningless.

The determinant in (15.3) can be expanded to form a polynomial in λ, normalized by multiplying through by $(-1)^n$ for convenience,[7] as

$$\lambda^n + a_{n-1}\lambda^{n-1} + a_{n-2}\lambda^{n-2} + \ldots + a_1\lambda^1 + a_0 = 0. \tag{15.4}$$

This polynomial in λ is called the characteristic equation, characteristic function or characteristic polynomial.

For example, if $n = 2$, then, the determinant in (15.3) is expanded as

$$\begin{aligned} |\mathbf{A} - \lambda\mathbf{I}| &= (a_{11}-\lambda)(a_{22}-\lambda) - a_{12}a_{21} \\ &= \lambda^2 - a_{11}\lambda - a_{22}\lambda + a_{11}a_{22} - a_{12}a_{21} \\ &= \lambda^2 + (-a_{11} - a_{22})\lambda + a_{11}a_{22} - a_{12}a_{21} \\ &= \lambda^2 + c_1\lambda + c_0 = 0. \end{aligned} \tag{15.5}$$

This characteristic equation is solved by the quadratic formula,

$$\lambda = \frac{-c_1 \pm \sqrt{c_1^2 - 4c_0}}{2}, \tag{15.6a}$$

[6] The addition of vectors can be done graphically by drawing each vector with a line from its origin to its tip, where its length is proportional to its magnitude and its angle reflects its direction. Then, each vector is added to the previous one by simply placing its origin on the tip of the previous vector. The sum is then given by the vector drawn from the origin (the location of the origin of the first vector in the sum) to the tip of the final vector – in this case, the final vector tip ends at the origin (i.e., we have just undone all of the additions to get back to where we started).

[7] This normalization, or simple scaling, is done to ensure that the polynomial is always written in the form of (15.4). If not done, for n equal to an odd number, the first term in the sum, λ^n, would be $-\lambda^n$.

or by factoring and solving for the *roots*:

$$(\lambda + b_1)(\lambda + b_2) = 0. \tag{15.6b}$$

The equation has two solutions: one, λ_1, when taking the "+" in the numerator of (15.6a); the other, λ_2, when taking the "–". These correspond to the two roots, $\lambda_1 = -b_1$ and $\lambda_2 = -b_2$ in (15.6b). So there are two different values of λ that satisfy this determinant and satisfy $\mathbf{Ax} = \lambda\mathbf{x}$.

Generalized, a polynomial of order n, and thus the determinant of (15.3), will always have n (and only n) roots that satisfy the equation (though all n roots needn't be unique).[8] These n roots, that is, the n values of λ that satisfy (15.1) are called the **eigenvalues**[9] of the matrix **A**.

For each eigenvalue, $\lambda_1, \lambda_2, \ldots, \lambda_n$, there exists an explicit system (15.2c) that is satisfied. So, for the case of $n = 2$, there exist

$$\begin{aligned} (a_{11} - \lambda_1)x_1 + a_{12}x_2 &= 0 \\ a_{21}x_1 + (a_{22} - \lambda_1)x_2 &= 0 \end{aligned} \tag{15.7a}$$

and

$$\begin{aligned} (a_{11} - \lambda_2)x_1 + a_{12}x_2 &= 0 \\ a_{21}x_1 + (a_{22} - \lambda_2)x_2 &= 0. \end{aligned} \tag{15.7b}$$

Given the values of λ_1 and λ_2, these two systems of equations (15.7a) and (15.7b) can be solved for the two **x** vectors, which are those that are simply scaled when multiplying the matrix **A** for which the eigenvalues correspond. These vectors that satisfy (15.1) and which exist for each eigenvalue, are called **eigenvectors**.

So, in (15.7a), the coefficients of the system, $(a_{11} - \lambda_1)$ and $(a_{22} - \lambda_1)$, yield values of x_{11} and x_{21}, forming the first eigenvector $\mathbf{x}_1$, that are different from the values of x_{12} and x_{22}, forming the second eigenvector, $\mathbf{x}_2$, resulting from $(a_{11} - \lambda_2)$ and $(a_{22} - \lambda_2)$.[10]

In the general case, with n eigenvalues, each eigenvalue, obtained by solving the determinant in (15.3), leads to an eigenvector, **e**, obtained by solving the system (15.2) for that eigenvalue:

$$\begin{aligned} \lambda_1 \text{ yields } \mathbf{e}_1 &= [x_{11}\ x_{21}\ x_{31}\ \ldots\ x_{n1}]^{\mathrm{T}} \\ \lambda_2 \text{ yields } \mathbf{e}_2 &= [x_{12}\ x_{22}\ x_{32}\ \ldots\ x_{n2}]^{\mathrm{T}} \\ &\vdots \\ \lambda_n \text{ yields } \mathbf{e}_n &= [x_{1n}\ x_{2n}\ x_{3n}\ \ldots\ x_{nn}]^{\mathrm{T}}. \end{aligned} \tag{15.8}$$

[8] The solution of an nth order polynomial for the roots is not a trivial task, and is best obtained using one of several numerical techniques, such as singular value decomposition (SVD), for example.

[9] Other names often associated with these are "characteristic values," "proper roots," "proper values," "λ-roots" or "latent roots." These names "proper," "latent" and "characteristic" also can be used in place of "eigen" in the previously defined terms.

[10] Actually, the two eigenvalues can be the same, but the point is that there will be a distinct eigenvector for each eigenvalue, regardless of whether they are the same or not.

Each eigenvalue, λ_j, allows the equation $\mathbf{Ax} = \lambda\mathbf{x}$ to be solved by the corresponding eigenvector, $\mathbf{x}_j$. Note that, since these eigenvalue-eigenvector pairs represent solutions to homogeneous equations, we can multiply any of the eigenvectors by a constant and still satisfy the equations. Therefore, this is consistent with the original premise that we are finding vectors that, when multiplied by a matrix, retain their original direction, but not necessarily their original length (magnitude). Thus, it is the *direction* of the eigenvectors, not their magnitude, that is unique for each eigenvector. Typically, we work with the eigenvectors after normalizing them to unit length: $\mathbf{e} = \mathbf{e}/(\mathbf{e}^T\mathbf{e})^{1/2}$.

Once we have all of the eigenvalues and corresponding eigenvectors, we can represent the complete **eigenstructure** – that is, the entire set of eigenvalues and eigenvectors – in a single equation of the form

$$\mathbf{AE} = \mathbf{E\Lambda}, \tag{15.9a}$$

where $\mathbf{A}$ is the original square matrix from which the eigenvalues and eigenvectors originate, $\mathbf{E}$ is the square ($n \times n$) matrix for which each column contains one eigenvector, and $\mathbf{\Lambda}$ (sometimes denoted as $\mathbf{L}$) is a square diagonal matrix ($n \times n$) that contains an eigenvalue on each diagonal element, such that the eigenvalues are aligned with their corresponding eigenvector in the $\mathbf{E}$ matrix,[11] as

$$\begin{pmatrix} a_{11} & a_{12} & \cdots & a_{1n} \\ a_{21} & a_{22} & & a_{2n} \\ \vdots & & & \\ a_{n1} & a_{n2} & & a_{nn} \end{pmatrix} \begin{bmatrix} e_1 & e_2 & \cdots & e_n \end{bmatrix} = \begin{bmatrix} e_1 & e_2 & \cdots & e_n \end{bmatrix} \begin{bmatrix} \lambda_1 & & & \\ & \lambda_2 & & \\ & & \ddots & \\ & & & \lambda_n \end{bmatrix}$$

$$= \begin{bmatrix} \lambda_1 e_1 & \lambda_2 e_2 & \cdots & \lambda_n e_n \end{bmatrix} \tag{15.9b}$$

15.3.2 Orthogonality for Symmetrical Matrices

Now consider a symmetrical matrix $\mathbf{A}$ and any two of its eigenvalues and corresponding (normalized) eigenvectors, say λ_i, $\mathbf{e}_i$ and λ_j, $\mathbf{e}_j$, such that

$$\mathbf{Ae}_i = \lambda_i \mathbf{e}_i \tag{15.10a}$$

$$\mathbf{Ae}_j = \lambda_j \mathbf{e}_j. \tag{15.10b}$$

[11] Post-multiplying by a diagonal matrix scales the columns of a matrix, whereas pre-multiplying scales the rows.

Box D15.1 Natural Eigenfunctions of Linear Systems

Consider the complex function $e^{i\omega t}$ in the context of linear systems, where a linear system has the property

$$L(a\mathbf{x} + b) = aL(\mathbf{x}) + b \tag{D15.1.1}$$

and L is a linear operator such as differentiation, integration, convolution, etc.

As an example, consider convolution:

$$h_t = \sum_{j=0}^{n-1} f_j g_{t-j}. \tag{D15.1.2a}$$

Here, f is the time-invariant impulse response function of a filter, h is the output from the filter and g is the input, both of which are functions of time (or lag), indexed by t. If g is given by the complex exponential, we have

$$\begin{aligned} h_t &= \sum_{j=0}^{n-1} f_j e^{i\omega(t-j)} \\ &= \sum_{j=0}^{n-1} f_j e^{i\omega t} e^{-i\omega j} \\ &= e^{i\omega t} \sum_{j=0}^{n-1} f_j e^{-i\omega j} \\ &= c(\omega)\, e^{i\omega t}, \end{aligned} \tag{D15.1.2b}$$

where

$$c(\omega) = \sum_{j=0}^{n-1} f_j e^{-i\omega j}, \tag{D15.1.2c}$$

which is the Fourier transform of the impulse response function, the **transfer function** of the filter (Chapter 13).

Therefore, the input function $e^{i\omega t}$, when operated upon by a linear filter, gives an output function that is proportional to the input function. In other words, *the complex exponential is a natural eigenfunction for this (and all) linear systems*. The function of eigenvalues, $c(\omega)$, is nothing more than the transfer function for the system.

Because we know that the complex exponential is the complex sum of a sine and cosine (as stated by Euler's law), this suggests that the sines and cosines used in Fourier analysis to fit time series are the natural eigenvectors to linear systems. This is one reason why they are excellent functions to use for fitting and decomposing time series (though they have some other generally beneficial properties as well).

Since $\mathbf{A}$ is symmetric, $\mathbf{A} = \mathbf{A}^{\mathrm{T}}$, and thus the transpose of one of the above equations is

$$(\mathbf{A}\mathbf{e}_i)^{\mathrm{T}} = \lambda_i \mathbf{e}_i^{\mathrm{T}}, \tag{15.11a}$$

or (recalling the reversal rule of transposed products)

$$\mathbf{e}_i^{\mathrm{T}}\mathbf{A} = \lambda_i \mathbf{e}_i^{\mathrm{T}}. \tag{15.11b}$$

Now consider pre-multiplying (15.10b) by $\mathbf{e}_\mathbf{i}^{\mathbf{T}}$ and post-multiplying (15.11b) by $\mathbf{e}_j$ to give

$$\mathbf{e}_i^{\mathrm{T}}\mathbf{A}\mathbf{e}_j = \lambda_j \mathbf{e}_i^{\mathrm{T}}\mathbf{e}_j \tag{15.12a}$$

$$\mathbf{e}_i^{\mathrm{T}}\mathbf{A}\mathbf{e}_j = \lambda_i \mathbf{e}_i^{\mathrm{T}}\mathbf{e}_j. \tag{15.12b}$$

The vector-products of both equations are the same (so the only differences are the eigenvalues), and (15.12) must hold for all i and j, including those (usually the majority) for which $\lambda_i \neq \lambda_j$, therefore the two equations (15.12) can only hold if

$$\mathbf{e}_i^{\mathrm{T}}\mathbf{e}_j = 0. \tag{15.13}$$

The condition that $\mathbf{e}_i^{\mathrm{T}}\mathbf{e}_j = 0$ *shows that (i.e., is the definition that) the eigenvectors, e, are orthogonal to one another.* Furthermore, we have already defined $\mathbf{e}$ to be unit length,

$$\mathbf{e}_i^{\mathrm{T}}\mathbf{e}_i = 1, \tag{15.14}$$

so the vectors as defined are in fact, **orthonormal**.

Empirical Orthogonal Functions (EOFs)

These orthogonal eigenvectors (*which result from all symmetric matrices A*), are often termed **empirical orthogonal functions (EOFs)**. Combining them into a single matrix $\mathbf{E}$ as in (15.9a) gives an orthonormal matrix. All non-diagonal elements in $\mathbf{E}^{\mathrm{T}}\mathbf{E}$ are the covariances between the vectors (zero for these orthogonal vectors) and the diagonals are the magnitude of the vectors, which were already normalized to equal one. So, $\mathbf{E}^{-1} = \mathbf{E}^{\mathrm{T}}$, $\mathbf{E}^{\mathrm{T}}\mathbf{E} = \mathbf{I} = \mathbf{E}\mathbf{E}^{\mathrm{T}}$, and thus,

$$\begin{aligned} \mathbf{AE} &= \mathbf{E\Lambda} \\ \mathbf{AEE}^{\mathrm{T}} &= \mathbf{E\Lambda E}^{\mathrm{T}} \\ \mathbf{A} &= \mathbf{E\Lambda E}^{\mathrm{T}} \end{aligned} \tag{15.15a}$$

or

$$\begin{aligned} \mathbf{AE} &= \mathbf{E\Lambda} \\ \mathbf{E}^{\mathrm{T}}\mathbf{AE} &= \mathbf{E}^{\mathrm{T}}\mathbf{E\Lambda} \\ \mathbf{E}^{\mathrm{T}}\mathbf{AE} &= \mathbf{\Lambda}, \end{aligned} \tag{15.15b}$$

where

$$\mathbf{E}^{\mathrm{T}}\mathbf{E} = \mathbf{E}\mathbf{E}^{\mathrm{T}} = \mathbf{I} = \begin{bmatrix} 1 & & & \\ & 1 & & \\ & & \ddots & \\ & & & 1 \end{bmatrix}. \tag{15.15c}$$

Therefore, any symmetric matrix A can be factored into the product of the diagonal matrix, Λ, containing the eigenvalues of the matrix which is pre- and post-multiplied by the orthogonal matrix, E, containing the eigenvectors (EOFs) of the matrix (15.15a). *Or, any symmetric matrix A can be reduced to a diagonal matrix, whose elements consist of its eigenvalues, by pre- and post-multiplying the A matrix with the orthogonal matrix containing its eigenvectors (this is called* **diagonalization***).*

Since the eigenvectors of any symmetrical matrix are orthogonal to one another, they can be combined with appropriate coefficients to produce any nonzero vector, $\mathbf{z}$ (where the order of the symmetric matrix is the same as that of the vector). That is, they can be used like any other orthogonal basis for interpolating, smoothing or any other purpose for which we have employed such functions previously. Specifically,

$$\begin{aligned} \mathbf{z} &= c_1\mathbf{e}_1 + c_2\mathbf{e}_2 + \ldots + c_n\mathbf{e}_n \\ &= \mathbf{EC}, \end{aligned} \tag{15.16a}$$

where $\mathbf{C}$ is the matrix containing the c_i constants.

Pre-multiplying (15.16a) through by $\mathbf{E}^{\mathrm{T}}$ gives

$$\begin{aligned} \mathbf{E}^{\mathrm{T}}\mathbf{z} &= \mathbf{E}^{\mathrm{T}}\mathbf{EC} \\ &= \mathbf{C}. \end{aligned} \tag{15.16b}$$

So, the coefficients in vector $\mathbf{C}$ are determined from $\mathbf{E}^{\mathrm{T}}\mathbf{z}$, from which $\mathbf{z}$ can be decomposed as $\mathbf{EC}$. Thus the eigenvectors form an orthogonal basis for $\mathbf{z}$, and *the details (i.e., structure or shape) of the basis changes with the composition of the matrix* ***A*** *from which it was derived.*

Alternatively, consider (15.15) for a zero-mean stationary/ergodic multidimensional data set,[12] containing a variable sampled at n locations and m times (containing an m-dimensional time series for each row), and stored in **data matrix X** whose sample covariance matrix $\hat{\boldsymbol{\Sigma}}_{\mathbf{X}}$ is proportional to $\mathbf{XX}^{\mathbf{T}}$.

$$\mathbf{X} = \overbrace{\begin{bmatrix} x_{11} & x_{12} & \cdots & x_{1m} \\ x_{21} & x_{22} & & x_{2m} \\ & \vdots & & \\ x_{n1} & x_{n2} & & x_{nm} \end{bmatrix}}^{\text{times } (j=1,\ j=2,\ \ldots,\ j=m)} \left.\begin{matrix} i=1 \\ i=2 \\ \\ i=n \end{matrix}\right\} \text{locations} \tag{15.17a}$$

[12] The stationarity/ergodic restriction is imposed so that we can estimate the covariance matrix directly from the major product moment $\mathbf{XX}^{\mathrm{T}}$. Otherwise, the covariance matrix is given as the expected value of the major product moment, as explained in Chapter 7. The standardization avoids the need for additional scaling of the major product moment to give variances and covariances.

$$\frac{\mathbf{X}\mathbf{X}^{\mathrm{T}}}{n-1} = \hat{\Sigma}_X = \begin{bmatrix} s_1^2 & s_{12} & \cdots & s_{1n} \\ s_{21} & s_2^2 & & s_{2n} \\ & & \vdots & \\ s_{n1} & s_{n2} & \cdots & s_n^2 \end{bmatrix} \qquad (15.17\text{b})$$

Where s_i^2 represents the sample temporal variance of the **X** *series over all sampled times at location 1, and s_{21} represents the sample covariance between the time series obtained at locations 2 and 1.* As constructed here, each *column* of the data matrix **X** contains the values of **X** at a time t = 1,2, . . ., or m across all n spatial locations, the second column contains all observations at time 2, etc.[13] In other words, the time series occupy the rows (ith row contains the time series for the ith location), and each column contains a *map* of values for time j (a "map" because it shows the values of the variable x at multiple locations all at the same time).

By placing time series in rows in the data matrix X (each row a different location) as many of us prefer, the operational order of the sample covariance matrix is the same for a single time series vector, x, as for a data matrix: E*[$\mathbf{x}\mathbf{x}^T$] and $\mathbf{X}\mathbf{X}^T$ (see "Variance Covariance Matrix" in* Appendix 1, §A.73).[14]

Rewriting (15.15b) using the sample covariance matrix of **X** for matrix **A** ($\hat{\boldsymbol{\Sigma}}_{\mathbf{X}}$) and substituting $\mathbf{X}\mathbf{X}^{\mathrm{T}}$ for $\hat{\boldsymbol{\Sigma}}_{\mathbf{X}}$ gives (**E**, $\hat{\boldsymbol{\Sigma}}_{\mathbf{X}}$ and $\boldsymbol{\Lambda}$ are all $n \times n$ matrices),

$$\mathbf{E}^{\mathrm{T}}\mathbf{X}\mathbf{X}^{\mathrm{T}}\mathbf{E} = \Lambda, \qquad (15.18\text{a})$$

and, defining

$$\mathbf{C}_{nm} = \mathbf{E}_{nn}^{\mathrm{T}}\mathbf{X}_{nm} \qquad (15.18\text{b})$$

$$\mathbf{X}_{nm} = \mathbf{E}_{nn}\mathbf{C}_{nm}, \qquad (15.18\text{c})$$

where **C** is now a fully populated ($n \times m$) matrix of coefficients. Equation (15.18c) states that the spatial series in **X** can be re-expressed as a linear sum of (spatial) eigenvectors as one map, j, per time: $i = 1, m$. Alternatively, in terms of the eigenvector maps and time-series coefficients (one complete time series for each map),

$$\mathbf{X} = \begin{bmatrix} & & & \\ \mathbf{e}_1 & \mathbf{e}_2 & \cdots & \mathbf{e}_n \\ & & & \end{bmatrix} \begin{bmatrix} c_{11} & c_{12} & c_{13} & \cdots & c_{1m} \\ c_{21} & c_{22} & c_{23} & & c_{2m} \\ \vdots & & & & \\ c_{n1} & c_{n2} & c_{n3} & \cdots & c_{nm} \end{bmatrix}. \qquad (15.18\text{d})$$

Thus, the data are represented as a linear sum of the EOF basis, as described in (15.16a), and consistent with any orthogonal interpolation problem. In (15.16), it is clear

[13] Some authors define a data matrix so that the time variation is given as column vectors (versus their presentation as row vectors, here). This is fine if you prefer to take this approach, but in doing so, it is important to be clear which way the data matrix is being set up, so that all of the ensuing manipulations are done in the proper arrangement.

[14] Single vectors are column vectors, unless otherwise stated.

that any vector can be described as a linear combination of *any n*-orthogonal eigenvectors; here, in (15.18), it is suggested that the eigenvectors arising from the sample covariance matrix of the data matrix may represent a natural basis for the data in question.

The coefficients, on the other hand, in (15.18b) look like

$$\mathbf{C} = \begin{bmatrix} \mathbf{e}_1^{\mathrm{T}} \\ \mathbf{e}_2^{\mathrm{T}} \\ \vdots \\ \mathbf{e}_n^{\mathrm{T}} \end{bmatrix} \begin{bmatrix} x_{11} & x_{12} & x_{13} & \cdots & x_{1m} \\ x_{21} & x_{22} & x_{23} & & x_{2m} \\ \vdots & & & & \\ x_{n1} & x_{n2} & x_{n3} & \cdots & x_{nm} \end{bmatrix}. \tag{15.18e}$$

So, each coefficient is simply the sum of the eigenvector elements with the corresponding element in $\mathbf{x}_j$ (i.e., the *projection* of the data on the eigenvectors). That is, it is the correlation between the eigenvector and data. That is exactly how the Fourier transform works. Consider the discrete Fourier transform, which gives the coefficient $\mathrm{J}_j = n^{-1}\sum_{p=1}^{n} x_p \mathrm{e}^{-i\omega_j \mathrm{t}_p}$, where the eigenvector is the $\mathrm{e}^{-i\omega\mathrm{t}}$ and each element of the data vector, $\mathbf{x}_j$, coincides with each time in x_t. Likewise, (15.18c) is of the same form as the inverse Fourier transform, converting the coefficients of the frequency domain (the orthogonal function domain) back to the original domain. Thus, the Fourier transform is just a specific case of the more general form that applies for any orthogonal basis, though as shown in (D15.1), the Fourier kernel ($\mathrm{e}^{-i\omega\mathrm{t}}$) is the natural eigenfunction of linear systems.

Note two important properties (though there are many more) of eigenvalues: (1) the trace of matrix $\hat{\boldsymbol{\Sigma}}_\mathbf{X} = \mathrm{tr}(\hat{\boldsymbol{\Sigma}}_\mathbf{X})$ = the sum of the λ_i = sum of the total variance in the data matrix over all time and space locations and (2) the determinant of matrix $\hat{\boldsymbol{\Sigma}}_\mathbf{X} = |\hat{\boldsymbol{\Sigma}}_\mathbf{X}| = \prod\lambda_i$. Also, a singular matrix must have at least one nonzero eigenvalue, and in fact the number of nonzero eigenvalues is equal to the rank of the matrix.

Singular Value Decomposition (SVD) Approach

Consider the relationship shown in (15.18b,c). Actually solving for eigenvalues typically requires a matrix method such as singular value decomposition (**SVD**). Recall from (A.51), the relationship defining SVD, that any $n \times m$ matrix $\mathbf{A}$ $(n \geq m)$ can be decomposed as

$$\mathbf{A}_{nm} = \mathbf{U}_{nn}\mathbf{S}_{nm}\mathbf{V}_{mm}^{\mathrm{T}}, \tag{15.19; A.51}$$

where $\mathbf{U}, \mathbf{V}^{\mathrm{T}}$ are orthogonal and $\mathbf{S}$ is a diagonal matrix. Therefore, if you perform SVD on your data matrix, $\mathbf{X}$, you obtain $\mathbf{X} = \mathbf{USV}^{\mathbf{T}}$. From this you can form the sample covariance matrix, $\hat{\boldsymbol{\Sigma}}_\mathbf{X}$, in the usual manner ($\mathbf{XX}^{\mathbf{T}}$), allowing

$$\mathbf{XX}^{\mathrm{T}} = \mathbf{USV}^{\mathrm{T}}\mathbf{VS}^{\mathrm{T}}\mathbf{U}^{\mathrm{T}} = \mathbf{USS}^{\mathrm{T}}\mathbf{U}^{\mathrm{T}}. \tag{15.20}$$

Calling $\mathbf{SS}^{\mathrm{T}}$, $\boldsymbol{\Lambda}$ (i.e., *the singular values in* $\boldsymbol{S}$ *are the square roots of the eigenvalues in* $\boldsymbol{\Lambda}$), gives

$$\mathbf{X}\mathbf{X}^{\mathrm{T}} = \mathbf{U}\boldsymbol{\Lambda}\mathbf{U}^{\mathrm{T}}. \tag{15.21}$$

This is of the same form as (15.15a), $\mathbf{A} = \mathbf{E}\boldsymbol{\Lambda}\mathbf{E}^{\mathrm{T}}$. It is a unique composition, thus: $\mathbf{S}\mathbf{S}^{\mathrm{T}} = \boldsymbol{\Lambda}$ represents the eigenvalues of $\mathbf{X}\mathbf{X}^{\mathrm{T}}$, and $\mathbf{U}$ (containing the **left singular vectors**) represent the eigenvectors (the $\mathbf{E}$ in (15.15a)) of $\mathbf{X}\mathbf{X}^{\mathrm{T}}$, our sample covariance matrix.

Likewise, if we form our covariance matrix between space series (instead of between time series), then, $\hat{\boldsymbol{\Sigma}}_{\mathbf{X}} = \mathbf{X}^{\mathrm{T}}\mathbf{X}$, and a repeat of the above procedure shows that

$$\mathbf{X}^{\mathrm{T}}\mathbf{X} = \mathbf{V}\mathbf{S}^{\mathrm{T}}\mathbf{U}^{\mathrm{T}}\mathbf{U}\mathbf{S}\mathbf{V}^{\mathrm{T}} = \mathbf{V}\mathbf{S}^{\mathrm{T}}\mathbf{S}\mathbf{V}^{\mathrm{T}}. \tag{15.22}$$

So in this case, $\mathbf{V}$ (the **right singular vectors**) contain the eigenvectors of the covariance matrix formed as $\mathbf{X}^{\mathrm{T}}\mathbf{X}$, and $\mathbf{S}^{\mathrm{T}}\mathbf{S}$ contain the eigenvalues.

*Thus, we can apply SVD analysis directly to the original (non-square) data matrix, **X**, and obtain the eigenvectors and eigenvalues without ever having to actually construct the sample covariance matrix*:

$$\mathbf{X}_{nm} = \mathbf{U}_{nn}\mathbf{S}_{nm}\mathbf{V}_{mm}^{\mathrm{T}}. \tag{15.23}$$

Furthermore, the SVD is a very robust decomposition method that typically provides a solution even if the covariance matrix is only marginally stable.

EOF Orthogonal Fitting of a Straight Line

In Chapter 5 (Smoothed Curve Fitting), the simple solution of the orthogonal fitting a straight line in a regression via EOFs was deferred to this chapter after explanation of the fundamentals of the EOF methodology. In §5.6 a more complicated approach was given where you rotate the axes to explain the most variance. In this section, the angle θ ($\hat{\theta}$ the amount remaining in §5.6) is found via EOF analysis. Here, I provide the solution as shown by Jackson (1991).

First, place your n (x,y) data pairs in a $n \times 2$ matrix $\mathbf{X}$:

$$\mathbf{X} = \begin{bmatrix} x_1 & y_1 \\ x_2 & y_2 \\ \vdots & \\ x_n & y_n \end{bmatrix}. \tag{15.24}$$

Now perform a singular-value decomposition on the $\mathbf{X}$ matrix, giving

$$\mathbf{X}_{nm} = \mathbf{U}_{nn}\mathbf{S}_{nm}\mathbf{V}_{mm}^{\mathrm{T}}. \tag{15.25}$$

The columns of $\mathbf{V}$ contain the eigenvectors ($\mathbf{e_1}$, $\mathbf{e_2}$) of the system, in this case, $m = 2$.

$$\mathbf{V} = \begin{bmatrix} e_{11} & e_{12} \\ e_{21} & e_{22} \end{bmatrix} \tag{15.26}$$

Now compute the direction cosines associated with each element of $\mathbf{e_1}$. So,

$$\cos \mathrm{e}_{11} = \theta_{11}. \tag{15.27}$$

θ_{11} gives the angle of rotation from the abscissa, and θ_{21} the angle of rotation from the ordinate ($\theta_{11} + \theta_{21} = 90°$). The orthogonal best-fit line is the line with the θ_{11} rotation.

15.4 Principal Components (PC)

Principal Components Analysis (PCA) is a specific case of EOF analysis and is one of the classic methods of multivariate statistical analysis. There is considerable material written about it in the multivariate statistics texts (e.g., Cooley and Lohnes, 1971; Anderson, 1984). It involves particular linear combinations of the original random variables such that they are orthogonal to one another and represent the greatest amount of variance (of the original set) in the fewest number of terms.[15] Thus, if we have *p* variables, the principal components are *p* new variables that sum together to interpolate the original data set, with the special characteristics that the new *p* variables are not correlated to one another, the first variable describes the most variance of the original dataset, the second variable describes the next-most variance (i.e., the most variance of the remaining undescribed variance), etc. Recall that in spectral analysis, the power of each fitted cosine was proportional to the amount of variance in the original time series that it described. The same is true here (as will be shown), only in this case the new variables are ordered so that each one describes less variance than the preceding ones. In this manner, it is hoped that we might find some small subset of new variables that describe most of the original variance in the data set, and that we only need to retain this minimal number of new variables (principal components) to retain the series signal.

For example, if we have measured, over some region of the ocean, the surface temperature, salinity, CO_2, freons, He^3, and eight other tracers, we might expect that the distribution of several of these are related to one another (i.e., are distributed via the same physical processes and have similar source/sink functions). Consequently, we might expect that some linear combination of the related variables will describe a significant amount of the variance displayed by the entire dataset, say 50 percent of the observed variance. Another combination may describe another large fraction of the variance, say 40 percent, while the remaining 11 combinations each describe only some small fraction (that we hope to be noise). Therefore, by retaining just the first two principal components, we are describing 90 percent of the total variance through only two orthogonal composite variables, instead of the original bulky 13 variables. Those variables were not independent, and thus were dragging along a lot of redundant information. Furthermore, since these two principal components are orthogonal, they may each represent different aspects of the local physics in the area, and by working strictly with them we might isolate the two physical processes that seem to dominate the distribution of all of these parameters in the region.[16]

[15] These represent the different variables that are related through a joint probability distribution, or in terms of the physical problem of interest, related through the phenomenon being studied.

[16] Often, factor analysis, another variant of EOF analysis, is more appropriate for this particular purpose.

Finally, we need not make any assumptions about the nature of the multivariate distribution between the various variables we will be examining. However, if the variables do obey a multivariate normal distribution, then they can be completely described by their means and variance-covariance structure (i.e., via a covariance matrix). Since the method of principal components makes use of this covariance matrix, for multivariate normal data, given zero mean (or normalizing as such), we are completely explaining the joint distribution of the variables through this analysis.

15.4.1 Definitions

Consider defining n new random variables, $\mathrm{P_i}$, as the linear combination, or composite, of n original random variables (or locations), each variable contained within a vector $\mathbf{X}_\mathrm{i}$.[17] Written in terms of random variables, $\mathrm{P_i}$ and X_i (i.e., not in terms of vectors, though each random variable can also be thought of as a vector),

$$\begin{aligned} \mathrm{P}_1 &= a_{11}X_1 + a_{12}X_2 + \ldots + a_{1n}X_m \\ \mathrm{P}_2 &= a_{21}X_1 + a_{22}X_2 + \ldots + a_{2n}X_m \\ &\vdots \\ \mathrm{P}_n &= a_{n1}X_1 + a_{n2}X_2 + \ldots + a_{nn}X_m. \end{aligned} \tag{15.28}$$

Each of these P_i is called a **principal component**. If the data represent a space-time field so that at different locations we have measurements at some specific times, then it is easy to see that $\mathrm{P_i}$ is the x value estimated to exist at one of the locations, given its multivariate correlation to the values taken at the same time at all of the other n space locations. In other words, we are saying that the value at any one location is correlated to the values observed at the other locations (the X_i), and we might use that covarying information to predict a single composite time series containing all of the covarying information. That is, *a principal component is a composite variable formed by simple multivariate linear regression that contains in this one composite time series all of the shared information, eliminating the need for all of the covarying time series carrying replicate information.* The various time elements in each principal component are given as

$$\begin{aligned} \mathrm{p}_{\ell 1} &= a_{\ell 1}x_{11} + a_{\ell 2}x_{21} + \ldots + a_{\ell n}x_{n1} \\ \mathrm{p}_{\ell 2} &= a_{\ell 1}x_{12} + a_{\ell 2}x_{22} + \ldots + a_{\ell n}x_{n2} \\ &\vdots \\ \mathrm{p}_{\ell m} &= a_{\ell 1}x_{1m} + a_{\ell 2}x_{2m} + \ldots + a_{\ell n}x_{nm} \end{aligned}$$

[17] This is essentially the reverse operation of what we did previously with the eigenvectors, in which case we combined eigenvectors to reproduce the data (15.18c). Here, we wish to construct p new variables as linear combinations of the original data. However, it is easily seen that the principal components defined here are comparable to the coefficients arising from the multiplication of the data matrix with the eigenvector matrix, shown in (15.18b).

$$\mathrm{p}_{\ell j} = \sum_{k=1}^{n} a_{\ell k} x_{kj}. \tag{15.29a}$$

In other words, the weights used to linearly combine each random variable change only with the random variable, not with the value of the random variable.

In matrix notation,

$$\begin{aligned} \mathbf{p}_1^{\mathrm{T}} &= \mathbf{a}_1^{\mathrm{T}}\mathbf{X} \\ \mathbf{p}_2^{\mathrm{T}} &= \mathbf{a}_2^{\mathrm{T}}\mathbf{X} \\ &\vdots \\ \mathbf{p}_n^{\mathrm{T}} &= \mathbf{a}_n^{\mathrm{T}}\mathbf{X}, \end{aligned} \tag{15.29b}$$

or

$$\mathbf{P} = \mathbf{AX}, \tag{15.29c}$$

where $\mathbf{a}_i$ and $\mathbf{x}_i$ are column vectors, with $\mathbf{p}_\mathbf{i}^\mathbf{T} = [\mathrm{p}_{i1}\ \mathrm{p}_{i2} \ldots \mathrm{p}_{in}]$, $\mathbf{a}_\mathbf{i}^\mathbf{T} = [\mathrm{a}_{i1}\ \mathrm{a}_{i2} \ldots \mathrm{a}_{in}]$ and $\mathbf{x}_\mathbf{i}^\mathbf{T} = [x_{i1}\ x_{i2} \ldots x_{in}]$, **P, A** and **X** are matrices with each row, i, containing the $\mathbf{p}_\mathrm{i}^\mathrm{T}$, $\mathbf{a}_\mathrm{i}^\mathrm{T}$ and $\mathbf{x}_\mathrm{i}^\mathrm{T}$, respectively.

Therefore, if the n data variables are $u_i, v_i \ldots, z_i$, then in matrix form,

$$\mathbf{X} = \begin{bmatrix} u_1 & u_2 & & u_m \\ v_1 & v_2 & \ldots & v_m \\ \vdots & & & \\ z_1 & z_2 & & z_m \end{bmatrix},$$

where each row represents a different variable and each column represents the time at which that variable was measured. Alternatively, if the data consist of only one variable, say temperature as a function of spatial location, then we might represent the data as before (m times and n spatial locations),

$$\mathbf{X} = \begin{bmatrix} x_{11} & x_{12} & & x_{1m} \\ x_{21} & x_{22} & \ldots & x_{2m} \\ \vdots & & & \\ x_{n1} & x_{n2} & & x_{nm} \end{bmatrix},$$

where each column is again time, but this time the rows represent different spatial locations. In either case, the data matrix can be represented via time-synchronous column vectors (each *element* in a vector $\mathbf{f}_j$ representing a different spatial location (= map) or a different variable, but always measured at the same times (j = 1 through m), in this example where time is the independent variable of interest), as

$$\mathbf{X} = \begin{bmatrix} \\ \mathbf{f}_1 & \mathbf{f}_2 & \ldots & \mathbf{f}_m \\ \\ \end{bmatrix}. \tag{15.29d}$$

For any of the representations of the data matrix, (15.29) can be written in terms of the principal components, i.e., the $\mathbf{P}_\ell$ vectors, as[18]

$$\mathbf{P}_\ell = a_{\ell 1}\mathbf{X_1} + a_{\ell 2}\mathbf{X_2} + \ldots + a_{\ell n}\mathbf{X}_n = \mathbf{X}^{\mathrm{T}}\mathbf{a}_\ell. \tag{15.30a}$$

Or,

$$\mathbf{P} = \begin{bmatrix} \mathbf{p}_1^{\mathrm{T}} \\ \mathbf{p}_2^{\mathrm{T}} \\ \vdots \\ \mathbf{p}_n^{\mathrm{T}} \end{bmatrix} = \begin{bmatrix} a_{11} & a_{12} & & a_{1n} \\ a_{21} & a_{22} & \ldots & a_{2n} \\ \vdots & & & \\ a_{n1} & a_{n2} & & a_{nn} \end{bmatrix} \begin{bmatrix} u_1 & u_2 & & u_m \\ v_1 & v_2 & \ldots & v_m \\ \vdots & & & \\ z_1 & z_2 & & z_m \end{bmatrix} \quad n \text{ variables over } m \text{ times}$$

$$= \begin{bmatrix} a_{11} & a_{12} & & a_{1n} \\ a_{21} & a_{22} & \ldots & a_{2n} \\ \vdots & & & \\ a_{n1} & a_{n2} & & a_{nn} \end{bmatrix} \begin{bmatrix} x_{11} & x_{12} & & x_{1m} \\ x_{21} & x_{22} & \ldots & x_{2m} \\ \vdots & & & \\ x_{n1} & x_{n2} & & x_{nm} \end{bmatrix} \quad \begin{array}{l} 1 \text{ variable measured at} \\ n \text{ locations over } m \text{ times} \end{array}$$

$$= \begin{bmatrix} a_{11} & a_{12} & & a_{1n} \\ a_{21} & a_{22} & \ldots & a_{2n} \\ \vdots & & & \\ a_{n1} & a_{n2} & & a_{nn} \end{bmatrix} \begin{bmatrix} \\ \mathbf{f}_1 & \mathbf{f}_2 & \ldots & \mathbf{f}_m \\ \\ \end{bmatrix} \quad \begin{array}{l} n - \text{length vectors} \\ \text{measured over } m \text{ times} \end{array}$$

which is seen by examining the elements of the ℓ th PC (each element representing a different time, j),

$$\mathbf{p}_\ell = \begin{bmatrix} p_{\ell 1} \\ p_{\ell 2} \\ p_{\ell 3} \\ \vdots \\ p_{\ell m} \end{bmatrix} = a_{\ell 1} \begin{bmatrix} x_{11} \\ x_{12} \\ x_{13} \\ \vdots \\ x_{1m} \end{bmatrix} + a_{\ell 2} \begin{bmatrix} x_{21} \\ x_{22} \\ x_{23} \\ \vdots \\ x_{2m} \end{bmatrix} + \ldots + a_{\ell n} \begin{bmatrix} x_{n1} \\ x_{n2} \\ x_{n3} \\ \vdots \\ x_{nm} \end{bmatrix}, \tag{15.30b}$$

directly giving the form of (15.30a) and (15.29a). In terms of multivariate series, $u, v, \ldots, z$, (15.30b) is

[18] This is a direct application of the reversal rule of transposed products to (15.29b). That is, if $\mathrm{p}^{\mathrm{T}} = \mathrm{a}^{\mathrm{T}}\mathrm{x}$, then the transpose of this is $\mathrm{p} = (\mathrm{a}^{\mathrm{T}}\mathrm{x})^{\mathrm{T}} = \mathrm{x}^{\mathrm{T}}\mathrm{a}$.

$$\mathbf{p}_\ell = \begin{bmatrix} p_{\ell 1} \\ p_{\ell 2} \\ p_{\ell 3} \\ \vdots \\ p_{\ell m} \end{bmatrix} = a_{\ell 1} \begin{bmatrix} u_1 \\ u_2 \\ u_3 \\ \vdots \\ u_m \end{bmatrix} + a_{\ell 2} \begin{bmatrix} v_1 \\ v_2 \\ v_3 \\ \vdots \\ v_m \end{bmatrix} + \ldots + a_{\ell n} \begin{bmatrix} z_1 \\ z_2 \\ z_3 \\ \vdots \\ z_m \end{bmatrix}. \tag{15.30c}$$

So, each principal component is an m-length vector containing the weighted linear sum of the n variables in the data set at each of m times. The weighting applied to each variable to form the new composite variable, the principal component, doesn't change with time as shown in Figure 15.2 (i.e., the variables are added as a weighted sum to form the principal component at time t = j = 1, using the same weights used to form the principal component at time t = j = m), so the weights are static. *This is a multivariate regression problem: the p_i are predicted as linearly related to the various x_i values. However, the principal components change in time, since the x_i variables themselves change with each time increment, even though the weights do not. The regression reduces the interdependencies between the variables to new variables that contain the interdependencies as the weighted sum in a single new variable.*

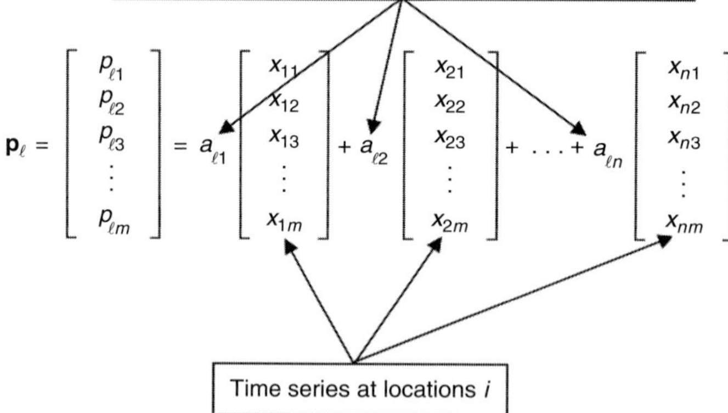

Figure 15.2 Description of each part of Principal Component. From this, it is clear that the PC has the same units as the x variable.

15.4.2 Solving for the PC Coefficients

Constrained Regression

We wish to determine those coefficient values in matrix **A** of (15.29c) such that the principal components in **P** sequentially describe the maximum amount of variance in **X** or its residual (i.e., that variance not described by the previous PCs). That is, the new variables, the PCs, are to be constructed so that they are uncorrelated to one another; the first one describes a maximum amount of variance in the original dataset; the second PC describes the most amount of variance in the data that were not described by the first PC (i.e., the residual variance). Each PC describes the maximum amount of the remaining variance until eventually the nth PC describes the final residual so that the n PCs together describe all of the variance in **X**. This solution is accomplished via constrained regression using the method of Lagrange multipliers, as presented in Chapter 6.

First, consider the variance of the principal components, P_i (which measures *variance in time, for the data matrix as defined here*). This is done directly from the formula developed earlier for the variance of a general linear function, where if

$$P_i = y(X_i) = \sum_{i=1}^{m} a_i X_i \tag{15.31a}$$

$$\mathbf{p}_i = \mathbf{X}^{\mathrm{T}}\mathbf{a}_i, \tag{15.31b}$$

then, according to (2.53a),

$$\mathrm{Var}[\mathrm{y}(\mathrm{X})] = \sum_{i=1}^{m}\sum_{j=1}^{m} a_i a_j \mathrm{Cov}[\mathrm{X}_i \mathrm{X}_j], \tag{15.32a}$$

or in matrix form,

$$= \mathbf{a}^{\mathrm{T}}\Sigma\mathbf{a}, \tag{15.32b}$$

where $\mathbf{\Sigma}$ is the (true) covariance matrix of X.[19]

Thus, for the ith PC,

$$\mathrm{Var}[\mathrm{P}_i] = \mathbf{a}_i^{\mathrm{T}}\Sigma\mathbf{a}_i. \tag{15.33a}$$

Similarly, the covariance between the different P_i is given as

$$\mathrm{Cov}[\mathrm{P}_i\mathrm{P}_j] = \mathbf{a}_i^{\mathrm{T}}\Sigma_{ij}\mathbf{a}_j. \tag{15.33b}$$

For principal components analysis, we wish to determine the a_{ij} coefficients in **A** such that $\mathrm{Cov}[\mathrm{P}_i\mathrm{P}_j] = 0$ when $i \neq j$ (i.e., making the principal components uncorrelated), and such that the variance described by P_1 is greater than that described by P_2, or any of the other PCs. The ith principal component describes the ith-largest variance, so $\mathrm{Var}[\mathrm{P}_1] >$

[19] This will be estimated by the sample covariance matrix, $\hat{\mathbf{\Sigma}}_{\mathbf{X}}$.

$\mathrm{Var}[P_2] > \mathrm{Var}[P_3]$, etc. We wish to maximize $\mathrm{Var}[P_1]$, or $\mathbf{a}_1^T\boldsymbol{\Sigma}\mathbf{a}_1$ Also, in order to preserve the total variance of the original system, we must normalize the coefficients such that their sum of squares (i.e., their scaling of the variance) is $\mathbf{a}_1^T\mathbf{a}_1 = 1$.

First PC

We can determine the coefficients of the first principal component, a_{1j}, via the same techniques we have previously used. Specifically, we wish to find those coefficients that maximize the variance, $\mathbf{a}_1^T\boldsymbol{\Sigma}\mathbf{a}_1$, subject to the constraint that $\mathbf{a}_1^T\mathbf{a}_1 - 1 = 0$.[20] This is done using the method of Lagrange multipliers, where the Lagrangian equation (to be maximized with the constraint) is given as[21]

$$\begin{aligned} L(\mathbf{a}_1) &= f(\mathbf{a}_1) - \lambda\theta(\mathbf{a}_1) \\ &= \mathbf{a}_1^T\boldsymbol{\Sigma}\mathbf{a}_1 - \lambda(\mathbf{a}_1^T\mathbf{a}_1 - 1) \\ &= \sum_{j=1}^{n}\sum_{i=1}^{n} a_{1i}a_{1j}\sigma_{\sigma ij} - \lambda\left(\sum_{j=1}^{n} a_{1j}^2 - 1\right). \end{aligned} \tag{15.34}$$

At the maximum, the derivative of the Lagrangian with respect to the unknown coefficient is zero,

$$\frac{\partial L}{\partial a_1} = \frac{\partial}{\partial a_{1j}}\left(\sum_{j=1}^{n}\sum_{i=1}^{n} a_{1i}a_{1j}\sigma_{ij}\right) - \lambda\frac{\partial}{\partial a_{1j}}\left(\sum_{j=1}^{n} a_{1j}^2 - 1\right) = 0, \tag{15.35a}$$

which, recalling that $\partial(\mathbf{x}^T\mathbf{A}\mathbf{x})/\partial\mathbf{x} = 2\mathbf{A}\mathbf{x}$, reduces to

$$2\boldsymbol{\Sigma}\mathbf{a}_1 - 2\lambda\mathbf{a}_1 = 0 \tag{15.35b}$$

or

$$(\boldsymbol{\Sigma} - \lambda I)\mathbf{a}_1 = 0. \tag{15.35c}$$

What great luck![22] *This potential mess just reduced to the standard form of the eigenvalue problem given in* (15.1b)*, where the eigenvalues correspond to the covariance matrix,* $\boldsymbol{\Sigma}$. That is, we can use this equation to solve for the Lagrange multiplier, or now clearly the eigenvalue λ, since the nontrivial solution only exists by making $(\boldsymbol{\Sigma} - \lambda \mathrm{I})$ singular.

The covariance matrix $\boldsymbol{\Sigma}$ must be estimated from the data via the sample covariance matrix, $\hat{\boldsymbol{\Sigma}}_X = \mathbf{X}\mathbf{X}^T$, and the eigenvalues determined from this, so the determinant is

$$|\hat{\boldsymbol{\Sigma}}_x - \lambda I|\mathbf{a}_1 = 0, \tag{15.36}$$

[20] This constraint, that the coefficients conserve variance, is simply that $\mathbf{a}_1^T\mathbf{a}_1 = 1$, but rewritten as required by the method of Lagrange multipliers: that is, $g(a) = 0$, or $\mathbf{a}_1^T\mathbf{a}_1 - 1 = 0$.

[21] In anticipation of results soon to follow, we define the Lagrange multiplier as $-\lambda$. Since this is an arbitrary constant to be solved for, this does nothing more than add later convenience.

[22] Or was it magic?

which is a polynomial in λ of order n, with n roots. Since $\hat{\boldsymbol{\Sigma}}_{\mathbf{X}}$ is positive definite, these roots, the eigenvalues, are all real and positive, and we order them such that

$$\lambda_1 \geq \lambda_2 \geq \lambda_3 \geq \ldots \geq \lambda_n \,. \tag{15.37}$$

We can now introduce the constraint equation to complete the constrained maximization problem by multiplying (15.35b) by $\mathbf{a}_1^{\mathrm{T}}$ to give

$$\mathbf{a}_1^{\mathrm{T}}\hat{\boldsymbol{\Sigma}}_X\mathbf{a}_1 \;-\; \boldsymbol{\lambda}\mathbf{a}_1^{\mathrm{T}}\mathbf{a}_1 \;=\; 0. \tag{15.38a}$$

Since $\mathbf{a}_1^{\mathrm{T}}\mathbf{a}_1 = 1$, introduction of this constraint gives

$$\mathbf{a}_1^{\mathrm{T}}\hat{\boldsymbol{\Sigma}}_X\mathbf{a}_1 \;=\; \lambda. \tag{15.38b}$$

As shown in (15.33a), $\mathrm{Var}[P_1] = \mathbf{a}_1^{\mathrm{T}}\boldsymbol{\Sigma}\mathbf{a}_1$, and since (15.38b) shows that $\mathbf{a}_1^{\mathrm{T}}\hat{\boldsymbol{\Sigma}}_X\mathbf{a}_{11} \;=\; \lambda$ then,

$$\mathrm{Var}_{sample}[\mathrm{P}_1] \;=\; \lambda. \tag{15.39}$$

The most variance is described by λ_1, since we ordered the eigenvalues in decreasing size (15.37). Therefore, the constraint effectively informs us to use the first eigenvalue in order to maximize the variance, though it explicitly forces us to work with normalized eigenvectors. So the first principal component, that which describes the maximum variance of linear combinations of the original variables, is given by

$$\mathbf{p}_1^{\mathrm{T}} \;=\; \mathbf{a}_1^{\mathrm{T}}\mathbf{X}, \tag{15.40}$$

where a_1 is that (normalized) eigenvector that satisfies $(\hat{\boldsymbol{\Sigma}}_{\mathbf{X}} - \lambda_1\mathbf{I})\mathbf{a}_1 \;=\; 0$. That is, the *coefficients* required to make this new composite variable, or PC are in fact given by the first *eigenvector* of $\hat{\boldsymbol{\Sigma}}_{\mathbf{X}}$.

Second PC

For the second principal component, we now wish to find that coefficient vector (i.e., those values of $\mathbf{a}_2^{\mathrm{T}}$) that maximizes the amount of remaining variance undescribed by the first principal component. The original constraint that the weights conserve variance must now be subjected to an additional constraint, that this principal component is uncorrelated to the first one. So,

$$\max\{\mathrm{Var}[\mathrm{P}_2](a_2)\} \;=\; \max\{\mathbf{a}_2^{\mathrm{T}}\boldsymbol{\Sigma}\mathbf{a}_2\}, \tag{15.41a}$$

subject to

$$\mathbf{a}_2^{\mathrm{T}}\mathbf{a}_2 \;=\; 1 \tag{15.41b}$$

$$\mathrm{Cov}[\mathrm{P}_1\mathrm{P}_2] \;=\; \mathbf{a}_2^{\mathrm{T}}\boldsymbol{\Sigma}_{12}\mathbf{a}_1 \;=\; 0. \tag{15.41c}$$

Here, the matrix form of $\mathrm{Cov}[P_1P_2]$ in (15.41c) is found by expanding the sample covariance sum in terms of these two components and substituting in the vector expression from (15.29b):

$$\begin{aligned}\mathrm{Cov}[\mathbf{P}_1\mathbf{P}_2] = \hat{\boldsymbol{\Sigma}}_{\mathbf{P}_{12}} &= \frac{1}{n-1}\sum_{i=1}^{n}\mathbf{P}_{2i}\mathbf{P}_{1i} \\ &= \frac{1}{n-1}\sum_{i=1}^{n}(\mathbf{a}_2^{\mathrm{T}}\mathbf{X}_i)(\mathbf{a}_1^{\mathrm{T}}\mathbf{X}_i)^{\mathrm{T}} \\ &= \frac{1}{n-1}\sum_{i=1}^{n}\mathbf{a}_2^{\mathrm{T}}\mathbf{X}_i\mathbf{X}_i^{\mathrm{T}}\mathbf{a}_1 \\ &= \mathbf{a}_2^{\mathrm{T}}\hat{\boldsymbol{\Sigma}}_{\mathbf{X}}\mathbf{a}_1.\end{aligned} \tag{15.42}$$

Further manipulation of this product[23] shows that it is equivalent to $\mathbf{a}_2^{\mathrm{T}}\mathbf{a}_1\lambda_1$, and thus, since this is equal to zero, reconfirms that the eigenvectors must be uncorrelated to one another.

Again we resort to constrained maximization using two Lagrange multipliers (λ and κ, the latter given as 2κ for convenience) to maximize the Lagrangian equation:

$$\begin{aligned}L(\mathbf{a}_2) &= \mathrm{f}(\mathbf{a}_2) - \lambda\theta(\mathbf{a}_2) - 2\kappa\theta(\mathbf{a}_2, \mathbf{a}_1) \\ &= \mathbf{a}_2^{\mathrm{T}}\boldsymbol{\Sigma}_{\mathbf{X}}\mathbf{a}_2 - \lambda(\mathbf{a}_2^{\mathrm{T}}\mathbf{a}_2 - 1) - 2\kappa\mathbf{a}_2^{\mathrm{T}}\boldsymbol{\Sigma}_{\mathbf{X}}\mathbf{a}_1 \\ &= \sum_{j=1}^{n}\sum_{i=1}^{n}a_{2i}a_{2j}\sigma_{ij} - \lambda\left(\sum_{j=1}^{n}a_{2j}^2 - 1\right) - 2\kappa\sum_{j=1}^{n}\sum_{i=1}^{n}a_{2i}a_{1j}\sigma_{ij}.\end{aligned} \tag{15.43}$$

At the maximum,

$$\frac{\partial L}{\partial \mathbf{a}} = \frac{\partial}{\partial a_{2j}}\left(\sum_{j=1}^{n}\sum_{i=1}^{n}a_{2i}a_{2j}\sigma_{ij}\right) - \lambda\frac{\partial}{\partial a_{2j}}\left(\sum_{j=1}^{n}a_{2j}^2 - 1\right) - 2\kappa\frac{\partial}{\partial a_{2j}}\left(\sum_{j=1}^{n}\sum_{i=1}^{n}a_{2i}a_{1j}\sigma_{ij}\right) = 0, \tag{15.44a}$$

which reduces to

$$2\boldsymbol{\Sigma}_{\mathbf{X}}\mathbf{a}_2 - 2\lambda\mathbf{a}_2 - 2\kappa\boldsymbol{\Sigma}_{\mathbf{X}}\mathbf{a}_1 = 0. \tag{15.44b}$$

Pre-multiplying through by $\mathbf{a}_1^{\mathrm{T}}$ gives

$$\mathbf{a}_1^{\mathrm{T}}\boldsymbol{\Sigma}_{\mathbf{X}}\mathbf{a}_2 - \lambda\mathbf{a}_1^{\mathrm{T}}\mathbf{a}_2 - \kappa\mathbf{a}_1^{\mathrm{T}}\boldsymbol{\Sigma}_{\mathbf{X}}\mathbf{a}_1 = 0, \tag{15.44c}$$

allowing application of the second constraint equation, (15.41c), which indicates that the first two terms are zero (see (15.33) and the statement immediately below it). Also, we know already from (15.38b) that the third term in (15.44c), $\mathbf{a}_1^{\mathrm{T}}\hat{\boldsymbol{\Sigma}}_{\mathbf{X}}\mathbf{a}_1 = \lambda_1$, is nonzero, so the only way (15.44c) can be satisfied, i.e., that

[23] This result comes almost directly from application of (15.15b), with the minor difference that here we need only examine one vector within the eigenvector matrix, **E**.

$$\kappa\lambda_1 = 0 \tag{15.44d}$$

is if $\kappa = 0$.

Finally, multiply (15.44b) by $\mathbf{a}_2^{\mathrm{T}}$, after setting $\kappa = 0$, to give

$$\mathbf{a}_2^{\mathrm{T}}\boldsymbol{\Sigma}_{\mathbf{X}}\mathbf{a}_2 - \lambda\mathbf{a}_2^{\mathrm{T}}\mathbf{a}_2 = 0. \tag{15.44e}$$

Application of the first constraint (second constraint has already been applied above), that $\mathbf{a}_2^{\mathrm{T}}\mathbf{a}_2 = 1$, yields

$$\mathbf{a}_2^{\mathrm{T}}\boldsymbol{\Sigma}_{\mathbf{X}}\mathbf{a}_2 = \lambda. \tag{15.44f}$$

As before, the variance of P_2 is equal to the above product, and thus λ_2 is that eigenvalue (the variance of P_2) that describes the maximum amount of as-yet undescribed variance. Therefore, as with the first principal component, the second one is simply determined in an analogous manner, where

$$\mathbf{p}_2^{\mathrm{T}} = \mathbf{a}_2^{\mathrm{T}}\mathbf{X}, \tag{15.45}$$

with $\mathbf{a}_2^{\mathrm{T}}$ being that normalized eigenvector that coincides with the λ_2 eigenvalue.

General Solution

This procedure is continued for each of the principal components, yielding each as simply products of the eigenvectors of the sample covariance matrix constructed from the original data matrix. The complete set of n principal components is represented as

$$\mathbf{P} = \mathbf{E}^{\mathrm{T}}\mathbf{X} \tag{15.46a}$$

or

$$\mathbf{p}_\ell^{\mathrm{T}} = \sum_{j=1}^{m} \mathbf{f}_j\mathbf{e}_\ell^{\mathrm{T}}, \tag{15.46b}$$

where the eigenvectors in $\mathbf{E}$ are derived from the sample covariance matrix, $\mathbf{X}\mathbf{X}^{\mathrm{T}}/(n-1)$. So,

$$\mathbf{P} = \begin{bmatrix} \mathbf{p}_1^{\mathrm{T}} \\ \mathbf{p}_2^{\mathrm{T}} \\ \vdots \\ \mathbf{p}_n^{\mathrm{T}} \end{bmatrix} = \begin{bmatrix} e_{11} & e_{12} & & e_{1n} \\ e_{21} & e_{22} & \cdots & e_{2n} \\ \vdots & & & \\ e_{n1} & e_{n2} & & e_{nn} \end{bmatrix} \begin{bmatrix} x_{11} & x_{12} & & x_{1m} \\ x_{21} & x_{22} & \cdots & x_{2m} \\ \vdots & & & \\ x_{n1} & x_{n2} & & x_{nm} \end{bmatrix}, \tag{15.46c}$$

where the rows of the coefficient matrix ($\mathbf{e}$, or $\mathbf{a}$ in (15.36)), are the eigenvectors, $\mathbf{e}_\ell$, derived from the covariance matrix,

$$\mathbf{P} = \begin{bmatrix} \mathbf{p}_1^{\mathrm{T}} \\ \mathbf{p}_2^{\mathrm{T}} \\ \vdots \\ \mathbf{p}_n^{\mathrm{T}} \end{bmatrix} = \begin{bmatrix} \mathbf{e}_1^{\mathrm{T}} \\ \mathbf{e}_2^{\mathrm{T}} \\ \vdots \\ \mathbf{e}_n^{\mathrm{T}} \end{bmatrix} \begin{bmatrix} x_{11} & x_{12} & & x_{1m} \\ x_{21} & x_{22} & \cdots & x_{2m} \\ \vdots & & & \\ x_{n1} & x_{n2} & & x_{nm} \end{bmatrix} = \begin{bmatrix} \mathbf{e}_1^{\mathrm{T}} \\ \mathbf{e}_2^{\mathrm{T}} \\ \vdots \\ \mathbf{e}_n^{\mathrm{T}} \end{bmatrix} \begin{bmatrix} \mathbf{f}_1 & \mathbf{f}_2 & \cdots & \mathbf{f}_m \end{bmatrix}. \tag{15.46d}$$

Here (15.46d) *is identical to* (15.18b), *but now we see that the coefficient vectors in C are the principal components, P.*

When decomposing the data matrix using SVD, where:

$$\mathbf{X} = \mathbf{USV}^{\mathrm{T}} \tag{15.46e}$$

this entire procedure collapses to being: eigenvector matrix is: $\mathbf{U}$, eigenvalues: $\mathbf{S}^{\mathbf{T}}\mathbf{S}$ (square of the singular values) and principal components: $\mathbf{SV}^{\mathbf{T}}$.

15.4.3 Interpretation and Use

Since *the variance of each principal component is equal to its corresponding eigenvalue, the total sample variance of the original data set, s_x^2, is equal to the sum of the eigenvalues*, and thus equal to the total variance of the principal components, i.e., $s_{\mathrm{X}}^2 = \mathrm{tr}(\hat{\boldsymbol{\Sigma}}_{\mathrm{X}}) = \mathrm{tr}(\boldsymbol{\Lambda})$. The relative fraction of variance described by each principal component is equal to $\mathrm{Var}[\mathrm{P}_i]/\mathrm{tr}(\boldsymbol{\Lambda})$, or $\lambda_i/s_{\mathrm{X}}^2$.

The principal components are nothing more than a new system of independent (uncorrelated) vectors that describe the complete dataset in the EOF domain, but in such a manner that most of the variance of the data is now contained in the first few principal components. Such a system is demonstrated in Figure 15.3. The similarity between the functions in (15.18d) and (15.46b), or their matrix equivalents in (15.18c) and (15.46a), are analogous to Fourier transform pairs. That is, (15.18c) acts like an inverse Fourier transform in which the original series can be recovered, given knowledge of the coefficients of the orthogonal basis function (i.e., this is the interpolation of the original data set, given the coefficients to the orthogonal interpolant). Equation (15.46b) serves the role of a Fourier transform, which converts the data from the standard time (or other, e.g., space) domain to the new, EOF domain. In a Fourier transform, the new domain was the frequency domain, given the EOFs that are harmonic sines and cosines of frequency f_i.

In this case, the analysis found the shape based on the data (i.e., didn't assume cycles) that shared a common time history. The shape sharing this common time history is the eigenvector (the EOF) and how its magnitude varies in time is given by the PC (multiply the eigenvector for any particular time by the value of the PC at that time, giving a map at that time, scaled to capture the shape at that time).

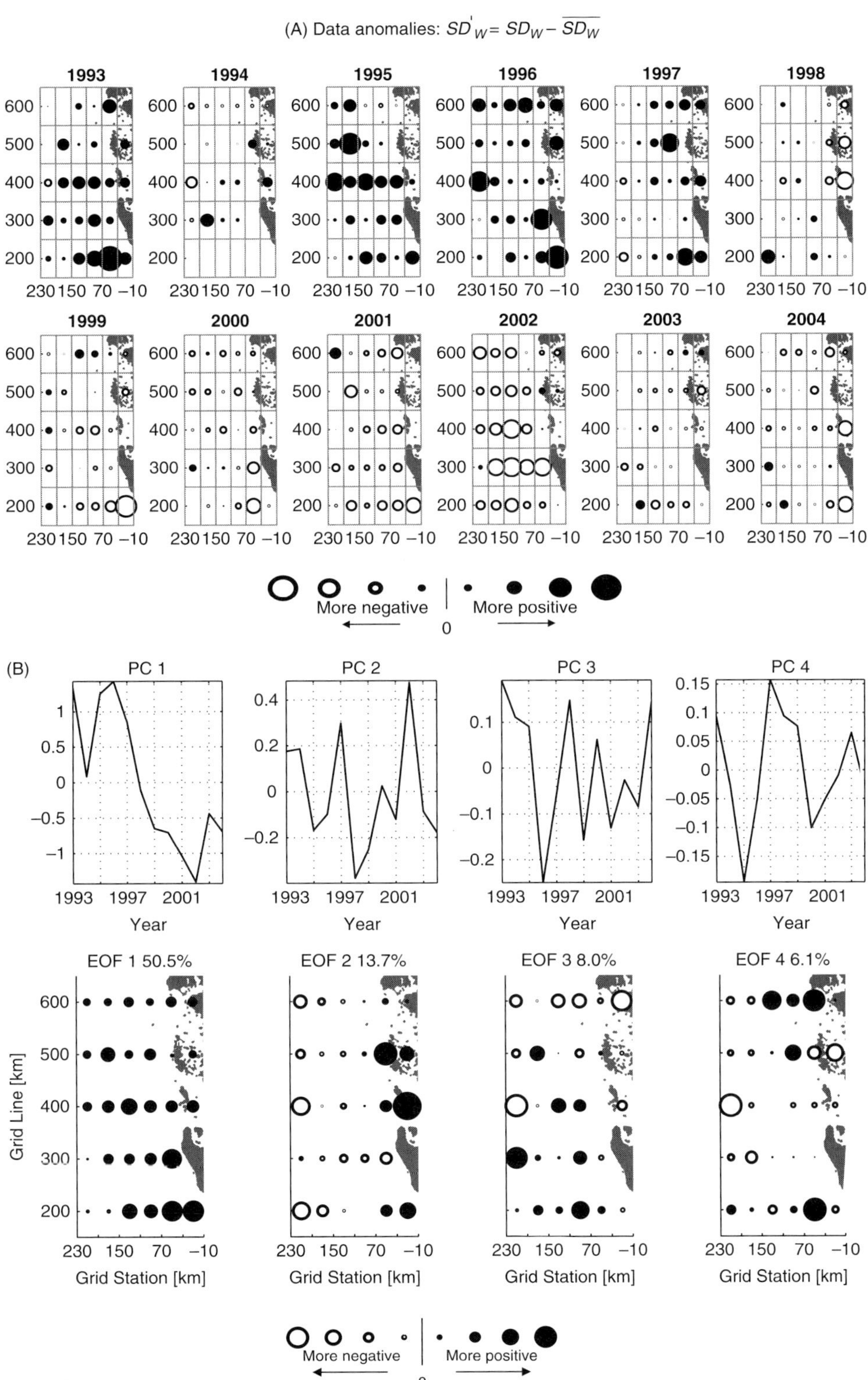

Figure 15.3 For an example of Figure 15.1, **(A)** anomaly maps for salt deficit, SD'_W ; **(B)** here are the first four modes of the EOF analysis. The first mode describes ~51 percent of the total temporal variance

Note that the principal components, i.e., the vectors in **P**, are orthogonal to each other, hence from (15.18a), $\mathbf{PP}^{\mathrm{T}} = \Lambda$, so they have completely eliminated any correlation between transformed variables. That is, both the eigenvectors and their time-varying coefficients, the PCs, are *orthogonal* sets.

Variance Decomposition

One can now consider the principal components in terms of their description of the variance in the original data, stored in **X**. That is, from (15.15a) and (15.18a) we see that the variance of the data, as described in the covariance matrix, $\mathbf{\Sigma}$, or the sample covariance matrix, $\hat{\mathbf{\Sigma}}_{\mathbf{X}}$, can be completely described by the eigenvectors and eigenvalue matrix as

$$\mathbf{XX}^{\mathrm{T}} = \hat{\mathbf{\Sigma}}_{\mathbf{X}} = \mathbf{S} = \mathbf{E\Lambda E}^{\mathrm{T}}, \tag{15.47a}$$

or, in summation form,

$$\begin{aligned} &= \sum_{i=1}^{n} \lambda_i \mathbf{e}_i \mathbf{e}_i^{\mathrm{T}} \\ &= \lambda_1 \mathrm{e}_1 \mathrm{e}_1^{\mathrm{T}} + \lambda_2 \mathrm{e}_2 \mathrm{e}_2^{\mathrm{T}} + \ldots + \lambda_n \mathrm{e}_n \mathrm{e}_n^{\mathrm{T}}, \end{aligned}$$

the sum of outer products, each giving a matrix allowing

$$= \mathbf{S}_1 + \mathbf{S}_2 + \ldots + \mathbf{S}_n \tag{15.47b}$$

$$\text{where, } \mathbf{S}_i = \lambda_i \mathbf{e}_i \mathbf{e}_i^{\mathrm{T}}.$$

Therefore, this same variance can be described (factored) as orthogonal, and thus additive, submatrices, $\mathbf{S}_i$, where each submatrix is the covariance structure of the particular eigenvector.

Equation (15.47) is known as a **spectral decomposition** of a matrix **A**.

If we determine the minimum number of components, p, that can be used to describe the maximum amount of variance in the dataset, then (15.47) can be written in terms of the sum of those components that contain most of the variance of the data, known as the **component theory, signal** or **information matrix**, $\hat{\mathbf{S}}$ and the sum of those remaining $n-p$ components describe little variance, and thus represent the **residual matrix, error** or **noise matrix**, $\widetilde{\mathbf{S}}$. In this case, the total variance is now described as

$$\begin{aligned} \hat{\mathbf{\Sigma}}_{\mathbf{X}} &= \hat{\mathbf{S}} + \widetilde{\mathbf{S}} \\ \mathbf{S}_i &= \lambda_i \mathbf{e}_i \mathbf{e}_i^{\mathrm{T}} \end{aligned} \tag{15.48}$$

Caption for Figure 15.3 (cont.)

(explaining why changes in PC1 were apparent from visual inspection of the anomaly). Second mode still captures a fair amount of variance, but third and fourth modes less so – in fact, according to North's rule of thumb (discussed below), these two modes are too similar to separate and should be combined into a single mode.

So, we are stating that the EOF basis determined, given the principal component approach, can be reduced from n vectors to p vectors with minimal loss of signal. That is, we have found those eigenvectors that maximize the variance in the least number of basis vectors.

How many modes (EOFs plus their principal components) you keep can be estimated in a variety of ways. One of the most general rules of thumb is to only keep those that describe more than $1/n$ of the total variance. That is, if there are 100 variables, then if the variance was evenly distributed throughout all of them, each variable would explain 1 percent of the total variance. After transforming to principal components, we only wish to preserve those components that explain more than the average variable would. This rule is typically prone to err on the side of keeping too many coefficients.

Alternatively, you may make use of a **scree plot**, in which the fraction of variance explained by each component is plotted as a function of the component. This is exactly analogous to making a power spectrum plot in spectral analysis, only in this case one only retains those components that lie above some obvious background "flat" level (or "noise floor") that the higher-order terms display. This test works well in those situations where the transition from the information matrix to the noise matrix is rather abrupt.

Finally, you can apply a more rigorous test in the form of a sphericity test. In this case, after extracting each eigenvalue the test checks to see if the remaining, unexplained covariance matrix is significantly different from singular. That is, is the determinant indistinguishable from zero? (Recall that the determinant is equal to the sum of the remaining eigenvalues.)

You can perform a similar type of test initially, before you do the principal components analysis, to determine whether the variables are not already orthogonal to one another. In this case, the test is to see if the original covariance matrix is significantly different from the identity matrix. If not, then the variables are already as independent as they can get, and additional decomposition into an EOF basis would be meaningless (and highly sensitive to small changes in the data).

As previously stated, this is essentially a generalized Fourier approach. In fact, it was shown earlier (§D15.1) that sines and cosines (at harmonic frequencies) are *natural* eigenvectors of systems undergoing translation of the origin and of time-invariant linear systems. Thus, we would decompose our data into a set of orthogonal sines and cosines (the eigenvectors, $\mathbf{E}^T$, in (15.18b)), and the variance would be spread over all frequencies, with the dominant variance captured by those harmonics that are closest to any true frequencies in the data. On the other hand, in principal components, we have found a different eigenvector basis, the one that is customized to the particular data set so that it captures the maximum variance in the fewest terms (which can be sines and cosines).

Finally, regarding the roots of the characteristic equation. If some of the eigenvalues are zero, it is an indication that the original matrix from which the eigenvalues are determined is singular. The true rank (number of independent data series) of the matrix is equal to the number of nonzero roots. For covariance matrices such as we are dealing with here, this should rarely if ever occur.

You may want to standardize the various data vectors going into the **A** matrix before you compute the EOFs. This leads to a correlation matrix instead of covariance matrix.

For the covariance matrix, the data with the largest absolute variance may simply dominate the entire decomposition, since the other vectors may not contribute much to the overall variance. For the correlation matrix, all data vectors contribute more equally to the overall variance. The latter is important if one wants to find the relative contributions of each data set. The former is important if the decomposition of the total variance is what is desired.

If there are multiple roots of the same values, the matrix is called **defective**, and the resulting eigenvectors will not be distinct. This is fairly uncommon, as it indicates that the variance of two different variables is identical. It is also uncommon computationally due to round-off errors. Fortunately, in symmetric matrices, this does not hurt us specifically, since the eigenvectors are still orthogonal, but in either case, the easiest solution is simply to perturb one of the numbers contributing to the multiple eigenvalue slightly, since the associated eigenvectors will most likely need to be ignored, given the fact that the uncertainties in the eigenvalues scale with the separation of the roots, and in this case the roots will be so close as to render the use of the associated eigenvectors meaningless. The number of equal eigenvalues is called the degree of **multiplicity**.

Uncertainties in eigenvalues are generally treated via a rule of thumb from North et al. (1982):

$$\Delta\lambda = \lambda(2/n)^{1/2}. \tag{15.49}$$

Using (15.49), if $\Delta\lambda \geq \lambda_\ell - \lambda_{\ell-1}$, the two eigenvalues are too close together to separate. That is, uncertainty in one overlaps the value of its neighbor, so they are not distinct, and should be combined into a single mode.

Another related technique, **factor analysis**, differs from principal components in that while the latter is mainly interested in finding the fewest components that describe the most variance, the former is interested in finding those components that show the strongest degrees of correlation between the variables of the dataset. For example, it might collapse a set of 50 variables in 50 new components for which each component represents a natural clustering (as determined by correlation) of some subset of variables.

Figure 15.4 summarizes the above operations and results for the original Antarctic example presented in the previous figures.

15.5 Singular Spectrum Analysis (SSA)

Singular Spectrum Analysis (**SSA**; or "Singular Systems Analysis," in some circles) deals with the application of EOF analysis to single time series through a transform known as the Karhunen–Loéve transform. Its strength, as with EOF analysis in general, is that it allows you to find modes that are data adaptive (i.e., that fit the shape and internal consistency of the time series). Vautard et al. (1992; hereafter **VYG**) have provided an excellent review/survey of this topic from the

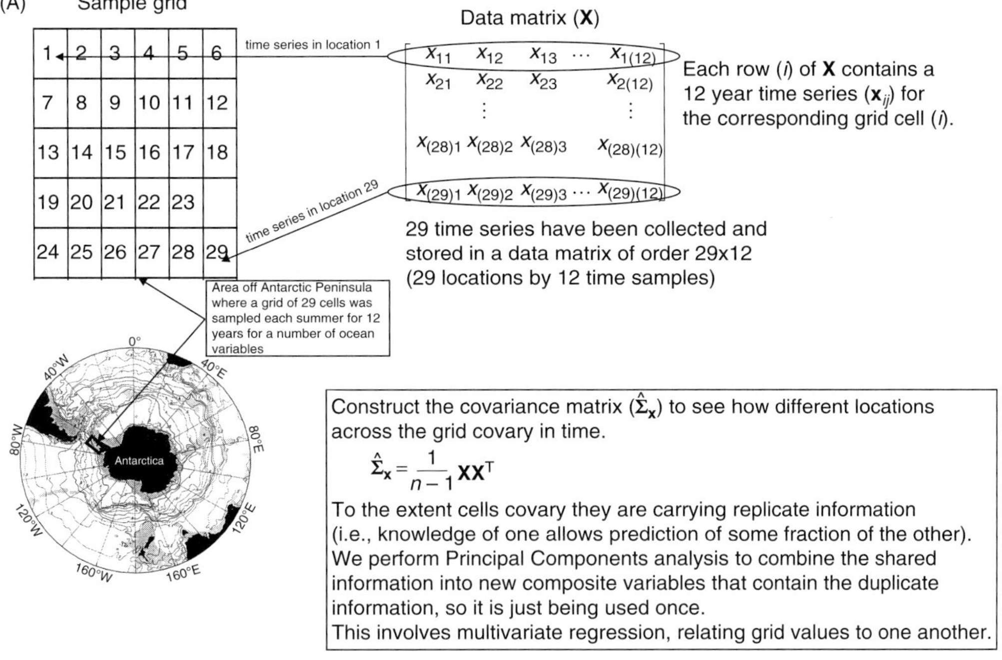

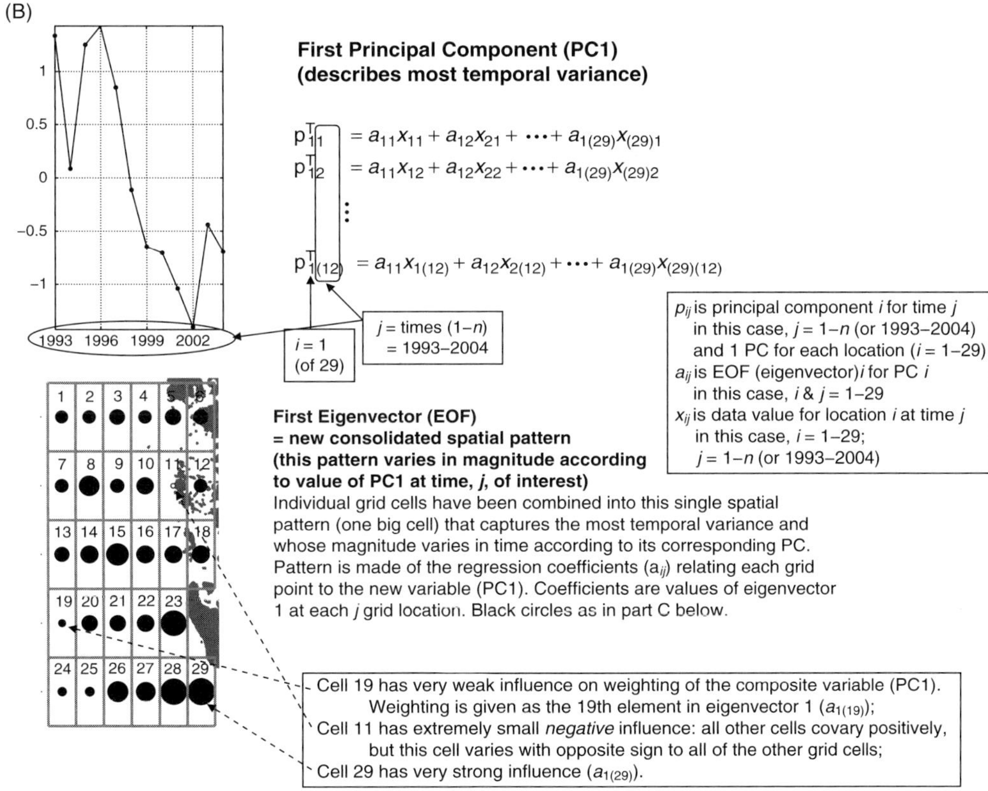

Figure 15.4 Overview of construction and interpretation of EOF analysis. **(A)** Data set of anomalies of a measure of upper-ocean salinity relative to the average (climatology), showing the sample grid and corresponding data matrix; **(B)** showing the first EOF and PC explaining the most variance;

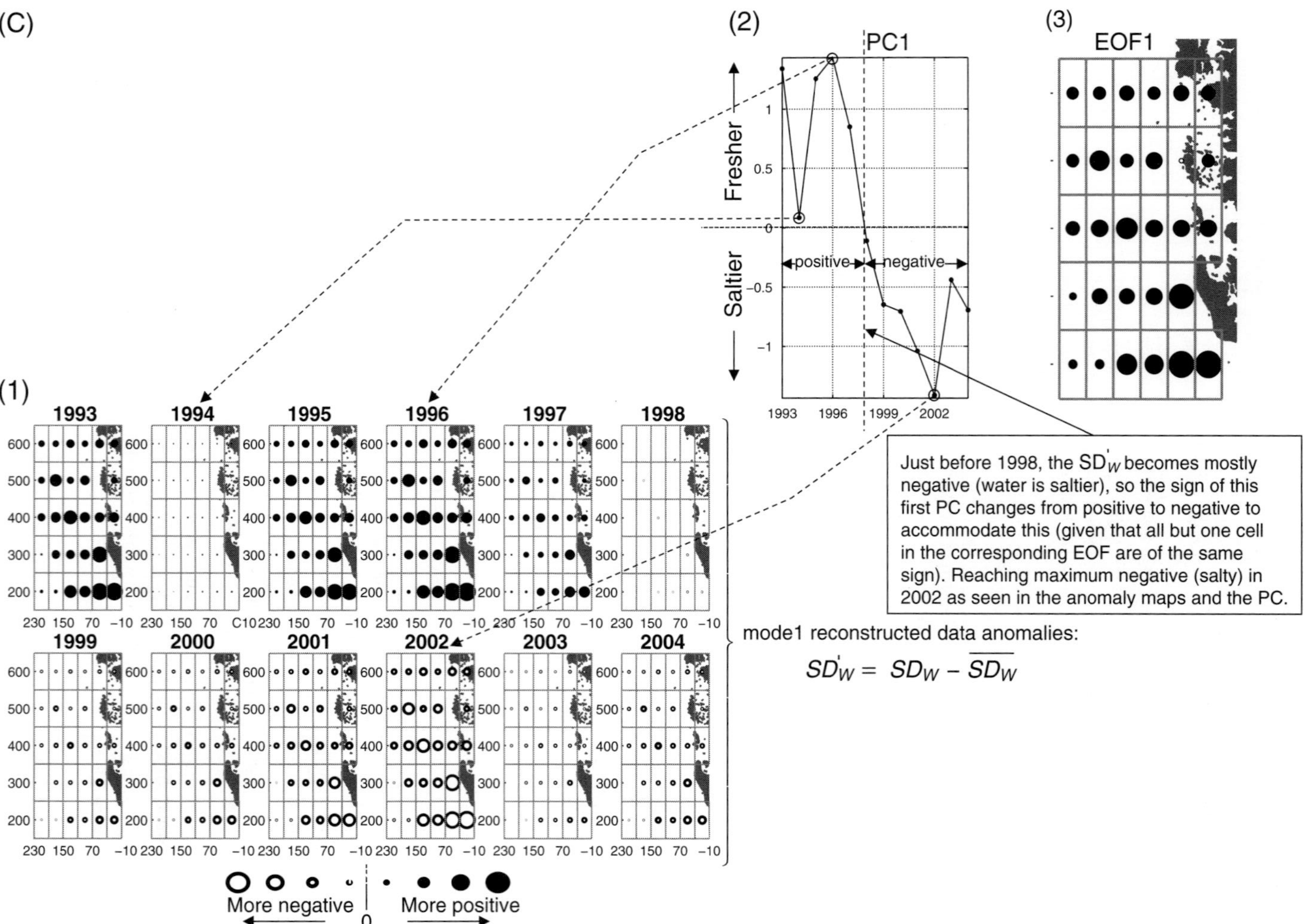

Figure 15.4 (cont.) (**C**) complete description of the parts for the first PC and EOF pattern, particularly showing that the EOF pattern is multiplied by the value of the PC to reconstruct each particular year for this mode (note that when the PC crosses 0, the open circles in EOF1 become solid and vice versa for the solid circles);

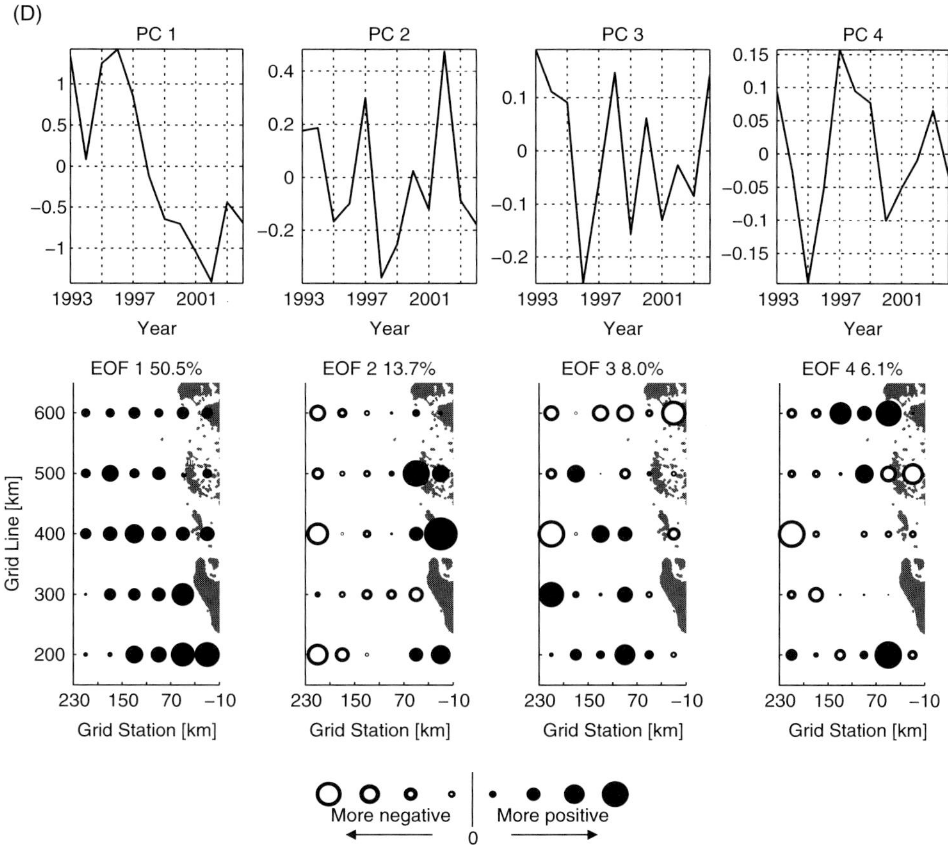

Figure 15.4 (cont.) **(D)** the first four principal components and eigenvectors and their percentage of the total variance described. This is a beautiful system: the *i*th eigenvector (EOF *i*) defines the pattern for that mode via the eigenvector elements, a_{ij} (the size of the circles in the figure are simply the a_{ij} values), and those elements are also the weights for the PC. Wow, we set up a potentially complicated constrained optimization problem and end up with an eigen-problem where everything is given by the eigenvectors!

standpoint of extracting signal (stochastic and deterministic) from noise in a data series. In keeping with our earlier attempt to decompose our data into a signal (or here, *intrinsic*) component and noise (*extrinsic*) component, VYG show how this is accomplished via SSA. Specifically, the lowest-order modes will represent the signal component, and how to determine this number of modes is discussed at length in VYG (though that concentrates more heavily on chaos theory than on what is relevant here).

Trajectory Matrix

A fundamental difference from EOF analysis is the construction of the sample covariance matrix to be decomposed for the eigenstructure. Dynamic system analysis suggests that we can segment the data into m offset lagged segments. As long as the time series is weakly stationary (mean and variance, or in this case, autocovariance, which actually implies it is ergodic), each segment will contain the same relationship between lagged values, and it will be this lagged covariance that the method will maximize in the PCs.

For the 2-dimensional fields used with EOF analysis, you had a data matrix, with each row containing a time series with as many rows as stations, or grid points, sampled (n, the number of EOF modes). For a single time series, that matrix is a row vector. Constructing the sample covariance matrix for this in the usual manner would clearly be inadequate. Instead, we form a **trajectory matrix** whereby we examine the time series as offset (lagged) copies of itself. To construct this, you must choose a "window length," or **embedding dimension**, m, equivalent to the number of spatial locations in the EOF example (i.e., the size of the maps and number of eigenvectors).

Assuming stationarity and ergodicity, you can produce a much better estimate of the covariance matrix through the autocovariance serial product. This provides a reasonable estimate through at least some lags. This resolution is dictated by the embedding dimension m, which determines the number of modes that will be constructed to fit the data. It also sets the duration in time of the window over which the time series will be examined, so if you have an idea of the time scale of the signal you seek, then m should be chosen to resolve this (this will become more obvious below). Figure 15.5 demonstrates the construction of a trajectory matrix.

For standard spatiotemporal data, each *column* in the data matrix represented a "map" for a single time (i.e., the value of the variable at a single time for each station location). Then the eigenvectors give new composite variables showing spatial patterns that contain the most shared variance, with the PCs showing how that new pattern changes in time. For SSA, the trajectory contains m rows, equivalent to m spatial locations, but in this case the "maps" are segments in time over which the new composite variables will contain the most shared variance, with the PCs showing how those segments vary in time over the full extent of the time series.

After choosing m, remove the last $m - 1$ points from the times series, giving an effective length $n = N - m + 1$, and then make an $m \times n$ matrix, where each row is an n-length segment of the time series, but each row shifts this series, beginning and ending one data point later. So, for a time series of $N = 20$ points that runs from 1 to 20, we choose for an example an embedding dimension of $m = 5$ and form our $m \times n$ data matrix as shown in the example of Figure 15.5.

$\mathbf{XX^T}/n$ gives an $m \times m$ sample autocovariance matrix ($\hat{\mathbf{\Sigma}}_X$), for which each diagonal, away from the principal diagonal, gives the lagged autocovariance, with the lag equal to the number of diagonals away from the principal diagonal.

(A) For a series of length N:

$x_t = x_1 x_2 x_3 \cdots x_N$

for an embedding dimension m, construct m single-lag offset segments (length $n = N - m + 1$):

$$x_{t1} = x_1 x_2 x_3 \cdots x_n$$
$$x_{t2} = x_2 x_3 x_4 \cdots x_{n+1}$$
$$x_{t3} = x_3 x_4 x_5 \cdots x_{n+2}$$
$$\vdots$$
$$x_{tm} = x_{69} x_{70} x_{71} \cdots x_{n+m-1(=N)}$$

$$X_M = \begin{bmatrix} x_1 & x_2 & x_3 & \cdots & x_{n+0} \\ x_2 & x_3 & & & x_{n+1} \\ x_3 & & & & x_{n+2} \\ \vdots & & & & \vdots \\ x_m & & & & x_{n+(m-1)} \end{bmatrix} \text{ } m \text{ offset segments}$$

N–m+1 length

$$\Sigma_{mm} = \mathbf{X}_{mn}\mathbf{X}_{nm}^{T} = \begin{bmatrix} \sigma_{11}^2 & \sigma_{12} & \cdots & \sigma_{1m} \\ & \sigma_{22}^2 & & \\ \vdots & & & \vdots \\ \sigma_{m1} & & & \sigma_{mm}^2 \end{bmatrix}$$

(B)

For a series of length $N = 20$:

$x_t = x_1\ x_2\ x_3 \cdots x_{17}\ x_{18}\ x_{19}\ x_{20}$

An embedding dimension $m = 5$, construct m single lag offset segments of length $n = N - m + 1$:

$$\begin{matrix} x_1 & x_2 & x_3 & x_4 & x_5 & x_6 & x_7 & x_8 & x_9 & x_{10} & x_{11} & x_{12} & x_{13} & x_{14} & x_{15} & x_{16} \\ x_2 & x_3 & x_4 & x_5 & x_6 & x_7 & x_8 & x_9 & x_{10} & x_{11} & x_{12} & x_{13} & x_{14} & x_{15} & x_{16} & x_{17} \\ x_3 & x_4 & x_5 & x_6 & x_7 & x_8 & x_9 & x_{10} & x_{11} & x_{12} & x_{13} & x_{14} & x_{15} & x_{16} & x_{17} & x_{18} \\ x_4 & x_5 & x_6 & x_7 & x_8 & x_9 & x_{10} & x_{11} & x_{12} & x_{13} & x_{14} & x_{15} & x_{16} & x_{17} & x_{18} & x_{19} \\ x_5 & x_6 & x_7 & x_8 & x_9 & x_{10} & x_{11} & x_{12} & x_{13} & x_{14} & x_{15} & x_{16} & x_{17} & x_{18} & x_{19} & x_{20} \end{matrix}$$

$$X_M = \begin{bmatrix} x_1 & x_2 & x_3 & \cdots & x_{(N-m)+1} \\ x_2 & x_3 & & & x_{(N-m)+2} \\ x_3 & & & & x_{(N-m)+3} \\ \vdots & & & & \vdots \\ x_m & & & & x_{(N-m)+m} \end{bmatrix} \text{ } m \text{ offset segments}$$

n (= $N - m + 1$) length

Trajectory matrix (X_M) is often defined as the transpose of what is shown here, which we prefer for consistency with our previous order of operations, so:

$$\mathbf{\Sigma_X} = \mathbf{XX}^{\mathrm{T}}, \text{ not } \mathbf{X}^{\mathrm{T}}\mathbf{X}$$

$$\mathbf{\Sigma}_{mm} = \mathbf{X}_{mn}\mathbf{X}_{nm}^{T} = \begin{bmatrix} \sigma_{11}^2 & \sigma_{12} & \cdots & \sigma_{1m} \\ & \sigma_{22}^2 & & \\ \vdots & & & \vdots \\ \sigma_{m1} & & & \sigma_{mm}^2 \end{bmatrix}$$

True covariance matrix is a **Toeplitz** matrix, showing the following symmetry:

$$\Sigma_{mm} = \begin{bmatrix} \sigma_X^2 & \gamma_1 & & \cdots & \gamma_m \\ \gamma_1 & \sigma_X^2 & \gamma_1 & & \\ \vdots & \gamma_1 & & & \vdots \\ & & & & \gamma_1 \\ \gamma_m & & & \gamma_1 & \sigma_X^2 \end{bmatrix}$$

(C)

For a series of length $N = 20$:

$x_t = x_1\ x_2\ x_3 \cdots x_{17}\ x_{18}\ x_{19}\ x_{20}$

An embedding dimension $m = 5$, construct m single lag offset segments of length $n = N - m + 1$:

$$\begin{matrix} x_1 & x_2 & x_3 & x_4 & x_5 & x_6 & x_7 & x_8 & x_9 & x_{10} & x_{11} & x_{12} & x_{13} & x_{14} & x_{15} & x_{16} \\ x_2 & x_3 & x_4 & x_5 & x_6 & x_7 & x_8 & x_9 & x_{10} & x_{11} & x_{12} & x_{13} & x_{14} & x_{15} & x_{16} & x_{17} \\ x_3 & x_4 & x_5 & x_6 & x_7 & x_8 & x_9 & x_{10} & x_{11} & x_{12} & x_{13} & x_{14} & x_{15} & x_{16} & x_{17} & x_{18} \\ x_4 & x_5 & x_6 & x_7 & x_8 & x_9 & x_{10} & x_{11} & x_{12} & x_{13} & x_{14} & x_{15} & x_{16} & x_{17} & x_{18} & x_{19} \\ x_5 & x_6 & x_7 & x_8 & x_9 & x_{10} & x_{11} & x_{12} & x_{13} & x_{14} & x_{15} & x_{16} & x_{17} & x_{18} & x_{19} & x_{20} \end{matrix}$$

↑ *Eigenvectors describe shape over segment*

↖ ↑ ↗ *PC gives amplitude of shape for each segment (i.e., captures local features)*

Each column of length m is effectively a "map," as was the case for the standard EOFs for our spatio-temporal example, but in this case each column represents a different segment of the full length time series from the first to last (N) point. So the eigenvectors will be patterns of length m, but in this case, not patterns in space, but rather patterns in time of length m. The PCs give how the amplitude of those patterns vary with the segment of time they represent.

Figure 15.5 **(A)** Generic example of trajectory matrix ($\mathbf{X}_{\mathrm{M}}$) construction and of covariance matrix for diagonalization. **(B)** Explicit example of trajectory matrix for $N = 20$ and $m = 5$. **(C)** Showing how embedding dimension relates to number of spatial locations (forming maps at particular times) in standard EOF analysis of spatio-temporal data, as described earlier.

You now diagonalize the covariance matrix as with EOFs, and then reconstruct the time series with any of the desired modes. Specifically,

$$\begin{aligned} \mathbf{E}^{\mathrm{T}}\boldsymbol{\Sigma}_X\mathbf{E} &= \boldsymbol{\Lambda} \\ \mathbf{E}^{\mathrm{T}}\mathbf{X}\mathbf{X}^{\mathrm{T}}\mathbf{E} &= \boldsymbol{\Lambda} \end{aligned} \tag{15.50a}$$

define

$$\begin{aligned} \mathbf{P}_{mn} &= \mathbf{E}_{mm}^{\mathrm{T}}\mathbf{X}_{mn} \\ \mathbf{X}_{mn} &= \mathbf{E}_{mm}\mathbf{P}_{mn}. \end{aligned} \tag{15.50b}$$

With SSA, you now have another difference from EOFs. If you simply project the eigenvectors on the PCs, you will reconstruct the time series of the trajectory matrix (i.e., missing m points and at the various different lags). In order to reconstruct the entire time series, VYG use least squares to fit the modes to the entire data set, allowing reconstruction of the entire time series using any set of modes, $R_{\mathfrak{I}}$ (e.g., if you wish to reconstruct the time series using only the signal modes, say, 1 through 4, then $R_{\mathfrak{I}}$ would be the first four eigenvectors):

$$\begin{aligned} (R_{\mathfrak{I}}X)_i &= \frac{1}{i}\sum_{k\in\mathfrak{I}}\sum_{j=1}^{i} p_{i-j+1}^{k}E_j^k & 1 \le i \le m-1 \\ (R_{\mathfrak{I}}X)_i &= \frac{1}{m}\sum_{k\in\mathfrak{I}}\sum_{j=1}^{m} p_{i-j+1}^{k}E_j^k & 1 \le i \le N-m+1. \\ (R_{\mathfrak{I}}X)_i &= N-i+1\sum_{k\in\mathfrak{I}}\;\sum_{j=1-N+m}^{m} p_{i-j+1}^{k}E_j^k & N-m+2 \le i \le N \end{aligned} \tag{15.51}$$

Picking Embedding Dimension

Picking the embedding dimension can proceed in a number of ways. EOFs provide for 2-dimensional data set patterns (EOFs) that share a common time history (PCs). Time one is showing the pattern of the entire spatial set (a map) at the first time (i.e., the first column of the data matrix), then the map at time two, etc. For SSA, the column length is dictated by m, so m serves to set the distance along the time series that will be fit as a pattern that is repeated in time with different amplitudes according to the PC. So, m should be chosen that is long enough to capture any changing features you wish to capture. Alternatively, you might simply use window closing – that is, choose an m that is likely too long, and one that is too short, and then some intermediate values, until you have a value where small changes in m yield similar modes.

We now apply this methodology to the LR04 record introduced previously in the text in Figure 15.6. First, this method should be ideal for identifying the stepwise trend displayed by the data, so perhaps after removing this trend we will be able to more definitively answer the original example question as to whether long-term climate shows

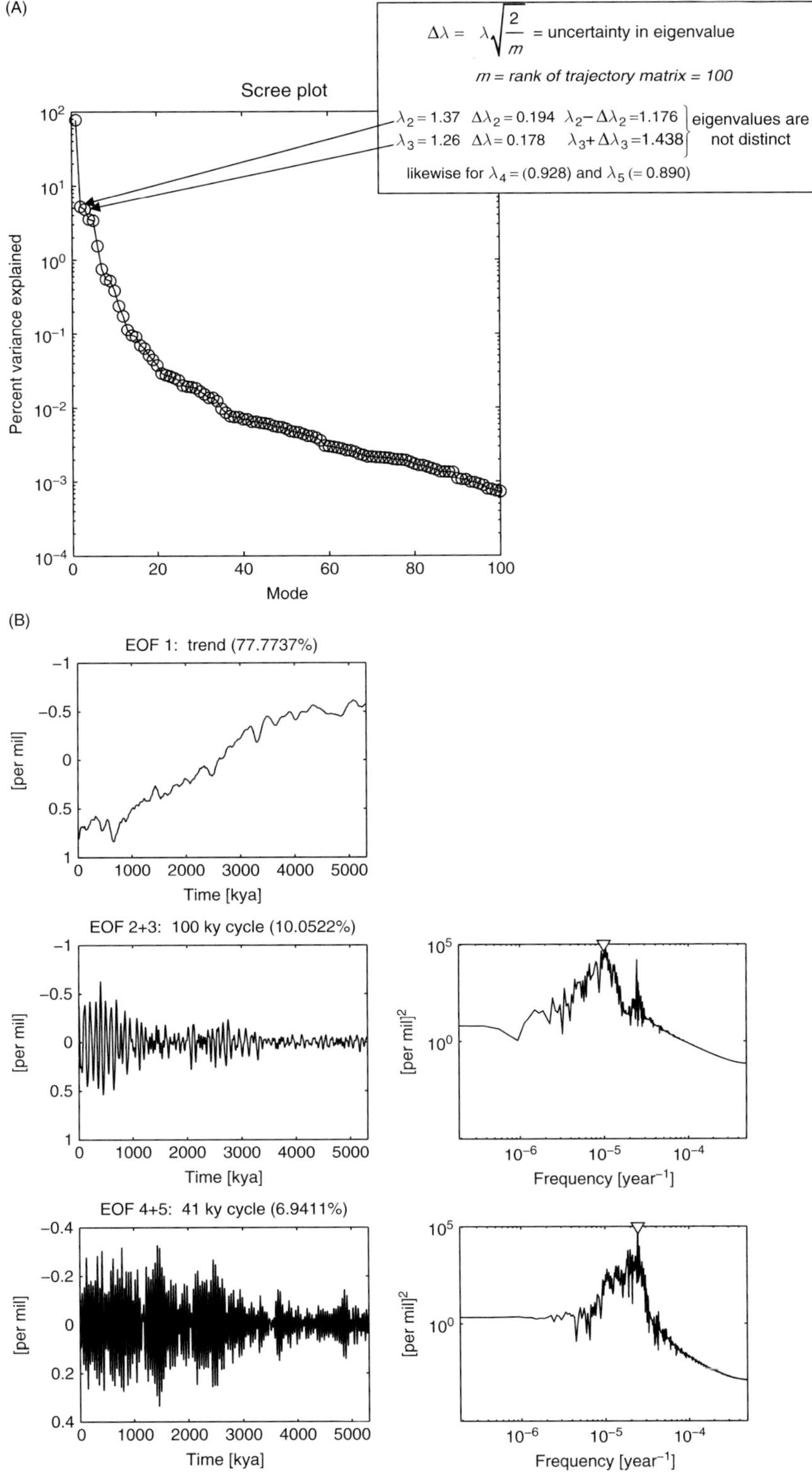

Figure 15.6 **(A)** Scree plot for LRO4 with $N = 5321$ and $m = 100$. Using North's rule of thumb, it is shown that eigenvalues two and three as well as four and five are too close together to separate, so those pairs are combined into a single reconstructed series. **(B)** First three reconstructed series. The first mode clearly captures the trend (describing 78 percent of the total LR04 variance), while the second series (modes two and three) appears to capture the "eccentricity" signal (highly variable, but centered on ~100 kyr cycle in the most recent million years) of the ice ages,

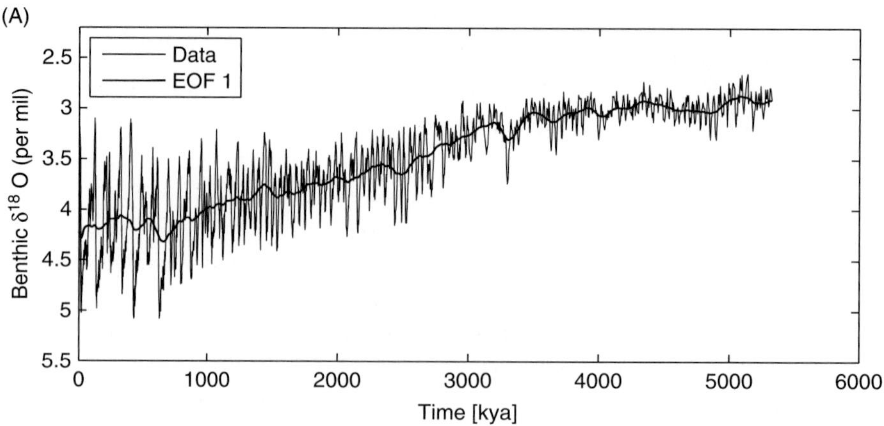

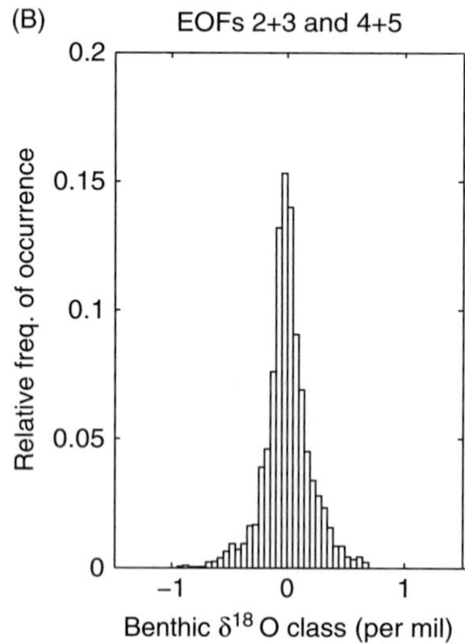

Figure 15.7 **(A)** Showing the fit of the first EOF to original data. It clearly captures the trend, as well as a few bumps here and there. **(B)** Probability mass function (PMF) showing distribution of glacial (cold; negative) and interglacial (warm; positive) periods of the series reconstructed from "signal modes" (two and three, and four and five), suggesting that natural climate variability is evenly distributed about a variable-mean state over the last 5 million years.

two modes (cold and warmer periods) or some sort of smoother continuum of climate. This curve contains $N = 5321$ data points, and we use $m = 100$. The eigenvalues are presented in a scree plot (Figure 15.6A) to gauge which ones dominate the variance.

Caption for Figure 15.6 (cont.)

describing ~10 percent of the variance (it is clearly more dominant in the most recent million years). The third series (modes four and five) captures the "obliquity" of Earth's axis (a ~41 kyr cycle) that captures only 7 percent of the variance (also, some of that signal is captured in the second series).

Clearly, the leading mode (mode 1) dominates, as one might expect from simple visual inspection of LR04. Modes 2 and 3 and modes 4 and 5 fail North's test (15.49) for distinct modes, and thus are combined into coupled reconstructed modes. Such pairs often represent regular oscillations in the data, the nonlinear equivalent of a Fourier sine-cosine pair.

When first introducing LR04 in Chapter 3 (example of PMF), we raised the question of whether climate was bimodal with two stable states: glacial (ice age) and interglacial (ice free). Our initial examination showed a hint of bimodality, but that was later seen to be a consequence of the warmer climate prior to 1 million years ago. We then investigated means of removing the cooling trend (climate cooling from 3 to 1 million years ago) by fitting various curves to LR04 to remove this long-term trend. With better fits to the curve, we starting seeing a loss of bimodality toward a more symmetrical distribution of climate. Here we have now found the best fit of the trend through the first SSA mode. When we remove this mode and reconstruct only the signal modes (two and three, and four and five) our PMF for this reconstructed signal version of LR04 show a nearly perfect Gaussian distribution (Figure 15.7).

15.6 Take-Home Points

1. EOF analysis is a means of reducing a large dependent dataset into a new set of variables that contain the information in a minimum of orthogonal variables, reducing all dependent information into these few new variables that are completely independent.
2. This information can be used for optimal interpolation, whereby if a data point is missing, the EOFs know how that point covaried with every other point in the grid, and that information can be used to estimate the most likely value of the missing datum.
3. By decomposing a data set into orthogonal functions with no covariability, one hopes that some of the empirical orthogonal functions (the EOFs) explain the underlying processes responsible for the random process (the times dataset).
4. An EOF decomposition of a time series (called Singular Spectrum Analysis; SSA) is functionally like a Fourier transform, but the orthogonal functions fitting the series are not harmonics or known mathematical functions; rather, they are functions that best fit the shape and internal consistency of the time series. This is a data-adaptive decomposition and is ideal for finding signal in noise.

15.7 Questions

Pencil and Paper Questions

1. a. Regarding EOF analysis:

    1. What does EOF mean?
    2. Describe your data matrix for such an analysis.

   b. Show the equations needed to do the EOF analysis, from manipulating the data matrix to making the PCs (don't need derivations; instead look for the fundamental matrix operations leading to and resulting in producing PCs).

2. a. Regarding SSA analysis:

    1. What does SSA mean?
    2. Describe your data matrix for such an analysis.

   b. Show the equations needed to do SSA, from manipulating the data matrix (give its name after describing) to making the PCs (don't need derivations; instead look for the fundamental matrix operations leading to and resulting in producing PCs).

Computer-Based Questions

1. Using LR04 and GISP2 truncated to an equal length in time, remove a nonstationary mean via SSA, then compute PSD for the first modes of each, and compare and contrast differences.

   Johnson and Wichern (1988) have an excellent discussion on the geometric interpretation of PCs (pages 49–50) and on spectral decomposition, just before that.

Appendix 1 Overview of Matrix Algebra

A1.1 Overview

Matrix algebra (also called *linear algebra*) is the branch of mathematics that deals with the algebraic manipulation of groups of numbers or equations stored in matrices. The advantage of matrix algebra lies in the fact that it provides concise and simple methods for manipulating large sets of numbers such as those frequently encountered in data analysis. As such, it is ideal for computers, and the compact form of matrices allows convenient notation for describing large tables of data. Matrix operations allow you to see complex relationships that would otherwise be obscured by the sheer size of the data (i.e., it aids clarification); most matrix manipulation involves just a few standard operations for which standard subroutines are readily available.

This review mainly introduces the terminology and tools available in matrix algebra. These will then be applied in numerous analysis techniques.

A1.2 Definitions

A **matrix** is a rectangular array of "elements" arranged in a series of m rows and n columns (like a table). A matrix is most commonly indicated by a boldface letter (e.g., **A, B, x**), an uppercase letter, or a letter with an underscore (e.g., $\underline{a}$ or $\tilde{a}$), in brackets (e.g., $[a]$, $[A]$), or with hat (e.g., $\hat{a}$). Here, matrices will consistently be indicated by a boldface letter.

Order of a matrix is the specification of the number of rows by the number of columns. Order for a matrix **A** is usually given as $n \times m$ (read, "n-by-m") or $\mathbf{A}_{n,m}$. Typically, the letters m, n and p are used to indicate order (row position is always given first).

Elements of a matrix **A** are given as a_{ij}, where the value of i specifies the row position and the value of j specifies the column position (the letters k and l are often used in place of i and j as well). An element can be a number (real or complex), algebraic expression or (with some restrictions) a matrix or matrix expression, or other component.

As a convention here (allowing all matrix manipulations to be presented in a common form), data matrices (i.e., matrices in which the elements represent data values) are constructed so that the rows of any one column contain the observations related to a particular variable or location. For example, the first column may contain all of the temperature observations, the second column the salinities, etc.; or the first column may contain the temperatures at location 1, the second column the temperatures from location 2, etc. If the data represent time or space series, typically the rows correspond to time or space. Since there are typically more observations than variables, such data matrices are usually rectangular, having more rows (n) than columns (m).

Column vector is a matrix containing only a single column of elements:

$$\tilde{a}_n = A = \begin{bmatrix} a_1 \\ a_2 \\ a_3 \\ \bullet \\ \bullet \\ \bullet \\ a_n \end{bmatrix}$$

Figure A1.1 Example of column vector containing n elements (dimension n).

Box A1.1 Example of Matrix Contents and Labeling

$$A := n\text{ rows} \overset{-\,m\text{ columns}\,-}{\begin{bmatrix} a_{11} & a_{12} & a_{13} \\ a_{21} & a_{22} & a_{23} \\ a_{31} & a_{32} & a_{33} \\ a_{41} & a_{42} & a_{43} \end{bmatrix}} = \overset{j=1\ j=2\ j=3}{\begin{bmatrix} 12 & 4 & 10 \\ 8 & 1 & 11 \\ 15 & 3 & 7 \\ 14 & 1 & 9 \end{bmatrix}} \begin{matrix} i=1 \\ i=2 \\ i=3 \\ i=4 \end{matrix}$$

This matrix, **A**, has order $n \times m = 4 \times 3$ (4 rows by 3 columns), the element $a_{23} = 11$, $a_{13} = 10$, etc.

Row vector is a matrix containing only a single row of elements:

$$\widetilde{a}_n = A = [a_1\ a_2\ a_3 \ldots a_n].$$

Vector size is simply the number of elements it contains (= n, in both examples above).

Note that some people do not explicitly differentiate between row and column vectors. Instead, they simply assume that the "vector" takes the form (row or column) required for the operation in which it is being used.

Summing vector, written as $\mathbf{1}_n$, is a vector of order n, in which all n elements are equal to the value 1. This vector is useful when the sum of a vector's elements is required.

Null matrix, written as $\mathbf{0}$ or $\mathbf{0}_{(n,m)}$, has all elements equal to 0. It plays the role of zero in matrix algebra.

Square matrix has the same numbers of rows as columns, so order is $n \times n$.

Diagonal matrix is a *square matrix* with zero in all positions except along the **principal diagonal** (or **lead diagonal**) as shown in (A1.1a):

$$\mathbf{D} = \begin{bmatrix} 3 & & \\ & 5 & \\ & & 5 \end{bmatrix} \tag{A1.1a}$$

or

$$\mathbf{A} = \mathbf{A}^T = \begin{bmatrix} 1 & 2 & 5 \\ 2 & 7 & 3 \\ 5 & 3 & 9 \end{bmatrix} = \text{symmetric}. \tag{A1.1b}$$

Those elements that do not lie on the principal diagonal are often called **off-diagonal** or nondiagonal elements.

Diagonal matrices are important for scaling rows or columns of other matrices. Note that some authors do not require that a diagonal matrix be square – they only require that the a_{ii} be nonzero and all other elements be zero.

Identity matrix (I) is a *diagonal matrix* with all of the nonzero elements – that is, the diagonal elements (the a_{ii}), equal to 1. Written as **I** or $\mathbf{I}_n$, it plays the role of 1 in matrix algebra. So,

$$\mathbf{AI} = \mathbf{IA} = \mathbf{A}. \tag{A1.2}$$

The order of **I** will always be taken to be conformable (this is discussed below) with that of the matrix it is multiplying.

Unit matrix is one in which *all* elements are equal to 1. This matrix is useful in some operations.

Lower-triangular matrix is a *square matrix* with all elements equal to zero *above* the principal diagonal:

$$\mathbf{A} = \begin{bmatrix} 1 & 0 & 0 \\ 3 & 7 & 0 \\ 2 & 3 & 9 \end{bmatrix} = \begin{bmatrix} 1 & & \\ 3 & 7 & \\ 2 & 3 & 9 \end{bmatrix} \tag{A1.3a}$$

or

$$a_{ij} = \begin{cases} 0 & i < j \\ \text{nonzero} & i \geq j. \end{cases} \tag{A1.3b}$$

Upper-triangular matrix is a *square matrix* with all elements equal to zero *below* the principal diagonal:

$$\mathbf{A} = \begin{bmatrix} 1 & 2 & 5 \\ 0 & 7 & 3 \\ 0 & 0 & 9 \end{bmatrix} = \begin{bmatrix} 1 & 2 & 5 \\ & 7 & 3 \\ & & 9 \end{bmatrix} \tag{A1.4a}$$

or

$$a_{ij} = \begin{cases} 0 & i > j \\ \text{nonzero} & i \leq j. \end{cases} \tag{A1.4b}$$

If one multiplies two triangular matrices of the same form, the result is a third matrix of the same form.

A **fully populated matrix** is a matrix with all of its elements nonzero.

A **sparse matrix** is a matrix with only a small proportion of its elements nonzero. From a computational standpoint, this type of matrix is often treated using techniques specially formulated to take advantage of the relatively few elements contained in the matrix.

A **scalar** is a number (or a matrix with a single element).

Matrix transpose (or transpose of a matrix) is obtained by interchanging the rows and columns of a matrix. So row i becomes column i and column j becomes row j; also, the order of the matrix is reversed from $n \times m$ to $m \times n$. The transpose of matrix $\mathbf{A}$ is denoted $\mathbf{A}^T$, or sometimes $\mathbf{A}'$:

$$A = \begin{bmatrix} 1 & 2 & 5 \\ 6 & 7 & 3 \\ 4 & 1 & 9 \end{bmatrix}; A^T = \begin{bmatrix} 1 & 6 & 4 \\ 2 & 7 & 1 \\ 5 & 3 & 9 \end{bmatrix}$$

So,

$$a_{ij} \rightarrow a_{ji}. \tag{A1.5b}$$

A diagonal matrix is its own transpose: $\mathbf{D}^T = \mathbf{D}$.

Symmetric matrix is a *square matrix* that is symmetrical about its principal diagonal, so

$$a_{ij} = a_{ji}. \tag{A1.6a}$$

A symmetrical matrix, $\mathbf{A}$, is equal to its own transpose:

$$\mathbf{A} = \mathbf{A}^T = \begin{bmatrix} 1 & 2 & 5 \\ 2 & 7 & 3 \\ 5 & 3 & 9 \end{bmatrix} = \text{symmetric.} \tag{A1.6b}$$

Skew symmetric matrix is a *square matrix* in which

$$a_{ij} = -a_{ij}, \tag{A1.7a}$$

so

$$\mathbf{A}^T = -\mathbf{A} \tag{A1.7b}$$

and

$$a_{ii} = 0. \tag{A1.7c}$$

That is, the principal diagonal elements are zero:

$$\mathbf{A} = \begin{bmatrix} 0 & -2 & 5 \\ 2 & 0 & 3 \\ -5 & -3 & 0 \end{bmatrix} = \text{skew symmetric.} \tag{A1.7d}$$

Any square matrix can be split into the sum and difference of a symmetric and skew symmetric matrix:

$$\mathbf{A} = \frac{1}{2}(\mathbf{A} + \mathbf{A}^{\mathrm{T}}) + \frac{1}{2}(\mathbf{A} - \mathbf{A}^{\mathrm{T}}). \tag{A1.8}$$

Though it looks a bit goofy, this operation is extremely important throughout matrix algebra.

A1.3 Basic Matrix Operations

Addition and subtraction require matrices of the same order, since this operation simply involves addition or subtraction of corresponding elements. This is shown in Figure A1.2.

Also,

$$\mathbf{A} + \mathbf{B} = \mathbf{B} + \mathbf{A} \tag{A1.9}$$

$$(\mathbf{A} + \mathbf{B}) + \mathbf{C} = \mathbf{A} + (\mathbf{B} + \mathbf{C}). \tag{A1.10}$$

Multiplication operations fall into several groups.

Scalar multiplication of a matrix involves multiplying a matrix by a constant (scalar). For a scalar β,

$$\beta\mathbf{A} = \beta \begin{bmatrix} a_{11} & a_{12} & a_{13} \\ a_{21} & a_{22} & a_{23} \\ a_{31} & a_{32} & a_{33} \end{bmatrix} = \begin{bmatrix} \beta a_{11} & \beta a_{12} & \beta a_{13} \\ \beta a_{21} & \beta a_{22} & \beta a_{23} \\ \beta a_{31} & \beta a_{32} & \beta a_{33} \end{bmatrix}, \tag{A1.11}$$

so the product is a matrix in which each element has been multiplied by β.

A + B = C

$$\mathbf{A} = \begin{bmatrix} a_{11} & a_{12} & a_{13} \\ a_{21} & a_{22} & a_{23} \\ a_{31} & a_{32} & a_{33} \end{bmatrix}; \qquad \mathbf{B} = \begin{bmatrix} b_{11} & b_{12} & b_{13} \\ b_{21} & b_{22} & b_{23} \\ b_{31} & b_{32} & b_{33} \end{bmatrix}$$

$$\mathbf{C} = \begin{bmatrix} a_{11}+b_{11} & a_{12}+b_{12} & a_{13}+b_{13} \\ a_{21}+b_{21} & a_{22}+b_{22} & a_{23}+b_{23} \\ a_{31}+b_{31} & a_{32}+b_{32} & a_{33}+b_{33} \end{bmatrix}$$

Figure A1.2 Example of adding two matrices together.

Scalar product (also **dot product** or **inner product**) is the (scalar) product of two *vectors* of the same size, defined as

$$\mathbf{A}\cdot\mathbf{B}=\beta, \tag{A1.12a}$$

where

$\mathbf{A}$ = a row vector (or the transpose of a column vector) of length n
$\mathbf{B}$ = a column vector (or the transpose of a row vector), also of length n
β = the *scalar* product of $\mathbf{A}\cdot\mathbf{B}$.

Specifically,

$$\mathbf{A}=[a_1 \quad a_2 \quad a_3]; \quad \mathbf{B}=\begin{bmatrix} b_1 \\ b_2 \\ b_3 \end{bmatrix} \tag{A1.12b}$$

and

$$\beta=a_1b_1+a_2b_2+a_3b_3, \tag{A1.12c}$$

so the product is *not* a matrix, but rather a scalar (i.e., a number).

Some people like to visualize this multiplication as shown in Figure A1.3, where the corresponding elements are multiplied (product indicated along the diagonal lines) and added to make the scalar product.

Conceptually, this product can be thought of as multiplying the length of one vector by the component of the other vector that is parallel to the first.

In terms of force vectors, $|\mathbf{B}|\cos\theta$ represents the component of force $\mathbf{B}$ acting to drive a displacement $\mathbf{A}$ in the direction indicated. The dot product gives the work in direction A. Alternatively, in terms of a decomposition, it gives that portion of the $\mathbf{B}$ vector that is represented along the $\mathbf{A}$ direction (the axis of one of the decomposing vectors). In this sense, the dot product is given as (see Figure A1.4)

$$\mathbf{A}\cdot\mathbf{B}=|\mathbf{A}||\mathbf{B}|\cos\theta=\beta, \tag{A1.12c}$$

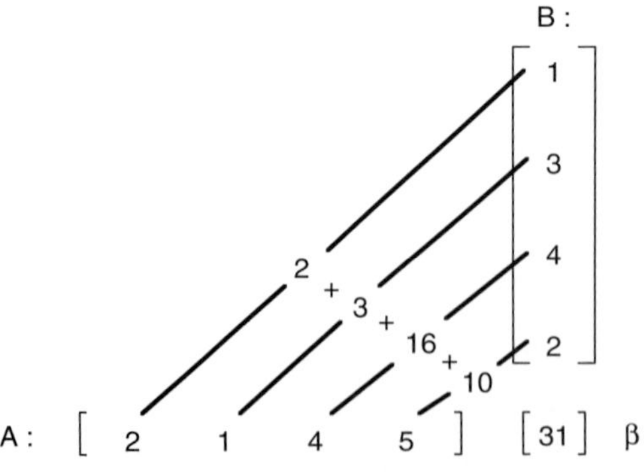

Figure A1.3 Schematic showing visualization of dot product, cell-by-cell multiplication and sum.

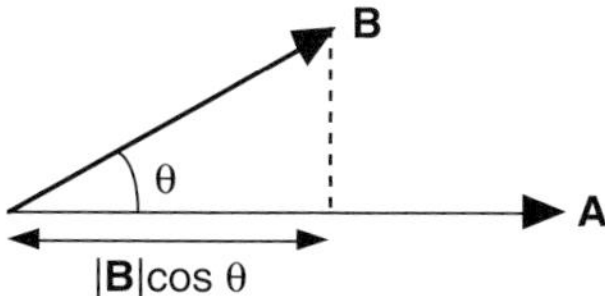

Figure A1.4 Dynamic schematic of dot product, where vector **B** is operating on vector **A**.

where $|\mathbf{x}|$ is the vector x **magnitude**, given as

$$|\mathbf{x}| = \sqrt{x_1^2 + x_2^2 + x_3^2 + \ldots + x_n^2}. \tag{A1.12d}$$

This is also referred to as a **projection** of **B** on **A**, which looks similar to a correlation (discussed later), and is given in summation form as

$$\mathbf{A} \cdot \mathbf{B} = \sum_{i=1}^{n} \mathrm{a}_i \mathrm{b}_i. \tag{A1.13}$$

Maximum principle says that the unit vector, **n** (a vector of magnitude unity), making **A**·**n** a maximum, is that unit vector pointing in the same direction as **A**. This is seen from the above definition of the scalar product, where, if

$$\mathbf{n} \parallel \mathbf{A}, \tag{A1.14a}$$

then

$$\cos\theta = \cos 0° = 1, \tag{A1.14b}$$

so

$$\mathbf{A} \cdot \mathbf{n} = |\mathbf{A}||\mathbf{n}|\cos\theta = |\mathbf{A}||\mathbf{n}| = |\mathbf{A}|. \tag{A1.14c}$$

For any other angle θ, the cos θ will be less than 1, yielding a smaller scalar product. This principle holds equally well for any vector **n** that is not a unit vector (note that **n** is usually used to denote a unit vector). So the vector that is parallel to **A** will give the largest dot product with **A**.

Parallel vectors (or **dependent vectors**) *geometrically* thus have cos θ = 1. So, if

$$\mathbf{A} \| \mathbf{B},$$

then

$$\mathbf{A} \cdot \mathbf{B} = |\mathbf{A}||\mathbf{B}|. \tag{A1.14}$$

Also,

$$\mathbf{A} = \beta\mathbf{B}, \tag{A1.15a}$$

This latter formula states that if two vectors are parallel, they are related by a simple constant:

$$\beta = \frac{|\mathbf{A}|}{|\mathbf{B}|}. \tag{A1.15b}$$

The fact that parallel vectors are related by a constant indicates that they are **dependent** vectors. This represents an important concept in matrix algebra.

Orthogonal vectors are *perpendicular*, and thus have

$$\cos\theta = \cos 90^\circ = 0,$$

so

$$\mathbf{A}\cdot\mathbf{B} = 0 \tag{A1.16}$$

when

$$\mathbf{A} \perp \mathbf{B}.$$

Orthogonal vectors represent a special limiting case of independent vectors. They have properties that make them ideal for a variety of mathematical operations.

Defining Independent, Orthogonal and Uncorrelated Vectors

For vectors, or time series: $\mathbf{X}_1, \mathbf{X}_2, \ldots, \mathbf{X}_n$ in terms of vector operations, the following material holds.

Vectors are only dependent when they are parallel (one being shorter, longer or reversed) relative to each other so that multiplication by a constant can make them identical. If they are not parallel, they are *independent*, satisfying for any constants (except 0), $a, b, \ldots, z$

$$a\mathbf{X}_1 + b\mathbf{X}_2 + \cdots + z\mathbf{X}_n \neq 0. \tag{A1.17a}$$

If they are at right angles to one another, as independent as can be, they are *orthogonal*, satisfying

$$\mathbf{X}_1^{\mathrm{T}}\mathbf{X}_2 = 0. \tag{A1.17b}$$

They are *uncorrelated* if

$$(\mathbf{X}_1 - \overline{\mathbf{X}_1})^T(\mathbf{X}_2 - \overline{\mathbf{X}_2}) = 0. \tag{A1.17c}$$

This says that the centered (i.e., vectors with means removed) are orthogonal, but that does not mean that the uncentered vectors are orthogonal (e.g., consider each vector as having a trend for a mean, which immediately forces a dependence).

Squaring vectors is simply

$$\text{For row vectors}: \mathbf{A}^2 = \mathbf{A}\mathbf{A}^{\mathrm{T}}$$

and

$$\text{For column vectors}: \mathbf{A}^2 = \mathbf{A}^{\mathrm{T}}\mathbf{A}. \tag{A1.18}$$

Matrix multiplication requires **conformable matrices**, which are those in which there are as many columns in the first as there are rows in the second, so

$$\mathbf{C}_{(n,m)} = \mathbf{A}_{(n,p)}\mathbf{B}_{(p,m)}, \tag{A1.19}$$

and the product matrix **C** is of order $n \times m$ with has elements c_{ij}. The elements c_{ij}, of **C**, are given by

$$c_{ij} = \sum_{k=1}^{p} a_{ik} b_{kj}. \tag{A1.18b}$$

This is an extension of the scalar product – in this case, each element of **C** represents the scalar product of a row vector in **A** and column vector in **B**,

$$\begin{bmatrix} c_{11} & c_{12} \\ c_{21} & c_{22} \end{bmatrix} = \begin{bmatrix} a_{11} & a_{12} & a_{13} \\ a_{21} & a_{22} & a_{23} \end{bmatrix} \begin{bmatrix} b_{11} & b_{12} \\ b_{21} & b_{22} \\ b_{31} & b_{32} \end{bmatrix}$$

or

$$c_{12} = a_{11}b_{12} + a_{12}b_{22} + a_{13}b_{32}.$$

In "box form," this product is shown by

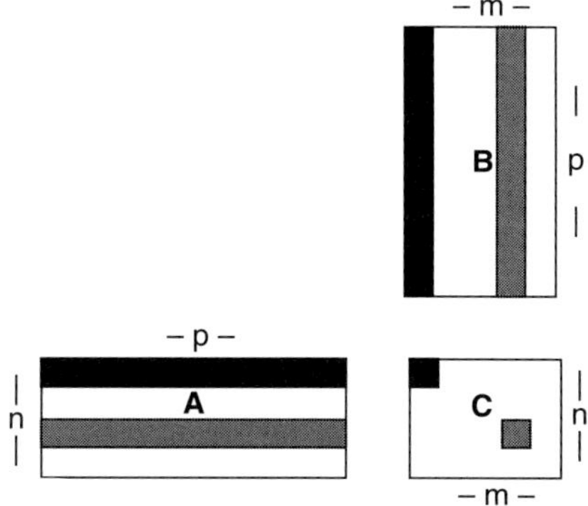

This form graphically illustrates that the elements of **C** correspond to the scalar products of the vectors indicated (i.e., consider the matrices to be composed of row and column vectors). The element locations in **C** occur where the extensions of the vectors intersect in the **C** matrix. Note that the order is shown for each matrix. **C** will have as many columns as **B** has (m) and as many rows as $\boldsymbol{A}$ has (n).

Order of multiplication is important – usually.

$$\mathbf{AB} \neq \mathbf{BA}, \tag{A1.20}$$

unless **A** and **B** are square matrices of the same order. Otherwise, one of the two products cannot even be formed (since the matrices will not be conformable in both directions).

Order is specified by stating that

A is **pre-multiplied** by **B** (for **BA**); or
A is **post-multiplied** by **B** (for **AB**).

Multiple products can be multiplied as

$$\mathbf{D} = (\mathbf{AB})\mathbf{C} = \mathbf{A}(\mathbf{BC}).$$

Computational Considerations

When examining a matrix multiplication, it is seen that the *order* in which pairs are multiplied in a multiple product can make a large computational difference. Consider the following example.

Box A1.2 Example of Matrix Indexing

$$\mathbf{C}_{(n,m)} = \mathbf{A}_{(n,p)}\mathbf{B}_{(p,m)}$$

This product involves $n \times m \times p$ multiplications ($n \times m$ vector products times p element multiplications in each product) and $n \times m \times (p - 1)$ additions (computed in a similar manner). If the matrix **E** is then the product,

$$\mathbf{E}_{(n,m)} = [\mathbf{A}_{(n,p)}\mathbf{B}_{(p,q)}]\mathbf{C}_{(q,m)},$$

the product in brackets involves mpq multiplications and the product of this with **C** then involves another nqm multiplication, for a total of $npq + nqm = nq(p + m)$ multiplications (and a nearly comparable number of additions).

Alternatively, if the product is multiplied as

$$\mathbf{E}_{(n,m)} = \mathbf{A}_{(n,p)}[\mathbf{B}_{(p,q)}\mathbf{C}_{(q,m)}],$$

the bracketed product involves pqm multiplications and the total product $pqm + npm = pm(q + n)$ multiplications.

Therefore,

1) $(\mathbf{AB})\mathbf{C} \rightarrow nq(p + m)$ multiplications;
2) $\mathbf{A}(\mathbf{BC}) \rightarrow pm(q + m)$ multiplications.

If **A** and **B** are both 100×100 order matrices and **C** is 100×1 order, then

$$n = 100$$
$$p = 100$$
$$q = 100$$
$$m = 1,$$

so the order of multiplication shown in 1) requires $\sim 10^6$ multiplications and 2) requires $\sim 10^4$ multiplications. Consequently, if one were to simply multiply **AB** and then this product by **C**, the computational cost would be nearly a million multiplications (and an almost equal number of additions) more than required by multiplying **BC** first and then pre-multiplying this product by **A**!

This is not just important for the obvious reason of computational time, but more importantly, the more the operations, the greater the accumulation of round-off error when dealing with very large matrices.

Transpose of a matrix product is simply multiplication by the transpose of the individual matrices in reverse order:

$$\mathbf{D} = \mathbf{ABC} \tag{A1.21a}$$

$$\mathbf{D}^T = \mathbf{C}^T\mathbf{B}^T\mathbf{A}^T. \tag{A1.21b}$$

A1.4 Special Matrix Products

Multiplication by I

$$\mathbf{AI} = \mathbf{IA} = \mathbf{A} \tag{A1.22}$$

as seen by

$$\begin{array}{cc} & \begin{bmatrix} 1 & 0 & 0 \\ 0 & 1 & 0 \\ 0 & 0 & 1 \end{bmatrix} \mathbf{I} \\ \mathbf{A}\begin{bmatrix} 3 & 6 & 9 \\ 2 & 8 & 7 \end{bmatrix} & \begin{bmatrix} 3 & 6 & 9 \\ 2 & 8 & 7 \end{bmatrix} \mathbf{AI} \end{array} .$$

Recall that the identity matrix, **I**, acts as 1 in matrix multiplications, and is always assumed to be conformable with the matrix being multiplied. Thus in (A1.22), if **A** has order $n \times m$, when post-multiplying **A** by **I, I** has order m (recall that **I** is a square matrix, so order n is equivalent to order n,n). When pre-multiplying **A** by **I, I** has order n.

Pre-multiplication by a diagonal matrix produces a matrix in which each *row* is scaled by a diagonal element of **D**. So, for

$$\mathbf{C} = \mathbf{DA}, \tag{A1.23}$$

where **D** is a diagonal matrix, and

$$D = \begin{bmatrix} d_{11} & & \\ & d_{22} & \\ & & d_{33} \end{bmatrix}; \quad A = \begin{bmatrix} a_{11} & a_{12} & a_{13} \\ a_{21} & a_{22} & a_{23} \\ a_{31} & a_{23} & a_{33} \end{bmatrix}$$

$$C = \begin{bmatrix} a_{11}d_{11} & a_{12}d_{11} & a_{13}d_{11} \\ a_{21}d_{22} & a_{22}d_{22} & a_{23}d_{22} \\ a_{31}d_{33} & a_{32}d_{33} & a_{33}d_{33} \end{bmatrix}$$

so the first row of **C** is scaled by d_{11}, the second row by d_{22}, etc.

Post-multiplication by a diagonal matrix produces a matrix in which each *column* has been scaled by a diagonal element of **D**. So, for

$$\mathbf{C} = \mathbf{AD}. \tag{A1.24}$$

Then,

$$\mathbf{C} = \begin{bmatrix} a_{11}d_{11} & a_{12}d_{22} & a_{13}d_{33} \\ a_{21}d_{11} & a_{22}d_{22} & a_{23}d_{33} \\ a_{31}d_{11} & a_{32}d_{22} & a_{33}d_{33} \end{bmatrix}$$

where the first column of **C** is scaled by d_{11}, the second by d_{22}, etc.

A1.4.1 Determinant of a Matrix

The determinant of a matrix is a single number representing a *property of a square matrix* (its exact meaning is dependent upon what the particular matrix represents). The main use here is to find the inverse of a matrix or solve simultaneous equations for which the determinant plays a major role. Symbolically, the determinant is usually given as det **A**, $|\mathbf{A}|$ or $\|\mathbf{A}\|$ (to differentiate it from magnitude).

Calculation of a 2×2 determinant is given by

$$|\mathbf{A}| = \begin{vmatrix} a_{11} & a_{12} \\ a_{21} & a_{22} \end{vmatrix} = a_{11}a_{22} - a_{12}a_{21}. \tag{A1.25}$$

This is the difference of the cross products (the determinant of a 1×1 matrix is just the particular element).

Calculation of an $n \times n$ determinant is given by

$$|\mathbf{A}| = a_{11}M_{11} - a_{12}M_{12} + a_{13}M_{13} - \ldots. - (-1)^n a_{1m}M_{1m}, \tag{A1.26}$$

where

M_{11} = the determinant, with the first row and column missing from the matrix;
M_{12} = the determinant, with the first row and second column missing; etc.

Generalizing, the determinant with the *i*th row and *j*th column removed is known as the **cofactor**, $(-1)^{i+j} M_{ij}$. This will be a useful concept later for inverting a matrix.

This formula can be used recursively. That is, once the first column and row of the matrix have been eliminated (to construct M_{11}), then the determinant of the remaining portion of the matrix can be computed by applying the above formula to this reduced matrix so that it, too, can be reduced until a manageable size is attained (e.g., 3×3 or smaller). But note that, in any case, the cofactors are only those that correspond to the first row of elements in the matrix. That is, as seen in (A1.26), the cofactors are multiplied by a_{11}, a_{12}, etc., never by a_{21}, a_{nn}. However, the cofactors themselves can involve the determinants in which these other rows are included if a recursive implementation is required. That is, the cofactor for M_{11} might involve cofactors of submatrices whose determinants involve a_{22}, a_{23}, etc.

For a 4×4 determinant, each M_{ij} would be an entire expansion like that given above for the 3×3 determinant – one quickly needs a computer. That is, the M_{11} term would eliminate the first row and first column of the matrix, leaving a 3×3 matrix. This individual 3×3 matrix would then be expanded as in the above example, allowing

Box A1.3 Example of a Determinant

Example of a 3×3 determinant:

$$|\mathbf{A}| = \begin{vmatrix} a_{11} & a_{12} & a_{13} \\ a_{21} & a_{22} & a_{23} \\ a_{31} & a_{32} & a_{33} \end{vmatrix}$$

$$M_{11} = \begin{vmatrix} a_{11} & a_{12} & a_{13} \\ a_{21} & a_{22} & a_{23} \\ a_{31} & a_{32} & a_{33} \end{vmatrix} = a_{22}\,a_{33} - a_{23}\,a_{32}$$

$$M_{12} = \begin{vmatrix} a_{11} & a_{12} & a_{13} \\ a_{21} & a_{22} & a_{23} \\ a_{31} & a_{32} & a_{33} \end{vmatrix} = a_{21}\,a_{33} - a_{23}\,a_{31}$$

$$M_{13} = \begin{vmatrix} a_{11} & a_{12} & a_{13} \\ a_{21} & a_{22} & a_{23} \\ a_{31} & a_{32} & a_{33} \end{vmatrix} = a_{21}\,a_{32} - a_{22}\,a_{31}$$

Applying the above formula gives

$$\begin{aligned} |\mathbf{A}| &= a_{11}M_{11} - a_{12}M_{12} + a_{23}M_{13} \\ &= a_{11}(a_{22}a_{33} - a_{23}a_{32}) - a_{12}(a_{21}a_{33} - a_{23}a_{31}) + a_{13}(a_{21}a_{32} - a_{22}a_{31}). \end{aligned}$$

solution for the M_{11} term. For the M_{12} term, the elimination of the first row and second column would also leave a 3×3 matrix, which would then be expanded to give M_{12}. This procedure is thus followed for each M_{1j} term until all of them have been determined.

Singular matrix is a *square matrix whose determinant is zero.* A determinant is zero if

1) any row or column in the matrix is zero;
2) any row or column, is equal to a linear combination of any other rows or columns.

Example:

$$\mathbf{A} = \begin{bmatrix} 1 & 6 & 4 \\ 2 & 1 & 0 \\ 5 & -3 & -4 \end{bmatrix},$$

where row 1 = 3(row 2) − row 3. So, computing the determinant gives

$$\begin{aligned} |\mathbf{A}| &= a_{11}(a_{22}a_{33} - a_{23}a_{32}) - a_{12}(a_{21}a_{33} - a_{23}a_{31}) + a_{13}(a_{21}a_{32} - a_{22}a_{31}) \\ &= 1[1(-4) - 0(-3)] - 6[2(-4) - 0(5)] + 4[2(-3) - 1(5)] \\ &= -4 + 48 - 44 = 0. \end{aligned}$$

Degree of clustering symmetrically about the principal diagonal is another (of many) properties of a determinant. The more clustering there is, the higher the value of the determinant.

A1.5 Matrix "Division": Inverse Matrix

Matrix division must be thought of as multiplying by the inverse, as is true for scalar division,

$$\frac{\mathbf{x}}{\mathbf{b}} = \mathbf{x}\mathbf{b}^{-1} \tag{A1.27a}$$

and

$$\mathbf{b}\mathbf{b}^{-1} = 1. \tag{A1.27b}$$

Matrices are divided by applying the above – i.e., by multiplying by the inverse matrix. *Nonsingular square matrices* may have a unique inverse symbolized as

$$\mathbf{A}^{-1}$$

and

$$\mathbf{A}\mathbf{A}^{-1} = \mathbf{I}. \tag{A1.28}$$

The most basic inverse involves the use of cofactors and the determinant. First, recall that the definition of a cofactor, C_{ij}, is

$$\mathrm{C}_{ij} = (-1)^{i+j}\mathrm{M}_{ij}, \tag{A1.29}$$

where M_{ij} is the determinant of the submatrix created by eliminating the ith row and jth column of the matrix. Given this, the determinant of a matrix $\mathbf{A}$ is given by

$$\det \mathbf{A} = \mathrm{a}_{i1}\mathrm{C}_{i1} + \mathrm{a}_{i2}\mathrm{C}_{i2} + \ldots + \mathrm{a}_{im}\mathrm{C}_{im}. \tag{A1.30}$$

Then, the *inverse is completely determined by knowledge of the cofactors*. In particular, $\mathbf{A}^{-1}$ is given as

$$\mathbf{A}^{-1} = \frac{\mathbf{C}^{\mathrm{T}}}{\det \mathbf{A}}, \tag{A1.31}$$

where $\mathbf{C}^{\mathrm{T}}$ is the transpose of the **adjoint matrix** (or **adjugate** or **cofactor matrix**), which is defined as

$$\mathbf{C} = \begin{bmatrix} \mathrm{c}_{11} & \mathrm{c}_{12} & \ldots & \mathrm{c}_{1m} \\ \mathrm{c}_{21} & \mathrm{c}_{22} & \ldots & \mathrm{c}_{2m} \\ \vdots & & & \\ \mathrm{c}_{n1} & \mathrm{c}_{n2} & \ldots & \mathrm{c}_{nm} \end{bmatrix}, \tag{A1.32}$$

and the determinant of the matrix $\mathbf{A}$ is also determined by combining the cofactors from the first row of the cofactor matrix, as seen in (A1.30).

Calculation of the inverse of a larger matrix is usually done using **elimination methods** (on the computer). These techniques (or methods used in place of

Box A1.4 Example of Inverse Using Adjoint

For a simple 2×2 matrix, the inverse is given by

$$\mathbf{A}^{-1} = \frac{1}{\det \ \mathbf{A}} \begin{bmatrix} a_{22} & -a_{12} \\ -a_{21} & a_{11} \end{bmatrix}$$

according to (A1.31). So for the **A** matrix

$$\mathbf{A} = \begin{bmatrix} 7 & 2 \\ 10 & 3 \end{bmatrix},$$

then

$$\mathbf{A}^{-1} = \frac{1}{21 - 20} \begin{bmatrix} 3 & -2 \\ -10 & 7 \end{bmatrix} = \begin{bmatrix} 3 & -2 \\ -10 & 7 \end{bmatrix}$$

and

$$\mathbf{A}\mathbf{A}^{-1} = \begin{bmatrix} 7 & 2 \\ 10 & 3 \end{bmatrix} \begin{bmatrix} 3 & -2 \\ -10 & 7 \end{bmatrix} = \begin{bmatrix} 1 & 0 \\ 0 & 1 \end{bmatrix} = \mathrm{I}.$$

inversion) will be presented when necessary, in context with specific data analysis techniques.

A1.5.1 Solution of Simultaneous Equations

A system of n simultaneous equations in n unknowns is easily handled by matrix algebra notation. For example, consider four equations in four unknowns:

$$\begin{aligned} a_{11}x_1 + a_{12}x_2 + a_{13}x_3 + a_{14}x_4 &= b_1 \\ a_{21}x_1 + a_{22}x_2 + a_{23}x_3 + a_{24}x_4 &= b_2 \\ a_{31}x_1 + a_{32}x_2 + a_{33}x_3 + a_{34}x_4 &= b_3 \\ a_{41}x_1 + a_{42}x_2 + a_{43}x_3 + a_{44}x_4 &= b_4. \end{aligned} \tag{A1.31a}$$

This is written in matrix form as

$$\mathbf{Ax} = \mathbf{b}, \tag{A1.31b}$$

where

$$\mathrm{A} = \begin{bmatrix} a_{11} & a_{12} & a_{13} & a_{14} \\ a_{21} & a_{22} & a_{23} & a_{24} \\ a_{31} & a_{32} & a_{33} & a_{34} \\ a_{41} & a_{42} & a_{43} & a_{44} \end{bmatrix} = \text{coefficient matrix}$$

$$x = \begin{bmatrix} x_1 \\ x_2 \\ x_3 \\ x_4 \end{bmatrix}; \; b = \begin{bmatrix} b_1 \\ b_2 \\ b_3 \\ b_4 \end{bmatrix}.$$

This is easily solved (symbolically) using the simple operations of matrix algebra as

$$\mathbf{A}^{-1}\mathbf{A}\mathbf{x} = \mathbf{A}^{-1}\mathbf{b} \tag{A1.32a}$$

(i.e., pre-multiply both sides by $\mathbf{A}^{-1}$; recall that order of multiplication is important). So,

$$\mathbf{I}\mathbf{x} = \mathbf{A}^{-1}\mathbf{b} \tag{A1.32b}$$

$$\mathbf{x} = \mathbf{A}^{-1}\mathbf{b}. \tag{A1.32c}$$

That is, the unknown elements of **x** are simply given by pre-multiplying the **b** column vector with the inverse of the coefficient matrix, $\mathbf{A}^{-1}$.

Computational Considerations

The case of matrix inverses suggests that, while the above approach is extremely elegant symbolically, sometimes the computational effort may be considerable when dealing with large systems and a direct solution by elimination methods is quicker. This is revealed by consideration of the number of multiplications involved in a matrix inversion approach.

Solution by the inversion (as shown above) requires m^3 multiplications for the inversion and m^2n more multiplications to finish the solution, where: m = number of equations per set and n = number of sets of equations (each of the same form, but different **b** matrix). Therefore, the total number of multiplications is: $m^3 + m^2n$.

Compare this to a direct solution by a typical elimination method (the standard way to solve a system of equations using algebra), which requires only $m^3/3 + m^2n$.

So, while the matrix form is easy to handle, one should not necessarily always use it blindly (we will consider many situations for which matrix solutions are ideal). Note that for sparse matrices, the above relationships may not hold; nor for symmetrical matrices or other special matrix forms in which case-special fast algorithms exist for the inversion. Also note that, for stability reasons, sometimes other special approaches are better suited for the solution of a large system. These will be introduced at the appropriate times.

A1.5.2 Additional Terms

Rank of a matrix is the number of linearly independent vectors the matrix contains (either row or column vectors).

Box A1.5 Example Solving Simple **Ax** = **b**

Consider solving the following two simultaneous equations (i.e., two equations in two unknowns):

$$5x_1 + 7x_2 = 19$$
$$3x_1 - 2x_2 = -1.$$

These represent the equations of lines, and the common solution represents the intersection of the two lines. In matrix form, these are given as

$$\begin{bmatrix} 5 & 7 \\ 3 & -2 \end{bmatrix} \begin{bmatrix} x_1 \\ x_2 \end{bmatrix} = \begin{bmatrix} 19 \\ -1 \end{bmatrix}$$
$$\mathbf{A} \quad \mathbf{x} \quad = \quad \mathbf{b}.$$

To solve this requires the inverse of the **A** matrix:

$$\mathbf{A}^{-1} = \frac{1}{-10-21}\begin{bmatrix} -2 & -7 \\ -3 & 5 \end{bmatrix} = \begin{bmatrix} \frac{2}{31} & \frac{7}{31} \\ \frac{3}{31} & \frac{-5}{31} \end{bmatrix}.$$

Then,

$$\mathbf{x} = \mathbf{A}^{-1}\mathbf{b}$$

and

$$\mathbf{A}^{-1}\mathbf{b} = \begin{bmatrix} \frac{2}{31} & \frac{7}{31} \\ \frac{3}{31} & \frac{-5}{31} \end{bmatrix} \begin{bmatrix} 19 \\ -1 \end{bmatrix} = \begin{bmatrix} \left(\frac{38}{31} - \frac{7}{31}\right) \\ \left(\frac{57}{31} + \frac{5}{31}\right) \end{bmatrix} = \begin{bmatrix} 1 \\ 2 \end{bmatrix}.$$

so the unknowns, $x_1 = 1$ and $x_2 = 2$ represent the solution and solve the system.

Rank has special meaning when dealing with systems of equations. Consider the common intersection of four lines of the standard form $y = mx + b$:

$$2 = m1 + b$$
$$3 = m2 + b$$
$$4 = m3 + b$$
$$5 = m4 + b,$$

or, in matrix form,

$$\begin{bmatrix} 1 & 1 \\ 2 & 1 \\ 3 & 1 \\ 4 & 1 \end{bmatrix} \begin{bmatrix} m \\ b \end{bmatrix} = \begin{bmatrix} 2 \\ 3 \\ 4 \\ 5 \end{bmatrix}.$$

Box A1.6 Example of Matrix Rank

Consider

$$\mathbf{A} = \begin{bmatrix} 1 & 4 & 0 & 2 \\ 1 & 0 & 1 & -1 \\ -3 & -4 & -2 & 0 \end{bmatrix}.$$

Since

$$\text{row 3} = -(\text{row 1}) - 2(\text{row2})$$

or

$$\text{col 3} = (\text{col 1}) - \frac{1}{4}(\text{col 2})$$

and

$$\text{col 4} = -(\text{col 1}) + \frac{3}{4}(\text{col 2}),$$

the matrix **A** has rank 2 (i.e., it has only two linearly independent vectors, independent of whether it is viewed by rows or columns).

Since the **A** matrix is order 4×2, the largest rank it can possibly have is 2 (obviously, there cannot be more linearly independent column or row vectors than the smallest number of columns or rows). This says that at least two of the rows (= equations) are dependent, and thus provide duplicate information. Inspection of the above matrix shows that row (4) = (row 2) + (row 3) − (row 1).

A matrix is **rank deficient** if its rank < min(n,m) and it is of **full rank** if its rank = min (n,m).

The **rank of a matrix product** must be less than or equal to the smallest rank of the matrices being multiplied, so

$$A_{(\text{rank }2)}B_{(\text{rank }1)} = C_{(\text{rank }1)}.$$

This seemingly esoteric fact is actually extremely helpful when considered from another angle. That is, *if a matrix has rank r, then any matrix factor of it must have rank of at least r*. Since the rank cannot be greater than the smallest of n or m, in a $n\times m$ matrix, this definition also limits the size (order) of factor matrices. So you cannot factor a matrix of rank 2 into two matrices of which either is of less than rank 2, so n and m of each factor must also be 2.

Rank of a matrix **A** is most easily determined by performing an SVD decomposition on the matrix and counting the number of nonzero singular values (= rank) in the diagonal matrix of the SVD decomposition. If there are no zeros, matrix **A** is of **full rank**.

Minor is the determinant of an equal number of rows and columns from a (encompassing) matrix. If a matrix has rank r, then there must be at least one nonzero minor of order r present in the matrix (and no nonzero minors of order $> r$). Remember that a determinant of a matrix is zero if its rank is not equal to its order (i.e., if any row or column in the matrix is linearly dependent upon any other rows or columns). Unfortunately, we cannot compute the rank of a matrix by examining for nonzero determinants due to machine round-off errors.

The **trace** of a square matrix is simply the sum of the elements along the principal diagonal. It is symbolized as tr**A.** This property is useful in calculating various quantities from matrices.

Submatrices are smaller matrix partitions of a larger, encompassing supermatrix. For example,

$$\begin{bmatrix} \\ \text{supermatrix} \\ \mathbf{F} \\ \\ \end{bmatrix} = \overset{\text{Submatrices :}}{\left[\begin{array}{c|c} \text{A} & \text{B} \\ \hline \text{C} & \text{D} \end{array}\right]} .$$

Such partitioning is frequently useful.

Major product moment is the *square matrix* resulting from the product $\mathbf{AA}^{\mathrm{T}}$ in which the diagonal elements are the sum of squares for each row of the matrix.

$$\begin{array}{cc} & \begin{bmatrix} 1 & 2 & 2 \\ -3 & 0 & 2 \end{bmatrix} \mathbf{A}^{\mathrm{T}} \\ \mathbf{A} \begin{bmatrix} 1 & -3 \\ 2 & 0 \\ 2 & 2 \end{bmatrix} & \begin{bmatrix} 10 & 2 & -4 \\ 2 & 4 & 4 \\ -4 & 4 & 8 \end{bmatrix} \mathbf{AA}^{\mathrm{T}} \end{array}$$

Minor product moment is the *square matrix* resulting from the product $\mathbf{A}^{\mathrm{T}}\mathbf{A}$ in which the diagonal elements are the sum of squares for each column of the matrix.

A1.6 Useful Properties

$$(\mathbf{A}^{\mathrm{T}})^{\mathrm{T}} = \mathbf{A} \tag{A1.33}$$

$$(\mathbf{A}^{-1})^{-1} = \mathbf{A} \tag{A1.34}$$

$$(\mathbf{A}^{-1})^{\mathrm{T}} = (\mathbf{A}^{\mathrm{T}})^{-1} = \mathbf{A}^{-\mathrm{T}} \tag{A1.35}$$

So, if

$$\mathbf{D} = \mathbf{ABC}, \tag{A1.36a}$$

then

$$\mathbf{D}^{-1} = \mathbf{C}^{-1}\mathbf{B}^{-1}\mathbf{A}^{-1} \tag{A1.36b}$$

$$\mathbf{D}^{\mathrm{T}} = \mathbf{C}^{\mathrm{T}}\mathbf{B}^{\mathrm{T}}\mathbf{A}^{\mathrm{T}}. \tag{A1.36c}$$

These last two "*reversal rules*" for inverse and transpose products are constantly needed for manipulation of matrix equations and are also useful for eliminating or minimizing the number of matrix inverses or transposes requiring calculation.

A1.7 More Advanced Topics

While the preceding gives an introduction to the manipulative capabilities of matrix algebra, the topic encompasses considerably more. In particular, linear algebra represents a different means through which most of higher mathematics can be viewed. Linear algebra provides a framework for thinking about the typical calculations we will encounter that tends to unify the otherwise diverse topics to be discussed. This section presents an overview of the elements of linear algebra that define this framework.

A1.7.1 Vector Space and Basis

Consider a set of *n-element* vectors $\{\mathbf{x}_1, \mathbf{x}_2, \ldots, \mathbf{x}_m\}$ belonging to **real vector space $\mathbf{R}^n$**. A collection of m vectors $\mathbf{x}_j \in \mathbf{R}^n$, along with every linear combination of those m vectors, forms a subspace defined as the **span** of $\mathbf{x}_j$. That is, they *span* the space because they can produce every *m-dimensional* vector in that space via some linear combination. If the set of vectors span the space and are independent, that is $\sum_{i=1}^{n} a_i\mathbf{x}_i \neq 0$ for all $a_1 \neq 0$, the set represents a *basis*. With this basis, every *n-element* vector in the space can be generated and the number of vectors in the basis is the dimension of the *space* (this **implies that** $n = m$). From our perspective, if we wish to describe (e.g., interpolate, generate, or otherwise form) any vector of n elements, we must do so using an *n-dimensional* basis, ensuring that we can fit *any* such vector, or, alternatively, allowing us to solve $\mathbf{Ax} = \mathbf{b}$ for any vector $\mathbf{b}$. So, with this set of basic vectors, we can describe any vector of the same dimension.

The **null space** of a matrix $\mathbf{A}$, denoted N($\mathbf{A}$), is defined as all vectors $\mathbf{x}$, such that $\mathbf{Ax} = \mathbf{0}$. In systems of equations, this represents the homogeneous system of equations (no source or sink terms). In linear algebra, it represents the following operation between the elements of $\mathbf{x}$ and the column vectors contained within $\mathbf{A}$:

$$x_1 \begin{bmatrix} a_{11} \\ a_{21} \\ a_{31} \\ \vdots \\ a_{n1} \end{bmatrix} + x_2 \begin{bmatrix} a_{12} \\ a_{22} \\ a_{32} \\ \vdots \\ a_{n2} \end{bmatrix} + \ldots + x_m \begin{bmatrix} a_{1m} \\ a_{2m} \\ a_{3m} \\ \vdots \\ a_{nm} \end{bmatrix} = \begin{bmatrix} 0 \\ 0 \\ 0 \\ \vdots \\ 0 \end{bmatrix}. \tag{A1.37}$$

This implies that the linear combination of column vectors (in **A**), with the appropriate constants (the elements of **x**), can be summed linearly to equal zero. For this to happen, the columns of **A** must be linearly dependent, unless the only vector **x** in the nullspace is the null vector – that is, the only way to add the columns of **A** together to get the zero (null) vector is when you multiply each column by zero (the trivial case).[1]

The **row space** of a matrix is the collection of all possible linear combinations of the row vectors in the matrix. Similarly, the **column space** of a matrix is the collection of all possible linear combinations of the column vectors in the matrix.

A1.7.2 Vector Differentiation

Direct differential manipulation of the vectors and matrix equations facilitates many computations. This is easily done, given the following convention for vector differentiation:

$$\frac{\partial}{\partial x}(\mathbf{Ax}) \equiv \frac{\partial}{\partial x}(\mathbf{Ax})^{\mathrm{T}} = \frac{\partial}{\partial x}\mathbf{x}^{\mathrm{T}}\mathbf{A}^{\mathrm{T}} = \mathbf{A}^{\mathrm{T}}. \tag{A1.38a}$$

For expressions often encountered in statistical applications, such as

$$\frac{\partial}{\partial x}(\mathrm{x}^{\mathrm{T}}\mathbf{Ax}) \equiv \frac{\partial}{\partial x}(\mathrm{x}^{\mathrm{T}}\mathbf{Ax})^{\mathrm{T}} = \frac{\partial}{\partial x}(\mathbf{x}^{\mathrm{T}}\mathbf{A}^{\mathrm{T}}x), \tag{A1.38b}$$

use the principle of differentiation by parts to give[2]

$$\begin{aligned}\frac{\partial}{\partial x}(\mathrm{x}^{\mathrm{T}}\mathbf{Ax}) &= \frac{\partial}{\partial x}\mathrm{x}^{\mathrm{T}}(\mathbf{Ax})^{\mathrm{T}} = \frac{\partial}{\partial x}\mathbf{x}^{\mathrm{T}}(\mathbf{A}^{\mathrm{T}}\mathrm{x}) \\ &= \mathbf{Ax} + \mathbf{A}^{\mathrm{T}}\mathbf{x}.\end{aligned} \tag{A1.38c}$$

A1.7.3 Quadratic Form

Up until now we have dealt with strictly linear forms of systems. Now we wish to deal with quadratic forms – in particular, $\mathbf{x}^{\mathrm{T}}\mathbf{Ax}$.

First we must establish some necessary terminology. Consider a vector product, $\mathbf{x}^{\mathrm{T}}\mathbf{Ax}$, where **A** is square. This product represents a quadratic form of the vector **x**. Basically, it

[1] Recall the definition of linear dependency: $a_1x_1 + a_2x_2 + \ldots + a_nx_n = 0$.

[2] This rule can be somewhat confusing. To help, set $\mathbf{x}^{\mathrm{T}} = \mathbf{y}$, then rewrite $\mathbf{x}^{\mathrm{T}}\mathbf{Ax} = \mathbf{yAx}$. Now, consider differentiation by parts for scalars: $\frac{\partial}{\partial x}(\mathbf{yAx}) = \mathbf{Ax}\frac{\partial}{\partial x}\mathbf{y} + \mathbf{yA}\frac{\partial}{\partial x}\mathbf{x}$. However, for vectors, the order of operations is important. Therefore, we take account of the equivalency given in (A.38), which states that the derivatives of $\mathbf{x}^{\mathrm{T}}\mathbf{Ax}$ and $\mathbf{x}^{\mathrm{T}}\mathbf{A}^{\mathrm{T}}\mathbf{x}$ are equal. We use the first form to treat **Ax** as being independent of **x**, so we form the counterpart to $\mathbf{Ax}\frac{\partial}{\partial x}\mathrm{y}$ as $(\frac{\partial}{\partial x}\mathrm{x}^{\mathrm{T}})\mathrm{A} = \mathrm{A}$. Having done this, we must now treat $\mathbf{x}^{\mathrm{T}}\mathbf{A}$ as being independent of x, which we do by using the second form, in which case the $\mathbf{x}^{\mathrm{T}}\mathbf{A}$ has been transposed and reversed to $\mathbf{A}^{\mathrm{T}}\mathbf{x}$. With this, the derivative can again be written in the proper form for vector differentiation, where the counterpart to $\mathbf{yA}\frac{\partial}{\partial x}\mathbf{x}$ is $(\frac{\partial}{\partial x}\mathbf{x}^{\mathrm{T}})\mathbf{B} = \mathbf{B}$, where $\mathbf{B} = \mathbf{A}^{\mathrm{T}}\mathbf{x}$. In this manner, we have accomplished the product expansion while satisfying the convention for the proper form of vector differentiation.

represents the weighted sum of squares and all cross-products (i.e., the $x_i x_j$, where $i \neq j$), where the weights are the various elements of matrix **A**. Explicitly,

$$\begin{aligned}
\mathbf{x}^T\mathbf{A}\mathbf{x} &= \sum_{i=1}^{n}\sum_{j=1}^{n} a_{ij}x_i x_j \\
&= \sum_{i=1}^{n} a_{ii}x_i^2 + \sum_{i=1}^{n}\sum_{\substack{j=1 \\ j\neq i}}^{n} a_{ij}x_i x_j \\
&= \sum_{i=1}^{n} a_{ii}x_i^2 + \sum_{i=1}^{n}\sum_{\substack{j=1 \\ j\neq i}}^{n} (a_{ij} + a_{ij})x_i x_j
\end{aligned} \tag{A1.39}$$

Inspection of (A1.38c), reveals that the exact same quadratic sum could be represented using a different square matrix, **A**. For example, $\mathbf{x}^T\mathbf{A}\mathbf{x} = \mathbf{x}^T\mathbf{B}\mathbf{x}$, as long as the diagonal elements and the sum of $(a_{ij} + a_{ij})$ (that is, the sum of symmetrical, off-diagonal pairs of both **A** and **B** are equal).[3] Since this can be satisfied by an infinite number of matrices, there is no single unique matrix **A** for this quadratic form of **x**.

For example,

$$A = \begin{bmatrix} 1 & 4 & 2 \\ 2 & 3 & 3 \\ 2 & 5 & 1 \end{bmatrix};\ B = \begin{bmatrix} 1 & 1 & 1 \\ 5 & 3 & 1 \\ 3 & 7 & 1 \end{bmatrix};\ \text{and } C = \begin{bmatrix} 1 & 3 & 2 \\ 3 & 3 & 4 \\ 2 & 4 & 1 \end{bmatrix} \tag{A1.40}$$

all yield the same quadratic sum, so the quadratic forms $\mathbf{x}^T\mathbf{A}\mathbf{x} = \mathbf{x}^T\mathbf{B}\mathbf{x} = \mathbf{x}^T\mathbf{C}\mathbf{x}$. More importantly, notice that **C** is a symmetrical matrix. Therefore, we are sacrificing nothing by stipulating that the coefficient matrix, **A**, in the quadratic form is symmetric. Since there is only *one* symmetrical matrix that satisfies this equality, it is unique. The symmetrical version of any matrix **A** satisfying the quadratic form can always be obtained as $(\mathbf{A} + \mathbf{A}^T)/2$.

Using the symmetric form, (A1.38c) can be rewritten as

$$\mathbf{x}^T\mathrm{Ax} = \sum_{i=1}^{n} a_{ii}x_i^2 + 2\sum_{i=1}^{n}\sum_{\substack{j=1 \\ j>1}}^{n} a_{ij}x_i x_j \tag{A1.41}$$

Because of the uniqueness of the symmetrical form, without loss of generality, and because of the special properties of symmetric matrices, we will always restrict the quadratic form to involve the unique symmetric coefficient matrix, **A** (so $\mathbf{A} = \mathbf{A}^T$).

Positive definite. In the simplest quadratic form case, where $\mathbf{A} = \mathbf{I}$, the quadratic sum reduces to $\mathbf{x}^T\mathbf{x}$, the sum of squares of the x values. This scalar product is a positive, nonzero number for all **x** except $\mathbf{x} = 0$. When the matrix **A** is not the identity matrix, its

[3] That is, these pairs consist of an element plus the element that occupies the same position of the current element in the transpose of the matrix. That is, the a_{ij} and a_{ji}.

elements now represent weights against which the squares and cross-product terms will be scaled before summing. In this case, the quadratic sum need not sum to positive values as it must when $\mathbf{A} = \mathbf{I}$. However, in those cases where

$$\mathbf{x}^T\mathbf{A}\mathbf{x} > 0 \text{ (all } \mathbf{x} \neq \mathbf{0}), \tag{A1.42a}$$

(that is, where the quadratic sum is greater than zero for all $\mathbf{x} > 0$), the quadratic form is called **positive definite**, and the corresponding (symmetric) coefficient matrix, $\mathbf{A}$, is a **positive definite matrix**.

If

$$\mathbf{x}^T\mathbf{A}\mathbf{x} \geq 0 \text{ (all } \mathbf{x} \neq \mathbf{0}\text{; } \mathbf{x}^T\mathbf{A}\mathbf{x} = 0 \text{ for some } \mathbf{x} \neq \mathbf{0}), \tag{A1.42b}$$

then the quadratic form is called **positive semidefinite** and the corresponding (symmetric) coefficient matrix, $\mathbf{A}$, is a **positive semidefinite matrix.**[4]

The quadratic form is most commonly encountered when dealing with variance and covariance matrices. For example, consider writing the sample variance[5] in its quadratic form. First, expand the sum of squared deviations about the mean:

$$\begin{aligned}
\sum_{i=1}^{n}(x_i - \bar{x})^2 &= \sum_{i=1}^{n} x_i^2 - n\bar{x}^2 \\
&= \sum_{i=1}^{n} x_i^2 - n\left[\frac{1}{n}\sum_{j=1}^{n} x_j\right]^2 \\
&= \sum_{i=1}^{n} x_i^2 - \sum_{j=1}^{n}\frac{1}{n}x_i^2 - \sum_{i=1}^{n}\sum_{\substack{j=1 \\ j\neq 1}}^{n}\frac{1}{n}x_i x_j \\
&= \sum_{i=1}^{n}\left(\frac{n-1}{n}x_i^2\right) - \sum_{i=1}^{n}\sum_{\substack{j=1 \\ j\neq 1}}^{n}\frac{1}{n}x_i x_j
\end{aligned} \tag{A1.43a}$$

Now multiply through by $1/(n - 1)$ to complete the operation:

$$= \left[\sum_{i=1}^{n}\left(\frac{1}{n}x_i^2\right) - \sum_{i=1}^{n}\sum_{\substack{j=1 \\ j\neq 1}}^{n}\left(\frac{1}{n(n-1)}x_i x_j\right)\right] \tag{A1.43b}$$

$$= \mathbf{x}^T\mathbf{C}\mathbf{x}. \tag{A1.43c}$$

The expanded sum (A1.43b) is in the form of (A1.38b) with the $a_{ii} = 1/n$ and the $a_{ij} = -1/([n(n-1)]$. Therefore, in the quadratic equivalent, $\mathbf{x}^T = (x_1\ x_2 \ldots x_n)$ and $\mathbf{C} = [(n+2)/n(n-1)]\mathbf{I} - [1/[n(n-1)]\mathbf{J}_n$, $\mathbf{J}_n$ is a square unity matrix (all elements equal to one). Because the matrix

[4] Note that some people like to abbreviate a positive definite and positive semidefinite matrix as **pd** and **psd**, respectively. I will avoid this notation, since I use the abbreviation psd to represent power spectral density. Regardless, the usage for the two representations of psd are so different that there should never be any confusion as to which meaning is intended. Also, some people use **non-negative definite** (**nnd**) for both positive definite and positive semidefinite, while others use non-negative definite to mean positive semidefinite only.

[5] The biased variance results when dividing the sum of squared perturbations about the mean by n, instead of $n - 1$ (which produces the unbiased estimate). Note that for the present example, the result would differ only in that the factor $1/n^2$ would become $1/[n(n-1)]$.

C serves to remove the mean (while also scaling), it is often referred to as the **centering matrix**.

Pseudo-inverse. Pseudo-inverses are the means for computing the "inverse" of a nonsquare matrix, **A**. Specifically, if **A** is overdetermined, having more rows than columns, then a pseudo-inverse known as the **left inverse** is obtained as

$$\mathbf{A}_{L}^{-1} = (\mathbf{A}^{T}\mathbf{A})^{-1}\mathbf{A}^{T}. \tag{A1.44a}$$

and for the case where **A** is underdetermined, having more columns than rows, then a pseudo-inverse known as the **right inverse** is obtained as

$$\mathbf{A}_{R}^{-1} = \mathbf{A}^{T}(\mathbf{A}\mathbf{A}^{T})^{-1}. \tag{A1.44b}$$

For equations of the form: $\mathbf{Ax} = \mathbf{b}$, these lead to the following solutions (discussed more in context later).

For overdetermined systems, where matrix **A** has more rows than columns, the inverse solution (in fact, the least-squares solution) is thus

$$\mathbf{x} = (\mathbf{A}^{T}\mathbf{A})^{-1}\mathbf{A}^{T}\mathbf{b} = \mathbf{A}_{L}^{-1}\mathbf{b}. \tag{A1.45a}$$

whereas for underdetermined systems, where matrix **A** has more columns than rows, the inverse solution is

$$\mathbf{x} = \mathbf{A}^{T}(\mathbf{A}\mathbf{A}^{T})^{-1}\mathbf{b}. \tag{A1.45b}$$

The implications of these are discussed more fully in the regression discussions. Using SVD (discussed below), the pseudo-inverse is:

$$\mathbf{VS^{*}U^{T}} \tag{A1.45c}$$

Where **S* is the original diagonal matrix** S, with each element inverted (i.e., σ_{11}becomes $1/\sigma_{11}$).

A1.7.4 Ill-Conditioning

Ill-conditioned systems are those in which a *small change* in one or more of the coefficient values (in the **A** matrix) or observed values (in the **b** vector) leads to a large change in the solution of the system unknowns. Ill-conditioning arises from three distinctly different sources:

- The physical problem itself may represent an unstable system, such as a bicycle balancing on a tightrope or a pencil balancing on its eraser.
- The basis chosen to represent the system may be poorly chosen, such as a polynomial basis over a range of 0 to 1, versus a basis that is orthogonal over this range.
- The method of solution may be sensitive to small changes, as are most matrix inversion techniques and pivotal methods based on elimination procedures under certain conditions.

Here we consider only the latter two sources, since they are the more common in curve fitting (interpolation and smoothing) problems. Conceptually, they can both be described in a similar manner.

Consider the solutions to the interpolation and least-squares problems that require inversion of **A** or the $\mathbf{A}^T\mathbf{A}$ matrix products. A square matrix, **A**, can only have an inverse if it is of full rank (i.e., there are as many *independent* vectors in the matrix as there are columns or rows). If this is not the case, the determinant of the matrix is zero, and **A** cannot be inverted – recall that the elements of $\mathbf{A}^{-1}$ are scaled by 1/(det **A**), which, for det **A** = 0, is singular (a square matrix that has its determinant equal to 0 is known as a *singular matrix*). If the vectors in **A** are *almost* dependent, then the value of det **A** is small in magnitude relative to certain of the matrix cofactors. This can result in a round-off problem, leading to the loss of accuracy.

The following two examples (from Jennings, 1977), demonstrate the problem. First, consider the system **Ax** = **b** for which the coefficient matrix has the following elements:

$$\mathbf{A} = \begin{bmatrix} .5 & (.5+\alpha) \\ .5 & (.5+\alpha) \\ .5 & (.5-\alpha) \\ .5 & (.5-\alpha) \end{bmatrix}. \tag{A1.46}$$

This matrix has rank 2 and the square matrix $\mathbf{A}^T\mathbf{A}$ is

$$\mathbf{A}^T\mathbf{A} = \begin{bmatrix} 1 & 1 \\ 1 & 1+4\alpha^2 \end{bmatrix}. \tag{A1.47}$$

This matrix product is square and of full rank, and thus has an inverse. However, if $\alpha \ll 1$, then $\alpha^2 \ll \alpha$. Machine round-off may force the computer to represent the small $4\alpha^2$ as 0. This will artificially make the matrix product in (A1.47) singular, with no matrix solution. Conceptually, the source of the problem is that the two rows (or columns) in $\mathbf{A}^T\mathbf{A}$ are nearly linearly dependent when α is small.

Note that the magnitude of the determinant is sometimes (erroneously) used as a measure of the **condition** of the matrix – the smaller the magnitude, the worse the conditioning, because the value of 1/(det **A**) will approach singularity. This can be shown to be invalid through a simple scaling experiment similar to the above example. That is, one can scale the above $\mathbf{A}^T\mathbf{A}$ matrix so that its determinant is changed tremendously, yet the condition of the matrix is not changed at all. There are actually more accurate measures of the condition of a matrix that reveal whether it is ill-conditioned or not, but they can be computationally intensive so people often avoid them.

Second, consider the exact fit to n data points using an nth-degree polynomial or a least-squares fit to these data of an $(m-1)$th-degree polynomial, where m is a large number. So,

$$y_i = a_1 + a_2x_i + a_3x_i^2 + a_4x_i^3 + \ldots + \tag{A1.48}$$

where the x_i are evenly spaced and span the range from 0 to 1.

If n is large, the coefficient matrix, **A**, can be very ill-conditioned (especially if scaled by $1/n$, in which case the matrix approximates the ill-conditioned *Hilbert matrix*). This is

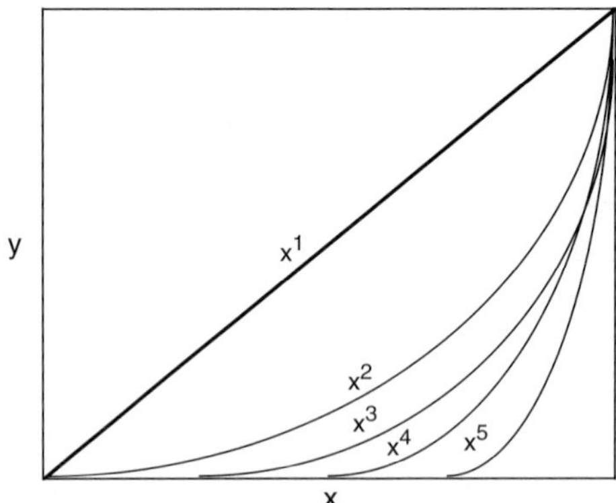

Figure A1.5 Polynomials of different degrees, showing how the higher degrees start to look similar at larger y values.

because the terms of the polynomial basis, x, x^2, … , x^n, have similar shape and thus display a strong degree of linear dependence, as seen in Figure A1.5.

This problem arises due to the choice of the basis for the given range of x. In this case, a simple solution exists if you choose another basis that has constituent functions that are strongly independent over the range of x – or better, that are orthogonal over the range (e.g., Chebyshev polynomials or orthogonal polynomials).

Alternatively, another solution is to convert the offending functions into a mutually orthogonal set through a set of simple operations on the system. This approach works equally well for both of the above two examples.

A1.7.5 Orthogonal Decomposition

Orthogonal decomposition is the name given to general procedures used to transform the variables into a set of mutually orthogonal functions. Since the orthogonal functions are "completely" independent, the above problems of ill-conditioning are largely circumvented.

Orthogonal decomposition relies on the identity that any $n \times m$ matrix **A** $(n \geq m)$ can be decomposed into the product of an $n \times m$ orthogonal matrix **Q** satisfying

$$\mathbf{Q}^{\mathrm{T}}\mathbf{Q} = \mathrm{I}, \tag{A1.49}$$

and another matrix or matrix product, such as an $m \times m$ upper-triangular matrix **R**, so

$$\mathbf{A}_{nm} = \mathbf{Q}_{nm}\mathbf{R}_{mm}. \tag{A1.50}$$

This is called **QR decomposition** and is one form of orthogonal decomposition.

Singular-value decomposition (**SVD**) is another popular form where

$$\mathbf{A}_{nm} = \mathbf{U}_{nn}\mathbf{S}_{nm}\mathbf{V}^{\mathbf{T}}{}_{mm} \tag{A1.51}$$

and **U**, $\mathbf{V}^{\mathrm{T}}$ are orthogonal and **S** is a diagonal matrix.

The matrices **Q**, **U** and **V** are all **orthonormal**. This means that each row (or column) has magnitude 1, so $||\mathrm{q}|| = 1$.

Both QR decomposition and SVD have particular properties that make them advantageous for a variety of matrix problems. Regarding curve fitting problems, the QR decomposition is often considered ideal for problems in which n $\lesssim$ 100, though Lawson and Hanson (1974) like SVD in most cases because it returns information regarding the coefficient and data matrices.

I will demonstrate QR decomposition here because it is the more straightforward.

Given the QR decomposition, a system,

$$\mathbf{Ax} = \mathbf{b}, \tag{A1.52}$$

is decomposed to

$$\mathbf{QRx} = \mathbf{b} \tag{A1.53a}$$

$$\mathbf{Q}^T\mathbf{QRx} = \mathbf{Q}^T\mathbf{b}, \tag{A1.53b}$$

and, since $\mathbf{Q}^T\mathbf{Q} = \mathbf{I}$,

$$\mathbf{Rx} = \mathbf{Q}^T\mathbf{b}. \tag{A1.54}$$

R is nonsingular, so this equation can always be solved (the implication of the solution when **A** is not square as discussed below). Furthermore, this matrix equation is easily solved using **backsubstitution** – a simple procedure solving each consecutive row of **Rx**.

The **Rx** system represents a system of the form

$$\begin{aligned} 2x_1 + 4x_2 - 3x_3 &= y_1 \\ 1x_2 + 2x_3 &= y_2 \\ x_3 &= y_3, \end{aligned} \tag{A1.55}$$

which is solved from the bottom up (by backsubstituting):

$$\begin{aligned} x_3 &= y_3 \\ x_2 &= (y_2 - 2x_3)/1 \\ x_1 &= (y_1 + 3x_3 - 4x_2)/2. \end{aligned} \tag{A1.56}$$

Now consider the QR decomposition given a nonsquare matrix **A**, so

$$\mathbf{Ax} = \mathbf{b} \tag{A1.57}$$

and the standard least-squares solution is given by

$$[\mathbf{A}^T\mathbf{A}]^{-1}\mathbf{A}^T\mathbf{Ax} = [\mathbf{A}^T\mathbf{A}]^{-1}\mathbf{A}^T\mathbf{b}. \tag{A1.58}$$

Multiplying (A1.57), by $\mathbf{A}^T$ to get

$$\mathbf{A}^T\mathbf{Ax} = \mathbf{A}^T\mathbf{b} \tag{A1.59}$$

and substituting $\mathbf{A} = \mathbf{QR}$ and $\mathbf{A}^T = \mathbf{R}^T\mathbf{Q}^T$ (employing the reversal rule of transposed products) gives

$$\mathbf{R}^T\mathbf{Q}^T\mathbf{QRx} = \mathbf{R}^T\mathbf{Q}^T\mathbf{b}, \tag{A1.60}$$

Box DA1.1 Derivation of Orthogonal Decomposition (Optional)

DA1.1.1 Gram–Schmidt Procedure

Orthogonal transformations (e.g., computing the **Q** and **R** matrices) are most easily demonstrated using a method known as the **Gram–Schmidt procedure**, though the **Householder transform** is more efficient (and popular), as is the **Givens transformation**.[6]

The Gram–Schmidt Procedure works as follows. Consider the product **QR** = **A**,

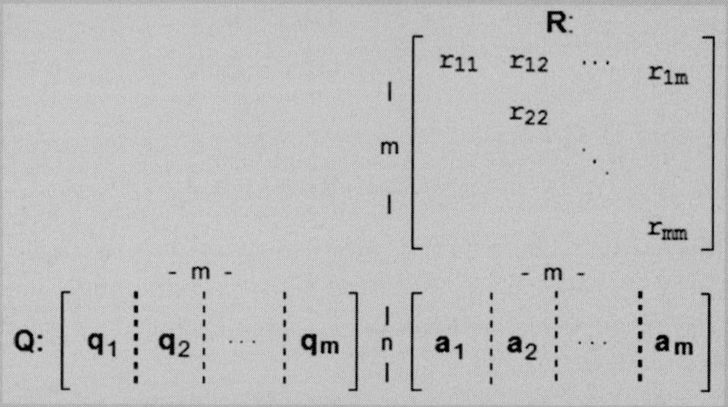

$$\mathbf{R}: \begin{bmatrix} r_{11} & r_{12} & \cdots & r_{1m} \\ & r_{22} & & \\ & & \ddots & \\ & & & r_{mm} \end{bmatrix} \quad (m \times m)$$

$$\mathbf{Q}: \left[\, \mathbf{q}_1 \;\vdots\; \mathbf{q}_2 \;\vdots\; \cdots \;\vdots\; \mathbf{q}_m \,\right] \quad (n \times m) \qquad \left[\, \mathbf{a}_1 \;\vdots\; \mathbf{a}_2 \;\vdots\; \cdots \;\vdots\; \mathbf{a}_m \,\right] \quad (n \times m)$$

where the **Q** and **A** matrices have been partitioned into m column vectors, so

$$a_1 = \begin{bmatrix} a_{11} \\ a_{21} \\ a_{31} \\ \vdots \\ a_{n1} \end{bmatrix}; \quad q_1 \begin{bmatrix} q_{11} \\ q_{21} \\ q_{31} \\ \vdots \\ q_{n1} \end{bmatrix}. \tag{DA1.1.2}$$

So, the multiplication **QR** can be done in terms of vectors. The first vector of **A** is

$$\mathbf{a}_1 = \mathbf{q}_1 r_{11} = r_{11}\mathbf{q}_1. \tag{DA1.1.3}$$

since r_{11} is a scalar.

Now consider the **orthonormal property** of **Q**. This states that

$$q_i^T q_j = \begin{cases} 1 & i = j \\ 0 & i \neq j. \end{cases} \tag{DA1.1.4}$$

In other words, (DA1.1.4) states that the magnitude (L_2 norm, or Euclidean length) of each vector within **Q** is unity. For example, $[q_{11}^2 + q_{12}^2 + \ldots + q_{n1}^2]^{1/2} = \mathbf{q}_1^T \mathbf{q}_1 = 1$. Also, each vector within **Q** is orthogonal to every other vector in **Q**, hence, by definition, the dot product between all of the vectors is zero.

[6] A popular method for least-squares (L2 norm) type problems involves Golub's method, using Householder transformations.

Box DA1.1 (Cont.)

Consider the case of three columns in $\mathbf{Q}$, so $m = 3$:

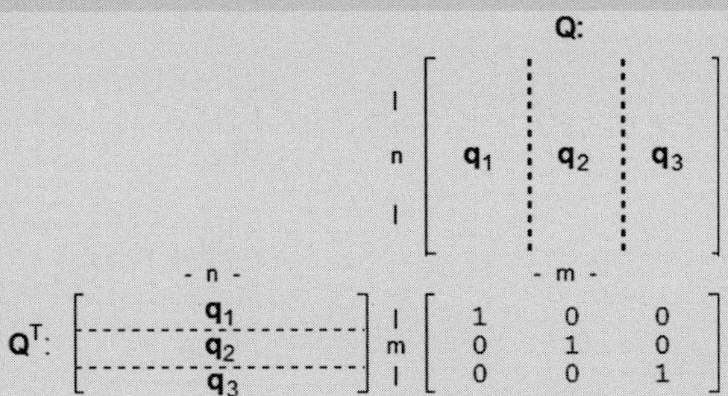

so $\mathbf{Q}^T\mathbf{Q} = \mathbf{I}$.

Multiplying (DA1.1.3) by a_1^T and then imposing the orthonormal constraint, $\mathbf{q}^T\mathbf{q}_1 = 1$ gives

$$\begin{aligned}\mathbf{a}_1^T\mathbf{a}_1 &= (\mathbf{q}_1 r_{11})^T(\mathbf{q}_1 r_{11}) \\ &= r_{11}\mathbf{q}_1^T\mathbf{q}_1 r_{11} \\ &= r_{11}^2.\end{aligned} \tag{DA1.1.6}$$

so

$$r_{11} = (\mathbf{a}_1^T\mathbf{a}_1)^{1/2} = \|\mathbf{a}_1\|, \tag{DA1.1.7}$$

which is the Euclidean norm of the vector $\mathbf{a}_1$. Given r_{11}, (DA1.1.3) can be solved for $\mathbf{q}_1$:

$$\mathbf{q}_1 = \frac{1}{r_{11}}\mathbf{a}_1 \tag{DA1.1.8}$$

So, $\mathbf{q}_1$ is simply the $\mathbf{a}_1$ vector normalized by its Euclidean norm (scaling a vector by its norm is known as **normalization** – it is a standard procedure used to make the norm of any vector unity).

The second vector of **A,** $\mathbf{a}_2$, is

$$\mathbf{a}_2 = \mathbf{q}_1 r_{12} + \mathbf{q}_2 r_{22}. \tag{DA1.1.9}$$

Pre-multiplying by $\mathbf{q}_1^T$,

$$\mathbf{q}_1^T\mathbf{a}_2 = \mathbf{q}_1^T\mathbf{q}_1 r_{12} + \mathbf{q}_1^T\mathbf{q}_2 r_{22}, \tag{DA1.1.10}$$

and imposing orthonormal conditions (i.e., $\mathbf{q}_1^T\mathbf{q}_1 = 1$ and $\mathbf{q}_1^T\mathbf{q}_2 = 0$) solves (DA1.1.10) for r_{12}:

$$\mathbf{q}_1^T\mathbf{a}_2 = r_{12}. \tag{DA1.1.11}$$

Rearranging equation (DA1.1.9) in terms of the two unknowns, $\mathbf{q}_2$ and r_{22}, gives

$$\mathbf{q}_2 r_{22} = \mathbf{a}_2 - \mathbf{q}_1 r_{12}. \tag{DA1.1.12}$$

This is abbreviated by defining the known quantity as

$$\widetilde{\mathbf{a}}_2 = \mathbf{a}_2 - \mathbf{q}_1 r_{12} \tag{DA1.1.13a}$$

Box DA1.1 (Cont.)

and, substituting (DA1.1.12) into (DA1.1.13a), gives

$$\widetilde{\mathbf{a}}_2 = \mathbf{q}_2 r_{22} \qquad \text{(DA1.1.13b)}$$

With this, we can follow a solution procedure similar to that used to solve (DA1.1.3):

$$\begin{aligned}\widetilde{\mathbf{a}}_2^{\mathrm{T}}\widetilde{\mathbf{a}}_2 &= r_{22}\mathbf{q}_2^{\mathrm{T}}\mathbf{q}_2 r_{22}\\ &= r_{22}^2\end{aligned} \qquad \text{(DA1.1.14)}$$

so

$$r_{22} = \left(\widetilde{\mathbf{a}}_2^{\mathrm{T}}\widetilde{\mathbf{a}}_2\right)^{1/2} = ||\widetilde{\mathbf{a}}_2|| \qquad \text{(DA1.1.15)}$$

Finally, substitution of (DA1.1.15) into (DA1.1.13b) gives $\mathbf{q}_2$:

$$\begin{aligned}\mathbf{q}_2 &= \frac{1}{r_{22}}\widetilde{\mathbf{a}}_2\\ &= \frac{1}{r_{22}}(\mathbf{a}_2 - \mathbf{q}_1 r_{12}).\end{aligned} \qquad \text{(DA1.1.16)}$$

This simple procedure is repeated until all elements of **R** and columns of **Q** are formed, thus transforming **A** to **QR**. Other approaches can also be used to efficiently obtain the QR decomposition besides the Gram–Schmidt procedure, which is relatively inefficient. (Note that the **R** matrix is identical to the upper-triangular matrix obtained by Cholesky decomposition, though the latter can suffer from round-off accumulation error.)

DA1.1 Overview of Householder Transformation

The **Householder transformation** is known as a **reflection transformation**. It is an efficient and popular method for actually computing the QR decomposition, though it serves a variety of other uses as well. It is a **unitary transformation,** meaning that it *preserves length* (magnitude), so if the transformation matrix M is unitary, then

$$\mathbf{M}^{\mathrm{T}}\mathbf{M} = \mathbf{I}, \qquad \text{(DA1.1.17)}$$

and **Mx** and **x** have the same length:

$$\begin{aligned}\mathbf{x}^{\mathrm{T}}\mathbf{x} &= (\mathbf{M}\mathbf{x})^{\mathrm{T}}\mathbf{M}x\\ &= \mathbf{x}^{\mathrm{T}}\mathbf{M}^{\mathrm{T}}\mathbf{M}\mathbf{x}\\ &= \mathbf{x}^{\mathrm{T}}\mathbf{x}.\end{aligned} \qquad \text{(DA1.1.18)}$$

The **Householder transformation matrix, H**, transforms the coefficient matrix, **A**, to an upper diagonal form (as all QR decompositions do).

Box DA1.1 (Cont.)

So,

$$\mathbf{A} = \mathbf{QR} \tag{DA1.1.19}$$

and, from condition (DA1.1.17),

$$\mathbf{Q}^{\mathrm{T}}\mathbf{A} = \mathbf{R}, \tag{DA1.1.20}$$

or, in terms of the Householder transform matrix, **H**,

$$\mathbf{HA} = \mathbf{R}. \tag{DA1.1.21}$$

Thus **H** (= $\mathbf{Q}^{\mathrm{T}}$) *transforms* **A** to **R** and the length of **A** is preserved.

The Householder transformation accomplishes the transform to **R** by transforming each column vector, $\mathbf{a}_j$, in **A**, one at a time,

$$\mathbf{Ha}_j = -\sigma||\mathbf{a}_j||\mathbf{e}_1, \tag{DA1.1.22}$$

where

$$\mathrm{e}_1 = \begin{bmatrix} 1 \\ 0 \\ \vdots \\ 0 \end{bmatrix}; \ \sigma = \begin{cases} +1 & \mathrm{a}_{1j} \geq 0 \\ -1 & \mathrm{a}_{1j} < 0 \end{cases}, \tag{DA1.1.23}$$

so the sign of the transformed column simply reflects the sign of the first element in the column vector being transformed.

The transformation of $\mathbf{a}_j$ is accomplished by defining an arbitrary column vector:

$$\mathbf{v} = \mathbf{a}_j + \sigma||\mathbf{a}_j||\mathbf{e}_1. \tag{DA1.1.24}$$

Then,

$$\mathbf{H} = \mathbf{I}_m - \frac{2\mathbf{v}\mathbf{v}^{\mathrm{T}}}{\mathbf{v}^{\mathrm{T}}\mathbf{v}}. \tag{DA1.1.25}$$

Note that while $\mathbf{v}^{\mathrm{T}}\mathbf{v}$ = scalar (the dot product); $\mathbf{vv}^{\mathrm{T}}$ is an $n \times n$ matrix (where n is the size of the **v** column vector):

$$\begin{matrix} & & [a_1 & a_2 & \dots & a_n] \\ \mathbf{vv}^{\mathrm{T}} = & \begin{bmatrix} a_1 \\ a_2 \\ \vdots \\ a_n \end{bmatrix} & \begin{bmatrix} a_1^2 & a_1 a_2 & \dots & a_1 a_n \\ a_2 a_1 & a_2^2 & & \\ \vdots & & \ddots & \\ a_n a_1 & a_n a_2 & & a_n^2 \end{bmatrix} \end{matrix}. \tag{DA1.1.26}$$

A **reflection matrix** such as **H,** above, is its own inverse, so

$$\mathbf{H} = \mathbf{H}^{-1} \tag{DA1.1.27}$$

Box DA1.1 (Cont.)

and

$$\mathbf{H}^{-1}\mathbf{H} = \mathbf{I} = \mathbf{H}^{\mathrm{T}}\mathbf{H}, \tag{DA1.1.28}$$

since

$$\begin{aligned}\mathbf{H}^{\mathrm{T}}\mathbf{H} &= \left(\mathbf{I} - \frac{2\mathbf{v}\mathbf{v}^{\mathrm{T}}}{\mathbf{v}^{\mathrm{T}}\mathbf{v}}\right)^2 \\ &= \mathbf{I} - \frac{4\mathbf{v}\mathbf{v}^{\mathrm{T}}}{\mathbf{v}^{\mathrm{T}}\mathbf{v}} + \frac{4\mathbf{v}(\mathbf{v}\mathbf{v}^{\mathrm{T}})\mathbf{v}^{\mathrm{T}}}{(\mathbf{v}^{\mathrm{T}}\mathbf{v})^2} \\ &= \mathbf{I}.\end{aligned} \tag{DA1.1.29}$$

Pre-multiplication by **H** converts to zeros all of the elements *below* the main diagonal of one column of the matrix being transformed. *Post-multiplication* zeros the elements *above* the main diagonal of one row of the matrix. Thus, depending upon the implementation, a succession of **H** transforms can transform a matrix to an upper-triangular, lower-triangular, tridiagonal or other useful form. In practice, the Householder transform can be implemented in a highly efficient manner.

and $\mathbf{Q}^{\mathrm{T}}\mathbf{Q} = \mathbf{I}$, so this reduces to

$$\mathbf{R}^{\mathrm{T}}\mathbf{R}\mathbf{x} = \mathbf{R}^{\mathrm{T}}\mathbf{Q}^{\mathrm{T}}\mathbf{b}. \tag{A1.61}$$

Multiplying through by $\mathbf{R}^{-\mathrm{T}}$ (i.e., the inverse of $\mathbf{R}^{\mathrm{T}}$) gives

$$\mathbf{R}\mathbf{x} = \mathbf{Q}^{\mathrm{T}}\mathbf{b}, \tag{A1.62}$$

which is identical to (A1.54). In fact, *the solution of an overdetermined system* $\boldsymbol{Ax} = \boldsymbol{b}$ *by QR decomposition leads directly to the least-squares solution*; (A1.58) is directly equivalent to the QR decomposition form of (A1.62).

Since **R** replaces $\mathbf{A}^{\mathrm{T}}\mathbf{A}$ and is upper triangular (i.e., solved by backsubstitution), we avoid computing $(\mathbf{A}^{\mathrm{T}}\mathbf{A})^{-1}$, which exacerbates problems of ill-conditioning. Orthogonal decomposition does not actually eliminate ill-conditioning – rather, by not computing the $(\mathbf{A}^{\mathrm{T}}\mathbf{A})^{-1}$, it reduces the sensitivity by reducing the loss of precision due to round-off error. In general, *orthogonal decomposition preserves about twice as many significant digits as does the comparable elimination (inverse) solution*, though the orthogonal decomposition requires more operations. Therefore, when a matrix inverse can be stably inverted by using double precision, an orthogonal decomposition will likely achieve the same level of stability using single precision, making "more" operations warranted.

If QR decomposition is used to solve weighted least-squares problems (done by modifying **A** and **b** prior to decomposition as discussed in Chapter 6), some row interchanges may be required if disparately weighted rows result. This may be necessary to avoid subtracting very small numbers from very large numbers, which can suffer from round-off errors during the backsubstitution.

A1.8 Statistical Topics

A1.8.1 Random Vectors and Matrices

We now consider the cases where we are working with vectors and matrices that are filled with random variables. Such vectors or matrices are themselves random vectors or matrices, since the actual elements occupying them can be any of a set of values representing the spread in each element. So, consider a random vector, **X**:

$$\mathbf{X} = \begin{bmatrix} X_1 \\ X_2 \\ X_3 \\ \vdots \\ X_n \end{bmatrix}, \qquad \text{(A1.63)}$$

where the elements each represent a random variable, X_i. Depending on which values are present in the vector, any one particular vector represents just one realization of an infinite (or finite, if each random variable has a finite number of possible values or events) number of possible vectors, **X**. Thus, the random vector is analogous to a scalar random variable.

Now consider the random matrix **Z**:

$$\mathbf{Z} = \begin{bmatrix} Z_{11} & Z_{12} & \dots & Z_{1m} \\ Z_{21} & Z_{22} & \dots & Z_{2m} \\ Z_{31} & Z_{32} & \dots & Z_{3m} \\ & & \vdots & \\ Z_{n1} & Z_{n2} & \dots & Z_{nm} \end{bmatrix}, \qquad \text{(A1.64)}$$

where again each element represents a random variable.

Random vectors and matrices represent an ideal means for manipulating multivariate random variables or situations in which we are dealing with systems of random variables.

A1.8.2 Statistical Moments of Random Matrices (Expectance Operations)

Mean

Now consider the expectance of a random matrix. As before, this is denoted E[**Z**], indicating

$$\mathrm{E}[Z] = \begin{bmatrix} \mathrm{E}[Z_{11}] & \mathrm{E}[Z_{12}] & \dots & \mathrm{E}[Z_{1m}] \\ \mathrm{E}[Z_{21}] & \mathrm{E}[Z_{22}] & \dots & \mathrm{E}[Z_{2m}] \\ \mathrm{E}[Z_{31}] & \mathrm{E}[Z_{32}] & \dots & \mathrm{E}[Z_{3m}] \\ & & \vdots & \\ \mathrm{E}[Z_{n1}] & \mathrm{E}[Z_{n2}] & \dots & \mathrm{E}[Z_{nm}] \end{bmatrix} = \begin{bmatrix} \mu_{11} & \mu_{12} & \cdots & \mu_{1m} \\ \mu_{21} & \mu_{22} & \cdots & \mu_{2m} \\ \mu_{31} & \mu_{32} & \cdots & \mu_{3m} \\ & & \vdots & \\ \mu_{n1} & \mu_{n2} & \cdots & \mu_{nm} \end{bmatrix}, \qquad \text{(A1.65a)}$$

or

$$= \boldsymbol{\mu}. \qquad \text{(A1.65b)}$$

So, the expectance of random matrix **Z** is the mean, or mean matrix, $\boldsymbol{\mu}$. Likewise, for the random vector, $E[\mathbf{X}] = \boldsymbol{\mu}$. The elements of $\boldsymbol{\mu}$ contain the expected value of each element of **Z** or **X**.

As shown previously, a scalar times a matrix is equal to the matrix with each individual element multiplied by the scalar. From this and (A1.65a), it is clear that

$$E[a\mathbf{X}] = aE[\mathbf{X}] = a\boldsymbol{\mu}. \tag{A1.66}$$

$$E[a\mathbf{X} + b] = aE[\mathbf{X}] + E[b] = a\boldsymbol{\mu} + b. \tag{A1.67}$$

Likewise for the vector products and sums, where **a, b** and **c** are nonrandom variable vectors, in which case,

$$E[\mathbf{aX}] = \mathbf{a}E[\mathbf{X}] = \mathbf{a}\boldsymbol{\mu}. \tag{A1.68}$$

$$E[\mathbf{aX} + \mathbf{b}] = \mathbf{a}E[\mathbf{X}] + E[\mathbf{b}] = \mathbf{a}\boldsymbol{\mu} + \mathbf{b}. \tag{A1.69}$$

$$E[\mathbf{aXb} + \mathbf{c}] = \mathbf{a}E[\mathbf{X}]\mathbf{b} + E[\mathbf{c}] = \mathbf{a}\boldsymbol{\mu}\mathbf{b} + \mathbf{c}. \tag{A1.70}$$

Variance (Covariance Matrix)

The variance of random vector, **X**, is analogous to the variance of a scalar random variable, **X**, where it was then the expectance of a function of the random variable $(\mathbf{X} - \boldsymbol{\mu})^2$, and in this case it is given as the expectance of a random matrix **Z** of the form

$$\mathbf{Z} = (\mathbf{X}-E[\mathbf{X}])(\mathbf{X}-E[\mathbf{X}])^T, \tag{A1.71}$$

so

$$\begin{aligned} \mathrm{Var}[\mathbf{X}] = \boldsymbol{\Sigma} &= E[(\mathbf{X} - E[\mathbf{X}])(\mathbf{X} - E[\mathbf{X}])^T] \\ &= E[(X - \mu)(X - \mu)^T], \end{aligned} \tag{A1.72a}$$

where $\boldsymbol{\Sigma}$ is the **covariance matrix** of **X** (some people refer to $\boldsymbol{\Sigma}$ as the **variance-covariance matrix**).

Equation (A1.72a) is expanded and reduced from (A1.68) as

$$\begin{aligned} &= E[(XX^T - \mu X^T - X\mu^T + \mu\mu^T] \\ &= E[XX^T] - E[\mu X^T] - E[X\mu^T] + E[\mu\mu^T] \\ &= E[XX^T] - \mu E[X^T] - E[X]\mu^T + \mu\mu^T \\ &= E[XX^T] - \mu\mu^T - \mu\mu^T + \mu\mu^T \\ &= E[XX^T] - \mu\mu^T. \end{aligned} \tag{A1.72b}$$

This is equivalent to the scalar operation

$$= E[(X_i - \mu_i)(X_j - \mu_j)] \tag{A1.72c}$$

for each element i,j, $i = 1,2,3, \ldots,n$, $j = 1,2,3, \ldots,m$. As a scalar operation for pairs of elements, it is clear (by considering the operation in (A1.72c) from Chapter 2 on Probability) that when $i \neq j$, this gives the covariance between the various pairs of random variables in **X**, and when $i = j$, it is giving the variance.

Consider the vector operation in (A1.72a). This involves the *outer product* $(\mathbf{X}-\boldsymbol{\mu})(\mathbf{X}-\boldsymbol{\mu})^T$. First, for the case where the mean vector is the null vector, i.e., the X random variables have zero mean, then this outer product is given as $\mathbf{XX}^T$, or

$$\mathbf{XX}^T = \begin{bmatrix} X_1 \\ X_2 \\ X_3 \\ \cdots \\ X_n \end{bmatrix} \begin{matrix} [X_1 & X_2 & \ldots & X_n] \\ \begin{bmatrix} X_1X_1 & X_1X_2 & \ldots & X_1X_n \\ X_2X_1 & X_2X_2 & \ldots & X_2X_n \\ X_3X_1 & X_3X_2 & \ldots & X_3X_n \\ & & \vdots & \\ X_nX_1 & X_nX_2 & \ldots & X_nX_n \end{bmatrix} \end{matrix}, \qquad \text{(A1.73a)}$$

and the expectance of this matrix, as indicated by (A1.65), is

$$E[\mathbf{XX}^T] = \begin{bmatrix} E[x_1x_1] & E[x_1x_2] & \ldots & E[x_1x_n] \\ E[x_2x_1] & E[x_2x_2] & \ldots & E[x_2x_n] \\ E[x_3x_1] & E[x_3x_2] & \ldots & E[x_3x_n] \\ & & \vdots & \\ E[x_nx_1] & E[x_nx_2] & \ldots & E[x_nx_n] \end{bmatrix}. \qquad \text{(A1.73b)}$$

If the mean is not zero, then the operation is the same, except that the mean for each random variable is subtracted off so that the above matrix product looks like this:

$$E[(\mathbf{X}-\boldsymbol{\mu})(\mathbf{X}-\boldsymbol{\mu})^T] =$$

$$\begin{bmatrix} E[(X_1-\mu_1)(X_1-\mu_1)] & E[(X_1-\mu_1)(X_2-\mu_2)] & \ldots & E[(X_1-\mu_1)(X_n-\mu_n)] \\ E[(X_2-\mu_2)(X_1-\mu_1)] & E[(X_2-\mu_2)(X_2-\mu_2)] & \ldots & E[(X_2-\mu_2)(X_n-\mu_n)] \\ E[(X_3-\mu_3)(X_1-\mu_1)] & E[(X_3-\mu_3)(X_2-\mu_2)] & \ldots & E[(X_3-\mu_3)(X_n-\mu_n)] \\ & & \vdots & \\ E[(X_n-\mu_n)(X_1-\mu_1)] & E[(X_n-\mu_n)(X_2-\mu_2)] & \ldots & E[(X_n-\mu_n)(X_n-\mu_n)] \end{bmatrix}$$
$$\text{(A1.73c)}$$

In either case, with $\boldsymbol{\mu}$ zero or not, when n equals the full population or ensemble size (N or ∞), (A1.73b,c) is

$$\mathrm{Var}[\mathbf{X}] = \boldsymbol{\Sigma} = \begin{bmatrix} \sigma_{11} & \sigma_{12} & \ldots & \sigma_{1n} \\ \sigma_{21} & \sigma_{22} & \ldots & \sigma_{2n} \\ \sigma_{31} & \sigma_{32} & \ldots & \sigma_{3n} \\ & & \vdots & \\ \sigma_{n1} & \sigma_{n2} & \ldots & \sigma_{nn} \end{bmatrix}$$

$$= \begin{bmatrix} \sigma_1^2 & \sigma_{12} & \ldots & \sigma_{1n} \\ \sigma_{21} & \sigma_2^2 & \ldots & \sigma_{2n} \\ \sigma_{31} & \sigma_{32} & \ldots & \sigma_{3n} \\ & & \vdots & \\ \sigma_{n1} & \sigma_{n2} & \ldots & \sigma_n^2 \end{bmatrix}, \qquad \text{(A1.73d)}$$

where the $\sigma_{ii} = \sigma_i^2$ = the true variance of the random variable $\mathbf{X}_i$ and the σ_{ij} = the true covariance between the random variable $\mathbf{X}_i$ and $\mathbf{X}_j$. When n is less than the full ensemble size, the above is the sample Var[**X**] = **S**, populated by sample variance (s_i^2) and covariance (s_{ij}) estimates, or lagged covariance estimates ($\gamma(k)$) for ergodic series.

Thus, the variance of random vector **X** is given by the covariance matrix, **Σ**.

Consider the variance of the following vector product and sum, where **a** and **b** are nonrandom variable vectors, in which case

$$\mathrm{Var}[\mathbf{aX} + \mathbf{b}] = \mathrm{E}[(\mathbf{aX} + \mathbf{b} - \mathrm{E}[\mathbf{aX} + \mathbf{b}])(\mathbf{aX} + \mathbf{b} - \mathrm{E}[\mathbf{aX} + \mathbf{b}])^{\mathrm{T}}]. \qquad \text{(A1.74a)}$$

The expectance term $\mathrm{E}[\mathrm{aX} + \mathrm{b}] = \mathrm{aE}[\mathrm{X}] + \mathrm{b} = \mathrm{a}\mu + \mathrm{b}$ from (A1.69), so (A1.74a) becomes

$$\begin{aligned}
&= \mathrm{E}[(\mathrm{a}\mathbf{X} + \mathbf{b} - (\mathrm{a}\mu + \mathbf{b}))(\mathrm{a}\mathbf{X} + \mathbf{b} - (\mathrm{a}\mu + \mathbf{b}))^{\mathrm{T}}] \\
&= \mathrm{E}[(\mathrm{a}\mathbf{X} + \mathbf{b} - \mathrm{a}\mu - \mathbf{b})(\mathrm{a}\mathbf{X} + \mathbf{b} - \mathrm{a}\mu - \mathbf{b})^{\mathrm{T}}] \\
&= \mathrm{E}[(\mathrm{a}\mathbf{X} - \mathrm{a}\mu)(\mathrm{a}\mathbf{X} - \mathrm{a}\mu)^{\mathrm{T}}] \\
&= \mathrm{E}[(\mathrm{a}\mathbf{X}\mathbf{X}^{\mathrm{T}}\mathrm{a}^{\mathrm{T}} - \mathrm{a}\mu\mathbf{X}^{\mathrm{T}}\mathrm{a}^{\mathrm{T}} - \mathrm{a}\mathbf{X}\mu^{\mathrm{T}}\mathrm{a}^{\mathrm{T}} + \mathrm{a}\mu\mu^{\mathrm{T}}\mathrm{a}^{\mathrm{T}}] \\
&= \mathrm{E}[\mathrm{a}\mathbf{X}\mathbf{X}^{\mathrm{T}}\,\mathrm{a}^{\mathrm{T}}] - \mathrm{E}\,[\mathrm{a}\mu\mathbf{X}^{\mathrm{T}}\mathrm{a}^{\mathrm{T}}] - \mathrm{E}[\mathrm{a}\mathbf{X}\mu^{\mathrm{T}}\mathrm{a}^{\mathrm{T}}] + \mathrm{E}[\mathrm{a}\mu\mu^{\mathrm{T}}\mathrm{a}^{\mathrm{T}}] \\
&= \mathrm{E}[\mathrm{a}\mathbf{X}\mathbf{X}^{\mathrm{T}}\mathrm{a}^{\mathrm{T}}] - \mathrm{a}\mu\mathrm{E}[\mathbf{X}^{\mathrm{T}}]\mathrm{a}^{\mathrm{T}} - \mathrm{aE}[\mathbf{X}]\mu^{\mathrm{T}}\mathrm{a}^{\mathrm{T}} + \mathrm{a}\mu\mu^{\mathrm{T}}\mathrm{a}^{\mathrm{T}} \\
&= \mathrm{aE}[\mathbf{X}\mathbf{X}^{\mathrm{T}}]\mathrm{a}^{\mathrm{T}} - \mathrm{a}\mu\mu^{\mathrm{T}}\mathrm{a}^{\mathrm{T}} - \mathrm{a}\mu\mu^{\mathrm{T}}\mathrm{a}^{\mathrm{T}} + \mathrm{a}\mu\mu^{\mathrm{T}}\mathrm{a}^{\mathrm{T}} \\
&= \mathrm{aE}[\mathbf{X}\mathbf{X}^{\mathrm{T}}]\mathrm{a}^{\mathrm{T}} - \mathrm{a}\mu\mu^{\mathrm{T}}\mathrm{a}^{\mathrm{T}},
\end{aligned} \qquad \text{(A1.74b)}$$

which, from (A1.72b), is seen to be

$$\begin{aligned}
&= \mathbf{a}\mathrm{Var}[\mathbf{X}]\mathbf{a}^{\mathrm{T}} \\
&= \mathbf{a}\boldsymbol{\Sigma}\mathbf{a}^{\mathrm{T}}.
\end{aligned} \qquad \text{(A1.74c)}$$

The sample covariance matrix – that is, the estimate of **Σ** for random vector **X** – is done by estimating the elements of **Σ** using the standard formulas for sample variance, given in (3.6), and sample covariance, given in (3.7). These estimates are then placed in the covariance matrix as indicated in (A1.73).

For stationary and ergodic time series, the covariance matrix comes directly from the autocovariance function discussed in Chapter 7 (equation (7.43a)). For that case, all variances (in the principle diagonal) are equal, and the non-principle diagonal covariance terms are simply a function of lag k (so $\sigma_{21} = \sigma_{32} = \gamma(k{=}1)$), as defined in Chapter 7 and shown in (7.44).

A1.9 Matrix References

Davis, John C., 1973. *Statistics and Data Analysis in Geology.* John Wiley and Sons, New York, NY. Chapter 4.

Good overview of the relevant basic matrix algebra, though he skips many important concepts. Good for a short introduction to the subject, though.

Householder, Alston S., 1975. *Theory of Matrices in Numerical Analysis*. Dover Publications Inc., New York, NY. 257 pp.

In most cases, I have only listed books that I like – this is the exception. Since this is an inexpensive Dover book written by the man for which the Householder transform is named, it may seem like an obvious reference – wrong. This book is incomprehensible unless you have already mastered all of the material he is presenting (which he does in a very abstruse manner). Other than that, it's a great reference.

Jennings, Alan, 1977. *Matrix Computation for Engineers and Scientists*. John Wiley and Sons, New York, NY. 329 pp.

Good book for matrix algebra on computers.

Press, William H., Brian P. Flannery, Saul A. Teukolsky and William T. Vetterling, 1986. *Numerical Recipes*. Cambridge University Press, Cambridge. 818 pp.

Superb general computational text with relevant theory, practical implementation tips and subroutines for everything. Not a review of basic linear algebra, but good stuff for matrix inversions and decompositions of all types.

Searle, Shayle R., 1982. *Matrix Algebra Useful for Statistics*. John Wiley and Sons, New York, NY. 438 pp.

Excellent book with a lot of insights provided that are not found elsewhere. This is not a statistics book, but it does show numerous statistical applications in matrix form. Good discussion on derivatives of matrices.

Strang, Gilbert, 2003. *Introduction to Linear Algebra and Its Applications*, 3rd ed. Wellesley-Cambridge Press, Wellesley, MA. 568 pp.

Excellent introduction to linear algebra.

Strang, Gilbert, 1988. *Linear Algebra and Its Applications*, 3rd ed. Harcourt Brace Jovanovich, Publishers, San Diego, CA. 505 pp.

Great book for advanced linear algebra (one of my very favorites). This is not just an introduction to the technical manipulations of matrices; rather, Strang approaches it as an applied mathematician and gives insights and reasons for the various (often, seemingly esoteric) aspects.

Appendix 2 Uncertainty Analysis

A2.1 Overview

Quantification of uncertainties associated with any particular study is the responsibility of the analyst. An intimidating number of tests have been developed for specific statistical testing, and if you desire, you can do endless research just finding the appropriate test for your particular statistic (assuming such exists). But there are two fundamental tools that can be applied to a great many analyses, regardless of whether they are standard analyses or something more specialized. These two are expectancy (first encountered in Chapter 2) and bootstrap. The former (expectancy) has the advantage of quantifying the contribution to the overall uncertainty of every contributing component, though it also has the disadvantage of being algebraically cumbersome or overly difficult for complex combinations of individual components. The bootstrap, on the other hand, speaks to one's logic, and that makes it understandable and easy to interpret. It also accounts for all hidden dangers, such as effective degrees of freedom, and easily provides the exact significance level that a hypothesis would be rejected (if approaching the uncertainty as a hypothesis test).

For convenient reference, this appendix combines material previously introduced in previous chapters, with elaboration here. Equation numbers are kept as their originals in the chapter where first presented, providing a pointer to the original discussion and development.

A2.2 Classification of Errors

Errors (uncertainties) in data are classified according to their source.

A2.2.1 Instrument Error

Under the best circumstances, the quality of the data is predominantly controlled by the capabilities of the recording device. Measurement capabilities are classified according to:

1) **Precision** – specifies how well a specific measurement of the same sample can be replicated. In statistical terms, precision is a measure of the variance (or standard deviation) of the sample.

For example, if a substance were repeatedly weighed 100 times, giving a mean weight of 100 kg but with a scatter about this mean of 0.1 kg, then 0.1 kg would represent the precision of the measurement.

2) **Accuracy** – specifies how well a specific measurement actually represents the true value of the measured quantity (often considered in terms of, say, a long-term instrument drift). In statistical terms, accuracy is often reported in terms of **bias**.
For example, if a scale repeatedly returns a weight of ~100.0 kg for a substance, but its true weight is 105.3 kg, the mismatch between the measured value and true value reflects the bias of the measurement. So, the scale is good to an accuracy of just over 5 kg, or the scale has a bias of ~5 kg.
3) **Resolution** – specifies the size of a discrete measurement interval of the recording instrument, or that used in the discretization process. In other words, it indicates how well the instrument (or digitized data) can *resolve* changes in the quantity being measured.
For example, if a thermometer only registers changes in temperature of 0.01°C, then it cannot distinguish changes in temperature smaller than this resolution. One would achieve the same resolution if, in the process of digitizing higher-resolution data, all values were rounded off to the nearest 0.01° C.
4) **Response time** – specifies how fast an instrument can respond to a change in the quantity being measured. This will limit the bandwidth (range of frequencies) of a measured time series (discussed in more detail in Chapter 10).

In general, accuracy reflects the degree of systematic errors, while precision reflects the degree of random errors. Statistics are well designed to treat the latter (precision), while they are not generally designed to address the former (accuracy). Accuracy must be estimated by the person who understands the data, using whatever means are practical and reasonable.

A2.2.2 Experimental/Observational Error

The experimental design, sampling program or observational methods may also lead to errors in precision and accuracy.

Precision. Consider estimating the precision of a specific brand of thermometer. If 100 of the thermometers were simultaneously used to measure the temperature of a well-mixed bath of water, the scatter about the mean temperature might typically be presented as the precision of the thermometers. However, this is really the precision of the estimated temperature, and it reflects the precision of the experimental design. Any one particular thermometer may have significantly better or worse precision than that suggested by the scatter achieved between 100 different thermometers. Also, the "well-mixed" bath may actually contain temperature gradients to some extent, which will also influence the scatter observed in the measurements. Repeating the above calibration, only this time making 100 replicate measurements using a single thermometer, may include some scatter due to subtle changes in the water bath between replicate measurements, so even that measured precision reflects some combination of instrument precision and experimental scatter.

In this respect, precision errors may well be attributable to a combination of both instrument and experimental error. This combination is responsible for the observed random scatter in replicate measurements that is often referred to as **noise** in data. Noise can also represent any portion of the data that does not conform to preconceived ideas concerning the nature of the data – recall the expression that "one person's noise is another person's signal."

One of the goals in data analysis is to detect signal in noise or reduce the degree of noise contamination. Noise is often classified according to its contribution relative to some more-stable (or non-fluctuating) component of the observations, referred to as the **signal**.

A2.3 Expectance as Variance

The following material using expectance shows the formal definitions in which the expectance of a random variable is given as μ, the true population mean, and σ_x^2, the true population variance. When you are *estimating* such parameters from sample-based data, you are not getting the true mean or variance, but rather the sample mean, $\bar{x}$, and sample variance, s_x^2.

A2.3.1 Variance of a Univariate Random Variable

For a function $Y(x)$, its expectance is given as

$$\mathrm{E}[X] = \mu = \frac{1}{N}\sum_{i=1}^{N} x_i. \qquad \text{(A2.1; 2.24b)}$$

So the mean of Y is resolved in terms of the distribution of x of which it is a function. When that function is $Y = (X - \mu)^2$, the expectance of this function gives the *variance* for X:

$$\mathrm{E}\left[(X-\mu)^2\right] = \sigma^2 = \sum_{j=1}^{h} (x_j - \mu)^2 \mathrm{p}_X(x_j) = \frac{1}{N}\sum_{i=1}^{N} (x-\mu)_i^2. \qquad \text{(A2.2; 2.31a)}$$

Expansion of the left side of (A2.2) is

$$\begin{aligned} \sigma^2 &= \mathrm{E}[(X-\mu)^2] \\ &= \mathrm{E}[X^2 - 2\mu X + \mu^2] \\ &= \mathrm{E}[X^2] - 2\mu\mathrm{E}[X] + \mu^2 \\ &= \mathrm{E}[X^2] - 2\mu\mu + \mu^2 \\ &= \mathrm{E}[X^2] - \mu^2. \end{aligned} \qquad \text{(A2.3a; 2.32a)}$$

So, the variance (spread) of a random variable is equal to the mean of the sum of squares of the random variable, minus the mean of the random variable squared. Since μ is nothing more than $E[X]$, $\mu^2 = E[X]^2$, and (A2.3) can be written as

$$\begin{aligned}\sigma^2 &= E[X^2] - E[X]^2 \\ &= \mathrm{Var}[X].\end{aligned} \qquad \text{(A2.3b; 2.32b)}$$

Because this is such a standard use of the expectance operator, $\mathbf{E}[(X-\mu)^2]$ *is often given directly as* Var[X]. Also, while we typically call the spread in a random variable the variance, and its root, the **standard deviation**, when estimating the standard deviation from a sample where the Central Limit Theorem has reduced the variance by $1/n$, it is instead called the **standard error (s.e.).**

For the present discussion, the point is that variance is the spread (uncertainty of a random variable). A function of a random variable is itself a random variable, because it, too, cannot be predicted precisely, given its construction using other variables that cannot be stated with absolute certainty. The result of data analysis is nearly always a random variable, given the inexact nature of the data. Therefore, the variance of that result gives the uncertainty in the result.

Variance of a Function of a Random Variable

One big advantage of the above definition for variance, using the expectance operator, is that it allows you to determine the variance of a function of the original random variable (e.g., the result of your analysis). This is invaluable for data analysis, since the analysis often involves the manipulation of your data (i.e., converting the random variable(s) to a new form through some mathematical relationship). Expectancy provides a means by which the uncertainty in the new function can be determined via knowledge of the uncertainty (i.e., the variance) of the original raw data.

The variance (uncertainty, as scatter about the mean, or precision) of a function $Y(X)$ is determined as

$$\begin{aligned}\sigma_Y^2 &= \mathrm{Var}[Y] \\ &= [(Y - E[Y])^2] \\ &= E[Y^2] - 2E[YE[X]] + E[Y]^2,\end{aligned} \qquad \text{(A2.4a; 2.34a)}$$

and, since the middle term is the expectance of Y times a constant (the constant being the mean of $Y = E[Y] = \mu_Y$), the linear property (2.25a) indicates that it is equal to the constant times the expectance, or $E[Y\,E[Y]] = E[Y]E[Y] = E[Y]^2$, so (A2.4a) reduces to

$$= E[Y^2] - E[Y]^2. \qquad \text{(A2.4b; 2.34b)}$$

Consider the specific case where $Y = aX$, where a is a constant. Then,

$$\begin{aligned} \sigma_Y^2 &= \text{Var}[Y] \\ &= \text{E}[Y^2] - \text{E}[Y]^2 \\ &= \text{E}[\text{a}^2X^2] - \text{E}[\text{a}X]^2 \\ &= \text{a}^2\text{E}[X^2] - \text{a}^2\text{E}[X]^2 \\ &= \text{a}^2(\text{E}[X^2] - \text{E}[X]^2). \end{aligned} \qquad (\text{A2.5a; 2.35a})$$

Therefore,

$$\begin{aligned} &= \text{a}^2\text{Var}[X] \\ &= \text{a}^2\sigma_X^2. \end{aligned} \qquad (\text{A2.5b; 2.35b})$$

Now consider a slightly different function: $Y = \text{a}X + \text{b}$ (a and b constants). Then,

$$\begin{aligned} \sigma_Y^2 &= \text{Var}[Y] \\ &= \text{E}[Y^2] - \text{E}[Y]^2 \\ &= \text{E}[(\text{a}X + \text{b})^2] - \text{E}[\text{a}X + \text{b}]^2. \end{aligned} \qquad (\text{A2.6a; 2.36a})$$

From the linear properties, we know that $\text{E}[\text{a}X+\text{b}] = \text{aE}[X] + \text{b}$, so the second term on the right-hand side is $\text{a}^2\text{E}[X]^2 + 2\text{abE}[X] + \text{b}^2$ and the first term on the right-hand side is $\text{E}[\text{a}^2 X^2 + 2\text{ab}X + \text{b}^2]$, so (A2.6a) is

$$\begin{aligned} &= \text{a}^2\text{E}[X^2] + 2\text{abE}[X] + \text{b}^2 - \text{a}^2\text{E}[X]^2 - 2\text{abE}[X] - \text{b}^2 \\ &= \text{a}^2\text{E}[X^2] - \text{a}^2\text{E}[X]^2 \\ &= \text{a}^2(\text{E}[X^2] - \text{E}[X^2]). \end{aligned} \qquad (\text{A2.6b; 2.36b})$$

Therefore,

$$\text{Var}[\text{a}X + \text{b}] = \text{a}^2\text{Var}[X] = \text{a}^2\sigma_X^2. \qquad (\text{A2.6c; 2.36c})$$

That is, the addition of a constant to a random variable does not influence the spread or uncertainty of the function of the random variable, so $\text{Var}[\text{a}X+\text{b}] = \text{Var}[\text{a}X] = \text{a}^2\text{Var}[X]$.

The above manipulations are independent of whether the random variable is continuous or discrete. This is because the expectance operator is linear in both cases.

A2.3.2 Multivariate Expectance

Variance of Multivariate Random Variables

Consider the general linear function $Y(X_j)$, $j = 2$ (i.e., for bivariate data),

$$Y(X_j) = \text{a}_1X_1 + \text{a}_2X_2, \qquad (\text{A2.7; 2.41})$$

where the a_1 and a_2 are constants. For example, Y may be the quantity of interest, say atmospheric pressure, that is computed from the “raw” observations, X_1 and X_2, temperature and humidity.

We can compute the variance of this function using expectance, as was done previously with univariate data, starting with those previous results that showed $\text{Var}[X] = E[X^2] - E[X]^2$, so

$$\begin{aligned}\sigma_Y^2 &= \text{Var}[a_1X_1 + a_2X_2] = \text{Var}[Y] = E[Y^2] - E[Y]^2 \\ &= E[(a_1X_1 + a_2X_2)^2] - E[(a_1X_1 + a_2X_2)]^2 \\ &= a_1^2E[X_1^2] + 2a_1a_2E[X_1X_2] + a_2^2E[X_2^2] - \{E[(a_1X_1) + E[(a_2X_2)]\}^2 \qquad \text{(A2.8a; 2.42a)} \\ &= a_1^2E[X_1^2] + 2a_1a_2E[X_1X_2] + a_2^2E[X_2^2] \\ &\quad - a_1^2E[X_1]^2 - 2a_1a_2E[X_1]E[X_2] - a_2^2E[X_2]^2.\end{aligned}$$

This expression can be reduced, since we know from (A2.5) $\text{Var}[aX] = a^2E[X^2] - a^2 E[X]^2$, so several terms in (A2.8a) can be combined to give

$$\begin{aligned}&= a_1^2\text{Var}[X_1] + a_2^2\text{Var}[X_2] + 2a_1a_2E[X_1X_2] - 2a_1a_2\mu_1\mu_2 \\ &= a_1^2\sigma_{X_1}^2 + a_2^2\sigma_{X_2}^2 + 2a_1a_2E[X_1X_2] - 2a_1a_2\mu_1\mu_2,\end{aligned} \qquad \text{(A2.8b; 2.42b)}$$

where $\mu_1 = E[X_1]$ and $\mu_2 = E[X_2]$.

The expression in (A2.8b) now contains a new term not present in the previous univariate cases, which involves $E[X_1X_2]$. This term provides information regarding the extent to which the two random variables X_1 and X_2 are dependent upon one another, or how they **covary**. The degree to which random variables covary involves the concept of **covariance**.

Covariance

Covariance is comparable to variance for a single random variable, only in this case the concept extends to include how two or more random variables vary together. Specifically, the covariance between two random variables X_1 and X_2, denoted by $\text{Cov}[X_1X_2]$ or σ_{12}^2, is given as the expected value of the product of their deviations about their means:

$$\begin{aligned}\sigma_{12}^2 &= \text{Cov}[X_1X_2] = E[(X_1 - E[X_1])(X_2 - E[X_2])] \\ &= E\Big[(X_1X_2 - X_1E[X_2] - E[X_1]X_2 + E[X_1]E[X_2]\Big] \\ &= E[(X_1X_2 - X_1\mu_2 - \mu_1X_2 + \mu_1\mu_2] \qquad \text{(A2.9a; 2.43a)} \\ &= E[X_1X_2] - E[X_1]\mu_2 - \mu_1E[X_2] + \mu_1\mu_2 \\ &= E[X_1X_2] - \mu_1\mu_2 - \mu_1\mu_2 + \mu_1\mu_2 \\ &= E[X_1X_2] - \mu_1\mu_2,\end{aligned}$$

or

$$= E[X_1X_2] - E[X_1]E[X_2]. \qquad \text{(A2.9b; 2.43b)}$$

From this definition, it is seen that (A2.8b) can be further reduced to

$$\begin{aligned}&= a_1^2\text{Var}[X_1] + a_2^2\text{Var}[X_2] + 2a_1a_2\text{Cov}[X_1X_2] \\ &= a_1^2\sigma_{X_1}^2 + a_2^2\sigma_{X_2}^2 + 2a_1a_2\sigma_{12}^2.\end{aligned} \qquad \text{(A2.10; 2.44)}$$

So, with respect to the previous example of estimating the atmospheric pressure, the uncertainty (i.e., the variance) in the computed pressure is the linear sum, weighted by the coefficients, of the variance of atmospheric temperature and humidity values, respectively, and the covariance between the temperature and humidity.

The similarity between variance of a univariate random variable and covariance is plainly seen when considering the case where $X_1 = X_2$. In that case, (A2.9) reduces to $E[X_1^2] - \mu^2$, which is the variance of X_1 (recall equation (A2.3a)), so

$$\mathrm{Cov}[X_1 X_1] = \mathrm{Var}[X_1]. \qquad \text{(A2.11; 2.45)}$$

Independent Random Variables (Definitions)

Chapter 2 presents the following terms for the different forms of independent variables:

linearly independent, defined for variables $y_1, y_2, \ldots y_n$:

$$a_1 y_1 + a_2 y_2 + a_3 y_3 + \ldots + a_n y_n \neq 0 \quad \text{for any nonzero values of } a_i$$

$$\mathbf{a}^T \mathbf{Y} \neq 0 \quad \text{(for vectors)}$$

orthogonal, defined for variables y and x:

$$x_1 y_1 + x_2 y_2 + x_3 y_3 + \ldots + x_n y_n = 0$$

$$\mathbf{X}^T \mathbf{Y} = 0 \quad \text{(for vectors)}$$

uncorrelated, defined for y and x:

$$(x - \bar{x})_1 (y - \bar{y})_1 + (x - \bar{x})_2 (y - \bar{y})_2 + (x - \bar{x})_3 (y - \bar{y})_3 + \ldots + (x - \bar{x})_n (y - \bar{y})_n = 0$$

$$\mathbf{X}'^T \mathbf{Y}' = 0 \quad \text{(for vectors)}$$

If two or more random variables are related in some manner, their joint distribution conveys this dependence. That is, the probability of getting any one value of a random variable is dependent upon the value of the other random variable. If the variables are not related, then they are considered to be **linearly independent** (or simply, **independent**) random variables (other statements of independence are presented later as well). In this case, the distribution of one random variable is given strictly by its own distribution, independent of the distribution of the other random variable(s).

When random variables are independent, then

$$P\{X_1 = x_{1i}, X_2 = x_{2j}\} = P\{X_1 = x_{1i}\} P\{X_2 = x_{2j}\} \qquad \text{(A2.12; 2.46)}$$

or,

$$p_{1,2}(x_1, x_2) = p_1(x_1) p_2(x_2). \qquad \text{(A2.13; 2.47a)}$$

Their joint probability table would show each value of x and y equal to their marginal probabilities. In fact, the joint probability table is unnecessary, since just the marginal PMFs show all that is needed for each variable.

Y / X	y_1	y_2	…	y_N	Marginal probabilities
x_1	$p(x_1)\,p(y_1)$	$p(x_1)\,p(y_2)$	…	$p(x_1)\,p(y_N)$	$p(x_1)$
x_2	$p(x_2)\,p(y_1)$	$p(x_2)\,p(y_2)$	…	$p(x_2)\,p(y_N)$	$p(x_2)$
⋮	⋮	⋮	…	⋮	⋮
x_M	$p(x_M)\,p(y_1)$	$p(x_M)\,p(y_2)$	…	$p(x_M)\,p(y_N)$	$p(x_M)$
Marginal probabilities	$p(y_1)$	$p(y_2)$	…	$p(y_N)$	1

Independence is also reflected in the cumulative distributions, so

$$F(x_1, x_2) = F_1(x_1)F_2(x_2) \quad \text{(A2.14; 2.48)}$$

for discrete and continuous variables.

When random variables are independent, the expectance of their product reduces to the product of their expectances, which is essentially a restatement of (A2.12), so

$$E[X_1X_2] = E[X_1]E[X_2] \quad (\text{if } X_1 \text{ and } X_2 \text{ are independent}), \quad \text{(A2.15; 2.49)}$$

and thus (A2.9) reduces to

$$\text{Cov}[X_1X_2] = 0 \quad (\text{if } X_1 \text{ and } X_2 \text{ are independent}). \quad \text{(A2.16; 2.50)}$$

Therefore, independent random variables do *not* covary. Conversely, if two or more random variables have zero covariance, then they are linearly independent, though they may still be dependent in a *nonlinear* fashion.

Following the example of atmospheric pressure, in this case, if the atmospheric temperature and humidity were independent of one another – that is, if the covariance between them were zero (the value of temperature showing no consistent dependency on the humidity, and vice versa) – then the variance of atmospheric pressure would be the linear sum, weighted by the squared coefficients, of the variance of temperature with the variance of humidity.

Variance of Multivariate Functions of Random Variables

The expectance operator can be applied to functions of multivariate random variables, with the most common application aimed at determining the variance of the function – that is, the *uncertainty* of the function stemming from its dependence on random variables, themselves uncertain.

Consider a more general expression of the function in (A2.7):

$$Y(X_j) = \mathrm{a}_1X_1 + \mathrm{a}_2X_2 + \mathrm{a}_3X_3 + \ldots + \mathrm{a}_kX_k = \sum_{j=1}^{k}\mathrm{a}_jX_j. \quad \text{(A2.17; 2.51)}$$

The variance of $Y(X_j)$ is given as

$$\begin{aligned}\sigma_Y^2 &= \mathrm{Var}\left[\sum_{j=1}^{k}\mathrm{a}_jX_j\right] \\ &= \mathrm{E}\left[\left(\sum_{j=1}^{k}\mathrm{a}_jX_j\right)^2\right] - \left\{\mathrm{E}\left[\left(\sum_{j=1}^{k}\mathrm{a}_jX_j\right)\right]\right\}^2 \\ &= \mathrm{E}\left[(\mathrm{a}_1X_1 + \mathrm{a}_2X_2 + \mathrm{a}_3X_3 + \ldots + \mathrm{a}_kX_k)^2\right] \\ &\quad -\{\mathrm{E}\,[\mathrm{a}_1X_1 + \mathrm{a}_2X_2 + \mathrm{a}_3X_3 + \ldots + \mathrm{a}_kX_k]\}^2 \\ &= \sum_{i=1}^{k}\sum_{j=1}^{k}\mathrm{a}_i\mathrm{a}_j\mathrm{Cov}[X_iX_j].\end{aligned} \quad \text{(A2.18a; 2.52a)}$$

The covariance between *independent* random variables is zero, so if the X_j in $Y(X_j)$ are independent, all of the terms in (A2.18a) equal zero *except* when $i = j$. So, expanding (A2.18a),

$$\begin{aligned}\sigma_Y^2 = {} & \mathrm{a}_1\mathrm{a}_1\mathrm{Cov}[X_1X_1] + \mathrm{a}_1\mathrm{a}_2\mathrm{Cov}[X_1X_2] + \ldots + \mathrm{a}_1\mathrm{a}_k\mathrm{Cov}[X_1X_k] + \ldots + \\ & \mathrm{a}_2\mathrm{a}_1\mathrm{Cov}[X_2X_1] + \mathrm{a}_2\mathrm{a}_2\mathrm{Cov}[X_2X_2] + \ldots + \mathrm{a}_2\mathrm{a}_k\mathrm{Cov}[X_2X_k] + \ldots \\ & + \mathrm{a}_k\mathrm{a}_1\mathrm{Cov}[X_kX_1] + \mathrm{a}_k\mathrm{a}_2\mathrm{Cov}[X_kX_2] + \ldots + \mathrm{a}_k\mathrm{a}_k\mathrm{Cov}[X_kX_k].\end{aligned}$$

(A2.18b; 2.52b)

All terms are eliminated (= 0) except for those where $i = j$, so for *independent* X_j, the variance of $Y(X_j)$ reduces to

$$\sigma_Y^2 = \sum_{j=1}^{k}\mathrm{a}_j^2\mathrm{Var}[X_j] \quad \text{(A2.19a; 2.53a)}$$

$$= \sum_{j=1}^{k}\mathrm{a}_j^2\sigma_j^2. \quad \text{(A2.19b; 2.53b)}$$

For cases in which all of the independent random variables, X_j, have the same variance, so $\mathrm{Var}[X_j] = \sigma^2$, then (A2.19) reduces to

$$\sigma_Y^2 = \sigma_j^2\sum_{j=1}^{k}\mathrm{a}_j^2. \quad \text{(A2.20; 2.54)}$$

If the function $Y(X_j)$ is given by

$$\begin{aligned} Y(X_j) &= \mathrm{a}_1X_1 + \mathrm{b}_1 + \mathrm{a}_2X_2 + \mathrm{b}_2 + \mathrm{a}_3X_3 + \mathrm{b}_3 + \ldots + \mathrm{a}_kX_k + \mathrm{b}_k \\ &= \mathrm{a}_1X_1 + \mathrm{a}_2X_2 + \mathrm{a}_3X_3 + \ldots + \mathrm{a}_kX_k + \mathrm{b} \\ &= \sum_{j=1}^{k}(\mathrm{a}_jX_j) + \mathrm{b}, \end{aligned} \qquad \text{(A2.21; 2.55)}$$

where the $\mathrm{b_j}$ are constants, and the constant b is the sum of the b_j. The addition of a constant b is comparable to the univariate case of (A2.6), where it was shown that an added constant does not influence the variance of a function. Consequently, the variance of (A2.21) is the same as that of (A2.17), so, again,

$$\sigma_Y^2 = \sum_{i=1}^{k}\sum_{j=1}^{k} \mathrm{a}_i\mathrm{a}_j\mathrm{Cov}[X_iX_j]. \qquad \text{(A2.22; 2.56)}$$

Note that one of the most valuable aspects of this application of the expectance operator to a function (e.g., a variable that is constructed from other variables or parameters) is that, if variability of the function with time or space is what is of interest (e.g., climate variability, seismic activity on a particular fault), then the variance of the function as determined from Var[X] reveals its *dependencies* on the variability of the parameters on which it depends. Therefore, you can determine the sensitivity of the function to these parameters – i.e., which parameters contribute most to the variability of the function of interest. This is most useful for interpretation, hypothesis testing, and future sampling strategies (e.g., sampling with better precision that parameter that causes the biggest error, or, defining the level of sample precision needed in each parameter or number of realizations required to achieve that precision by the Central Limit Theorem to be able to estimate desired variable within desired precision).

A2.3.3 Moments of Nonlinear Functions of Random Variables

Heretofore, the functions $Y(X_j)$ have been linear in the random variables. When this is not the case, the simple reductions within the linear expectance operator that were previously applied often cannot be adopted. In these cases, the expectance used to compute the first two central moments of the nonlinear function (i.e., the mean and variance) can be approximated by expanding the nonlinear function in a Taylor series about the mean, and then truncating this expansion to first order, providing a linear approximation to the function. The expectance operator is easily applied to this truncated, linear approximation to provide an approximation of the moments.[1]

For a nonlinear function $Y(X_j)$,

$$Y(X_1, X_2, \ldots, X_k), \qquad \text{(A2.23; 2.57)}$$

[1] And, I must add, beautifully!

the **Taylor series expansion** of this function about the mean, μ (for this case expand about the mean, not some other point), truncated to first order (that is, after the first derivative term), is given by

$$Y(X_1, X_2, \ldots, X_K) \approx Y(\mu_1, \mu_2, \ldots, \mu_K) + \sum_{j=1}^{K} \left. \frac{\partial Y(X_j)}{\partial X_j} \right|_{\mu_j} (X_j - \mu_j), \qquad \text{(A2.24; 2.58)}$$

where the derivatives are evaluated at the μ_j.

Mean of Nonlinear Functions of Random Variables

The expected value of the first-order *approximation* to $\mu(X_j)$, in (A2.24), is given as

$$Y(X_1, X_2, \ldots, X_k) \approx Y(\mu_1, \mu_2, \ldots, \mu_k) + \sum_{j=1}^{k} \left. \frac{\partial Y(X_j)}{\partial X_j} \right|_{\mu_j} (\mathrm{E}[X_j] - \mu_j)$$
$$\approx Y(\mu_1, \mu_2, \ldots, \mu_k). \qquad \text{(A2.25; 2.59)}$$

The second term on the right-hand side is eliminated, since the derivatives evaluated at the points μ_j yield constants, in which case the expectance operator operates on the only random variables within the sum, which are $\mathrm{E}[X_j]$. Since $\mathrm{E}[X_j] = \mu_j$, the difference within the sum, $\mathrm{E}[X_j] - \mu_j = \mu_j - \mu_j$, eliminates the term, leaving only the first term.

Therefore, the mean of a nonlinear multivariate function of a random variable, $Y(X_j)$, is approximated by the function evaluated at the mean of each of the individual random variables of which $Y(X_j)$ is a function.

Variance of Nonlinear Functions of Random Variables

The variance of $Y(X_j)$ is given as

$$\mathrm{Var}[Y(X_1, X_2, \ldots, X_K)] = \mathrm{E}[\{Y(X_1, X_2, \ldots, X_K) - \mathrm{E}[Y(X_1, X_2, \ldots, X_K)]\}^2] \quad \text{(A2.26)}$$

but, using (A2.25) for the second term in brackets on the right-hand side, then (A2.24)

$$Y(X_1, X_2, \ldots, X_k) - Y(\mu_1, \mu_2, \ldots, \mu_k) \approx \sum_{j=1}^{k} \left. \frac{\partial Y(X_j)}{\partial X_j} \right|_{\mu_j} (X_j - \mu_j), \qquad \text{(A2.27; 2.61)}$$

and, from (A2.25), where $\mathrm{E}[Y(X_1, X_2, \ldots, X_K)] \approx Y(\mu_1, \mu_2, \ldots, \mu_K)$, (A2.27) is approximately equivalent to

$$Y(X_1, X_2, \ldots, X_k) - \mathrm{E}[Y(X_1, X_2, \ldots, X_k)] \approx \sum_{j=1}^{k} \left. \frac{\partial Y(X_j)}{\partial X_j} \right|_{\mu_j} (X_j - \mu_j), \quad \text{(A2.28a; 2.62a)}$$

where the left-hand side is equivalent to the term in braces in (A2.26).

Because the derivatives on the right-hand side of (A2.28a) are constants and the means, where these derivatives are evaluated at μ_j, are constants, the right-hand side can be rewritten, changing the notation only (using a_j and b_j for the constants), as

$$Y(X_1, X_2, \ldots, X_k) - E[Y(X_1, X_2, \ldots, X_k)] \approx \sum_{j=1}^{k} (a_j X_j) - b,$$

with the summation form being equal to

$$= \sum_{j=1}^{k} (a_j X_j) - b, \qquad \text{(A2.28b; 2.61)}$$

where b is the sum of the b_j, the b_j being the products of the derivatives, a_j (evaluated at the means) and means, μ_j.

This function, (A2.28b), is linear in the random variables. Furthermore, it is of the same form as (A2.21), which means that the variance of this function is given by (A2.22), or

$$\mathrm{Var}[Y(X_1, X_2, \ldots, X_k)] = \sigma_Y^2 \approx \sum_{i=1}^{k} \sum_{j=1}^{k} a_i a_j \mathrm{Cov}[X_i X_j] \qquad \text{(A2.29a; 2.63a)}$$

or,

$$\sigma_Y^2 \approx \sum_{i=1}^{k} \left. \frac{\partial Y(X_i)}{\partial X_i} \right|_{\mu_i} \sum_{j=1}^{k} \left. \frac{\partial Y(X_j)}{\partial X_j} \right|_{\mu_j} \mathrm{Cov}\left[X_i X_j\right]. \qquad \text{(A2.29b; 2.63b)}$$

If the X_j are independent, then the above sum is reduced to become equivalent in form with (2.53); if the independent random variables all have the same variance, it reduces to a form equivalent to (2.54).

Moments of Univariate Nonlinear Functions

When $K = 1$ in the preceding nonlinear functions, the moments are reduced to those for the univariate case where the sums are replaced in the above expressions by single terms.

For example, consider $K = 1$ and $Y(X) = X^2$. In this case, the mean is given by (A2.25), or specifically,

$$\begin{aligned} E[Y] &= \mu_Y \approx Y(\mu_X) \\ &= \mu_X^2. \end{aligned} \qquad \text{(A2.30; 2.63)}$$

The variance will then be given by (A2.29), or specifically,

$$\begin{aligned}\sigma_Y^2 &\approx \sum_{i=1}^{1}\sum_{j=1}^{1} \frac{\partial Y(X_i)}{\partial X_i}\frac{\partial Y(X_j)}{\partial X_j}\mathrm{Cov}\left[X_i X_j\right] \\ &\approx \left.\frac{\partial Y(X_1)}{\partial X_1}\right|_{\mu_1}^{2} \mathrm{Cov}[X_1 X_1] \\ &\approx 4\mu_X^2 \mathrm{Var}[X] \\ &\approx 4\mu_X^2\sigma_X^2.\end{aligned} \tag{A2.31; 2.64}$$

Box A2.1 Example of First-Order Linear Approximation to Error

Consider the simple situation where $X = 2, 3, 4, 5$, so $\mu_X = 3.5$, and $\sigma_X^2 = 1.25$; whereas $X^2 = 4,9,16,25$ has a true mean of 13.5 and true variance of 62.25. The first-order linear approximations to X^2 using (A2.30) for the mean, and (A2.31) for the variance, give $\mu_Y = \mu_X^2 = 12.25$ ($Y = X^2$, evaluated at the X mean, which is the square of μ_X) for the mean, and $4\mu_X^2\sigma_X^2 = 61.25$ for the variance. Thus, in this case, the mean of the first-order approximation to X^2 is better than 92 percent of the true value, while the variance is better than 98 percent.

Obviously, the degree of nonlinearity of a function will determine just how well or how poorly the linear approximation works. From a practical perspective, the approximations do provide reasonable estimates of the lower-order moments. If more precise values are required, more elaborate numerical techniques must be employed.

Moments of Nonlinear Functions of Independent Random Variables

If the random variables of the nonlinear function are independent, then the covariance terms in the variance approximation of (A2.29) are zero whenever $i \neq j$, so (A2.29) reduces to

$$\begin{aligned}\mathrm{Var}[Y(X_1, X_2, \ldots, X_k)] &= \sigma_Y^2 \approx \sum_{j=1}^{k}\left(\left.\frac{\partial Y(X_j)}{\partial X_j}\right|_{\mu_1}^{2} \mathrm{Var}\left[X_j\right]\right) \\ &\approx \sum_{j=1}^{k}\left(\left.\frac{\partial Y(X_j)}{\partial X_j}\right|_{\mu_1}^{2} \sigma_j^2\right).\end{aligned} \tag{A2.32; 2.65}$$

If the variances of the different independent X_j are equivalent, the approximation of (A2.32) reduces to

$$\mathrm{Var}[Y(X_1, X_2, \ldots, X_k)] = \sigma_Y^2 \approx \sigma_j^2 \sum_{j=1}^{k} \left(\frac{\partial Y(X_j)}{\partial X_j} \bigg|_{\mu_1}^2 \right). \tag{A2.33; 2.66}$$

A2.3.4 Expectance with Random Vectors and Matrices

Consider working with vectors and matrices that are filled with random variables. Such vectors or matrices are themselves random vectors or matrices, since the actual elements occupying them can be any of a series of values, representing the spread in each element. Random vector, **X**, is an example

$$\mathbf{X} = \begin{bmatrix} X_1 \\ X_2 \\ X_3 \\ \vdots \\ X_n \end{bmatrix} \tag{A2.34}$$

where the elements each represent a random variable X_i. Depending on which values are present in the vector, any one particular vector represents just one realization of an infinite (or finite, if each random variable has a finite number of possible values or events) number of possible vectors, **X**. Thus, the random vector is analogous to a scalar random variable.

Now consider the random matrix **Z**,

$$\mathbf{Z} = \begin{bmatrix} Z_{11} & Z_{12} & & Z_{1m} \\ Z_{21} & Z_{22} & & Z_{2m} \\ Z_{31} & Z_{32} & & Z_{3m} \\ & & & \\ Z_{n1} & Z_{n2} & & Z_{nm} \end{bmatrix}, \tag{A2.35}$$

where again each element represents a random variable.

Random vectors and matrices represent an ideal means for manipulating multivariate random variables or situations in which you are dealing with systems of random variables.

Mean

The expectance of a random matrix, as before, is denoted E[**Z**], indicating

$$\mathrm{E}[\mathbf{Z}] = \begin{bmatrix} \mathrm{E}[Z_{11}] & \mathrm{E}[Z_{12}] & \mathrm{E}[Z_{1m}] \\ \mathrm{E}[Z_{21}] & \mathrm{E}[Z_{22}] & \mathrm{E}[Z_{2m}] \\ \mathrm{E}[Z_{31}] & \mathrm{E}[Z_{32}] & \mathrm{E}[Z_{3m}] \\ & & \\ \mathrm{E}[Z_{n1}] & \mathrm{E}[Z_{n2}] & \mathrm{E}[Z_{nm}] \end{bmatrix} = \begin{bmatrix} \mu_{11} & \mu_{12} & \mu_{1m} \\ \mu_{21} & \mu_{22} & \mu_{2m} \\ \mu_{31} & \mu_{32} & \mu_{3m} \\ & & \\ \mu_{n1} & \mu_{n2} & \mu_{nm} \end{bmatrix} \tag{A2.36a}$$

or

$$= \boldsymbol{\mu}. \tag{A2.36b}$$

So, the expectance of random matrix **Z** is the mean, or the mean matrix, $\boldsymbol{\mu}$. Likewise, for the random vector, E[**X**] = $\boldsymbol{\mu}$. The elements of $\boldsymbol{\mu}$ contain the expected value of each element of **Z** or **X**.

As shown previously, a scalar multiplied by a matrix is equal to the matrix with each individual element multiplied by the scalar. From this and (A2.36a), it is clear that

$$\mathrm{E}[\mathbf{aX}] = \mathbf{a}\mathrm{E}[\mathbf{X}] = \mathbf{a}\mu \tag{A2.37}$$

$$\mathrm{E}[\mathbf{aX} + \mathrm{b}] = \mathbf{a}\mathrm{E}[\mathbf{X}] + \mathrm{E}[\mathrm{b}] = \mathbf{a}\mu + \mathrm{b}. \tag{A2.38}$$

The same is true for the vector products and sums, where **a, b** and **c** are nonrandom variable vectors, in which case

$$\mathrm{E}[\mathbf{aX}] = \mathbf{a}\mathrm{E}[\mathbf{X}] = \mathbf{a}\mu \tag{A2.39}$$

$$\mathrm{E}[\mathbf{aX} + \mathbf{b}] = \mathbf{a}\mathrm{E}[\mathbf{X}] + \mathrm{E}[\mathbf{b}] = \mathbf{a}\mu + \mathbf{b} \tag{A2.40}$$

$$\mathrm{E}[\mathbf{aXb} + c] = \mathbf{a}\mathrm{E}[\mathbf{X}]\mathbf{b} + \mathrm{E}[c] = \mathbf{a}\mu\mathbf{b} + c. \tag{A2.41}$$

Variance (Covariance Matrix)

The variance of random vector **X** is analogous to the variance of a random variable, X : the expectance of a function of the random variable $(X - \mu)^2$, and in this case it is given as the expectance of a random matrix **Z** of the form

$$\mathbf{Z} = (\mathbf{X} - \mathrm{E}[\mathbf{X}])(\mathbf{X} - \mathrm{E}[\mathbf{X}])^{\mathrm{T}}, \tag{A2.42}$$

so

$$\begin{aligned} \mathrm{Var}[X] = \Sigma = \mathrm{E}[\mathbf{Z}] &= \mathrm{E}\left[(\mathbf{X} - \mathrm{E}[\mathbf{X}])(\mathbf{X} - \mathrm{E}[\mathbf{X}])^{\mathrm{T}}\right] \\ &= (\mathbf{X} - \mu)(\mathbf{X} - \mu)^{\mathrm{T}}, \end{aligned} \tag{A2.43a}$$

where $\boldsymbol{\Sigma}$ is the **covariance matrix** of **X** (often referred to as the **variance-covariance matrix**).

Equation (A2.43a) is expanded and reduced from (A2.39) as

$$\begin{aligned} &= \mathrm{E}[(\mathbf{X}\mathbf{X}^{\mathrm{T}} - \mu\mathbf{X}^{\mathrm{T}} - \mathbf{X}\mu^{\mathrm{T}} + \mu\mu^{\mathrm{T}})] \\ &= \mathrm{E}[\mathbf{X}\mathbf{X}^{\mathrm{T}}] - \mu\mathrm{E}[\mathbf{X}^{\mathrm{T}}] - \mathrm{E}[\mathbf{X}]\mu^{\mathrm{T}} + \mu\mu^{\mathrm{T}} \\ &= \mathrm{E}[\mathbf{X}\mathbf{X}^{\mathrm{T}}] - \mu\mu^{\mathrm{T}} - \mu\mu^{\mathrm{T}} + \mu\mu^{\mathrm{T}} \\ &= \mathrm{E}[\mathbf{X}\mathbf{X}^{\mathrm{T}}] - \mu\mu^{\mathrm{T}}. \end{aligned} \tag{A2.43b}$$

This is equivalent to the operation,

$$= \mathrm{E}\ [(X_i - \mu_i)(X_j - \mu_j)], \tag{A2.43c}$$

for each element i, j: $i = 1,2,3, \ldots, n$, $j = 1,2,3, \ldots, m$. As a scalar operation for pairs of elements, it is clear (by considering the operation in (A2.43c) from Chapter 2 on Probability) that when $i \neq j$, this gives the covariance between the various pairs of random variables in **X**, and when $i = j$, it is giving the variance.

Consider the vector operation in (A2.43a). This involves the *outer product* $(\mathbf{X} - \boldsymbol{\mu})(\mathbf{X} - \boldsymbol{\mu})^{\mathrm{T}}$. First, consider the case where the mean vector is the null vector, i.e., the X random variables have zero mean. Then this outer product is given as $\mathbf{X}\mathbf{X}^{\mathrm{T}}$, or

$$\mathbf{X}\mathbf{X}^{\mathrm{T}} = \begin{bmatrix} X_1 \\ X_2 \\ X_3 \\ \vdots \\ X_n \end{bmatrix} \begin{matrix} [X_1 & X_2 & \ldots & X_n\ ] \\ \begin{bmatrix} X_1X_1 & X_1X_2 & & X_1X_n \\ X_2X_1 & X_2X_2 & \ldots & X_2X_n \\ X_3X_1 & X_3X_2 & & X_3X_n \\ & \vdots & & \\ X_nX_1 & X_nX_2 & & X_nX_n \end{bmatrix} \end{matrix}, \tag{A2.44a}$$

and the expectance of this matrix, as indicated by (A2.36) is

$$\mathrm{E}[\mathbf{X}\mathbf{X}^{\mathrm{T}}] = \begin{bmatrix} \mathrm{E}[X_1X_1] & \mathrm{E}[X_1X_2] & \dots & \mathrm{E}[X_1X_n] \\ \mathrm{E}[X_2X_1] & \mathrm{E}[X_2X_2] & & \mathrm{E}[X_2X_n] \\ \mathrm{E}[X_3X_1] & \mathrm{E}[X_3X_2] & & \mathrm{E}[X_3X_n] \\ & \vdots & \ddots & \\ \mathrm{E}[X_nX_1] & \mathrm{E}[X_nX_2] & & \mathrm{E}[X_nX_n] \end{bmatrix}. \tag{A2.44b}$$

If the mean is not zero, then the operation is the same, except that the mean for each random variable is subtracted off prior to forming the outer product of (A2.44b).

In either case, with $\boldsymbol{\mu}$ zero or not, when n equals the full population or ensemble size (N or ∞), (A2.44b) is

$$\begin{aligned} \mathrm{Var}[\mathbf{X}] = \boldsymbol{\Sigma} &= \begin{bmatrix} \sigma_{11} & \sigma_{12} & & \sigma_{1n} \\ \sigma_{21} & \sigma_{22} & \dots & \sigma_{2n} \\ \sigma_{31} & \sigma_{32} & & \sigma_{3n} \\ & \vdots & & \\ \sigma_{n1} & \sigma_{n2} & & \sigma_{nn} \end{bmatrix} \\ &= \begin{bmatrix} \sigma_1^2 & \sigma_{12} & \dots & \sigma_{1n} \\ \sigma_{21} & \sigma_2^2 & & \sigma_{2n} \\ \sigma_{31} & \sigma_{32} & & \sigma_{3n} \\ & \vdots & & \\ \sigma_{n1} & \sigma_{n2} & & \sigma_n^2 \end{bmatrix}, \end{aligned} \tag{A2.44c}$$

where the $\sigma_{ii} = \sigma_i^2 =$ the true variance of the random variable $\mathbf{X}_i$ and the $\sigma_{ij} =$ the true covariance between the random variables $\mathbf{X}_i$ and $\mathbf{X}_j$. When n is less than the full ensemble size (i.e., for a sample), the above is the sample Var[**X**] = **S**, populated by sample variance (s_i^2) and covariance (s_{ij}) estimates, or lagged covariance estimates, $\gamma(k)$, for ergodic series.

Thus, the variance of random vector **X** is given by the covariance matrix, **Σ**.

Consider the variance of the following vector product and sum where **a** and **b** are nonrandom variable vectors, in which case

$$\mathrm{Var}[\mathbf{aX}+\mathbf{b}] = \mathrm{E}\left[(\mathbf{aX}+\mathbf{b}-\mathrm{E}[\mathbf{aX}+\mathbf{b}])(\mathbf{aX}+\mathbf{b}-\mathrm{E}[\mathbf{aX}+\mathbf{b}])^{\mathrm{T}}\right]. \tag{A2.45a}$$

The expectance term E[**aX** + **b**] = **a**E[**X**] + **b** = **aμ** + **b** from (A2.40), so (A2.45a) becomes

$$
\begin{aligned}
&= \mathrm{E}[(\mathbf{aX}+\mathbf{b}-\mathbf{a}\mu-\mathbf{b})(\mathbf{aX}+\mathbf{b}-\mathbf{a}\mu-\mathbf{b})^{\mathrm{T}}] \\
&= \mathrm{E}[(\mathbf{aX}-\mathbf{a}\mu)(\mathbf{aX}-\mathbf{a}\mu)^{\mathrm{T}}] \\
&= \mathrm{E}[\mathbf{aXX}^{\mathrm{T}}\mathbf{a}^{\mathrm{T}}-\mathbf{a}\mu\mathbf{X}^{\mathrm{T}}\mathbf{a}^{\mathrm{T}}-\mathbf{aX}\,\mu^{\mathrm{T}}\mathbf{a}^{\mathrm{T}}+\mathbf{a}\mu\mu^{\mathrm{T}}\mathbf{a}^{\mathrm{T}}] \\
&= \mathrm{E}[\mathbf{aXX}^{\mathrm{T}}\mathbf{a}^{\mathrm{T}}]-\mathrm{E}[\mathbf{a}\mu\mathbf{X}^{\mathrm{T}}\mathbf{a}^{\mathrm{T}}]-\mathrm{E}[\mathbf{aX}\,\mu^{\mathrm{T}}\mathbf{a}^{\mathrm{T}}]+\mathrm{E}[\mathbf{a}\mu\mu^{\mathrm{T}}\mathbf{a}^{\mathrm{T}}] \qquad \text{(A2.45b)} \\
&= \mathbf{a}\mathrm{E}[\mathbf{XX}^{\mathrm{T}}]\mathbf{a}^{\mathrm{T}}-\mathbf{a}\mu\mathrm{E}[\mathbf{X}^{\mathrm{T}}]\mathbf{a}^{\mathrm{T}}-\mathbf{a}\mathrm{E}[\mathbf{X}\,]\mu^{\mathrm{T}}\mathbf{a}^{\mathrm{T}}+\mathbf{a}\mu\mu^{\mathrm{T}}\mathbf{a}^{\mathrm{T}} \\
&= \mathbf{a}\mathrm{E}[\mathbf{XX}^{\mathrm{T}}]\mathbf{a}^{\mathrm{T}}-\mathbf{a}\mu\mu^{\mathrm{T}}\mathbf{a}^{\mathrm{T}}-\mathbf{a}\mu\mu^{\mathrm{T}}\mathbf{a}^{\mathrm{T}}+\mathbf{a}\mu\mu^{\mathrm{T}}\mathbf{a}^{\mathrm{T}} \\
&= \mathbf{a}\mathrm{E}[\mathbf{XX}^{\mathrm{T}}]\mathbf{a}^{\mathrm{T}}-\mathbf{a}\mu\mu^{\mathrm{T}}\mathbf{a}^{\mathrm{T}},
\end{aligned}
$$

which, from (A2.43b) is seen to be

$$
\begin{aligned}
&= \mathbf{a}\mathrm{Var}[\mathbf{X}]\mathbf{a}^{\mathrm{T}} \\
&= \mathbf{a}\Sigma\mathbf{a}^{\mathrm{T}}.
\end{aligned} \qquad \text{(A2.45c)}
$$

The sample covariance matrix – that is, the estimate of $\mathbf{\Sigma}$ for random vector $\mathbf{X}$ – is done by estimating the elements of $\mathbf{\Sigma}$ using the standard formulas for sample variance, given in (3.6), and sample covariance, given in (3.7). These estimates are then placed in the covariance matrix as indicated in (A2.44). For stationary and ergodic time series, the estimates come directly from the outer product of $\mathrm{E}[(\mathrm{X}-\overline{\mathrm{x}})(\mathrm{X}-\overline{\mathrm{x}})^{\mathrm{T}}]$, since this produces the *serial product* estimates appropriate for ergodic series as discussed in Chapter 7. For that case, all variances (in the principal diagonal) are equal, and the non-principal diagonal covariance terms are simply a function of lag k (so $\sigma_{21} = \sigma_{32} = \gamma(k = 1)$), as defined in Chapter 7).

A2.4 Bootstrap

There is a relatively new means of estimating uncertainties that has proliferated because of the power of computers. This is presented as "resampling" statistics by Simon (1992), or the **bootstrap** method. The bootstrap, formally developed by Efron (1981), provides a rigorous foundation for the method, but the theoretical development is a bit difficult to follow for the casual user. Simon has presented the general concept as resampling statistics and suggests in his book that the technique is ideal for teaching most of statistics, as it tends to demystify statistics and probability, replacing the black-magic formulations with an intuitively obvious approach.

From my perspective, the basic concept of this technique is similar to that of maximum likelihood, where for the latter, with knowledge of the PDF, you use the method of maximum likelihood to determine the best estimates of the population parameters, in the sense that they would be those parameters most likely to yield the sample you did indeed draw. A logical extension of this is to ignore the mathematical form of the distribution, but rather use the distribution of the sample as being representative of the true population, and then repeat your own sampling from the sample population multiple times to determine the likelihood of achieving any particular statistic or test result. You can do this multiple times to compute, say, the likelihood of achieving a particular ratio or other result by drawing multiple random samples from your distribution and then computing the parameter or statistic of interest for each sample drawn, and then, from those, compute the PMF of the quantity and use that to estimate the likelihood of achieving the answer you indeed got. I will provide some examples of this later, particularly with time series analysis.

Fundamental to the evaluation of your data is the ability to assess the uncertainty of your result(s). We have dealt with that formally with expectance, but there are many instances in which expectance cannot be applied, or its application would be too cumbersome or difficult. Also, we often wish to access the results of a statistical test (recall the null hypothesis and hypothesis testing). Simple formulas exist for simple statistical tests such as the t-test. Those formulas usually require nothing more than the degrees of freedom (or equivalent degrees of freedom (EDOF)) and satisfaction of some fundamental properties (e.g., the data show Gaussian distribution). Great, but what about those awkward cases where it is not clear what the EDOF are, or where the data are not Gaussian? In all cases, the bootstrap provides a simple means for using your own data to construct similar (but random) data with which to repeat the test and determine the statistical uncertainty directly (by simulating the random data to preserve the fundamental characteristics of your data, your tests will naturally account for the EDOF and the data distribution).

This sounds too good to be true, and it was until powerful computers came along and gave us the power to do this.

Application of the bootstrap typically involves the following steps:

1) For sequential data, determine the fundamental (lowest-order) statistical moments of your data that must be preserved; for nonsequential data, preserve the sample PMF.
2) Generate random data (*bootstrap data*) that preserve those statistical moments and the number of data points (discussed below).
3) Repeat whatever test or analysis you wish to test with these bootstrap data.
4) Repeat steps 2 and 3 multiple times (easily generating thousands of repeat tests).
5) Build a bootstrap PMF of the results for the multiple bootstrap tests.
6) From the bootstrap PMF, determine the probability of attaining the value you attained from the real data in order to get its significance (e.g., the random data only give a test score as high as you achieved in 4 out of 100 tests, so your real results are significant at $\alpha = 0.04$; i.e., you are very unlikely to have achieved this result by random chance).

As used here, "bootstrap" is a general term. Here, it includes variants that have more specific names such as jackknife, resampling, Monte Carlo or other colorful phrases. In the end, the common denominator in all cases is that random data are simulated and the tests repeated to access the significance of your findings. Specifically, you are preserving those statistical moments that are fundamental characteristics of the population or ensemble from which your sample was drawn. That way, the "random" data you generate are effectively more samples from the same process that generated your data. You will be testing the probability of getting the value you did obtain, compared to what you might expect to have obtained, given an unusually large number of samples collected from the population. In some cases the bootstrap is the only game in town, and in most cases, it is the easiest game.

A2.4.1 Generating Your Random Data

Nonsequential Data

For this case, it is best to preserve the PMF itself (i.e., a histogram of the values assumed to represent the data population). You will then resample from this histogram to "draw" another sample. There are many ways to resample your PMF. For example, assume that you have a sample size of $N = 100$ and your binning shows the relative frequency of each discrete value (or range of values). From this, you might generate an array of 10^4 (or so) numbers that show the same relative frequency as the discrete numbers. Then randomly select N realizations from that array as a second sample, and continue until you have a huge number of samples.

If your original sample size is so small that it doesn't seem reasonable to make a PMF, you can still make it and sample it. Or you can get inventive, and make a number of slight variants of the PMF and sample those (testing in the end to see if using these subtly different PMFs changes your results – if so, examine the results closely to determine why, and then go with the one you think is most representative and defensible, or even consider presenting the different results, and stating how they are different and which one or ones you choose to use).

Even for something as simple as a coin toss, you have two results, each having a 0.5 chance of occurring. So, you can make an array of thousands of numbers being 1 (heads) and equal number of 2 (tails), and then sample randomly. Or run a random number generator, giving values between 0 and 1; above 0.5 is heads, below is tails (a perfect 0.5 is the coin landing on its edge, which you just ignore).

Sequential (Time Series) Data

For time series, you need to preserve the mean and variance (two lowest order univariate moments). But a time series is characterized by the order in which the values occur. That order is captured by the multivariate moments, the most basic of which is the lowest bivariate moment: the autocovariance function (acvf). Fortunately, as you recall (if not, look at Chapter 9), the Fourier transform of the acvf is the power spectrum. The 0-frequency power is the mean, the integral of the PSD is the variance, and the PSD is itself the acvf in the frequency domain. So, you are in luck – the PSD preserves everything you want to simulate – a time series with the same "color" of noise, including any periodic components superimposed on the noise background. Alternatively, you can fit the PSD with a simple function to generate a more pure random series preserving just the color, with no periodicity appearing in your data set, or you can add some amount of noise to this smooth background noise spectrum before inverse transforming back to produce the simulated time series.

To generate new simulated realizations (y_s), you compute the PSD (Fourier transform of data, which gives amplitude spectrum (A_j), that is, the square root of the PSD). The PSD (transform of the acf) does not give a phase spectrum, so generate random phase, ε_j, and then invert the amplitude spectrum (A_j) with the random phase. This produces a time series with the same mean, variance and acf of the original data (i.e., a time series of colored noise, colored as if your real-time series were simply noise). *The key here is to do the inverse transform in polar coordinates* as

$$y_{si} = \sum_{j>-n/2}^{\leq n/2} A_j e^{i\varepsilon_j t_i} \quad i = 1, \ldots, N. \tag{A2.46}$$

Otherwise, you must work harder to preserve the symmetry relationships of the real and imaginary components of the phase; recall that the real component (the a_j) is even and the imaginary (b_j) is odd, so you would have to make the random phase, $\varphi_j = \tan^{-1}(b_j/a_j)$, obey that symmetry while preserving the PSD, $A_j^2 = a_j^2 + b_j^2$. But when inverting using the polar form, that relationship is automatically accounted for.

Box A2.3 Example: Evaluating Significance of r-Values on a Correlation Map

We are interested in finding those locations on Earth that are related to changes in the ocean-atmosphere-ice (OAI) interactions in the polar oceans around Antarctica. We expect that such interactions play a role in climate, and here are looking for global linkages between dominant EOF modes of these interactions with the surface air temperature anomalies (SAT*) measured over time around the globe. We do this via what is known as a heterogeneous correlation map, where we correlate the PCs of the OAI EOF modes to time series of a different variable elsewhere. The OAI interactions are characterized by two meaningful integrated characteristics of the upper ocean, referred to as *bulk stability* and *ocean heat flux*. The former represents how much sea ice would have to form to "destabilize" the water column (this is the ocean equivalent to forming deep (very high) cumulus convection clouds, but in this case it implies transporting a considerable amount of very cold and fresh water to the ocean bottom). The latter, as the name implies, indicates the amount of heat ventilating from the ocean into the overlying sea ice and/or atmosphere in winter. This analysis is applied to waters of the Antarctic polar ocean at the southernmost extent of the Atlantic Ocean (the Weddell region).[2]

The Earth's surface has been divided into ~1000 grid boxes, and the annual (detrended) SAT anomalies computed for each grid box.[3] The Weddell OAI parameters fields are dominated by the first three EOF modes, displayed in Figure A2.1.

[2] Published in Martinson, D. G. and R. A. Iannuzzi, 2003. Spatial/Temporal Patterns in Weddell Gyre Characteristics and Their Relationship to Global Climate, Journal of Geophysical Research, 108(C4), doi:10.1029/2000JC000538.

[3] Detrended near-surface temperature anomalies (SAT*) around the globe determined and supplied by National Centers for Environmental Prediction (NCEP) and the National Center for Atmospheric Research (NCAR).

Box A2.3 (Cont.)

Now we search for "teleconnections." These are links to other regions around the Earth; something like finding that when there is an El Niño, numerous weather anomalies are initiated elsewhere (e.g., floods in Ecuador and Peru, droughts in Australia and Southern Africa, and many more). We search by correlating the PCs of the Weddell OAI indicators to SAT* around the globe and find patterns of r-values, as shown in Figure A2.1.

To interpret these results, we need the ability to evaluate the likelihood of getting artificially high r-values (i.e., what value of r constitutes a correlation that is not likely to occur by pure chance). You might consider applying the standard significance test for each individual grid box (i.e., determine the EDOF for the time series being correlated, choose the level of significance desired, α, and then compute that value of r, r_α, above which the correlation is significant).

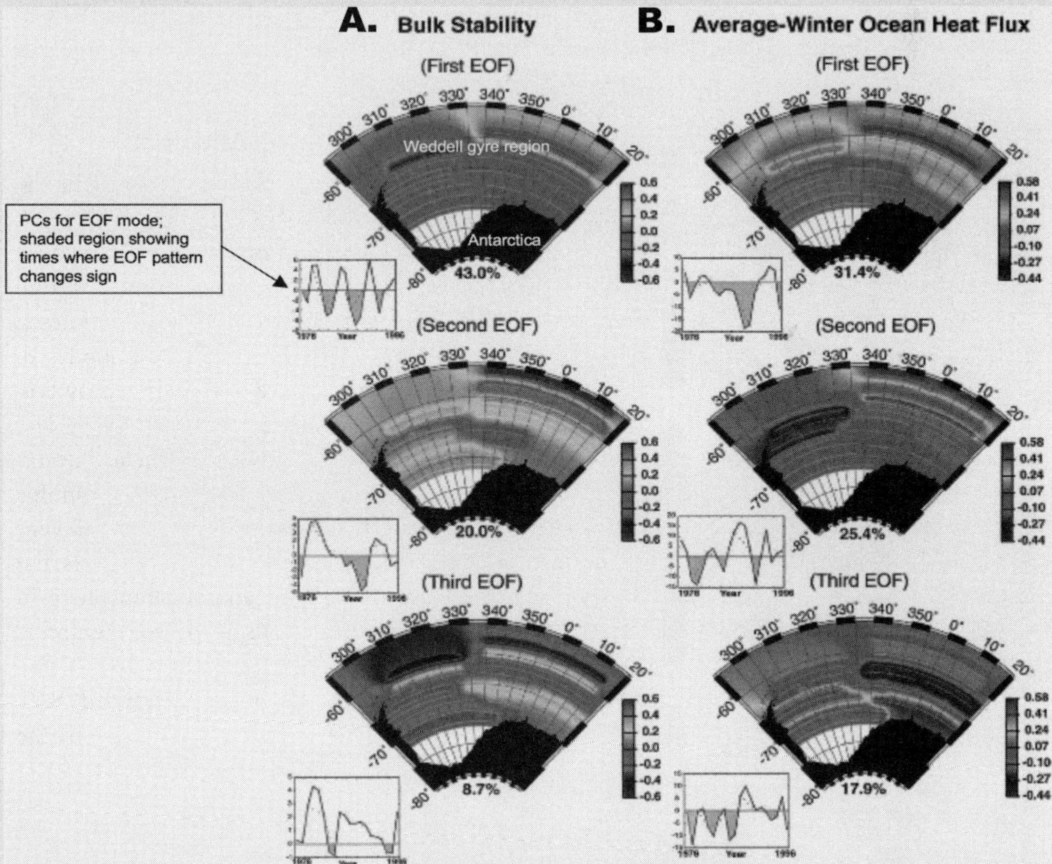

Figure A2.1 Dominant modes for two ocean-atmosphere-ice (OAI) parameters in the Weddell gyre region (indicated): **(A)** bulk stability, **(B)** winter ocean heat flux.

In this case, such a test is not correct, and there are several reasons why. The most important of these is something known as **multiplicity**. Assume for the moment that we

Box A2.3 (Cont.)

feel that $\alpha = 0.1$ is most appropriate for this particular study (though, actually, I would prefer to simply contour α values). First of all, if we had 1000 grid cells and were correlating pure noise, by definition we would expect that 100(!) of the grid cells would be significant at $\alpha = 0.1$. In other words, this significance level indicates that 10 cases in every 100 correlations of noise would be expected to actually exceed r_α by pure chance. Another effect that leads to more elevated r-values than might otherwise be expected is known sometimes as **local effects**. In that case, if an EOF mode of the OAI index is well correlated with some area strongly related to the OAI state, we would expect a similarly high value everywhere within a region related to it. We need to determine what values actually are expected by pure chance from a lot of correlations so that we can establish values that only occur rarely when correlating noise (rarely, as in 100α percent of the time for a significance level of α). For this, we rely on the bootstrap to determine the distribution of r-values expected when correlating noise with the same coloration (same PSD) as that of the actual OAI EOF time series (the PCs). Results of that are presented in Figure A2.2.

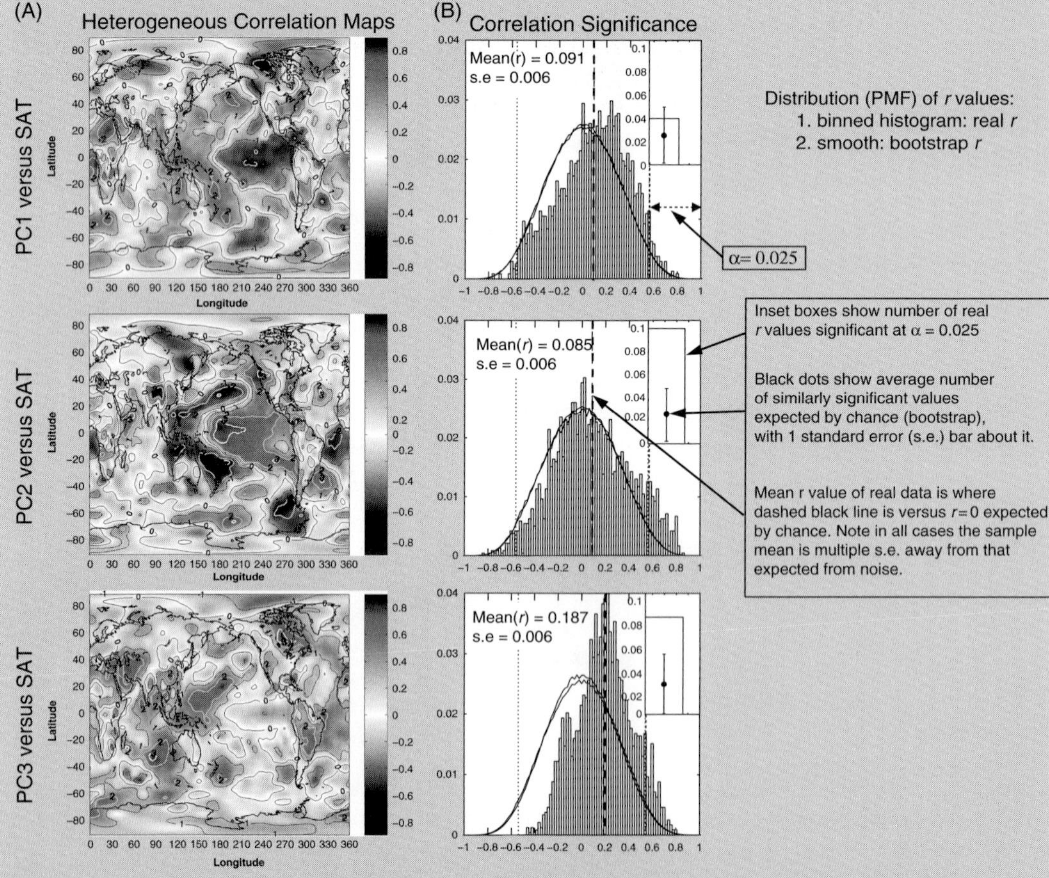

Figure A2.2 **(A)** Heterogeneous correlation maps for PCs 1 through 3 in Figure A2.1, correlating time series of detrended NCEP/NCAR near-surface temperature anomalies (SAT*) around the

Box A2.3 (Cont.)

Caption for Figure A2.2 (cont.)

globe with PCs 1 through 3 of Weddell bulk stability (column A of Figure A2.1). Grayscale reveals r value. **(B)** Significance of correlations via bootstrap. Correlations relative to that expected from correlation with colored noise, identical in lowest two bivariate statistical moments of PCs, is accessed by comparison of sample PMF for sample correlation map (histogram) to bootstrap (noise) PDF (smooth Gaussian-like curves) for each mode. Solid dots in insets near PDFs show average frequency of occurrences of correlations in the positive upper 2.5 percent of the bootstrap (noise) and one-sigma standard error bar about this average value, based on the accumulated values achieved in 1000 bootstrap noise correlation maps. Box in inset is a combined histogram class for the upper 2.5 percent, or of all of the classes found in the actual frequency of occurrence achieved from the observations. In all cases, the data show considerably more strong (rare) correlations than would be expected from correlations involving quasi-periodic colored noise time series (based on PSD of PCs). This includes that number appearing in the upper 2.5 percent as well as a highly significant (several s.e. from expected mean) shift of the mean toward higher r-values in the real data. Finally, in panel **A**, confidence intervals showing significance of correlations achieved and displayed in the correlation maps are labeled in terms of the number of standard deviations above or below that expected from noise. Number of standard errors confidence is given by labeled contours (1 through 3). 3σ indicates 99.98 percent confidence level or 0.02 significance level. Note the considerable number of highly significant correlations. Significance accounts for autocorrelations in space, time and multiplicity.

A2.5 Expectance Versus Bootstrap

The bootstrap is easy to use, understand and apply, so you might question why anyone would want to use expectance. The advantage of expectance is that it quantitatively shows the contribution of each term contributing to your overall uncertainty. With that, you can target specific terms for improvement to achieve some desired precision in the final result. On the other hand, the bootstrap is simple to understand and apply. It can handle much more complicated aspects of an analysis (e.g., multiplicity, effective degrees of freedom, etc.). In complicated cases, you must give considerable thought to the construction of the bootstrap so that you are testing exactly what it is you want to test.

References

Anderson, T. W., 1984. *An Introduction to Multivariate Statistical Analysis*, John Wiley and Sons, New York, NY, 704 pp.

Bartlett, M. S., 1946. On the theoretical specification and sampling properties of autocorrelated time series. *Journal of the Royal Statistical Society Supplement*, **V8** p1, pp. 27–41.

Blackman, R. B. and J. W. Tukey, 1958. *The Measurement of Power Spectra*, Dover Publications Inc., New York, NY, 190 pp.

Bloomfield, Peter, 1976. *Fourier Analysis of Time Series: An Introduction*, John Wiley and Sons, New York, NY, 258 pp.

Box, George E. P. and Gwilym M. Jenkins, 1976. *Time Series Analysis Forecasting and Control*, Holden-Day, Oakland, CA, 575 pp.

Bracewell, Ronald N., 1965. *The Fourier Transform and Its Applications*, McGraw Hill, New York, NY, 444 pp.

Brigham, E. Oran, 1974. *The Fast Fourier Transform*, Prentice-Hall, Inc., Englewood Cliffs, NJ, 252 pp.

Chatfield, C., 1989. *The Analysis of Time Series*, Chapman and Hall Ltd., New York, NY, 241 pp.

Cooley, W. W. and P. R. Lohnes, 1971. *Multivariate Data Analysis*, John Wiley and Sons, New York, NY, 364 pp.

Davis, John C., 1973. *Statistics and Data Analysis in Geology*. John Wiley and Sons, New York, NY, chapter 4.

Dougherty, E. R., 1990. *Probability and Statistics for the Engineering, Computing and Physical Sciences*. Prentice-Hall Inc., Englewood Cliffs, NJ, 800 pp.

Draper, N. and H. Smith, 1981. *Applied Regression Analysis*, second edition, Wiley-Interscience, New York, NY, 709 pp.

Draper, N. and H. Smith, 1998. *Applied Regression Analysis*, 3rd edition, Wiley-Interscience, New York, NY, 706 pp.

Eddy, S. R., 2004. What is Bayesian statistics? *Nature Biotechnology*, **22**, 9, pp. 1177–1178.

Edwards, A. L., 1976. *An Introduction to Linear Regression and Correlation*, W. H. Freeman and Co. Ltd., 213 pp.

Efron, B., 1981. Nonparametric estimates of standard error: The jackknife, the bootstrap and other methods. *Biometrika*, **68**, pp. 589–599.

Elliot, Douglas F., 1987. *Handbook of Digital Signal Processing, Engineering Applications*, Academic Press, New York, NY, 999 pp.

Gelman, A., 2008. Objections to Bayesian statistics. *International Society of Bayesian Analysis*, **3**, pp. 445–450.

Hamming, R. W., 1989. *Digital Filters*, Prentice-Hall, Inc, Englewood Cliffs, NJ, 284 pp.

Householder, Alston S., 1975. *Theory of Matrices in Numerical Analysis*, Dover Publications Inc., New York, NY, 257 pp.

Jackson, J. E., 1991. *A Users Guide to Principal Components*. Wiley-Interscience, New York, NY, 569 pp.

Jeffries, Harold, 1998. *Theory of Probability*, Oxford Classic Texts, London, 470 pp.

Jenkins, Gwilym M. and Donald G. Watts, 1968. *Spectral Analysis and Its Applications*, Holden-Day, Inc., San Francisco, CA, 525 pp.

Jennings, Alan, 1977. *Matrix Computation for Engineers and Scientists*, John Wiley and Sons, New York, NY, 329 pp.

Johnson, R. A., D. W. Wichern, 1988. *Applied Multivariate Statistical Analysis*, second edition, Prentice Hall, Englewood Cliffs, NJ, 607 pp.

Kanasewich, E. R., 1981. *Time Sequence Analysis in Geophysics*, University of Alberta Press, Winnipeg, Alta., 480 pp.

Katz, R. W. and B. G. Brown, 1991. The problem of multiplicity in research on teleconnections, *Inter. J. Climatol.*, **11**, 505–513.

Kay, Stephen M., 1988. *Modern Spectral Estimation, Theory and Application*, Prentice-Hall, Inc., Englewood Cliffs, NJ, 543 pp.

Kolmogorov, A. N., 1956. *Foundations of the Theory of Probability* (second English translation, by N. Morrison, of 1933 German monograph), Chelsea Publishing Company, New York, NY, 84 pp.

Koopmans, L. H., 1974. *The Spectral Analysis of Time Series*, Academic Press, New York, NY, 366 pp.

Lavine, M., 2000. What is Bayesian statistics and why everything else is wrong. Published online at: www2.stat.duke.edu/courses/Spring06/sta114/whatisbayes.pdf.

Lisiecki, L. E. and M. E. Raymo, 2005. A Pliocene-Pleistocene stack of 57 globally distributed benthic d18O records. *Paleoceanography*, **20**, PA1003.

Marple Jr., S. Lawrence, 1987. *Digital Spectral Analysis with Application*, Prentice-Hall, Inc., Englewood Cliffs, NJ, 492 pp.

Menke, W., 1984. *Discrete Inverse Theory*, Academic Press, New York, 260 pp.

Miller, R. G., 1986. *Beyond ANOVA: Basics of Applied Statistics*, John Wiley and Sons, New York, NY, 317 pp.

Nelson, Charles R., 1973. *Applied Time Series Analysis for Managerial Forecasting*, Holden-Day, San Francisco, CA, 231 pp.

North, G. R., T. L. Bell, R. F. Cahalan, and F. J. Moeng, 1982. Sampling errors in the estimation of empirical orthogonal functions. *Monthly Weather Review*, July 1982, pp. 699–706.

Percival, Donald B. and Andrew T. Walden, 1993. *Spectral Analysis for Physical Applications*, Cambridge University Press, Cambridge, 583 pp.

Press, William H., Brian P. Flannery, Saul A. Teukolsky, and William T. Vetterling, 1986. *Numerical Recipes*, Cambridge University Press, Cambridge, 818 pp.

Priestley, M. B., 1981. *Spectral Analysis and Time Series*, Academic Press, New York, NY, 890+ pp.

Robinson, E. A. and S. Treitel, 1980. *Geophysical Signal Analysis*, Prentice-Hall, Englewood Cliffs, NJ, 466 pp.

Rodgers, J. L., W. A. Nicewander and L. Toothaker, 1984. Linear independent, orthogonal, and uncorrelated variables. *The American Statistician*, **V38**, #2, pp 133–134.

Rousseeuw, P. J. and A. M. Leroy, 1987. *Robust Regression and Outlier Detection*, John Wiley and Sons, New York, NY, 329 pp.

Schama, S., 1989. Citizens*: A Chronicle of the French Revolution*, Knopf, New York, NY, 948 pp.

Schroeder, M., 1990. *Fractals, Chaos and Power Laws*, W. H. Freeman and Company, 429 pp.

Searle, Shayle R., 1982. *Matrix Algebra Useful for Statistics*, John Wiley and Sons, New York, NY, 438 pp.

Simon, J. L., 1992. *Resampling: The New Statistics*, Resampling Stats Inc., 279 pp.

Slepian, D., 1978. Prolate spheroidal wave functions, Fourier analysis, and uncertainty -V: The discrete case. *Bell Systems Technical Journal*, **V57**, 1371–1429.

Strang, Gilbert, 1988. *Linear Algebra and Its Applications*, third edition, Harcourt Brace Jovanovich, Publishers, San Diego, CA, 505 pp.

Strang, Gilbert, 2003. *Introduction to Linear Algebra and Its Applications*, third edition, Wellesley-Cambridge Press, Wellesley, MA, 568 pp.

Thomson, D. J., 1982. Spectrum estimation and harmonic analysis. *Proceedings of the IEEE*, **V70**, N9, 1055–1096.

Von Storch, H. and F. Zwiers, 2002. *Statistical Analysis in Climate Research*, Cambridge University Press, London, 496 pp.

Welch, P. D., 1967. The use of fast Fourier transform for the estimation of power spectra: A method based on time averaging over short, modified periodograms. *IEEE Transactions on Audio and Electroacoustics*, **V15**, #2, 70–73.

Index